MULTIVARIABLE MATHEMATICS
Linear Algebra, Multivariable Calculus, and Manifolds

MULTIVARIABLE MATHEMATICS

Linear Algebra, Multivariable Calculus, and Manifolds

THEODORE SHIFRIN
University of Georgia

John Wiley & Sons, Inc.

Associate Publisher   *Laurie Rosatone*
Editorial Assistant   *Kelly Boyle*
Executive Marketing Manager   *Julie Lindstrom*
Senior Production Editor   *Sujin Hong*
Senior Designer   *Madelyn Lesure*

This book was set in *Times Roman* by *Techsetters, Inc.* and printed and bound by *Malloy Lithograph*. The cover was printed by *Phoenix Color*.

This book is printed on acid free paper. ∞

Copyright © 2005 John Wiley & Sons, Inc. All rights reserved.

No part of this publication may be reproduced, stored in a retrieval system or transmitted in any form or by any means, electronic, mechanical, photocopying, recording, scanning or otherwise, except as permitted under Sections 107 or 108 of the 1976 United States Copyright Act, without either the prior written permission of the Publisher, or authorization through payment of the appropriate per-copy fee to the Copyright Clearance Center, Inc., 222 Rosewood Drive, Danvers, MA 01923, (978)750-8400, Fax: (978)750-4470. Requests to the Publisher for permission should be addressed to the Permissions Department, John Wiley & Sons, Inc., 111 River Street, Hoboken, NJ 07030, (201) 748-6011, Fax: (201) 748-6008, E-Mail: PERMREQ@WILEY.COM. To order books or for customer service please call 1-800-CALL WILEY (225-5945).

ISBN 0-471-52638-X (Domestic)
ISBN 0-471-63160-4 (WIE)

Printed in the United States of America

10 9 8 7 6 5 4 3 2 1

CONTENTS

Preface vii

CHAPTER 1
VECTORS AND MATRICES 1

1. Vectors in $\mathbb{R}^n$ 1
2. Dot Product 8
3. Subspaces of $\mathbb{R}^n$ 16
4. Linear Transformations and Matrix Algebra 23
5. Introduction to Determinants and the Cross Product 43

CHAPTER 2
FUNCTIONS, LIMITS, AND CONTINUITY 53

1. Scalar- and Vector-Valued Functions 53
2. A Bit of Topology in $\mathbb{R}^n$ 64
3. Limits and Continuity 72

CHAPTER 3
THE DERIVATIVE 81

1. Partial Derivatives and Directional Derivatives 81
2. Differentiability 87
3. Differentiation Rules 97
4. The Gradient 104
5. Curves 109
6. Higher-Order Partial Derivatives 120

CHAPTER 4
IMPLICIT AND EXPLICIT SOLUTIONS OF LINEAR SYSTEMS 127

1. Gaussian Elimination and the Theory of Linear Systems 127
2. Elementary Matrices and Calculating Inverse Matrices 147
3. Linear Independence, Basis, and Dimension 156
4. The Four Fundamental Subspaces 171
5. The Nonlinear Case: Introduction to Manifolds 186

CHAPTER 5
EXTREMUM PROBLEMS 196

1. Compactness and the Maximum Value Theorem 196
2. Maximum/Minimum Problems 202
3. Quadratic Forms and the Second Derivative Test 208
4. Lagrange Multipliers 216
5. Projections, Least Squares, and Inner Product Spaces 225

CHAPTER 6
SOLVING NONLINEAR PROBLEMS 244

1. The Contraction Mapping Principle 244
2. The Inverse and Implicit Function Theorems 251
3. Manifolds Revisited 261

CHAPTER 7
INTEGRATION 267

1. Multiple Integrals 267
2. Iterated Integrals and Fubini's Theorem 276
3. Polar, Cylindrical, and Spherical Coordinates 288
4. Physical Applications 298
5. Determinants and n-Dimensional Volume 309
6. Change of Variables Theorem 324

CHAPTER 8
DIFFERENTIAL FORMS AND INTEGRATION ON MANIFOLDS 333

1. Motivation 333
2. Differential Forms 335
3. Line Integrals and Green's Theorem 348
4. Surface Integrals and Flux 367
5. Stokes's Theorem 379
6. Applications to Physics 393
7. Applications to Topology 403

CHAPTER 9
EIGENVALUES, EIGENVECTORS, AND APPLICATIONS   413

1. Linear Transformations and Change of Basis   413
2. Eigenvalues, Eigenvectors, and Diagonalizability   422
3. Difference Equations and Ordinary Differential Equations   436
4. The Spectral Theorem   455

GLOSSARY OF NOTATIONS AND RESULTS FROM SINGLE-VARIABLE CALCULUS   467

FOR FURTHER READING   473

ANSWERS TO SELECTED EXERCISES   474

INDEX   488

PREFACE

I began writing this text as I taught a brand-new course combining linear algebra and a rigorous approach to multivariable calculus. Some of the students had already taken a proof-oriented single-variable calculus course (using Spivak's beautiful book, *Calculus*), but many had not: There were sophomores who wanted a more challenging entrée to higher-level mathematics, as well as freshmen who'd scored a 5 on the Advanced Placement Calculus BC exam. My goal was to include all the standard computational material found in the usual linear algebra and multivariable calculus courses *and more*, interweaving the material as effectively as possible, and include complete proofs.

Although there have been a number of books that include both the linear algebra and the calculus material, they have tended to segregate the material. Advanced calculus books treat the rigorous multivariable calculus, but presume the students have already mastered linear algebra. I wanted to integrate the material so as to emphasize the recurring theme of *implicit* versus *explicit* that persists in linear algebra and analysis. In every linear algebra course we should learn how to go back and forth between a system of equations and a parametrization of its solution set. But the same problem occurs, in principle, in calculus: To solve constrained maximum/minimum problems we must either parametrize the constraint set or use Lagrange multipliers; to integrate over a curve or surface, we need a parametric representation. Of course, in the linear case one can globally go back and forth; it's not so easy in the nonlinear case, but, as we'll learn, it should at least be possible in principle *locally*.

The prerequisites for this book are a solid background in single-variable calculus and, if not some experience writing proofs, a strong interest in grappling with them. In presenting the material, I have included plenty of examples, clear proofs, and significant motivation for the crucial concepts. We all know that to learn (and enjoy?) mathematics one must work lots of problems, from the routine to the more challenging. To this end, I have provided numerous exercises of varying levels of difficulty, both computational and more proof-oriented. Some of the proof exercises require the student "merely" to understand and modify a proof in the text; others may require a good deal of ingenuity. I also ask students for lots of examples and counterexamples. Generally speaking, exercises are arranged in order of increasing difficulty. To offer a bit more guidance, I have marked with an asterisk (*) those problems to which short answers, hints, or—in some cases—complete solutions are given at the back of the book. As a guide to the new teacher, I have marked with a sharp ($^\sharp$) some important exercises to which reference is made later. An Instructor's Solutions Manual (ISBN 0-471-64915-5) is available from the publisher.

▶ COMMENTS ON CONTENTS

The linear algebraic material with which we begin the course in Chapter 1 is concrete, establishes the link with geometry, and is a good self-contained setting for working on proofs. We introduce vectors, dot products, subspaces, and linear transformations and matrix computations. At this early stage we emphasize the two interpretations of multiplying a matrix A by a vector $\mathbf{x}$: the linear equations viewpoint (considering the dot products of the *rows* of A with $\mathbf{x}$) and the linear combinations viewpoint (taking the linear combination of the *columns* of A weighted by the coordinates of $\mathbf{x}$). We end the chapter with a discussion of 2×2 and 3×3 determinants, area, volume, and the cross product.

In Chapter 2 we begin to make the transition to calculus, introducing scalar functions of a vector variable—their graphs and their level sets—and vector-valued functions. We introduce the requisite language of open and closed sets, sequences, and limits and continuity, including the proofs of the usual limit theorems. (Generally, however, I give these short shrift in lecture, as I don't have the time to emphasize δ-ε arguments.)

We come to the concepts of differential calculus in Chapter 3. We quickly introduce partial and directional derivatives as immediate to calculate, and then come to the definition of differentiability, the characterization of differentiable functions, and the standard differentiation rules. We give the gradient vector its own brief section, in which we emphasize its geometric meaning. Then comes a section on curves, in which we mention Kepler's laws (the second is proved in the text and the other two are left as an exercise), arclength, and curvature of a space curve.

In the first four sections of Chapter 4 we give an accelerated treatment of Gaussian elimination (including a proof of uniqueness of reduced echelon form) and the theory of linear systems, the standard material on linear independence and dimension (including a brief mention of abstract vector spaces), and the four fundamental subspaces associated to a matrix. In the last section, we begin our assault on the nonlinear case, introducing (with no proofs) the implicit function theorem and the notion of a manifold.

Chapter 5 is a blend of topology, calculus, and linear algebra—quadratic forms and projections. We start with the topological notion of compactness and prove the maximum value theorem in higher dimensions. We then turn to the calculus of applied maximum/minimum problems and then to the analysis of the second-derivative test and the Hessian. Then comes one of the most important topics in applications, Lagrange multipliers (with a rigorous proof). In the last section, we return to linear algebra, to discuss projections (from both the explicit and the implicit approaches), least-squares solutions of inconsistent systems, the Gram-Schmidt process, and a brief discussion of abstract inner product spaces (including a nice proof of Lagrange interpolation).

Chapter 6 is a brief, but sophisticated, introduction to the inverse and implicit function theorems. We present our favorite proof using the contraction mapping principle (which is both more elegant and works just fine in the infinite-dimensional setting). In the last section we prove that all three definitions of a manifold are (locally) equivalent: the implicit representation, the parametric representation, and the representation as a graph. (In the year-long course that I teach, I find I have time to treat this chapter only lightly.)

In Chapter 7 we study the multidimensional (Riemann) integral. In the first two sections we deal predominantly with the theory of the multiple integral and, then, Fubini's Theorem and the computation of iterated integrals. Then we introduce (as is customary in a typical multivariable calculus course) polar, cylindrical, and spherical coordinates and various physical applications. We conclude the chapter with a careful treatment of determinants (which will play a crucial role in Chapters 8 and 9) and a proof of the Change of Variables Theorem.

In single-variable calculus, one of the truly impressive results is the Fundamental Theorem of Calculus. In Chapter 8 we start by laying the groundwork for the analogous multidimensional result, introducing differential forms in a very explicit fashion. We then parallel a traditional vector calculus course, introducing line integrals and Green's Theorem, surface integrals and flux, and, then, finally stating and proving the general Stokes's Theorem for compact oriented manifolds. We do not skimp on concrete and nontrivial examples throughout. In Section 8.6 we introduce the standard terminology of divergence and curl and give the "classical" versions of Stokes's and the Divergence Theorems, along with some applications to physics. In Section 8.7 we begin to illustrate the power of Stokes's Theorem by proving the Fundamental Theorem of Algebra, a special case of the argument principle, and the "hairy ball theorem" from topology.

In Chapter 9 we complete our study of linear algebra, including standard material on change of basis (with a geometric slant), eigenvalues, eigenvectors and discussion of diagonalizability. The remainder of the chapter is devoted to applications: difference and differential equations, and a brief discussion of flows and their relation to the Divergence Theorem of Chapter 8. We close with the Spectral Theorem. (With the exception of Section 3.3, which relies on Chapter 8, and the proof of the Spectral Theorem, which relies on Section 4 of Chapter 5, topics in this chapter can be covered at any time after completing Chapter 4.)

We have included a glossary of notations and a quick compilation of relevant results from trigonometry and single-variable calculus (including a short table of integrals), along with a much-requested list of the Greek alphabet.

There are over 800 exercises in the text, many with multiple parts. Here are a few particularly interesting (and somewhat unusual) exercises included in this text:

- Exercises 1.2.22–26 and Exercises 1.5.19 and 1.5.20 on the geometry of triangles, and Exercise 1.5.17, a nice glimpse of affine geometry
- Exercise 2.1.12, a parametrization of a hyperboloid of one sheet in which the parameter curves are the two families of rulings
- Exercises 2.3.15–17, 3.1.10, and 3.2.18–19, exploring the infamous sorts of discontinuous and nondifferentiable functions
- Example 3.4.3 introducing the reflectivity property of the ellipse via the gradient, with follow-ups in Exercises 3.4.8, 3.4.9, and 3.4.13, and then Kepler's first and third laws in Exercise 3.5.15.
- Exercise 3.5.14, the famous fact (due to Huygens) that the evolute of a cycloid is a congruent cycloid
- Exercise 4.5.13, in which we discover that the lines passing through three pairwise-skew lines generate a saddle surface
- Exercises 5.1.5, 5.1.7, 9.4.11, exploring the (operator) norm of a matrix
- Exercise 5.2.15, introducing the Fermat/Steiner point of a triangle
- Exercises 5.3.2 and 5.3.4, pointing out a local minimum along every line need not be a local minimum (an issue that is mishandled in surprisingly many multivariable calculus texts) and that a lone critical point that is a local minimum may not be a global minimum

- Exercises 5.4.32, 5.4.34, and 9.4.21, giving the interpretation of the Lagrange multiplier, introducing the bordered Hessian, and giving a proof that the bordered Hessian gives a sufficient test for constrained critical points
- Exercises 6.1.8 and 6.1.10, giving Kantarovich's Theorem (first in one dimension and then in higher), a sufficient condition for Newton's method to converge (a beautiful result I learned from Hubbard and Hubbard)
- Exercise 6.2.13, introducing the envelope of a family of curves
- Exercise 7.3.24, my favorite triple integral challenge problem
- Exercises 7.4.27 and 7.4.28
- Exercises 7.5.25–27, some nice applications of the determinant
- Exercises 8.3.23, 8.3.25, and 8.3.26, some interesting applications of line integration and Green's Theorem
- Exercise 8.5.22, giving a calibrations proof that the minimal surface equation gives surfaces of least area
- The discussion in Chapter 8, Section 7, of counting roots (reminiscent of the treatment of winding numbers and Gauss's Law in earlier sections) and Exercises 8.7.9 and 9.4.22, in which we prove that the roots of a complex polynomial depend continuously on its coefficients, and then derive Sylvester's Law of Inertia as a corollary
- Exercises 9.1.12 and 9.1.13, some interesting applications of the change-of-basis framework
- Exercises 9.2.19, 9.2.20, 9.2.23, and 9.2.24, some more standard but more challenging linear algebra exercises

▶ POSSIBLE WAYS TO USE THIS BOOK

I have been using the text for a number of years in a course for highly motivated freshmen and sophomores. Since this is the first "serious" course in mathematics for many of them, because of time limitations, I must give somewhat short shrift to many of the complicated analytic proofs. For example, I only have time to talk about the Inverse and Implicit Function Theorems and to sketch the proof of the Change of Variables Theorem, and do not include all the technical aspects of the proof of Stokes's Theorem. On the other hand, I cover most of the linear algebra material thoroughly. I do plenty of examples and assign a broad range of homework problems, from the computational to the more challenging proofs.

It would also be quite appropriate to use the text in courses in advanced calculus or multivariable analysis. Depending on the students' background, I might bypass the linear algebra material or assign some of it as review reading and highlight a few crucial results. I would spend more time on the analytic material (especially in Chapters 3, 6, and 7) and treat Stokes's Theorem from the differential form viewpoint very carefully, including the applications in Section 8.7. The approach of the text will give the students a very hands-on understanding of rather abstract material. In such courses, I would spend more time in class on proofs and assign a greater proportion of theoretical homework problems.

▶ ACKNOWLEDGMENTS

I would like to thank my students of the past years for enduring preliminary versions of this text and for all their helpful comments and suggestions. I would like to acknowledge helpful conversations with my colleagues Malcolm Adams and Jason Cantarella. I would also like to thank the following reviewers, along with several anonymous referees, who offered many helpful comments:

Michael T. Anderson	SUNY, Stony Brook
Quo-Shin Chi	Washington University
Mohamed Elhamdadi	University of South Florida
Nathaniel Emerson	University of California, Los Angeles
Greg Friedman	Yale University
Stephen Sperber	University of Minnesota
John Stalker	Princeton University
Philip B. Yasskin	Texas A&M University

I am very grateful to my editor, Laurie Rosatone, for her enthusiastic support, encouragement, and guidance.

I welcome any comments and suggestions. Please address any e-mail correspondence to

shifrin@math.uga.edu

and please keep an eye on

www.math.uga.edu/~shifrin/Multivariable.html

or

www.wiley.com/college/shifrin

for the latest in typos and corrections.

Theodore Shifrin

CHAPTER

1

VECTORS AND MATRICES

Linear algebra provides a beautiful example of the interplay between two branches of mathematics, geometry and algebra. Moreover, it provides the foundations for all of our upcoming work with calculus, which is based on the idea of approximating the general function locally by a linear one. In this chapter, we introduce the basic language of vectors, linear functions, and matrices. We emphasize throughout the symbiotic relation between geometric and algebraic calculations and interpretations. This is true also of the last section, where we discuss the determinant in two and three dimensions and define the cross product.

▶ 1 VECTORS IN $\mathbb{R}^n$

A point in $\mathbb{R}^n$ is an ordered n-tuple of real numbers, written $(x_1, \ldots, x_n)$. To it we may associate the vector $\mathbf{x} = \begin{bmatrix} x_1 \\ x_2 \\ \vdots \\ x_n \end{bmatrix}$, which we visualize *geometrically* as the arrow pointing from the origin to the point. We shall (purposely) use the boldface letter $\mathbf{x}$ to denote both the point and the corresponding vector, as illustrated in Figure 1.1. We denote by $\mathbf{0}$ the vector all of whose coordinates are 0, called the *zero vector*.

More generally, any two points A and B in space determine the arrow pointing from A to B, as shown in Figure 1.2, again specifying a vector that we denote $\overrightarrow{AB}$. We often refer

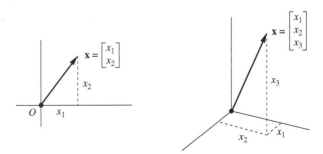

Figure 1.1

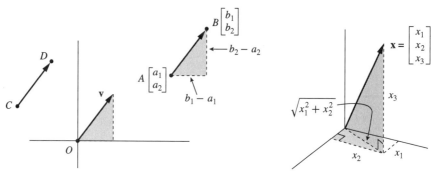

Figure 1.2   **Figure 1.3**

to A as the "tail" of the vector $\overrightarrow{AB}$ and B as its "head." If $A = \begin{bmatrix} a_1 \\ \vdots \\ a_n \end{bmatrix}$ and $B = \begin{bmatrix} b_1 \\ \vdots \\ b_n \end{bmatrix}$, then $\overrightarrow{AB}$ is equal to the vector $\mathbf{v} = \begin{bmatrix} b_1 - a_1 \\ \vdots \\ b_n - a_n \end{bmatrix}$, whose tail is at the origin, as indicated in Figure 1.2.

The Pythagorean Theorem tells us that when $n = 2$ the length of the vector $\mathbf{x}$ is $\sqrt{x_1^2 + x_2^2}$. A repeated application of the Pythagorean Theorem, as indicated in Figure 1.3, leads to the following

Definition We define the *length* of the vector

$$\mathbf{x} = \begin{bmatrix} x_1 \\ x_2 \\ \vdots \\ x_n \end{bmatrix} \in \mathbb{R}^n \quad \text{to be} \quad \|\mathbf{x}\| = \sqrt{x_1^2 + x_2^2 + \cdots + x_n^2}.$$

We say $\mathbf{x}$ is a *unit vector* if it has length 1, i.e., if $\|\mathbf{x}\| = 1$.

There are two crucial algebraic operations one can perform on vectors, both of which have clear geometric interpretations.

Scalar multiplication: If c is a real number and $\mathbf{x} = \begin{bmatrix} x_1 \\ x_2 \\ \vdots \\ x_n \end{bmatrix}$ is a vector, then we define $c\mathbf{x}$ to be the vector $\begin{bmatrix} cx_1 \\ cx_2 \\ \vdots \\ cx_n \end{bmatrix}$. Note that $c\mathbf{x}$ points in either the same direction as $\mathbf{x}$ or the opposite direction, depending on whether $c > 0$ or $c < 0$, respectively. Thus, multiplication by the real number c simply stretches (or shrinks) the vector by a factor of $|c|$ and reverses

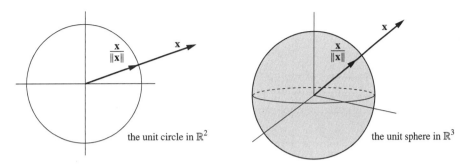

Figure 1.4

its direction when c is negative. Since this is a geometric "change of scale," we refer to the real number c as a *scalar* and the multiplication $c\mathbf{x}$ as *scalar multiplication*.

Note that whenever $\mathbf{x} \neq \mathbf{0}$ we can find a unit vector with the same direction by taking

$$\frac{\mathbf{x}}{\|\mathbf{x}\|} = \frac{1}{\|\mathbf{x}\|}\mathbf{x},$$

as shown in Figure 1.4.

Given a nonzero vector $\mathbf{x}$, any scalar multiple $c\mathbf{x}$ lies on the line through the origin and passing through the head of the vector $\mathbf{x}$. For this reason, we make the following

Definition We say two vectors $\mathbf{x}$ and $\mathbf{y}$ are *parallel* if one is a scalar multiple of the other, i.e., if there is a scalar c so that $\mathbf{y} = c\mathbf{x}$ or $\mathbf{x} = c\mathbf{y}$. We say $\mathbf{x}$ and $\mathbf{y}$ are *nonparallel* if they are not parallel.

Vector addition: If $\mathbf{x} = \begin{bmatrix} x_1 \\ \vdots \\ x_n \end{bmatrix}$ and $\mathbf{y} = \begin{bmatrix} y_1 \\ \vdots \\ y_n \end{bmatrix}$, then we define $\mathbf{x} + \mathbf{y} = \begin{bmatrix} x_1 + y_1 \\ \vdots \\ x_n + y_n \end{bmatrix}$.

To understand this geometrically, we move the vector $\mathbf{y}$ so that its tail is at the head of $\mathbf{x}$, and draw the arrow from the origin to its head. This is the so-called *parallelogram law* for vector addition, for, as we see in Figure 1.5, $\mathbf{x} + \mathbf{y}$ is the "long" diagonal of the

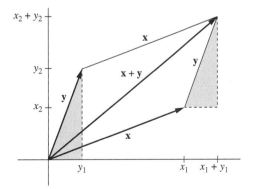

Figure 1.5

parallelogram spanned by **x** and **y**. Notice that the picture makes it clear that vector addition is commutative; i.e.,

$$\mathbf{x} + \mathbf{y} = \mathbf{y} + \mathbf{x}.$$

This also follows immediately from the algebraic definition because addition of real numbers is commutative. (See Exercise 12 for an exhaustive list of the properties of vector addition and scalar multiplication.)

Remark We emphasize here that the notions of vector addition and scalar multiplication make sense geometrically for vectors in the form $\overrightarrow{AB}$ which do not necessarily have their tails at the origin. If we wish to add $\overrightarrow{AB}$ to $\overrightarrow{CD}$, we simply recall that $\overrightarrow{CD}$ is equal to *any* vector with the same length and direction, so we just translate $\overrightarrow{CD}$ so that C and B coincide; then the arrow from A to the point D in its new position is the sum $\overrightarrow{AB} + \overrightarrow{CD}$.

Subtraction of one vector from another is easy to define algebraically. If **x** and **y** are as above, then we set

$$\mathbf{x} - \mathbf{y} = \begin{bmatrix} x_1 - y_1 \\ \vdots \\ x_n - y_n \end{bmatrix}.$$

As is the case with real numbers, we have the following interpretation of the difference $\mathbf{x} - \mathbf{y}$: It is the vector we add to **y** in order to obtain **x**; i.e.,

$$(\mathbf{x} - \mathbf{y}) + \mathbf{y} = \mathbf{x}.$$

Pictorially, we see that $\mathbf{x} - \mathbf{y}$ is drawn, as shown in Figure 1.6, by putting its tail at **y** and its head at **x**, thereby resulting in the other diagonal of the parallelogram determined by **x** and **y**. Note that if A and B are points in space and we set $\mathbf{x} = \overrightarrow{OA}$ and $\mathbf{y} = \overrightarrow{OB}$, then $\mathbf{y} - \mathbf{x} = \overrightarrow{AB}$. Moreover, as Figure 1.6 also suggests, we have $\mathbf{x} - \mathbf{y} = \mathbf{x} + (-\mathbf{y})$.

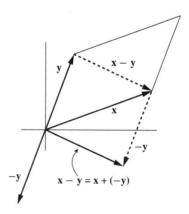

Figure 1.6

▶ **EXAMPLE 1**

Let A and B be points in $\mathbb{R}^n$. The *midpoint* M of the line segment joining them is the point halfway from A to B; that is, $\overrightarrow{AM} = \frac{1}{2}\overrightarrow{AB}$. Using the notation as above, we set $\mathbf{x} = \overrightarrow{OA}$ and $\mathbf{y} = \overrightarrow{OB}$, and we have

(∗) $$\overrightarrow{OM} = \mathbf{x} + \overrightarrow{AM} = \mathbf{x} + \tfrac{1}{2}(\mathbf{y} - \mathbf{x}) = \tfrac{1}{2}(\mathbf{x} + \mathbf{y}).$$

In particular, the vector from the origin to the midpoint of $\overline{AB}$ is the average of the vectors $\mathbf{x}$ and $\mathbf{y}$. See Exercise 8 for a generalization to three vectors and Section 4 of Chapter 7 for more.

From this formula follows one of the classic results from high school geometry: The diagonals of a parallelogram bisect one another. We've seen that the midpoint M of $\overline{AB}$ is, by virtue of the formula (∗), also the midpoint of diagonal $\overline{OC}$. (See Figure 1.7.) ◀

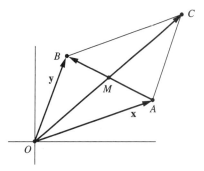

Figure 1.7

It should now be evident that vector methods provide a great tool for translating theorems from Euclidean geometry into simple algebraic statements. Here is another example. Recall that a *median* of a triangle is a line segment from a vertex to the midpoint of the opposite side.

Proposition 1.1 *The medians of a triangle intersect at a point that is two-thirds of the way from each vertex to the opposite side.*

Proof We may put one of the vertices of the triangle at the origin, so that the picture is as shown in Figure 1.8(a). Let $\mathbf{x} = \overrightarrow{OA}$, $\mathbf{y} = \overrightarrow{OB}$, and let L, M, and N be the midpoints of $\overline{OA}$, $\overline{AB}$, and $\overline{OB}$, respectively. The battle plan is the following: We let P denote the point $2/3$ of the way from B to L, Q the point $2/3$ of the way from O to M, and R the point $2/3$ of the way from A to N. Although we've indicated P, Q, and R as distinct points in Figure 1.8(b), our goal is to prove that $P = Q = R$; we do this by expressing all the vectors $\overrightarrow{OP}$, $\overrightarrow{OQ}$, and $\overrightarrow{OR}$ in terms of $\mathbf{x}$ and $\mathbf{y}$.

$$\overrightarrow{OP} = \overrightarrow{OB} + \overrightarrow{BP} = \overrightarrow{OB} + \tfrac{2}{3}\overrightarrow{BL} = \mathbf{y} + \tfrac{2}{3}(\tfrac{1}{2}\mathbf{x} - \mathbf{y})$$
$$= \tfrac{1}{3}\mathbf{x} + \tfrac{1}{3}\mathbf{y};$$
$$\overrightarrow{OQ} = \tfrac{2}{3}\overrightarrow{OM} = \tfrac{2}{3}\left(\tfrac{1}{2}(\mathbf{x} + \mathbf{y})\right) = \tfrac{1}{3}(\mathbf{x} + \mathbf{y}); \text{ and}$$

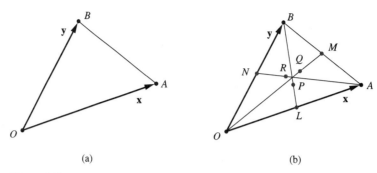

Figure 1.8

$$\overrightarrow{OR} = \overrightarrow{OA} + \overrightarrow{AR} = \overrightarrow{OA} + \tfrac{2}{3}\overrightarrow{AN} = \mathbf{x} + \tfrac{2}{3}(\tfrac{1}{2}\mathbf{y} - \mathbf{x}) = \tfrac{1}{3}\mathbf{x} + \tfrac{1}{3}\mathbf{y}.$$

We conclude that, as desired, $\overrightarrow{OP} = \overrightarrow{OQ} = \overrightarrow{OR}$, and so $P = Q = R$. That is, if we go 2/3 of the way down any of the medians, we end up at the same point; this is, of course, the point of intersection of the three medians. ∎

The astute reader might notice that we could have been more economical in the last proof. Suppose we merely check that the points 2/3 of the way down *two* of the medians (say P and Q) agree. It would then follow (say, by relabeling the triangle slightly) that the same is true of a different pair of medians (say P and R). But since any two pairs must have a point in common, we may now conclude that all three points are equal.

▶ EXERCISES 1.1

1. Given $\mathbf{x} = \begin{bmatrix} 2 \\ 3 \end{bmatrix}$ and $\mathbf{y} = \begin{bmatrix} -1 \\ 1 \end{bmatrix}$, calculate the following both algebraically and geometrically.

   (a) $\mathbf{x} + \mathbf{y}$
   (b) $\mathbf{x} - \mathbf{y}$
   (c) $\mathbf{x} + 2\mathbf{y}$
   (d) $\tfrac{1}{2}\mathbf{x} + \tfrac{1}{2}\mathbf{y}$
   (e) $\mathbf{y} - \mathbf{x}$
   (f) $2\mathbf{x} - \mathbf{y}$
   (g) $\|\mathbf{x}\|$
   (h) $\mathbf{x}/\|\mathbf{x}\|$

*2. Three vertices of a parallelogram are $\begin{bmatrix} 1 \\ 2 \\ 1 \end{bmatrix}$, $\begin{bmatrix} 2 \\ 4 \\ 3 \end{bmatrix}$, and $\begin{bmatrix} 3 \\ 1 \\ 5 \end{bmatrix}$. What are all the possible positions of the fourth vertex? Give your reasoning.

3. The origin is at the center of a regular polygon.
   (a) What is the sum of the vectors to each of the vertices of the polygon? Give your reasoning. (Hint: What are the symmetries of the polygon?)
   (b) What is the sum of the vectors from one fixed vertex to each of the remaining vertices? Give your reasoning.

4. Given $\triangle ABC$, let M and N be the midpoints of $\overline{AB}$ and $\overline{AC}$, respectively. Prove that $\overrightarrow{MN} = \tfrac{1}{2}\overrightarrow{BC}$.

5. Let $ABCD$ be an arbitrary quadrilateral. Let P, Q, R, and S be the midpoints of $\overline{AB}$, $\overline{BC}$, $\overline{CD}$, and $\overline{DA}$, respectively. Use vector methods to prove that $PQRS$ is a parallelogram. (Hint: Use Exercise 4.)

*6. In $\triangle ABC$ pictured in Figure 1.9, $\|\overrightarrow{AD}\| = \frac{2}{3}\|\overrightarrow{AB}\|$ and $\|\overrightarrow{CE}\| = \frac{2}{5}\|\overrightarrow{CB}\|$. Let Q denote the midpoint of $\overline{CD}$; show that $\overrightarrow{AQ} = c\overrightarrow{AE}$ for some scalar c and determine the ratio $c = \|\overrightarrow{AQ}\|/\|\overrightarrow{AE}\|$. In what ratio does $\overline{CD}$ divide $\overline{AE}$?

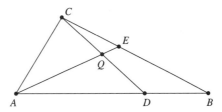

Figure 1.9

7. Consider parallelogram $ABCD$. Suppose $\overrightarrow{AE} = \frac{1}{3}\overrightarrow{AB}$ and $\overrightarrow{DP} = \frac{3}{4}\overrightarrow{DE}$. Show that P lies on the diagonal $\overline{AC}$. (See Figure 1.10.)

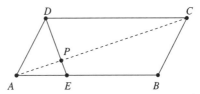

Figure 1.10

8. Let A, B, and C be vertices of a triangle in $\mathbb{R}^3$. Let $\mathbf{x} = \overrightarrow{OA}$, $\mathbf{y} = \overrightarrow{OB}$, and $\mathbf{z} = \overrightarrow{OC}$. Show that the head of the vector $\mathbf{v} = \frac{1}{3}(\mathbf{x} + \mathbf{y} + \mathbf{z})$ lies on each median of $\triangle ABC$ (and thus is the point of intersection of the three medians). It follows (see Section 4 of Chapter 7) that when we put equal masses at A, B, and C, the center of mass of that system is given by the intersections of the medians of the triangle.

9. (a) Let $\mathbf{u}, \mathbf{v} \in \mathbb{R}^2$. Describe the vectors $\mathbf{x} = s\mathbf{u} + t\mathbf{v}$, where $s + t = 1$. Pay particular attention to the location of $\mathbf{x}$ when $s \geq 0$ and when $t \geq 0$.

(b) Let $\mathbf{u}, \mathbf{v}, \mathbf{w} \in \mathbb{R}^3$. Describe the vectors $\mathbf{x} = r\mathbf{u} + s\mathbf{v} + t\mathbf{w}$, where $r + s + t = 1$. Pay particular attention to the location of $\mathbf{x}$ when each of r, s, and t is positive.

10. Suppose $\mathbf{x}, \mathbf{y} \in \mathbb{R}^n$ are nonparallel vectors. (Recall the definition on p. 3.)
(a) Prove that if $s\mathbf{x} + t\mathbf{y} = \mathbf{0}$, then $s = t = 0$. (Hint: Show that neither $s \neq 0$ nor $t \neq 0$ is possible.)
(b) Prove that if $a\mathbf{x} + b\mathbf{y} = c\mathbf{x} + d\mathbf{y}$, then $a = c$ and $b = d$.

11. "Discover" the fraction 2/3 that appears in Proposition 1.1 by finding the intersection of two medians. (Hint: A point on the line $\overleftrightarrow{OM}$ can be written in the form $t(\mathbf{x} + \mathbf{y})$ for some scalar t, and a point on the line $\overleftrightarrow{AN}$ can be written in the form $\mathbf{x} + s(\frac{1}{2}\mathbf{y} - \mathbf{x})$ for some scalar s. You will need to use the result of Exercise 10.)

12. Verify both algebraically and geometrically that the following properties of vector arithmetic hold. (Do so for $n = 2$ if the general case is too intimidating.)
(a) For all $\mathbf{x}, \mathbf{y} \in \mathbb{R}^n$, $\mathbf{x} + \mathbf{y} = \mathbf{y} + \mathbf{x}$.
(b) For all $\mathbf{x}, \mathbf{y}, \mathbf{z} \in \mathbb{R}^n$, $(\mathbf{x} + \mathbf{y}) + \mathbf{z} = \mathbf{x} + (\mathbf{y} + \mathbf{z})$.
(c) $\mathbf{0} + \mathbf{x} = \mathbf{x}$ for all $\mathbf{x} \in \mathbb{R}^n$.

(d) For each $\mathbf{x} \in \mathbb{R}^n$, there is a vector $-\mathbf{x}$ so that $\mathbf{x} + (-\mathbf{x}) = \mathbf{0}$.
(e) For all $c, d \in \mathbb{R}$ and $\mathbf{x} \in \mathbb{R}^n$, $c(d\mathbf{x}) = (cd)\mathbf{x}$.
(f) For all $c \in \mathbb{R}$ and $\mathbf{x}, \mathbf{y} \in \mathbb{R}^n$, $c(\mathbf{x} + \mathbf{y}) = c\mathbf{x} + c\mathbf{y}$.
(g) For all $c, d \in \mathbb{R}$ and $\mathbf{x} \in \mathbb{R}^n$, $(c + d)\mathbf{x} = c\mathbf{x} + d\mathbf{x}$.
(h) For all $\mathbf{x} \in \mathbb{R}^n$, $1\mathbf{x} = \mathbf{x}$.

♯13. (a) Using only the properties listed in Exercise 12, prove that for any $\mathbf{x} \in \mathbb{R}^n$, we have $0\mathbf{x} = \mathbf{0}$. (It often surprises students that this is a consequence of the properties in Exercise 12.)

(b) Using the result of part a, prove that $(-1)\mathbf{x} = -\mathbf{x}$. (Be sure that you didn't use this fact in your proof of part a!)

▶ 2 DOT PRODUCT

We discuss next one of the crucial constructions in linear algebra, the dot product $\mathbf{x} \cdot \mathbf{y}$ of two vectors $\mathbf{x}, \mathbf{y} \in \mathbb{R}^n$. By way of motivation, let's recall some basic results from plane geometry. Let $P = \begin{bmatrix} x_1 \\ x_2 \end{bmatrix}$ and $Q = \begin{bmatrix} y_1 \\ y_2 \end{bmatrix}$ be points in the plane, as pictured in Figure 2.1.

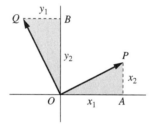

Figure 2.1

Then we observe that when $\angle POQ$ is a right angle, $\triangle OAP$ is similar to $\triangle OBQ$, and so $x_2/x_1 = -y_1/y_2$, whence $x_1 y_1 + x_2 y_2 = 0$. This leads us to make the following

Definition Given vectors $\mathbf{x}, \mathbf{y} \in \mathbb{R}^2$, define their *dot product*
$$\mathbf{x} \cdot \mathbf{y} = x_1 y_1 + x_2 y_2.$$
More generally, given vectors $\mathbf{x}, \mathbf{y} \in \mathbb{R}^n$, define their dot product
$$\mathbf{x} \cdot \mathbf{y} = x_1 y_1 + x_2 y_2 + \cdots + x_n y_n.$$

We know that when the vectors $\mathbf{x}$ and $\mathbf{y} \in \mathbb{R}^2$ are perpendicular, their dot product is 0. By starting with the algebraic properties of the dot product, we are able to get a great deal of geometry out of it.

Proposition 2.1 *The dot product has the following properties:*

1. $\mathbf{x} \cdot \mathbf{y} = \mathbf{y} \cdot \mathbf{x}$ *for all* $\mathbf{x}, \mathbf{y} \in \mathbb{R}^n$ *(dot product is* commutative*);*
2. $\mathbf{x} \cdot \mathbf{x} = \|\mathbf{x}\|^2 \geq 0$ *and* $\mathbf{x} \cdot \mathbf{x} = 0 \iff \mathbf{x} = \mathbf{0}$;
3. $(c\mathbf{x}) \cdot \mathbf{y} = c(\mathbf{x} \cdot \mathbf{y})$ *for all* $\mathbf{x}, \mathbf{y} \in \mathbb{R}^n$ *and* $c \in \mathbb{R}$;
4. $\mathbf{x} \cdot (\mathbf{y} + \mathbf{z}) = \mathbf{x} \cdot \mathbf{y} + \mathbf{x} \cdot \mathbf{z}$ *for all* $\mathbf{x}, \mathbf{y}, \mathbf{z} \in \mathbb{R}^n$ *(the* distributive *property).*

Proof In order to simplify the notation, we give the proof with $n = 2$. Since multiplication of real numbers is commutative, we have
$$\mathbf{x} \cdot \mathbf{y} = x_1 y_1 + x_2 y_2 = y_1 x_1 + y_2 x_2 = \mathbf{y} \cdot \mathbf{x}.$$

The square of a real number is nonnegative and the sum of nonnegative numbers is nonnegative, so $\mathbf{x} \cdot \mathbf{x} = x_1^2 + x_2^2 \geq 0$ and is equal to 0 only when $x_1 = x_2 = 0$. The next property follows from the associative and distributive properties of real numbers:
$$(c\mathbf{x}) \cdot \mathbf{y} = (cx_1)y_1 + (cx_2)y_2 = c(x_1 y_1) + c(x_2 y_2) = c(x_1 y_1 + x_2 y_2) = c(\mathbf{x} \cdot \mathbf{y}).$$

The last result follows from the commutative, associative, and distributive properties of real numbers:
$$\begin{aligned}\mathbf{x} \cdot (\mathbf{y} + \mathbf{z}) &= x_1(y_1 + z_1) + x_2(y_2 + z_2) = x_1 y_1 + x_1 z_1 + x_2 y_2 + x_2 z_2 \\ &= (x_1 y_1 + x_2 y_2) + (x_1 z_1 + x_2 z_2) = \mathbf{x} \cdot \mathbf{y} + \mathbf{x} \cdot \mathbf{z}. \blacksquare\end{aligned}$$

Corollary 2.2 $\|\mathbf{x} + \mathbf{y}\|^2 = \|\mathbf{x}\|^2 + 2\mathbf{x} \cdot \mathbf{y} + \|\mathbf{y}\|^2$.

Proof Using the properties repeatedly, we have
$$\begin{aligned}\|\mathbf{x} + \mathbf{y}\|^2 &= (\mathbf{x} + \mathbf{y}) \cdot (\mathbf{x} + \mathbf{y}) = \mathbf{x} \cdot \mathbf{x} + \mathbf{x} \cdot \mathbf{y} + \mathbf{y} \cdot \mathbf{x} + \mathbf{y} \cdot \mathbf{y} \\ &= \|\mathbf{x}\|^2 + 2\mathbf{x} \cdot \mathbf{y} + \|\mathbf{y}\|^2,\end{aligned}$$
as desired. $\blacksquare$

The geometric meaning of this result comes from the Pythagorean Theorem: When $\mathbf{x}$ and $\mathbf{y}$ are perpendicular vectors in $\mathbb{R}^2$, then we have $\|\mathbf{x} + \mathbf{y}\|^2 = \|\mathbf{x}\|^2 + \|\mathbf{y}\|^2$, and so, by Corollary 2.2, it must be the case that $\mathbf{x} \cdot \mathbf{y} = 0$. (And the converse follows, too, from the converse of the Pythagorean Theorem.) That is, two vectors in $\mathbb{R}^2$ are perpendicular if and only if their dot product is 0.

Motivated by this, we use the algebraic definition of dot product of vectors in $\mathbb{R}^n$ to bring in the geometry. In keeping with current use of the terminology and falling prey to the penchant to have several names for the same thing, we make the following

Definition We say vectors $\mathbf{x}$ and $\mathbf{y}$ are *orthogonal* if $\mathbf{x} \cdot \mathbf{y} = 0$.

Armed with this definition, we proceed to a construction that will be important in much of our future work. Starting with two vectors $\mathbf{x}, \mathbf{y} \in \mathbb{R}^n$, where $\mathbf{y} \neq \mathbf{0}$, Figure 2.2 suggests that we should be able to write $\mathbf{x}$ as the sum of a vector, $\mathbf{x}^{\|}$, that is parallel to $\mathbf{y}$ and a vector, $\mathbf{x}^{\perp}$, that is orthogonal to $\mathbf{y}$. Let's suppose we have such an equation:
$$\mathbf{x} = \mathbf{x}^{\|} + \mathbf{x}^{\perp}, \quad \text{where}$$
$\mathbf{x}^{\|}$ is a scalar multiple of $\mathbf{y}$ and $\mathbf{x}^{\perp}$ is orthogonal to $\mathbf{y}$.

To say that $\mathbf{x}^{\|}$ is a scalar multiple of $\mathbf{y}$ means that we can write $\mathbf{x}^{\|} = c\mathbf{y}$ for some scalar c. Now, assuming such an expression exists, we can determine c by taking the dot product of both sides of the equation with $\mathbf{y}$:
$$\mathbf{x} \cdot \mathbf{y} = (\mathbf{x}^{\|} + \mathbf{x}^{\perp}) \cdot \mathbf{y} = (\mathbf{x}^{\|} \cdot \mathbf{y}) + (\mathbf{x}^{\perp} \cdot \mathbf{y}) = \mathbf{x}^{\|} \cdot \mathbf{y} = (c\mathbf{y}) \cdot \mathbf{y} = c\|\mathbf{y}\|^2.$$

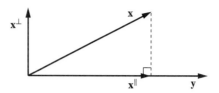

Figure 2.2

This means that

$$c = \frac{\mathbf{x} \cdot \mathbf{y}}{\|\mathbf{y}\|^2}, \quad \text{and so} \quad \mathbf{x}^\| = \frac{\mathbf{x} \cdot \mathbf{y}}{\|\mathbf{y}\|^2}\mathbf{y}.$$

The vector $\mathbf{x}^\|$ is called the *projection of* $\mathbf{x}$ *onto* $\mathbf{y}$, written $\text{proj}_\mathbf{y}\mathbf{x}$.

The fastidious reader may be puzzled by the logic here. We have apparently assumed that we can write $\mathbf{x} = \mathbf{x}^\| + \mathbf{x}^\perp$ in order to prove that we can do so. Of course, as it stands, this is not fair. Here's how we fix it. We now *define*

$$\mathbf{x}^\| = \frac{\mathbf{x} \cdot \mathbf{y}}{\|\mathbf{y}\|^2}\mathbf{y}$$

$$\mathbf{x}^\perp = \mathbf{x} - \frac{\mathbf{x} \cdot \mathbf{y}}{\|\mathbf{y}\|^2}\mathbf{y}.$$

Obviously, $\mathbf{x}^\| + \mathbf{x}^\perp = \mathbf{x}$ and $\mathbf{x}^\|$ is a scalar multiple of $\mathbf{y}$. All we need to check is that $\mathbf{x}^\perp$ is in fact orthogonal to $\mathbf{y}$. Well,

$$\mathbf{x}^\perp \cdot \mathbf{y} = \left(\mathbf{x} - \frac{\mathbf{x} \cdot \mathbf{y}}{\|\mathbf{y}\|^2}\mathbf{y}\right) \cdot \mathbf{y} = \mathbf{x} \cdot \mathbf{y} - \frac{\mathbf{x} \cdot \mathbf{y}}{\|\mathbf{y}\|^2}\mathbf{y} \cdot \mathbf{y} = \mathbf{x} \cdot \mathbf{y} - \frac{\mathbf{x} \cdot \mathbf{y}}{\|\mathbf{y}\|^2}\|\mathbf{y}\|^2 = \mathbf{x} \cdot \mathbf{y} - \mathbf{x} \cdot \mathbf{y} = 0,$$

as required. Note, moreover, that $\mathbf{x}^\|$ is the *unique* multiple of $\mathbf{y}$ that satisfies the equation $(\mathbf{x} - \mathbf{x}^\|) \cdot \mathbf{y} = 0$.

▶ **EXAMPLE 1**

Let $\mathbf{x} = \begin{bmatrix} 2 \\ 3 \\ 1 \end{bmatrix}$ and $\mathbf{y} = \begin{bmatrix} -1 \\ 1 \\ 1 \end{bmatrix}$. Then

$$\mathbf{x}^\| = \frac{\mathbf{x} \cdot \mathbf{y}}{\|\mathbf{y}\|^2}\mathbf{y} = \frac{\begin{bmatrix} 2 \\ 3 \\ 1 \end{bmatrix} \cdot \begin{bmatrix} -1 \\ 1 \\ 1 \end{bmatrix}}{\left\| \begin{bmatrix} -1 \\ 1 \\ 1 \end{bmatrix} \right\|^2} \begin{bmatrix} -1 \\ 1 \\ 1 \end{bmatrix} = \frac{2}{3}\begin{bmatrix} -1 \\ 1 \\ 1 \end{bmatrix} \quad \text{and}$$

$$\mathbf{x}^\perp = \begin{bmatrix} 2 \\ 3 \\ 1 \end{bmatrix} - \frac{2}{3}\begin{bmatrix} -1 \\ 1 \\ 1 \end{bmatrix} = \begin{bmatrix} \frac{8}{3} \\ \frac{7}{3} \\ \frac{1}{3} \end{bmatrix}.$$

To double-check, we compute $\mathbf{x}^\perp \cdot \mathbf{y} = \begin{bmatrix} 8/3 \\ 7/3 \\ 1/3 \end{bmatrix} \cdot \begin{bmatrix} -1 \\ 1 \\ 1 \end{bmatrix} = 0$, as it should be.

Suppose $\mathbf{x}, \mathbf{y} \in \mathbb{R}^2$. We shall see next that the formula for the projection of $\mathbf{x}$ onto $\mathbf{y}$ enables us to calculate the *angle* between the vectors $\mathbf{x}$ and $\mathbf{y}$. Consider the right triangle in Figure 2.3; let θ denote the angle between the vectors $\mathbf{x}$ and $\mathbf{y}$. Remembering that the cosine of an angle is the ratio of the *signed* length of the adjacent side to the length of the hypotenuse, we see that

$$\cos\theta = \frac{\text{signed length of } \mathbf{x}^\|}{\text{length of } \mathbf{x}} = \frac{c\|\mathbf{y}\|}{\|\mathbf{x}\|} = \frac{\frac{\mathbf{x}\cdot\mathbf{y}}{\|\mathbf{y}\|^2}\|\mathbf{y}\|}{\|\mathbf{x}\|} = \frac{\mathbf{x}\cdot\mathbf{y}}{\|\mathbf{x}\|\|\mathbf{y}\|}.$$

This, then, is the geometric interpretation of the dot product:

$$\boxed{\mathbf{x}\cdot\mathbf{y} = \|\mathbf{x}\|\|\mathbf{y}\|\cos\theta.}$$

Will this formula still make sense even when $\mathbf{x}, \mathbf{y} \in \mathbb{R}^n$? Geometrically, we simply restrict our attention to the plane spanned by $\mathbf{x}$ and $\mathbf{y}$ and measure the angle θ in that plane, and so we blithely make the

Definition Let $\mathbf{x}$ and $\mathbf{y}$ be nonzero vectors in $\mathbb{R}^n$. We define the *angle* between them to be the unique θ satisfying $0 \le \theta \le \pi$ so that

$$\cos\theta = \frac{\mathbf{x}\cdot\mathbf{y}}{\|\mathbf{x}\|\|\mathbf{y}\|}.$$

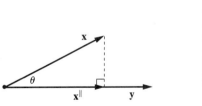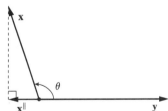

Figure 2.3

Since our geometric intuition may be misleading in $\mathbb{R}^n$, we should check *algebraically* that this definition makes sense. Since $|\cos\theta| \le 1$, the following result gives us what is needed.

Proposition 2.3 (Cauchy-Schwarz Inequality) *If* $\mathbf{x}, \mathbf{y} \in \mathbb{R}^n$, *then*

$$|\mathbf{x}\cdot\mathbf{y}| \le \|\mathbf{x}\|\|\mathbf{y}\|.$$

Moreover, equality holds if and only if one of the vectors is a scalar multiple of the other.

Proof If $\mathbf{y} = \mathbf{0}$, then there's nothing to prove. If $\mathbf{y} \neq \mathbf{0}$, then we observe that the quadratic function of t given by

$$g(t) = \|\mathbf{x} + t\mathbf{y}\|^2 = \|\mathbf{x}\|^2 + 2t\mathbf{x} \cdot \mathbf{y} + t^2\|\mathbf{y}\|^2$$

takes its minimum at $t_0 = -\dfrac{\mathbf{x} \cdot \mathbf{y}}{\|\mathbf{y}\|^2}$. The minimum value

$$g(t_0) = \|\mathbf{x}\|^2 - 2\frac{(\mathbf{x} \cdot \mathbf{y})^2}{\|\mathbf{y}\|^2} + \frac{(\mathbf{x} \cdot \mathbf{y})^2}{\|\mathbf{y}\|^2} = \|\mathbf{x}\|^2 - \frac{(\mathbf{x} \cdot \mathbf{y})^2}{\|\mathbf{y}\|^2}$$

is necessarily nonnegative, so

$$(\mathbf{x} \cdot \mathbf{y})^2 \le \|\mathbf{x}\|^2 \|\mathbf{y}\|^2,$$

and, since square root preserves inequality,

$$|\mathbf{x} \cdot \mathbf{y}| \le \|\mathbf{x}\| \|\mathbf{y}\|,$$

as desired. Equality holds if and only if $\mathbf{x} + t\mathbf{y} = \mathbf{0}$ for some scalar t. (See Exercise 9 for a discussion of how this proof relates to our formula for $\mathrm{proj}_\mathbf{y}\mathbf{x}$ above.) ∎

One of the most useful applications of this result is the famed *triangle inequality*, which tells us that the sum of the lengths of two sides of a triangle cannot be less than the length of the third.

Corollary 2.4 (Triangle Inequality) *For any vectors* $\mathbf{x}, \mathbf{y} \in \mathbb{R}^n$, *we have* $\|\mathbf{x} + \mathbf{y}\| \le \|\mathbf{x}\| + \|\mathbf{y}\|$.

Proof By Corollary 2.2 and Proposition 2.3 we have

$$\|\mathbf{x} + \mathbf{y}\|^2 = \|\mathbf{x}\|^2 + 2\mathbf{x} \cdot \mathbf{y} + \|\mathbf{y}\|^2 \le \|\mathbf{x}\|^2 + 2\|\mathbf{x}\|\|\mathbf{y}\| + \|\mathbf{y}\|^2 = (\|\mathbf{x}\| + \|\mathbf{y}\|)^2.$$

Since square root preserves inequality, we conclude that $\|\mathbf{x} + \mathbf{y}\| \le \|\mathbf{x}\| + \|\mathbf{y}\|$, as desired. ∎

Remark The dot product also arises in situations removed from geometry. The economist introduces the *commodity vector* $\mathbf{x}$, whose entries are the quantities of various commodities that happen to be of interest and the *price vector* $\mathbf{p}$. For example, we might consider

$$\mathbf{x} = \begin{bmatrix} x_1 \\ x_2 \\ x_3 \\ x_4 \\ x_5 \end{bmatrix} \quad \text{and} \quad \mathbf{p} = \begin{bmatrix} p_1 \\ p_2 \\ p_3 \\ p_4 \\ p_5 \end{bmatrix} \in \mathbb{R}^5,$$

where x_1 represents the number of pounds of flour, x_2 the number of dozens of eggs, x_3 the number of pounds of chocolate chips, x_4 the number of pounds of walnuts, and x_5 the number of pounds of butter needed to produce a certain massive quantity of chocolate chip cookies, and p_i is the price (in dollars) of a unit of the ith commodity (e.g., p_2 is the price of a dozen eggs). Then it is easy to see that

$$\mathbf{p} \cdot \mathbf{x} = p_1 x_1 + p_2 x_2 + p_3 x_3 + p_4 x_4 + p_5 x_5$$

is the total cost of producing the massive quantity of cookies. (To be realistic, we might also want to include x_6 as the number of hours of labor, with corresponding hourly wage p_6.) We will return to this interpretation in Section 4.

▶ EXERCISES 1.2

1. For each of the following pairs of vectors $\mathbf{x}$ and $\mathbf{y}$, calculate $\mathbf{x} \cdot \mathbf{y}$ and the angle θ between the vectors.

(a) $\mathbf{x} = \begin{bmatrix} 2 \\ 5 \end{bmatrix}, \mathbf{y} = \begin{bmatrix} -5 \\ 2 \end{bmatrix}$

(b) $\mathbf{x} = \begin{bmatrix} 2 \\ 1 \end{bmatrix}, \mathbf{y} = \begin{bmatrix} -1 \\ 1 \end{bmatrix}$

*(c) $\mathbf{x} = \begin{bmatrix} 1 \\ 8 \end{bmatrix}, \mathbf{y} = \begin{bmatrix} 7 \\ -4 \end{bmatrix}$

(d) $\mathbf{x} = \begin{bmatrix} 1 \\ 4 \\ -3 \end{bmatrix}, \mathbf{y} = \begin{bmatrix} 5 \\ 1 \\ 3 \end{bmatrix}$

(e) $\mathbf{x} = \begin{bmatrix} 1 \\ -1 \\ 6 \end{bmatrix}, \mathbf{y} = \begin{bmatrix} 5 \\ 3 \\ 2 \end{bmatrix}$

*(f) $\mathbf{x} = \begin{bmatrix} 3 \\ -4 \\ 5 \end{bmatrix}, \mathbf{y} = \begin{bmatrix} -1 \\ 0 \\ 1 \end{bmatrix}$

(g) $\mathbf{x} = \begin{bmatrix} 1 \\ 1 \\ 1 \\ 1 \end{bmatrix}, \mathbf{y} = \begin{bmatrix} 1 \\ -3 \\ -1 \\ 5 \end{bmatrix}$

*2. For each pair of vectors in Exercise 1, calculate $\text{proj}_\mathbf{y} \mathbf{x}$ and $\text{proj}_\mathbf{x} \mathbf{y}$.

*3. Find the angle between the long diagonal of a cube and a face diagonal.

4. Find the angle that the long diagonal of a $3 \times 4 \times 5$ rectangular box makes with the longest edge.

5. Suppose $\mathbf{x}, \mathbf{y} \in \mathbb{R}^n$, $\|\mathbf{x}\| = 2$, $\|\mathbf{y}\| = 1$, and the angle θ between $\mathbf{x}$ and $\mathbf{y}$ is $\theta = \arccos(1/4)$. Prove that the vectors $\mathbf{x} - 3\mathbf{y}$ and $\mathbf{x} + \mathbf{y}$ are orthogonal.

6. Suppose $\mathbf{x}, \mathbf{y}, \mathbf{z} \in \mathbb{R}^2$ are unit vectors satisfying $\mathbf{x} + \mathbf{y} + \mathbf{z} = \mathbf{0}$. What can you say about the angles between each pair?

7. Let $\mathbf{e}_1 = \begin{bmatrix} 1 \\ 0 \\ 0 \end{bmatrix}, \mathbf{e}_2 = \begin{bmatrix} 0 \\ 1 \\ 0 \end{bmatrix}$, and $\mathbf{e}_3 = \begin{bmatrix} 0 \\ 0 \\ 1 \end{bmatrix}$ be the so-called *standard basis vectors* for $\mathbb{R}^3$. Let $\mathbf{x} \in \mathbb{R}^3$ be a nonzero vector. For $i = 1, 2, 3$, let θ_i denote the angle between $\mathbf{x}$ and $\mathbf{e}_i$. Compute $\cos^2 \theta_1 + \cos^2 \theta_2 + \cos^2 \theta_3$.

*8. Let $\mathbf{x} = \begin{bmatrix} 1 \\ 1 \\ 1 \\ \vdots \\ 1 \end{bmatrix}$ and $\mathbf{y} = \begin{bmatrix} 1 \\ 2 \\ 3 \\ \vdots \\ n \end{bmatrix} \in \mathbb{R}^n$. Let θ_n be the angle between $\mathbf{x}$ and $\mathbf{y}$ in $\mathbb{R}^n$. Find $\lim_{n \to \infty} \theta_n$.

(Hint: You may need to recall the formulas for $1 + 2 + \cdots + n$ and $1^2 + 2^2 + \cdots + n^2$ from your beginning calculus course.)

9. With regard to the proof of Proposition 2.3, how is $t_0\mathbf{y}$ related to $\mathbf{x}^{\|}$? What does this say about $\text{proj}_\mathbf{y}\mathbf{x}$?

10. Use vector methods to prove that a parallelogram is a rectangle if and only if its diagonals have the same length.

11. Use the fundamental properties of the dot product to prove that
$$\|\mathbf{x}+\mathbf{y}\|^2 + \|\mathbf{x}-\mathbf{y}\|^2 = 2\left(\|\mathbf{x}\|^2 + \|\mathbf{y}\|^2\right).$$
Interpret the result geometrically.

*12. Use the dot product to prove the law of cosines: As shown in Figure 2.4,
$$c^2 = a^2 + b^2 - 2ab\cos\theta.$$

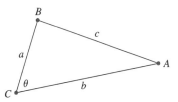

Figure 2.4

13. Use vector methods to prove that the diagonals of a parallelogram are orthogonal if and only if the parallelogram is a rhombus (i.e., has all sides of equal length).

♯14. Use vector methods to prove that a triangle inscribed in a circle and having a diameter as one of its sides must be a right triangle. (Hint: See Figure 2.5.)

   Geometric challenge: More generally, given two points A and B in the plane, what is the locus of points X so that $\angle AXB$ has a fixed measure?

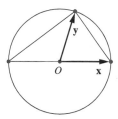

Figure 2.5

15. (a) Let $\mathbf{y} \in \mathbb{R}^n$. If $\mathbf{x}\cdot\mathbf{y} = 0$ for *all* $\mathbf{x} \in \mathbb{R}^n$, then prove that $\mathbf{y} = \mathbf{0}$.
    (b) Suppose $\mathbf{y}, \mathbf{z} \in \mathbb{R}^n$ and $\mathbf{x}\cdot\mathbf{y} = \mathbf{x}\cdot\mathbf{z}$ for all $\mathbf{x} \in \mathbb{R}^n$. What can you conclude?

16. If $\mathbf{x} = \begin{bmatrix} x_1 \\ x_2 \end{bmatrix} \in \mathbb{R}^2$, set $\rho(\mathbf{x}) = \begin{bmatrix} -x_2 \\ x_1 \end{bmatrix}$.
    (a) Check that $\rho(\mathbf{x})$ is orthogonal to $\mathbf{x}$; indeed, $\rho(\mathbf{x})$ is obtained by rotating $\mathbf{x}$ an angle $\pi/2$ counterclockwise.
    (b) Given $\mathbf{x}, \mathbf{y} \in \mathbb{R}^2$, prove that $\mathbf{x}\cdot\rho(\mathbf{y}) = -\rho(\mathbf{x})\cdot\mathbf{y}$. Interpret this statement geometrically.

♯17. Prove that for any vectors $\mathbf{x}, \mathbf{y} \in \mathbb{R}^n$, we have $\|\mathbf{x}\| - \|\mathbf{y}\| \leq \|\mathbf{x}-\mathbf{y}\|$. Deduce that $\big|\|\mathbf{x}\| - \|\mathbf{y}\|\big| \leq \|\mathbf{x}-\mathbf{y}\|$. (Hint: Apply the result of Corollary 2.4 directly.)

18. Use the Cauchy-Schwarz inequality to solve the following max/min problem: If the (long) diagonal of a rectangular box has length c, what is the greatest the sum of the length, width, and height of the box can be? For what shape box does the maximum occur?

19. Give an alternative proof of the Cauchy-Schwarz inequality, as follows. Let $a = \|\mathbf{x}\|$, $b = \|\mathbf{y}\|$, and deduce from $\|b\mathbf{x} - a\mathbf{y}\|^2 \geq 0$ that $\mathbf{x} \cdot \mathbf{y} \leq ab$. Now how do you show that $|\mathbf{x} \cdot \mathbf{y}| \leq ab$? When does equality hold?

♯20. (a) Let $\mathbf{x}$ and $\mathbf{y}$ be vectors with $\|\mathbf{x}\| = \|\mathbf{y}\|$. Prove that the vector $\mathbf{x} + \mathbf{y}$ bisects the angle between $\mathbf{x}$ and $\mathbf{y}$.

(b) More generally, if $\mathbf{x}$ and $\mathbf{y}$ are arbitrary nonzero vectors, let $a = \|\mathbf{x}\|$ and $b = \|\mathbf{y}\|$. Prove that the vector $b\mathbf{x} + a\mathbf{y}$ bisects the angle between $\mathbf{x}$ and $\mathbf{y}$.

21. Use vector methods to prove that the diagonals of a parallelogram bisect the vertex angles if and only if the parallelogram is a rhombus.

22. Given $\triangle ABC$ with D on $\overline{BC}$ as shown in Figure 2.6. Prove that if $\overrightarrow{AD}$ bisects $\angle BAC$, then $\|\overrightarrow{BD}\|/\|\overrightarrow{CD}\| = \|\overrightarrow{AB}\|/\|\overrightarrow{AC}\|$. (Hint: Use Exercise 20b. Let $\mathbf{x} = \overrightarrow{AB}$ and $\mathbf{y} = \overrightarrow{AC}$; give two expressions for $\overrightarrow{AD}$ in terms of $\mathbf{x}$ and $\mathbf{y}$ and use Exercise 1.1.10.)

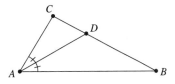

Figure 2.6

23. Use vector methods to prove that the angle bisectors of a triangle have a common point. (Hint: Given $\triangle OAB$, let $\mathbf{x} = \overrightarrow{OA}$, $\mathbf{y} = \overrightarrow{OB}$, $a = \|\overrightarrow{OA}\|$, $b = \|\overrightarrow{OB}\|$, and $c = \|\overrightarrow{AB}\|$. If we define the point P by $\overrightarrow{OP} = \frac{1}{a+b+c}(b\mathbf{x} + a\mathbf{y})$, use Exercise 20b to show that P lies on all three angle bisectors.)

24. Use vector methods to prove that the altitudes of a triangle have a common point. Recall that altitudes of a triangle are the lines passing through a vertex and perpendicular to the opposite side. (Hint: See Figure 2.7. Let C be the point of intersection of the altitude from B to $\overleftrightarrow{OA}$ and the altitude from A to $\overleftrightarrow{OB}$. Prove that $\overrightarrow{OC}$ is orthogonal to $\overrightarrow{AB}$.)

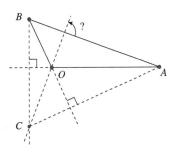

Figure 2.7

25. Use vector methods to prove that the perpendicular bisectors of the sides of a triangle intersect in a point, as follows. Assume the triangle OAB has one vertex at the origin, and let $\mathbf{x} = \overrightarrow{OA}$ and $\mathbf{y} = \overrightarrow{OB}$.

(a) Let **z** be the point of intersection of the perpendicular bisectors of $\overline{OA}$ and $\overline{OB}$. Prove that (using the notation of Exercise 16)

$$\mathbf{z} = \tfrac{1}{2}\mathbf{x} + c\rho(\mathbf{x}), \quad \text{where} \quad c = \frac{\|\mathbf{y}\|^2 - \mathbf{x} \cdot \mathbf{y}}{2\rho(\mathbf{x}) \cdot \mathbf{y}}.$$

(b) Show that **z** lies on the perpendicular bisector of $\overline{AB}$. (Hint: What is the dot product of $\mathbf{z} - \tfrac{1}{2}(\mathbf{x}+\mathbf{y})$ with $\mathbf{y}-\mathbf{x}$?)

26. Let P be the intersection of the medians of $\triangle OAB$ (see Proposition 1.1), Q the intersection of its altitudes (see Exercise 24), and R the intersection of the perpendicular bisectors of its sides (see Exercise 25). Show that P, Q, and R are collinear and that P is two-thirds of the way from Q to R. Does the intersection of the angle bisectors (see Exercise 23) lie on this line as well?

3 SUBSPACES OF $\mathbb{R}^n$

As we proceed in our study of "linear objects," it is fundamental to concentrate on subsets of $\mathbb{R}^n$ that are generalizations of lines and planes through the origin.

Definition A set $V \subset \mathbb{R}^n$ (a *subset* of $\mathbb{R}^n$) is called a *subspace* of $\mathbb{R}^n$ if it satisfies the following properties:

1. $\mathbf{0} \in V$ (the zero vector belongs to V);
2. whenever $\mathbf{v} \in V$ and $c \in \mathbb{R}$, we have $c\mathbf{v} \in V$ (V is closed under scalar multiplication);
3. whenever $\mathbf{v}, \mathbf{w} \in V$, we have $\mathbf{v} + \mathbf{w} \in V$ (V is closed under addition).

EXAMPLE 1

Let's begin with some familiar examples.

a. The *trivial subspace* consisting of just the zero vector $\mathbf{0} \in \mathbb{R}^n$ is a subspace since $c\mathbf{0} = \mathbf{0}$ for any scalar c and $\mathbf{0} + \mathbf{0} = \mathbf{0}$.

b. $\mathbb{R}^n$ itself is likewise a subspace of $\mathbb{R}^n$.

c. Fix a nonzero vector $\mathbf{u} \in \mathbb{R}^n$, and consider

$$\ell = \{\mathbf{x} \in \mathbb{R}^n : \mathbf{x} = t\mathbf{u} \text{ for some } t \in \mathbb{R}\}.$$

We check that the three criteria hold:

1. Setting $t = 0$, we see that $\mathbf{0} \in \ell$.
2. If $\mathbf{v} \in \ell$ and $c \in \mathbb{R}$, then $\mathbf{v} = t\mathbf{u}$ for some $t \in \mathbb{R}$, and so $c\mathbf{v} = c(t\mathbf{u}) = (ct)\mathbf{u}$, which is again a scalar multiple of $\mathbf{u}$ and hence an element of ℓ.
3. If $\mathbf{v}, \mathbf{w} \in \ell$, this means that $\mathbf{v} = s\mathbf{u}$ and $\mathbf{w} = t\mathbf{u}$ for some scalars s and t. Then $\mathbf{v} + \mathbf{w} = s\mathbf{u} + t\mathbf{u} = (s+t)\mathbf{u}$, so $\mathbf{v} + \mathbf{w} \in \ell$, as needed.

ℓ is called a *line* through the origin.

d. Fix two nonparallel vectors $\mathbf{u}$ and $\mathbf{v} \in \mathbb{R}^n$. Set

$$\mathcal{P} = \{\mathbf{x} \in \mathbb{R}^n : \mathbf{x} = s\mathbf{u} + t\mathbf{v} \text{ for some } s, t \in \mathbb{R}\},$$

as shown in Figure 3.1. $\mathcal{P}$ is called a *plane* through the origin. To see that $\mathcal{P}$ is a subspace, we do the obligatory checks:

1. Setting s and $t = 0$, we see that $\mathbf{0} = 0\mathbf{u} + 0\mathbf{v}$, so $\mathbf{0} \in \mathcal{P}$.
2. Suppose $\mathbf{x} \in \mathcal{P}$ and $c \in \mathbb{R}$. Then $\mathbf{x} = s\mathbf{u} + t\mathbf{v}$ for some scalars s and t, and $c\mathbf{x} = c(s\mathbf{u} + t\mathbf{v}) = (cs)\mathbf{u} + (ct)\mathbf{v}$, so $c\mathbf{x} \in \mathcal{P}$ as well.
3. Suppose $\mathbf{x}, \mathbf{y} \in \mathcal{P}$. This means that $\mathbf{x} = s\mathbf{u} + t\mathbf{v}$ for some scalars s and t, and $\mathbf{y} = s'\mathbf{u} + t'\mathbf{v}$ for some scalars s' and t'. Then
$$\mathbf{x} + \mathbf{y} = (s\mathbf{u} + t\mathbf{v}) + (s'\mathbf{u} + t'\mathbf{v}) = (s + s')\mathbf{u} + (t + t')\mathbf{v},$$
so $\mathbf{x} + \mathbf{y} \in \mathcal{P}$, as required.

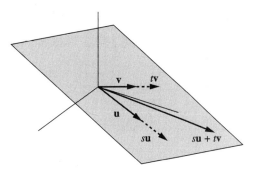

Figure 3.1

e. Fix a nonzero vector $\mathbf{A} \in \mathbb{R}^n$, and consider
$$V = \{\mathbf{x} \in \mathbb{R}^n : \mathbf{A} \cdot \mathbf{x} = 0\}.$$

V consists of all vectors orthogonal to the given vector $\mathbf{A}$, as pictured in Figure 3.2. We check once again that the three criteria hold:

1. Since $\mathbf{A} \cdot \mathbf{0} = 0$, we know that $\mathbf{0} \in V$.
2. Suppose $\mathbf{v} \in V$ and $c \in \mathbb{R}$. Then $\mathbf{A} \cdot (c\mathbf{v}) = c(\mathbf{A} \cdot \mathbf{v}) = 0$, so $c\mathbf{v} \in V$.

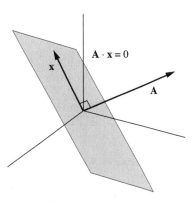

Figure 3.2

3. Suppose $\mathbf{v}, \mathbf{w} \in V$. Then $\mathbf{A} \cdot (\mathbf{v} + \mathbf{w}) = (\mathbf{A} \cdot \mathbf{v}) + (\mathbf{A} \cdot \mathbf{w}) = 0 + 0 = 0$, so $\mathbf{v} + \mathbf{w} \in V$, as required.

Thus, V is a subspace of $\mathbb{R}^n$. We call V a *hyperplane* in $\mathbb{R}^n$, having *normal vector* $\mathbf{A}$. More generally, given any collection of vectors $\mathbf{A}_1, \ldots, \mathbf{A}_m \in \mathbb{R}^n$, the set of solutions of the homogeneous system of linear equations

$$\mathbf{A}_1 \cdot \mathbf{x} = 0, \quad \mathbf{A}_2 \cdot \mathbf{x} = 0, \quad \ldots, \quad \mathbf{A}_m \cdot \mathbf{x} = 0$$

forms a subspace of $\mathbb{R}^n$. ◀

▶ **EXAMPLE 2**

Let's consider next a few subsets of $\mathbb{R}^2$, as pictured in Figure 3.3, that are *not* subspaces.

a. $S = \left\{ \begin{bmatrix} x_1 \\ x_2 \end{bmatrix} \in \mathbb{R}^2 : x_2 = 2x_1 + 1 \right\}$ is not a subspace. All three criteria fail, but it suffices to point out $\mathbf{0} \notin S$.

b. $S = \left\{ \begin{bmatrix} x_1 \\ x_2 \end{bmatrix} \in \mathbb{R}^2 : x_1 x_2 = 0 \right\}$ is not a subspace. Each of the vectors $\mathbf{v} = \begin{bmatrix} 1 \\ 0 \end{bmatrix}$ and $\mathbf{w} = \begin{bmatrix} 0 \\ 1 \end{bmatrix}$ lies in S, and yet their sum $\mathbf{v} + \mathbf{w} = \begin{bmatrix} 1 \\ 1 \end{bmatrix}$ does not.

c. $S = \left\{ \begin{bmatrix} x_1 \\ x_2 \end{bmatrix} \in \mathbb{R}^2 : x_2 \geq 0 \right\}$ is not a subspace. The vector $\mathbf{v} = \begin{bmatrix} 0 \\ 1 \end{bmatrix}$ lies in S, and yet any negative scalar multiple of it, e.g., $(-2)\mathbf{v} = \begin{bmatrix} 0 \\ -2 \end{bmatrix}$, does not. ◀

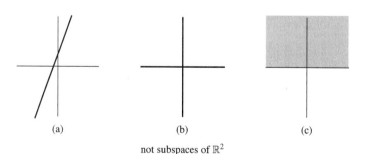

(a) (b) (c)

not subspaces of $\mathbb{R}^2$

Figure 3.3

Given a collection of vectors in $\mathbb{R}^n$, it is natural to try to "build" a subspace from them. We begin with some crucial definitions.

Definition Let $\mathbf{v}_1, \ldots, \mathbf{v}_k \in \mathbb{R}^n$. If $c_1, \ldots, c_k \in \mathbb{R}$, the vector

$$\mathbf{v} = c_1 \mathbf{v}_1 + c_2 \mathbf{v}_2 + \cdots + c_k \mathbf{v}_k$$

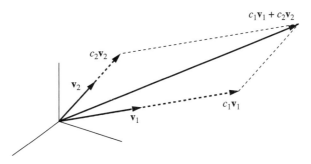

Figure 3.4

(as illustrated in Figure 3.4) is called a *linear combination* of $\mathbf{v}_1, \ldots, \mathbf{v}_k$. The set of all linear combinations of $\mathbf{v}_1, \ldots, \mathbf{v}_k$ is called their *span*, denoted $\mathrm{Span}(\mathbf{v}_1, \ldots, \mathbf{v}_k)$.

Every vector in $\mathbb{R}^n$ can be written as a linear combination of the vectors

$$\mathbf{e}_1 = \begin{bmatrix} 1 \\ 0 \\ \vdots \\ 0 \\ 0 \end{bmatrix}, \quad \mathbf{e}_2 = \begin{bmatrix} 0 \\ 1 \\ \vdots \\ 0 \\ 0 \end{bmatrix}, \quad \ldots, \quad \mathbf{e}_n = \begin{bmatrix} 0 \\ 0 \\ \vdots \\ 0 \\ 1 \end{bmatrix}.$$

The vectors $\mathbf{e}_1, \ldots, \mathbf{e}_n$ are often called the *standard basis* vectors for $\mathbb{R}^n$. Obviously, given the vector

$$\mathbf{x} = \begin{bmatrix} x_1 \\ \vdots \\ x_n \end{bmatrix}, \quad \text{we have} \quad \mathbf{x} = x_1 \mathbf{e}_1 + x_2 \mathbf{e}_2 + \cdots + x_n \mathbf{e}_n.$$

Proposition 3.1 *Let $\mathbf{v}_1, \ldots, \mathbf{v}_k \in \mathbb{R}^n$. Then $V = \mathrm{Span}(\mathbf{v}_1, \ldots, \mathbf{v}_k)$ is a subspace of $\mathbb{R}^n$.*

Proof We check that all three criteria hold.

1. To see that $\mathbf{0} \in V$, we merely take $c_1 = c_2 = \cdots = c_k = 0$. Then (by Exercise 1.1.13) $c_1 \mathbf{v}_1 + c_2 \mathbf{v}_2 + \cdots + c_k \mathbf{v}_k = 0\mathbf{v}_1 + \cdots + 0\mathbf{v}_k = \mathbf{0} + \cdots + \mathbf{0} = \mathbf{0}$.
2. Suppose $\mathbf{v} \in V$ and $c \in \mathbb{R}$. By definition, there are scalars $c_1, \ldots, c_k$ so that $\mathbf{v} = c_1 \mathbf{v}_1 + c_2 \mathbf{v}_2 + \cdots + c_k \mathbf{v}_k$. Thus,

$$c\mathbf{v} = c(c_1 \mathbf{v}_1 + c_2 \mathbf{v}_2 + \cdots + c_k \mathbf{v}_k) = (cc_1)\mathbf{v}_1 + (cc_2)\mathbf{v}_2 + \cdots + (cc_k)\mathbf{v}_k,$$

which is again a linear combination of $\mathbf{v}_1, \ldots, \mathbf{v}_k$, so $c\mathbf{v} \in V$, as desired.

3. Suppose $\mathbf{v}, \mathbf{w} \in V$. This means there are scalars $c_1, \ldots, c_k$ and $d_1, \ldots, d_k$ so that

$$\mathbf{v} = c_1\mathbf{v}_1 + \cdots + c_k\mathbf{v}_k \text{ and}$$
$$\mathbf{w} = d_1\mathbf{v}_1 + \cdots + d_k\mathbf{v}_k;$$

adding, we obtain

$$\mathbf{v} + \mathbf{w} = (c_1\mathbf{v}_1 + \cdots + c_k\mathbf{v}_k) + (d_1\mathbf{v}_1 + \cdots + d_k\mathbf{v}_k)$$
$$= (c_1 + d_1)\mathbf{v}_1 + \cdots + (c_k + d_k)\mathbf{v}_k,$$

which is again a linear combination of $\mathbf{v}_1, \ldots, \mathbf{v}_k$, hence an element of V.

This completes the verification that V is a subspace of $\mathbb{R}^n$. ∎

Remark Let $V \subset \mathbb{R}^n$ be a subspace and let $\mathbf{v}_1, \ldots, \mathbf{v}_k \in V$. We say that $\mathbf{v}_1, \ldots, \mathbf{v}_k$ *span* V if $\text{Span}(\mathbf{v}_1, \ldots, \mathbf{v}_k) = V$. (The point here is that every vector in V must be a linear combination of the vectors $\mathbf{v}_1, \ldots, \mathbf{v}_k$.) As we shall see in Chapter 4, it takes at least n vectors to span $\mathbb{R}^n$; the smallest number of vectors required to span a given subspace will be a measure of its "size" or "dimension."

▶ **EXAMPLE 3**

The plane

$$\mathcal{P}_1 = \left\{ s \begin{bmatrix} 1 \\ -1 \\ 2 \end{bmatrix} + t \begin{bmatrix} 2 \\ 0 \\ 1 \end{bmatrix} : s, t \in \mathbb{R} \right\}$$

is the span of the vectors

$$\mathbf{v}_1 = \begin{bmatrix} 1 \\ -1 \\ 2 \end{bmatrix} \text{ and } \mathbf{v}_2 = \begin{bmatrix} 2 \\ 0 \\ 1 \end{bmatrix}$$

and is therefore a subspace of $\mathbb{R}^3$. On the other hand, the plane

$$\mathcal{P}_2 = \left\{ \begin{bmatrix} 1 \\ 0 \\ 0 \end{bmatrix} + s \begin{bmatrix} 1 \\ -1 \\ 2 \end{bmatrix} + t \begin{bmatrix} 2 \\ 0 \\ 1 \end{bmatrix} : s, t \in \mathbb{R} \right\}$$

is not a subspace. This is most easily verified by checking that $\mathbf{0} \notin \mathcal{P}_2$, for $\mathbf{0} \in \mathcal{P}_2$ precisely when we can find values of s and t so that

$$\begin{bmatrix} 0 \\ 0 \\ 0 \end{bmatrix} = \begin{bmatrix} 1 \\ 0 \\ 0 \end{bmatrix} + s \begin{bmatrix} 1 \\ -1 \\ 2 \end{bmatrix} + t \begin{bmatrix} 2 \\ 0 \\ 1 \end{bmatrix}.$$

This amounts to the system of equations:

$$s + 2t = -1$$
$$-s = 0$$
$$2s + t = 0,$$

which we easily see has no solution.

A word of warning here: We might have expressed $\mathcal{P}_1$ in the form

$$\left\{ \begin{bmatrix} 1 \\ 1 \\ -1 \end{bmatrix} + s \begin{bmatrix} 1 \\ -1 \\ 2 \end{bmatrix} + t \begin{bmatrix} 2 \\ 0 \\ 1 \end{bmatrix} : s, t \in \mathbb{R} \right\},$$

so that, despite the presence of the "shifting" term, the plane may still pass through the origin. ◂

There are really two different ways in which subspaces of $\mathbb{R}^n$ arise: as being the span of a collection of vectors (the "parametric" approach) or as being the set of solutions of a (homogeneous) system of linear equations (the "implicit" approach). We shall study the connections between the two in detail in Chapter 4.

▶ **EXAMPLE 4**

As the reader can verify, the vector $\mathbf{A} = \begin{bmatrix} -1 \\ 3 \\ 2 \end{bmatrix}$ is orthogonal to both the vectors that span the plane $\mathcal{P}_1$ given in Example 3 above. Thus, every vector in $\mathcal{P}_1$ is orthogonal to $\mathbf{A}$, and we suspect that

$$\mathcal{P}_1 = \{\mathbf{x} \in \mathbb{R}^3 : \mathbf{A} \cdot \mathbf{x} = 0\} = \{\mathbf{x} \in \mathbb{R}^3 : -x_1 + 3x_2 + 2x_3 = 0\}.$$

Strictly speaking, we only know that every vector in $\mathcal{P}_1$ is a solution of this equation. But note that if $\mathbf{x}$ is a solution, then

$$\mathbf{x} = \begin{bmatrix} x_1 \\ x_2 \\ x_3 \end{bmatrix} = \begin{bmatrix} 3x_2 + 2x_3 \\ x_2 \\ x_3 \end{bmatrix} = x_2 \begin{bmatrix} 3 \\ 1 \\ 0 \end{bmatrix} + x_3 \begin{bmatrix} 2 \\ 0 \\ 1 \end{bmatrix} = (-x_2) \begin{bmatrix} 1 \\ -1 \\ 2 \end{bmatrix} + (2x_2 + x_3) \begin{bmatrix} 2 \\ 0 \\ 1 \end{bmatrix},$$

so $\mathbf{x} \in \mathcal{P}_1$ and the two sets are equal.[1] Thus, the discussion of Example 1e gives another justification that $\mathcal{P}_1$ is a subspace of $\mathbb{R}^3$.

On the other hand, one can check, analogously, that

$$\mathcal{P}_2 = \{\mathbf{x} \in \mathbb{R}^3 : -x_1 + 3x_2 + 2x_3 = -1\},$$

and so clearly $\mathbf{0} \notin \mathcal{P}_2$ and $\mathcal{P}_2$ is not a subspace. It is an *affine* plane parallel to $\mathcal{P}_1$. ◂

Definition Let V and W be subspaces of $\mathbb{R}^n$. We say they are *orthogonal subspaces* if every element of V is orthogonal to every element of W, i.e., if

$$\mathbf{v} \cdot \mathbf{w} = 0 \quad \text{for every } \mathbf{v} \in V \text{ and every } \mathbf{w} \in W.$$

---

[1] Ordinarily, the easiest way to establish that two sets are equal is to show that each is a subset of the other.

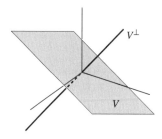

Figure 3.5

As indicated in Figure 3.5, given a subspace $V \subset \mathbb{R}^n$, define
$$V^\perp = \{\mathbf{x} \in \mathbb{R}^n : \mathbf{x} \cdot \mathbf{v} = 0 \text{ for every } \mathbf{v} \in V\}.$$
$V^\perp$ (read "V perp") is called the *orthogonal complement* of V.[2]

Proposition 3.2 *$V^\perp$ is also a subspace of $\mathbb{R}^n$.*

Proof We leave this to the reader in Exercise 4. ∎

▶ **EXAMPLE 5**

Let $V = \mathrm{Span}\left(\begin{bmatrix} 1 \\ 2 \\ 1 \end{bmatrix}\right)$. Then $V^\perp$ is the plane $W = \{x \in \mathbb{R}^3 : x_1 + 2x_2 + x_3 = 0\}$. Now what is the orthogonal complement of W? We suspect it is just the line V, but we will have to wait until Chapter 4 to have the appropriate tools. ◀

If V and W are orthogonal subspaces of $\mathbb{R}^n$, then certainly $W \subset V^\perp$ (why?). Of course, W need not be equal to $V^\perp$: Consider, for example, the x_1-axis and the x_2-axis in $\mathbb{R}^3$.

▶ **EXERCISES 1.3**

*1. Which of the following are subspaces? Justify your answer in each case.
  (a) $\{\mathbf{x} \in \mathbb{R}^2 : x_1 + x_2 = 1\}$
  (b) $\{\mathbf{x} \in \mathbb{R}^3 : \mathbf{x} = \begin{bmatrix} a \\ b \\ a+b \end{bmatrix} \text{ for some } a, b \in \mathbb{R}\}$
  (c) $\{\mathbf{x} \in \mathbb{R}^3 : x_1 + 2x_2 < 0\}$
  (d) $\{\mathbf{x} \in \mathbb{R}^3 : x_1^2 + x_2^2 + x_3^2 = 1\}$
  (e) $\{\mathbf{x} \in \mathbb{R}^3 : x_1^2 + x_2^2 + x_3^2 = 0\}$
  (f) $\{\mathbf{x} \in \mathbb{R}^3 : x_1^2 + x_2^2 + x_3^2 = -1\}$

---

[2] In fact, both this definition and Proposition 3.2 work just fine for any sub*set* $V \subset \mathbb{R}^n$.

(g) $\{\mathbf{x} \in \mathbb{R}^3 : \mathbf{x} = s\begin{bmatrix}2\\1\\1\end{bmatrix} + t\begin{bmatrix}1\\2\\1\end{bmatrix}$ for some $s, t \in \mathbb{R}\}$

(h) $\{\mathbf{x} \in \mathbb{R}^3 : \mathbf{x} = \begin{bmatrix}3\\0\\1\end{bmatrix} + s\begin{bmatrix}2\\1\\1\end{bmatrix} + t\begin{bmatrix}1\\2\\1\end{bmatrix}$ for some $s, t \in \mathbb{R}\}$

(i) $\{\mathbf{x} \in \mathbb{R}^3 : \mathbf{x} = \begin{bmatrix}2\\4\\-1\end{bmatrix} + s\begin{bmatrix}2\\1\\1\end{bmatrix} + t\begin{bmatrix}1\\2\\-1\end{bmatrix}$ for some $s, t \in \mathbb{R}\}$

*2. Criticize the following argument: By Exercise 1.1.13, for any vector $\mathbf{v}$, we have $0\mathbf{v} = \mathbf{0}$. So the first criterion for subspaces is, in fact, a consequence of the second criterion and could therefore be omitted.

♯3. Suppose $\mathbf{x}, \mathbf{v}_1, \ldots, \mathbf{v}_k \in \mathbb{R}^n$ and $\mathbf{x}$ is orthogonal to each of the vectors $\mathbf{v}_1, \ldots, \mathbf{v}_k$. Prove that $\mathbf{x}$ is orthogonal to any linear combination $c_1\mathbf{v}_1 + c_2\mathbf{v}_2 + \cdots + c_k\mathbf{v}_k$.

4. Prove Proposition 3.2.

5. Given vectors $\mathbf{v}_1, \ldots, \mathbf{v}_k \in \mathbb{R}^n$, prove that $V = \mathrm{Span}(\mathbf{v}_1, \ldots, \mathbf{v}_k)$ is the *smallest* subspace containing them all. That is, prove that if $W \subset \mathbb{R}^n$ is a subspace and $\mathbf{v}_1, \ldots, \mathbf{v}_k \in W$, then $V \subset W$.

♯6. (a) Let U and V be subspaces of $\mathbb{R}^n$. Define
$$U \cap V = \{\mathbf{x} \in \mathbb{R}^n : \mathbf{x} \in U \text{ and } \mathbf{x} \in V\}.$$
Prove that $U \cap V$ is a subspace of $\mathbb{R}^n$. Give two examples.

(b) Is $U \cup V = \{\mathbf{x} \in \mathbb{R}^n : \mathbf{x} \in U \text{ or } \mathbf{x} \in V\}$ a subspace of $\mathbb{R}^n$? Give a proof or counterexample.

(c) Let U and V be subspaces of $\mathbb{R}^n$. Define
$$U + V = \{\mathbf{x} \in \mathbb{R}^n : \mathbf{x} = \mathbf{u} + \mathbf{v} \text{ for some } \mathbf{u} \in U \text{ and } \mathbf{v} \in V\}.$$
Prove that $U + V$ is a subspace of $\mathbb{R}^n$. Give two examples.

7. Let $\mathbf{v}_1, \ldots, \mathbf{v}_k \in \mathbb{R}^n$ and let $\mathbf{v} \in \mathbb{R}^n$. Prove that
$$\mathrm{Span}(\mathbf{v}_1, \ldots, \mathbf{v}_k) = \mathrm{Span}(\mathbf{v}_1, \ldots, \mathbf{v}_k, \mathbf{v}) \iff \mathbf{v} \in \mathrm{Span}(\mathbf{v}_1, \ldots, \mathbf{v}_k).$$

♯*8. Let $V \subset \mathbb{R}^n$ be a subspace. Prove that $V \cap V^\perp = \{\mathbf{0}\}$.

♯9. Suppose $U, V \subset \mathbb{R}^n$ are subspaces and $U \subset V$. Prove that $V^\perp \subset U^\perp$.

♯10. Let $V \subset \mathbb{R}^n$ be a subspace. Prove that $V \subset (V^\perp)^\perp$. Do you think more is true?

♯11. Suppose $V = \mathrm{Span}(\mathbf{v}_1, \ldots, \mathbf{v}_k) \subset \mathbb{R}^n$. Show that there are vectors $\mathbf{w}_1, \ldots, \mathbf{w}_k \in V$ that are mutually orthogonal (i.e., $\mathbf{w}_i \cdot \mathbf{w}_j = 0$ whenever $i \neq j$) that also span V. (Hint: Let $\mathbf{w}_1 = \mathbf{v}_1$. Using techniques of Section 2, define $\mathbf{w}_2$ so that $\mathrm{Span}(\mathbf{w}_1, \mathbf{w}_2) = \mathrm{Span}(\mathbf{v}_1, \mathbf{v}_2)$ and $\mathbf{w}_1 \cdot \mathbf{w}_2 = 0$. Continue.)

12. Suppose U and V are subspaces of $\mathbb{R}^n$. Prove that $(U + V)^\perp = U^\perp \cap V^\perp$. (See the footnote on p. 21.)

4 LINEAR TRANSFORMATIONS AND MATRIX ALGEBRA

We are heading toward calculus and the study of functions. As we learned in the case of one variable, differential calculus is based on the idea of the best (affine) linear approximation of a function. Thus, our first brush with functions is with those that are linear.

First we introduce a bit of notation. If X and Y are sets, a *function* $f: X \to Y$ is a rule that assigns to *each* element $x \in X$ a *single* element $y \in Y$; we write $y = f(x)$. We call X the *domain* of f and Y the *range*. The *image* of f is the set of all its values; i.e., $\{y \in Y : y = f(x) \text{ for some } x \in X\}$.

Definition A function $T: \mathbb{R}^n \to \mathbb{R}^m$ is called a *linear transformation* or *linear map* if it satisfies

i. $T(\mathbf{u} + \mathbf{v}) = T(\mathbf{u}) + T(\mathbf{v})$ for all $\mathbf{u}, \mathbf{v} \in \mathbb{R}^n$;

ii. $T(c\mathbf{v}) = cT(\mathbf{v})$ for all $\mathbf{v} \in \mathbb{R}^n$ and scalars c.

If we think visually of T as mapping $\mathbb{R}^n$ to $\mathbb{R}^m$, then we have a diagram like Figure 4.1. The main point of the linearity properties is that the values of T on the standard basis vectors $\mathbf{e}_1, \ldots, \mathbf{e}_n$ completely determine the function T: For suppose $\mathbf{x} = x_1\mathbf{e}_1 + \cdots + x_n\mathbf{e}_n \in \mathbb{R}^n$; then

$$(*) \quad \begin{aligned} T(\mathbf{x}) &= T(x_1\mathbf{e}_1 + \cdots + x_n\mathbf{e}_n) \\ &= T(x_1\mathbf{e}_1) + \cdots + T(x_n\mathbf{e}_n) = x_1 T(\mathbf{e}_1) + \ldots x_n T(\mathbf{e}_n). \end{aligned}$$

In particular, let

$$T(\mathbf{e}_j) = \begin{bmatrix} a_{1j} \\ a_{2j} \\ \vdots \\ a_{mj} \end{bmatrix} \in \mathbb{R}^m;$$

then to T we can naturally associate the $m \times n$ array

$$A = \begin{bmatrix} a_{11} & \cdots & a_{1n} \\ a_{21} & \cdots & a_{2n} \\ \vdots & \ddots & \vdots \\ a_{m1} & \cdots & a_{mn} \end{bmatrix},$$

which we call the *standard matrix* for T. (We will often denote this by $[T]$.) To emphasize: The jth column of A is the vector in $\mathbb{R}^m$ obtained by applying T to the jth standard basis vector, $\mathbf{e}_j$.

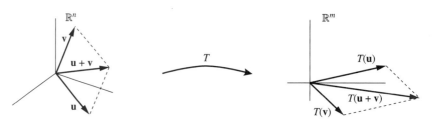

Figure 4.1

EXAMPLE 1

The most basic example of a linear map is the following. Fix $\mathbf{a} \in \mathbb{R}^n$, and define $T: \mathbb{R}^n \to \mathbb{R}$ by $T(\mathbf{x}) = \mathbf{a} \cdot \mathbf{x}$. By Proposition 2.1, we have

$$T(\mathbf{u} + \mathbf{v}) = \mathbf{a} \cdot (\mathbf{u} + \mathbf{v}) = (\mathbf{a} \cdot \mathbf{u}) + (\mathbf{a} \cdot \mathbf{v}) = T(\mathbf{u}) + T(\mathbf{v}), \quad \text{and}$$
$$T(c\mathbf{v}) = \mathbf{a} \cdot (c\mathbf{v}) = c(\mathbf{a} \cdot \mathbf{v}) = cT(\mathbf{v}),$$

as required. Moreover, it is easy to see that

$$\text{if} \quad \mathbf{a} = \begin{bmatrix} a_1 \\ a_2 \\ \vdots \\ a_n \end{bmatrix}, \quad \text{then} \quad [T] = \begin{bmatrix} a_1 & a_2 & \cdots & a_n \end{bmatrix}.$$

EXAMPLE 2

a. Consider the function $T: \mathbb{R}^2 \to \mathbb{R}^2$ defined by rotating vectors in the plane counterclockwise by $90°$. Then it is easy to see from the geometry in Figure 4.2 that

$$T\left(\begin{bmatrix} x_1 \\ x_2 \end{bmatrix}\right) = \begin{bmatrix} -x_2 \\ x_1 \end{bmatrix}.$$

Now the linearity properties can be checked algebraically: If

$$\mathbf{x} = \begin{bmatrix} x_1 \\ x_2 \end{bmatrix} \quad \text{and} \quad \mathbf{y} = \begin{bmatrix} y_1 \\ y_2 \end{bmatrix}$$

are vectors, then

$$T(\mathbf{x} + \mathbf{y}) = T\left(\begin{bmatrix} x_1 + y_1 \\ x_2 + y_2 \end{bmatrix}\right) = \begin{bmatrix} -(x_2 + y_2) \\ x_1 + y_1 \end{bmatrix} = \begin{bmatrix} -x_2 \\ x_1 \end{bmatrix} + \begin{bmatrix} -y_2 \\ y_1 \end{bmatrix} = T(\mathbf{x}) + T(\mathbf{y}),$$

and, even easier,

$$T(c\mathbf{x}) = T\left(c\begin{bmatrix} x_1 \\ x_2 \end{bmatrix}\right) = T\left(\begin{bmatrix} cx_1 \\ cx_2 \end{bmatrix}\right) = \begin{bmatrix} -cx_2 \\ cx_1 \end{bmatrix} = c\begin{bmatrix} -x_2 \\ x_1 \end{bmatrix} = cT(\mathbf{x}),$$

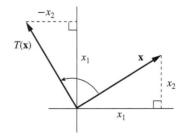

Figure 4.2

as required. The standard matrix for T is

$$[T] = \begin{bmatrix} 0 & -1 \\ 1 & 0 \end{bmatrix}.$$

Better yet, since rotation carries lines through the origin to lines through the origin and triangles to congruent triangles, it is clear on geometric grounds that T must satisfy properties (i) and (ii).

b. Consider the function $T: \mathbb{R}^2 \to \mathbb{R}^2$ defined by reflecting vectors across the line $x_1 = x_2$, as shown in Figure 4.3. (Visualize this as looking at vectors through a mirror along that line.) Once again, we see from the geometry that

$$T\left(\begin{bmatrix} x_1 \\ x_2 \end{bmatrix}\right) = \begin{bmatrix} x_2 \\ x_1 \end{bmatrix},$$

and linearity is obvious algebraically. But it should also be clear on geometric grounds that stretching a vector and then looking at it in the mirror is the same as stretching its mirror image, and likewise for addition of vectors. The standard matrix for T is

$$[T] = \begin{bmatrix} 0 & 1 \\ 1 & 0 \end{bmatrix}.$$

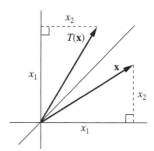

Figure 4.3

c. Consider the linear transformation $T: \mathbb{R}^2 \to \mathbb{R}^2$, whose standard matrix is

$$A = \begin{bmatrix} 1 & 1 \\ 0 & 1 \end{bmatrix}.$$

The effect of T is pictured in Figure 4.4. One might slide a deck of cards in this fashion, and such a motion is called a *shear*.

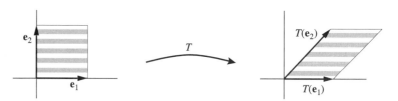

Figure 4.4

d. Consider the function $T\colon \mathbb{R}^3 \to \mathbb{R}^3$ defined by reflecting across the plane $x_3 = 0$. Then $T(\mathbf{e}_1) = \mathbf{e}_1$, $T(\mathbf{e}_2) = \mathbf{e}_2$, and $T(\mathbf{e}_3) = -\mathbf{e}_3$, so the standard matrix for T is

$$\begin{bmatrix} 1 & 0 & 0 \\ 0 & 1 & 0 \\ 0 & 0 & -1 \end{bmatrix}.$$

e. Generalizing part a, we consider rotation of $\mathbb{R}^2$ through the angle θ (given in radians). By the same geometric argument we suggested earlier (see Figure 4.5), this is a linear transformation of $\mathbb{R}^2$. Now, as we can see from Figure 4.6, the standard matrix has as its first column

$$T(\mathbf{e}_1) = \begin{bmatrix} \cos\theta \\ \sin\theta \end{bmatrix}$$

(by the usual definition of $\cos\theta$ and $\sin\theta$, in fact) and as its second

$$T(\mathbf{e}_2) = \begin{bmatrix} -\sin\theta \\ \cos\theta \end{bmatrix}$$

(since $\mathbf{e}_2$ is obtained by rotating $\mathbf{e}_1$ through $\pi/2$, then so is $T(\mathbf{e}_2)$ obtained by rotating $T(\mathbf{e}_1)$ through $\pi/2$). Thus, the standard matrix for T is

$$A_\theta = \begin{bmatrix} \cos\theta & -\sin\theta \\ \sin\theta & \cos\theta \end{bmatrix}.$$

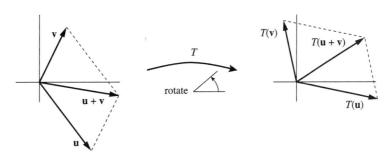

Figure 4.5

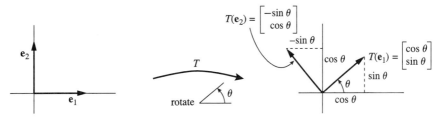

Figure 4.6

f. If $\ell \subset \mathbb{R}^2$ is the line spanned by $\begin{bmatrix} 1 \\ 2 \end{bmatrix}$, then we can consider the linear maps $S, T: \mathbb{R}^2 \to \mathbb{R}^2$ given respectively by projection onto, and reflection across, the line ℓ. Their standard matrices are

$$A = \begin{bmatrix} \frac{1}{5} & \frac{2}{5} \\ \frac{2}{5} & \frac{4}{5} \end{bmatrix} \quad \text{and} \quad B = \begin{bmatrix} -\frac{3}{5} & \frac{4}{5} \\ \frac{4}{5} & \frac{3}{5} \end{bmatrix}.$$

If we consider larger matrices, e.g.,

$$C = \begin{bmatrix} \frac{1}{6} & \frac{1}{3} + \frac{\sqrt{6}}{6} & \frac{1}{6} - \frac{\sqrt{6}}{3} \\ \frac{1}{3} - \frac{\sqrt{6}}{6} & \frac{2}{3} & \frac{1}{3} + \frac{\sqrt{6}}{6} \\ \frac{1}{6} + \frac{\sqrt{6}}{3} & \frac{1}{3} - \frac{\sqrt{6}}{6} & \frac{1}{6} \end{bmatrix},$$

then it seems impossible to discern the geometric nature of the linear map represented by such a matrix.[3] In these examples, the standard "coordinate system" built into matrices just masks the geometry, and as we shall see, the solution is to change our coordinate system. This we do in Chapter 9. ◀

Let $T: \mathbb{R}^n \to \mathbb{R}^m$ be a linear map, and let A be its standard matrix. We want to define the product of the $m \times n$ matrix A with the vector $\mathbf{x} \in \mathbb{R}^n$ in such a way that the vector $T(\mathbf{x}) \in \mathbb{R}^m$ is equal to $A\mathbf{x}$. (We will occasionally denote the linear map defined in this way by μ_A.) In accordance with the formula (∗) on p. 24, we have

$$A\mathbf{x} = T(\mathbf{x}) = \sum_{i=1}^{n} x_i T(\mathbf{e}_i) = \sum_{i=1}^{n} x_i \mathbf{a}_i,$$

where

$$\mathbf{a}_1 = \begin{bmatrix} a_{11} \\ \vdots \\ a_{m1} \end{bmatrix}, \quad \mathbf{a}_2 = \begin{bmatrix} a_{12} \\ \vdots \\ a_{m2} \end{bmatrix}, \dots, \quad \mathbf{a}_n = \begin{bmatrix} a_{1n} \\ \vdots \\ a_{mn} \end{bmatrix} \in \mathbb{R}^m$$

are the column vectors of the matrix A. That is, $A\mathbf{x}$ is the linear combination of the vectors $\mathbf{a}_1, \dots, \mathbf{a}_n$, weighted according to the coordinates of the vector $\mathbf{x}$.

There is, however, an alternative interpretation. Let

$$\mathbf{A}_1 = \begin{bmatrix} a_{11} \\ \vdots \\ a_{1n} \end{bmatrix}, \quad \mathbf{A}_2 = \begin{bmatrix} a_{21} \\ \vdots \\ a_{2n} \end{bmatrix}, \dots, \quad \mathbf{A}_m = \begin{bmatrix} a_{m1} \\ \vdots \\ a_{mn} \end{bmatrix} \in \mathbb{R}^n$$

---

[3] For the curious among you, multiplication by C gives a rotation of $\mathbb{R}^3$ through an angle of $\pi/2$ about the line spanned by $\begin{bmatrix} 1 \\ 2 \\ 1 \end{bmatrix}$. See Exercise 9.2.21.

be the row vectors of the matrix A. Then

$$Ax = \begin{bmatrix} a_{11}x_1 + \cdots + a_{1n}x_n \\ a_{21}x_1 + \cdots + a_{2n}x_n \\ \vdots \\ a_{m1}x_1 + \cdots + a_{mn}x_n \end{bmatrix} = \begin{bmatrix} \mathbf{A}_1 \cdot \mathbf{x} \\ \mathbf{A}_2 \cdot \mathbf{x} \\ \vdots \\ \mathbf{A}_m \cdot \mathbf{x} \end{bmatrix}.$$

As we shall study in great detail in Chapter 4, this allows us to interpret the equation $A\mathbf{x} = \mathbf{y}$ as a system of m linear equations in the variables $x_1, \ldots, x_n$.

4.1 Algebra of Linear Functions

Denote by $\mathcal{M}_{m \times n}$ the set of all $m \times n$ matrices. In an obvious way this set can be identified with $\mathbb{R}^{mn}$ (how?). Indeed, we begin by observing that we can add $m \times n$ matrices and multiply them by scalars, just as we did with vectors.

For future reference, we call a matrix *square* if $m = n$ (i.e., it has equal numbers of rows and columns). We refer to the entries a_{ii}, $i = 1, \ldots, n$, as *diagonal* entries. We call the (square) matrix a *diagonal matrix* if $a_{ij} = 0$ whenever $i \neq j$, i.e., if every nondiagonal entry is 0. A square matrix all of whose entries below the diagonal are 0 is called *upper triangular*; one all of whose entries above the diagonal are 0 is called *lower triangular*.

If $S, T: \mathbb{R}^n \to \mathbb{R}^m$ are linear maps and $c \in \mathbb{R}$, then we can obviously form the linear maps $cT: \mathbb{R}^n \to \mathbb{R}^m$ and $S + T: \mathbb{R}^n \to \mathbb{R}^m$, defined, respectively, by

$$(cT)(\mathbf{x}) = c(T(\mathbf{x}))$$
$$(S + T)(\mathbf{x}) = S(\mathbf{x}) + T(\mathbf{x}).$$

The corresponding algebraic manipulations with matrices are clear: If $A = [a_{ij}]$, $i = 1, \ldots, m$, $j = 1, \ldots, n$, then cA is the matrix whose entries are ca_{ij}:

$$cA = c \begin{bmatrix} a_{11} & \cdots & a_{1n} \\ a_{21} & \cdots & a_{2n} \\ \vdots & \ddots & \vdots \\ a_{m1} & \cdots & a_{mn} \end{bmatrix} = \begin{bmatrix} ca_{11} & \cdots & ca_{1n} \\ ca_{21} & \cdots & ca_{2n} \\ \vdots & \ddots & \vdots \\ ca_{m1} & \cdots & ca_{mn} \end{bmatrix}.$$

Given two matrices A and $B \in \mathcal{M}_{m \times n}$, we define their *sum* entry by entry. In symbols, when

$$A = \begin{bmatrix} a_{11} & \cdots & a_{1n} \\ a_{21} & \cdots & a_{2n} \\ \vdots & \ddots & \vdots \\ a_{m1} & \cdots & a_{mn} \end{bmatrix} \quad \text{and} \quad B = \begin{bmatrix} b_{11} & \cdots & b_{1n} \\ b_{21} & \cdots & b_{2n} \\ \vdots & \ddots & \vdots \\ b_{m1} & \cdots & b_{mn} \end{bmatrix},$$

we define

$$A + B = \begin{bmatrix} a_{11} + b_{11} & \cdots & a_{1n} + b_{1n} \\ a_{21} + b_{21} & \cdots & a_{2n} + b_{2n} \\ \vdots & \ddots & \vdots \\ a_{m1} + b_{m1} & \cdots & a_{mn} + b_{mn} \end{bmatrix}.$$

▶ **EXAMPLE 3**

Let $c = -2$ and

$$A = \begin{bmatrix} 1 & 2 & 3 \\ 2 & 1 & -2 \\ 4 & -1 & 3 \end{bmatrix}, \quad B = \begin{bmatrix} 6 & 4 & -1 \\ -3 & 1 & 1 \\ 0 & 0 & 0 \end{bmatrix}, \quad C = \begin{bmatrix} 1 & 2 & 1 \\ 2 & 1 & 1 \end{bmatrix}.$$

Then

$$cA = \begin{bmatrix} -2 & -4 & -6 \\ -4 & -2 & 4 \\ -8 & 2 & -6 \end{bmatrix}, \quad A + B = \begin{bmatrix} 7 & 6 & 2 \\ -1 & 2 & -1 \\ 4 & -1 & 3 \end{bmatrix},$$

and neither sum $A + C$ nor $B + C$ makes sense since C has a different shape from A and B. (One should not expect to be able to add functions with different domains or ranges.) ◀

Denote by O the *zero matrix*, the $m \times n$ matrix all of whose entries are 0. As the reader can easily check, scalar multiplication of matrices and matrix addition satisfy the same properties as scalar multiplication of vectors and vector addition (see Exercise 1.1.12). We list them here for reference.

Proposition 4.1 Let $A, B, C \in \mathcal{M}_{m \times n}$ and let $c, d \in \mathbb{R}$.

1. $A + B = B + A$.
2. $(A + B) + C = A + (B + C)$.
3. $O + A = A$.
4. There is a matrix $-A$ so that $A + (-A) = O$.
5. $c(dA) = (cd)A$.
6. $c(A + B) = cA + cB$.
7. $(c + d)A = cA + dA$.
8. $1A = A$.

Of all the operations one performs on functions, probably the most powerful is composition. Recall that when $g(x)$ is in the domain of f, we define $(f \circ g)(x) = f(g(x))$. So, suppose we have linear maps $S \colon \mathbb{R}^p \to \mathbb{R}^n$ and $T \colon \mathbb{R}^n \to \mathbb{R}^m$. Then we define $T \circ S \colon \mathbb{R}^p \to \mathbb{R}^m$

by $(T \circ S)(\mathbf{x}) = T(S(\mathbf{x}))$. It is well known that composition of functions is not commutative[4] but *is* associative, inasmuch as

$$((f \circ g) \circ h)(x) = f(g(h(x))) = (f \circ (g \circ h))(x).$$

We want to define matrix multiplication so that it corresponds to the composition of linear maps. Let A be the $m \times n$ matrix representing T and let B be the $n \times p$ matrix representing S. We expect that the $m \times p$ matrix C representing $T \circ S$ can be expressed in terms of A and B. The jth column of C is the vector $(T \circ S)(\mathbf{e}_j) \in \mathbb{R}^m$. Now,

$$T(S(\mathbf{e}_j)) = T \begin{pmatrix} b_{1j} \\ b_{2j} \\ \vdots \\ b_{nj} \end{pmatrix} = b_{1j}\mathbf{a}_1 + b_{2j}\mathbf{a}_2 + \cdots + b_{nj}\mathbf{a}_n,$$

where $\mathbf{a}_1, \ldots, \mathbf{a}_n$ are the column vectors of A. That is, the jth column of C is the product of the matrix A with the vector $\mathbf{b}_j$. So we now make the definition:

Definition Let A be an $m \times n$ matrix and B an $n \times p$ matrix. Their product AB is the $m \times p$ matrix whose j^{th} column is the product of A with the jth column of B. That is, its ij-entry is

$$(AB)_{ij} = a_{i1}b_{1j} + a_{i2}b_{2j} + \cdots + a_{in}b_{nj},$$

i.e., the dot product of the ith row vector of A and the jth column vector of B, both of which are vectors in $\mathbb{R}^n$. Graphically, we have

$$\begin{bmatrix} a_{11} & a_{12} & \cdots & a_{1n} \\ & \vdots & & \\ \boxed{a_{i1} \;\; a_{i2} \;\; \cdots \;\; a_{in}} \\ & \vdots & & \\ a_{m1} & a_{m2} & \cdots & a_{mn} \end{bmatrix} \begin{bmatrix} b_{11} & & b_{1j} & & b_{1p} \\ b_{21} & & b_{2j} & & b_{2p} \\ \vdots & \cdots & \vdots & \cdots & \vdots \\ b_{n1} & & b_{nj} & & b_{np} \end{bmatrix}$$

$$= \begin{bmatrix} \cdots & \cdots & \vdots & \cdots & \cdots \\ \cdots & \cdots & \boxed{(AB)_{ij}} & \cdots & \cdots \\ \cdots & \cdots & \vdots & \cdots & \cdots \end{bmatrix}.$$

We reiterate that in order for the product AB to be defined, the number of *columns* of A must equal the number of *rows* of B.

---

[4]E.g., $\sin(x^2) \neq \sin^2 x$.

▶ **EXAMPLE 4**

If
$$A = \begin{bmatrix} 1 & 3 \\ 2 & -1 \\ 1 & 1 \end{bmatrix} \quad \text{and} \quad B = \begin{bmatrix} 4 & 1 & 0 & -2 \\ -1 & 1 & 5 & 1 \end{bmatrix},$$

then
$$AB = \begin{bmatrix} 1 & 3 \\ 2 & -1 \\ 1 & 1 \end{bmatrix} \begin{bmatrix} 4 & 1 & 0 & -2 \\ -1 & 1 & 5 & 1 \end{bmatrix} = \begin{bmatrix} 1 & 4 & 15 & 1 \\ 9 & 1 & -5 & -5 \\ 3 & 2 & 5 & -1 \end{bmatrix}.$$

Notice also that the product BA does not make sense: B is a 2×4 matrix and A is 3×2, and $4 \ne 3$. ◀

The preceding example brings out an important point about the nature of matrix multiplication: It can happen that the matrix product AB is defined and the product BA is not. Now if A is an $m \times n$ matrix and B is an $n \times m$ matrix, then both products AB and BA make sense: AB is $m \times m$ and BA is $n \times n$. Notice that these are both square matrices, but of different sizes. But even if we start with both A and B as $n \times n$ matrices, the products AB and BA need not be equal.

▶ **EXAMPLE 5**

Let
$$A = \begin{bmatrix} 1 & 2 \\ -3 & 1 \end{bmatrix} \quad \text{and} \quad B = \begin{bmatrix} -1 & 0 \\ 1 & 0 \end{bmatrix}.$$

Then
$$AB = \begin{bmatrix} 1 & 0 \\ 4 & 0 \end{bmatrix}, \quad \text{whereas} \quad BA = \begin{bmatrix} -1 & -2 \\ 1 & 2 \end{bmatrix}. \quad ◀$$

When—and only when—A is a square matrix (i.e., $m = n$), we can multiply A by itself, obtaining $A^2 = AA$, $A^3 = A^2A = AA^2$, etc. If we think of $A\mathbf{x}$ as resulting from $\mathbf{x}$ by performing some geometric procedure, then $(A^2)\mathbf{x}$ should result from performing that procedure twice, $(A^3)\mathbf{x}$ thrice, and so on.

▶ **EXAMPLE 6**

Let
$$A = \begin{bmatrix} \frac{1}{5} & \frac{2}{5} \\ \frac{2}{5} & \frac{4}{5} \end{bmatrix}.$$

Then it is easy to check that $A^2 = A$, so $A^n = A$ for all positive integers n (why?). What is the geometric explanation? Note that

$$A \begin{bmatrix} 1 \\ 0 \end{bmatrix} = \begin{bmatrix} \frac{1}{5} \\ \frac{2}{5} \end{bmatrix} = \frac{1}{5}\begin{bmatrix} 1 \\ 2 \end{bmatrix} \quad \text{and} \quad A \begin{bmatrix} 0 \\ 1 \end{bmatrix} = \begin{bmatrix} \frac{2}{5} \\ \frac{4}{5} \end{bmatrix} = \frac{2}{5}\begin{bmatrix} 1 \\ 2 \end{bmatrix},$$

so that for every $\mathbf{x} \in \mathbb{R}^2$, we see that $A\mathbf{x}$ lies on the line spanned by $\begin{bmatrix} 1 \\ 2 \end{bmatrix}$. Indeed, we can tell more:

$$A \begin{bmatrix} x_1 \\ x_2 \end{bmatrix} = \begin{bmatrix} \frac{1}{5}(x_1 + 2x_2) \\ \frac{2}{5}(x_1 + 2x_2) \end{bmatrix} = \frac{x_1 + 2x_2}{5}\begin{bmatrix} 1 \\ 2 \end{bmatrix} = \frac{\mathbf{x} \cdot \begin{bmatrix} 1 \\ 2 \end{bmatrix}}{\left\| \begin{bmatrix} 1 \\ 2 \end{bmatrix} \right\|^2}\begin{bmatrix} 1 \\ 2 \end{bmatrix}$$

is the projection of $\mathbf{x}$ onto the line spanned by $\begin{bmatrix} 1 \\ 2 \end{bmatrix}$. This explains why $A^2\mathbf{x} = A\mathbf{x}$ for every $\mathbf{x} \in \mathbb{R}^2$: $A^2\mathbf{x} = A(A\mathbf{x})$, and once we've projected the vector $\mathbf{x}$ onto the line, it stays exactly the same.

▶ **EXAMPLE 7**

There is an interesting way to interpret matrix powers in terms of directed graphs. Starting with the matrix

$$A = \begin{bmatrix} 0 & 2 & 1 \\ 1 & 1 & 1 \\ 1 & 0 & 1 \end{bmatrix},$$

we draw a graph with three nodes (vertices) and a_{ij} directed edges (paths) from node i to node j, as shown in Figure 4.7. For example, there are two edges from node 1 to node 2 and none from node 3 to node 2.

We calculate

$$A^2 = \begin{bmatrix} 0 & 2 & 1 \\ 1 & 1 & 1 \\ 1 & 0 & 1 \end{bmatrix} \begin{bmatrix} 0 & 2 & 1 \\ 1 & 1 & 1 \\ 1 & 0 & 1 \end{bmatrix} = \begin{bmatrix} 3 & 2 & 3 \\ 2 & 3 & 3 \\ 1 & 2 & 2 \end{bmatrix},$$

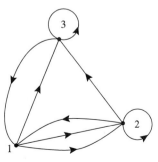

Figure 4.7

$$A^3 = \begin{bmatrix} 5 & 8 & 8 \\ 6 & 7 & 8 \\ 4 & 4 & 5 \end{bmatrix}, \quad \text{and} \quad \ldots$$

$$A^7 = \begin{bmatrix} 272 & 338 & 377 \\ 273 & 337 & 377 \\ 169 & 208 & 233 \end{bmatrix}.$$

For example, the 13-entry of A^2 is

$$(A^2)_{13} = a_{11}a_{13} + a_{12}a_{23} + a_{13}a_{33} = (0)(1) + (2)(1) + (1)(1) = 3.$$

With a bit of thought, the reader will convince herself that the ij-entry of A^2 is the number of "two-step" directed paths from node i to node j. Similarly, the ij-entry of A^n is the number of n-step directed paths from node i to node j. ◀

We have seen that, in general, matrix multiplication is *not* commutative. However, it does have the following crucial properties. Let I_n denote the $n \times n$ matrix with 1's on the diagonal and 0's elsewhere.

Proposition 4.2 *Let A and A' be $m \times n$ matrices; let B and B' be $n \times p$ matrices; let C be a $p \times q$ matrix, and let c be a scalar. Then*

1. $AI_n = A = I_m A$. For this reason, I_n is called the $n \times n$ identity matrix.
2. $(A + A')B = AB + A'B$ and $A(B + B') = AB + AB'$. This is the distributive *property of matrix multiplication over matrix addition.*
3. $(cA)B = c(AB) = A(cB)$.
4. $(AB)C = A(BC)$. This is the associative *property of matrix multiplication.*

Proof These are all immediate from the linear map viewpoint. ■

One of the important concepts is that of the inverse of a function.

Definition Let A be an $n \times n$ matrix. We say A is *invertible* if there is an $n \times n$ matrix B so that

$$AB = BA = I_n.$$

We call B the *inverse* of the matrix A and denote this by $B = A^{-1}$.

If A is the matrix representing the linear transformation $T : \mathbb{R}^n \to \mathbb{R}^n$, then A^{-1} represents the inverse function T^{-1}, which must then also be a linear transformation.

▶ **EXAMPLE 8**

Let

$$A = \begin{bmatrix} 2 & 5 \\ 1 & 3 \end{bmatrix} \quad \text{and} \quad B = \begin{bmatrix} 3 & -5 \\ -1 & 2 \end{bmatrix}.$$

Then $AB = I_2$ and $BA = I_2$, so B is the inverse matrix of A. ◀

EXAMPLE 9

It will be convenient for our future work to have the inverse of a 2×2 matrix

$$A = \begin{bmatrix} a & b \\ c & d \end{bmatrix}.$$

Provided $ad - bc \neq 0$, if we set

$$A^{-1} = \frac{1}{ad-bc}\begin{bmatrix} d & -b \\ -c & a \end{bmatrix},$$

then an easy calculation shows that $AA^{-1} = A^{-1}A = I_2$, as needed. ◂

EXAMPLE 10

It follows immediately from Example 9 that for our rotation matrix

$$A_\theta = \begin{bmatrix} \cos\theta & -\sin\theta \\ \sin\theta & \cos\theta \end{bmatrix}, \quad \text{we have} \quad A_\theta^{-1} = \begin{bmatrix} \cos\theta & \sin\theta \\ -\sin\theta & \cos\theta \end{bmatrix}.$$

Since $\cos(-\theta) = \cos\theta$ and $\sin(-\theta) = -\sin\theta$, we see that this is the matrix $A_{(-\theta)}$. If we think about the corresponding linear maps, this result becomes obvious: To invert (or "undo") a rotation through angle θ, we must rotate through angle $-\theta$. ◂

EXAMPLE 11

As an application of Example 9, we can now show that any two nonparallel vectors $\mathbf{u},\mathbf{v} \in \mathbb{R}^2$ must span $\mathbb{R}^2$. It is easy to check that if $\mathbf{u} = \begin{bmatrix} u_1 \\ u_2 \end{bmatrix}$ and $\mathbf{v} = \begin{bmatrix} v_1 \\ v_2 \end{bmatrix}$ are nonparallel, then $u_1 v_2 - u_2 v_1 \neq 0$, so the matrix

$$A = \begin{bmatrix} u_1 & v_1 \\ u_2 & v_2 \end{bmatrix}$$

is invertible. Given $\mathbf{x} \in \mathbb{R}^2$, define $\mathbf{c} = \begin{bmatrix} c_1 \\ c_2 \end{bmatrix}$ by $\mathbf{c} = A^{-1}\mathbf{x}$. Then we have

$$\mathbf{x} = A(A^{-1}\mathbf{x}) = A\mathbf{c} = c_1\mathbf{u} + c_2\mathbf{v},$$

thereby establishing that an arbitrary $\mathbf{x} \in \mathbb{R}^2$ is a linear combination of $\mathbf{u}$ and $\mathbf{v}$. Indeed, more is true: That linear combination is *unique* since $\mathbf{x} = A\mathbf{c}$ if and only if $\mathbf{c} = A^{-1}\mathbf{x}$. We shall study the generalization of this result to higher dimensions in great detail in Chapter 4. ◂

We shall learn in Chapter 4 how to calculate the inverse of a matrix in a straightforward fashion. We end the present discussion of inverses with a very important observation.

Proposition 4.3 *Suppose A and B are invertible $n \times n$ matrices. Then their product AB is invertible, and*

$$(AB)^{-1} = B^{-1}A^{-1}.$$

Remark Some people refer to this result rather endearingly as the "shoe-sock theorem," for to undo (invert) the process of putting on one's socks and then one's shoes, one must first remove the shoes and then remove the socks.

Proof To prove the matrix AB is invertible, we need only check that the candidate for the inverse works. That is, we need to check that

$$(AB)(B^{-1}A^{-1}) = I_n \quad \text{and} \quad (B^{-1}A^{-1})(AB) = I_n.$$

But these follow immediately from associativity:

$$(AB)(B^{-1}A^{-1}) = A(BB^{-1})A^{-1} = AI_n A^{-1} = AA^{-1} = I_n, \text{ and}$$
$$(B^{-1}A^{-1})(AB) = B^{-1}(A^{-1}A)B = B^{-1}I_n B = B^{-1}B = I_n. \quad \blacksquare$$

4.2 The Transpose

The final matrix operation we will discuss in this chapter is the *transpose*. When A is an $m \times n$ matrix with entries a_{ij}, the matrix A^T (read "A transpose") is the $n \times m$ matrix whose ij-entry is a_{ji}; i.e., the ith row of A^T is the ith column of A. We say a square matrix A is *symmetric* if $A^\mathsf{T} = A$ and *skew-symmetric* if $A^\mathsf{T} = -A$.

▶ **EXAMPLE 12**

Suppose

$$A = \begin{bmatrix} 1 & 2 & 1 \\ 3 & -1 & 0 \end{bmatrix}, \quad B = \begin{bmatrix} 1 & 3 \\ 2 & -1 \\ 1 & 0 \end{bmatrix}, \quad C = \begin{bmatrix} 1 \\ 2 \\ -3 \end{bmatrix}, \quad \text{and} \quad D = \begin{bmatrix} 1 & 2 & -3 \end{bmatrix}.$$

Then $A^\mathsf{T} = B$, $B^\mathsf{T} = A$, $C^\mathsf{T} = D$, and $D^\mathsf{T} = C$. Note, in particular, that the transpose of a column vector, i.e., an $n \times 1$ matrix, is a row vector, i.e., a $1 \times n$ matrix. An example of a symmetric matrix is

$$S = \begin{bmatrix} 1 & 2 & 3 \\ 2 & 0 & -1 \\ 3 & -1 & 7 \end{bmatrix} \quad \text{since} \quad S^\mathsf{T} = \begin{bmatrix} 1 & 2 & 3 \\ 2 & 0 & -1 \\ 3 & -1 & 7 \end{bmatrix} = S. \quad \triangleleft$$

The basic properties of the transpose operation are as follows:

Proposition 4.4 *Let A and A' be $m \times n$ matrices, let B be an $n \times p$ matrix, and let c be a scalar. Then*

1. $(A^\mathsf{T})^\mathsf{T} = A$;
2. $(cA)^\mathsf{T} = cA^\mathsf{T}$;
3. $(A + A')^\mathsf{T} = A^\mathsf{T} + A'^\mathsf{T}$;
4. $(AB)^\mathsf{T} = B^\mathsf{T} A^\mathsf{T}$.

Proof The first is obvious since we swap rows and columns and then swap again, returning to our original matrix. The second and third can be immediately checked. The last result is more interesting, and we will use it to derive a crucial result in a moment. Note, first, that AB is an $m \times p$ matrix, so $(AB)^\mathsf{T}$ will be a $p \times m$ matrix; $B^\mathsf{T} A^\mathsf{T}$ is the product of a $p \times n$ matrix and an $n \times m$ matrix and hence will be $p \times m$ as well, so the shapes agree. Now, the ji-entry of AB is the dot product of the jth row vector of A and the ith column vector of B; i.e., the ij-entry of $(AB)^\mathsf{T}$ is

$$\left((AB)^\mathsf{T}\right)_{ij} = (AB)_{ji} = \mathbf{A}_j \cdot \mathbf{b}_i.$$

On the other hand, the ij-entry of $B^\mathsf{T} A^\mathsf{T}$ is the dot product of the ith row vector of B^T and the jth column vector of A^T; but this is, by definition, the dot product of the ith column vector of B and the jth row vector of A. That is,

$$(B^\mathsf{T} A^\mathsf{T})_{ij} = \mathbf{b}_i \cdot \mathbf{A}_j,$$

and, since dot product is commutative, the two formulas agree. ∎

The transpose matrix will be very important to us because of the interplay between dot product and transpose. If $\mathbf{x}$ and $\mathbf{y}$ are vectors in $\mathbb{R}^n$, then by virtue of our very definition of matrix multiplication,

$$\boxed{\mathbf{x} \cdot \mathbf{y} = \mathbf{x}^\mathsf{T} \mathbf{y},}$$

provided we agree to think of a 1×1 matrix as a scalar. Now we have the highly useful

Proposition 4.5 *Let A be an $m \times n$ matrix, $\mathbf{x} \in \mathbb{R}^n$, and $\mathbf{y} \in \mathbb{R}^m$. Then*

$$A\mathbf{x} \cdot \mathbf{y} = \mathbf{x} \cdot A^\mathsf{T} \mathbf{y}.$$

(On the left, we take the dot product of vectors in $\mathbb{R}^m$; on the right, of vectors in $\mathbb{R}^n$.)

Remark You might remember this: To move the matrix "across the dot product," you must transpose it.

Proof We just calculate, using the formula for the transpose of a product and, as usual, associativity:

$$A\mathbf{x} \cdot \mathbf{y} = (A\mathbf{x})^\mathsf{T} \mathbf{y} = (\mathbf{x}^\mathsf{T} A^\mathsf{T})\mathbf{y} = \mathbf{x}^\mathsf{T}(A^\mathsf{T} \mathbf{y}) = \mathbf{x} \cdot A^\mathsf{T} \mathbf{y}. \quad \blacksquare$$

▶ **EXAMPLE 13**

We return to the economic interpretation of dot product given in the remark on p. 12. Suppose that m different ingredients are required to manufacture n different products. To manufacture the product vector $\mathbf{x} = \begin{bmatrix} x_1 \\ \vdots \\ x_n \end{bmatrix}$ requires the ingredient vector $\mathbf{y} = \begin{bmatrix} y_1 \\ \vdots \\ y_m \end{bmatrix}$, and we suppose $\mathbf{x}$ and $\mathbf{y}$ are related by the equation $\mathbf{y} = A\mathbf{x}$ for some $m \times n$ matrix A. If each unit of ingredient j costs a price p_j, then the cost of producing $\mathbf{x}$ is

$$\sum_{j=1}^{m} p_j y_j = \mathbf{y} \cdot \mathbf{p} = A\mathbf{x} \cdot \mathbf{p} = \mathbf{x} \cdot A^\mathsf{T}\mathbf{p} = \sum_{i=1}^{n} q_i x_i,$$

where $\mathbf{q} = A^\mathsf{T}\mathbf{p}$. Notice then that q_i is the amount it costs to produce a unit of the ith product. Our fundamental formula, Proposition 4.5, tells us that the total cost of the ingredients should equal the total worth of the products we manufacture. ◀

▶ EXERCISES 1.4

1. Let $A = \begin{bmatrix} 1 & 2 \\ 3 & 4 \end{bmatrix}$, $B = \begin{bmatrix} 2 & 1 \\ 4 & 3 \end{bmatrix}$, $C = \begin{bmatrix} 1 & 2 & 1 \\ 0 & 1 & 2 \end{bmatrix}$, and $D = \begin{bmatrix} 0 & 1 \\ 1 & 0 \\ 2 & 3 \end{bmatrix}$. Calculate each of the following expressions or explain why it is not defined.

(a) $A + B$ (e) AB (i) BD
*(b) $2A - B$ *(f) BA (j) DB
(c) $A - C$ *(g) AC *(k) CD
(d) $C + D$ *(h) CA *(l) DC

2. (a) If A is an $m \times n$ matrix and $A\mathbf{x} = \mathbf{0}$ for all $\mathbf{x} \in \mathbb{R}^n$, prove that $A = O$.
   (b) If A and B are $m \times n$ matrices and $A\mathbf{x} = B\mathbf{x}$ for all $\mathbf{x} \in \mathbb{R}^n$, prove that $A = B$.

♯3. Let A be an $m \times n$ matrix. Show that $V = \{\mathbf{x} \in \mathbb{R}^n : A\mathbf{x} = \mathbf{0}\}$ is a subspace of $\mathbb{R}^n$.

♯4. Let A be an $m \times n$ matrix.
   (a) Show that $V = \left\{ \begin{bmatrix} \mathbf{x} \\ A\mathbf{x} \end{bmatrix} : \mathbf{x} \in \mathbb{R}^n \right\} \subset \mathbb{R}^{m+n}$ is a subspace of $\mathbb{R}^{m+n}$.
   (b) When $m = 1$, show that $V \subset \mathbb{R}^{n+1}$ is a hyperplane (see Example 1e in Section 3) by finding a vector $\mathbf{b} \in \mathbb{R}^{n+1}$ so that $V = \{\mathbf{z} \in \mathbb{R}^{n+1} : \mathbf{b} \cdot \mathbf{z} = 0\}$.

5. Give 2×2 matrices A so that for any $\mathbf{x} \in \mathbb{R}^2$ we have, respectively,
   (a) $A\mathbf{x}$ is the vector whose components are, respectively, the sum and difference of the components of $\mathbf{x}$;
   *(b) $A\mathbf{x}$ is the vector obtained by projecting $\mathbf{x}$ onto the line $x_1 = x_2$ in $\mathbb{R}^2$;
   (c) $A\mathbf{x}$ is the vector obtained by first reflecting $\mathbf{x}$ across the line $x_1 = 0$ and then reflecting the resulting vector across the line $x_2 = 0$;
   (d) $A\mathbf{x}$ is the vector obtained by projecting $\mathbf{x}$ onto the line $2x_1 - x_2 = 0$;
   *(e) $A\mathbf{x}$ is the vector obtained by first projecting $\mathbf{x}$ onto the line $2x_1 - x_2 = 0$ and then rotating the resulting vector $\pi/2$ counterclockwise;
   (f) $A\mathbf{x}$ is the vector obtained by first rotating $\mathbf{x}$ an angle of $\pi/2$ counterclockwise and then projecting the resulting vector onto the line $2x_1 - x_2 = 0$.

6. (a) Calculate $A_\theta A_\phi$ and $A_\phi A_\theta$. (Recall the definition of the rotation matrix on p. 27.)
   (b) Use your answer to part a to derive the addition formulas for cos and sin.

7. Let A_θ be the rotation matrix defined on p. 27, $0 \le \theta \le \pi$. Prove that
(a) $\|A_\theta \mathbf{x}\| = \|\mathbf{x}\|$ for all $\mathbf{x} \in \mathbb{R}^2$;
(b) the angle between $\mathbf{x}$ and $A_\theta \mathbf{x}$ is θ.
These properties should characterize a rotation of the plane through angle θ.

8. Prove or give a counterexample. Assume all relevant matrices are square and of the same size.
(a) If $AB = CB$ and $B \ne O$, then $A = C$.
(b) If $A^2 = A$, then $A = O$ or $A = I$.
(c) $(A + B)(A - B) = A^2 - B^2$.
(d) If $AB = BC$ and B is invertible, then $A = C$.

9. Find all 2×2 matrices $A = \begin{bmatrix} a & b \\ c & d \end{bmatrix}$ satisfying
(a) $A^2 = I_2$;
*(b) $A^2 = O$;
(c) $A^2 = -I_2$.

10. (a) Show that the matrix giving *reflection* across the line spanned by $\begin{bmatrix} \cos\theta \\ \sin\theta \end{bmatrix}$ is

$$R = \begin{bmatrix} \cos 2\theta & \sin 2\theta \\ \sin 2\theta & -\cos 2\theta \end{bmatrix}.$$

(b) Letting A_θ be the rotation matrix defined on p. 27, check that

$$A_{2\theta} \begin{bmatrix} 1 & 0 \\ 0 & -1 \end{bmatrix} = R = A_\theta \begin{bmatrix} 1 & 0 \\ 0 & -1 \end{bmatrix} A_{(-\theta)}.$$

11. For each of the following matrices A, find a formula for A^n. (If you know how to give an inductive proof, please do so.)

(a) $A = \begin{bmatrix} 1 & 1 \\ 0 & 1 \end{bmatrix}$

(b) $A = \begin{bmatrix} d_1 & & & \\ & d_2 & & \\ & & \ddots & \\ & & & d_m \end{bmatrix}$ (all nondiagonal entries are 0)

12. Suppose A and A' are $m \times m$ matrices, B and B' are $m \times n$ matrices, C and C' are $n \times m$ matrices, and D and D' are $n \times n$ matrices. Check the following formula for the product of "block" matrices:

$$\left[\begin{array}{c|c} A & B \\ \hline C & D \end{array}\right] \left[\begin{array}{c|c} A' & B' \\ \hline C' & D' \end{array}\right] = \left[\begin{array}{c|c} AA' + BC' & AB' + BD' \\ \hline CA' + DC' & CB' + DD' \end{array}\right].$$

*13. Let $T \colon \mathbb{R}^2 \to \mathbb{R}^2$ be the linear transformation defined by rotating the plane $\pi/2$ counterclockwise; let $S \colon \mathbb{R}^2 \to \mathbb{R}^2$ be the linear transformation defined by reflecting the plane across the line $x_1 + x_2 = 0$.
(a) Give the standard matrices representing S and T.
(b) Give the standard matrix representing $T \circ S$.
(c) Give the standard matrix representing $S \circ T$.

14. Calculate the standard matrix for each of the following linear transformations T:
(a) $T \colon \mathbb{R}^2 \to \mathbb{R}^2$ given by rotating $-\pi/4$ about the origin and then reflecting across the line $x_1 - x_2 = 0$.
(b) $T \colon \mathbb{R}^3 \to \mathbb{R}^3$ given by rotating $\pi/2$ about the x_1-axis (as viewed from the positive side) and then reflecting across the plane $x_2 = 0$.

(c) $T: \mathbb{R}^3 \to \mathbb{R}^3$ given by rotating $-\pi/2$ about the x_1-axis (as viewed from the positive side) and then rotating $\pi/2$ about the x_3-axis.

15. Consider the cube with vertices $\begin{bmatrix} \pm 1 \\ \pm 1 \\ \pm 1 \end{bmatrix}$, pictured in Figure 4.8. (Note that the coordinate axes pass through the centers of the various faces.) Give the standard matrices for each of the following symmetries of the cube.

(a) 90° rotation about the x_3-axis (viewed from high above)

(b) 180° rotation about the line joining $\begin{bmatrix} -1 \\ 0 \\ 1 \end{bmatrix}$ and $\begin{bmatrix} 1 \\ 0 \\ -1 \end{bmatrix}$

(c) 120° rotation about the line joining $\begin{bmatrix} -1 \\ -1 \\ -1 \end{bmatrix}$ and $\begin{bmatrix} 1 \\ 1 \\ 1 \end{bmatrix}$ (viewed from high above)

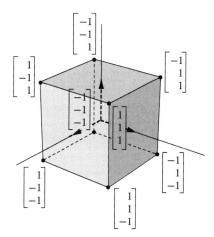

Figure 4.8

16. Consider the tetrahedron with vertices $\begin{bmatrix} 1 \\ 1 \\ 1 \end{bmatrix}, \begin{bmatrix} -1 \\ -1 \\ 1 \end{bmatrix}, \begin{bmatrix} 1 \\ -1 \\ -1 \end{bmatrix}$, and $\begin{bmatrix} -1 \\ 1 \\ -1 \end{bmatrix}$, pictured in Figure 4.9. Give the standard matrices for each of the following symmetries of the tetrahedron.

(a) 120° rotation counterclockwise (as viewed from high above) about the line joining $\begin{bmatrix} 0 \\ 0 \\ 0 \end{bmatrix}$ and the vertex $\begin{bmatrix} 1 \\ 1 \\ 1 \end{bmatrix}$

(b) 180° rotation about the line joining $\begin{bmatrix} 0 \\ 0 \\ 1 \end{bmatrix}$ and $\begin{bmatrix} 0 \\ 0 \\ -1 \end{bmatrix}$

(c) reflection across the plane containing one edge and the midpoint of the opposite edge
(Hint: Note where the coordinate axes intersect the tetrahedron.)

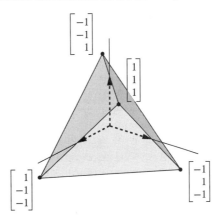

Figure 4.9

*17. Suppose A is an $n \times n$ matrix and B is an invertible $n \times n$ matrix. Calculate the following.
   (a) $(BAB^{-1})^2$
   (b) $(BAB^{-1})^n$ (n a positive integer)
   (c) $(BAB^{-1})^{-1}$ (what additional assumption is required here?)

18. Find matrices A so that
   (a) $A \neq O$, but $A^2 = O$; (b) $A^2 \neq O$, but $A^3 = O$.
   Can you make a conjecture about matrices satisfying $A^{n-1} \neq O$ but $A^n = O$?

*19. Suppose A is an invertible $n \times n$ matrix and $\mathbf{x} \in \mathbb{R}^n$ satisfies $A\mathbf{x} = 7\mathbf{x}$. Calculate $A^{-1}\mathbf{x}$.

20. Suppose A is a square matrix satisfying the equation $A^3 - 3A + 2I = 0$. Show that A is invertible. (Hint: Can you give an explicit formula for A^{-1}?)

21. Suppose A is an $n \times n$ matrix satisfying $A^{10} = O$. Prove that the matrix $I_n - A$ is invertible. (Hint: As a warm-up, try assuming $A^2 = O$.)

22. Define the *trace* of an $n \times n$ matrix A (denoted $\mathrm{tr}\,A$) to be the sum of its diagonal entries:
$$\mathrm{tr}\,A = \sum_{i=1}^{n} a_{ii}.$$

   (a) Prove that $\mathrm{tr}\,A = \mathrm{tr}(A^\mathsf{T})$.
   (b) Prove that $\mathrm{tr}(A + B) = \mathrm{tr}\,A + \mathrm{tr}\,B$ and $\mathrm{tr}(cA) = c\,\mathrm{tr}\,A$ for any scalar c.
   (c) Prove that $\mathrm{tr}(AB) = \mathrm{tr}(BA)$. (Hint: $\sum_{k=1}^{n}\sum_{\ell=1}^{n} c_{k\ell} = \sum_{\ell=1}^{n}\sum_{k=1}^{n} c_{k\ell}$.)

23. Let
$$A = \begin{bmatrix} 1 & 2 \\ 3 & 4 \end{bmatrix}, \quad B = \begin{bmatrix} 2 & 1 \\ 4 & 3 \end{bmatrix}, \quad C = \begin{bmatrix} 1 & 2 & 1 \\ 0 & 1 & 2 \end{bmatrix}, \quad \text{and} \quad D = \begin{bmatrix} 0 & 1 \\ 1 & 0 \\ 2 & 3 \end{bmatrix}.$$

Calculate each of the following expressions or explain why it is not defined.
   (a) A^T
   *(b) $2A - B^\mathsf{T}$
   (c) C^T
   (d) $C^\mathsf{T} + D$
   *(e) $A^\mathsf{T} C$
   (f) AC^T

*(g) $C^T A^T$ (i) $D^T B$ *(k) $C^T C$
(h) $B D^T$ *(j) $C C^T$ (l) $C^T D^T$

*24. Suppose A and B are symmetric. Prove that AB is symmetric if and only if $AB = BA$.

25. Let A be an arbitrary $m \times n$ matrix. Prove that $A^T A$ is symmetric.

26. Suppose A is invertible. Check that $(A^{-1})^T A^T = I$ and $A^T (A^{-1})^T = I$, and deduce that A^T is likewise invertible.

*27. Let A_θ be the rotation matrix defined on p. 27. Explain why $A_\theta^{-1} = A_\theta^T$.

28. An $n \times n$ matrix is called a *permutation matrix* if it has a single 1 in each row and column and all its remaining entries are 0.
   (a) Write down all the 2×2 permutation matrices. How many are there?
   (b) Write down all the 3×3 permutation matrices. How many are there?
   (c) Prove that the product of two permutation matrices is again a permutation matrix. Do they commute?
   (d) Prove that every permutation matrix is invertible and $P^{-1} = P^T$.
   (e) If A is an $n \times n$ matrix and P is an $n \times n$ permutation matrix, describe the matrices PA and AP.

♯29. Let A be an $m \times n$ matrix and let $\mathbf{x}, \mathbf{y} \in \mathbb{R}^n$. Prove that if $A\mathbf{x} = \mathbf{0}$ and $\mathbf{y} = A^T \mathbf{b}$ for some $\mathbf{b} \in \mathbb{R}^m$, then $\mathbf{x} \cdot \mathbf{y} = 0$.

♯30. Suppose A is a symmetric $n \times n$ matrix. Let $V \subset \mathbb{R}^n$ be a subspace with the property that $A\mathbf{x} \in V$ for every $\mathbf{x} \in V$. Prove that $A\mathbf{y} \in V^\perp$ for all $\mathbf{y} \in V^\perp$.

*31. Given the matrix

$$A = \begin{bmatrix} 1 & 2 & 1 \\ 1 & 3 & 1 \\ 0 & 1 & -1 \end{bmatrix} \quad \text{and its inverse matrix} \quad A^{-1} = \begin{bmatrix} 4 & -3 & 1 \\ -1 & 1 & 0 \\ -1 & 1 & -1 \end{bmatrix},$$

find (with no computation) the inverse of

(a) $\begin{bmatrix} 1 & 1 & 0 \\ 2 & 3 & 1 \\ 1 & 1 & -1 \end{bmatrix}$,   (b) $\begin{bmatrix} 1 & 2 & 1 \\ 0 & 1 & -1 \\ 1 & 3 & 1 \end{bmatrix}$,   (c) $\begin{bmatrix} 1 & 2 & 1 \\ 1 & 3 & 1 \\ 0 & 2 & -2 \end{bmatrix}$.

♯*32. Suppose A is an $m \times n$ matrix and $\mathbf{x} \in \mathbb{R}^n$ satisfies $(A^T A)\mathbf{x} = \mathbf{0}$. Prove that $A\mathbf{x} = \mathbf{0}$. (Hint: What is $\|A\mathbf{x}\|$?)

33. Suppose A is a symmetric matrix satisfying $A^2 = O$. Prove that $A = O$. Give an example to show that the hypothesis of symmetry is required.

♯34. We say an $n \times n$ matrix A is *orthogonal* if $A^T A = I_n$.
   (a) Prove that the column vectors $\mathbf{a}_1, \ldots, \mathbf{a}_n$ of an orthogonal matrix A are unit vectors that are orthogonal to one another; i.e.,

$$\mathbf{a}_i \cdot \mathbf{a}_j = \begin{cases} 1, & i = j \\ 0, & i \neq j \end{cases}.$$

   (b) Fill in the missing columns in the following matrices to make them orthogonal:

$$\begin{bmatrix} \frac{\sqrt{3}}{2} & ? \\ -\frac{1}{2} & ? \end{bmatrix}, \quad \begin{bmatrix} 1 & 0 & ? \\ 0 & -1 & ? \\ 0 & 0 & ? \end{bmatrix}, \quad \begin{bmatrix} \frac{1}{3} & ? & \frac{2}{3} \\ \frac{2}{3} & ? & -\frac{2}{3} \\ \frac{2}{3} & ? & \frac{1}{3} \end{bmatrix}.$$

(c) Prove that any 2×2 orthogonal matrix A must be of the form
$$\begin{bmatrix} \cos\theta & -\sin\theta \\ \sin\theta & \cos\theta \end{bmatrix} \quad \text{or} \quad \begin{bmatrix} \cos\theta & \sin\theta \\ \sin\theta & -\cos\theta \end{bmatrix}$$
for some real number θ. (Hint: Use part a, rather than the original definition.)

*(d) Prove that if A is an orthogonal 2×2 matrix, then $\mu_A \colon \mathbb{R}^2 \to \mathbb{R}^2$ is either a rotation or the composition of a rotation and a reflection.

(e) Assume for now that $A^\mathsf{T} = A^{-1}$ when A is orthogonal (this is a consequence of Corollary 2.2 of Chapter 4). Prove that the row vectors $\mathbf{A}_1, \ldots, \mathbf{A}_n$ of an orthogonal matrix A are unit vectors that are orthogonal to one another.

35. (Recall the definition of orthogonal matrices from Exercise 34.)
(a) Prove that if A and B are orthogonal $n \times n$ matrices, then so is AB.
*(b) Prove that if A is an orthogonal matrix, then so is A^{-1}.

36. (a) Prove that the only matrix that is both symmetric and skew-symmetric is O.
(b) Given any square matrix A, prove that $S = \tfrac{1}{2}(A + A^\mathsf{T})$ is symmetric and $K = \tfrac{1}{2}(A - A^\mathsf{T})$ is skew-symmetric.
(c) Prove that any square matrix A can be written in the form $A = S + K$, where S is symmetric and K is skew-symmetric.
(d) Prove that the expression in part c is unique: If $A = S + K$ and $A = S' + K'$ (where S and S' are symmetric and K and K' are skew-symmetric), then $S = S'$ and $K = K'$. (Hint: Use part a.)

37. Suppose A is an $n \times n$ matrix that commutes with all $n \times n$ matrices; i.e., $AB = BA$ for all $B \in \mathcal{M}_{n \times n}$. What can you say about A?

5 INTRODUCTION TO DETERMINANTS AND THE CROSS PRODUCT

Let $\mathbf{x}$ and $\mathbf{y}$ be vectors in $\mathbb{R}^2$ and consider the parallelogram $\mathcal{P}$ they span. The area of $\mathcal{P}$ is nonzero as long as $\mathbf{x}$ and $\mathbf{y}$ are not collinear. We want to express the area of $\mathcal{P}$ in terms of the coordinates of $\mathbf{x}$ and $\mathbf{y}$. First notice that the area of the parallelogram pictured in Figure 5.1 is the same as the area of the rectangle obtained by moving the shaded triangle from the right side to the left. This rectangle has area $A = bh$, where $b = \|\mathbf{x}\|$ is the base and $h = \|\mathbf{y}\| \sin\theta$ is the height. We could calculate $\sin\theta$ from the formula
$$\cos\theta = \frac{\mathbf{x} \cdot \mathbf{y}}{\|\mathbf{x}\| \|\mathbf{y}\|},$$
but instead we note (see Figure 5.2) that
$$\|\mathbf{x}\| \|\mathbf{y}\| \sin\theta = \|\mathbf{x}\| \|\mathbf{y}\| \cos\left(\tfrac{\pi}{2} - \theta\right) = \rho(\mathbf{x}) \cdot \mathbf{y},$$
where $\rho(\mathbf{x})$ is the vector obtained by rotating $\mathbf{x}$ an angle $\pi/2$ counterclockwise (see Exercise 1.2.16). If $\mathbf{x} = \begin{bmatrix} x_1 \\ x_2 \end{bmatrix}$ and $\mathbf{y} = \begin{bmatrix} y_1 \\ y_2 \end{bmatrix}$, then we have
$$\text{area}(\mathcal{P}) = \rho(\mathbf{x}) \cdot \mathbf{y} = \begin{bmatrix} -x_2 \\ x_1 \end{bmatrix} \cdot \begin{bmatrix} y_1 \\ y_2 \end{bmatrix} = x_1 y_2 - x_2 y_1.$$

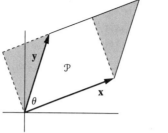
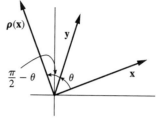

Figure 5.1   **Figure 5.2**

▶ EXAMPLE 1

If $\mathbf{x} = \begin{bmatrix} 3 \\ 1 \end{bmatrix}$ and $\mathbf{y} = \begin{bmatrix} 4 \\ 3 \end{bmatrix}$, then the area of the parallelogram spanned by $\mathbf{x}$ and $\mathbf{y}$ is $x_1 y_2 - x_2 y_1 = 3 \cdot 3 - 1 \cdot 4 = 5$. On the other hand, if we interchange the two, letting $\mathbf{x} = \begin{bmatrix} 4 \\ 3 \end{bmatrix}$ and $\mathbf{y} = \begin{bmatrix} 3 \\ 1 \end{bmatrix}$, then we get $x_1 y_2 - x_2 y_1 = 4 \cdot 1 - 3 \cdot 3 = -5$. Certainly the parallelogram hasn't changed; nor does it make sense to have negative area. What is the explanation? In deriving our formula for the area above, we assumed $0 < \theta < \pi$; but if we must turn clockwise to get from $\mathbf{x}$ to $\mathbf{y}$, this means that θ is negative, resulting in a sign discrepancy in the area calculation. ◀

So we should amend our earlier result. We define the *signed area* of the parallelogram $\mathcal{P}$ to be the area of $\mathcal{P}$ when one turns counterclockwise from $\mathbf{x}$ to $\mathbf{y}$ and to be *negative* the area of $\mathcal{P}$ when one turns clockwise from $\mathbf{x}$ to $\mathbf{y}$, as illustrated in Figure 5.3. Then we have

$$\text{signed area}(\mathcal{P}) = x_1 y_2 - x_2 y_1.$$

Because of its geometric significance, we consider the function[5]

(∗) $$\mathcal{D}(\mathbf{x}, \mathbf{y}) = x_1 y_2 - x_2 y_1;$$

this is the function that associates to each ordered pair of vectors $\mathbf{x}, \mathbf{y} \in \mathbb{R}^2$ the signed area of the parallelogram they span.

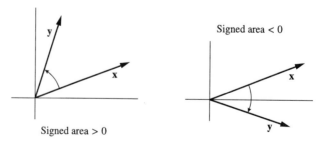

Figure 5.3

---

[5] Here, since $\mathbf{x}$ and $\mathbf{y}$ are themselves vectors, we use the customary notation for functions.

Next, let's explore the properties of the signed area function $\mathcal{D}$ on $\mathbb{R}^2 \times \mathbb{R}^2$.[6]

Property 1 If $\mathbf{x}, \mathbf{y} \in \mathbb{R}^2$, then $\mathcal{D}(\mathbf{y}, \mathbf{x}) = -\mathcal{D}(\mathbf{x}, \mathbf{y})$.

Algebraically, we have
$$\mathcal{D}(\mathbf{y}, \mathbf{x}) = y_1 x_2 - y_2 x_1 = -(x_1 y_2 - x_2 y_1) = -\mathcal{D}(\mathbf{x}, \mathbf{y}).$$
Geometrically, this was the point of our introducing the notion of signed area.

Property 2 If $\mathbf{x}, \mathbf{y} \in \mathbb{R}^2$ and $c \in \mathbb{R}$, then
$$\mathcal{D}(c\mathbf{x}, \mathbf{y}) = c\mathcal{D}(\mathbf{x}, \mathbf{y}) = \mathcal{D}(\mathbf{x}, c\mathbf{y}).$$

This follows immediately from the formula (∗):
$$\mathcal{D}(c\mathbf{x}, \mathbf{y}) = (cx_1)y_2 - (cx_2)y_1 = c(x_1 y_2 - x_2 y_1) = c\mathcal{D}(\mathbf{x}, \mathbf{y}).$$
Geometrically, if we stretch one of the edges of the parallelogram by a factor of $c > 0$, then the area is multiplied by a factor of c. And if $c < 0$, the area is multiplied by a factor of $|c|$ and the signed area changes sign (why?).

Property 3 If $\mathbf{x}, \mathbf{y}, \mathbf{z} \in \mathbb{R}^2$, then
$$\mathcal{D}(\mathbf{x} + \mathbf{y}, \mathbf{z}) = \mathcal{D}(\mathbf{x}, \mathbf{z}) + \mathcal{D}(\mathbf{y}, \mathbf{z}) \quad \text{and} \quad \mathcal{D}(\mathbf{x}, \mathbf{y} + \mathbf{z}) = \mathcal{D}(\mathbf{x}, \mathbf{y}) + \mathcal{D}(\mathbf{x}, \mathbf{z}).$$

We can check this explicitly in coordinates (but the clever reader should try to use properties of the dot product to give a better algebraic proof): If $\mathbf{x} = \begin{bmatrix} x_1 \\ x_2 \end{bmatrix}$, $\mathbf{y} = \begin{bmatrix} y_1 \\ y_2 \end{bmatrix}$, and $\mathbf{z} = \begin{bmatrix} z_1 \\ z_2 \end{bmatrix}$, then

$$\mathcal{D}(\mathbf{x} + \mathbf{y}, \mathbf{z}) = (x_1 + y_1)z_2 - (x_2 + y_2)z_1$$
$$= (x_1 z_2 - x_2 z_1) + (y_1 z_2 - y_2 z_1) = \mathcal{D}(\mathbf{x}, \mathbf{z}) + \mathcal{D}(\mathbf{y}, \mathbf{z}),$$

as required. (The formula for $\mathcal{D}(\mathbf{x}, \mathbf{y} + \mathbf{z})$ can now be deduced by using Property 1.) Geometrically, we can deduce the result from Figure 5.4: The area of parallelogram $OBCD$ ($\mathcal{D}(\mathbf{x} + \mathbf{y}, \mathbf{z})$) is equal to the sum of the areas of parallelograms $OAED$ ($\mathcal{D}(\mathbf{x}, \mathbf{z})$) and $ABCE$ ($\mathcal{D}(\mathbf{y}, \mathbf{z})$). The proof of this, in turn, follows from the fact that $\triangle OAB$ is congruent to $\triangle DEC$.

---

[6]Recall that, given two sets X and Y, their *product* $X \times Y$ consists of all ordered pairs (x, y), where $x \in X$ and $y \in Y$.

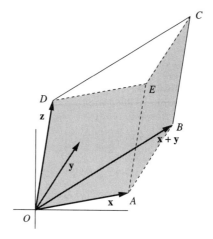

Figure 5.4

Property 4 For the standard basis vectors e_1, e_2, we have $\mathcal{D}(e_1, e_2) = 1$.

The expression $\mathcal{D}(x, y)$ is a 2×2 *determinant*, often written $\begin{vmatrix} x & y \end{vmatrix}$. Indeed, given a 2×2 matrix A with column vectors $a_1, a_2 \in \mathbb{R}^2$, we define

$$\det A = \mathcal{D}(a_1, a_2) = \begin{vmatrix} a_1 & a_2 \end{vmatrix}.$$

As we ask the reader to check in Exercise 4, one can deduce from the four properties above and the geometry of linear maps the fact that the determinant represents the signed area of the parallelogram.

We next turn to the case of 3×3 determinants. The general case will wait until Chapter 7. Given three vectors,

$$x = \begin{bmatrix} x_1 \\ x_2 \\ x_3 \end{bmatrix}, \quad y = \begin{bmatrix} y_1 \\ y_2 \\ y_3 \end{bmatrix}, \quad \text{and} \quad z = \begin{bmatrix} z_1 \\ z_2 \\ z_3 \end{bmatrix} \in \mathbb{R}^3,$$

we define

$$\mathcal{D}(x, y, z) = \begin{vmatrix} | & | & | \\ x & y & z \\ | & | & | \end{vmatrix} = x_1 \begin{vmatrix} y_2 & z_2 \\ y_3 & z_3 \end{vmatrix} - x_2 \begin{vmatrix} y_1 & z_1 \\ y_3 & z_3 \end{vmatrix} + x_3 \begin{vmatrix} y_1 & z_1 \\ y_2 & z_2 \end{vmatrix}.$$

Multiplying this out, we get three positive terms and three negative terms; a handy mnemonic device for this formula is depicted in Figure 5.5.

This function $\mathcal{D}$ of three vectors in $\mathbb{R}^3$ has properties quite analogous to those in the two-dimensional case. In particular, it follows immediately from the latter that if $x, y, z,$ and w are vectors in $\mathbb{R}^3$ and c is a scalar, then

$$\mathcal{D}(x, z, y) = -\mathcal{D}(x, y, z),$$
$$\mathcal{D}(x, cy, z) = c\mathcal{D}(x, y, z) = \mathcal{D}(x, y, cz),$$

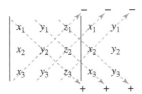

Figure 5.5

$$\mathcal{D}(\mathbf{x}, \mathbf{y} + \mathbf{w}, \mathbf{z}) = \mathcal{D}(\mathbf{x}, \mathbf{y}, \mathbf{z}) + \mathcal{D}(\mathbf{x}, \mathbf{w}, \mathbf{z}), \quad \text{and}$$
$$\mathcal{D}(\mathbf{x}, \mathbf{y}, \mathbf{z} + \mathbf{w}) = \mathcal{D}(\mathbf{x}, \mathbf{y}, \mathbf{z}) + \mathcal{D}(\mathbf{x}, \mathbf{y}, \mathbf{w}).$$

It is also immediately obvious from the definition that if $\mathbf{x}, \mathbf{y}, \mathbf{z}$, and $\mathbf{w}$ are vectors in $\mathbb{R}^3$ and c is a scalar, then

$$\mathcal{D}(c\mathbf{x}, \mathbf{y}, \mathbf{z}) = c\mathcal{D}(\mathbf{x}, \mathbf{y}, \mathbf{z}),$$
$$\mathcal{D}(\mathbf{x} + \mathbf{w}, \mathbf{y}, \mathbf{z}) = \mathcal{D}(\mathbf{x}, \mathbf{y}, \mathbf{z}) + \mathcal{D}(\mathbf{w}, \mathbf{y}, \mathbf{z}).$$

Least elegant is the verification that $\mathcal{D}(\mathbf{y}, \mathbf{x}, \mathbf{z}) = -\mathcal{D}(\mathbf{x}, \mathbf{y}, \mathbf{z})$:

$$\mathcal{D}(\mathbf{y}, \mathbf{x}, \mathbf{z}) = y_1 \begin{vmatrix} x_2 & z_2 \\ x_3 & z_3 \end{vmatrix} - y_2 \begin{vmatrix} x_1 & z_1 \\ x_3 & z_3 \end{vmatrix} + y_3 \begin{vmatrix} x_1 & z_1 \\ x_2 & z_2 \end{vmatrix}$$
$$= y_1(x_2 z_3 - x_3 z_2) + y_2(x_3 z_1 - x_1 z_3) + y_3(x_1 z_2 - x_2 z_1)$$
$$= -x_1(y_2 z_3 - y_3 z_2) + x_2(y_1 z_3 - y_3 z_1) - x_3(y_1 z_2 - y_2 z_1)$$
$$= -x_1 \begin{vmatrix} y_2 & z_2 \\ y_3 & z_3 \end{vmatrix} + x_2 \begin{vmatrix} y_1 & z_1 \\ y_3 & z_3 \end{vmatrix} - x_3 \begin{vmatrix} y_1 & z_2 \\ y_2 & z_2 \end{vmatrix}$$
$$= -\mathcal{D}(\mathbf{x}, \mathbf{y}, \mathbf{z}).$$

Summarizing, we have

Property 1 If $\mathbf{x}, \mathbf{y}, \mathbf{z} \in \mathbb{R}^3$, then

$$\mathcal{D}(\mathbf{y}, \mathbf{x}, \mathbf{z}) = \mathcal{D}(\mathbf{x}, \mathbf{z}, \mathbf{y}) = \mathcal{D}(\mathbf{z}, \mathbf{y}, \mathbf{x}) = -\mathcal{D}(\mathbf{x}, \mathbf{y}, \mathbf{z}).$$

Note that, as a consequence, whenever two of $\mathbf{x}, \mathbf{y}$, and $\mathbf{z}$ are the same, we have $\mathcal{D}(\mathbf{x}, \mathbf{y}, \mathbf{z}) = 0$.

Property 2 If $\mathbf{x}, \mathbf{y}, \mathbf{z} \in \mathbb{R}^3$ and $c \in \mathbb{R}$, then

$$\mathcal{D}(c\mathbf{x}, \mathbf{y}, \mathbf{z}) = \mathcal{D}(\mathbf{x}, c\mathbf{y}, \mathbf{z}) = \mathcal{D}(\mathbf{x}, \mathbf{y}, c\mathbf{z}) = c\mathcal{D}(\mathbf{x}, \mathbf{y}, \mathbf{z}).$$

Property 3 If $\mathbf{x}, \mathbf{y}, \mathbf{z} \in \mathbb{R}^3$, then

$$\mathcal{D}(\mathbf{x} + \mathbf{w}, \mathbf{y}, \mathbf{z}) = \mathcal{D}(\mathbf{x}, \mathbf{y}, \mathbf{z}) + \mathcal{D}(\mathbf{w}, \mathbf{y}, \mathbf{z}),$$
$$\mathcal{D}(\mathbf{x}, \mathbf{y} + \mathbf{w}, \mathbf{z}) = \mathcal{D}(\mathbf{x}, \mathbf{y}, \mathbf{z}) + \mathcal{D}(\mathbf{x}, \mathbf{w}, \mathbf{z}), \quad \text{and}$$
$$\mathcal{D}(\mathbf{x}, \mathbf{y}, \mathbf{z} + \mathbf{w}) = \mathcal{D}(\mathbf{x}, \mathbf{y}, \mathbf{z}) + \mathcal{D}(\mathbf{x}, \mathbf{y}, \mathbf{w}).$$

Property 4 For the standard basis vectors e_1, e_2, e_3, we have $\mathcal{D}(e_1, e_2, e_3) = 1$.

If we let $y' = y - \text{proj}_x y$ and $z' = z - \text{proj}_x z - \text{proj}_{y'} z$, then it follows from the properties of $\mathcal{D}$ that $\mathcal{D}(x, y, z) = \mathcal{D}(x, y', z')$. Moreover, we shall see when we study determinants in Chapter 7 that the results of Exercise 4 hold in three dimensions as well, so that the latter value is not changed by rotating $\mathbb{R}^3$ to make $x = \alpha e_1$, $y' = \beta e_2$, and $z' = \gamma e_3$. Since rotation doesn't change signed volume, we deduce that $\mathcal{D}(x, y, z)$ equals the signed volume of the parallelepiped spanned by x, y, and z, as suggested in Figure 5.6. For an alternative argument, see Exercise 18.

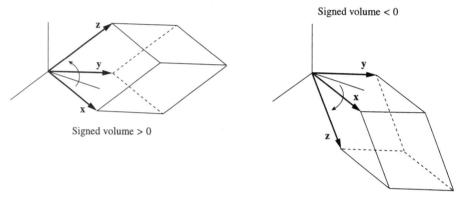

Figure 5.6

Given two vectors $x, y \in \mathbb{R}^3$, define a vector, called their *cross product*, by

$$x \times y = (x_2 y_3 - x_3 y_2) e_1 + (x_3 y_1 - x_1 y_3) e_2 + (x_1 y_2 - x_2 y_1) e_3$$

$$= \begin{vmatrix} e_1 & x_1 & y_1 \\ e_2 & x_2 & y_2 \\ e_3 & x_3 & y_3 \end{vmatrix},$$

where the latter is to be interpreted "formally." The geometric interpretation of the cross product, as indicated in Figure 5.7, is the content of the following

Proposition 5.1 *The cross product $x \times y$ of two vectors $x, y \in \mathbb{R}^3$ is orthogonal to both x and y and $\|x \times y\|$ is the area of the parallelogram $\mathcal{P}$ spanned by x and y. Moreover,*

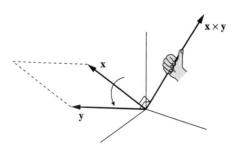

Figure 5.7

when **x** and **y** are nonparallel, the vectors **x**, **y**, **x** × **y** determine a parallelepiped of positive signed volume.

Remark More colloquially, if you curl the fingers of your *right* hand from **x** toward **y**, your thumb points in the direction of **x** × **y**.

Proof The orthogonality is an immediate consequence of the properties once we realize that the formula for the cross product guarantees that

$$\mathbf{z} \cdot (\mathbf{x} \times \mathbf{y}) = \mathcal{D}(\mathbf{z}, \mathbf{x}, \mathbf{y}).$$

In particular, $\mathbf{x} \cdot (\mathbf{x} \times \mathbf{y}) = \mathcal{D}(\mathbf{x}, \mathbf{x}, \mathbf{y}) = 0$.

Now, $\mathcal{D}(\mathbf{x}, \mathbf{y}, \mathbf{x} \times \mathbf{y})$ is the signed volume of the parallelepiped spanned by **x**, **y**, and **x** × **y**. Since **x** × **y** is orthogonal to the plane spanned by **x** and **y**, that volume is the product of the area of $\mathcal{P}$ and $\|\mathbf{x} \times \mathbf{y}\|$. On the other hand,

$$\mathcal{D}(\mathbf{x}, \mathbf{y}, \mathbf{x} \times \mathbf{y}) = \mathcal{D}(\mathbf{x} \times \mathbf{y}, \mathbf{x}, \mathbf{y}) = (\mathbf{x} \times \mathbf{y}) \cdot (\mathbf{x} \times \mathbf{y}) = \|\mathbf{x} \times \mathbf{y}\|^2.$$

Setting the two expressions equal, we infer that

$$\|\mathbf{x} \times \mathbf{y}\| = \text{area}(\mathcal{P}).$$

When **x** and **y** are nonparallel, we have $\mathcal{D}(\mathbf{x}, \mathbf{y}, \mathbf{x} \times \mathbf{y}) = \|\mathbf{x} \times \mathbf{y}\|^2 > 0$, so the vectors span a parallelepiped of positive signed volume, as desired. ∎

▶ **EXAMPLE 2**

We can use the cross product to find the equation of the subspace $\mathcal{P}$ spanned by the vectors $\mathbf{u} = \begin{bmatrix} 1 \\ 1 \\ -1 \end{bmatrix}$

and $\mathbf{v} = \begin{bmatrix} 1 \\ 2 \\ 1 \end{bmatrix}$. For the normal vector to $\mathcal{P}$ is

$$\mathbf{A} = \mathbf{u} \times \mathbf{v} = \begin{vmatrix} \mathbf{e}_1 & 1 & 1 \\ \mathbf{e}_2 & 1 & 2 \\ \mathbf{e}_3 & -1 & 1 \end{vmatrix} = \begin{bmatrix} 3 \\ -2 \\ 1 \end{bmatrix},$$

and so

$$\mathcal{P} = \{\mathbf{x} \in \mathbb{R}^3 : \mathbf{A} \cdot \mathbf{x} = 0\} = \{\mathbf{x} \in \mathbb{R}^3 : 3x_1 - 2x_2 + x_3 = 0\}.$$

Moreover, as depicted schematically in Figure 5.8, the affine plane $\mathcal{P}_1$ parallel to $\mathcal{P}$ and passing through the point $\mathbf{x}_0 = \begin{bmatrix} 2 \\ 0 \\ 1 \end{bmatrix}$ is given by

$$\mathcal{P}_1 = \{\mathbf{x} \in \mathbb{R}^3 : \mathbf{A} \cdot (\mathbf{x} - \mathbf{x}_0) = 0\} = \{\mathbf{x} \in \mathbb{R}^3 : \mathbf{A} \cdot \mathbf{x} = \mathbf{A} \cdot \mathbf{x}_0\}$$
$$= \{\mathbf{x} \in \mathbb{R}^3 : 3x_1 - 2x_2 + x_3 = 7\}. \quad \triangleleft$$

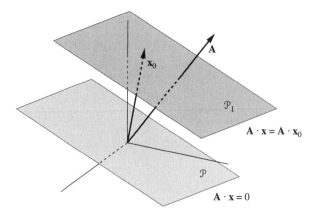

Figure 5.8

EXERCISES 1.5

1. Give a geometric proof that $\mathcal{D}(\mathbf{x}, \mathbf{y} + c\mathbf{x}) = \mathcal{D}(\mathbf{x}, \mathbf{y})$ for any scalar c.

2. Show that if a function $\mathcal{D}\colon \mathbb{R}^2 \times \mathbb{R}^2 \to \mathbb{R}$ satisfies Properties 1–4, then $\mathcal{D}(\mathbf{x}, \mathbf{y}) = x_1 y_2 - x_2 y_1$.

3. Suppose a polygon in the plane has vertices $\begin{bmatrix} x_1 \\ y_1 \end{bmatrix}, \begin{bmatrix} x_2 \\ y_2 \end{bmatrix}, \ldots, \begin{bmatrix} x_n \\ y_n \end{bmatrix}$. Give a formula for its area.

4. (a) Check that when A and B are 2×2 matrices, we have $\det(AB) = \det A \det B$.
   (b) Let $A = A_\theta$ be a rotation matrix. Check that $\det(A_\theta B) = \det B$ for any 2×2 matrix B.
   (c) Use the result of part b and the properties of determinants to give an alternative proof that $\mathcal{D}(\mathbf{x}, \mathbf{y})$ is the signed area of the parallelogram spanned by $\mathbf{x}$ and $\mathbf{y}$.

5. Calculate the cross product of the given vectors $\mathbf{x}$ and $\mathbf{y}$.

   *(a) $\mathbf{x} = \begin{bmatrix} 1 \\ 0 \\ -1 \end{bmatrix}, \mathbf{y} = \begin{bmatrix} 1 \\ 2 \\ 1 \end{bmatrix}$
   (b) $\mathbf{x} = \begin{bmatrix} 1 \\ -2 \\ 1 \end{bmatrix}, \mathbf{y} = \begin{bmatrix} 7 \\ 1 \\ -5 \end{bmatrix}$

6. Find the area of the triangle with the given vertices.

   *(a) $A = \begin{bmatrix} 0 \\ 0 \\ 0 \end{bmatrix}, B = \begin{bmatrix} 1 \\ 0 \\ -1 \end{bmatrix}, C = \begin{bmatrix} 1 \\ 2 \\ 1 \end{bmatrix}$
   (c) $A = \begin{bmatrix} 0 \\ 0 \\ 0 \end{bmatrix}, B = \begin{bmatrix} 1 \\ -2 \\ 1 \end{bmatrix}, C = \begin{bmatrix} 7 \\ 1 \\ -5 \end{bmatrix}$

   (b) $A = \begin{bmatrix} 1 \\ -1 \\ 1 \end{bmatrix}, B = \begin{bmatrix} 2 \\ -1 \\ 0 \end{bmatrix}, C = \begin{bmatrix} 2 \\ 1 \\ 2 \end{bmatrix}$
   (d) $A = \begin{bmatrix} 1 \\ 1 \\ 1 \end{bmatrix}, B = \begin{bmatrix} 2 \\ -1 \\ 2 \end{bmatrix}, C = \begin{bmatrix} 8 \\ 2 \\ -4 \end{bmatrix}$

7. Find the equation of the (affine) plane containing the three points

   *(a) $A = \begin{bmatrix} 0 \\ 0 \\ 0 \end{bmatrix}, B = \begin{bmatrix} 1 \\ 0 \\ -1 \end{bmatrix}, C = \begin{bmatrix} 1 \\ 2 \\ 1 \end{bmatrix}$,
   (c) $A = \begin{bmatrix} 0 \\ 0 \\ 0 \end{bmatrix}, B = \begin{bmatrix} 1 \\ -2 \\ 1 \end{bmatrix}, C = \begin{bmatrix} 7 \\ 1 \\ -5 \end{bmatrix}$,

(b) $A = \begin{bmatrix} 1 \\ -1 \\ 1 \end{bmatrix}$, $B = \begin{bmatrix} 2 \\ -1 \\ 0 \end{bmatrix}$, $C = \begin{bmatrix} 2 \\ 1 \\ 2 \end{bmatrix}$, *(d) $A = \begin{bmatrix} 1 \\ 1 \\ 1 \end{bmatrix}$, $B = \begin{bmatrix} 2 \\ -1 \\ 2 \end{bmatrix}$, $C = \begin{bmatrix} 8 \\ 2 \\ -4 \end{bmatrix}$.

*8. Find the equation of the (affine) plane containing the points $\begin{bmatrix} 1 \\ -1 \\ 2 \end{bmatrix}$ and $\begin{bmatrix} 0 \\ 1 \\ 3 \end{bmatrix}$ and parallel to the vector $\begin{bmatrix} 1 \\ 0 \\ 1 \end{bmatrix}$.

9. Find the intersection of the two planes $x_1 + x_2 - 2x_3 = 0$ and $2x_1 + x_2 + x_3 = 0$.

10. Given the nonzero vector $\mathbf{a} \in \mathbb{R}^3$, $\mathbf{a} \cdot \mathbf{x} = b \in \mathbb{R}$, and $\mathbf{a} \times \mathbf{x} = \mathbf{c} \in \mathbb{R}^3$, can you determine the vector $\mathbf{x} \in \mathbb{R}^3$? If so, give a geometric construction for $\mathbf{x}$.

*11. Find the distance between the given skew lines in $\mathbb{R}^3$:

$$\ell : \begin{bmatrix} 2 \\ 1 \\ 1 \end{bmatrix} + t \begin{bmatrix} 0 \\ 1 \\ -1 \end{bmatrix} \quad \text{and} \quad m : \begin{bmatrix} 1 \\ 1 \\ 0 \end{bmatrix} + s \begin{bmatrix} 1 \\ 1 \\ 1 \end{bmatrix}.$$

*12. Find the volume of the parallelepiped spanned by

$$\mathbf{x} = \begin{bmatrix} 1 \\ 2 \\ 1 \end{bmatrix}, \quad \mathbf{y} = \begin{bmatrix} 2 \\ 3 \\ 1 \end{bmatrix}, \quad \text{and} \quad \mathbf{z} = \begin{bmatrix} -1 \\ 0 \\ 3 \end{bmatrix}.$$

13. Let $\mathcal{P}$ be a parallelogram in $\mathbb{R}^3$. Let $\mathcal{P}_1$ be its projection on the x_2x_3-plane, $\mathcal{P}_2$ be its projection on the x_1x_3-plane, and $\mathcal{P}_3$ be its projection on the x_1x_2-plane. Prove that
$$\bigl(\text{area}(\mathcal{P})\bigr)^2 = \bigl(\text{area}(\mathcal{P}_1)\bigr)^2 + \bigl(\text{area}(\mathcal{P}_2)\bigr)^2 + \bigl(\text{area}(\mathcal{P}_3)\bigr)^2.$$
(How's that for a generalization of the Pythagorean Theorem?)

14. Let $\mathbf{x}, \mathbf{y}, \mathbf{z} \in \mathbb{R}^3$.
(a) Show that $\mathbf{x} \times \mathbf{y} = -\mathbf{y} \times \mathbf{x}$ and $\mathbf{x} \times (\mathbf{y} + \mathbf{z}) = \mathbf{x} \times \mathbf{y} + \mathbf{x} \times \mathbf{z}$.
(b) Show that cross product is *not* associative; i.e., give specific vectors so that $(\mathbf{x} \times \mathbf{y}) \times \mathbf{z} \neq \mathbf{x} \times (\mathbf{y} \times \mathbf{z})$.

*15. Given $\mathbf{a} = \begin{bmatrix} a \\ b \\ c \end{bmatrix} \in \mathbb{R}^3$, define $T: \mathbb{R}^3 \to \mathbb{R}^3$ by $T(\mathbf{x}) = \mathbf{a} \times \mathbf{x}$. Prove that T is a linear transformation and give its standard matrix. Explain in the context of Proposition 4.5 why $[T]$ is skew-symmetric.

16. Let $\mathbf{x}, \mathbf{y}, \mathbf{z}, \mathbf{w} \in \mathbb{R}^3$. Show that $(\mathbf{x} \times \mathbf{y}) \cdot (\mathbf{z} \times \mathbf{w}) = \begin{vmatrix} \mathbf{x} \cdot \mathbf{z} & \mathbf{x} \cdot \mathbf{w} \\ \mathbf{y} \cdot \mathbf{z} & \mathbf{y} \cdot \mathbf{w} \end{vmatrix}$.

17. Suppose $\mathbf{u}, \mathbf{v}, \mathbf{w} \in \mathbb{R}^2$ are noncollinear points, and let $\mathbf{x} \in \mathbb{R}^2$.
(a) Show that we can write $\mathbf{x}$ uniquely in the form $\mathbf{x} = r\mathbf{u} + s\mathbf{v} + t\mathbf{w}$, where $r + s + t = 1$. (Hint: The vectors $\mathbf{v} - \mathbf{u}$ and $\mathbf{w} - \mathbf{u}$ must be nonparallel. Now apply the result of Example 11 of Section 4.)
(b) Show that r is the ratio of the signed area of the triangle with vertices $\mathbf{x}, \mathbf{v}$, and $\mathbf{w}$ to the signed area of the triangle with vertices $\mathbf{u}, \mathbf{v}$, and $\mathbf{w}$. Give corresponding formulas for s and t.

(c) Suppose **x** is the the intersection of the medians of the triangle with vertices **u**, **v**, and **w**. Compare the areas of the three triangles formed by joining **x** with any pair of the vertices. (Cf. Exercise 1.1.8.)

(d) Let $r = \mathcal{D}(\mathbf{v}, \mathbf{w})$, $s = \mathcal{D}(\mathbf{w}, \mathbf{u})$, and $t = \mathcal{D}(\mathbf{u}, \mathbf{v})$. Show that $r\mathbf{u} + s\mathbf{v} + t\mathbf{w} = \mathbf{0}$. Give a physical interpretation of this result.

18. In this exercise, we give a self-contained derivation of the geometric interpretation of the 3×3 determinant as signed volume.

(a) By direct algebraic calculation, show that $\|\mathbf{x} \times \mathbf{y}\|^2 = \|\mathbf{x}\|^2 \|\mathbf{y}\|^2 - (\mathbf{x} \cdot \mathbf{y})^2$. Deduce that $\|\mathbf{x} \times \mathbf{y}\|$ is the area of the parallelogram spanned by **x** and **y**.

(b) Show that $\mathbf{z} \cdot (\mathbf{x} \times \mathbf{y})$ is the signed volume of the parallelepiped spanned by **x**, **y**, and **z**.

(c) Conclude that $\mathcal{D}(\mathbf{x}, \mathbf{y}, \mathbf{z})$ equals the signed volume of that parallelepiped.

19. (**Heron's formula**) Given $\triangle OAB$, let $\overrightarrow{OA} = \mathbf{x}$ and $\overrightarrow{OB} = \mathbf{y}$, and set $\|\mathbf{x}\| = a$, $\|\mathbf{y}\| = b$, and $\|\mathbf{x} - \mathbf{y}\| = c$. Let $s = \frac{1}{2}(a + b + c)$ be the semiperimeter of the triangle. Use the formulas

$$\|\mathbf{x} \times \mathbf{y}\|^2 = \|\mathbf{x}\|^2 \|\mathbf{y}\|^2 - (\mathbf{x} \cdot \mathbf{y})^2 \quad \text{(see Exercise 18),}$$
$$\|\mathbf{x} - \mathbf{y}\|^2 = \|\mathbf{x}\|^2 + \|\mathbf{y}\|^2 - 2\mathbf{x} \cdot \mathbf{y}$$

to prove that the area $\mathcal{A}$ of $\triangle OAB$ satisfies

$$\mathcal{A}^2 = \frac{1}{4}\left(a^2 b^2 - \frac{1}{4}(c^2 - a^2 - b^2)^2\right) = s(s-a)(s-b)(s-c).$$

20. Let $\triangle ABC$ have sides a, b, and c. Let $s = \frac{1}{2}(a + b + c)$ be its semiperimeter. Prove that the *inradius* of the triangle (i.e., the radius of its inscribed circle) is $r = \sqrt{(s-a)(s-b)(s-c)/s}$.

CHAPTER 2

FUNCTIONS, LIMITS, AND CONTINUITY

In this brief chapter we introduce examples of nonlinear functions, their graphs, and their level sets. As usual in calculus, the notion of limit is a cornerstone on which calculus is built. To discuss "nearness," we need the concepts of open and closed sets and of convergent sequences. We then give the usual theorems on limits of functions and several equivalent ways of thinking about continuity. All of this will be the foundation for our work on differential calculus, which comes next.

1 SCALAR- AND VECTOR-VALUED FUNCTIONS

In first-year calculus we studied real-valued functions defined on intervals in $\mathbb{R}$ (or perhaps on all of $\mathbb{R}$). In Chapter 1 we began our study of *linear* functions from $\mathbb{R}^n$ to $\mathbb{R}^m$. There are three steps we might imagine to understand more complicated vector-valued functions of a vector variable.

1.1 Parametrized Curves

First, we might study a vector-valued function of a single variable. If we think of the independent variable t as time, then we can visualize $\mathbf{f}\colon (a,b) \to \mathbb{R}^n$ as a *parametrized curve*—we can imagine a particle moving in $\mathbb{R}^n$ as time varies, and $\mathbf{f}(t)$ gives its position at time t. At this point, we just give an assortment of examples. The careful analysis, including the associated differential calculus and physical interpretations, will come in the next chapter.

▶ **EXAMPLE 1**

The easiest examples, perhaps, are linear. Imagine a particle starting at position $\mathbf{x}_0$ and moving with constant velocity $\mathbf{v}$. Then its position at time t is evidently $\mathbf{f}(t) = \mathbf{x}_0 + t\mathbf{v}$ and its trajectory is a line passing through $\mathbf{x}_0$ and having *direction vector* $\mathbf{v}$, as shown in Figure 1.1. We refer to the vector-valued function $\mathbf{f}$ as a *parametrization* of the line. Here t is free to vary over all of $\mathbb{R}$. When we wish to parametrize the line passing through two points A and B, it is natural to use one of those points, say A, as $\mathbf{x}_0$ and the vector $\overrightarrow{AB}$ as the direction vector $\mathbf{v}$, as indicated in Figure 1.2. ◀

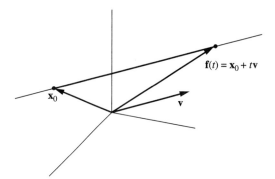

Figure 1.1

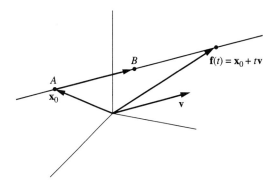

Figure 1.2

▶ **EXAMPLE 2**

The next curve with which every mathematics student is familiar is the circle. Essentially by the very definition of the trigonometric functions cos and sin, we obtain a very natural parametrization of a circle of radius a, as pictured in Figure 1.3(a):

$$\mathbf{f}(t) = a \begin{bmatrix} \cos t \\ \sin t \end{bmatrix} = \begin{bmatrix} a \cos t \\ a \sin t \end{bmatrix}, \quad 0 \le t \le 2\pi.$$

Now, if $a, b > 0$ and we apply the linear map

$$T \colon \mathbb{R}^2 \to \mathbb{R}^2, \quad T\begin{pmatrix} x \\ y \end{pmatrix} = \begin{bmatrix} ax \\ by \end{bmatrix},$$

we see that the unit circle $x^2 + y^2 = 1$ maps to the ellipse $\dfrac{x^2}{a^2} + \dfrac{y^2}{b^2} = 1$. Since $T\begin{pmatrix} \cos t \\ \sin t \end{pmatrix} = \begin{bmatrix} a \cos t \\ b \sin t \end{bmatrix}$, the latter gives a natural parametrization of the ellipse, as shown in Figure 1.3(b). Be warned, however: Here t is *not* the angle between the position vector and the positive x-axis, as Figure 1.3(c) indicates. ◀

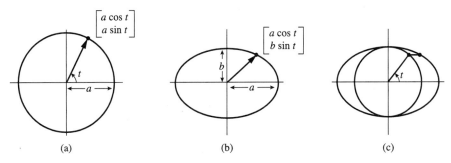

Figure 1.3

Now we come to some more interesting examples.

▶ **EXAMPLE 3**

Consider the two cubic curves in $\mathbb{R}^2$, illustrated in Figure 1.4. On the left is the *cuspidal cubic* $y^2 = x^3$, and on the right is the *nodal cubic* $y^2 = x^3 + x^2$. These can be parametrized, respectively, by the functions

$$\mathbf{f}(t) = \begin{bmatrix} t^2 \\ t^3 \end{bmatrix} \quad \text{and} \quad \mathbf{f}(t) = \begin{bmatrix} t^2 - 1 \\ t(t^2 - 1) \end{bmatrix},$$

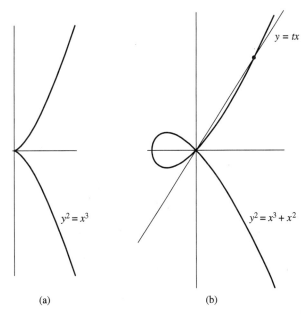

Figure 1.4

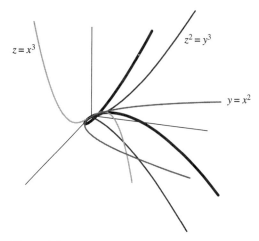

Figure 1.5

as the reader can verify.[1] Now consider the *twisted cubic* in $\mathbb{R}^3$, illustrated in Figure 1.5, given by

$$\mathbf{f}(t) = \begin{bmatrix} t \\ t^2 \\ t^3 \end{bmatrix}, \quad t \in \mathbb{R}.$$

Its projections in the xy-, xz-, and yz-coordinate planes are, respectively, $y = x^2$, $z = x^3$, and $z^2 = y^3$ (the cuspidal cubic).

▶ EXAMPLE 4

Our last example is a classic called the *cycloid*: It is the trajectory of a dot on a rolling wheel (circle). Consider the illustration in Figure 1.6. Assuming the wheel rolls without slipping, we see that the distance it travels along the ground is equal to the length of the circular arc subtended by the angle through which it has turned. That is, if the radius of the circle is a and it has turned through angle t, then the point of contact with the x-axis, Q, is at units to the right. The vector from the origin to the point P can be expressed as the sum of the three vectors $\overrightarrow{OQ}$, $\overrightarrow{QC}$, and $\overrightarrow{CP}$ (see Figure 1.7):

$$\overrightarrow{OP} = \overrightarrow{OQ} + \overrightarrow{QC} + \overrightarrow{CP}$$
$$= \begin{bmatrix} at \\ 0 \end{bmatrix} + \begin{bmatrix} 0 \\ a \end{bmatrix} + \begin{bmatrix} -a \sin t \\ -a \cos t \end{bmatrix},$$

and hence the function

$$\mathbf{f}(t) = \begin{bmatrix} at - a \sin t \\ a - a \cos t \end{bmatrix} = a \begin{bmatrix} t - \sin t \\ 1 - \cos t \end{bmatrix}, \quad t \in \mathbb{R}$$

gives a parametrization of the cycloid. ◀

---

[1] To see where the latter came from, as suggested by Figure 1.4(b), we substitute $y = tx$ in the equation and solve for x.

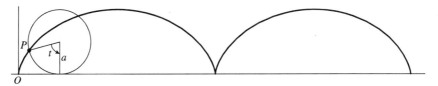

Figure 1.6

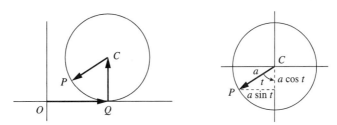

Figure 1.7

1.2 Scalar Functions of Several Variables

Next, we might study a scalar-valued function of several variables. For example, we might study elevation of the earth as a function of position on the surface of the earth; temperature at noon as a function of position in space; or, indeed, temperature as a function of both position and time. If we have a function of n variables, to avoid cumbersome notation, we will typically write

$$f\begin{pmatrix} x_1 \\ \vdots \\ x_n \end{pmatrix} \quad \text{rather than} \quad f\left(\begin{bmatrix} x_1 \\ \vdots \\ x_n \end{bmatrix}\right).$$

It would be typographically more pleasant and economical to suppress the vector notation and write merely $f(x_1, \ldots, x_n)$, as do most mathematicians. We hope our choice will make it easier for the reader to keep vectors in columns and not confuse rows and columns of matrices.

When $n = 1$ or $n = 2$, such functions are often best visualized by their graphs

$$\text{graph}(f) = \left\{ \begin{bmatrix} \mathbf{x} \\ f(\mathbf{x}) \end{bmatrix} : \mathbf{x} \in \mathbb{R}^n \right\} \subset \mathbb{R}^{n+1},$$

as pictured, for example, in Figure 1.8. There are two ways to try to visualize functions and their graphs, as we shall see in further detail in Chapter 3. One is to fix all of the coordinates of $\mathbf{x}$ but one, and see how f varies with each of $x_1, \ldots, x_n$ individually. This corresponds to taking slices of the graph, as shown in Figure 1.9. The other is to think of a topographical map, in which we see curves representing points at the same elevation. One then can lift each of these up to the appropriate height and imagine the surface interpolating among them, as illustrated in Figure 1.10. These curves are called *level curves* or *contour curves* of the function.

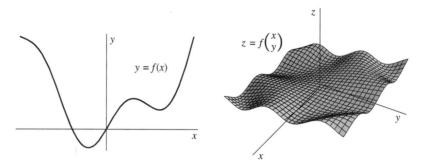

Figure 1.8

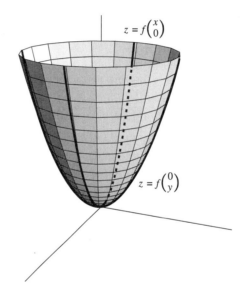

Figure 1.9

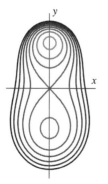 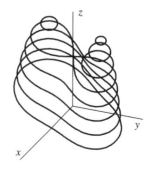 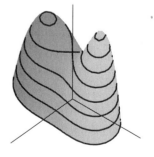

Figure 1.10

EXAMPLE 5

Suppose we see families of concentric circles as the level curves, as shown in Figure 1.11. We see that in (a) the circles are evenly spaced, whereas in (b) they grow closer together as we move outward. This tells us that in (a) the value of f grows linearly with the distance from the origin and in (b) it grows more quickly. Indeed, it is not surprising to see the corresponding graphs in Figure 1.12: The respective functions are $f(\mathbf{x}) = \|\mathbf{x}\|$ and $f(\mathbf{x}) = \|\mathbf{x}\|^2$.

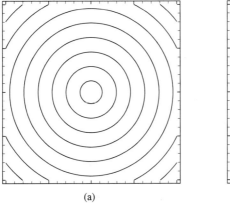
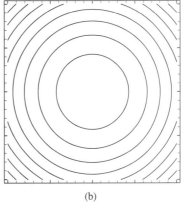

(a)   (b)

Figure 1.11

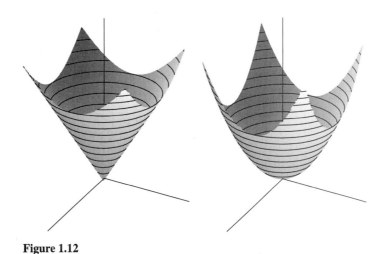

Figure 1.12

1.3 Vector Functions of Several Variables

Last, we think of vector-valued functions of several variables. Of course, the linear versions arise in the study of linear maps, as we've already seen, and a good deal more in the solution of systems of linear equations. Sometimes it is easiest to think of a vector-valued function

60 ▶ Chapter 2. Functions, Limits, and Continuity

$\mathbf{f}: \mathbb{R}^n \to \mathbb{R}^m$ as merely a collection of m scalar functions of n variables:

$$\mathbf{f}\begin{pmatrix} x_1 \\ \vdots \\ x_n \end{pmatrix} = \begin{bmatrix} f_1(\mathbf{x}) \\ \vdots \\ f_m(\mathbf{x}) \end{bmatrix}.$$

But in other instances, we really want to think of the values as geometrically defined vectors; fundamental examples are parametrized surfaces and vector fields (both of which we shall study a good deal in Chapter 8). Note that we will indicate a vector-valued function by boldface type.

▶ **EXAMPLE 6**

Consider the mapping

$$\mathbf{f}: (0, \infty) \times [0, 2\pi) \to \mathbb{R}^2 - \{\mathbf{0}\}$$

$$\mathbf{f}\begin{pmatrix} r \\ \theta \end{pmatrix} = \begin{bmatrix} r\cos\theta \\ r\sin\theta \end{bmatrix},$$

as illustrated in Figure 1.13. This is a one-to-one mapping onto $\mathbb{R}^2 - \{\mathbf{0}\}$. The coordinates $\begin{bmatrix} r \\ \theta \end{bmatrix}$ are often called the *polar coordinates* of the point $\begin{bmatrix} x \\ y \end{bmatrix} = \begin{bmatrix} r\cos\theta \\ r\sin\theta \end{bmatrix}$. ◀

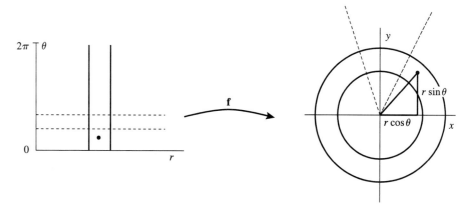

Figure 1.13

▶ **EXAMPLE 7**

Consider the mapping

$$\mathbf{f}\begin{pmatrix} u \\ v \end{pmatrix} = \begin{bmatrix} u\cos v \\ u\sin v \\ u \end{bmatrix}, \quad u > 0, \ 0 \le v \le 2\pi.$$

When we fix $u = u_0$, the image is a circle of radius u_0 at height u_0; when we fix $v = v_0$, the image is a ray making an angle of $\pi/4$ with the z-axis and whose projection into the xy-plane makes an angle of v_0 with the positive x-axis. Thus, the image of $\mathbf{f}$ is a cone, as pictured in Figure 1.14. ◀

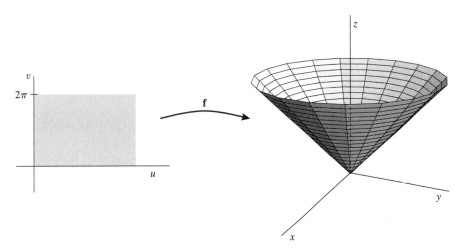

Figure 1.14

▶ **EXERCISES 2.1**

1. Find parametrizations of each of the following lines:
   (a) $3x_1 + 4x_2 = 6$,

   *(b) the line with slope $1/3$ that passes through $\begin{bmatrix} -1 \\ 2 \end{bmatrix}$,

   (c) the line through $A = \begin{bmatrix} 1 \\ 2 \\ 1 \end{bmatrix}$ and $B = \begin{bmatrix} 2 \\ 1 \\ 0 \end{bmatrix}$,

   (d) the line through $A = \begin{bmatrix} -2 \\ 1 \end{bmatrix}$ perpendicular to $\begin{bmatrix} 3 \\ 5 \end{bmatrix}$,

   *(e) the line through $\begin{bmatrix} 1 \\ 1 \\ 0 \\ -1 \end{bmatrix}$ parallel to $\mathbf{g}(t) = \begin{bmatrix} 2+t \\ 1-2t \\ 3t \\ 4-t \end{bmatrix}$.

2. (a) Give parametric equations for the circle $x^2 + y^2 = 1$ in terms of the length t pictured in Figure 1.15. (Hint: Use similar triangles and algebra.)

   (b) Use your answer to part a to produce infinitely many positive integer solutions[2] of $X^2 + Y^2 = Z^2$ with distinct ratios Y/X.

---

[2] These are called *Pythagorean triples*. Fermat asked whether there were any nonzero integer solutions of the corresponding equations $X^n + Y^n = Z^n$ for $n \geq 3$. In 1995, Andrew Wiles proved in a tour de force of algebraic number theory that there can be none.

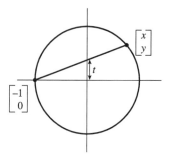

Figure 1.15

3. A string is unwound from a circular reel of radius a, being pulled taut at each instant. Give parametric equations for the tip of the string P in terms of the angle θ, as pictured in Figure 1.16.

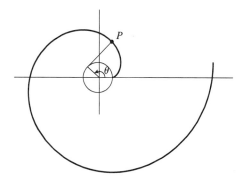

Figure 1.16

4. A wheel of radius a (perhaps belonging to a train) rolls along the x-axis. If a point P (on the wheel) is located a distance b from the center of the wheel, what are the parametric equations of its locus as the wheel rolls? (Note that when $b = a$ we obtain a cycloid.) See Figure 1.17.

Figure 1.17

5. *(a) A circle of radius b rolls without slipping outside a circle of radius $a > b$. Give the parametric equations of a point P on the circumference of the rolling circle (in terms of the angle θ of the line joining the centers of the two circles). (See Figure 1.18(a).)

(b) Now it rolls inside. Do the same as for part a.

These curves are called, respectively, an *epicycloid* and a *hypocycloid*.

6. A coin of radius $1''$ is rolled (without slipping) around the outside of a coin of radius $2''$. How many complete revolutions does its "head" make? Now explain the correct answer! (There is a famous story that the Educational Testing Service screwed this one up and was challenged by a precocious high school student who knew that he had done the problem correctly.)

*7. A dog buries a bone at $\begin{bmatrix} 0 \\ 1 \end{bmatrix}$. He is at the end of a 1-unit long leash, and his master walks down the positive x-axis, dragging the dog along. Since the dog wants to get back to the bone, he pulls

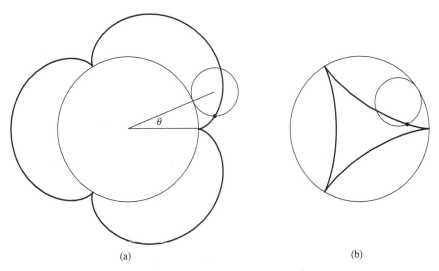

Figure 1.18

the leash taut. (It was pointed out to me by some students a few years ago that the realism of this model leaves something to be desired.) The curve the dog travels is called a *tractrix* (why?). Give parametric equations of the curve in terms of the parameters

(a) θ, (b) t,

as pictured in Figure 1.19. (Hint: The fact that the leash is pulled taut means that the leash is tangent to the curve. Show that $\theta'(t) = \sin \theta(t)$.)

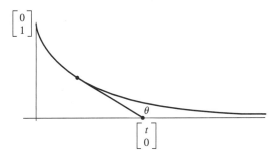

Figure 1.19

8. Prove that the twisted cubic (given in Example 3) has the property that any three distinct points on it determine a plane; i.e., no three distinct points are collinear.

9. Sketch families of level curves and the graphs of the following functions f.

(a) $f\begin{pmatrix} x \\ y \end{pmatrix} = 1 - y$ \quad (c) $f\begin{pmatrix} x \\ y \end{pmatrix} = x^2 - y^2$

(b) $f\begin{pmatrix} x \\ y \end{pmatrix} = x^2 - y$ \quad (d) $f\begin{pmatrix} x \\ y \end{pmatrix} = xy$

10. Consider the surfaces

$$X = \left\{ \begin{bmatrix} x \\ y \\ z \end{bmatrix} : x^2 + y^2 - z^2 = 1 \right\} \quad \text{and} \quad Y = \left\{ \begin{bmatrix} x \\ y \\ z \end{bmatrix} : x^2 + y^2 - z^2 = -1 \right\}.$$

(a) Sketch the surfaces.

(b) Give a rigorous argument (not merely based on your pictures) that every pair of points of X can be joined by a curve in X but that the same is not true of Y.

11. Consider the function

$$\mathbf{g}\begin{pmatrix} s \\ t \end{pmatrix} = \begin{bmatrix} (2 + \cos t) \cos s \\ (2 + \cos t) \sin s \\ \sin t \end{bmatrix}, \quad 0 \le s, t \le 2\pi.$$

(a) Sketch the image, X, of $\mathbf{g}$.

*(b) Find an algebraic equation satisfied by all the points of X.

12. Consider the function (defined wherever $st \ne 1$)

$$\mathbf{g}\begin{pmatrix} s \\ t \end{pmatrix} = \begin{bmatrix} \frac{st+1}{st-1} \\ \frac{s-t}{st-1} \\ \frac{s+t}{st-1} \end{bmatrix}.$$

(a) Show that every point in the image of $\mathbf{g}$ lies on the hyperboloid $x^2 + y^2 - z^2 = 1$.

(b) Show that the curves $\mathbf{g}\begin{pmatrix} s_0 \\ t \end{pmatrix}$ and $\mathbf{g}\begin{pmatrix} s \\ t_0 \end{pmatrix}$ (for s_0 and t_0 constants) are (subsets of) lines. (See Figure 1.20.)

(c) More challenging: What is the image of $\mathbf{g}$?

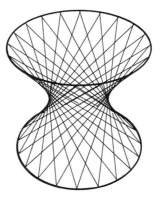

Figure 1.20

▶ 2 A BIT OF TOPOLOGY IN $\mathbb{R}^n$

Having introduced functions, we must next decide what it means for a function to be continuous. In one-variable calculus, we study functions defined on intervals and come to appreciate the difference between open and closed intervals. For example, the notion of limit is couched in terms of open intervals, whereas the maximum value theorem for continuous functions depends crucially on closed intervals. Matters are somewhat more subtle in higher dimensions, and we begin our assault on the analogous notions in $\mathbb{R}^n$.

Definition Let $\mathbf{a} \in \mathbb{R}^n$ and let $\delta > 0$. The *ball* of radius δ centered at $\mathbf{a}$ is
$$B(\mathbf{a}, \delta) = \{\mathbf{x} \in \mathbb{R}^n : \|\mathbf{x} - \mathbf{a}\| < \delta\}.$$
This is often called a *neighborhood* of $\mathbf{a}$.

Note that if $|x_i - a_i| < \dfrac{\delta}{\sqrt{n}}$ for all $i = 1, \ldots, n$, then
$$\|\mathbf{x} - \mathbf{a}\| = \sqrt{\sum_{i=1}^{n}(x_i - a_i)^2} < \sqrt{n\left(\frac{\delta}{\sqrt{n}}\right)^2} = \delta,$$
so $\mathbf{x} \in B(\mathbf{a}, \delta)$. And if $\mathbf{x} \in B(\mathbf{a}, \delta)$, then $|x_i - a_i| \le \|\mathbf{x} - \mathbf{a}\| < \delta$ for all $i = 1, \ldots, n$. Figure 2.1 illustrates these relationships. If $a_i < b_i$ for $i = 1, \ldots, n$, we can consider the *rectangle*
$$R = [a_1, b_1] \times [a_2, b_2] \times \cdots \times [a_n, b_n] = \{\mathbf{x} \in \mathbb{R}^n : a_i \le x_i \le b_i, \ i = 1, \ldots, n\}.$$
(Strictly speaking, we should call this a rectangular parallelepiped, but that's too much of a mouthful.) For reasons that will be obvious in a moment, when we construct the rectangle from *open* intervals, viz.,
$$S = (a_1, b_1) \times (a_2, b_2) \times \cdots \times (a_n, b_n) = \{\mathbf{x} \in \mathbb{R}^n : a_i < x_i < b_i, \ i = 1, \ldots, n\},$$
we call it an *open rectangle*.

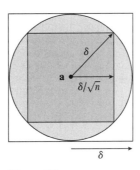

Figure 2.1

Definition We say a subset $U \subset \mathbb{R}^n$ is *open* if for every $\mathbf{a} \in U$ there is some ball centered at $\mathbf{a}$ that is completely contained in U; that is, there is $\delta > 0$ so that $B(\mathbf{a}, \delta) \subset U$.

▶ **EXAMPLE 1**

a. First of all, an open interval $(a, b) \subset \mathbb{R}$ is an open subset. Given any $c \in (a, b)$, choose $\delta < \min(c - a, b - c)$. Then $B(c, \delta) \subset (a, b)$. However, suppose we view this interval as a subset of $\mathbb{R}^2$; namely, $S = \left\{ \begin{bmatrix} x \\ 0 \end{bmatrix} : a < x < b \right\}$. Then it is no longer an open subset because no *ball* in $\mathbb{R}^2$ centered at $\begin{bmatrix} c \\ 0 \end{bmatrix}$ is contained in S, as Figure 2.2 plainly indicates.

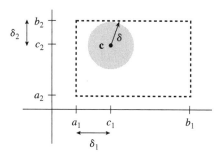

Figure 2.2

b. An open rectangle is an open set. As indicated in Figure 2.3, suppose $\mathbf{c} \in S = (a_1, b_1) \times (a_2, b_2) \times \cdots \times (a_n, b_n)$. Let $\delta_i = \min(c_i - a_i, b_i - c_i)$, $i = 1, \ldots, n$, and set $\delta = \min(\delta_1, \ldots, \delta_n)$. Then we claim that $B(\mathbf{c}, \delta) \subset S$. For if $\mathbf{x} \in B(\mathbf{c}, \delta)$, then $|x_i - c_i| \le \|\mathbf{x} - \mathbf{c}\| < \delta \le \delta_i$, so $a_i < x_i < b_i$, as required.

Figure 2.3

c. Consider $S = \left\{ \begin{bmatrix} x \\ y \end{bmatrix} : 0 < xy < 1 \right\}$. We want to show that S is open, so we choose $\mathbf{c} = \begin{bmatrix} a \\ b \end{bmatrix} \in S$. Without loss of generality, we may assume that $0 < b \le a$, as shown in Figure 2.4. We claim that the ball of radius

$$\delta = \frac{1}{\frac{1}{2}(a + \frac{1}{b})} - b = b\left(\frac{1 - ab}{1 + ab}\right)$$

centered at $\mathbf{c}$ is wholly contained in the region S. We consider the open rectangle centered at $\mathbf{c}$ with base $\frac{1}{b} - a$ and height 2δ; by construction, this rectangle is contained in S. Since $b \le a$ and $ab < 1$, it easy to check that the height is smaller than the length, and so the ball of radius δ centered at $\mathbf{c}$ is contained in the rectangle, hence in S. ◀

As we shall see in the next section, the concept of open sets is integral to the notion of continuity of a function.

We turn next to a discussion of *sequences*. The connections to open sets will become clear.

Definition A *sequence* of vectors (or points) in $\mathbb{R}^n$ is a function from the set of natural numbers, $\mathbb{N}$, to $\mathbb{R}^n$, i.e., an assignment of a vector $\mathbf{x}_k \in \mathbb{R}^n$ to each natural number $k \in \mathbb{N}$. We refer to $\mathbf{x}_k$ as the kth term of the sequence. We often abuse notation and write $\{\mathbf{x}_k\}$ for such a sequence, even though we are thinking of the actual function and not the set of its values.

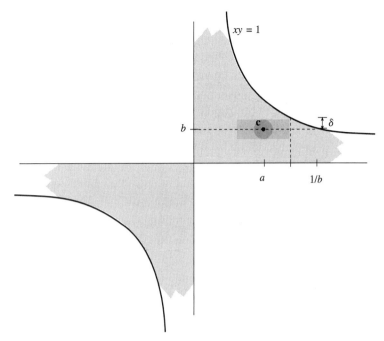

Figure 2.4

We say the sequence $\{\mathbf{x}_k\}$ *converges* to $\mathbf{a}$ (denoted $\mathbf{x}_k \to \mathbf{a}$ or $\lim_{k\to\infty} \mathbf{x}_k = \mathbf{a}$) if for all $\varepsilon > 0$, there is $K \in \mathbb{N}$ such that

$$\|\mathbf{x}_k - \mathbf{a}\| < \varepsilon \quad \text{whenever } k > K.$$

(That is, given any neighborhood of $\mathbf{a}$, "eventually"—past some K—all the elements $\mathbf{x}_k$ of the sequence lie inside.) We say the sequence $\{\mathbf{x}_k\}$ is *convergent* if it converges to some $\mathbf{a}$.

▶ **EXAMPLE 2**

Here are a few examples of sequences, both convergent and nonconvergent.

a. Let $x_k = \frac{k}{k+1}$. We suspect that $x_k \to 1$. To prove this, note that, given any $\varepsilon > 0$,

$$|x_k - 1| = \left|\frac{k}{k+1} - 1\right| = \frac{1}{k+1} < \varepsilon$$

whenever $k + 1 > 1/\varepsilon$. If we let $K = [1/\varepsilon]$ (the greatest integer less than or equal to $1/\varepsilon$), then it is easy to see that $k > K \implies k + 1 > 1/\varepsilon$, as required.

b. The sequence $\{x_k = (1 + \frac{1}{k})^k\}$ of real numbers is a famous one (think of compound interest) and converges to e, as the reader can check by taking logs and applying Proposition 3.6.

c. The sequence $1, -1, 1, -1, 1, \ldots$, i.e., $\{x_k = (-1)^{k+1}\}$, is not convergent. Since its consecutive terms are two units apart, no matter what $a \in \mathbb{R}$ and $K \in \mathbb{N}$ we pick, whenever $\varepsilon < 1$, we cannot have $|x_k - a| < \varepsilon$ whenever $k > K$. For if we did, we would have (by the triangle inequality)

$$2 = |x_{k+1} - x_k| \leq |x_{k+1} - a + a - x_k| \leq |x_{k+1} - a| + |x_k - a| < 2\varepsilon < 2$$

whenever $k > K$, which is, of course, impossible.

d. Let $\mathbf{x}_0 \in \mathbb{R}^n$ be a fixed vector. Define a sequence (recursively) by $\mathbf{x}_k = \frac{1}{2}\mathbf{x}_{k-1}$, $k \geq 1$. This means, of course, that $\mathbf{x}_k = \left(\frac{1}{2}\right)^k \mathbf{x}_0$, and so we suspect that $\mathbf{x}_k \to \mathbf{0}$. If $\mathbf{x}_0 = \mathbf{0}$, there is nothing to prove. Suppose $\mathbf{x}_0 \neq \mathbf{0}$ and $\varepsilon > 0$. Then we will have

$$\|\mathbf{x}_k - \mathbf{0}\| = \|\mathbf{x}_k\| = \left(\frac{1}{2}\right)^k \|\mathbf{x}_0\| < \varepsilon$$

whenever $k > \log_2(\|\mathbf{x}_0\|/\varepsilon) = \log(\|\mathbf{x}_0\|/\varepsilon)/\log 2$. So, if we take $K = [\log(\|\mathbf{x}_0\|/\varepsilon)/\log 2] + 1$, then it follows that whenever $k > K$ we have $\|\mathbf{x}_k\| < \varepsilon$, as required.

e. Let $A = \begin{bmatrix} 2 & 0 \\ 0 & 1 \end{bmatrix}$ and $\mathbf{x}_0 = \begin{bmatrix} 1 \\ 1 \end{bmatrix}$. Define a sequence of vectors in $\mathbb{R}^2$ recursively by

$$\mathbf{x}_k = \frac{A\mathbf{x}_{k-1}}{\|A\mathbf{x}_{k-1}\|}, \quad k \geq 1.$$

As the reader can easily prove by induction, we have $\mathbf{x}_k = \frac{1}{\sqrt{2^{2k}+1}} \begin{bmatrix} 2^k \\ 1 \end{bmatrix}$, and it follows that $\lim_{k \to \infty} \mathbf{x}_k = \begin{bmatrix} 1 \\ 0 \end{bmatrix}$.

▶ EXAMPLE 3

Suppose $\mathbf{x}_k, \mathbf{y}_k \in \mathbb{R}^n$, $\mathbf{x}_k \to \mathbf{a}$, and $\mathbf{y}_k \to \mathbf{b}$. Then it seems quite plausible that $\mathbf{x}_k + \mathbf{y}_k \to \mathbf{a} + \mathbf{b}$. Given $\varepsilon > 0$, we are to find $K \in \mathbb{N}$ so that whenever $k > K$ we have $\|(\mathbf{x}_k + \mathbf{y}_k) - (\mathbf{a} + \mathbf{b})\| < \varepsilon$. Rewriting, we observe that (by the triangle inequality)

$$\|(\mathbf{x}_k + \mathbf{y}_k) - (\mathbf{a} + \mathbf{b})\| = \|(\mathbf{x}_k - \mathbf{a}) + (\mathbf{y}_k - \mathbf{b})\| \leq \|\mathbf{x}_k - \mathbf{a}\| + \|\mathbf{y}_k - \mathbf{b}\|,$$

and so we can make $\|(\mathbf{x}_k + \mathbf{y}_k) - (\mathbf{a} + \mathbf{b})\| < \varepsilon$ by making $\|\mathbf{x}_k - \mathbf{a}\| < \varepsilon/2$ and $\|\mathbf{y}_k - \mathbf{b}\| < \varepsilon/2$. To this end, we use the definition of convergence of the sequences $\{\mathbf{x}_k\}$ and $\{\mathbf{y}_k\}$ as follows: There are $K_1, K_2 \in \mathbb{N}$ so that

$$\text{whenever } k > K_1, \text{ we have } \|\mathbf{x}_k - \mathbf{a}\| < \varepsilon/2$$

and

$$\text{whenever } k > K_2, \text{ we have } \|\mathbf{y}_k - \mathbf{b}\| < \varepsilon/2.$$

Thus, if we take $K = \max(K_1, K_2)$, whenever $k > K$, we will have $k > K_1$ and $k > K_2$, and so

$$\|(\mathbf{x}_k + \mathbf{y}_k) - (\mathbf{a} + \mathbf{b})\| \leq \|\mathbf{x}_k - \mathbf{a}\| + \|\mathbf{y}_k - \mathbf{b}\| < \frac{\varepsilon}{2} + \frac{\varepsilon}{2} = \varepsilon,$$

as was required. ◀

A crucial topological property of $\mathbb{R}$ is the *least upper bound property*:

> A subset $S \subset \mathbb{R}$ is *bounded above* if there is some $b \in \mathbb{R}$ so that $a \le b$ for all $a \in S$. Such a real number b is called an *upper bound* of S. Then every nonempty set S that is bounded above has a *least* upper bound, denoted $\sup S$. That is, $a \le \sup S$ for all $a \in S$ and $\sup S \le b$ for every upper bound b of S.

▶ **EXAMPLE 4**

    a. Let $S = [0, 1]$. Then S is bounded above (e.g., by 2) and $\sup S = 1$.

    b. Let $S = \{x \in \mathbb{Q} : x^2 < 2\}$. Then S is bounded above (e.g., by 2), and $\sup S = \sqrt{2}$. (Note that $\sqrt{2} \notin \mathbb{Q}$. The point is that the irrational numbers fill in all the "holes" among the rationals.)

    c. Suppose $\{x_k\}$ is a sequence of real numbers that is both bounded above and nondecreasing (i.e., $x_k \le x_{k+1}$ for all $k \in \mathbb{N}$). Then the sequence must converge. Since the sequence is bounded above, there is a least upper bound, α, for the set of its values. Now we claim that $x_k \to \alpha$. Given $\varepsilon > 0$, there is $K \in \mathbb{N}$ so that $\alpha - x_K < \varepsilon$ (for otherwise α would not be the *least* upper bound). But then the fact that the sequence is nondecreasing tells us that whenever $k > K$ we have $0 \le \alpha - x_k \le \alpha - x_K < \varepsilon$, as required. ◀

Definition Suppose $S \subset \mathbb{R}^n$. If S has the property that every *convergent* sequence of points *in* S converges to a point *in* S, then we say S is *closed*. That is, S is closed if the following is true: Whenever a convergent sequence $\mathbf{x}_k \to \mathbf{a}$ has the property that $\mathbf{x}_k \in S$ for all $k \in \mathbb{N}$, then $\mathbf{a} \in S$ as well.

▶ **EXAMPLE 5**

$S = \mathbb{R} - \{0\} = \{x \in \mathbb{R} : x \ne 0\} \subset \mathbb{R}$ is not closed, for if we take the sequence $x_k = 1/k \in S$, clearly $x_k \to 0$, and $0 \notin S$. ◀

This definition seems a bit strange, but it is exactly what we will need for many applications to come. In the meantime, if we need to decide whether or not a set is closed, it is easiest to use the following.

Proposition 2.1 *The subset $S \subset \mathbb{R}^n$ is closed if and only if its complement, $\mathbb{R}^n - S = \{\mathbf{x} \in \mathbb{R}^n : \mathbf{x} \notin S\}$, is open.*

Proof Suppose $\mathbb{R}^n - S$ is open and $\{\mathbf{x}_k\}$ is a convergent sequence with $\mathbf{x}_k \in S$ and limit $\mathbf{a}$. Suppose that $\mathbf{a} \notin S$. Then there is a neighborhood $B(\mathbf{a}, \varepsilon)$ of $\mathbf{a}$ wholly contained in $\mathbb{R}^n - S$, which means *no* element of the sequence $\{\mathbf{x}_k\}$ lies in that neighborhood, contradicting the fact that $\mathbf{x}_k \to \mathbf{a}$. Therefore, $\mathbf{a} \in S$, as desired.

Suppose S is closed and $\mathbf{b} \notin S$. We claim that there is a neighborhood of $\mathbf{b}$ lying entirely in $\mathbb{R}^n - S$. Suppose not. Then for every $k \in \mathbb{N}$, the ball $B(\mathbf{b}, 1/k)$ intersects S; that is, we can find a point $\mathbf{x}_k \in S$ with $\|\mathbf{x}_k - \mathbf{b}\| < 1/k$. Then $\{\mathbf{x}_k\}$ is a sequence of points in S converging to the point $\mathbf{b} \notin S$, contradicting the hypothesis that S is closed. ∎

EXAMPLE 6

It now follows easily that the closed interval $[a, b] = \{x \in \mathbb{R} : a \leq x \leq b\}$ is a closed subset of $\mathbb{R}$, inasmuch as its complement is the union of two open intervals. Similarly, the *closed ball* $\overline{B}(\mathbf{a}, r) = \{\mathbf{x} \in \mathbb{R}^n : \|\mathbf{x} - \mathbf{a}\| \leq r\}$ is a closed subset of $\mathbb{R}^n$, as we ask the reader to check in Exercise 5. In summary, our choice of terminology is felicitous indeed. ◄

Note that most sets are neither open nor closed. For example, the interval $S = (0, 1] \subset \mathbb{R}$ is not open because there is no neighborhood of the point 1 contained in S, and it is not closed because of the reasoning in Example 5. Be careful not to make a common mistake here: Just because a set isn't open, it need not be closed, and vice versa.

For future use, we make the following

Definition Suppose $S \subset \mathbb{R}^n$. We define the *closure* of S to be the smallest closed set containing S. It is denoted by $\overline{S}$.

We should think of $\overline{S}$ as containing all the points of S and all points that can be obtained as limits of convergent sequences of points of S. A slightly different formulation of this notion is given in Exercise 8.

EXERCISES 2.2

*1. Which of the following subsets of $\mathbb{R}^n$ is open? closed? neither? Prove your answer.
   (a) $\{x : 0 < x \leq 2\} \subset \mathbb{R}$
   (b) $\{x : x = 2^{-k} \text{ for some } k \in \mathbb{N} \text{ or } x = 0\} \subset \mathbb{R}$
   (c) $\left\{ \begin{bmatrix} x \\ y \end{bmatrix} : y > 0 \right\} \subset \mathbb{R}^2$
   (d) $\left\{ \begin{bmatrix} x \\ y \end{bmatrix} : y \geq 0 \right\} \subset \mathbb{R}^2$
   (e) $\left\{ \begin{bmatrix} x \\ y \end{bmatrix} : y > x \right\} \subset \mathbb{R}^2$
   (f) $\left\{ \begin{bmatrix} x \\ y \end{bmatrix} : xy \neq 0 \right\} \subset \mathbb{R}^2$
   (g) $\left\{ \begin{bmatrix} x \\ y \end{bmatrix} : y = x \right\} \subset \mathbb{R}^2$
   (h) $\{\mathbf{x} : 0 < \|\mathbf{x}\| < 1\} \subset \mathbb{R}^n$
   (i) $\{\mathbf{x} : \|\mathbf{x}\| > 1\} \subset \mathbb{R}^n$
   (j) $\{\mathbf{x} : \|\mathbf{x}\| \leq 1\} \subset \mathbb{R}^n$
   (k) the set of rational numbers, $\mathbb{Q} \subset \mathbb{R}$
   (l) $\left\{ \mathbf{x} : \|\mathbf{x}\| < 1 \text{ or } \left\|\mathbf{x} - \begin{bmatrix} 1 \\ 0 \end{bmatrix}\right\| < 1 \right\} \subset \mathbb{R}^2$
   (m) $\emptyset$ (the empty set)

♯2. Let $\{\mathbf{x}_k\}$ be a sequence of points in $\mathbb{R}^n$. For $i = 1, \ldots, n$, let $x_{k,i}$ denote the ith coordinate of the vector $\mathbf{x}_k$. Prove that $\mathbf{x}_k \to \mathbf{a}$ if and only if $x_{k,i} \to a_i$ for all $i = 1, \ldots, n$.

3. Suppose $\{\mathbf{x}_k\}$ is a sequence of points (vectors) in $\mathbb{R}^n$ converging to $\mathbf{a}$.
   (a) Prove that $\|\mathbf{x}_k\| \to \|\mathbf{a}\|$. (Hint: See Exercise 1.2.17.)
   (b) Prove that if $\mathbf{b} \in \mathbb{R}^n$ is any vector, then $\mathbf{b} \cdot \mathbf{x}_k \to \mathbf{b} \cdot \mathbf{a}$.

♯4. Prove that a rectangle $R = [a_1, b_1] \times \cdots \times [a_n, b_n] \subset \mathbb{R}^n$ is closed.

*5. Prove that the closed ball $\overline{B}(\mathbf{a}, r) = \{\mathbf{x} \in \mathbb{R}^n : \|\mathbf{x} - \mathbf{a}\| \leq r\} \subset \mathbb{R}^n$ is closed.

♯6. Given a sequence $\{\mathbf{x}_k\}$ of points in $\mathbb{R}^n$, a *subsequence* is formed by taking $\mathbf{x}_{k_1}, \mathbf{x}_{k_2}, \ldots, \mathbf{x}_{k_j}, \ldots$, where $k_1 < k_2 < k_3 < \cdots$.

(a) Prove that if the sequence $\{\mathbf{x}_k\}$ converges to $\mathbf{a}$, then any subsequence $\{\mathbf{x}_{k_j}\}$ converges to $\mathbf{a}$ as well.

(b) Is the converse valid? Give a proof or counterexample.

♯**7.** (a) Suppose U and V are open subsets of $\mathbb{R}^n$. Prove that $U \cup V$ and $U \cap V$ are open as well. (Recall that $U \cup V = \{\mathbf{x} \in \mathbb{R}^n : \mathbf{x} \in U \text{ or } \mathbf{x} \in V\}$ and $U \cap V = \{\mathbf{x} \in \mathbb{R}^n : \mathbf{x} \in U \text{ and } \mathbf{x} \in V\}$.)

(b) Suppose C and D are closed subsets of $\mathbb{R}^n$. Prove that $C \cup D$ and $C \cap D$ are closed as well.

♯**8.** Let $S \subset \mathbb{R}^n$. We say $\mathbf{a} \in S$ is an *interior point* of S if some neighborhood of $\mathbf{a}$ is contained in S. We say $\mathbf{a} \in \mathbb{R}^n$ is a *frontier point* of S if every neighborhood of $\mathbf{a}$ contains both points in S and points not in S.

(a) Show that every point of S is either an interior point or a frontier point, but give examples to show that a frontier point of S may or may not belong to S.

(b) Give an example of a set S every point of which is a frontier point.

(c) Prove that the set of frontier points of S is always a closed set.

(d) Let S' be the union of S and the set of frontier points of S. Prove that S' is closed.

(e) Suppose C is a closed set containing S. Prove that $S' \subset C$. Thus, S' is the smallest closed set containing S, which we have earlier called $\overline{S}$, the closure of S. (Hint: Show that $\mathbb{R}^n - C \subset \mathbb{R}^n - S'$.)

9. Continuing Exercise 8:

(a) Is it true that all the interior points of $\overline{S}$ are points of S? Is this true if S is open? (Give proofs or counterexamples.)

(b) Let $S \subset \mathbb{R}^n$ and let F be the set of the frontier points of S. Is it true that the set of frontier points of F is F itself? (Give a proof or counterexample.)

♯*__**10.** (a) Suppose $I_0 = [a, b]$ is a closed interval, and for each $k \in \mathbb{N}$, I_k is a *closed* interval with the property that $I_k \subset I_{k-1}$. Prove that there is a point $x \in \mathbb{R}$ so that $x \in I_k$ for all $k \in \mathbb{N}$.

(b) Give an example to show that the result of part a is false if the intervals are not closed.

11. Prove that the only subsets of $\mathbb{R}$ that are both open and closed are the empty set and $\mathbb{R}$ itself. (Hint: Suppose S is such a nonempty subset that is not equal to $\mathbb{R}$. Then there are some points $a \in S$ and $b \notin S$. Without loss of generality (how?), assume $a < b$. Let $\alpha = \sup\{x \in \mathbb{R} : [a, x] \subset S\}$. Show that neither $\alpha \in S$ nor $\alpha \notin S$ is possible.)

12. A sequence $\{\mathbf{x}_k\}$ of points in $\mathbb{R}^n$ is called a *Cauchy sequence* if for all $\varepsilon > 0$ there is $K \in \mathbb{N}$ so that whenever $k, \ell > K$, we have $\|\mathbf{x}_k - \mathbf{x}_\ell\| < \varepsilon$.

(a) Prove that any convergent sequence is Cauchy.

(b) Prove that if a subsequence of a Cauchy sequence converges, then the sequence itself must converge. (Hint: Suppose $\varepsilon > 0$. If $\mathbf{x}_{k_j} \to \mathbf{a}$, then there is $J \in \mathbb{N}$ so that whenever $j > J$, we have $\|\mathbf{x}_{k_j} - \mathbf{a}\| < \varepsilon/2$. There is also $K \in \mathbb{N}$ so that whenever $k, \ell > K$, we have $\|\mathbf{x}_k - \mathbf{x}_\ell\| < \varepsilon/2$. Choose $j > J$ so that $k_j > K$.)

*__**13.** Prove that if $\{\mathbf{x}_k\}$ is a Cauchy sequence, then all the points lie in some ball centered at the origin.

14. (a) Suppose $\{x_k\}$ is a sequence of points in $\mathbb{R}$ satisfying $a \le x_k \le b$ for all $k \in \mathbb{N}$. Prove that $\{x_k\}$ has a convergent subsequence (see Exercise 6). (Hint: If there are only finitely many distinct terms in the sequence, this should be easy. If there are infinitely many distinct terms in the sequence, then there must be infinitely many either in the left half-interval $[a, \frac{a+b}{2}]$ or in the right half-interval $[\frac{a+b}{2}, b]$. Let $[a_1, b_1]$ be such a half-interval. Continue the process, and apply Exercise 10.)

(b) Use the results of Exercises 12 and 13 to prove that any Cauchy sequence in $\mathbb{R}$ is convergent.

(c) Now prove that any Cauchy sequence in $\mathbb{R}^n$ is convergent. (Hint: Use Exercise 2.)

♯**15.** Suppose $S \subset \mathbb{R}^n$ is a closed set that is a subset of the rectangle $[a_1, b_1] \times \cdots \times [a_n, b_n]$. Prove that any sequence of points in S has a convergent subsequence. (Hint: Use repeatedly the idea of Exercise 14a.)

Chapter 2. Functions, Limits, and Continuity

3 LIMITS AND CONTINUITY

The concept on which all of calculus is founded is that of the *limit*. Limits are rather more subtle when we consider functions of more than one variable. We begin with the obligatory definition and some standard properties of limits.

Definition Let $U \subset \mathbb{R}^n$ be an open subset containing a neighborhood of $\mathbf{a} \in \mathbb{R}^n$, except perhaps for the point $\mathbf{a}$ itself. Suppose $\mathbf{f}: U \to \mathbb{R}^m$. We say that

$$\lim_{\mathbf{x} \to \mathbf{a}} \mathbf{f}(\mathbf{x}) = \boldsymbol{\ell}$$

($\mathbf{f}(\mathbf{x})$ approaches $\boldsymbol{\ell} \in \mathbb{R}^m$ as $\mathbf{x}$ approaches $\mathbf{a}$) if for every $\varepsilon > 0$ there is $\delta > 0$ so that

$$\|\mathbf{f}(\mathbf{x}) - \boldsymbol{\ell}\| < \varepsilon \quad \text{whenever} \quad 0 < \|\mathbf{x} - \mathbf{a}\| < \delta.$$

(Note that even if $\mathbf{f}(\mathbf{a})$ is defined, we say nothing whatsoever about its relation to $\boldsymbol{\ell}$.)

We begin by observing that for a vector-*valued* function, calculating a limit may be done component by component. As is customary by now, we denote the components of $\mathbf{f}$ by $f_1, \ldots, f_m$.

Proposition 3.1 $\lim_{\mathbf{x} \to \mathbf{a}} \mathbf{f}(\mathbf{x}) = \boldsymbol{\ell}$ *if and only if* $\lim_{\mathbf{x} \to \mathbf{a}} f_j(\mathbf{x}) = \ell_j$ *for all* $j = 1, \ldots, m$.

Proof The proof is based on Figure 2.1. Suppose $\lim_{\mathbf{x} \to \mathbf{a}} \mathbf{f}(\mathbf{x}) = \boldsymbol{\ell}$. We must show that for any $j = 1, \ldots, m$, we have $\lim_{\mathbf{x} \to \mathbf{a}} f_j(\mathbf{x}) = \ell_j$. Given $\varepsilon > 0$, there is $\delta > 0$ so that whenever $0 < \|\mathbf{x} - \mathbf{a}\| < \delta$, we have $\|\mathbf{f}(\mathbf{x}) - \boldsymbol{\ell}\| < \varepsilon$. But since we have

$$|f_j(\mathbf{x}) - \ell_j| \le \|\mathbf{f}(\mathbf{x}) - \boldsymbol{\ell}\|,$$

we see that whenever $0 < \|\mathbf{x} - \mathbf{a}\| < \delta$, we have $|f_j(\mathbf{x}) - \ell_j| < \varepsilon$, as required.

Now, suppose that $\lim_{\mathbf{x} \to \mathbf{a}} f_j(\mathbf{x}) = \ell_j$ for $j = 1, \ldots, m$. Given $\varepsilon > 0$, there are $\delta_1, \ldots, \delta_m > 0$ so that

$$|f_j(\mathbf{x}) - \ell_j| < \frac{\varepsilon}{\sqrt{m}} \quad \text{whenever} \quad 0 < \|\mathbf{x} - \mathbf{a}\| < \delta_j.$$

Let $\delta = \min(\delta_1, \ldots, \delta_m)$. Then whenever $0 < \|\mathbf{x} - \mathbf{a}\| < \delta$, we have

$$\|\mathbf{f}(\mathbf{x}) - \boldsymbol{\ell}\| = \sqrt{\sum_{j=1}^{m}(f_j(\mathbf{x}) - \ell_j)^2} < \sqrt{m\left(\frac{\varepsilon}{\sqrt{m}}\right)^2} = \varepsilon,$$

as required. ∎

EXAMPLE 1

Fix a nonzero vector $\mathbf{b} \in \mathbb{R}^n$. Let $f: \mathbb{R}^n \to \mathbb{R}$ be defined by $f(\mathbf{x}) = \mathbf{b} \cdot \mathbf{x}$. We claim that $\lim_{\mathbf{x} \to \mathbf{a}} f(\mathbf{x}) = \mathbf{b} \cdot \mathbf{a}$ because

$$|f(\mathbf{x}) - \mathbf{b} \cdot \mathbf{a}| = |\mathbf{b} \cdot \mathbf{x} - \mathbf{b} \cdot \mathbf{a}| = |\mathbf{b} \cdot (\mathbf{x} - \mathbf{a})| \le \|\mathbf{b}\| \|\mathbf{x} - \mathbf{a}\|,$$

by the Cauchy-Schwarz Inequality, Proposition 2.3 of Chapter 1. Thus, given $\varepsilon > 0$, if we take $\delta = \varepsilon/\|\mathbf{b}\|$, then whenever $0 < \|\mathbf{x} - \mathbf{a}\| < \delta$, we have $|f(\mathbf{x}) - \mathbf{b} \cdot \mathbf{a}| < \|\mathbf{b}\|\dfrac{\varepsilon}{\|\mathbf{b}\|} = \varepsilon$, as needed.

Note, moreover, that as a consequence of Proposition 3.1, for any linear map $T: \mathbb{R}^n \to \mathbb{R}^m$ it is the case that $\lim_{\mathbf{x} \to \mathbf{a}} T(\mathbf{x}) = T(\mathbf{a})$. ◀

▶ **EXAMPLE 2**

Let $f: \mathbb{R}^n \to \mathbb{R}$ be defined by $f(\mathbf{x}) = \|\mathbf{x}\|^2$. Then we claim that $\lim_{\mathbf{x} \to \mathbf{a}} f(\mathbf{x}) = \|\mathbf{a}\|^2$.

1. Suppose first that $\mathbf{a} = \mathbf{0}$. Since $r^2 \le r$ whenever $0 \le r \le 1$, we know that when $0 < \varepsilon \le 1$, we can choose $\delta = \varepsilon$ and then

$$0 < \|\mathbf{x}\| < \delta = \varepsilon \implies |f(\mathbf{x})| = \|\mathbf{x}\|^2 < \varepsilon^2 \le \varepsilon,$$

as required. But what if some (admittedly, silly) person hands us an $\varepsilon > 1$? The trick to take care of this is to let $\delta = \min(1, \varepsilon)$. Should ε be bigger than 1, then $\delta = 1$, and so when $0 < \|\mathbf{x}\| < \delta$, we know that $\|\mathbf{x}\| < 1$ and, once again, $|f(\mathbf{x})| < 1 < \varepsilon$, as required.

2. Now suppose $\mathbf{a} \ne \mathbf{0}$. Given $\varepsilon > 0$, let $\delta = \min\left(\|\mathbf{a}\|, \dfrac{\varepsilon}{3\|\mathbf{a}\|}\right)$. Now suppose $0 < \|\mathbf{x} - \mathbf{a}\| < \delta$. Then, in particular, we have $\|\mathbf{x}\| < \|\mathbf{a}\| + \delta \le 2\|\mathbf{a}\|$, so that $\|\mathbf{x} + \mathbf{a}\| \le \|\mathbf{x}\| + \|\mathbf{a}\| < 3\|\mathbf{a}\|$. Then

$$|f(\mathbf{x}) - \|\mathbf{a}\|^2| = |\mathbf{x} \cdot \mathbf{x} - \mathbf{a} \cdot \mathbf{a}| = |(\mathbf{x} + \mathbf{a}) \cdot (\mathbf{x} - \mathbf{a})|$$
$$\le \|\mathbf{x} + \mathbf{a}\|\|\mathbf{x} - \mathbf{a}\| < 3\|\mathbf{a}\| \cdot \dfrac{\varepsilon}{3\|\mathbf{a}\|} = \varepsilon,$$

as required.

Such sleight of hand (and more) is often required when the function is nonlinear. ◀

▶ **EXAMPLE 3**

Define $f: \mathbb{R}^2 - \{\mathbf{0}\} \to \mathbb{R}$ by $f\begin{pmatrix} x \\ y \end{pmatrix} = \dfrac{x^2 y}{x^2 + y^2}$. Does $\lim_{\mathbf{x} \to \mathbf{0}} f(\mathbf{x})$ exist? Since $|x| \le \sqrt{x^2 + y^2}$ and $|y| \le \sqrt{x^2 + y^2}$, we have (writing $\mathbf{x} = \begin{bmatrix} x \\ y \end{bmatrix}$)

$$|f(\mathbf{x})| \le \dfrac{\|\mathbf{x}\|^3}{\|\mathbf{x}\|^2} = \|\mathbf{x}\|,$$

and so $f(\mathbf{x}) \to 0$ as $\mathbf{x} \to \mathbf{0}$. (In particular, taking $\delta = \varepsilon$ will work.) An alternative approach, which will be useful later, is this:

$$|f(\mathbf{x})| = |y|\dfrac{x^2}{x^2 + y^2} \le |y|$$

since $0 \le \dfrac{x^2}{x^2 + y^2} \le 1$. Once again, $|y| \le \|\mathbf{x}\|$ and hence approaches 0 as $\mathbf{x} \to \mathbf{0}$. Thus, so does $|f(\mathbf{x})|$. (See Figure 3.1(a).) ◀

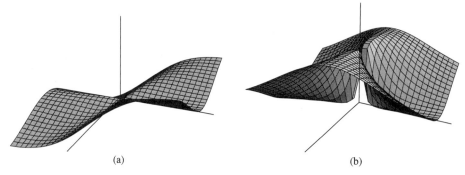

(a)   (b)

Figure 3.1

▶ **EXAMPLE 4**

Let's modify the previous example slightly. Define $f : \mathbb{R}^2 - \{\mathbf{0}\} \to \mathbb{R}$ by $f\begin{pmatrix} x \\ y \end{pmatrix} = \dfrac{x^2}{x^2 + y^2}$. We ask again whether $\lim_{\mathbf{x} \to \mathbf{0}} f(\mathbf{x})$ exists. Note that

$$\lim_{h \to 0} f\begin{pmatrix} h \\ 0 \end{pmatrix} = \lim_{h \to 0} \frac{h^2}{h^2} = 1, \quad \text{whereas}$$

$$\lim_{k \to 0} f\begin{pmatrix} 0 \\ k \end{pmatrix} = \lim_{k \to 0} \frac{0}{k^2} = 0.$$

Thus, $\lim_{\mathbf{x} \to \mathbf{0}} f(\mathbf{x})$ cannot exist (there is no number ℓ so that both 1 and 0 are less than ε away from ℓ when $0 < \varepsilon < 1/2$). (See Figure 3.1(b).) Now, what about $f\begin{pmatrix} x \\ y \end{pmatrix} = \dfrac{xy}{x^2 + y^2}$? In this case we have

$$\lim_{h \to 0} f\begin{pmatrix} h \\ 0 \end{pmatrix} = \lim_{k \to 0} f\begin{pmatrix} 0 \\ k \end{pmatrix} = 0,$$

so we might surmise that the limit exists and equals 0. But consider what happens if $\mathbf{x}$ approaches $\mathbf{0}$ along the line $y = x$:

$$\lim_{h \to 0} f\begin{pmatrix} h \\ h \end{pmatrix} = \lim_{h \to 0} \frac{h^2}{2h^2} = \frac{1}{2}.$$

Once again, the limit does not exist. ◀

The fundamental properties of limits with which every calculus student is familiar generalize in an obvious way to the multivariable setting.

Theorem 3.2 *Suppose $\mathbf{f}$ and $\mathbf{g}$ map a neighborhood of $\mathbf{a} \in \mathbb{R}^n$ (with the possible exception of the point $\mathbf{a}$ itself) to $\mathbb{R}^m$ and k maps the same neighborhood to $\mathbb{R}$. Suppose*

$$\lim_{\mathbf{x} \to \mathbf{a}} \mathbf{f}(\mathbf{x}) = \boldsymbol{\ell}, \quad \lim_{\mathbf{x} \to \mathbf{a}} \mathbf{g}(\mathbf{x}) = \mathbf{m}, \quad \text{and} \quad \lim_{\mathbf{x} \to \mathbf{a}} k(\mathbf{x}) = c.$$

Then

$$\lim_{x \to a} f(x) + g(x) = \ell + m,$$

$$\lim_{x \to a} f(x) \cdot g(x) = \ell \cdot m,$$

$$\lim_{x \to a} k(x)f(x) = c\ell.$$

Proof Given $\varepsilon > 0$, there are $\delta_1, \delta_2 > 0$ so that

$$\|f(x) - \ell\| < \frac{\varepsilon}{2} \quad \text{whenever} \quad 0 < \|x - a\| < \delta_1$$

and

$$\|g(x) - m\| < \frac{\varepsilon}{2} \quad \text{whenever} \quad 0 < \|x - a\| < \delta_2.$$

Let $\delta = \min(\delta_1, \delta_2)$. Whenever $0 < \|x - a\| < \delta$, we have

$$\|(f(x) + g(x)) - (\ell + m)\| \leq \|f(x) - \ell\| + \|g(x) - m\| < \frac{\varepsilon}{2} + \frac{\varepsilon}{2} = \varepsilon,$$

as required.

Given $\varepsilon > 0$, there are (different) $\delta_1, \delta_2 > 0$ so that

$$\|f(x) - \ell\| < \min\left(\frac{\varepsilon}{2(\|m\| + 1)}, 1\right) \quad \text{whenever} \quad 0 < \|x - a\| < \delta_1$$

and

$$\|g(x) - m\| < \frac{\varepsilon}{2(\|\ell\| + 1)} \quad \text{whenever} \quad 0 < \|x - a\| < \delta_2.$$

Note that when $0 < \|x - a\| < \delta_1$, we have (by the triangle inequality) $\|f(x)\| < \|\ell\| + 1$. Now, let $\delta = \min(\delta_1, \delta_2)$. Whenever $0 < \|x - a\| < \delta$, we have

$$|f(x) \cdot g(x) - \ell \cdot m| = |f(x) \cdot (g(x) - m) + (f(x) - \ell) \cdot m|$$
$$\leq \|f(x)\| \|g(x) - m\| + \|f(x) - \ell\| \|m\|$$
$$< (\|\ell\| + 1)\|g(x) - m\| + \|f(x) - \ell\| \|m\|$$
$$< (\|\ell\| + 1)\frac{\varepsilon}{2(\|\ell\| + 1)} + \frac{\varepsilon}{2(\|m\| + 1)}\|m\| < \frac{\varepsilon}{2} + \frac{\varepsilon}{2} = \varepsilon,$$

as required.

The proof of the last equality is left to the reader in Exercise 4. ∎

Once we have the concept of limit, the definition of continuity is quite straightforward.

Definition Let $U \subset \mathbb{R}^n$ be an open subset containing a neighborhood of $a \in \mathbb{R}^n$, and let $f: U \to \mathbb{R}^m$. We say f is *continuous at* a if

$$\lim_{x \to a} f(x) = f(a).$$

That is, **f** is continuous at **a** if, given any $\varepsilon > 0$, there is $\delta > 0$ so that

$$\|\mathbf{f}(\mathbf{x}) - \mathbf{f}(\mathbf{a})\| < \varepsilon \quad \text{whenever} \quad \|\mathbf{x} - \mathbf{a}\| < \delta.$$

We say **f** is *continuous* if it is continuous at every point of its domain.

As an immediate consequence of Theorem 3.2 and this definition we have

Corollary 3.3 *Suppose* **f** *and* **g** *map a neighborhood of* $\mathbf{a} \in \mathbb{R}^n$ *to* $\mathbb{R}^m$ *and* k *maps the same neighborhood to* $\mathbb{R}$. *If each function is continuous at* **a**, *then so are* $\mathbf{f} + \mathbf{g}$, $\mathbf{f} \cdot \mathbf{g}$, *and* $k\mathbf{f}$.

It is perhaps a bit more interesting to relate the definition of continuity to our notions of open and closed sets from the previous section. Let's first introduce a bit of standard notation: If $f\colon X \to Y$ is a function and $Z \subset Y$, we write $f^{-1}(Z) = \{x \in X : f(x) \in Z\}$, as illustrated in Figure 3.2. This is called the *preimage* of Z under the mapping f; be careful to remember that f may not be one-to-one and hence may well have no inverse function.

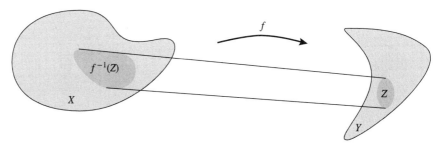

Figure 3.2

Proposition 3.4 *Let* $U \subset \mathbb{R}^n$ *be an open set. The function* $\mathbf{f}\colon U \to \mathbb{R}^m$ *is continuous if and only if for every open subset* $V \subset \mathbb{R}^m$, $\mathbf{f}^{-1}(V)$ *is an open set* (*i.e., the preimage of every open set is open*).

Proof $\Longleftarrow$: Suppose $\mathbf{a} \in U$ and we wish to prove **f** is continuous at **a**. Given $\varepsilon > 0$, we must find $\delta > 0$ (chosen small enough so that $B(\mathbf{a}, \delta) \subset U$) so that whenever $\|\mathbf{x} - \mathbf{a}\| < \delta$, we have $\|\mathbf{f}(\mathbf{x}) - \mathbf{f}(\mathbf{a})\| < \varepsilon$. Take $V = B(\mathbf{f}(\mathbf{a}), \varepsilon)$. Since $\mathbf{f}^{-1}(V)$ is open and $\mathbf{a} \in \mathbf{f}^{-1}(V)$, there is $\delta > 0$ so that $B(\mathbf{a}, \delta) \subset \mathbf{f}^{-1}(V)$. We then know that whenever $\|\mathbf{x} - \mathbf{a}\| < \delta$, we have $\mathbf{f}(\mathbf{x}) \in V = B(\mathbf{f}(\mathbf{a}), \varepsilon)$, and so we're done. (See Figure 3.3.)

$\Longrightarrow$: Suppose now that **f** is continuous and $V \subset \mathbb{R}^m$ is open. Let $\mathbf{a} \in \mathbf{f}^{-1}(V)$ be arbitrary. Since $\mathbf{f}(\mathbf{a}) \in V$ and V is open, there is $\varepsilon > 0$ so that $B(\mathbf{f}(\mathbf{a}), \varepsilon) \subset V$. Since **f** is continuous at **a**, there is $\delta > 0$ so that whenever $\|\mathbf{x} - \mathbf{a}\| < \delta$, we have $\|\mathbf{f}(\mathbf{x}) - \mathbf{f}(\mathbf{a})\| < \varepsilon$. So, whenever $\mathbf{x} \in B(\mathbf{a}, \delta)$, we have $\mathbf{f}(\mathbf{x}) \in B(\mathbf{f}(\mathbf{a}), \varepsilon) \subset V$. This means that $B(\mathbf{a}, \delta) \subset \mathbf{f}^{-1}(V)$, and so $\mathbf{f}^{-1}(V)$ is open. ■

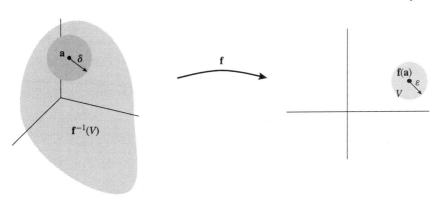

Figure 3.3

Proposition 3.5 *Suppose $U \subset \mathbb{R}^n$ and $W \subset \mathbb{R}^p$ are open, $\mathbf{f}: U \to \mathbb{R}^m$, $\mathbf{g}: W \to \mathbb{R}^n$, and the composition of functions $\mathbf{f} \circ \mathbf{g}$ is defined (i.e., $\mathbf{g}(\mathbf{x}) \in U$ for all $\mathbf{x} \in W$). Then if $\mathbf{f}$ and $\mathbf{g}$ are continuous, so is $\mathbf{f} \circ \mathbf{g}$.*

Proof Let $V \subset \mathbb{R}^m$ be open. We need to see that $(\mathbf{f} \circ \mathbf{g})^{-1}(V)$ is an open subset of $\mathbb{R}^p$. By the definition of composition, $(\mathbf{f} \circ \mathbf{g})(\mathbf{x}) = \mathbf{f}(\mathbf{g}(\mathbf{x})) \in V$ if and only if $\mathbf{g}(\mathbf{x}) \in \mathbf{f}^{-1}(V)$ if and only if $\mathbf{x} \in \mathbf{g}^{-1}(\mathbf{f}^{-1}(V))$. By the continuity of $\mathbf{f}$, we know that $\mathbf{f}^{-1}(V) \subset \mathbb{R}^n$ is open, and then by the continuity of $\mathbf{g}$, we deduce that $\mathbf{g}^{-1}(\mathbf{f}^{-1}(V)) \subset \mathbb{R}^p$ is open. That is, $(\mathbf{f} \circ \mathbf{g})^{-1}(V) \subset \mathbb{R}^p$ is open, as required. ∎

▶ **EXAMPLE 5**

Consider the function

$$f\begin{pmatrix} x \\ y \end{pmatrix} = \frac{x^2 y}{x^4 + y^2}, \quad \mathbf{x} \neq \mathbf{0}, \quad f(\mathbf{0}) = 0,$$

whose graph is shown in Figure 3.4. We ask whether f is continuous. Since the denominator vanishes only at the origin, it follows from Corollary 3.3 that f is continuous away from the origin. Now, since $f\begin{pmatrix} h \\ 0 \end{pmatrix} = f\begin{pmatrix} 0 \\ k \end{pmatrix} = 0$ for all h and k, we are encouraged. What's more, the restriction of f to any line $y = mx$ through the origin is continuous since

$$f\begin{pmatrix} x \\ mx \end{pmatrix} = \frac{mx^3}{x^4 + m^2 x^2} = \frac{mx}{m^2 + x^2} \quad \text{for all } x.$$

On the other hand, if we consider the restriction of f to the parabola $y = x^2$, we find that

$$f\begin{pmatrix} x \\ x^2 \end{pmatrix} = \begin{cases} \frac{x^4}{2x^4} = \frac{1}{2}, & x \neq 0 \\ 0, & x = 0 \end{cases},$$

which is definitely not a continuous function. Thus, f cannot be continuous. (If it were, according to Proposition 3.5, letting $\mathbf{g}(x) = \begin{bmatrix} x \\ x^2 \end{bmatrix}$, $f \circ \mathbf{g}$ would have to be continuous.) ◀

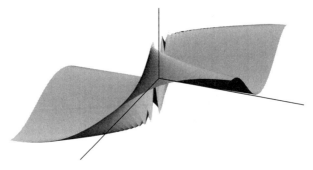

Figure 3.4

Next we come to the relation between continuity and convergent sequences.

Proposition 3.6 *Suppose $U \subset \mathbb{R}^n$ is open and $\mathbf{f}: U \to \mathbb{R}^m$. Then $\mathbf{f}$ is continuous at $\mathbf{a}$ if and only if for every sequence $\{\mathbf{x}_k\}$ of points in U converging to $\mathbf{a}$ the sequence $\{\mathbf{f}(\mathbf{x}_k)\}$ converges to $\mathbf{f}(\mathbf{a})$.*

Proof Suppose $\mathbf{f}$ is continuous at $\mathbf{a}$. Given $\varepsilon > 0$, there is $\delta > 0$ so that whenever $\|\mathbf{x} - \mathbf{a}\| < \delta$, we have $\|\mathbf{f}(\mathbf{x}) - \mathbf{f}(\mathbf{a})\| < \varepsilon$. Suppose $\mathbf{x}_k \to \mathbf{a}$. There is $K \in \mathbb{N}$ so that whenever $k > K$, we have $\|\mathbf{x}_k - \mathbf{a}\| < \delta$, and hence $\|\mathbf{f}(\mathbf{x}_k) - \mathbf{f}(\mathbf{a})\| < \varepsilon$. Thus, $\mathbf{f}(\mathbf{x}_k) \to \mathbf{f}(\mathbf{a})$, as required.

The converse is a bit trickier. We proceed by proving the contrapositive. Suppose $\mathbf{f}$ is *not* continuous at $\mathbf{a}$. This means that for *some* $\varepsilon_0 > 0$, it is the case that for *every* $\delta > 0$ there is *some* $\mathbf{x}$ with $\|\mathbf{x} - \mathbf{a}\| < \delta$ and $\|\mathbf{f}(\mathbf{x}) - \mathbf{f}(\mathbf{a})\| \geq \varepsilon_0$. So, for each $k \in \mathbb{N}$, there is a point $\mathbf{x}_k$ so that $\|\mathbf{x}_k - \mathbf{a}\| < 1/k$ and $\|\mathbf{f}(\mathbf{x}_k) - \mathbf{f}(\mathbf{a})\| \geq \varepsilon_0$. But this means that the sequence $\{\mathbf{x}_k\}$ converges to $\mathbf{a}$ and yet clearly the sequence $\{\mathbf{f}(\mathbf{x}_k)\}$ cannot converge to $\mathbf{f}(\mathbf{a})$. ∎

Corollary 3.7 *Suppose $\mathbf{f}: \mathbb{R}^n \to \mathbb{R}^m$ is continuous. Then for any $\mathbf{c} \in \mathbb{R}^m$, the level set $\mathbf{f}^{-1}(\{\mathbf{c}\}) = \{\mathbf{x} \in \mathbb{R}^n : \mathbf{f}(\mathbf{x}) = \mathbf{c}\}$ is a closed set.*

Proof Suppose $\{\mathbf{x}_k\}$ is a convergent sequence of points in $\mathbf{f}^{-1}(\{\mathbf{c}\})$, and let $\mathbf{a}$ be its limit. By Proposition 3.6, $\mathbf{f}(\mathbf{x}_k) \to \mathbf{f}(\mathbf{a})$. Since $\mathbf{f}(\mathbf{x}_k) = \mathbf{c}$ for all k, it follows that $\mathbf{f}(\mathbf{a}) = \mathbf{c}$ as well, and so $\mathbf{a} \in \mathbf{f}^{-1}(\{\mathbf{c}\})$, as we needed to show. ∎

▶ **EXAMPLE 6**

By Example 2, the function $f: \mathbb{R}^n \to \mathbb{R}$, $f(\mathbf{x}) = \|\mathbf{x}\|^2$, is continuous. The level sets of f are spheres centered at the origin. It follows that these spheres are closed sets. ◀

▶ **EXERCISES 2.3**

1. Prove that if $\lim_{\mathbf{x} \to \mathbf{a}} \mathbf{f}(\mathbf{x})$ exists, it must be unique. (Hint: If $\boldsymbol{\ell}$ and $\boldsymbol{m}$ are two putative limits, choose $\varepsilon = \|\boldsymbol{\ell} - \boldsymbol{m}\|/2$.)

♯2. Prove that $f: \mathbb{R}^n \to \mathbb{R}$, $f(\mathbf{x}) = \|\mathbf{x}\|$ is continuous. (Hint: Use Exercise 1.2.17.)

‡3. **(Squeeze Principle)** Suppose f, g, and h are real-valued functions on a neighborhood of $\mathbf{a}$ (perhaps not including the point $\mathbf{a}$ itself). Suppose $f(\mathbf{x}) \le g(\mathbf{x}) \le h(\mathbf{x})$ for all $\mathbf{x}$ and $\lim_{\mathbf{x} \to \mathbf{a}} f(\mathbf{x}) = \ell = \lim_{\mathbf{x} \to \mathbf{a}} h(\mathbf{x})$. Prove that $\lim_{\mathbf{x} \to \mathbf{a}} g(\mathbf{x}) = \ell$. (Hint: Given $\varepsilon > 0$, show that there is $\delta > 0$ so that whenever $0 < \|\mathbf{x} - \mathbf{a}\| < \delta$, we have $-\varepsilon < f(\mathbf{x}) - \ell \le g(\mathbf{x}) - \ell \le h(\mathbf{x}) - \ell < \varepsilon$.)

4. Suppose $\lim_{\mathbf{x} \to \mathbf{a}} \mathbf{f}(x) = \boldsymbol{\ell}$ and $\lim_{\mathbf{x} \to \mathbf{a}} k(x) = c$. Prove that $\lim_{\mathbf{x} \to \mathbf{a}} k(x)\mathbf{f}(x) = c\boldsymbol{\ell}$.

‡5. Suppose $U \subset \mathbb{R}^n$ is open and $f\colon U \to \mathbb{R}$ is continuous. If $\mathbf{a} \in U$ and $f(\mathbf{a}) > 0$, prove that there is $\delta > 0$ so that $f(\mathbf{x}) > 0$ for all $\mathbf{x} \in B(\mathbf{a}, \delta)$. (That is, a continuous function that is positive at a point must be positive on a neighborhood of that point.) Can you state a somewhat stronger result?

6. Let $U \subset \mathbb{R}^n$ be open. Suppose $g\colon U \to \mathbb{R}$ is continuous and $g(\mathbf{a}) \neq 0$. Prove that $1/g$ is continuous on some neighborhood of $\mathbf{a}$. (Hint: Apply Proposition 3.5.)

‡7. Suppose $T\colon \mathbb{R}^n \to \mathbb{R}^m$ is a linear map.
(a) Prove that T is continuous. (See Example 1.)
(b) Deduce the result of part a an alternative way by showing that for any $m \times n$ matrix A, we have
$$\|A\mathbf{x}\| \le \left(\sum_{i,j} a_{ij}^2\right)^{1/2} \|\mathbf{x}\|.$$

*8. Using Theorem 3.2 whenever possible (and standard facts from one-variable calculus), decide in each case whether $\lim_{\mathbf{x} \to \mathbf{0}} f(\mathbf{x})$ exists. Provide appropriate justification.

(a) $f\begin{pmatrix}x\\y\end{pmatrix} = \dfrac{xy}{x+y+1}$

(b) $f\begin{pmatrix}x\\y\end{pmatrix} = \dfrac{\sin(x^2+y^2)}{x^2+y^2}$

(c) $f\begin{pmatrix}x\\y\end{pmatrix} = \dfrac{x^2-y^2}{x-y}$, $x \neq y$, $f\begin{pmatrix}x\\x\end{pmatrix} = 0$

(d) $f\begin{pmatrix}x\\y\end{pmatrix} = e^{x^2+y^2}$

(e) $f\begin{pmatrix}x\\y\end{pmatrix} = e^{-1/(x^2+y^2)}$

(f) $f\begin{pmatrix}x\\y\end{pmatrix} = \dfrac{x^2+y^2}{x}$, $x \neq 0$, $f\begin{pmatrix}0\\y\end{pmatrix} = 0$

(g) $f\begin{pmatrix}x\\y\end{pmatrix} = \dfrac{x^3}{x^2+y^2}$

(h) $f\begin{pmatrix}x\\y\end{pmatrix} = \dfrac{x\sin^2 y}{x^2+y^2}$

(i) $f\begin{pmatrix}x\\y\end{pmatrix} = \dfrac{xy}{x^3-y^3}$, $x \neq y$, $f\begin{pmatrix}x\\x\end{pmatrix} = 0$

(j) $f\begin{pmatrix}x\\y\end{pmatrix} = \dfrac{x^2+y^2}{x+y}$, $x \neq -y$, $f\begin{pmatrix}x\\-x\end{pmatrix} = 0$

9. Suppose $\mathbf{f}\colon \mathbb{R}^n \to \mathbb{R}^n$ is continuous and $\mathbf{x}_0$ is arbitrary. Define a sequence by $\mathbf{x}_k = \mathbf{f}(\mathbf{x}_{k-1})$, $k = 1, 2, \ldots$. Prove that if $\mathbf{x}_k \to \mathbf{a}$, then $\mathbf{f}(\mathbf{a}) = \mathbf{a}$. We say $\mathbf{a}$ is a *fixed point* of $\mathbf{f}$.

10. Use Exercise 9 to find the limit of each of the following sequences of points in $\mathbb{R}$, presuming it exists.

*(a) $x_0 = 1$, $x_k = \sqrt{2x_{k-1}}$

(b) $x_0 = 5$, $x_k = \dfrac{x_{k-1}}{2} + \dfrac{2}{x_{k-1}}$

(c) $x_0 = 1$, $x_k = 1 + \dfrac{1}{x_{k-1}}$

*(d) $x_0 = 1$, $x_k = 1 + \dfrac{1}{1+x_{k-1}}$

11. Give an example of a discontinuous function $f\colon \mathbb{R} \to \mathbb{R}$ having the property that for every $c \in \mathbb{R}$ the level set $f^{-1}(\{c\})$ is closed.

12. If $f\colon X \to Y$ is a function and $U \subset X$, recall that the *image* of U is the set $f(U) = \{y \in Y : y = f(x) \text{ for some } x \in U\}$. Prove or give a counterexample: If $\mathbf{f}$ is continuous, then the image of every open set is open. (Cf. Proposition 3.4.)

13. Prove that if $\mathbf{f}$ is continuous, then the preimage of every closed set is closed.

14. Identify $\mathcal{M}_{m\times n}$, the set of $m \times n$ matrices, with $\mathbb{R}^{mn}$ in the obvious way.
(a) Prove that when $n = 2$ or 3, the set of $n \times n$ matrices with nonzero determinant is an open subset of $\mathcal{M}_{n\times n}$.
(b) Prove that the set of $n \times n$ matrices A satisfying $A^\mathsf{T} A = I_n$ is a closed subset of $\mathcal{M}_{n\times n}$.

15. (a) Let
$$f\begin{pmatrix} x \\ y \end{pmatrix} = \begin{cases} 0, & |y| > x^2 \text{ or } y = 0 \\ 1, & \text{otherwise} \end{cases}.$$
Show that f is continuous at $\mathbf{0}$ on every line through the origin but is not continuous at $\mathbf{0}$.

(b) Give a function that is continuous at $\mathbf{0}$ along every line and every parabola $y = kx^2$ through the origin but is not continuous at $\mathbf{0}$.

16. Give a function $f: \mathbb{R}^2 \to \mathbb{R}$ that is
(a) continuous at $\mathbf{0}$ along every line through the origin but unbounded in every neighborhood of $\mathbf{0}$;
(b) continuous at $\mathbf{0}$ along every line through the origin, unbounded in every neighborhood of $\mathbf{0}$, and discontinuous only at the origin.

17. Generalizing Example 5, determine for what positive values of α, β, γ, and δ the analogous function
$$f\begin{pmatrix} x \\ y \end{pmatrix} = \frac{|x|^\alpha |y|^\beta}{|x|^\gamma + |y|^\delta}, \quad \mathbf{x} \neq \mathbf{0}, \qquad f(\mathbf{0}) = 0$$
is continuous at $\mathbf{0}$.

18. (a) Suppose A is an invertible $n \times n$ matrix. Show that the solution of $A\mathbf{x} = \mathbf{b}$ varies continuously with $\mathbf{b} \in \mathbb{R}^n$.

(b) Show that the solution of $A\mathbf{x} = \mathbf{b}$ varies continuously as a function of $\begin{bmatrix} A \\ \mathbf{b} \end{bmatrix}$, as A varies over all invertible matrices and $\mathbf{b}$ over $\mathbb{R}^n$. (You should be able to get the cases $n = 1$ and $n = 2$. What do you need for $n > 2$?)

CHAPTER 3

THE DERIVATIVE

In this chapter we start in earnest on calculus. The immediate goal is to define the tangent plane at a point to the graph of a function, which should be the suitable generalization of the tangent lines in single-variable calculus. The fundamental computational tool is the partial derivative, a direct application of single-variable calculus tools. But the actual definition of a *differentiable* function immediately involves linear algebra. We establish various differentiation rules and then introduce the gradient, which, as common parlance has come to suggest, tells us in which direction a scalar function increases the fastest; thus, it is highly important for physical and mathematical applications. We conclude the chapter with a discussion of Kepler's laws, the geometry of curves, and higher-order derivatives.

▶ 1 PARTIAL DERIVATIVES AND DIRECTIONAL DERIVATIVES

Whenever possible it is desirable to reduce problems in multivariable calculus to those in single-variable calculus. It is reasonable to think that we should understand a function by knowing how it varies with each variable, fixing all the others. (A physical analogy is this: To find the change in energy of a gas as we change its volume and temperature, we imagine that we can first fix the volume and change the temperature and then, fixing the temperature, change the volume.)

We begin by considering a real-valued function f of two variables, x and y.

Definition We define the *partial derivatives* $\dfrac{\partial f}{\partial x}$ and $\dfrac{\partial f}{\partial y}$ as follows:

$$\frac{\partial f}{\partial x}\begin{pmatrix}a\\b\end{pmatrix} = \lim_{h\to 0} \frac{f\begin{pmatrix}a+h\\b\end{pmatrix} - f\begin{pmatrix}a\\b\end{pmatrix}}{h}$$

$$\frac{\partial f}{\partial y}\begin{pmatrix}a\\b\end{pmatrix} = \lim_{k\to 0} \frac{f\begin{pmatrix}a\\b+k\end{pmatrix} - f\begin{pmatrix}a\\b\end{pmatrix}}{k}.$$

Very simply, if we fix b, then $\dfrac{\partial f}{\partial x}\begin{pmatrix}a\\b\end{pmatrix}$ is the derivative at a (or slope) of the function $F(x) = f\begin{pmatrix}x\\b\end{pmatrix}$, as indicated in Figure 1.1. There is an analogous interpretation of $\dfrac{\partial f}{\partial y}\begin{pmatrix}a\\b\end{pmatrix}$.

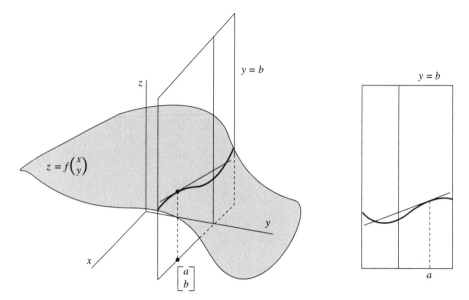

Figure 1.1

More generally, if $U \subset \mathbb{R}^n$ is open and $\mathbf{a} \in U$, we define the jth partial derivative of $\mathbf{f} \colon U \to \mathbb{R}^m$ at $\mathbf{a}$ to be

$$\frac{\partial \mathbf{f}}{\partial x_j}(\mathbf{a}) = \lim_{t \to 0} \frac{\mathbf{f}(\mathbf{a} + t\mathbf{e}_j) - \mathbf{f}(\mathbf{a})}{t}, \quad j = 1, \ldots, n$$

(provided this limit exists). Many authors use the alternative notation $D_j f(\mathbf{a})$ to represent the jth partial derivative of f at $\mathbf{a}$.

▶ **EXAMPLE 1**

Let $f\begin{pmatrix} x \\ y \end{pmatrix} = x^3 y^5 + e^{xy} \sin(2x + 3y)$. Then

$$\frac{\partial f}{\partial x}\begin{pmatrix} x \\ y \end{pmatrix} = 3x^2 y^5 + e^{xy}\bigl(y \sin(2x + 3y) + 2 \cos(2x + 3y)\bigr) \quad \text{and}$$

$$\frac{\partial f}{\partial y}\begin{pmatrix} x \\ y \end{pmatrix} = 5x^3 y^4 + e^{xy}\bigl(x \sin(2x + 3y) + 3 \cos(2x + 3y)\bigr). \quad \triangleleft$$

The partial derivatives of f measure the rate of change of f in the directions of the coordinate axes, i.e., in the directions of the standard basis vectors $\mathbf{e}_1, \ldots, \mathbf{e}_n$. Given any nonzero vector $\mathbf{v}$, it is natural to consider the rate of change of f in the direction of $\mathbf{v}$.

Definition Let $U \subset \mathbb{R}^n$ be open and $\mathbf{a} \in U$. We define the *directional derivative* of $\mathbf{f}: U \to \mathbb{R}^m$ at $\mathbf{a}$ *in the direction* $\mathbf{v}$ to be

$$D_{\mathbf{v}}\mathbf{f}(\mathbf{a}) = \lim_{t \to 0} \frac{\mathbf{f}(\mathbf{a} + t\mathbf{v}) - \mathbf{f}(\mathbf{a})}{t},$$

provided this limit exists.

Note that the jth partial derivative of $\mathbf{f}$ at $\mathbf{a}$ is just $D_{\mathbf{e}_j}\mathbf{f}(\mathbf{a})$. When $n = 2$ and $m = 1$, as we see from Figure 1.2, if $\|\mathbf{v}\| = 1$, the directional derivative $D_{\mathbf{v}}f(\mathbf{a})$ is just the slope at $\mathbf{a}$ of the graph we obtain by restricting to the line through $\mathbf{a}$ with direction $\mathbf{v}$.

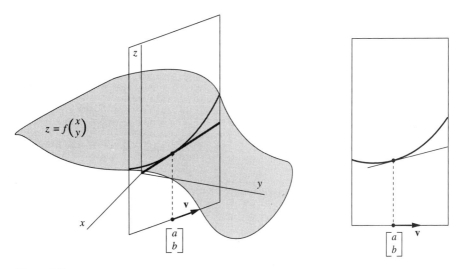

Figure 1.2

Remark Our terminology might be a bit misleading. Note that since

$$\begin{aligned}
D_{c\mathbf{v}}\mathbf{f}(\mathbf{a}) &= \lim_{t \to 0} \frac{\mathbf{f}(\mathbf{a} + t(c\mathbf{v})) - \mathbf{f}(\mathbf{a})}{t} = \lim_{t \to 0} \frac{\mathbf{f}(\mathbf{a} + (ct)\mathbf{v}) - \mathbf{f}(\mathbf{a})}{t} \\
&= c \lim_{t \to 0} \frac{\mathbf{f}(\mathbf{a} + (ct)\mathbf{v}) - \mathbf{f}(\mathbf{a})}{ct} = c \lim_{s \to 0} \frac{\mathbf{f}(\mathbf{a} + s\mathbf{v}) - \mathbf{f}(\mathbf{a})}{s} \\
&= c D_{\mathbf{v}}\mathbf{f}(\mathbf{a}),
\end{aligned}$$

the directional derivative depends not only on the direction of $\mathbf{v}$, but also on its magnitude. It is for this reason that many calculus books require that one specify a *unit* vector $\mathbf{v}$. It makes more sense to think of $D_{\mathbf{v}}\mathbf{f}(\mathbf{a})$ as the rate of change of $\mathbf{f}$ as experienced by an observer moving with instantaneous velocity $\mathbf{v}$. We shall return to this interpretation in Section 3.

EXAMPLE 2

Let $f : \mathbb{R}^2 \to \mathbb{R}$ be defined by

$$f\begin{pmatrix} x \\ y \end{pmatrix} = \frac{|x|y}{\sqrt{x^2 + y^2}}, \quad \mathbf{x} \neq \mathbf{0}, \quad f(\mathbf{0}) = 0,$$

whose graph is shown in Figure 1.3. Then the directional derivative of f at $\mathbf{0}$ in the direction of the unit vector $\mathbf{v} = \begin{bmatrix} v_1 \\ v_2 \end{bmatrix}$ is

$$D_{\mathbf{v}} f(\mathbf{0}) = \lim_{t \to 0} \frac{f(t\mathbf{v}) - f(\mathbf{0})}{t} = \lim_{t \to 0} \frac{\frac{|tv_1|(tv_2)}{|t|}}{t} = |v_1| v_2.$$

Note that both partial derivatives of f at $\mathbf{0}$ are 0, and yet the remaining directional derivatives are nonzero. ◀

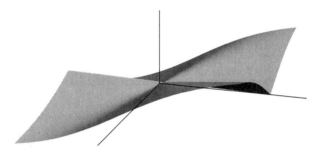

Figure 1.3

EXAMPLE 3

Let $f: \mathbb{R}^n \to \mathbb{R}$ be defined by $f(\mathbf{x}) = \|\mathbf{x}\|$. Let $\mathbf{a} \neq \mathbf{0}$ be arbitrary. Let $\mathbf{v} = \mathbf{a}/\|\mathbf{a}\|$ be a unit vector pointing radially outward at $\mathbf{a}$. Then

$$D_{\mathbf{v}} f(\mathbf{a}) = \lim_{t \to 0} \frac{f(\mathbf{a} + t\mathbf{v}) - f(\mathbf{a})}{t} = \lim_{t \to 0} \frac{\|\mathbf{a} + t\frac{\mathbf{a}}{\|\mathbf{a}\|}\| - \|\mathbf{a}\|}{t} = \lim_{t \to 0} \frac{(\|\mathbf{a}\| + t) - \|\mathbf{a}\|}{t} = 1.$$

On the other hand, if $\mathbf{v} \cdot \mathbf{a} = 0$, then

$$D_{\mathbf{v}} f(\mathbf{a}) = \lim_{t \to 0} \frac{\|\mathbf{a} + t\mathbf{v}\| - \|\mathbf{a}\|}{t} = 0,$$

inasmuch as $t = 0$ is a global minimum of the function $g(t) = \|\mathbf{a} + t\mathbf{v}\|$. (Why?) ◀

EXAMPLE 4

Let $f\begin{pmatrix} x \\ y \\ z \end{pmatrix} = x^2 y + e^{3x+y-z}$; let $\mathbf{a} = \begin{bmatrix} 1 \\ -1 \\ 2 \end{bmatrix}$ and $\mathbf{v} = \begin{bmatrix} 2 \\ 3 \\ -1 \end{bmatrix}$. What is the directional derivative $D_{\mathbf{v}} f(\mathbf{a})$?

We define $\varphi: \mathbb{R} \to \mathbb{R}$ by $\varphi(t) = f(\mathbf{a} + t\mathbf{v})$. Note that

$$\varphi'(0) = \lim_{t \to 0} \frac{\varphi(t) - \varphi(0)}{t} = \lim_{t \to 0} \frac{f(\mathbf{a} + t\mathbf{v}) - f(\mathbf{a})}{t} = D_\mathbf{v} f(\mathbf{a}).$$

So we just calculate φ and compute its derivative at 0:

$$\varphi(t) = f\begin{pmatrix} 1+2t \\ -1+3t \\ 2-t \end{pmatrix} = (1+2t)^2(-1+3t) + e^{(3(1+2t)+(-1+3t)-(2-t))}$$

$$= (1+2t)^2(-1+3t) + e^{10t}, \quad \text{so}$$

$$\varphi'(t) = 4(1+2t)(-1+3t) + 3(1+2t)^2 + 10e^{10t},$$

from which we conclude that $D_\mathbf{v} f(\mathbf{a}) = \varphi'(0) = 9$.

This approach is usually more convenient than calculating the limit directly when we are given commonplace functions. ◀

▶ EXERCISES 3.1

1. Calculate the partial derivatives of the following functions.

   *(a) $f\begin{pmatrix} x \\ y \end{pmatrix} = x^3 + 3xy^2 - 2y + 7$

   (b) $f\begin{pmatrix} x \\ y \end{pmatrix} = \sqrt{x^2 + y^2}$

   *(c) $f\begin{pmatrix} x \\ y \end{pmatrix} = \arctan\left(\frac{y}{x}\right)$

   (d) $f\begin{pmatrix} x \\ y \end{pmatrix} = e^{-(x^2+y^2)}$

   (e) $f\begin{pmatrix} x \\ y \end{pmatrix} = (x + y^2)\log x$

   (f) $f\begin{pmatrix} x \\ y \\ z \end{pmatrix} = e^{xy}z^2 - xy\sin(\pi yz)$

2. Calculate the directional derivative of the given function f at the given point $\mathbf{a}$ in the direction of the given vector $\mathbf{v}$.

   *(a) $f\begin{pmatrix} x \\ y \end{pmatrix} = x^2 + xy$, $\mathbf{a} = \begin{bmatrix} 2 \\ 1 \end{bmatrix}$, $\mathbf{v} = \begin{bmatrix} 1 \\ -1 \end{bmatrix}$

   *(b) $f\begin{pmatrix} x \\ y \end{pmatrix} = x^2 + xy$, $\mathbf{a} = \begin{bmatrix} 2 \\ 1 \end{bmatrix}$, $\mathbf{v} = \frac{1}{\sqrt{2}}\begin{bmatrix} 1 \\ -1 \end{bmatrix}$

   (c) $f\begin{pmatrix} x \\ y \end{pmatrix} = ye^{-x}$, $\mathbf{a} = \begin{bmatrix} 0 \\ 1 \end{bmatrix}$, $\mathbf{v} = \begin{bmatrix} 3 \\ 4 \end{bmatrix}$

   (d) $f\begin{pmatrix} x \\ y \end{pmatrix} = ye^{-x}$, $\mathbf{a} = \begin{bmatrix} 0 \\ 1 \end{bmatrix}$, $\mathbf{v} = \frac{1}{5}\begin{bmatrix} 3 \\ 4 \end{bmatrix}$

3. For each of the following functions f and points $\mathbf{a}$, find the *unit* vector $\mathbf{v}$ with the property that $D_\mathbf{v} f(\mathbf{a})$ is as large as possible.

   *(a) $f\begin{pmatrix} x \\ y \end{pmatrix} = x^2 + xy$, $\mathbf{a} = \begin{bmatrix} 2 \\ 1 \end{bmatrix}$

   (b) $f\begin{pmatrix} x \\ y \end{pmatrix} = ye^{-x}$, $\mathbf{a} = \begin{bmatrix} 0 \\ 1 \end{bmatrix}$

   (c) $f\begin{pmatrix} x \\ y \\ z \end{pmatrix} = \frac{1}{x} + \frac{1}{y} + \frac{1}{z}$, $\mathbf{a} = \begin{bmatrix} 1 \\ -1 \\ 1 \end{bmatrix}$

4. Suppose $D_{\mathbf{v}}f(\mathbf{a})$ exists. Prove that $D_{-\mathbf{v}}f(\mathbf{a})$ exists and calculate it in terms of the former.

5. (a) Show that there can be no function $f\colon \mathbb{R}^n \to \mathbb{R}$ so that for some point $\mathbf{a} \in \mathbb{R}^n$ we have $D_{\mathbf{v}}f(\mathbf{a}) > 0$ for all nonzero vectors $\mathbf{v} \in \mathbb{R}^n$.

(b) Show that there can, however, be a function $f\colon \mathbb{R}^n \to \mathbb{R}$ so that for some vector $\mathbf{v} \in \mathbb{R}^n$ we have $D_{\mathbf{v}}f(\mathbf{a}) > 0$ for all points $\mathbf{a} \in \mathbb{R}^n$.

6. Consider the ideal gas law $pV = nRT$. (Here p is pressure, V is volume, n is the number of moles of gas present, R is the universal gas constant, and T is temperature.) Assume n is fixed. Solve for each of p, V, and T as functions of the others, viz.,

$$p = f\begin{pmatrix} V \\ T \end{pmatrix}, \quad V = g\begin{pmatrix} p \\ T \end{pmatrix}, \quad \text{and} \quad T = h\begin{pmatrix} p \\ V \end{pmatrix}.$$

Compute the partial derivatives of f, g, and h. What is

$$\frac{\partial f}{\partial V} \cdot \frac{\partial g}{\partial T} \cdot \frac{\partial h}{\partial p}, \quad \text{or, more colloquially,} \quad \frac{\partial p}{\partial V} \cdot \frac{\partial V}{\partial T} \cdot \frac{\partial T}{\partial p}?$$

7. Suppose $f\colon \mathbb{R} \to \mathbb{R}$ is differentiable, and let $g\begin{pmatrix} x \\ y \end{pmatrix} = f\left(\dfrac{x}{y}\right)$. Show that

$$x\frac{\partial g}{\partial x} + y\frac{\partial g}{\partial y} = 0.$$

8. Suppose $f\colon \mathbb{R} \to \mathbb{R}$ is differentiable, and let $g\begin{pmatrix} x \\ y \end{pmatrix} = f(\sqrt{x^2 + y^2})$ for $\mathbf{x} \neq \mathbf{0}$. Show that

$$y\frac{\partial g}{\partial x} = x\frac{\partial g}{\partial y}.$$

9. Let $f\colon \mathbb{R}^2 \to \mathbb{R}$ be defined by

$$f\begin{pmatrix} x \\ y \end{pmatrix} = \frac{xy}{x^2 + y^2}, \quad \mathbf{x} \neq \mathbf{0}, \quad f(\mathbf{0}) = 0.$$

Show that the partial derivatives of f exist at $\mathbf{0}$ and yet f is not continuous at $\mathbf{0}$. Do other directional derivatives of f exist at $\mathbf{0}$?

10. (a) Let $f\colon \mathbb{R}^2 \to \mathbb{R}$ be the function defined in Example 5 of Chapter 2, Section 3. Calculate $D_{\mathbf{v}}f(\mathbf{0})$ for any $\mathbf{v} \in \mathbb{R}^2$.

(b) Give an example of a function $f\colon \mathbb{R}^2 \to \mathbb{R}$ all of whose directional derivatives at $\mathbf{0}$ are 0 but is, nevertheless, discontinuous at $\mathbf{0}$.

*11. Suppose $T\colon \mathbb{R}^n \to \mathbb{R}^m$ is a linear map. Show that the directional derivative $D_{\mathbf{v}}T(\mathbf{a})$ exists for all $\mathbf{a} \in \mathbb{R}^n$ and all $\mathbf{v} \in \mathbb{R}^n$ and calculate it.

12. Identify the set $\mathcal{M}_{n \times n}$ of $n \times n$ matrices with $\mathbb{R}^{n^2}$.
(a) Define $\mathbf{f}\colon \mathcal{M}_{n \times n} \to \mathcal{M}_{n \times n}$ by $\mathbf{f}(A) = A^{\mathsf{T}}$. For any $A, B \in \mathcal{M}_{n \times n}$, prove that $D_B\mathbf{f}(A) = B^{\mathsf{T}}$.
(b) Define $\mathbf{f}\colon \mathcal{M}_{n \times n} \to \mathcal{M}_{n \times n}$ by $\mathbf{f}(A) = \operatorname{tr} A$. For any $A, B \in \mathcal{M}_{n \times n}$, prove that $D_B\mathbf{f}(A) = \operatorname{tr} B$. (For the definition of trace, see Exercise 1.4.22.)

13. Identify the set $\mathcal{M}_{n \times n}$ of $n \times n$ matrices with $\mathbb{R}^{n^2}$.
(a) Define $\mathbf{f}\colon \mathcal{M}_{n \times n} \to \mathcal{M}_{n \times n}$ by $\mathbf{f}(A) = A^2$. For any $A, B \in \mathcal{M}_{n \times n}$, prove that $D_B\mathbf{f}(A) = AB + BA$.
(b) Define $\mathbf{f}\colon \mathcal{M}_{n \times n} \to \mathcal{M}_{n \times n}$ by $\mathbf{f}(A) = A^{\mathsf{T}}A$. Calculate $D_B\mathbf{f}(A)$.

2 DIFFERENTIABILITY

For a function $f\colon \mathbb{R} \to \mathbb{R}$, one of the fundamental consequences of being differentiable (at a) is that the function must be continuous (at a). We have already seen that for a function $f\colon \mathbb{R}^n \to \mathbb{R}$, having partial derivatives (or, indeed, all directional derivatives) at $\mathbf{a}$ need not guarantee continuity at $\mathbf{a}$. We now seek the appropriate definition.

Recall that the derivative is defined to be
$$f'(a) = \lim_{h \to 0} \frac{f(a+h) - f(a)}{h};$$
alternatively, if it exists, it is the unique number m with the property that
$$\lim_{h \to 0} \frac{f(a+h) - f(a) - mh}{h} = 0.$$
That is, the tangent line—the line passing through $\begin{bmatrix} a \\ f(a) \end{bmatrix}$ with slope $m = f'(a)$—is the best (affine) linear approximation to the graph of f at a, in the sense that the error goes to 0 faster than h as $h \to 0$. (See Figure 2.1.) Generalizing the latter notion, we make this

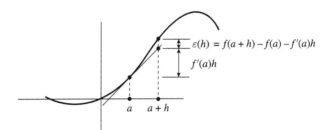

Figure 2.1

Definition Let $U \subset \mathbb{R}^n$ be open, and let $\mathbf{a} \in U$. A function $\mathbf{f}\colon U \to \mathbb{R}^m$ is *differentiable* at $\mathbf{a}$ if there is a linear map $D\mathbf{f}(\mathbf{a})\colon \mathbb{R}^n \to \mathbb{R}^m$ so that
$$\lim_{\mathbf{h} \to 0} \frac{\mathbf{f}(\mathbf{a}+\mathbf{h}) - \mathbf{f}(\mathbf{a}) - D\mathbf{f}(\mathbf{a})\mathbf{h}}{\|\mathbf{h}\|} = \mathbf{0}.$$

This says that $D\mathbf{f}(\mathbf{a})$ is the best linear approximation to the function $\mathbf{f} - \mathbf{f}(\mathbf{a})$ at $\mathbf{a}$, in the sense that the difference $\mathbf{f}(\mathbf{a}+\mathbf{h}) - \mathbf{f}(\mathbf{a}) - D\mathbf{f}(\mathbf{a})\mathbf{h}$ is small compared to $\mathbf{h}$. See Figure 2.2 and compare Figure 2.1. Equivalently, writing $\mathbf{x} = \mathbf{a} + \mathbf{h}$, the function $\mathbf{g}(\mathbf{x}) = \mathbf{f}(\mathbf{a}) + D\mathbf{f}(\mathbf{a})(\mathbf{x} - \mathbf{a})$ is the best affine linear approximation to $\mathbf{f}$ near $\mathbf{a}$. Indeed, the graph of $\mathbf{g}$ is called the *tangent plane* of the graph at $\mathbf{a}$. The tangent plane is obtained by translating the graph of $D\mathbf{f}(\mathbf{a})$, a subspace of $\mathbb{R}^n \times \mathbb{R}^m$, so that it passes through $\begin{bmatrix} \mathbf{a} \\ \mathbf{f}(\mathbf{a}) \end{bmatrix}$.

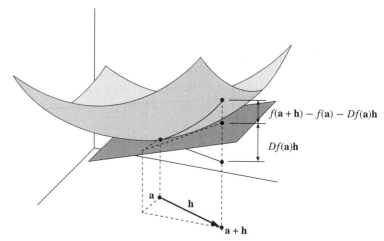

Figure 2.2

Remark The derivative $D\mathbf{f}(\mathbf{a})$, if it exists, must be unique. If there were two linear maps $T, T'\colon \mathbb{R}^n \to \mathbb{R}^m$ satisfying

$$\lim_{\mathbf{h}\to 0} \frac{\mathbf{f}(\mathbf{a}+\mathbf{h}) - \mathbf{f}(\mathbf{a}) - T(\mathbf{h})}{\|\mathbf{h}\|} = \mathbf{0} \quad \text{and} \quad \lim_{\mathbf{h}\to 0} \frac{\mathbf{f}(\mathbf{a}+\mathbf{h}) - \mathbf{f}(\mathbf{a}) - T'(\mathbf{h})}{\|\mathbf{h}\|} = \mathbf{0},$$

then we would have

$$\lim_{\mathbf{h}\to 0} \frac{(T - T')(\mathbf{h})}{\|\mathbf{h}\|} = \mathbf{0}.$$

In particular, letting $\mathbf{h} = t\mathbf{e}_i$ for any $i = 1, \ldots, n$, we see that

$$\lim_{t\to 0^+} \frac{(T - T')(t\mathbf{e}_i)}{t} = (T - T')(\mathbf{e}_i) = \mathbf{0} \quad \text{for } i = 1, \ldots, n,$$

and so $T = T'$.

It is worth observing that a vector-valued function $\mathbf{f}$ is differentiable at $\mathbf{a}$ if and only if each of its coordinate functions f_i is differentiable at $\mathbf{a}$. (See Exercise 6 for a proof.)

Proposition 2.1 *If $\mathbf{f}\colon \mathbb{R}^n \to \mathbb{R}^m$ is differentiable at $\mathbf{a}$, then the partial derivatives $\dfrac{\partial f_i}{\partial x_j}(\mathbf{a})$ exist and*

$$[D\mathbf{f}(\mathbf{a})] = \left[\frac{\partial f_i}{\partial x_j}(\mathbf{a})\right].$$

The latter matrix is often called the Jacobian matrix *of $\mathbf{f}$ at $\mathbf{a}$.*

Proof Since we assume $\mathbf{f}$ is differentiable at $\mathbf{a}$, we know there is a linear map $D\mathbf{f}(\mathbf{a})$ with the property that

$$\lim_{\mathbf{h}\to 0} \frac{\mathbf{f}(\mathbf{a}+\mathbf{h}) - \mathbf{f}(\mathbf{a}) - D\mathbf{f}(\mathbf{a})\mathbf{h}}{\|\mathbf{h}\|} = \mathbf{0}.$$

As we did in the remark above, for any $j = 1, \ldots, n$, we consider $\mathbf{h} = t\mathbf{e}_j$ and let $t \to 0$. Then we have
$$0 = \lim_{t \to 0} \frac{\mathbf{f}(\mathbf{a} + t\mathbf{e}_j) - \mathbf{f}(\mathbf{a}) - D\mathbf{f}(\mathbf{a})(t\mathbf{e}_j)}{|t|} = \mathbf{0}.$$

Considering separately the cases $t > 0$ and $t < 0$, we find that
$$0 = \lim_{t \to 0^+} \frac{\mathbf{f}(\mathbf{a} + t\mathbf{e}_j) - \mathbf{f}(\mathbf{a}) - D\mathbf{f}(\mathbf{a})(t\mathbf{e}_j)}{t} = \lim_{t \to 0^+} \frac{\mathbf{f}(\mathbf{a} + t\mathbf{e}_j) - \mathbf{f}(\mathbf{a})}{t} - D\mathbf{f}(\mathbf{a})(\mathbf{e}_j),$$
$$0 = \lim_{t \to 0^-} \frac{\mathbf{f}(\mathbf{a} + t\mathbf{e}_j) - \mathbf{f}(\mathbf{a}) - D\mathbf{f}(\mathbf{a})(t\mathbf{e}_j)}{-t} = -\left(\lim_{t \to 0^-} \frac{\mathbf{f}(\mathbf{a} + t\mathbf{e}_j) - \mathbf{f}(\mathbf{a})}{t} - D\mathbf{f}(\mathbf{a})(\mathbf{e}_j)\right),$$

and so $D\mathbf{f}(\mathbf{a})(\mathbf{e}_j) = \lim_{t \to 0} \frac{\mathbf{f}(\mathbf{a} + t\mathbf{e}_j) - \mathbf{f}(\mathbf{a})}{t} = \frac{\partial \mathbf{f}}{\partial x_j}(\mathbf{a})$, as required. ∎

▶ **EXAMPLE 1**

When $n = 1$, we have parametric equations of a curve in $\mathbb{R}^m$. We see that if $\mathbf{f}$ is differentiable at a, then
$$D\mathbf{f}(a) = \begin{bmatrix} f_1'(a) \\ f_2'(a) \\ \vdots \\ f_m'(a) \end{bmatrix},$$

and we can think of $D\mathbf{f}(a) = D\mathbf{f}(a)(1)$ as the velocity vector of the parametrized curve at the point $\mathbf{f}(a)$, which we will usually denote by the (more) familiar $\mathbf{f}'(a)$. See Section 5 for further discussion of this topic. ◀

▶ **EXAMPLE 2**

Let $f\begin{pmatrix} x \\ y \end{pmatrix} = xy$. To prove that f is differentiable at $\mathbf{a} = \begin{bmatrix} a \\ b \end{bmatrix}$, we must exhibit a linear map $Df(\mathbf{a})$ with the requisite property. By Proposition 2.1, we know the only candidate is
$$Df\begin{pmatrix} a \\ b \end{pmatrix} = \begin{bmatrix} b & a \end{bmatrix},$$

so now we just prove that the appropriate limit is really 0:
$$\lim_{\begin{bmatrix} h \\ k \end{bmatrix} \to 0} \frac{f\begin{pmatrix} a+h \\ b+k \end{pmatrix} - f\begin{pmatrix} a \\ b \end{pmatrix} - \begin{bmatrix} b & a \end{bmatrix}\begin{bmatrix} h \\ k \end{bmatrix}}{\left\|\begin{bmatrix} h \\ k \end{bmatrix}\right\|} = \lim_{\begin{bmatrix} h \\ k \end{bmatrix} \to 0} \frac{(ab + bh + ak + hk) - ab - (bh + ak)}{\sqrt{h^2 + k^2}}$$
$$= \lim_{\begin{bmatrix} h \\ k \end{bmatrix} \to 0} \frac{hk}{\sqrt{h^2 + k^2}} = \lim_{\begin{bmatrix} h \\ k \end{bmatrix} \to 0} \frac{h}{\sqrt{h^2 + k^2}} k = 0$$

because $\frac{|h|}{\sqrt{h^2 + k^2}} \leq 1$ and $k \to 0$ as $\begin{bmatrix} h \\ k \end{bmatrix} \to \mathbf{0}$. ◀

EXAMPLE 3

The tangent plane of the graph $z = f\begin{pmatrix} x \\ y \end{pmatrix} = xy$ at $\mathbf{a} = \begin{bmatrix} 2 \\ 1 \end{bmatrix}$ is

$$z = f\begin{pmatrix} 2 \\ 1 \end{pmatrix} + Df\begin{pmatrix} 2 \\ 1 \end{pmatrix}\begin{bmatrix} x-2 \\ y-1 \end{bmatrix}$$

$$= 2 + \begin{bmatrix} 1 & 2 \end{bmatrix}\begin{bmatrix} x-2 \\ y-1 \end{bmatrix} = 2 + (x-2) + 2(y-1) = x + 2y - 2.\ \triangleleft$$

EXAMPLE 4

Let $f\begin{pmatrix} x \\ y \end{pmatrix} = \dfrac{x}{y}$. First, we claim that f is differentiable at $\mathbf{a} = \begin{bmatrix} a \\ b \end{bmatrix}$, provided $b \neq 0$. The putative derivative is

$$Df\begin{pmatrix} a \\ b \end{pmatrix} = \begin{bmatrix} \dfrac{1}{b} & -\dfrac{a}{b^2} \end{bmatrix},$$

and we check that

$$\lim_{\mathbf{h} \to 0} \frac{\mathbf{f}(\mathbf{a}+\mathbf{h}) - \mathbf{f}(\mathbf{a}) - D\mathbf{f}(\mathbf{a})\mathbf{h}}{\|\mathbf{h}\|} = \mathbf{0}.$$

Well,

$$f\begin{pmatrix} a+h \\ b+k \end{pmatrix} - f\begin{pmatrix} a \\ b \end{pmatrix} - \begin{bmatrix} \dfrac{1}{b} & -\dfrac{a}{b^2} \end{bmatrix}\begin{bmatrix} h \\ k \end{bmatrix} = \frac{a+h}{b+k} - \frac{a}{b} - \frac{h}{b} + \frac{ak}{b^2}$$

$$= \frac{(a+h)(-bk) + ak(b+k)}{b^2(b+k)} = \frac{k(ak-bh)}{b^2(b+k)},$$

and so

$$\lim_{\begin{bmatrix} h \\ k \end{bmatrix} \to 0} \frac{f\begin{pmatrix} a+h \\ b+k \end{pmatrix} - f\begin{pmatrix} a \\ b \end{pmatrix} - \begin{bmatrix} \dfrac{1}{b} & -\dfrac{a}{b^2} \end{bmatrix}\begin{bmatrix} h \\ k \end{bmatrix}}{\left\|\begin{bmatrix} h \\ k \end{bmatrix}\right\|} = \lim_{\begin{bmatrix} h \\ k \end{bmatrix} \to 0} \frac{\frac{k(ak-bh)}{b^2(b+k)}}{\sqrt{h^2+k^2}} = 0$$

since $\dfrac{|k|}{\sqrt{h^2+k^2}} \leq 1$ and $\dfrac{ak-bh}{b^2(b+k)} \to 0$ as $\begin{bmatrix} h \\ k \end{bmatrix} \to \mathbf{0}$.

Now, as a (not totally facetious) application, consider the problem of calculating one's gas mileage, having used y gallons of gas to travel x miles. For example, without having a calculator on hand, we can use linear approximation afforded us by the derivative to estimate our gas mileage if we've used 10.8 gallons to drive 344 miles. Using $a = 350$ and $b = 10$, we have

$$f\begin{pmatrix} 344 \\ 10.8 \end{pmatrix} \approx f\begin{pmatrix} a \\ b \end{pmatrix} + Df\begin{pmatrix} a \\ b \end{pmatrix}\begin{bmatrix} 344-a \\ 10.8-b \end{bmatrix}$$

$$= 35 + \begin{bmatrix} 0.1 & -3.5 \end{bmatrix}\begin{bmatrix} -6 \\ 0.8 \end{bmatrix} = 35 - 0.6 - 2.8 = 31.6.$$

(The actual value, to two decimal places, is 31.85.) $\triangleleft$

▶ **EXAMPLE 5**

As we said earlier, a function $\mathbf{f}\colon \mathbb{R}^2 \to \mathbb{R}^2$ is differentiable if and only if both its component functions $f_i\colon \mathbb{R}^2 \to \mathbb{R}$ are differentiable. It follows from Examples 2 and 4 that the function $\mathbf{f}\colon \mathbb{R}^2 \to \mathbb{R}^2$ given by

$$\mathbf{f}\begin{pmatrix} x \\ y \end{pmatrix} = \begin{bmatrix} xy \\ x/y \end{bmatrix}$$

is differentiable, and the Jacobian matrix of $\mathbf{f}$ at $\mathbf{a}$ is

$$D\mathbf{f}\begin{pmatrix} a \\ b \end{pmatrix} = \begin{bmatrix} b & a \\ 1/b & -a/b^2 \end{bmatrix}. \quad \blacktriangleleft$$

One indication that we have the correct definition is the following.

Proposition 2.2 *If $\mathbf{f}\colon \mathbb{R}^n \to \mathbb{R}^m$ is differentiable at $\mathbf{a}$, then $\mathbf{f}$ is continuous at $\mathbf{a}$.*

Proof Suppose $\mathbf{f}$ is differentiable at $\mathbf{a}$; we must show that $\lim_{\mathbf{x}\to\mathbf{a}} \mathbf{f}(\mathbf{x}) = \mathbf{f}(\mathbf{a})$ or, equivalently, that $\lim_{\mathbf{h}\to\mathbf{0}} \mathbf{f}(\mathbf{a}+\mathbf{h}) = \mathbf{f}(\mathbf{a})$. We have a linear map $D\mathbf{f}(\mathbf{a})\colon \mathbb{R}^n \to \mathbb{R}^m$ so that

$$\lim_{\mathbf{h}\to\mathbf{0}} \frac{\mathbf{f}(\mathbf{a}+\mathbf{h}) - \mathbf{f}(\mathbf{a}) - D\mathbf{f}(\mathbf{a})\mathbf{h}}{\|\mathbf{h}\|} = \mathbf{0}.$$

This means that

$$\lim_{\mathbf{h}\to\mathbf{0}} \mathbf{f}(\mathbf{a}+\mathbf{h}) - \mathbf{f}(\mathbf{a}) - D\mathbf{f}(\mathbf{a})\mathbf{h} = \lim_{\mathbf{h}\to\mathbf{0}} \left(\frac{\mathbf{f}(\mathbf{a}+\mathbf{h}) - \mathbf{f}(\mathbf{a}) - D\mathbf{f}(\mathbf{a})\mathbf{h}}{\|\mathbf{h}\|} \|\mathbf{h}\| \right)$$

$$= \lim_{\mathbf{h}\to\mathbf{0}} \frac{\mathbf{f}(\mathbf{a}+\mathbf{h}) - \mathbf{f}(\mathbf{a}) - D\mathbf{f}(\mathbf{a})\mathbf{h}}{\|\mathbf{h}\|} \lim_{\mathbf{h}\to\mathbf{0}} \|\mathbf{h}\| = \mathbf{0}.$$

By Exercise 2.3.7, $\lim_{\mathbf{h}\to\mathbf{0}} D\mathbf{f}(\mathbf{a})\mathbf{h} = \mathbf{0}$, and so

$$\lim_{\mathbf{h}\to\mathbf{0}} \mathbf{f}(\mathbf{a}+\mathbf{h}) - \mathbf{f}(\mathbf{a}) = \lim_{\mathbf{h}\to\mathbf{0}} (\mathbf{f}(\mathbf{a}+\mathbf{h}) - \mathbf{f}(\mathbf{a}) - D\mathbf{f}(\mathbf{a})\mathbf{h}) + \lim_{\mathbf{h}\to\mathbf{0}} D\mathbf{f}(\mathbf{a})\mathbf{h} = \mathbf{0},$$

as required. ∎

Let's now study a few examples to see just how subtle the issue of differentiability is.

▶ **EXAMPLE 6**

Define $f\colon \mathbb{R}^2 \to \mathbb{R}$ by

$$f\begin{pmatrix} x \\ y \end{pmatrix} = \frac{xy}{x^2 + y^2}, \quad \mathbf{x} \neq \mathbf{0}, \qquad f(\mathbf{0}) = 0.$$

Since $f\begin{pmatrix} x \\ 0 \end{pmatrix} = 0$ for all x and $f\begin{pmatrix} 0 \\ y \end{pmatrix} = 0$ for all y, certainly

$$\frac{\partial f}{\partial x}\begin{pmatrix} 0 \\ 0 \end{pmatrix} = \frac{\partial f}{\partial y}\begin{pmatrix} 0 \\ 0 \end{pmatrix} = 0.$$

However, we have already seen in Exercise 3.1.9 that f is discontinuous, so it cannot be differentiable. For practice, we check directly: If $Df(\mathbf{0})$ existed, by Proposition 2.1 we would have $Df(\mathbf{0}) = \mathbf{0}$. Now let's consider

$$\lim_{\mathbf{h} \to \mathbf{0}} \frac{f(\mathbf{h}) - f(\mathbf{0}) - Df(\mathbf{0})\mathbf{h}}{\|\mathbf{h}\|} = \lim_{\mathbf{h} \to \mathbf{0}} \frac{f(\mathbf{h})}{\|\mathbf{h}\|} = \lim_{\begin{bmatrix} h \\ k \end{bmatrix} \to \mathbf{0}} \frac{hk}{(h^2 + k^2)^{3/2}}.$$

Like many of the limits we considered in Chapter 2, this one obviously does not exist; indeed, as $\mathbf{h} \to \mathbf{0}$ along the line $h = k$, this fraction becomes

$$\frac{h^2}{(2h^2)^{3/2}} = \frac{1}{2\sqrt{2}|h|},$$

which is clearly unbounded as $h \to 0$. What's more, as the reader can check, f has directional derivatives at $\mathbf{0}$ only in the directions of the axes. ◀

▶ **EXAMPLE 7**

Define $f : \mathbb{R}^2 \to \mathbb{R}$ by

$$f\begin{pmatrix} x \\ y \end{pmatrix} = \frac{x^2 y}{x^2 + y^2}, \quad \mathbf{x} \neq \mathbf{0}, \quad f(\mathbf{0}) = 0.$$

As in Example 6, both partial derivatives of this function at $\mathbf{0}$ are 0. This function, as we saw in Example 5 of Chapter 2, Section 3, *is* continuous, so differentiability is a bit more unclear. But we just try to calculate:

$$\lim_{\mathbf{h} \to \mathbf{0}} \frac{f(\mathbf{h}) - f(\mathbf{0}) - Df(\mathbf{0})\mathbf{h}}{\|\mathbf{h}\|} = \lim_{\mathbf{h} \to \mathbf{0}} \frac{f(\mathbf{h})}{\|\mathbf{h}\|} = \lim_{\begin{bmatrix} h \\ k \end{bmatrix} \to \mathbf{0}} \frac{h^2 k}{(h^2 + k^2)^{3/2}}.$$

When $\mathbf{h} \to \mathbf{0}$ along either coordinate axis, the limit is obviously 0; however, when $\mathbf{h} \to \mathbf{0}$ along the line $h = k$, the limit does not exist, as the expression is equal to $+\frac{1}{2\sqrt{2}}$ when $h > 0$ and $-\frac{1}{2\sqrt{2}}$ when $h < 0$. Thus, f is not differentiable at $\mathbf{0}$. ◀

Proposition 2.3 *When* $\mathbf{f}$ *is differentiable at* $\mathbf{a}$, *for any* $\mathbf{v} \in \mathbb{R}^n$, *the directional derivative of* $\mathbf{f}$ *at* $\mathbf{a}$ *in the direction* $\mathbf{v}$ *is given by*

$$D_\mathbf{v} \mathbf{f}(\mathbf{a}) = D\mathbf{f}(\mathbf{a})\mathbf{v}.$$

Proof Since $\mathbf{f}$ is differentiable at $\mathbf{a}$, we know that its derivative, $D\mathbf{f}(\mathbf{a})$, has the property that

$$\lim_{\mathbf{h} \to \mathbf{0}} \frac{\mathbf{f}(\mathbf{a} + \mathbf{h}) - \mathbf{f}(\mathbf{a}) - D\mathbf{f}(\mathbf{a})\mathbf{h}}{\|\mathbf{h}\|} = \mathbf{0}.$$

Substituting $\mathbf{h} = t\mathbf{v}$ and letting $t \to 0$, we have

$$\lim_{t \to 0} \frac{\mathbf{f}(\mathbf{a} + t\mathbf{v}) - \mathbf{f}(\mathbf{a}) - D\mathbf{f}(\mathbf{a})(t\mathbf{v})}{|t|} = \mathbf{0}.$$

Since $D\mathbf{f}(\mathbf{a})$ is a linear map, $D\mathbf{f}(\mathbf{a})(t\mathbf{v}) = tD\mathbf{f}(\mathbf{a})\mathbf{v}$. Proceeding as in the proof of Proposition 2.1, letting t approach 0 through positive values, we have

$$\lim_{t \to 0^+} \frac{\mathbf{f}(\mathbf{a}+t\mathbf{v}) - \mathbf{f}(\mathbf{a}) - tD\mathbf{f}(\mathbf{a})\mathbf{v}}{t} = \mathbf{0}, \quad \text{and so}$$

$$\lim_{t \to 0^+} \frac{\mathbf{f}(\mathbf{a}+t\mathbf{v}) - \mathbf{f}(\mathbf{a})}{t} = D\mathbf{f}(\mathbf{a})\mathbf{v}.$$

Similarly, when t approaches 0 through negative values, we have $|t| = -t$ and

$$\lim_{t \to 0^-} \frac{\mathbf{f}(\mathbf{a}+t\mathbf{v}) - \mathbf{f}(\mathbf{a}) - tD\mathbf{f}(\mathbf{a})\mathbf{v}}{-t} = \mathbf{0}, \quad \text{so}$$

$$\lim_{t \to 0^-} \frac{\mathbf{f}(\mathbf{a}+t\mathbf{v}) - \mathbf{f}(\mathbf{a}) - tD\mathbf{f}(\mathbf{a})\mathbf{v}}{t} = \mathbf{0}, \quad \text{and, as before,}$$

$$\lim_{t \to 0^-} \frac{\mathbf{f}(\mathbf{a}+t\mathbf{v}) - \mathbf{f}(\mathbf{a})}{t} = D\mathbf{f}(\mathbf{a})\mathbf{v}.$$

Thus,

$$D_\mathbf{v}\mathbf{f}(\mathbf{a}) = \lim_{t \to 0} \frac{\mathbf{f}(\mathbf{a}+t\mathbf{v}) - \mathbf{f}(\mathbf{a})}{t} = D\mathbf{f}(\mathbf{a})\mathbf{v},$$

as required. ∎

Remark Let's consider the case of a function $f \colon \mathbb{R}^2 \to \mathbb{R}$, as we pictured in Figures 1.1 and 1.2. As a consequence of Proposition 2.3, the tangent plane of the graph of f at $\mathbf{a}$ contains the tangent lines at $\mathbf{a}$ of the slices by all vertical planes. The function f given in Example 2 of Section 1 cannot be differentiable at $\mathbf{0}$, as it is clear from Figure 1.3 that the tangent lines to the various vertical slices at the origin do not lie in a plane.

Since it is so tedious to determine from the definition whether a function is differentiable, the following proposition is useful indeed.

Proposition 2.4 *If* $\mathbf{f} \colon U \to \mathbb{R}^m$ *has continuous partial derivatives, then* $\mathbf{f}$ *is differentiable.*

A function with continuous partial derivatives is said to be $\mathcal{C}^1$ or *continuously differentiable*. The reason for this notation will become clear when we study partial derivatives of higher order.

Proof By Exercise 6, it suffices to treat the case $m = 1$. For clarity, we give the proof in the case $n = 2$, although the general case is not conceptually any harder. As usual, we write $\mathbf{a} = \begin{bmatrix} a \\ b \end{bmatrix}$ and $\mathbf{h} = \begin{bmatrix} h \\ k \end{bmatrix}$.

As usual, if f is to be differentiable, we know that $Df(\mathbf{a})$ must be given by the Jacobian matrix of f at $\mathbf{a}$. To prove that f is differentiable at $\mathbf{a} \in U$, we need to estimate

$$f(\mathbf{a}+\mathbf{h}) - f(\mathbf{a}) - Df(\mathbf{a})\mathbf{h} = f(\mathbf{a}+\mathbf{h}) - f(\mathbf{a}) - \left(\frac{\partial f}{\partial x}(\mathbf{a})h + \frac{\partial f}{\partial y}(\mathbf{a})k \right).$$

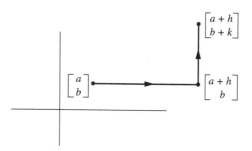

Figure 2.3

Now, here is the new twist: As Figure 2.3 indicates, we calculate $f(\mathbf{a} + \mathbf{h}) - f(\mathbf{a})$ by taking a two-step route:

$$f(\mathbf{a} + \mathbf{h}) - f(\mathbf{a}) = f\begin{pmatrix} a+h \\ b+k \end{pmatrix} - f\begin{pmatrix} a \\ b \end{pmatrix}$$

$$= \left(f\begin{pmatrix} a+h \\ b \end{pmatrix} - f\begin{pmatrix} a \\ b \end{pmatrix} \right) + \left(f\begin{pmatrix} a+h \\ b+k \end{pmatrix} - f\begin{pmatrix} a+h \\ b \end{pmatrix} \right),$$

and so, regrouping in a clever fashion and using the Mean Value Theorem twice, we obtain

$$f(\mathbf{a} + \mathbf{h}) - f(\mathbf{a}) - Df(\mathbf{a})\mathbf{h}$$

$$= \left(f\begin{pmatrix} a+h \\ b \end{pmatrix} - f\begin{pmatrix} a \\ b \end{pmatrix} - \frac{\partial f}{\partial x}(\mathbf{a})h \right) + \left(f\begin{pmatrix} a+h \\ b+k \end{pmatrix} - f\begin{pmatrix} a+h \\ b \end{pmatrix} - \frac{\partial f}{\partial y}(\mathbf{a})k \right)$$

$$= \left(\frac{\partial f}{\partial x}\begin{pmatrix} a+\xi \\ b \end{pmatrix} h - \frac{\partial f}{\partial x}(\mathbf{a})h \right) + \left(\frac{\partial f}{\partial y}\begin{pmatrix} a+h \\ b+\eta \end{pmatrix} k - \frac{\partial f}{\partial y}(\mathbf{a})k \right)$$

for some ξ between 0 and h and some η between 0 and k

$$= \left(\frac{\partial f}{\partial x}\begin{pmatrix} a+\xi \\ b \end{pmatrix} - \frac{\partial f}{\partial x}(\mathbf{a}) \right) h + \left(\frac{\partial f}{\partial y}\begin{pmatrix} a+h \\ b+\eta \end{pmatrix} - \frac{\partial f}{\partial y}(\mathbf{a}) \right) k.$$

Now, observe that $\frac{|h|}{\|\mathbf{h}\|} \leq 1$ and $\frac{|k|}{\|\mathbf{h}\|} \leq 1$; as $\mathbf{h} \to \mathbf{0}$, continuity of the partial derivatives guarantees that

$$\lim_{\mathbf{h} \to \mathbf{0}} \left(\frac{\partial f}{\partial x}\begin{pmatrix} a+\xi \\ b \end{pmatrix} - \frac{\partial f}{\partial x}(\mathbf{a}) \right) = \lim_{\mathbf{h} \to \mathbf{0}} \left(\frac{\partial f}{\partial y}\begin{pmatrix} a+h \\ b+\eta \end{pmatrix} - \frac{\partial f}{\partial y}(\mathbf{a}) \right) = 0$$

since $\xi \to 0$ and $\eta \to 0$ as $\mathbf{h} \to \mathbf{0}$. Thus,

$$\frac{|f(\mathbf{a}+\mathbf{h}) - f(\mathbf{a}) - Df(\mathbf{a})\mathbf{h}|}{\|\mathbf{h}\|}$$

$$\leq \left| \frac{\partial f}{\partial x}\begin{pmatrix} a+\xi \\ b \end{pmatrix} - \frac{\partial f}{\partial x}(\mathbf{a}) \right| \frac{|h|}{\|\mathbf{h}\|} + \left| \frac{\partial f}{\partial y}\begin{pmatrix} a+h \\ b+\eta \end{pmatrix} - \frac{\partial f}{\partial y}(\mathbf{a}) \right| \frac{|k|}{\|\mathbf{h}\|},$$

and therefore indeed approaches 0 as $\mathbf{h} \to \mathbf{0}$. ∎

EXAMPLE 8

We know that the function f given in Example 7 is not differentiable. It follows from Proposition 2.4 that f cannot be $\mathcal{C}^1$ at $\mathbf{0}$. Let's verify this directly.

It is obvious that $\dfrac{\partial f}{\partial x}(\mathbf{0}) = \dfrac{\partial f}{\partial y}(\mathbf{0}) = 0$, and for $\mathbf{x} \neq \mathbf{0}$, we have

$$\frac{\partial f}{\partial x}\begin{pmatrix} x \\ y \end{pmatrix} = \frac{2xy^3}{(x^2+y^2)^2} \quad \text{and} \quad \frac{\partial f}{\partial y}\begin{pmatrix} x \\ y \end{pmatrix} = \frac{x^2(x^2-y^2)}{(x^2+y^2)^2}.$$

So we see that when $x \neq 0$, $\dfrac{\partial f}{\partial x}\begin{pmatrix} x \\ x \end{pmatrix} = \dfrac{1}{2}$ and $\dfrac{\partial f}{\partial y}\begin{pmatrix} x \\ 0 \end{pmatrix} = 1$, neither of which approaches 0 as $x \to 0$.

Thus, f is not $\mathcal{C}^1$ at $\mathbf{0}$. ◀

EXAMPLE 9

To see that the sufficient condition for differentiability given by Proposition 2.4 is not necessary, we consider the classic example of the function $f: \mathbb{R} \to \mathbb{R}$ defined by

$$f(x) = \begin{cases} x^2 \sin \dfrac{1}{x}, & x \neq 0 \\ 0, & x = 0 \end{cases}.$$

Then it is easy to check that $f'(0) = 0$, and yet $f'(x) = 2x \sin \dfrac{1}{x} - \cos \dfrac{1}{x}$ has no limit as $x \to 0$.

Thus, f is differentiable on all of $\mathbb{R}$ but is not $\mathcal{C}^1$. ◀

EXERCISES 3.2

1. Find the equation of the tangent plane of the graph of f at the indicated point.

   *(a) $f\begin{pmatrix} x \\ y \end{pmatrix} = e^{xy}$, $\mathbf{a} = \begin{bmatrix} -1 \\ 2 \end{bmatrix}$

   (b) $f\begin{pmatrix} x \\ y \end{pmatrix} = x^2 + y^2$, $\mathbf{a} = \begin{bmatrix} -1 \\ 2 \end{bmatrix}$

   (c) $f\begin{pmatrix} x \\ y \end{pmatrix} = \sqrt{x^2+y^2}$, $\mathbf{a} = \begin{bmatrix} 3 \\ 4 \end{bmatrix}$

   (d) $f\begin{pmatrix} x \\ y \end{pmatrix} = \sqrt{4-x^2-y^2}$, $\mathbf{a} = \begin{bmatrix} 1 \\ 1 \end{bmatrix}$

   (e) $f\begin{pmatrix} x \\ y \\ z \end{pmatrix} = xyz$, $\mathbf{a} = \begin{bmatrix} 1 \\ 2 \\ 3 \end{bmatrix}$

   *(f) $f\begin{pmatrix} x \\ y \\ z \end{pmatrix} = \sin(xy)z^2 + e^{xz+1}$, $\mathbf{a} = \begin{bmatrix} 1 \\ 0 \\ -1 \end{bmatrix}$

2. Calculate the directional derivative of f at $\mathbf{a}$ in the given direction $\mathbf{v}$.

   *(a) $f\begin{pmatrix} x \\ y \end{pmatrix} = e^x \cos y$, $\mathbf{a} = \begin{bmatrix} 0 \\ \pi/4 \end{bmatrix}$, $\mathbf{v} = \begin{bmatrix} 1 \\ 1 \end{bmatrix}$

   (b) $f\begin{pmatrix} x \\ y \end{pmatrix} = e^x \cos y$, $\mathbf{a} = \begin{bmatrix} 0 \\ \pi/4 \end{bmatrix}$, $\mathbf{v} = \begin{bmatrix} 1 \\ -1 \end{bmatrix}$

   (c) $f\begin{pmatrix} x \\ y \end{pmatrix} = xy^2$, $\mathbf{a} = \begin{bmatrix} 3 \\ 1 \end{bmatrix}$, $\mathbf{v} = \begin{bmatrix} 1 \\ 2 \end{bmatrix}$

(d) $f\begin{pmatrix}x\\y\end{pmatrix} = x^2 + y^2$, $\mathbf{a} = \begin{bmatrix}2\\1\end{bmatrix}$, $\mathbf{v} = \frac{1}{\sqrt{5}}\begin{bmatrix}2\\1\end{bmatrix}$

*(e) $f\begin{pmatrix}x\\y\end{pmatrix} = \sqrt{x^2 + y^2}$, $\mathbf{a} = \begin{bmatrix}2\\1\end{bmatrix}$, $\mathbf{v} = \frac{1}{\sqrt{5}}\begin{bmatrix}2\\1\end{bmatrix}$

(f) $f\begin{pmatrix}x\\y\\z\end{pmatrix} = e^{xyz}$, $\mathbf{a} = \begin{bmatrix}1\\-1\\-1\end{bmatrix}$, $\mathbf{v} = \begin{bmatrix}2\\2\\1\end{bmatrix}$

*3. Give the derivative matrix of each of the following vector-valued functions.

(a) $\mathbf{f}: \mathbb{R}^2 \to \mathbb{R}^2$, $\mathbf{f}\begin{pmatrix}x\\y\end{pmatrix} = \begin{bmatrix}xy\\x^2 + y^2\end{bmatrix}$

(d) $\mathbf{f}: \mathbb{R}^3 \to \mathbb{R}^2$, $\mathbf{f}\begin{pmatrix}x\\y\\z\end{pmatrix} = \begin{bmatrix}xyz\\x+y+z^2\end{bmatrix}$

(b) $\mathbf{f}: \mathbb{R} \to \mathbb{R}^3$, $\mathbf{f}(t) = \begin{bmatrix}\cos t\\ \sin t\\ e^t\end{bmatrix}$

(e) $\mathbf{f}: \mathbb{R}^2 \to \mathbb{R}^3$, $\mathbf{f}\begin{pmatrix}x\\y\end{pmatrix} = \begin{bmatrix}x\cos y\\ x\sin y\\ y\end{bmatrix}$

(c) $\mathbf{f}: \mathbb{R}^2 \to \mathbb{R}^2$, $\mathbf{f}\begin{pmatrix}s\\t\end{pmatrix} = \begin{bmatrix}s\cos t\\ s\sin t\end{bmatrix}$

*4. Use the technique of Example 4 to estimate your gas mileage if you used 6.5 gallons to drive 224 miles.

5. Two sides of a triangle are $x = 3$ and $y = 4$, and the included angle is $\theta = \pi/3$. To a small change in which of these three variables is the area of the triangle most sensitive? Why?

6. Let $U \subset \mathbb{R}^n$ be an open set, and let $\mathbf{a} \in U$. Suppose $m > 1$. Prove that the function $\mathbf{f}: U \to \mathbb{R}^m$ is differentiable at $\mathbf{a}$ if and only if each component function f_i, $i = 1, \ldots, m$, is differentiable at $\mathbf{a}$. (Hint: Review the proof of Proposition 3.1 of Chapter 2.)

7. Show that any linear map is differentiable and is its own derivative (at an arbitrary point).

8. Show that the tangent plane of the cone $z^2 = x^2 + y^2$ at $\begin{bmatrix}a\\b\\c\end{bmatrix} \neq \mathbf{0}$ intersects the cone in a line.

9. Show that the tangent plane of the saddle surface $z = xy$ at any point intersects the surface in a pair of lines.

10. Find the derivative of the map $\mathbf{f}\begin{pmatrix}x\\y\end{pmatrix} = \begin{bmatrix}x^2 - y^2\\ 2xy\end{bmatrix}$ at the point $\mathbf{a}$. Show that whenever $\mathbf{a} \neq \mathbf{0}$, the linear map $D\mathbf{f}(\mathbf{a})$ is a scalar multiple of a rotation matrix.

11. Prove from the definition that the following functions are differentiable.

(a) $f\begin{pmatrix}x\\y\end{pmatrix} = x^2 + y^2$
(b) $f\begin{pmatrix}x\\y\end{pmatrix} = xy^2$
(c) $f: \mathbb{R}^n \to \mathbb{R}$, $f(\mathbf{x}) = \|\mathbf{x}\|^2$

12. Let

$$f\begin{pmatrix}x\\y\end{pmatrix} = \frac{x^2 y}{x^4 + y^2}, \quad \mathbf{x} \neq \mathbf{0}, \qquad f(\mathbf{0}) = 0.$$

Show directly that f fails to be $\mathcal{C}^1$ at the origin. (Of course, this follows from Example 5 of Section 3 of Chapter 2 and Propositions 2.2 and 2.4.)

13. Use the results of Exercise 3.1.13 to show that $\mathbf{f}(A) = A^2$ and $\mathbf{f}(A) = A^\mathsf{T} A$ are differentiable functions mapping $\mathcal{M}_{n\times n}$ to $\mathcal{M}_{n\times n}$.

♯14. Let A be an $n \times n$ matrix. Define $f \colon \mathbb{R}^n \to \mathbb{R}$ by $f(\mathbf{x}) = A\mathbf{x} \cdot \mathbf{x} = \mathbf{x}^\mathsf{T} A\mathbf{x}$.
  (a) Show that f is differentiable and $Df(\mathbf{a})\mathbf{h} = A\mathbf{a} \cdot \mathbf{h} + A\mathbf{h} \cdot \mathbf{a}$.
  (b) Deduce that when A is symmetric, $Df(\mathbf{a})\mathbf{h} = 2A\mathbf{a} \cdot \mathbf{h}$.

15. Let $\mathbf{a} \in \mathbb{R}^n$, $\delta > 0$, and suppose $f \colon B(\mathbf{a}, \delta) \to \mathbb{R}$ is differentiable at $\mathbf{a}$. Suppose $f(\mathbf{a}) \geq f(\mathbf{x})$ for all $\mathbf{x} \in B(\mathbf{a}, \delta)$. Prove that $Df(\mathbf{a}) = O$.

16. Let $\mathbf{a} \in \mathbb{R}^2$, $\delta > 0$, and suppose $f \colon B(\mathbf{a}, \delta) \to \mathbb{R}$ is differentiable and $Df(\mathbf{x}) = O$ for all $\mathbf{x} \in B(\mathbf{a}, \delta)$. Prove that $f(\mathbf{x}) = f(\mathbf{a})$ for all $\mathbf{x} \in B(\mathbf{a}, \delta)$. (Hint: Start with the proof of Proposition 2.4.)

17. Let
$$f\begin{pmatrix}x\\y\end{pmatrix} = \begin{cases} y, & x \neq 0 \\ 0, & x = 0 \end{cases}.$$
  (a) Prove that f is continuous at $\mathbf{0}$.
  (b) Determine whether f is differentiable at $\mathbf{0}$. Give a careful proof.

18. Let
$$f\begin{pmatrix}x\\y\end{pmatrix} = \frac{xy^6}{x^4 + y^8}, \quad \mathbf{x} \neq \mathbf{0}, \quad f(\mathbf{0}) = 0.$$
  (a) Find all the directional derivatives of f at $\mathbf{0}$.
  (b) Is f continuous at $\mathbf{0}$?
  (c) Is f differentiable at $\mathbf{0}$?

19. (a) Let $f \colon \mathbb{R}^2 \to \mathbb{R}$ be the function defined in Example 5 of Chapter 2, Section 3. Show that f has directional derivatives at $\mathbf{0}$ in every direction but is not differentiable at $\mathbf{0}$.
  (b) Find a function all of whose directional derivatives at $\mathbf{0}$ are 0 but which, nevertheless, is not differentiable at $\mathbf{0}$.
  (c) Find a function all of whose directional derivatives at $\mathbf{0}$ are 0 but which is *unbounded* in any neighborhood of $\mathbf{0}$.
  (d) Find a function all of whose directional derivatives at $\mathbf{0}$ are 0, all of whose directional derivatives exist at every point, and which is *unbounded* in any neighborhood of $\mathbf{0}$.

▶ 3 DIFFERENTIATION RULES

In practice, most of the time Proposition 2.4 is sufficient for us to calculate explicit derivatives. However, it is reassuring to know that the sum, product, and quotient rules from elementary calculus pertain to the multivariable case. We shall come to the chain rule shortly.

For the next proofs, we need the notion of the *norm* of a linear map $T \colon \mathbb{R}^n \to \mathbb{R}^m$. We set
$$\|T\| = \max_{\|\mathbf{x}\|=1} \|T(\mathbf{x})\|.$$

(In Section 1 of Chapter 5 we will prove the maximum value theorem, which states that a continuous function on a closed and bounded subset of $\mathbb{R}^n$ achieves its maximum value.

Since the unit sphere in $\mathbb{R}^n$ is closed and bounded, this maximum exists.) When $\mathbf{x} \neq \mathbf{0}$, we have $\left\| T\left(\frac{\mathbf{x}}{\|\mathbf{x}\|}\right)\right\| \leq \|T\|$, and so, by linearity, the following formula follows immediately:

$$\|T(\mathbf{x})\| \leq \|T\|\|\mathbf{x}\|.$$

Proposition 3.1 *Suppose $U \subset \mathbb{R}^n$ is open and $\mathbf{f}: U \to \mathbb{R}^m$, $\mathbf{g}: U \to \mathbb{R}^m$, and $k: U \to \mathbb{R}$. Suppose $\mathbf{a} \in U$ and $\mathbf{f}$, $\mathbf{g}$, and k are differentiable at $\mathbf{a}$. Then*

1. $\mathbf{f} + \mathbf{g}: U \to \mathbb{R}^m$ *is differentiable at $\mathbf{a}$ and $D(\mathbf{f}+\mathbf{g})(\mathbf{a}) = D\mathbf{f}(\mathbf{a}) + D\mathbf{g}(\mathbf{a})$.*
2. $k\mathbf{f}: U \to \mathbb{R}^m$ *is differentiable at $\mathbf{a}$ and $D(k\mathbf{f})(\mathbf{a})\mathbf{v} = \bigl(Dk(\mathbf{a})\mathbf{v}\bigr)\mathbf{f}(\mathbf{a}) + k(\mathbf{a})D\mathbf{f}(\mathbf{a})\mathbf{v}$ for any $\mathbf{v} \in \mathbb{R}^n$.*
3. $\mathbf{f} \cdot \mathbf{g}: U \to \mathbb{R}$ *is differentiable at $\mathbf{a}$ and $D(\mathbf{f} \cdot \mathbf{g})(\mathbf{a})\mathbf{v} = \bigl(D\mathbf{f}(\mathbf{a})\mathbf{v}\bigr) \cdot \mathbf{g}(\mathbf{a}) + \mathbf{f}(\mathbf{a}) \cdot \bigl(D\mathbf{g}(\mathbf{a})\mathbf{v}\bigr)$ for any $\mathbf{v} \in \mathbb{R}^n$.*

Proof These are much like the proofs of the corresponding results in single-variable calculus. Here, however, we insert the candidate for the derivative in the definition and check that the limit is indeed $\mathbf{0}$:

1. $$\lim_{\mathbf{h}\to\mathbf{0}} \frac{(\mathbf{f}+\mathbf{g})(\mathbf{a}+\mathbf{h}) - (\mathbf{f}+\mathbf{g})(\mathbf{a}) - \bigl(D\mathbf{f}(\mathbf{a}) + D\mathbf{g}(\mathbf{a})\bigr)\mathbf{h}}{\|\mathbf{h}\|}$$
    $$= \lim_{\mathbf{h}\to\mathbf{0}} \frac{\bigl(\mathbf{f}(\mathbf{a}+\mathbf{h}) - \mathbf{f}(\mathbf{a}) - D\mathbf{f}(\mathbf{a})\mathbf{h}\bigr) + \bigl(\mathbf{g}(\mathbf{a}+\mathbf{h}) - \mathbf{g}(\mathbf{a}) - D\mathbf{g}(\mathbf{a})\mathbf{h}\bigr)}{\|\mathbf{h}\|}$$
    $$= \lim_{\mathbf{h}\to\mathbf{0}} \frac{\mathbf{f}(\mathbf{a}+\mathbf{h}) - \mathbf{f}(\mathbf{a}) - D\mathbf{f}(\mathbf{a})\mathbf{h}}{\|\mathbf{h}\|} + \lim_{\mathbf{h}\to\mathbf{0}} \frac{\mathbf{g}(\mathbf{a}+\mathbf{h}) - \mathbf{g}(\mathbf{a}) - D\mathbf{g}(\mathbf{a})\mathbf{h}}{\|\mathbf{h}\|} = \mathbf{0} + \mathbf{0} = \mathbf{0}.$$

2. We proceed much as in the proof of the limit of the product in Theorem 3.2 of Chapter 2.
    $$\lim_{\mathbf{h}\to\mathbf{0}} \frac{(k\mathbf{f})(\mathbf{a}+\mathbf{h}) - (k\mathbf{f})(\mathbf{a}) - \bigl((Dk(\mathbf{a})\mathbf{h})\mathbf{f}(\mathbf{a}) + k(\mathbf{a})D\mathbf{f}(\mathbf{a})\mathbf{h}\bigr)}{\|\mathbf{h}\|}$$
    $$= \lim_{\mathbf{h}\to\mathbf{0}} \frac{k(\mathbf{a}+\mathbf{h})\mathbf{f}(\mathbf{a}+\mathbf{h}) - k(\mathbf{a})\mathbf{f}(\mathbf{a}) - \bigl((Dk(\mathbf{a})\mathbf{h})\mathbf{f}(\mathbf{a}) + k(\mathbf{a})D\mathbf{f}(\mathbf{a})\mathbf{h}\bigr)}{\|\mathbf{h}\|}$$
    $$= \lim_{\mathbf{h}\to\mathbf{0}} \frac{\bigl((k(\mathbf{a}+\mathbf{h}) - k(\mathbf{a}))\mathbf{f}(\mathbf{a}+\mathbf{h}) + k(\mathbf{a})(\mathbf{f}(\mathbf{a}+\mathbf{h}) - \mathbf{f}(\mathbf{a}))\bigr) - \bigl((Dk(\mathbf{a})\mathbf{h})\mathbf{f}(\mathbf{a}) + k(\mathbf{a})D\mathbf{f}(\mathbf{a})\mathbf{h}\bigr)}{\|\mathbf{h}\|}$$
    $$= \lim_{\mathbf{h}\to\mathbf{0}} \frac{(k(\mathbf{a}+\mathbf{h}) - k(\mathbf{a}))\mathbf{f}(\mathbf{a}+\mathbf{h}) - (Dk(\mathbf{a})\mathbf{h})\mathbf{f}(\mathbf{a})}{\|\mathbf{h}\|} + \lim_{\mathbf{h}\to\mathbf{0}} \frac{k(\mathbf{a})(\mathbf{f}(\mathbf{a}+\mathbf{h}) - \mathbf{f}(\mathbf{a})) - k(\mathbf{a})D\mathbf{f}(\mathbf{a})\mathbf{h}}{\|\mathbf{h}\|}$$
    $$= \lim_{\mathbf{h}\to\mathbf{0}} \frac{(k(\mathbf{a}+\mathbf{h}) - k(\mathbf{a}))\mathbf{f}(\mathbf{a}+\mathbf{h}) - (Dk(\mathbf{a})\mathbf{h})\mathbf{f}(\mathbf{a})}{\|\mathbf{h}\|} + k(\mathbf{a})\lim_{\mathbf{h}\to\mathbf{0}} \frac{\mathbf{f}(\mathbf{a}+\mathbf{h}) - \mathbf{f}(\mathbf{a}) - D\mathbf{f}(\mathbf{a})\mathbf{h}}{\|\mathbf{h}\|}.$$

    Now, the second term clearly approaches $\mathbf{0}$. To handle the first term, we have to use continuity in a rather subtle way, remembering that if $\mathbf{f}$ is differentiable at $\mathbf{a}$, then it is necessarily continuous at $\mathbf{a}$ (Proposition 2.2):

$$\frac{\bigl(k(\mathbf{a}+\mathbf{h})-k(\mathbf{a})\bigr)\mathbf{f}(\mathbf{a}+\mathbf{h})-\bigl(Dk(\mathbf{a})\mathbf{h}\bigr)\mathbf{f}(\mathbf{a})}{\|\mathbf{h}\|}$$

$$=\frac{\bigl(k(\mathbf{a}+\mathbf{h})-k(\mathbf{a})-\bigl(Dk(\mathbf{a})\mathbf{h}\bigr)\bigr)\mathbf{f}(\mathbf{a}+\mathbf{h})+\bigl(Dk(\mathbf{a})\mathbf{h}\bigr)\bigl(\mathbf{f}(\mathbf{a}+\mathbf{h})-\mathbf{f}(\mathbf{a})\bigr)}{\|\mathbf{h}\|}$$

$$=\frac{k(\mathbf{a}+\mathbf{h})-k(\mathbf{a})-Dk(\mathbf{a})\mathbf{h}}{\|\mathbf{h}\|}\mathbf{f}(\mathbf{a}+\mathbf{h})+\frac{\bigl(Dk(\mathbf{a})\mathbf{h}\bigr)\bigl(\mathbf{f}(\mathbf{a}+\mathbf{h})-\mathbf{f}(\mathbf{a})\bigr)}{\|\mathbf{h}\|}.$$

Now here the first term clearly approaches $\mathbf{0}$, but the second term is a bit touchy. The length of the second term is

$$\frac{|Dk(\mathbf{a})\mathbf{h}|\,\|\mathbf{f}(\mathbf{a}+\mathbf{h})-\mathbf{f}(\mathbf{a})\|}{\|\mathbf{h}\|}\le\|Dk(\mathbf{a})\|\,\|\mathbf{f}(\mathbf{a}+\mathbf{h})-\mathbf{f}(\mathbf{a})\|,$$

which in turn goes to 0 as $\mathbf{h}\to\mathbf{0}$ by continuity of $\mathbf{f}$ at $\mathbf{a}$. This concludes the proof of (2).

The proof of (3) is virtually identical to that of (2) and is left to the reader in Exercise 9. ∎

Theorem 3.2 (The Chain Rule) *Suppose* $\mathbf{g}\colon\mathbb{R}^n\to\mathbb{R}^m$ *and* $\mathbf{f}\colon\mathbb{R}^m\to\mathbb{R}^\ell$, $\mathbf{g}$ *is differentiable at* $\mathbf{a}$, *and* $\mathbf{f}$ *is differentiable at* $\mathbf{g}(\mathbf{a})$. *Then* $\mathbf{f}\circ\mathbf{g}$ *is differentiable at* $\mathbf{a}$ *and*

$$D(\mathbf{f}\circ\mathbf{g})(\mathbf{a})=D\mathbf{f}(\mathbf{g}(\mathbf{a}))\circ D\mathbf{g}(\mathbf{a}).$$

Proof We must show that

$$\lim_{\mathbf{h}\to\mathbf{0}}\frac{\mathbf{f}(\mathbf{g}(\mathbf{a}+\mathbf{h}))-\mathbf{f}(\mathbf{g}(\mathbf{a}))-D\mathbf{f}(\mathbf{g}(\mathbf{a}))D\mathbf{g}(\mathbf{a})\mathbf{h}}{\|\mathbf{h}\|}=\mathbf{0}.$$

Letting $\mathbf{b}=\mathbf{g}(\mathbf{a})$, we know that

$$\lim_{\mathbf{h}\to\mathbf{0}}\frac{\mathbf{g}(\mathbf{a}+\mathbf{h})-\mathbf{g}(\mathbf{a})-D\mathbf{g}(\mathbf{a})\mathbf{h}}{\|\mathbf{h}\|}=\mathbf{0}\quad\text{and}$$

$$\lim_{\mathbf{k}\to\mathbf{0}}\frac{\mathbf{f}(\mathbf{b}+\mathbf{k})-\mathbf{f}(\mathbf{b})-D\mathbf{f}(\mathbf{b})\mathbf{k}}{\|\mathbf{k}\|}=\mathbf{0}.$$

Given $\varepsilon>0$, this means that there are $\delta_1>0$ and $\eta>0$ so that

(∗) $\quad 0<\|\mathbf{h}\|<\delta_1 \implies \|\mathbf{g}(\mathbf{a}+\mathbf{h})-\mathbf{g}(\mathbf{a})-D\mathbf{g}(\mathbf{a})\mathbf{h}\|<\varepsilon\|\mathbf{h}\|$ and

(∗∗) $\quad \|\mathbf{k}\|<\eta \implies \|\mathbf{f}(\mathbf{b}+\mathbf{k})-\mathbf{f}(\mathbf{b})-D\mathbf{f}(\mathbf{b})\mathbf{k}\|\le\varepsilon\|\mathbf{k}\|.$

Setting $\mathbf{k}=\mathbf{g}(\mathbf{a}+\mathbf{h})-\mathbf{g}(\mathbf{a})$ and rewriting (∗), we conclude that whenever $0<\|\mathbf{h}\|<\delta_1$, we have

$$\|\mathbf{k}-D\mathbf{g}(\mathbf{a})\mathbf{h}\|<\varepsilon\|\mathbf{h}\|,$$

and so

$$\|\mathbf{k}\|<\|D\mathbf{g}(\mathbf{a})\mathbf{h}\|+\varepsilon\|\mathbf{h}\|\le\bigl(\|D\mathbf{g}(\mathbf{a})\|+\varepsilon\bigr)\|\mathbf{h}\|.$$

Let $\delta_2=\eta/\bigl(\|D\mathbf{g}(\mathbf{a})\|+\varepsilon\bigr)$ and set $\delta=\min(\delta_1,\delta_2)$.

Finally, we start with the numerator of the fraction whose limit we seek.

$$\mathbf{f}(\mathbf{b}+\mathbf{k}) - \mathbf{f}(\mathbf{b}) - D\mathbf{f}(\mathbf{b})D\mathbf{g}(\mathbf{a})\mathbf{h}$$
$$= [\mathbf{f}(\mathbf{b}+\mathbf{k}) - \mathbf{f}(\mathbf{b}) - D\mathbf{f}(\mathbf{b})\mathbf{k}] + [D\mathbf{f}(\mathbf{b})\mathbf{k} - D\mathbf{f}(\mathbf{b})D\mathbf{g}(\mathbf{a})\mathbf{h}]$$
$$= [\mathbf{f}(\mathbf{b}+\mathbf{k}) - \mathbf{f}(\mathbf{b}) - D\mathbf{f}(\mathbf{b})\mathbf{k}] + D\mathbf{f}(\mathbf{b})(\mathbf{k} - D\mathbf{g}(\mathbf{a})\mathbf{h}).$$

Therefore, whenever $0 < \|\mathbf{h}\| < \delta$, we have

$$\|\mathbf{f}(\mathbf{b}+\mathbf{k}) - \mathbf{f}(\mathbf{b}) - D\mathbf{f}(\mathbf{b})D\mathbf{g}(\mathbf{a})\mathbf{h}\|$$
$$\leq \|\mathbf{f}(\mathbf{b}+\mathbf{k}) - \mathbf{f}(\mathbf{b}) - D\mathbf{f}(\mathbf{b})\mathbf{k}\| + \|D\mathbf{f}(\mathbf{b})\| \|\mathbf{k} - D\mathbf{g}(\mathbf{a})\mathbf{h}\|$$
$$< \varepsilon \|\mathbf{k}\| + \|D\mathbf{f}(\mathbf{b})\| \varepsilon \|\mathbf{h}\| < \varepsilon \|\mathbf{h}\| (\|D\mathbf{g}(\mathbf{a})\| + \varepsilon + \|D\mathbf{f}(\mathbf{b})\|).$$

Thus, whenever $0 < \|\mathbf{h}\| < \delta$, we have

$$\frac{\|\mathbf{f}(\mathbf{g}(\mathbf{a}+\mathbf{h})) - \mathbf{f}(\mathbf{g}(\mathbf{a})) - D\mathbf{f}(\mathbf{g}(\mathbf{a}))D\mathbf{g}(\mathbf{a})\mathbf{h}\|}{\|\mathbf{h}\|} < \varepsilon(\|D\mathbf{g}(\mathbf{a})\| + \varepsilon + \|D\mathbf{f}(\mathbf{b})\|).$$

Since $\varepsilon > 0$ is arbitrary, this shows that

$$\lim_{\mathbf{h}\to 0} \frac{\|\mathbf{f}(\mathbf{g}(\mathbf{a}+\mathbf{h})) - \mathbf{f}(\mathbf{g}(\mathbf{a})) - D\mathbf{f}(\mathbf{g}(\mathbf{a}))D\mathbf{g}(\mathbf{a})\mathbf{h}\|}{\|\mathbf{h}\|} = 0,$$

as required. ∎

Remark Those who wish to end with a perfect ε at the end may replace the ε in (∗) with $\dfrac{\varepsilon}{2(\|D\mathbf{f}(\mathbf{b})\| + 1)}$ and that in (∗∗) with $\dfrac{\varepsilon}{2(\|D\mathbf{g}(\mathbf{a})\| + \varepsilon)}$.

▶ **EXAMPLE 1**

Suppose the temperature in space is given by $f\begin{pmatrix} x \\ y \\ z \end{pmatrix} = xyz^2 + e^{3xy-2z}$ and the position of a bumblebee is given as a function of time t by $\mathbf{g} \colon \mathbb{R} \to \mathbb{R}^3$. If at time $t = 0$ the bumblebee is at $\mathbf{a} = \begin{bmatrix} 1 \\ 2 \\ 3 \end{bmatrix}$ and her velocity vector is $\mathbf{v} = \begin{bmatrix} -1 \\ 1 \\ 2 \end{bmatrix}$, as indicated in Figure 3.1, then we might ask at what rate she perceives the temperature to be changing at that instant. The temperature she measures at time t is $(f \circ \mathbf{g})(t)$, and so she wants to calculate $(f \circ \mathbf{g})'(0) = D(f \circ \mathbf{g})(0)$. We have

$$Df\begin{pmatrix} x \\ y \\ z \end{pmatrix} = \begin{bmatrix} yz^2 + 3ye^{3xy-2z} & xz^2 + 3xe^{3xy-2z} & 2xyz - 2e^{3xy-2z} \end{bmatrix},$$

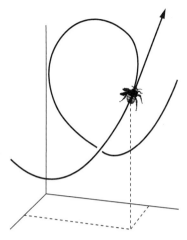

Figure 3.1

so $Df(\mathbf{a}) = \begin{bmatrix} 24 & 12 & 10 \end{bmatrix}$. Then

$$(f \circ \mathbf{g})'(0) = Df(\mathbf{g}(0))\mathbf{g}'(0) = Df(\mathbf{a})\mathbf{v} = \begin{bmatrix} 24 & 12 & 10 \end{bmatrix} \begin{bmatrix} -1 \\ 1 \\ 2 \end{bmatrix} = 8.$$

Note that in order to apply the chain rule, we need to know only her position and velocity vector at that instant, not even what her path near $\mathbf{a}$ might be. ◀

Remark Suppose $\mathbf{f} \colon \mathbb{R}^n \to \mathbb{R}^m$ is differentiable at $\mathbf{a}$ and we wish to evaluate $D_\mathbf{v}\mathbf{f}(\mathbf{a})$ for some $\mathbf{v} \in \mathbb{R}^n$. Define $\mathbf{g} \colon \mathbb{R} \to \mathbb{R}^n$ by $\mathbf{g}(t) = \mathbf{a} + t\mathbf{v}$, and consider $\varphi(t) = (\mathbf{f} \circ \mathbf{g})(t)$. By definition, we have $D_\mathbf{v}\mathbf{f}(\mathbf{a}) = \varphi'(0)$. Then, by the chain rule, we have

$$D_\mathbf{v}\mathbf{f}(\mathbf{a}) = \varphi'(0) = (\mathbf{f} \circ \mathbf{g})'(0) = D\mathbf{f}(\mathbf{a})\mathbf{g}'(0) = D\mathbf{f}(\mathbf{a})\mathbf{v}.$$

This is an alternative derivation of the result of Proposition 2.3. (Cf. Example 4 of Section 1.)

Indeed, if $\mathbf{g}$ is *any* differentiable function with $\mathbf{g}(0) = \mathbf{a}$ and $\mathbf{g}'(0) = \mathbf{v}$, we see that $D_\mathbf{v}\mathbf{f}(\mathbf{a}) = (\mathbf{f} \circ \mathbf{g})'(0)$, so this shows, as we suggested in the remark on p. 83, that we should think of the directional derivative as the rate of change perceived by an observer at $\mathbf{a}$ moving with instantaneous velocity $\mathbf{v}$.

▶ **EXAMPLE 2**

Let

$$\mathbf{f}\begin{pmatrix} x \\ y \end{pmatrix} = \begin{bmatrix} x^2 - y^2 \\ 2xy \end{bmatrix} \quad \text{and} \quad \mathbf{g}\begin{pmatrix} u \\ v \end{pmatrix} = \begin{bmatrix} u \cos v \\ u \sin v \end{bmatrix}.$$

Since

$$D\mathbf{f}\begin{pmatrix}x\\y\end{pmatrix} = \begin{bmatrix} 2x & -2y \\ 2y & 2x \end{bmatrix} \quad \text{and} \quad D\mathbf{g}\begin{pmatrix}u\\v\end{pmatrix} = \begin{bmatrix} \cos v & -u \sin v \\ \sin v & u \cos v \end{bmatrix},$$

we have

$$D(\mathbf{f} \circ \mathbf{g})\begin{pmatrix}u\\v\end{pmatrix} = D\mathbf{f}\left(\mathbf{g}\begin{pmatrix}u\\v\end{pmatrix}\right) D\mathbf{g}\begin{pmatrix}u\\v\end{pmatrix}$$

$$= \begin{bmatrix} 2u \cos v & -2u \sin v \\ 2u \sin v & 2u \cos v \end{bmatrix} \begin{bmatrix} \cos v & -u \sin v \\ \sin v & u \cos v \end{bmatrix} = 2 \begin{bmatrix} u \cos 2v & -u^2 \sin 2v \\ u \sin 2v & u^2 \cos 2v \end{bmatrix}.$$

On the other hand, as the reader can verify, $(\mathbf{f} \circ \mathbf{g})\begin{pmatrix}u\\v\end{pmatrix} = \begin{bmatrix} u^2 \cos 2v \\ u^2 \sin 2v \end{bmatrix}$, and so we can double-check the calculation of the derivative directly. ◀

▶ EXERCISES 3.3

*1. Suppose $f: \mathbb{R}^3 \to \mathbb{R}$ is differentiable and

$$\mathbf{g}(t) = \begin{bmatrix} \cos t + \sin t \\ t+1 \\ t^2 + 4t - 1 \end{bmatrix} \quad \text{and} \quad Df\begin{pmatrix}1\\1\\-1\end{pmatrix} = \begin{bmatrix} 2 & 1 & -1 \end{bmatrix}.$$

Find $(f \circ \mathbf{g})'(0)$.

*2. Suppose

$$\mathbf{f}\begin{pmatrix}x\\y\end{pmatrix} = \begin{bmatrix} 2y - \sin x \\ e^{x+3y} \\ xy + y^3 \end{bmatrix} \quad \text{and} \quad \mathbf{g}\begin{pmatrix}x\\y\\z\end{pmatrix} = \begin{bmatrix} 3x + y - z \\ x + yz + 1 \end{bmatrix}.$$

Calculate $D(\mathbf{f} \circ \mathbf{g})(\mathbf{0})$ and $D(\mathbf{g} \circ \mathbf{f})(\mathbf{0})$.

3. Suppose $\mathbf{g}(t) = \begin{bmatrix} \cos t \\ \sin t \\ 2\sin(t/2) \end{bmatrix}$ and $f\begin{pmatrix}x\\y\\z\end{pmatrix} = x^2 + y^2 + z^2 + 2x$. Use the chain rule to calculate $(f \circ \mathbf{g})'(t)$. What do you conclude?

4. An ant moves along a helical path with trajectory $\mathbf{g}(t) = \begin{bmatrix} 3\cos t \\ 3\sin t \\ 5t \end{bmatrix}$.

(a) At what rate is his distance from the origin changing at $t = 2\pi$?

(b) The temperature in space is given by the function $f: \mathbb{R}^3 \to \mathbb{R}$, $f\begin{pmatrix}x\\y\\z\end{pmatrix} = xy + z^2$. At what rate does the ant detect the temperature to be changing at $t = 3\pi/4$?

*5. An airplane is flying near a radar tower. At the instant it is exactly 3 miles due west of the tower, it is 4 miles high and flying with a ground speed of 450 mph and climbing at a rate of 5 mph. If at that instant it is flying

(a) due east, (b) northeast,
at what rate is it approaching the radar tower at that instant?

*6. An ideal gas obeys the law $pV = nRT$, where p is pressure, V is volume, n is the number of moles, R is the universal gas constant, and T is temperature. Suppose for a certain quantity of ideal gas, $nR = 1$ l-atm/°K. At a given instant, the volume is 10 l and is increasing at the rate of 1 l/min; the temperature is 300°K and is increasing at the rate of 5°K/min. At what rate is the pressure increasing at that instant?

7. Ohm's law tells us that $V = IR$, where V is the voltage in an electric circuit, I is the current flow (in amps), and R is the resistance (in ohms). Suppose that as time passes, the voltage decreases as the battery wears out and the resistance increases as the resistor heats up. Assuming V and R vary as differentiable functions of t, at what rate is the current flow changing at the instant t_0 if $R(t_0) = 100$ ohm, $R'(t_0) = 0.5$ ohm/sec, $I(t_0) = 0.1$ amp, and $V'(t_0) = -0.1$ volt/sec?

8. Let $U \subset \mathbb{R}^n$ be open. Suppose $g \colon U \to \mathbb{R}$ is differentiable at $\mathbf{a} \in U$ and $g(\mathbf{a}) \neq 0$. Prove that $1/g$ is differentiable at $\mathbf{a}$ and $D(1/g)(\mathbf{a}) = -1/(g(\mathbf{a}))^2 Dg(\mathbf{a})$.

9. Prove (3) in Proposition 3.1. (One approach is to mimic the proof given of (2). Another is to apply (1) and (2) appropriately.)

♯10. Suppose $U \subset \mathbb{R}^n$ is open and $\mathbf{a} \in U$. Let $\mathbf{f}, \mathbf{g} \colon U \to \mathbb{R}^3$ be differentiable at $\mathbf{a}$. Prove that $\mathbf{f} \times \mathbf{g}$ is differentiable at $\mathbf{a}$ and $D(\mathbf{f} \times \mathbf{g})(\mathbf{a})\mathbf{v} = (D\mathbf{f}(\mathbf{a})\mathbf{v}) \times \mathbf{g}(\mathbf{a}) + \mathbf{f}(\mathbf{a}) \times (D\mathbf{g}(\mathbf{a})\mathbf{v})$ for any $\mathbf{v} \in \mathbb{R}^n$. (Hint: Follow the proof of part (2) of Proposition 3.1, and use Exercise 1.5.14.)

11. (**Euler's Theorem on Homogeneous Functions**) We say $f \colon \mathbb{R}^n - \{0\} \to \mathbb{R}$ is *homogeneous of degree k* if $f(t\mathbf{x}) = t^k f(\mathbf{x})$ for all $t > 0$. Prove that f is homogeneous of degree k if and only if $Df(\mathbf{x})\mathbf{x} = kf(\mathbf{x})$ for all nonzero $\mathbf{x} \in \mathbb{R}^n$. (Hint: Fix $\mathbf{x}$ and consider $h(t) = t^{-k} f(t\mathbf{x})$.)

♯12. Suppose $U \subset \mathbb{R}^n$ is open and convex (i.e., given any points $\mathbf{a}, \mathbf{b} \in U$, the line segment joining them lies in U as well). If $\mathbf{f} \colon U \to \mathbb{R}^m$ is differentiable and $D\mathbf{f}(\mathbf{x}) = O$ for all $\mathbf{x} \in U$, prove that $\mathbf{f}$ is constant. Can you prove this when U is open and *connected* (i.e., any pair of points can be joined by a piecewise-$\mathcal{C}^1$ path)?

13. Suppose $f \colon \mathbb{R} \to \mathbb{R}$ is differentiable and let $h\begin{pmatrix} x \\ y \end{pmatrix} = f(\sqrt{x^2 + y^2})$ for $\mathbf{x} \neq \mathbf{0}$. Letting $r = \sqrt{x^2 + y^2}$, show that
$$x \frac{\partial h}{\partial x} + y \frac{\partial h}{\partial y} = rf'(r).$$

*14. Suppose $h \colon \mathbb{R} \to \mathbb{R}$ is continuous and $u, v \colon (a, b) \to \mathbb{R}$ are differentiable. Prove that the function $F \colon (a, b) \to \mathbb{R}$ given by
$$F(t) = \int_{u(t)}^{v(t)} h(s)\,ds$$
is differentiable and calculate F'. (Hint: Recall that the Fundamental Theorem of Calculus tells you how to differentiate functions such as $H(x) = \int_a^x h(s)\,ds$.)

15. If $f \colon \mathbb{R}^2 \to \mathbb{R}$ is differentiable and $F\begin{pmatrix} u \\ v \end{pmatrix} = f\begin{pmatrix} u+v \\ u-v \end{pmatrix}$, show that
$$\frac{\partial F}{\partial u} \frac{\partial F}{\partial v} = \left(\frac{\partial f}{\partial x}\right)^2 - \left(\frac{\partial f}{\partial y}\right)^2,$$
where the functions on the right-hand side are evaluated at $\begin{bmatrix} u+v \\ u-v \end{bmatrix}$.

*16. Suppose $f\colon \mathbb{R}^2 \to \mathbb{R}$ is differentiable and let $F\begin{pmatrix} r \\ \theta \end{pmatrix} = f\begin{pmatrix} r\cos\theta \\ r\sin\theta \end{pmatrix}$. Calculate

$$\left(\frac{\partial F}{\partial r}\right)^2 + \frac{1}{r^2}\left(\frac{\partial F}{\partial \theta}\right)^2$$

in terms of the partial derivatives of f.

17. Suppose $f\colon \mathbb{R}^2 \to \mathbb{R}$ is differentiable and $\dfrac{\partial f}{\partial t} = c\dfrac{\partial f}{\partial x}$ for some nonzero constant c. Prove that $f\begin{pmatrix} x \\ t \end{pmatrix} = h(x+ct)$ for some function h. (Hint: Let $\begin{bmatrix} u \\ v \end{bmatrix} = \begin{bmatrix} x \\ x+ct \end{bmatrix}$.)

4 THE GRADIENT

To develop physical intuition, it is important to recast Proposition 2.3 in more geometric terms when f is a scalar-valued function.

Definition Let $f\colon \mathbb{R}^n \to \mathbb{R}$ be differentiable at $\mathbf{a}$. We define the *gradient* of f at $\mathbf{a}$ to be the vector

$$\nabla f(\mathbf{a}) = \bigl(Df(\mathbf{a})\bigr)^{\mathsf{T}} = \begin{bmatrix} \dfrac{\partial f}{\partial x_1}(\mathbf{a}) \\ \dfrac{\partial f}{\partial x_2}(\mathbf{a}) \\ \vdots \\ \dfrac{\partial f}{\partial x_n}(\mathbf{a}) \end{bmatrix}.$$

Now we can interpret the directional derivative of a differentiable function as a dot product:

(*) $\qquad D_{\mathbf{v}}f(\mathbf{a}) = Df(\mathbf{a})\mathbf{v} = \nabla f(\mathbf{a}) \cdot \mathbf{v}.$

If we consider the directional derivative in the direction of various *unit* vectors $\mathbf{v}$, we infer from the Cauchy-Schwarz inequality, Proposition 2.3 of Chapter 1, that

$$D_{\mathbf{v}}f(\mathbf{a}) \le \|\nabla f(\mathbf{a})\|,$$

with equality holding if and only if $\nabla f(\mathbf{a})$ is a positive scalar multiple of $\mathbf{v}$.

As a consequence, we have

Proposition 4.1 *Suppose f is differentiable at $\mathbf{a}$. Then $\nabla f(\mathbf{a})$ points in the direction in which f increases at the greatest rate, and $\|\nabla f(\mathbf{a})\|$ is that greatest possible rate of change; i.e.,*

$$\|\nabla f(\mathbf{a})\| = \max_{\|\mathbf{v}\|=1} D_{\mathbf{v}}f(\mathbf{a}).$$

▶ **EXAMPLE 1**

Let $f\colon \mathbb{R}^n \to \mathbb{R}$ be defined by $f(\mathbf{x}) = \|\mathbf{x}\|$. It is simple enough to calculate partial derivatives of f, but we'd rather use the geometric meaning of the gradient to figure out $\nabla f(\mathbf{a})$ for any $\mathbf{a} \ne \mathbf{0}$. Clearly, if we are at $\mathbf{a}$, the direction in which distance from the origin increases most rapidly is in the direction of **a** itself (i.e., to move away from the origin as fast as possible, we should move radially outward). Moreover, we saw in Example 3 of Section 1 that the directional derivative $D_{\mathbf{v}} f(\mathbf{a}) = 1$ when $\mathbf{v} = \mathbf{a}/\|\mathbf{a}\|$. Therefore, we infer from Proposition 4.1 that $\nabla f(\mathbf{a})$ is a vector pointing radially outward and having length 1. That is,

$$\nabla f(\mathbf{a}) = \frac{\mathbf{a}}{\|\mathbf{a}\|}.$$

As corroboration, we observe that if we move orthogonal to **a**, then instantaneously our distance from the origin is not changing, so $D_{\mathbf{v}} f(\mathbf{a}) = \nabla f(\mathbf{a}) \cdot \mathbf{v} = 0$ when $\mathbf{v} \cdot \mathbf{a} = 0$, as it should. ◀

An equally important interpretation, which will emerge in a significant rôle in Section 5 of Chapter 4 and then in Chapter 5, is this: Suppose $f : \mathbb{R}^2 \to \mathbb{R}$ is a $\mathcal{C}^1$ function, c is a constant, and $C = \{\mathbf{x} \in \mathbb{R}^2 : f(\mathbf{x}) = c\}$ is a *level curve* of f. We shall prove later that for any $\mathbf{a} \in C$, provided $\nabla f(\mathbf{a}) \neq \mathbf{0}$, C has a tangent line at **a** and $\nabla f(\mathbf{a})$ is orthogonal to that tangent line. Intuitively, this is quite plausible: If **v** is tangent to C at **a**, then since f does not change as we move along C, it therefore does not change instantaneously as we move in the direction of **v**, and so $D_{\mathbf{v}} f(\mathbf{a}) = 0$. Therefore, by (∗), $\nabla f(\mathbf{a})$ is orthogonal to **v**. (See also Exercise 6.) More generally, if $f : \mathbb{R}^n \to \mathbb{R}$ is differentiable and $\nabla f(\mathbf{a}) \neq \mathbf{0}$, then $\nabla f(\mathbf{a})$ is orthogonal to the level set $\{\mathbf{x} \in \mathbb{R}^n : f(\mathbf{x}) = c\}$ of f passing through **a**.

▶ **EXAMPLE 2**

Consider the surface M defined by $f\begin{pmatrix} x \\ y \\ z \end{pmatrix} = e^{x+2y} \cos z - xz + y = 2$. Note that the point $\mathbf{a} = \begin{bmatrix} -2 \\ 1 \\ 0 \end{bmatrix}$ lies on M. We want to find the equation of the tangent plane to M at **a**. We know that $\nabla f(\mathbf{a})$ gives the normal to the plane, so we calculate

$$\nabla f \begin{pmatrix} x \\ y \\ z \end{pmatrix} = \begin{bmatrix} -z + e^{x+2y} \cos z \\ 1 + 2e^{x+2y} \cos z \\ -x - e^{x+2y} \sin z \end{bmatrix}, \quad \text{and therefore} \quad \nabla f \begin{pmatrix} -2 \\ 1 \\ 0 \end{pmatrix} = \begin{bmatrix} 1 \\ 3 \\ 2 \end{bmatrix}.$$

Thus, the equation of the tangent plane of M at **a** is $1(x+2) + 3(y-1) + 2(z-0) = 0$ or $x + 3y + 2z = 1$. ◀

Remark Be sure to distinguish between a *level surface* of f and the *graph* of f (which, in this case, would reside in $\mathbb{R}^4$).

▶ **EXAMPLE 3**

As a beautiful application of this principle, we use the results of Example 1 to derive a fundamental physical property of the ellipse. Given two points F_1 and F_2 in the plane, an ellipse (as pictured in Figure 4.1) is the locus of points P so that

$$\|\overrightarrow{F_1 P}\| + \|\overrightarrow{F_2 P}\| = 2a$$

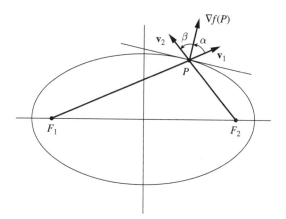

Figure 4.1

for some positive constant a. Write $f_i(\mathbf{x}) = \|\overrightarrow{F_i \mathbf{x}}\|$, $i = 1, 2$, and set $f(\mathbf{x}) = f_1(\mathbf{x}) + f_2(\mathbf{x})$. Then, by the results of Example 1, we have

$$\nabla f(P) = \nabla f_1(P) + \nabla f_2(P) = \underbrace{\frac{\overrightarrow{F_1 P}}{\|\overrightarrow{F_1 P}\|}}_{\mathbf{v}_1} + \underbrace{\frac{\overrightarrow{F_2 P}}{\|\overrightarrow{F_2 P}\|}}_{\mathbf{v}_2}.$$

Both $\mathbf{v}_1$ and $\mathbf{v}_2$ are unit vectors pointing radially away from F_1 and F_2, respectively, and therefore $\nabla f(P)$ bisects the angle between them (see Exercise 1.2.20). Thus, $\alpha = \beta$, and so the tangent line to the ellipse at P makes equal angles with the lines $\overleftrightarrow{F_1 P}$ and $\overleftrightarrow{F_2 P}$. Thus, a light ray emanating from one focus reflects off the ellipse back to the other focus. ◀

▶ EXERCISES 3.4

1. Give the equation of the tangent line of the given level curve at the prescribed point $\mathbf{a}$.

   *(a) $x^3 + y^3 = 9$, $\mathbf{a} = \begin{bmatrix} 1 \\ 2 \end{bmatrix}$

   (b) $3xy^2 + e^{xy} - \sin(\pi y) = 1$, $\mathbf{a} = \begin{bmatrix} 0 \\ 1 \end{bmatrix}$

   (c) $x^3 + xy^2 - y^4 = 1$, $\mathbf{a} = \begin{bmatrix} 1 \\ -1 \end{bmatrix}$

2. Give the equation of the tangent plane of the given level surface at the prescribed point $\mathbf{a}$.

   (a) $x^2 + y^2 + z^2 = 5$, $\mathbf{a} = \begin{bmatrix} 1 \\ 0 \\ 2 \end{bmatrix}$

   *(b) $yz^2 + 2e^{xy}z^5 = 4$, $\mathbf{a} = \begin{bmatrix} 0 \\ 2 \\ 1 \end{bmatrix}$

   (c) $x^3 + xz^2 + y^2z + y^3 = 0$, $\mathbf{a} = \begin{bmatrix} -1 \\ 1 \\ 0 \end{bmatrix}$

   (d) $e^{2x+z} \cos(3y) - xy + z = 3$, $\mathbf{a} = \begin{bmatrix} -1 \\ 0 \\ 2 \end{bmatrix}$

3. Given the topographical map in Figure 4.2, sketch on the map an approximate route of steepest ascent from P to Q, the top of the mountain. What about from R?

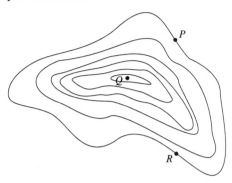

Figure 4.2

*4. Suppose a hillside is given by $z = f(\mathbf{x})$, $\mathbf{x} \in U \subset \mathbb{R}^2$. Suppose $f(\mathbf{a}) = c$ and $Df(\mathbf{a}) = \begin{bmatrix} 3 & -4 \end{bmatrix}$.

(a) Find a vector tangent to the curve of steepest ascent on the hill at $\begin{bmatrix} \mathbf{a} \\ c \end{bmatrix}$.

(b) Find the angle that a stream makes with the horizontal at $\begin{bmatrix} \mathbf{a} \\ c \end{bmatrix}$ if it flows in the $\mathbf{e}_2$ direction at that point.

5. As shown in Figure 4.3, at a certain moment, a ladybug is at position $\mathbf{x}_0$ and moving with velocity vector $\mathbf{v}$. At that moment, the angle $\angle \mathbf{a}\mathbf{x}_0\mathbf{b} = \pi/2$, her velocity bisects that angle, and her speed is 5 units/sec. At what rate is the *sum* of her distances from $\mathbf{a}$ and $\mathbf{b}$ decreasing at that moment? Give your reasoning clearly.

Figure 4.3

6. Suppose that, in a neighborhood of the point $\mathbf{a}$, the level curve $C = \{\mathbf{x} \in \mathbb{R}^2 : f(\mathbf{x}) = c\}$ can be parametrized by a differentiable function $\mathbf{g}: (-\varepsilon, \varepsilon) \to \mathbb{R}^2$, with $\mathbf{g}(0) = \mathbf{a}$. Use the chain rule to prove that $\nabla f(\mathbf{a})$ is orthogonal to the tangent vector to C at $\mathbf{a}$.

7. Check that the definition of an ellipse given in Example 3 gives the usual Cartesian equation of the form
$$\frac{x^2}{a^2} + \frac{y^2}{b^2} = 1$$
when the foci are at $\begin{bmatrix} \pm c \\ 0 \end{bmatrix}$. (Hint: You should find that $a^2 = b^2 + c^2$.)

8. By analogy with Example 3, prove that light emanating from the focus of a parabola reflects off the parabola in the direction of the axis of the parabola. This is why automobile headlights use parabolic

reflectors. (A convenient definition of a parabola is this: It is the locus of points equidistant from a point (the *focus*) and a line (the *directrix*), as pictured in Figure 4.4.)

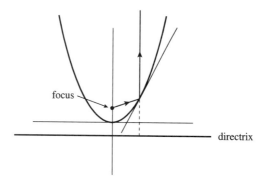

Figure 4.4

9. Using Figure 4.5 as a guide, complete Dandelin's proof (dating from 1822) that the appropriate conic section is an ellipse. Find spheres that are inscribed in the cone and tangent to the plane of the ellipse. Letting F_1 and F_2 be the points of tangency and P a point of the ellipse, let Q_1 and Q_2 be points where the generator of the cone through P intersects the respective spheres. Show that $\|\overrightarrow{Q_i P}\| = \|\overrightarrow{F_i P}\|$, $i = 1, 2$, and deduce that $\|\overrightarrow{F_1 P}\| + \|\overrightarrow{F_2 P}\| = $ const. (What happens when we tilt the plane to obtain a parabola or hyperbola?)

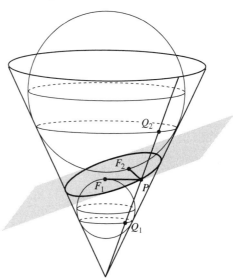

Figure 4.5

10. Suppose $f: \mathbb{R}^2 \to \mathbb{R}$ is a differentiable function whose gradient is nowhere $\mathbf{0}$ and that satisfies
$$\frac{\partial f}{\partial x} = 2 \frac{\partial f}{\partial y}$$
everywhere.
(a) Find (with proof) the level curves of f.
(b) Show that there is a differentiable function $F: \mathbb{R} \to \mathbb{R}$ so that $f\begin{pmatrix} x \\ y \end{pmatrix} = F(2x + y)$.

11. Suppose $f\colon \mathbb{R}^2 - \{\mathbf{0}\} \to \mathbb{R}$ is a differentiable function whose gradient is nowhere $\mathbf{0}$ and that satisfies
$$-y\frac{\partial f}{\partial x} + x\frac{\partial f}{\partial y} = 0$$
everywhere.
(a) Find (with proof) the level curves of f.
(b) Show that there is a differentiable function F defined on the set of positive real numbers so that $f(\mathbf{x}) = F(\|\mathbf{x}\|)$.

*12. Find all constants c for which the surfaces
$$x^2 + y^2 + z^2 = 1 \quad \text{and} \quad z = x^2 + y^2 + c$$

(a) intersect tangentially at each point and (b) intersect orthogonally at each point

13. Prove the so-called *pedal property* of the ellipse: If $\mathbf{n}$ is the unit normal to the ellipse at P, then $(\overrightarrow{F_1 P} \cdot \mathbf{n})(\overrightarrow{F_2 P} \cdot \mathbf{n}) = $ constant.

14. The height of land in the vicinity of a hill is given in terms of horizontal coordinates x and y by $h\begin{pmatrix} x \\ y \end{pmatrix} = \dfrac{40}{4 + x^2 + 3y^2}$. A stream passes through the point $\begin{bmatrix} 1 \\ 1 \\ 5 \end{bmatrix}$ and follows a path of "steepest descent." Find the equation of the path of the stream on a map of the region.

15. A drop of water falls onto a football and rolls down, following the path of steepest descent; that is, it moves in the direction tangent to the football most nearly vertically downward. Find the path the water drop follows if the surface of the football is ellipsoidal and given by the equation
$$4x^2 + y^2 + 4z^2 = 9$$
and the drop starts at the point $\begin{bmatrix} 1 \\ 1 \\ 1 \end{bmatrix}$.

5 CURVES

In this section, we return to the study of (parametrized) curves with which we began Chapter 2. Now we bring in the appropriate differential calculus to discuss velocity, acceleration, some basic principles from physics, and the notion of curvature.

If $\mathbf{g}\colon (a,b) \to \mathbb{R}^n$ is a twice-differentiable vector-valued function, we can visualize $\mathbf{g}(t)$ as denoting the position of a particle at time t, and hence the image of $\mathbf{g}$ represents its trajectory as time passes. Then $\mathbf{g}'(t)$ is the *velocity vector* of the particle at time t and $\mathbf{g}''(t)$ is its *acceleration vector* at time t. The length of the velocity vector, $\|\mathbf{g}'(t)\|$, is called the *speed* of the particle. In physics, a particle of mass m is said to have *kinetic energy*
$$\text{K.E.} = \tfrac{1}{2}m(\text{speed})^2,$$
and acceleration looms large because of *Newton's second law of motion*, which says that a *force* acting on an object imparts an acceleration according to the equation
$$\overrightarrow{\text{force}} = (\text{mass})\overrightarrow{\text{acceleration}}, \quad \text{or, in other words,} \quad \mathbf{F} = m\mathbf{a}.$$

As a quick application of some vector calculus, let's discuss a few properties of motion in a *central force field*. We call a force field $\mathbf{F}\colon U \to \mathbb{R}^3$ on an open subset $U \subset \mathbb{R}^3$ *central* if $\mathbf{F}(\mathbf{x}) = \psi(\mathbf{x})\mathbf{x}$ for some continuous function $\psi\colon U \to \mathbb{R}$; that is, $\mathbf{F}$ is everywhere a scalar multiple of the position vector.

Newton discovered that the gravitational field of a point mass M is an *inverse square* force directed toward the point mass. If we assume the point mass is at the origin, then the force exerted on a unit test mass at position $\mathbf{x}$ is

$$\mathbf{F}(\mathbf{x}) = -\frac{GM}{\|\mathbf{x}\|^2}\frac{\mathbf{x}}{\|\mathbf{x}\|} = -\frac{GM}{\|\mathbf{x}\|^3}\mathbf{x},$$

where G is the universal gravitational constant. Newton published his laws of motion in 1687 in his *Philosophiae Naturalis Principia Mathematica*. Interestingly, Kepler had published his empirical observations almost a century earlier, in 1596.[1]

Kepler's first law: Planets move in ellipses with the sun at one focus.

Kepler's second law: The position vector from the sun to the planet sweeps out area at a constant rate.

Kepler's third law: The square of the period of a planet is proportional to the cube of the semimajor axis of its elliptical orbit.

For the first and third laws we refer the reader to Exercise 15, but here we prove a generalization of the second.

Proposition 5.1 *Let $\mathbf{F}$ be a central force field on $\mathbb{R}^3$. Then the trajectory of any particle lies in a plane; assuming the trajectory is not a line, the position vector sweeps out area at a constant rate.*

Proof Let the trajectory of the particle be given by $\mathbf{g}(t)$, and let its mass be m. Consider the vector function $\mathbf{A}(t) = \mathbf{g}(t) \times \mathbf{g}'(t)$. By Exercise 3.3.10 and by Newton's second law of motion, we have

$$\mathbf{A}'(t) = \mathbf{g}'(t) \times \mathbf{g}'(t) + \mathbf{g}(t) \times \mathbf{g}''(t) = \mathbf{g}(t) \times \tfrac{1}{m}\psi(\mathbf{g}(t))\mathbf{g}(t) = \mathbf{0}$$

since the cross product of any vector with a scalar multiple of itself is $\mathbf{0}$. Thus, $\mathbf{A}(t) = \mathbf{A}_0$ is a constant. If $\mathbf{A}_0 = \mathbf{0}$, the particle moves on a line (why?). If $\mathbf{A}_0 \neq \mathbf{0}$, then note that $\mathbf{g}$ lies on the plane

$$\mathbf{A}_0 \cdot \mathbf{x} = 0,$$

as $\mathbf{A}_0 \cdot \mathbf{g}(t) = \mathbf{A}(t) \cdot \mathbf{g}(t) = 0$ for all t.

Assume now the trajectory is not linear. Let $\mathcal{A}(t)$ denote the area swept out by the position vector $\mathbf{g}(t)$ from time t_0 to time t. Since $\mathcal{A}(t+h) - \mathcal{A}(t)$ equals the area subtended

---

[1] Somewhat earlier he had surmised that the positions of the six known planets were linked to the famous five regular polyhedra.

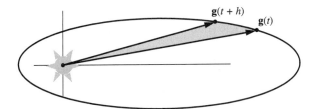

Figure 5.1

by the position vectors $\mathbf{g}(t)$ and $\mathbf{g}(t+h)$ (see Figure 5.1), for h small, this is approximately the area of the triangle determined by the pair of vectors or, equivalently, by the vectors $\mathbf{g}(t)$ and $\mathbf{g}(t+h) - \mathbf{g}(t)$. According to Proposition 5.1 of Chapter 1, this area is $\frac{1}{2}\|\mathbf{g}(t) \times (\mathbf{g}(t+h) - \mathbf{g}(t))\|$, so that

$$\begin{aligned}\mathcal{A}'(t) &= \lim_{h \to 0} \frac{\mathcal{A}(t+h) - \mathcal{A}(t)}{h} \\ &= \lim_{h \to 0^+} \frac{1}{2} \frac{\|\mathbf{g}(t) \times (\mathbf{g}(t+h) - \mathbf{g}(t))\|}{h} \\ &= \lim_{h \to 0^+} \frac{1}{2} \left\| \mathbf{g}(t) \times \frac{\mathbf{g}(t+h) - \mathbf{g}(t)}{h} \right\| \\ &= \tfrac{1}{2}\|\mathbf{g}(t) \times \mathbf{g}'(t)\| = \tfrac{1}{2}\|\mathbf{A}_0\|.\end{aligned}$$

That is, the position vector sweeps out area at a constant rate. ■

One of the most useful (yet intuitively quite apparent) results about curves is the following.

Proposition 5.2 *Suppose* $\mathbf{g}\colon (a,b) \to \mathbb{R}^n$ *is a differentiable parametrized curve with the property that* $\mathbf{g}$ *has constant length (i.e., the curve lies on a sphere centered at the origin). Then* $\mathbf{g}(t) \cdot \mathbf{g}'(t) = 0$ *for all t; i.e., the velocity vector is everywhere orthogonal to the position vector.*

Proof By (3) of Proposition 3.1, we differentiate the equation

$$\mathbf{g}(t) \cdot \mathbf{g}(t) = \text{const}$$

to obtain

$$\mathbf{g}'(t) \cdot \mathbf{g}(t) + \mathbf{g}(t) \cdot \mathbf{g}'(t) = 2\mathbf{g}(t) \cdot \mathbf{g}'(t) = 0,$$

as required. ■

Physically, one should think of it this way: If the velocity vector had a nonzero projection on the position vector, that would mean that the particle's distance from the center of the sphere would be changing. Analogously, as we ask the reader to show in Exercise 2, if a particle moves with constant speed, then its acceleration must be orthogonal to its velocity.

Now we leave physics behind for a while and move on to discuss some geometry. We begin with a generalization of the triangle inequality, Corollary 2.4 of Chapter 1.

Lemma 5.3 *Suppose* $\mathbf{g}\colon [a,b] \to \mathbb{R}^n$ *is continuous (except perhaps at finitely many points). Then, defining the integral of* $\mathbf{g}$ *component by component, i.e.,*

$$\int_a^b \mathbf{g}(t)dt = \begin{bmatrix} \int_a^b g_1(t)dt \\ \vdots \\ \int_a^b g_n(t)dt \end{bmatrix},$$

we have

$$\left\| \int_a^b \mathbf{g}(t)dt \right\| \le \int_a^b \|\mathbf{g}(t)\|dt.$$

Proof Let $\mathbf{v} = \int_a^b \mathbf{g}(t)dt$. If $\mathbf{v} = \mathbf{0}$, there is nothing to prove. By the Cauchy-Schwarz inequality, Proposition 2.3 of Chapter 1, $|\mathbf{v} \cdot \mathbf{g}(t)| \le \|\mathbf{v}\| \|\mathbf{g}(t)\|$, so

$$\|\mathbf{v}\|^2 = \mathbf{v} \cdot \int_a^b \mathbf{g}(t)dt = \int_a^b \mathbf{v} \cdot \mathbf{g}(t)dt \le \int_a^b \|\mathbf{v}\| \|\mathbf{g}(t)\|dt = \|\mathbf{v}\| \int_a^b \|\mathbf{g}(t)\|dt.$$

Assuming $\mathbf{v} \ne \mathbf{0}$, we now infer that $\|\mathbf{v}\| \le \int_a^b \|\mathbf{g}(t)\|dt$, as required. ∎

Definition Let $\mathbf{g}\colon [a,b] \to \mathbb{R}^n$ be a (continuous) parametrized curve. Given a partition $\mathcal{P} = \{a = t_0 < t_1 < \cdots < t_k = b\}$ of the interval $[a,b]$, let

$$\ell(\mathbf{g}, \mathcal{P}) = \sum_{i=1}^k \|\mathbf{g}(t_i) - \mathbf{g}(t_{i-1})\|.$$

That is, $\ell(\mathbf{g}, \mathcal{P})$ is the length of the inscribed polygon with vertices at $\mathbf{g}(t_i)$, $i = 0, \ldots, k$, as indicated in Figure 5.2. We define the *arclength* of $\mathbf{g}$ to be

$$\ell(\mathbf{g}) = \sup\{\ell(\mathbf{g}, \mathcal{P}) : \mathcal{P} \text{ partition of } [a,b]\},$$

provided the set of polygonal lengths is bounded above.

The following result is not in the least surprising: The distance a particle travels is the integral of its speed.

Proposition 5.4 *Let* $\mathbf{g}\colon [a,b] \to \mathbb{R}^n$ *be a piecewise-$\mathcal{C}^1$ parametrized curve. Then*

$$\ell(\mathbf{g}) = \int_a^b \|\mathbf{g}'(t)\|dt.$$

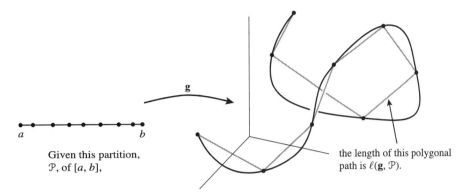

Figure 5.2

Given this partition, $\mathcal{P}$, of $[a, b]$,

the length of this polygonal path is $\ell(\mathbf{g}, \mathcal{P})$.

Proof For any partition $\mathcal{P}$ of $[a, b]$, we have, by Lemma 5.3,

$$\ell(\mathbf{g}, \mathcal{P}) = \sum_{i=1}^{k} \|\mathbf{g}(t_i) - \mathbf{g}(t_{i-1})\|$$
$$= \sum_{i=1}^{k} \left\| \int_{t_{i-1}}^{t_i} \mathbf{g}'(t)\,dt \right\| \leq \sum_{i=1}^{k} \int_{t_{i-1}}^{t_i} \|\mathbf{g}'(t)\|\,dt = \int_{a}^{b} \|\mathbf{g}'(t)\|\,dt,$$

so $\ell(\mathbf{g}) \leq \int_{a}^{b} \|\mathbf{g}'(t)\|\,dt$. The same holds on any interval.

Now, for $a \leq t \leq b$, define $s(t)$ to be the arclength of the curve $\mathbf{g}$ on the interval $[a, t]$. Then for $h > 0$ we have

$$\frac{\|\mathbf{g}(t+h) - \mathbf{g}(t)\|}{h} \leq \frac{s(t+h) - s(t)}{h} \leq \frac{1}{h} \int_{t}^{t+h} \|\mathbf{g}'(u)\|\,du$$

since $s(t+h) - s(t)$ is the arclength of the curve $\mathbf{g}$ on the interval $[t, t+h]$. Now

$$\lim_{h \to 0^+} \frac{\|\mathbf{g}(t+h) - \mathbf{g}(t)\|}{h} = \|\mathbf{g}'(t)\| = \lim_{h \to 0^+} \frac{1}{h} \int_{t}^{t+h} \|\mathbf{g}'(u)\|\,du.$$

Therefore, by the squeeze principle (see Exercise 2.3.3),

$$\lim_{h \to 0^+} \frac{s(t+h) - s(t)}{h} = \|\mathbf{g}'(t)\|.$$

A similar argument works for $h < 0$, and we conclude that $s'(t) = \|\mathbf{g}'(t)\|$. Therefore,

$$s(t) = \int_{a}^{t} \|\mathbf{g}'(u)\|\,du, \quad a \leq t \leq b,$$

and, in particular, $s(b) = \ell(\mathbf{g}) = \int_{a}^{b} \|\mathbf{g}'(t)\|\,dt$, as desired. ∎

Chapter 3. The Derivative

EXAMPLE 1

Consider the *helix*

$$\mathbf{g}(t) = \begin{bmatrix} a \cos t \\ a \sin t \\ bt \end{bmatrix}, \quad t \in \mathbb{R},$$

as pictured in Figure 5.3. Note that it twists around the cylinder of radius a, heading "uphill" at a constant pitch. If we take one "coil" of the helix, letting t run from 0 to 2π, then the arclength of that portion is

$$\ell(\mathbf{g}) = \int_0^{2\pi} \|\mathbf{g}'(t)\| dt = \int_0^{2\pi} \left\| \begin{bmatrix} -a \sin t \\ a \cos t \\ b \end{bmatrix} \right\| dt = \int_0^{2\pi} \sqrt{a^2 + b^2} \, dt = 2\pi \sqrt{a^2 + b^2}.$$

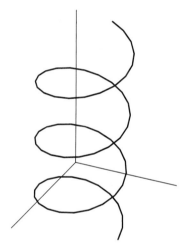

Figure 5.3

We say the parametrized curve is *arclength-parametrized* if $\|\mathbf{g}'(t)\| = 1$ for all t, so $s(t) = t + c$ for some constant c. Typically, when the curve is arclength-parametrized, we use s as the parameter.

EXAMPLE 2

The following curves are arclength-parametrized.

a. Consider the following parametrization of a circle of radius a:

$$\mathbf{g}(s) = \begin{bmatrix} a \cos(s/a) \\ a \sin(s/a) \end{bmatrix}, \quad 0 \le s \le 2\pi a.$$

Then note that

$$\mathbf{g}'(s) = \begin{bmatrix} -\sin(s/a) \\ \cos(s/a) \end{bmatrix}, \quad \text{and} \quad \|\mathbf{g}'(s)\| = 1.$$

b. Consider the curve

$$\mathbf{g}(s) = \begin{bmatrix} \frac{1}{3}(1+s)^{3/2} \\ \frac{1}{3}(1-s)^{3/2} \\ \frac{s}{\sqrt{2}} \end{bmatrix}, \quad -1 < s < 1.$$

Then

$$\mathbf{g}'(s) = \begin{bmatrix} \frac{1}{2}\sqrt{1+s} \\ \frac{1}{2}\sqrt{1-s} \\ \frac{1}{\sqrt{2}} \end{bmatrix}, \quad \text{and} \quad \|\mathbf{g}'(s)\| = 1. \blacktriangleleft$$

If $\mathbf{g}$ is arclength-parametrized, then the velocity vector $\mathbf{g}'(s)$ is the *unit tangent vector* at each point, which we denote by $\mathbf{T}(s)$. Let's assume now that $\mathbf{g}$ is twice differentiable. Since $\|\mathbf{T}(s)\| = 1$ for all s, it follows from Proposition 5.2 that $\mathbf{T}(s) \cdot \mathbf{T}'(s) = 0$. Define the *curvature* of the curve to be $\kappa(s) = \|\mathbf{T}'(s)\|$; assuming $\mathbf{T}'(s) \neq \mathbf{0}$, define the *principal normal vector* $\mathbf{N}(s) = \mathbf{T}'(s)/\|\mathbf{T}'(s)\|$. (See Figure 5.4.)

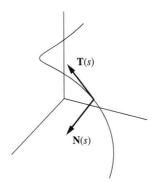

Figure 5.4

▶ **EXAMPLE 3**

If $\mathbf{g}$ is a line, then $\mathbf{T}$ is constant and $\kappa = 0$ (and conversely). If we start with a circle of radius a, then from Example 2a we have

$$\mathbf{T}(s) = \begin{bmatrix} -\sin(s/a) \\ \cos(s/a) \end{bmatrix}$$

from which we compute that

$$\mathbf{T}'(s) = \frac{1}{a}\begin{bmatrix} -\cos(s/a) \\ -\sin(s/a) \end{bmatrix}.$$

In particular, we see that $\mathbf{N}(s)$ is centripetal (pointing toward the center of the circle) and $\kappa(s) = 1/a$ for all s. ◀

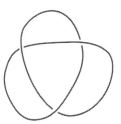

Figure 5.5

Remark If the arclength-parametrized curve $\mathbf{g}\colon [0, L] \to \mathbb{R}^3$ is *closed* (meaning that $\mathbf{g}(0) = \mathbf{g}(L)$), then it is interesting to consider its *total curvature*, $\int_0^L \kappa(s)\,ds$. For a circle or, indeed, for any convex plane curve, this integral is 2π. Not surprisingly, at least that much total curvature is required for the curve to "close up." A famous theorem in differential geometry, called the Farý-Milnor Theorem, states that total curvature at least 4π is required to make a *knot*, a closed curve that cannot be continuously deformed into a circle without crossing itself. A trefoil knot is pictured in Figure 5.5. See, e.g., doCarmo, *Differential Geometry of Curves and Surfaces*, §5.7.

Note that as long as $\mathbf{g}'(t)$ never vanishes, the arclength s is a differentiable function of t with positive derivative everywhere; thus, it has a differentiable inverse function, which we write $t(s)$. We can "reparametrize by arclength" by considering the composition $\mathbf{h}(s) = \mathbf{g}(t(s))$, and then, of course, $\mathbf{g}(t) = \mathbf{h}(s(t))$. Writing[2] $\upsilon(t) = s'(t) = \|\mathbf{g}'(t)\|$ for the speed, we have by the chain rule

$$\mathbf{g}'(t) = \mathbf{h}'(s(t))s'(t) = \upsilon(t)\mathbf{T}(s(t)) \quad\text{and}$$
(†) $\quad \mathbf{g}''(t) = \upsilon'(t)\mathbf{T}(s(t)) + \upsilon(t)^2\mathbf{T}'(s(t)) = \upsilon'(t)\mathbf{T}(s(t)) + \kappa(s(t))\upsilon(t)^2\mathbf{N}(s(t)).$

▶ **EXAMPLE 4**

Consider the parametrized curve

$$\mathbf{g}(t) = \begin{bmatrix} \cos^3 t \\ \sin^3 t \end{bmatrix}, \quad 0 < t < \pi/2.$$

Then we have

$$\mathbf{g}'(t) = 3\cos t \sin t \begin{bmatrix} -\cos t \\ \sin t \end{bmatrix}, \quad \text{so} \quad \upsilon(t) = 3\cos t \sin t \quad \text{and} \quad \mathbf{T}(s(t)) = \begin{bmatrix} -\cos t \\ \sin t \end{bmatrix}.$$

Then, by the chain rule, we have

$$\begin{bmatrix} \sin t \\ \cos t \end{bmatrix} = (\mathbf{T} \circ s)'(t) = \mathbf{T}'(s(t))s'(t) = \kappa(s(t))\upsilon(t)\mathbf{N}(s(t)),$$

---

[2] For those who might not know, υ is the Greek letter *upsilon*, not to be confused with ν, the Greek letter *nu*.

from which we conclude that

$$\kappa(s(t))v(t) = \left\|\begin{bmatrix} \sin t \\ \cos t \end{bmatrix}\right\| = 1, \quad \text{and} \quad \kappa(s(t)) = \frac{1}{v(t)} = \frac{1}{3\cos t \sin t}.$$

EXERCISES 3.5

1. Suppose $\mathbf{g}: (a, b) \to \mathbb{R}^n$ is a differentiable parametrized curve with the property that at each t, the position and velocity vectors are orthogonal. Prove that $\mathbf{g}$ lies on a sphere centered at the origin.

2. Suppose $\mathbf{g}: (a, b) \to \mathbb{R}^n$ is a twice-differentiable parametrized curve. Prove that $\mathbf{g}$ has constant speed if and only if the velocity and acceleration vectors are orthogonal at every t.

3. Suppose $\mathbf{f}, \mathbf{g}: (a, b) \to \mathbb{R}^n$ are differentiable and $\mathbf{f} \cdot \mathbf{g} = \text{const}$. Prove that $\mathbf{f}' \cdot \mathbf{g} = -\mathbf{g}' \cdot \mathbf{f}$. Interpret the result geometrically in the event that $\mathbf{f}$ and $\mathbf{g}$ are always unit vectors.

4. Suppose a particle moves in a central force field in $\mathbb{R}^3$ with constant speed. What can you say about its trajectory? (Proof?)

5. Suppose $\mathbf{g}: (a, b) \to \mathbb{R}^n$ is nowhere zero and $\mathbf{g}'(t) = \lambda(t)\mathbf{g}(t)$ for some scalar function λ. Prove (rigorously) that $\mathbf{g}/\|\mathbf{g}\|$ is constant. (Hint: Set $\mathbf{h} = \mathbf{g}/\|\mathbf{g}\|$, write $\mathbf{g} = \|\mathbf{g}\|\mathbf{h}$, and differentiate.)

6. Suppose $\mathbf{g}: (a, b) \to \mathbb{R}^n$ is a differentiable parametrized curve and that for some point $\mathbf{p} \in \mathbb{R}^n$ we have $\|\mathbf{g}(t_0) - \mathbf{p}\| \le \|\mathbf{g}(t) - \mathbf{p}\|$ for all $t \in (a, b)$. Prove that $\mathbf{g}'(t_0) \cdot (\mathbf{g}(t_0) - \mathbf{p}) = 0$. Give a geometric explanation.

7. Find the arclength of the following parametrized curves.

*(a) $\mathbf{g}(t) = \begin{bmatrix} e^t \cos t \\ e^t \sin t \\ e^t \end{bmatrix}$, $a \le t \le b$

*(c) $\mathbf{g}(t) = \begin{bmatrix} t \\ 3t^2 \\ 6t^3 \end{bmatrix}$, $0 \le t \le 1$

(b) $\mathbf{g}(t) = \begin{bmatrix} \frac{1}{2}(e^t + e^{-t}) \\ \frac{1}{2}(e^t - e^{-t}) \\ t \end{bmatrix}$, $-1 \le t \le 1$

(d) $\mathbf{g}(t) = \begin{bmatrix} a(t - \sin t) \\ a(1 - \cos t) \end{bmatrix}$, $0 \le t \le 2\pi$

8. Calculate the unit tangent vector and curvature of the following curves.

*(a) $\mathbf{g}(t) = \begin{bmatrix} \frac{1}{\sqrt{3}}\cos t + \frac{1}{\sqrt{2}}\sin t \\ \frac{1}{\sqrt{3}}\cos t \\ \frac{1}{\sqrt{3}}\cos t - \frac{1}{\sqrt{2}}\sin t \end{bmatrix}$

(c) $\mathbf{g}(t) = \begin{bmatrix} t \\ t^2 \\ t^3 \end{bmatrix}$

*(b) $\mathbf{g}(t) = \begin{bmatrix} e^{-t} \\ e^t \\ \sqrt{2}t \end{bmatrix}$

9. Prove that for a parametrized curve $\mathbf{g}: (a, b) \to \mathbb{R}^3$, we have $\kappa = \|\mathbf{g}' \times \mathbf{g}''\|/v^3$.

10. Using the formula (†) for acceleration, explain how engineers might decide at what angle to bank a road that is a circle of radius $1/4$ mile and around which cars wish to drive safely at 40 mph.

11. **(Frenet Formulas)** Let $\mathbf{g}: [0, L] \to \mathbb{R}^3$ be a three-times differentiable arclength-parametrized curve with $\kappa > 0$, and let $\mathbf{T}$ and $\mathbf{N}$ be defined as above. Define the *binormal* $\mathbf{B} = \mathbf{T} \times \mathbf{N}$.
(a) Show that $\|\mathbf{B}\| = 1$. Assuming the result of Exercise 1.4.34e, show that every vector in $\mathbb{R}^3$ can be expressed as a linear combination of $\mathbf{T}(s)$, $\mathbf{N}(s)$, and $\mathbf{B}(s)$. (Hint: See Example 11 of Chapter 1, Section 4.)

(b) Show that $\mathbf{B}' \cdot \mathbf{T} = \mathbf{B}' \cdot \mathbf{B} = 0$, and deduce that $\mathbf{B}'(s)$ is a scalar multiple of $\mathbf{N}(s)$ for every s. (Hint: See Exercise 3.)

(c) Define the *torsion* τ of the curve by $\mathbf{B}' = -\tau \mathbf{N}$. Show that $\mathbf{g}$ is a planar curve if and only if $\tau(s) = 0$ for all s.

(d) Show that $\mathbf{N}' = -\kappa \mathbf{T} + \tau \mathbf{B}$.

The equations

$$\mathbf{T}' = \kappa \mathbf{N}, \qquad \mathbf{N}' = -\kappa \mathbf{T} + \tau \mathbf{B}, \qquad \mathbf{B}' = -\tau \mathbf{N}$$

are called the *Frenet formulas* for the arclength-parametrized curve $\mathbf{g}$.

*12. (See Exercise 11 for the definition of torsion.) Calculate the curvature and torsion of the helix presented in Example 1. Explain the meaning of the sign of the torsion.

13. (See Exercise 11 for the definition of torsion.) Calculate the curvature and torsion of the curve

$$\mathbf{g}(t) = \begin{bmatrix} e^t \cos t \\ e^t \sin t \\ e^t \end{bmatrix}.$$

14. A pendulum is made, as pictured in Figure 5.6, by hanging from the cusp where two arches of a cycloid meet a length of string equal to the length of one of the arches. As it swings, the string wraps around the cycloid and extends tangentially to the bob at the end. Given the equation

$$\mathbf{f}(t) = \begin{bmatrix} t + \sin t \\ 1 - \cos t \end{bmatrix}, \qquad 0 \le t \le 2\pi,$$

for the cycloid, find the parametric equation of the bob, P, of the pendulum.[3]

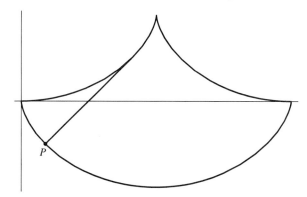

Figure 5.6

15. Assuming that the force field is inverse square, prove Kepler's first and third laws, as follows. Without loss of generality, we may assume that the planet has mass 1 and moves in the xy-plane. (You will need to use polar coordinates, as introduced in Example 6 of Chapter 2, Section 1.)

(a) Suppose $a, b > 0$ and $a^2 = b^2 + c^2$. Show that the polar coordinates equation of the ellipse

$$\frac{(x-c)^2}{a^2} + \frac{y^2}{b^2} = 1 \quad \text{is} \quad r\left(1 - \frac{c}{a}\cos\theta\right) = \frac{b^2}{a}.$$

This is an ellipse with semimajor axis a and semiminor axis b, with one focus at the origin. (Hint: Expand the left-hand side in polar coordinates and express the result as a difference of squares.)

---

[3]This phenomenon was originally discovered by the Dutch mathematician Huygens in an effort to design a pendulum whose period would not depend on the amplitude of its motion, hence one ideal for an accurate clock.

(b) Let $r(t)$ and $\theta(t)$ be the polar coordinates of $\mathbf{g}(t)$, and let

$$\mathbf{e}_r(t) = \begin{bmatrix} \cos\theta(t) \\ \sin\theta(t) \end{bmatrix} \quad \text{and} \quad \mathbf{e}_\theta(t) = \begin{bmatrix} -\sin\theta(t) \\ \cos\theta(t) \end{bmatrix},$$

as pictured in Figure 5.7. Show that

$$\mathbf{g}'(t) = r'(t)\mathbf{e}_r(t) + r(t)\theta'(t)\mathbf{e}_\theta(t),$$
$$\mathbf{g}''(t) = \big(r''(t) - r(t)\theta'(t)^2\big)\mathbf{e}_r(t) + \big(2r'(t)\theta'(t) + r(t)\theta''(t)\big)\mathbf{e}_\theta(t).$$

(c) Let $\mathbf{A}_0$ be as in the proof of Proposition 5.1. Show that $\mathbf{g}''(t) \times \mathbf{A}_0 = GM\theta'(t)\mathbf{e}_\theta(t) = GM\mathbf{e}_r'(t)$, and deduce that $\mathbf{g}'(t) \times \mathbf{A}_0 = GM(\mathbf{e}_r(t) + \mathbf{c})$ for some constant vector $\mathbf{c}$.

(d) Dot the previous equation with $\mathbf{g}(t)$ and use the fact that $\mathbf{g}(t) \times \mathbf{g}'(t) = \mathbf{A}_0$ to deduce that $GMr(t)(1 - \|\mathbf{c}\|\cos\theta(t)) = \|\mathbf{A}_0\|^2$ if we assume $\mathbf{c}$ is a negative scalar multiple of $\mathbf{e}_1$. Deduce that when $\|\mathbf{c}\| \geq 1$ the path of the planet is unbounded and that when $\|\mathbf{c}\| < 1$ the orbit of the planet is an ellipse with one focus at the origin.

(e) As we shall see in Chapter 7, the area of an ellipse with semimajor axis a and semiminor axis b is πab; show that the period $T = 2\pi ab/\|\mathbf{A}_0\|$. Now prove that $T^2 = \dfrac{4\pi^2}{GM}a^3$.

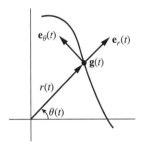

Figure 5.7

16. (Pilfered from *Which Way did the Bicycle Go ...and Other Intriguing Mathematical Mysteries*, published by the M.A.A. Copyright The Mathematical Association of America, Washington, DC, 1996. All rights reserved.)

> "This track, as you perceive, was made by a rider who was going from the direction of the school."
>
> "Or towards it?"
>
> "No, no, my dear Watson.... It was undoubtedly heading away from the school."

So spoke Sherlock Holmes.[4] Imagine a 20-foot wide mud patch through which a bicycle has just passed, with its front and rear tires leaving tracks as illustrated in Figure 5.8. (We have taken the liberty of helping you in your capacity as sleuth by using dashes for the path of one of the wheels.) In which direction was the bicyclist traveling? Explain your answer.

---

[4] "The Adventure of the Priory School," *The Return of Sherlock Holmes*.

Figure 5.8

6 HIGHER-ORDER PARTIAL DERIVATIVES

Suppose $U \subset \mathbb{R}^n$ is open and $\mathbf{f}: U \to \mathbb{R}^m$ is a vector-valued function on U. Recall that we said $\mathbf{f}$ is $\mathcal{C}^1$ (on U) if the partial derivatives $\dfrac{\partial \mathbf{f}}{\partial x_i}$ exist and are continuous on U. Suppose this is the case. Then we can ask whether they in turn have partial derivatives, i.e., whether the functions

$$\frac{\partial^2 \mathbf{f}}{\partial x_j \partial x_i} \underset{\text{def}}{=} \frac{\partial}{\partial x_j}\left(\frac{\partial \mathbf{f}}{\partial x_i}\right)$$

are defined. These functions are, for obvious reasons, called *second-order partial derivatives* of $\mathbf{f}$. We say $\mathbf{f}$ is $\mathcal{C}^2$ (on U) if all its first- and second-order partial derivatives exist and are continuous (on U). More generally, we say $\mathbf{f}$ is $\mathcal{C}^k$ (on U) if all its first-, second-, ..., and k^{th}-order partial derivatives

$$\frac{\partial^k \mathbf{f}}{\partial x_{i_k} \partial x_{i_{k-1}} \cdots \partial x_{i_2} \partial x_{i_1}}, \quad 1 \leq i_1, i_2, \ldots, i_k \leq n,$$

exist and are continuous (on U). We say $\mathbf{f}$ is $\mathcal{C}^\infty$ (or *smooth*) if all its partial derivatives of *all* orders exist.

▶ **EXAMPLE 1**

Let $f\begin{pmatrix} x \\ y \\ z \end{pmatrix} = e^{xy} \sin z + xy^3 z^4$. Then f is smooth and

$$\frac{\partial f}{\partial x} = y e^{xy} \sin z + y^3 z^4,$$

$$\frac{\partial^2 f}{\partial z \partial x} = y e^{xy} \cos z + 4 y^3 z^3,$$

$$\frac{\partial^3 f}{\partial z^2 \partial x} = -y e^{xy} \sin z + 12 y^3 z^2, \quad \text{and}$$

$$\frac{\partial^3 f}{\partial y \partial z \partial x} = e^{xy}(xy + 1) \cos z + 12 y^2 z^3. \quad \blacktriangleleft$$

It is a hassle to keep track of the order in which we calculate higher-order partial derivatives. Luckily, the following result tells us that for smooth functions, the order in which we calculate the partial derivatives does not matter. This is an intuitively obvious result, but the proof is quite subtle.

Theorem 6.1 *Let $U \subset \mathbb{R}^n$ be open, and suppose $\mathbf{f} \colon U \to \mathbb{R}^m$ is a $\mathcal{C}^2$ function. Then for any i and j we have*

$$\frac{\partial^2 \mathbf{f}}{\partial x_i \partial x_j} = \frac{\partial^2 \mathbf{f}}{\partial x_j \partial x_i}.$$

Proof It suffices to prove the result when $m = 1$. For ease of notation, we take $n = 2$, $i = 1$, and $j = 2$. Introduce the function

$$\Delta \begin{pmatrix} h \\ k \end{pmatrix} = f \begin{pmatrix} a+h \\ b+k \end{pmatrix} - f \begin{pmatrix} a+h \\ b \end{pmatrix} - f \begin{pmatrix} a \\ b+k \end{pmatrix} + f \begin{pmatrix} a \\ b \end{pmatrix},$$

as indicated schematically in Figure 6.1. Letting $q(s) = f \begin{pmatrix} s \\ b+k \end{pmatrix} - f \begin{pmatrix} s \\ b \end{pmatrix}$ and applying the Mean Value Theorem, we have

$$\Delta \begin{pmatrix} h \\ k \end{pmatrix} = q(a+h) - q(a) = hq'(\xi) \quad \text{for some } \xi \text{ between } a \text{ and } a+h$$

$$= h \left(\frac{\partial f}{\partial x} \begin{pmatrix} \xi \\ b+k \end{pmatrix} - \frac{\partial f}{\partial x} \begin{pmatrix} \xi \\ b \end{pmatrix} \right)$$

$$= hk \frac{\partial^2 f}{\partial y \partial x} \begin{pmatrix} \xi \\ \eta \end{pmatrix} \quad \text{for some } \eta \text{ between } b \text{ and } b+k.$$

On the other hand, letting $r(t) = f \begin{pmatrix} a+h \\ t \end{pmatrix} - f \begin{pmatrix} a \\ t \end{pmatrix}$, we have

$$\Delta \begin{pmatrix} h \\ k \end{pmatrix} = r(b+k) - r(k) = kr'(\tau) \quad \text{for some } \tau \text{ between } b \text{ and } b+k$$

$$= k \left(\frac{\partial f}{\partial y} \begin{pmatrix} a+h \\ \tau \end{pmatrix} - \frac{\partial f}{\partial y} \begin{pmatrix} a \\ \tau \end{pmatrix} \right)$$

$$= hk \frac{\partial^2 f}{\partial x \partial y} \begin{pmatrix} \sigma \\ \tau \end{pmatrix} \quad \text{for some } \sigma \text{ between } a \text{ and } a+h.$$

Therefore, we have

$$\frac{1}{hk} \Delta \begin{pmatrix} h \\ k \end{pmatrix} = \frac{\partial^2 f}{\partial y \partial x} \begin{pmatrix} \xi \\ \eta \end{pmatrix} = \frac{\partial^2 f}{\partial x \partial y} \begin{pmatrix} \sigma \\ \tau \end{pmatrix}.$$

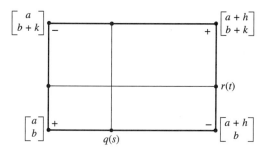

Figure 6.1

Now $\xi, \sigma \to a$ and $\eta, \tau \to b$ as $h, k \to 0$, and since the functions $\dfrac{\partial^2 f}{\partial x \partial y}$ and $\dfrac{\partial^2 f}{\partial y \partial x}$ are continuous, we have

$$\frac{\partial^2 f}{\partial x \partial y}\begin{pmatrix}a\\b\end{pmatrix} = \frac{\partial^2 f}{\partial y \partial x}\begin{pmatrix}a\\b\end{pmatrix},$$

as required. ∎

To see why the $\mathcal{C}^2$ hypothesis is necessary, see Exercise 1.

Second-order derivatives appear in the study of the local behavior of functions near critical points and, more importantly, in differential equations and physics—as we've seen, Newton's second law of motion tells us that forces induce acceleration. At this juncture, we give a few examples of higher-order partial derivatives and *partial differential equations* that arise in the further study of mathematics and physics.

▶ **EXAMPLE 2**

(**Harmonic Functions**) If f is a $\mathcal{C}^2$ function on (an open subset of) $\mathbb{R}^n$, the expression

$$\nabla^2 f = \frac{\partial^2 f}{\partial x_1^2} + \frac{\partial^2 f}{\partial x_2^2} + \cdots + \frac{\partial^2 f}{\partial x_n^2}$$

is called the *Laplacian* of f. A solution of the equation $\nabla^2 f = 0$ is called a *harmonic function*. As we shall see in Chapter 8, the Laplacian and harmonic functions play an important role in physical applications. For example, the gravitational (resp., electrostatic) potential is a harmonic function in mass-free (resp., charge-free) space. ◀

▶ **EXAMPLE 3**

(**Wave Equation**) The equation

(∗) $$\frac{\partial^2 f}{\partial t^2} = c^2 \frac{\partial^2 f}{\partial x^2}$$

models the displacement of a one-dimensional vibrating string (with "wave velocity" c) from its equilibrium position. By a clever use of the chain rule, we can find an explicit formula for its general solution, assuming f is $\mathcal{C}^2$. Let

$$\begin{bmatrix} x \\ t \end{bmatrix} = \mathbf{g}\begin{pmatrix} u \\ v \end{pmatrix} = \begin{bmatrix} \frac{1}{2}(u+v) \\ \frac{1}{2c}(u-v) \end{bmatrix}$$

(so that $u = x + ct$ and $v = x - ct$), and set $F\begin{pmatrix} u \\ v \end{pmatrix} = f\left(\mathbf{g}\begin{pmatrix} u \\ v \end{pmatrix}\right)$. Then by the chain rule, we have

$$DF\begin{pmatrix} u \\ v \end{pmatrix} = Df\left(\mathbf{g}\begin{pmatrix} u \\ v \end{pmatrix}\right) D\mathbf{g}\begin{pmatrix} u \\ v \end{pmatrix}$$

$$= \begin{bmatrix} \frac{\partial f}{\partial x}\left(\mathbf{g}\begin{pmatrix} u \\ v \end{pmatrix}\right) & \frac{\partial f}{\partial t}\left(\mathbf{g}\begin{pmatrix} u \\ v \end{pmatrix}\right) \end{bmatrix} \begin{bmatrix} \frac{1}{2} & \frac{1}{2} \\ \frac{1}{2c} & -\frac{1}{2c} \end{bmatrix},$$

so

$$\frac{\partial F}{\partial v} = \begin{bmatrix} \frac{\partial f}{\partial x}\left(\mathbf{g}\begin{pmatrix} u \\ v \end{pmatrix}\right) & \frac{\partial f}{\partial t}\left(\mathbf{g}\begin{pmatrix} u \\ v \end{pmatrix}\right) \end{bmatrix} \begin{bmatrix} \frac{1}{2} \\ -\frac{1}{2c} \end{bmatrix} = \frac{1}{2}\frac{\partial f}{\partial x}\left(\mathbf{g}\begin{pmatrix} u \\ v \end{pmatrix}\right) - \frac{1}{2c}\frac{\partial f}{\partial t}\left(\mathbf{g}\begin{pmatrix} u \\ v \end{pmatrix}\right).$$

Now, differentiating with respect to u, we have to apply the chain rule to each of the functions $\frac{\partial f}{\partial x}\left(\mathbf{g}\begin{pmatrix} u \\ v \end{pmatrix}\right)$ and $\frac{\partial f}{\partial t}\left(\mathbf{g}\begin{pmatrix} u \\ v \end{pmatrix}\right)$:

$$\frac{\partial^2 F}{\partial u \partial v} = \frac{1}{2}\begin{bmatrix} \frac{\partial^2 f}{\partial x^2}\left(\mathbf{g}\begin{pmatrix} u \\ v \end{pmatrix}\right) & \frac{\partial^2 f}{\partial t \partial x}\left(\mathbf{g}\begin{pmatrix} u \\ v \end{pmatrix}\right) \end{bmatrix}\begin{bmatrix} \frac{1}{2} \\ \frac{1}{2c} \end{bmatrix} - \frac{1}{2c}\begin{bmatrix} \frac{\partial^2 f}{\partial x \partial t}\left(\mathbf{g}\begin{pmatrix} u \\ v \end{pmatrix}\right) & \frac{\partial^2 f}{\partial t^2}\left(\mathbf{g}\begin{pmatrix} u \\ v \end{pmatrix}\right) \end{bmatrix}\begin{bmatrix} \frac{1}{2} \\ \frac{1}{2c} \end{bmatrix}$$

$$= \frac{1}{4}\left\{\frac{\partial^2 f}{\partial x^2}\left(\mathbf{g}\begin{pmatrix} u \\ v \end{pmatrix}\right) + \frac{1}{c}\frac{\partial^2 f}{\partial t \partial x}\left(\mathbf{g}\begin{pmatrix} u \\ v \end{pmatrix}\right) - \frac{1}{c}\frac{\partial^2 f}{\partial x \partial t}\left(\mathbf{g}\begin{pmatrix} u \\ v \end{pmatrix}\right) - \frac{1}{c^2}\frac{\partial^2 f}{\partial t^2}\left(\mathbf{g}\begin{pmatrix} u \\ v \end{pmatrix}\right)\right\}$$

$$= \frac{1}{4}\left\{\frac{\partial^2 f}{\partial x^2}\left(\mathbf{g}\begin{pmatrix} u \\ v \end{pmatrix}\right) - \frac{1}{c^2}\frac{\partial^2 f}{\partial t^2}\left(\mathbf{g}\begin{pmatrix} u \\ v \end{pmatrix}\right)\right\} = 0,$$

where at the last step we use Theorem 6.1. Now what can we say about the general solution of the equation $\frac{\partial^2 F}{\partial u \partial v} = 0$? On any rectangle in the uv-plane, we can infer that

$$F\begin{pmatrix} u \\ v \end{pmatrix} = \phi(u) + \psi(v)$$

for some differentiable functions ϕ and ψ. (For $\frac{\partial}{\partial u}\left(\frac{\partial F}{\partial v}\right) = 0$ tells us that $\frac{\partial F}{\partial v}$ is independent of u, hence a function of v only, whose antiderivative we call $\psi(v)$. But the constant of integration can be an arbitrary function of u. To examine this argument a bit more carefully, we recommend that the reader consider Exercise 11.)

In conclusion, on a suitable domain, the general solution of the wave equation ($*$) can be written in the form

$$f\begin{pmatrix} x \\ t \end{pmatrix} = \phi(x + ct) + \psi(x - ct)$$

for arbitrary $\mathcal{C}^2$ functions ϕ and ψ. The physical interpretation is this: The general solution is the superposition of two traveling waves, one moving to the right along the string with speed c, the other moving to the left with speed c. ◀

EXAMPLE 4

(**Minimal Surfaces**) When you dip a piece of wire shaped in the form of a closed curve C into soap film, the resulting surface you see is called a *minimal surface*, so called because in principle surface tension dictates that the surface should have least area among all those surfaces having that curve C as boundary. If the minimal surface is in the form of a graph $z = f\begin{pmatrix} x \\ y \end{pmatrix}$, then it is shown in a differential geometry course that f must be a solution of the *minimal surface equation*

$$\left(1 + \left(\frac{\partial f}{\partial y}\right)^2\right)\frac{\partial^2 f}{\partial x^2} - 2\frac{\partial f}{\partial x}\frac{\partial f}{\partial y}\frac{\partial^2 f}{\partial x \partial y} + \left(1 + \left(\frac{\partial f}{\partial x}\right)^2\right)\frac{\partial^2 f}{\partial y^2} = 0.$$

(See also Exercise 8.5.22.) Examples of minimal surfaces are

  a. a plane;
  b. a helicoid—the spiral surface obtained by joining points of a helix "horizontally" to its vertical axis, as pictured in Figure 6.2(a);
  c. a catenoid—the surface of revolution obtained by rotating a catenary $y = \frac{1}{2c}(e^{cx} + e^{-cx})$ (for any $c > 0$) about the x-axis, as pictured in Figure 6.2(b).

(See Exercise 10.)

(a)   (b)

Figure 6.2

EXERCISES 3.6

1. Define $f : \mathbb{R}^2 \to \mathbb{R}$ by

$$f\begin{pmatrix} x \\ y \end{pmatrix} = xy\frac{x^2 - y^2}{x^2 + y^2}, \quad \mathbf{x} \neq \mathbf{0}, \qquad f(\mathbf{0}) = 0.$$

(a) Show that $\dfrac{\partial f}{\partial x}\begin{pmatrix} 0 \\ y \end{pmatrix} = -y$ for all y and $\dfrac{\partial f}{\partial y}\begin{pmatrix} x \\ 0 \end{pmatrix} = x$ for all x.

(b) Deduce that $\dfrac{\partial^2 f}{\partial x \partial y}(\mathbf{0}) = 1$ but $\dfrac{\partial^2 f}{\partial y \partial x}(\mathbf{0}) = -1$.

(c) Conclude that f is not $\mathcal{C}^2$ at $\mathbf{0}$.

2. Check that the following are harmonic functions.

(a) $f\begin{pmatrix} x \\ y \end{pmatrix} = 3x^2 - 5xy - 3y^2$

(b) $f\begin{pmatrix} x \\ y \end{pmatrix} = \log(x^2 + y^2)$

(c) $f\begin{pmatrix} x \\ y \\ z \end{pmatrix} = x^2 + xy + 2y^2 - 3z^2 + xyz$

(d) $f\begin{pmatrix} x \\ y \\ z \end{pmatrix} = (x^2 + y^2 + z^2)^{-1/2}$

3. Check that the following functions are solutions of the one-dimensional wave equation given in Example 3.

(a) $f\begin{pmatrix} x \\ t \end{pmatrix} = \cos(x + ct)$

(b) $f\begin{pmatrix} x \\ t \end{pmatrix} = \sin 5x \cos 5ct$

4. Let $f\begin{pmatrix} x \\ t \end{pmatrix} = t^{-1/2} e^{-x^2/4kt}$. Show that f is a solution of the one-dimensional heat equation $\dfrac{\partial f}{\partial t} = k\dfrac{\partial^2 f}{\partial x^2}$.

*5. Suppose we are given a solution f of the one-dimensional wave equation, with initial position $f\begin{pmatrix} x \\ 0 \end{pmatrix} = h(x)$ and initial velocity $\dfrac{\partial f}{\partial t}\begin{pmatrix} x \\ 0 \end{pmatrix} = k(x)$. Express the functions ϕ and ψ in the solution of Example 3 in terms of h and k.

6. Suppose $f: \mathbb{R}^2 \to \mathbb{R}$ and $\mathbf{g}: \mathbb{R}^2 \to \mathbb{R}^2$ are C^2, and let $F\begin{pmatrix} u \\ v \end{pmatrix} = f\left(\mathbf{g}\begin{pmatrix} u \\ v \end{pmatrix}\right)$. Writing $g_1\begin{pmatrix} u \\ v \end{pmatrix} = x\begin{pmatrix} u \\ v \end{pmatrix}$ and $g_2\begin{pmatrix} u \\ v \end{pmatrix} = y\begin{pmatrix} u \\ v \end{pmatrix}$, show that

$$\frac{\partial^2 F}{\partial u \partial v} = \frac{\partial f}{\partial x}\frac{\partial^2 x}{\partial u \partial v} + \frac{\partial f}{\partial y}\frac{\partial^2 y}{\partial u \partial v} + \frac{\partial^2 f}{\partial x^2}\frac{\partial x}{\partial u}\frac{\partial x}{\partial v} + \frac{\partial^2 f}{\partial x \partial y}\left(\frac{\partial x}{\partial u}\frac{\partial y}{\partial v} + \frac{\partial x}{\partial v}\frac{\partial y}{\partial u}\right) + \frac{\partial^2 f}{\partial y^2}\frac{\partial y}{\partial u}\frac{\partial y}{\partial v},$$

where the partial derivatives of f are evaluated at $\mathbf{g}\begin{pmatrix} u \\ v \end{pmatrix}$.

7. Suppose $f: \mathbb{R}^2 \to \mathbb{R}$ is C^2. Let $F\begin{pmatrix} r \\ \theta \end{pmatrix} = f\begin{pmatrix} r\cos\theta \\ r\sin\theta \end{pmatrix}$. Show that

$$\frac{\partial^2 F}{\partial r^2} + \frac{1}{r}\frac{\partial F}{\partial r} + \frac{1}{r^2}\frac{\partial^2 F}{\partial \theta^2} = \frac{\partial^2 f}{\partial x^2} + \frac{\partial^2 f}{\partial y^2},$$

where the lefthand side is evaluated at $\begin{bmatrix} r \\ \theta \end{bmatrix}$ and the righthand side is evaluated at $\begin{bmatrix} r\cos\theta \\ r\sin\theta \end{bmatrix}$. (This is the formula for the Laplacian in polar coordinates.)

8. Use the result of Exercise 7 to show that for any integer n, the functions $F\begin{pmatrix} r \\ \theta \end{pmatrix} = r^n \cos n\theta$ and $F\begin{pmatrix} r \\ \theta \end{pmatrix} = r^n \sin n\theta$ are harmonic.

*9. Use the result of Exercise 7 to find all radially symmetric harmonic functions on the plane. (This means that $F\begin{pmatrix} r \\ \theta \end{pmatrix}$ is independent of θ, so we can call it $h(r)$.)

10. Check that the following functions $f\colon \mathbb{R}^2 \to \mathbb{R}$ are indeed solutions of the minimal surface equation given in Example 4.

    (a) $f\begin{pmatrix} x \\ y \end{pmatrix} = c$

    (b) $f\begin{pmatrix} x \\ y \end{pmatrix} = \arctan(y/x)$

    (c) $f\begin{pmatrix} x \\ y \end{pmatrix} = \sqrt{(\frac{1}{2}(e^x + e^{-x}))^2 - y^2}$ (For this one, a computer algebra system is recommended.)

11. Define $F\begin{pmatrix} u \\ v \end{pmatrix} = \begin{cases} 0, & u < 0 \text{ or } v < 0 \\ u^3, & u \geq 0 \text{ and } v > 0 \end{cases}$. Show that F is $\mathcal{C}^2$ and $\dfrac{\partial^2 F}{\partial u \partial v} = 0$, and yet F cannot be written in the form prescribed by the discussion of Example 3. Resolve this paradox.

CHAPTER 4

IMPLICIT AND EXPLICIT SOLUTIONS OF LINEAR SYSTEMS

We have seen that we can view the unit circle $\{\mathbf{x} \in \mathbb{R}^2 : \|\mathbf{x}\| = 1\}$ either as the set of solutions of an equation or in terms of a parametric representation $\left\{ \begin{bmatrix} \cos t \\ \sin t \end{bmatrix} : t \in [0, 2\pi) \right\}$. These are, respectively, the *implicit* and *explicit* representations of this subset of $\mathbb{R}^2$. Similarly, any subspace $V \subset \mathbb{R}^n$ can be represented in two ways:

   i. $V = \text{Span}(\mathbf{v}_1, \ldots, \mathbf{v}_k)$ for appropriate vectors $\mathbf{v}_1, \ldots, \mathbf{v}_k \in \mathbb{R}^n$—this is the explicit or parametric representation;

   ii. $V = \{\mathbf{x} \in \mathbb{R}^n : A\mathbf{x} = \mathbf{0}\}$ for an appropriate $m \times n$ matrix A—this is the implicit representation, viewing V as the intersection of the hyperplanes defined by $\mathbf{A}_i \cdot \mathbf{x} = 0$.

In this chapter we will see how to go back and forth between these two approaches. The central tool is Gaussian elimination, with which we deal in depth in the first two sections. We then come to the central notion of dimension and some useful applications. In the last section, we will begin to investigate to what extent we can relate implicit and explicit descriptions in the nonlinear setting.

▶ 1 GAUSSIAN ELIMINATION AND THE THEORY OF LINEAR SYSTEMS

In this section we give an explicit algorithm for solving a system of m linear equations in n variables:

$$\begin{aligned} a_{11}x_1 + a_{12}x_2 + \ldots + a_{1n}x_n &= b_1 \\ a_{21}x_1 + a_{22}x_2 + \ldots + a_{2n}x_n &= b_2 \\ &\vdots \\ a_{m1}x_1 + a_{m2}x_2 + \ldots + a_{mn}x_n &= b_m. \end{aligned}$$

Chapter 4. Implicit and Explicit Solutions of Linear Systems

Of course, we can write this in the form $A\mathbf{x} = \mathbf{b}$, where

$$A = \begin{bmatrix} a_{11} & a_{12} & \cdots & a_{1n} \\ a_{21} & a_{22} & \cdots & a_{2n} \\ \vdots & \vdots & \ddots & \vdots \\ a_{m1} & a_{m2} & \cdots & a_{mn} \end{bmatrix}, \quad \mathbf{x} = \begin{bmatrix} x_1 \\ x_2 \\ \vdots \\ x_n \end{bmatrix}, \quad \text{and} \quad \mathbf{b} = \begin{bmatrix} b_1 \\ \vdots \\ b_m \end{bmatrix}.$$

Geometrically, a solution of the system $A\mathbf{x} = \mathbf{b}$ is a vector $\mathbf{x}$ having the requisite dot products with the row vectors $\mathbf{A}_i$ of the matrix A:

$$\mathbf{A}_i \cdot \mathbf{x} = b_i \quad \text{for all } i = 1, 2, \ldots, m.$$

That is, the system of equations describes the intersection of the m hyperplanes with normal vectors $\mathbf{A}_i$ and at (signed) distance $b_i / \|\mathbf{A}_i\|$ from the origin.

To solve a system of linear equations, we want to give an explicit *parametric* description of the general solution. Some systems are relatively simple to solve. For example, taking the system

$$\begin{aligned} x_1 \phantom{{}+x_2} - x_3 &= 1 \\ x_2 + 2x_3 &= 2, \end{aligned}$$

we see that these equations allow us to determine x_1 and x_2 in terms of x_3; in particular, we can write $x_1 = 1 + x_3$ and $x_2 = 2 - 2x_3$, where x_3 is *free* to take on any real value. Thus, any solution of this system is of the form $\mathbf{x} = \begin{bmatrix} 1+t \\ 2-2t \\ t \end{bmatrix}$ for some $t \in \mathbb{R}$. (It is easily checked that every vector of this form is in fact a solution, as $(1+t) - t = 1$ and $(2-2t) + 2t = 2$ for every $t \in \mathbb{R}$.) Thus, we see that the intersection of the two given planes is the line in $\mathbb{R}^3$ passing through $\begin{bmatrix} 1 \\ 2 \\ 0 \end{bmatrix}$ with direction vector $\begin{bmatrix} 1 \\ -2 \\ 1 \end{bmatrix}$.

More complicated systems of equations require some algebraic manipulations before we can easily read off the general solution in parametric form. There are three basic operations we can perform on systems of equations that will not affect the solution set. They are the following *elementary operations*:

i. interchange any pair of equations;
ii. multiply any equation by a nonzero real number;
iii. replace any equation by its sum with a multiple of any other equation.

▶ **EXAMPLE 1**

Consider the system of linear equations

$$\begin{aligned} 3x_1 - 2x_2 + 9x_4 &= 4 \\ 2x_1 + 2x_2 - 4x_4 &= 6. \end{aligned}$$

We can use operation (i) to replace this system with

$$2x_1 + 2x_2 - 4x_4 = 6$$
$$3x_1 - 2x_2 + 9x_4 = 4;$$

then we use operation (ii), multiplying the first equation by 1/2, to get

$$x_1 + x_2 - 2x_4 = 3$$
$$3x_1 - 2x_2 + 9x_4 = 4;$$

now we use operation (iii), adding -3 times the first equation to the second:

$$x_1 + x_2 - 2x_4 = 3$$
$$-5x_2 + 15x_4 = -5.$$

Next we use operation (ii) again, multiplying the second equation by $-1/5$, to obtain

$$x_1 + x_2 - 2x_4 = 3$$
$$x_2 - 3x_4 = 1;$$

finally, we use operation (iii), adding -1 times the second equation to the first:

$$x_1 + x_4 = 2$$
$$x_2 - 3x_4 = 1.$$

From this we see that x_1 and x_2 are determined by x_4, whereas x_3 and x_4 are free to take on any values. Thus, we read off the general solution of the system of equations:

$$x_1 = 2 \quad - x_4$$
$$x_2 = 1 \quad + 3x_4$$
$$x_3 = \quad x_3$$
$$x_4 = \quad x_4$$

Thus, the general solution is

$$\mathbf{x} = \begin{bmatrix} x_1 \\ x_2 \\ x_3 \\ x_4 \end{bmatrix} = \begin{bmatrix} 2 \\ 1 \\ 0 \\ 0 \end{bmatrix} + x_3 \begin{bmatrix} 0 \\ 0 \\ 1 \\ 0 \end{bmatrix} + x_4 \begin{bmatrix} -1 \\ 3 \\ 0 \\ 1 \end{bmatrix},$$

which is a parametric representation of a plane in $\mathbb{R}^4$. ◀

We now describe a systematic technique, using the three allowable elementary operations, for solving systems of m equations in n variables. Before going any further, we should make the official observation that performing elementary operations on a system of equations does not change its solutions.

Proposition 1.1 *If a system of equations* $A\mathbf{x} = \mathbf{b}$ *is changed into the new system* $C\mathbf{x} = \mathbf{d}$ *by elementary operations, then the systems have the same set of solutions.*

Proof Left to the reader in Exercise 1. ■

We introduce one further piece of shorthand notation, the *augmented matrix*

$$[A \mid \mathbf{b}] = \begin{bmatrix} a_{11} & \cdots & a_{1n} & b_1 \\ a_{21} & \cdots & a_{2n} & b_2 \\ \vdots & \ddots & \vdots & \vdots \\ a_{m1} & \cdots & a_{mn} & b_m \end{bmatrix}.$$

Notice that the augmented matrix contains all of the information of the original system of equations since we can recover the latter by filling in the x_i's, $+$'s, and $=$'s as needed.

The elementary operations on a system of equations become operations on the rows of the augmented matrix; in this setting, we refer to them as *elementary row operations* of the corresponding three types:

  i. interchange any pair of rows;
  ii. multiply all the entries of any row by a nonzero real number;
  iii. replace any row by its sum with a multiple of any other row.

Since we have established that elementary operations do not affect the solution set of a system of equations, we can freely perform elementary row operations on the augmented matrix of a system of equations with the goal of finding an "equivalent" augmented matrix from which we can easily read off the general solution.

▶ **EXAMPLE 2**

We revisit Example 1 in the notation of augmented matrices. To solve

$$3x_1 - 2x_2 + 9x_4 = 4$$
$$2x_1 + 2x_2 - 4x_4 = 6,$$

we begin by forming the appropriate augmented matrix

$$\begin{bmatrix} 3 & -2 & 0 & 9 & 4 \\ 2 & 2 & 0 & -4 & 6 \end{bmatrix}.$$

We denote the process of performing row operations by the symbol $\rightsquigarrow$ and (in this example) we indicate above it the type of operation we are performing:

$$\begin{bmatrix} 3 & -2 & 0 & 9 & 4 \\ 2 & 2 & 0 & -4 & 6 \end{bmatrix} \overset{(i)}{\rightsquigarrow} \begin{bmatrix} 2 & 2 & 0 & -4 & 6 \\ 3 & -2 & 0 & 9 & 4 \end{bmatrix} \overset{(ii)}{\rightsquigarrow} \begin{bmatrix} 1 & 1 & 0 & -2 & 3 \\ 3 & -2 & 0 & 9 & 4 \end{bmatrix}$$

$$\overset{(iii)}{\rightsquigarrow} \begin{bmatrix} 1 & 1 & 0 & -2 & 3 \\ 0 & -5 & 0 & 15 & -5 \end{bmatrix} \overset{(ii)}{\rightsquigarrow} \begin{bmatrix} 1 & 1 & 0 & -2 & 3 \\ 0 & 1 & 0 & -3 & 1 \end{bmatrix} \overset{(iii)}{\rightsquigarrow} \begin{bmatrix} 1 & 0 & 0 & 1 & 2 \\ 0 & 1 & 0 & -3 & 1 \end{bmatrix}.$$

From the final augmented matrix we are able to recover the simpler form of the equations,

$$x_1 \quad\quad\quad + x_4 = 2$$
$$x_2 - 3x_4 = 1,$$

and read off the general solution just as before. ◀

Definition We call the first *nonzero* entry of a row (reading left to right) its *leading entry*. A matrix is in *echelon*[1] *form* if

1. the leading entries move to the right in successive rows;
2. the entries of the column *below* each leading entry are all 0;[2]
3. all rows of zeroes are at the bottom of the matrix.

A matrix is in *reduced echelon form* if it is in echelon form and, in addition,

4. every leading entry is 1;
5. all the entries of the column *above* each leading entry are 0 as well.

We call the leading entry of a certain row of a matrix a *pivot* if there is no leading entry *above* it in the *same* column. When a matrix is in echelon form, we refer to the columns in which a pivot appears as *pivot columns* and to the corresponding variables (in the original system of equations) as *pivot variables*. The remaining variables are called *free variables*.

The augmented matrices

$$\begin{bmatrix} 1 & 2 & 0 & -1 & | & 1 \\ 0 & 0 & 1 & 2 & | & 2 \end{bmatrix}, \quad \begin{bmatrix} 1 & 2 & 1 & 1 & | & 3 \\ 0 & 0 & 1 & 2 & | & 2 \end{bmatrix}, \quad \begin{bmatrix} 1 & 2 & 0 & | & 3 \\ 1 & 0 & -1 & | & 2 \end{bmatrix}$$

are, respectively, in reduced echelon form, in echelon form, and in neither. The key point is this: When the matrix is in reduced echelon form, we are able to determine the general solution by expressing each of the *pivot* variables in terms of the *free* variables.

▶ **EXAMPLE 3**

The augmented matrix

$$\begin{bmatrix} 1 & 2 & 0 & 0 & 4 & | & 1 \\ 0 & 0 & 1 & 0 & -2 & | & 2 \\ 0 & 0 & 0 & 1 & 1 & | & 1 \end{bmatrix}$$

is in reduced echelon form. The corresponding system of equations is

$$\begin{aligned} x_1 + 2x_2 \quad\quad\quad\quad + 4x_5 &= 1 \\ x_3 \quad\quad - 2x_5 &= 2 \\ x_4 + x_5 &= 1. \end{aligned}$$

---

[1] The word *echelon* derives from the French *échelle*, "ladder." Although we don't usually draw the rungs of the ladder, they are there: $\begin{bmatrix} 1 & 2 & 3 & 4 \\ 0 & 0 & 1 & 2 \\ 0 & 0 & 0 & 3 \end{bmatrix}$.

[2] Condition (2) is actually a consequence of (1), but we state it anyway for clarity.

Notice that the pivot variables, x_1, x_3, and x_4, are completely determined by the free variables x_2 and x_5. As usual, we can write the general solution in terms of the free variables only:

$$\mathbf{x} = \begin{bmatrix} x_1 \\ x_2 \\ x_3 \\ x_4 \\ x_5 \end{bmatrix} = \begin{bmatrix} 1-2x_2-4x_5 \\ x_2 \\ 2 \ +2x_5 \\ 1 \ - x_5 \\ x_5 \end{bmatrix} = \begin{bmatrix} 1 \\ 0 \\ 2 \\ 1 \\ 0 \end{bmatrix} + x_2 \begin{bmatrix} -2 \\ 1 \\ 0 \\ 0 \\ 0 \end{bmatrix} + x_5 \begin{bmatrix} -4 \\ 0 \\ 2 \\ -1 \\ 1 \end{bmatrix}.$$

In this last example, we see that the general solution is the sum of a *particular solution*—obtained by setting all the free variables equal to 0—and a linear combination of vectors, one for each free variable—obtained by setting that free variable equal to 1 and the remaining free variables equal to 0 and ignoring the particular solution. In other words, if x_k is a free variable, the corresponding vector in the general solution has k^{th} coordinate equal to 1 and j^{th} coordinate equal to 0 for all the other free variables x_j. Concentrate on the circled entries in the vectors from Example 3:

$$x_2 \begin{bmatrix} -2 \\ \boxed{1} \\ 0 \\ 0 \\ 0 \end{bmatrix} + x_5 \begin{bmatrix} -4 \\ 0 \\ 2 \\ -1 \\ \boxed{1} \end{bmatrix}.$$

We refer to this as the *standard form* of the general solution. The general solution of any system in *reduced echelon form* can be presented in this manner.

Our strategy now is to transform the augmented matrix of any system of linear equations into echelon form by performing a sequence of elementary row operations. The algorithm goes by the name of *Gaussian elimination*.

The first step is to identify the first column (starting at the left) that does not consist only of 0's; usually this is the first column, but it may not be. Pick a row whose entry in this column is nonzero—usually the uppermost such row, but you may choose another if it helps with the arithmetic—and interchange this with the first row; now the first entry of the first nonzero column is nonzero. This will be our first *pivot*. Next, we add the appropriate multiple of the top row to all the remaining rows to make all the entries below the pivot equal to 0. To consider two examples, if we begin with the matrices

$$A = \begin{bmatrix} 3 & -1 & 2 & 7 \\ 2 & 1 & 3 & 3 \\ 2 & 2 & 4 & 2 \end{bmatrix} \quad \text{and} \quad B = \begin{bmatrix} 0 & 2 & 4 & 3 \\ 0 & 1 & 2 & -1 \\ 0 & 2 & 3 & 3 \end{bmatrix},$$

then we begin by switching the first and third rows of A and the first and second rows of B (to avoid fractions). After clearing out the first pivot column we have

$$A' = \begin{bmatrix} \boxed{2} & 2 & 4 & 2 \\ 0 & -1 & -1 & 1 \\ 0 & -4 & -4 & 4 \end{bmatrix} \quad \text{and} \quad B' = \begin{bmatrix} 0 & \boxed{1} & 2 & -1 \\ 0 & 0 & 0 & 5 \\ 0 & 0 & -1 & 5 \end{bmatrix}.$$

We have circled the pivots for emphasis. (If we are headed for the reduced echelon form, we might replace the first row of A' by $\begin{bmatrix} 1 & 1 & 2 & 1 \end{bmatrix}$.)

The next step is to find the first column (again, starting at the left) in the *new* matrix having a nonzero entry *below the first row*. Pick a row below the first that has a nonzero entry in this column, and, if necessary, interchange it with the second row. Now the second entry of this column is nonzero; this is our second pivot. (Once again, if we're calculating the reduced echelon form, we multiply by the reciprocal of this entry to make the pivot 1.) We then add appropriate multiples of the second row to the rows beneath it to make all the entries beneath the pivot equal to 0. Continuing with our examples, we obtain

$$A'' = \begin{bmatrix} 2 & 2 & 4 & 2 \\ 0 & -1 & -1 & 1 \\ 0 & 0 & 0 & 0 \end{bmatrix} \quad \text{and} \quad B'' = \begin{bmatrix} 0 & 1 & 2 & -1 \\ 0 & 0 & -1 & 5 \\ 0 & 0 & 0 & 5 \end{bmatrix}.$$

At this point, both A'' and B'' are in echelon form; note that the zero row of A'' is at the bottom, and that the pivots move toward the right and down.

The process continues until we can find no more pivots—either because we have a pivot in each row or because we're left with nothing but rows of zeroes. At this stage, if we are interested in finding the reduced echelon form, we clear out the entries in the pivot columns *above* the pivots and then make all the pivots equal to 1. (Two words of advice here: If we start at the *right* and work our way up and to the left, we in general minimize the amount of arithmetic that must be done. Also, we always do our best to avoid fractions.) Continuing with our examples, we find the reduced echelon forms of A and B, respectively:

$$A'' = \begin{bmatrix} 2 & 2 & 4 & 2 \\ 0 & -1 & -1 & 1 \\ 0 & 0 & 0 & 0 \end{bmatrix} \rightsquigarrow \begin{bmatrix} 1 & 1 & 2 & 1 \\ 0 & 1 & 1 & -1 \\ 0 & 0 & 0 & 0 \end{bmatrix} \rightsquigarrow \begin{bmatrix} 1 & 0 & 1 & 2 \\ 0 & 1 & 1 & -1 \\ 0 & 0 & 0 & 0 \end{bmatrix} = R_A$$

$$B'' = \begin{bmatrix} 0 & 1 & 2 & -1 \\ 0 & 0 & -1 & 5 \\ 0 & 0 & 0 & 5 \end{bmatrix} \rightsquigarrow \begin{bmatrix} 0 & 1 & 2 & -1 \\ 0 & 0 & 1 & -5 \\ 0 & 0 & 0 & 1 \end{bmatrix} \rightsquigarrow \begin{bmatrix} 0 & 1 & 2 & 0 \\ 0 & 0 & 1 & 0 \\ 0 & 0 & 0 & 1 \end{bmatrix}$$

$$\rightsquigarrow \begin{bmatrix} 0 & 1 & 0 & 0 \\ 0 & 0 & 1 & 0 \\ 0 & 0 & 0 & 1 \end{bmatrix} = R_B.$$

We must be careful from now on to distinguish between the symbols "$=$" and "$\rightsquigarrow$"; when we convert one matrix to another by performing one or more row operations, we do **not** have equal matrices.

Here is one last example:

▶ **EXAMPLE 4**

Give the general solution of the following system of linear equations:

$$\begin{aligned} x_1 + x_2 + 3x_3 - x_4 &= 0 \\ -x_1 + x_2 + x_3 + x_4 + 2x_5 &= -4 \\ x_2 + 2x_3 + 2x_4 - x_5 &= 0 \\ 2x_1 - x_2 + x_4 - 6x_5 &= 9. \end{aligned}$$

We begin with the augmented matrix of coefficients and put it in reduced echelon form:

$$\begin{bmatrix} 1 & 1 & 3 & -1 & 0 & | & 0 \\ -1 & 1 & 1 & 1 & 2 & | & -4 \\ 0 & 1 & 2 & 2 & -1 & | & 0 \\ 2 & -1 & 0 & 1 & -6 & | & 9 \end{bmatrix} \rightsquigarrow \begin{bmatrix} 1 & 1 & 3 & -1 & 0 & | & 0 \\ 0 & 2 & 4 & 0 & 2 & | & -4 \\ 0 & 1 & 2 & 2 & -1 & | & 0 \\ 0 & -3 & -6 & 3 & -6 & | & 9 \end{bmatrix}$$

$$\rightsquigarrow \begin{bmatrix} 1 & 1 & 3 & -1 & 0 & | & 0 \\ 0 & 1 & 2 & 0 & 1 & | & -2 \\ 0 & 0 & 0 & 2 & -2 & | & 2 \\ 0 & 0 & 0 & 3 & -3 & | & 3 \end{bmatrix} \rightsquigarrow \begin{bmatrix} 1 & 1 & 3 & -1 & 0 & | & 0 \\ 0 & 1 & 2 & 0 & 1 & | & -2 \\ 0 & 0 & 0 & 1 & -1 & | & 1 \\ 0 & 0 & 0 & 0 & 0 & | & 0 \end{bmatrix}$$

$$\rightsquigarrow \begin{bmatrix} 1 & 0 & 1 & 0 & -2 & | & 3 \\ 0 & 1 & 2 & 0 & 1 & | & -2 \\ 0 & 0 & 0 & 1 & -1 & | & 1 \\ 0 & 0 & 0 & 0 & 0 & | & 0 \end{bmatrix}$$

Thus, the system of equations is given in reduced echelon form by

$$\begin{aligned} x_1 \quad\quad + x_3 \quad\quad - 2x_5 &= 3 \\ x_2 + 2x_3 \quad\quad + x_5 &= -2 \\ x_4 - x_5 &= 1, \end{aligned}$$

from which we read off

$$\begin{aligned} x_1 &= 3 - x_3 + 2x_5 \\ x_2 &= -2 - 2x_3 - x_5 \\ x_3 &= x_3 \\ x_4 &= 1 + x_5 \\ x_5 &= x_5, \end{aligned}$$

and so the general solution is

$$\mathbf{x} = \begin{bmatrix} x_1 \\ x_2 \\ x_3 \\ x_4 \\ x_5 \end{bmatrix} = \begin{bmatrix} 3 \\ -2 \\ 0 \\ 1 \\ 0 \end{bmatrix} + x_3 \begin{bmatrix} -1 \\ -2 \\ 1 \\ 0 \\ 0 \end{bmatrix} + x_5 \begin{bmatrix} 2 \\ -1 \\ 0 \\ 1 \\ 1 \end{bmatrix}. \blacktriangleleft$$

When we reduce a matrix to echelon form, we must make a number of choices along the way, and the echelon form may well depend on the choices. But we shall now prove (using an inductive argument) that any two echelon forms of the same matrix must have pivots in the same columns, and from this it will follow that the *reduced* echelon form must be unique.

Theorem 1.2 *Suppose A and B are echelon forms of the same nonzero matrix M. Then all of their pivots appear in the same positions. As a consequence, if they are in reduced echelon form, then they are equal.*

Proof We begin by noting that we can transform M to both A and B by sequences of elementary row operations. It follows that we can proceed from A to B by a sequence of elementary row operations: The inverse of an elementary row operation is itself an elementary row operation, so we can first transform A to M and then transform M to B.

Suppose the ith column of A is its first pivot column; this column vector is the standard basis vector $\mathbf{e}_1 \in \mathbb{R}^m$ and all previous columns are zero. If we perform any elementary row operation on A, the first $i-1$ columns remain zero and the ith column remains nonzero. Thus, the ith column is the first nonzero column of B; i.e., it is B's first pivot column.

Next we prove that all the pivots must be in the same locations. We do this by induction on m, the number of rows. We've already established that this must be the case for $m=1$. Now assume that the statement is true for $m=k$ and consider $(k+1) \times n$ matrices A and B satisfying the hypotheses. By what we've already said, A and B have the same first pivot column; by using an elementary row operation of type (ii) appropriately, we may assume those respective first pivot entries in the first row are equal. Now, the $k \times n$ matrices A' and B' obtained from A and B by deleting their first rows are also in echelon form. Furthermore, any sequence of elementary row operations that transforms A to B cannot involve the first row in a nontrivial way (if we add a multiple of the first row to any other row, we must later subtract it again). Thus, A' can be transformed to B' by a sequence of elementary row operations. By the induction hypothesis we can now conclude that A' and B' have pivots in the same locations and, thus, so do A and B.

Last, we prove that if A and B are in *reduced* echelon form, then they are equal. Again we proceed by induction on m. The case $m=1$ is trivial. Assume that the statement is true for $m=k$ and consider the case $m=k+1$. If the matrix A has a row of zeroes, then so must the matrix B; we delete these rows and apply the induction hypothesis to conclude that $A=B$. Now, if the last row of A is nonzero, it must contain the last pivot of A (say, in the jth column). Then we know that the last pivot of B must be in the jth column as well. Since the matrices are in reduced echelon form, their j^{th} columns must be the last standard basis vector $\mathbf{e}_m \in \mathbb{R}^m$. Because of this, the sequence of elementary row operations that transforms A to B cannot involve the last row in a nontrivial way. Thus, if we let A' and B' be the matrices obtained from A and B by deleting the last row, we see that A' can be transformed to B' by a sequence of elementary row operations and that A' and B' are both in reduced echelon form. The induction hypothesis applies to A' and B', so we conclude that $A'=B'$. Finally, we need to argue that the bottom rows of A and B are identical. But any elementary row operation that would alter the last row would also have to make some change in the first j entries. Since the last rows of A and B are known to agree in the first j entries, we conclude that they must agree everywhere. ∎

1.1 Consistency

We recall from Chapter 1 that the product $A\mathbf{x}$ can be expressed as

$$(*) \quad A\mathbf{x} = \begin{bmatrix} a_{11}x_1 + \cdots + a_{1n}x_n \\ a_{21}x_1 + \cdots + a_{2n}x_n \\ \vdots \\ a_{m1}x_1 + \cdots + a_{mn}x_n \end{bmatrix} = x_1 \begin{bmatrix} a_{11} \\ a_{21} \\ \vdots \\ a_{m1} \end{bmatrix} + x_2 \begin{bmatrix} a_{12} \\ a_{22} \\ \vdots \\ a_{m2} \end{bmatrix} + \cdots + x_n \begin{bmatrix} a_{1n} \\ a_{2n} \\ \vdots \\ a_{mn} \end{bmatrix}$$

$$= x_1 \mathbf{a}_1 + x_2 \mathbf{a}_2 + \cdots + x_n \mathbf{a}_n,$$

where $\mathbf{a}_1, \ldots, \mathbf{a}_n \in \mathbb{R}^m$ are the column vectors of the matrix A. Thus, a solution $\mathbf{c} = \begin{bmatrix} c_1 \\ \vdots \\ c_n \end{bmatrix}$ of the linear system $A\mathbf{x} = \mathbf{b}$ provides scalars $c_1, \ldots, c_n$ so that

$$\mathbf{b} = c_1 \mathbf{a}_1 + \cdots + c_n \mathbf{a}_n;$$

i.e., a solution gives a representation of the vector $\mathbf{b}$ as a linear combination, $c_1 \mathbf{a}_1 + \cdots + c_n \mathbf{a}_n$, of the column vectors of A.

▶ **EXAMPLE 5**

Consider the four vectors

$$\mathbf{b} = \begin{bmatrix} 4 \\ 3 \\ 1 \\ 2 \end{bmatrix}, \quad \mathbf{v}_1 = \begin{bmatrix} 1 \\ 0 \\ 1 \\ 2 \end{bmatrix}, \quad \mathbf{v}_2 = \begin{bmatrix} 1 \\ 1 \\ 1 \\ 1 \end{bmatrix}, \quad \text{and} \quad \mathbf{v}_3 = \begin{bmatrix} 2 \\ 1 \\ 1 \\ 2 \end{bmatrix}.$$

Suppose we want to express the vector $\mathbf{b}$ as a linear combination of the vectors $\mathbf{v}_1$, $\mathbf{v}_2$, and $\mathbf{v}_3$. Writing out the expression

$$x_1 \mathbf{v}_1 + x_2 \mathbf{v}_2 + x_3 \mathbf{v}_3 = x_1 \begin{bmatrix} 1 \\ 0 \\ 1 \\ 2 \end{bmatrix} + x_2 \begin{bmatrix} 1 \\ 1 \\ 1 \\ 1 \end{bmatrix} + x_3 \begin{bmatrix} 2 \\ 1 \\ 1 \\ 2 \end{bmatrix} = \begin{bmatrix} 4 \\ 3 \\ 1 \\ 2 \end{bmatrix},$$

we obtain the system of equations

$$\begin{aligned} x_1 + x_2 + 2x_3 &= 4 \\ x_2 + x_3 &= 3 \\ x_1 + x_2 + x_3 &= 1 \\ 2x_1 + x_2 + 2x_3 &= 2. \end{aligned}$$

In matrix notation, we must solve $A\mathbf{x} = \mathbf{b}$, where

$$A = \begin{bmatrix} 1 & 1 & 2 \\ 0 & 1 & 1 \\ 1 & 1 & 1 \\ 2 & 1 & 2 \end{bmatrix}.$$

So we take the augmented matrix to reduced echelon form:

$$[A \mid \mathbf{b}] = \begin{bmatrix} 1 & 1 & 2 & | & 4 \\ 0 & 1 & 1 & | & 3 \\ 1 & 1 & 1 & | & 1 \\ 2 & 1 & 2 & | & 2 \end{bmatrix} \rightsquigarrow \begin{bmatrix} 1 & 1 & 2 & | & 4 \\ 0 & 1 & 1 & | & 3 \\ 0 & 0 & -1 & | & -3 \\ 0 & -1 & -2 & | & -6 \end{bmatrix}$$

$$\rightsquigarrow \begin{bmatrix} 1 & 1 & 2 & | & 4 \\ 0 & 1 & 1 & | & 3 \\ 0 & 0 & 1 & | & 3 \\ 0 & 0 & 0 & | & 0 \end{bmatrix} \rightsquigarrow \begin{bmatrix} 1 & 0 & 0 & | & -2 \\ 0 & 1 & 0 & | & 0 \\ 0 & 0 & 1 & | & 3 \\ 0 & 0 & 0 & | & 0 \end{bmatrix}.$$

This tells us that the solution is

$$\mathbf{x} = \begin{bmatrix} -2 \\ 0 \\ 3 \end{bmatrix}, \quad \text{so} \quad \mathbf{b} = -2\mathbf{v}_1 + 0\mathbf{v}_2 + 3\mathbf{v}_3,$$

which, as the reader can check, works. ◂

Now we modify the preceding example slightly.

▶ **EXAMPLE 6**

We would like to express the vector

$$\mathbf{b} = \begin{bmatrix} 1 \\ 1 \\ 0 \\ 1 \end{bmatrix}$$

as a linear combination of the same vectors $\mathbf{v}_1$, $\mathbf{v}_2$, and $\mathbf{v}_3$. This then leads analogously to the system of equations

$$\begin{aligned} x_1 + x_2 + 2x_3 &= 1 \\ x_2 + x_3 &= 1 \\ x_1 + x_2 + x_3 &= 0 \\ 2x_1 + x_2 + 2x_3 &= 1 \end{aligned}$$

and to the augmented matrix

$$\begin{bmatrix} 1 & 1 & 2 & | & 1 \\ 0 & 1 & 1 & | & 1 \\ 1 & 1 & 1 & | & 0 \\ 2 & 1 & 2 & | & 1 \end{bmatrix},$$

whose echelon form is

$$\begin{bmatrix} 1 & 1 & 2 & | & 1 \\ 0 & 1 & 1 & | & 1 \\ 0 & 0 & 1 & | & 1 \\ 0 & 0 & 0 & | & 1 \end{bmatrix}.$$

The last row of the augmented matrix corresponds to the equation

$$0x_1 + 0x_2 + 0x_3 = 1,$$

which obviously has no solution. Thus, the original system of equations has no solution: The vector $\mathbf{b}$ in this example cannot be written as a linear combination of $\mathbf{v}_1$, $\mathbf{v}_2$, and $\mathbf{v}_3$.

These examples lead us to make the following

Definition If the system of equations $A\mathbf{x} = \mathbf{b}$ has no solutions, the system is said to be *inconsistent*; if it has at least one solution, then it is said to be *consistent*.

A system of equations is consistent precisely when a solution *exists*. We see that the system of equations in Example 6 is inconsistent and the system of equations in Example 5 is consistent. It is easy to recognize an inconsistent system of equations from the echelon form of its augmented matrix: The system is inconsistent only when there is an equation that reads

$$0x_1 + 0x_2 + \cdots + 0x_n = c$$

for some nonzero scalar c, i.e., when there is a row in the echelon form of the augmented matrix where all but the rightmost entry are 0.

Turning this around a bit, let $[U \mid \mathbf{c}]$ denote the echelon form of the augmented matrix $[A \mid \mathbf{b}]$. The system $A\mathbf{x} = \mathbf{b}$ is consistent if and only if any zero row in U corresponds to a zero entry in the vector $\mathbf{c}$.

There are two geometric interpretations of consistency. From the standpoint of row vectors, the system $A\mathbf{x} = \mathbf{b}$ is consistent precisely when the intersection of the hyperplanes

$$\mathbf{A}_1 \cdot \mathbf{x} = b_1, \quad \ldots, \quad \mathbf{A}_m \cdot \mathbf{x} = b_m$$

is nonempty. From the point of view of column vectors, the system $A\mathbf{x} = \mathbf{b}$ is consistent precisely when the vector $\mathbf{b}$ can be written as a linear combination of the column vectors $\mathbf{a}_1, \ldots, \mathbf{a}_n$ of A.

In the next example, we characterize those vectors $\mathbf{b} \in \mathbb{R}^4$ that can be expressed as a linear combination of the three vectors $\mathbf{v}_1$, $\mathbf{v}_2$, and $\mathbf{v}_3$ from Examples 5 and 6.

▶ **EXAMPLE 7**

For what vectors

$$\mathbf{b} = \begin{bmatrix} b_1 \\ b_2 \\ b_3 \\ b_4 \end{bmatrix}$$

will the system of equations

$$x_1 + x_2 + 2x_3 = b_1$$
$$x_2 + x_3 = b_2$$
$$x_1 + x_2 + x_3 = b_3$$
$$2x_1 + x_2 + 2x_3 = b_4$$

have a solution? We form the augmented matrix $[A \mid \mathbf{b}]$ and determine its echelon form:

$$\begin{bmatrix} 1 & 1 & 2 & b_1 \\ 0 & 1 & 1 & b_2 \\ 1 & 1 & 1 & b_3 \\ 2 & 1 & 2 & b_4 \end{bmatrix} \rightsquigarrow \begin{bmatrix} 1 & 1 & 2 & b_1 \\ 0 & 1 & 1 & b_2 \\ 0 & 0 & -1 & b_3 - b_1 \\ 0 & -1 & -2 & b_4 - 2b_1 \end{bmatrix} \rightsquigarrow \begin{bmatrix} 1 & 1 & 2 & b_1 \\ 0 & 1 & 1 & b_2 \\ 0 & 0 & 1 & b_1 - b_3 \\ 0 & 0 & 0 & -b_1 + b_2 - b_3 + b_4 \end{bmatrix}.$$

We infer from the last row of the latter matrix that the original system of equations will have a solution if and only if

(†) $$-b_1 + b_2 - b_3 + b_4 = 0.$$

That is, the vector $\mathbf{b}$ can be written as a linear combination of $\mathbf{v}_1$, $\mathbf{v}_2$, and $\mathbf{v}_3$ precisely when $\mathbf{b}$ satisfies the *constraint equation* (†). ◀

▶ **EXAMPLE 8**

Given

$$A = \begin{bmatrix} 1 & -1 & 1 \\ 3 & 2 & -1 \\ 1 & 4 & -3 \\ 3 & -3 & 3 \end{bmatrix},$$

we wish to find all vectors $\mathbf{b} \in \mathbb{R}^4$ so that $A\mathbf{x} = \mathbf{b}$ is consistent, i.e., all vectors $\mathbf{b}$ that can be expressed as a linear combination of the columns of A.

We consider the augmented matrix $[A \mid \mathbf{b}]$ and determine its echelon form $[U \mid \mathbf{c}]$. In order for the system to be consistent, every entry of $\mathbf{c}$ corresponding to a row of zeroes in U must be 0 as well:

$$[A \mid \mathbf{b}] = \begin{bmatrix} 1 & -1 & 1 & b_1 \\ 3 & 2 & -1 & b_2 \\ 1 & 4 & -3 & b_3 \\ 3 & 3 & -3 & b_4 \end{bmatrix} \rightsquigarrow \begin{bmatrix} 1 & -1 & 1 & b_1 \\ 0 & 5 & -4 & b_2 - 3b_1 \\ 0 & 5 & -4 & b_3 - b_1 \\ 0 & 0 & 0 & b_4 - 3b_1 \end{bmatrix}$$

$$\rightsquigarrow \begin{bmatrix} 1 & -1 & 1 & b_1 \\ 0 & 5 & -4 & b_2 - 3b_1 \\ 0 & 0 & 0 & b_3 - b_2 + 2b_1 \\ 0 & 0 & 0 & b_4 - 3b_1 \end{bmatrix}.$$

Thus, we conclude that $A\mathbf{x} = \mathbf{b}$ is consistent if and only if $\mathbf{b}$ satisfies the *constraint equations*

$$2b_1 - b_2 + b_3 = 0 \quad \text{and} \quad -3b_1 + b_4 = 0.$$

These equations describe the intersection of two hyperplanes through the origin in $\mathbb{R}^4$ with respective

normal vectors $\begin{bmatrix} 2 \\ -1 \\ 1 \\ 0 \end{bmatrix}$ and $\begin{bmatrix} -3 \\ 0 \\ 0 \\ 1 \end{bmatrix}$. ◂

Notice that here we have reversed the process at the beginning of this section. There we expressed the general solution of a system of linear equations as a linear combination of certain vectors. Here, starting with the column vectors of the matrix A, we have found the *constraint equations* a vector **b** must satisfy in order to be a linear combination of them (that is, to be in the plane they span). This is the process of determining Cartesian equations for a space defined parametrically.

1.2 Existence and Uniqueness of Solutions

In general, given an $m \times n$ matrix, we might wonder how many conditions a vector $\mathbf{b} \in \mathbb{R}^m$ must satisfy in order to be a linear combination of the columns of A. From the procedure we've just followed, the answer is quite clear: Each row of zeroes in the echelon form of A contributes one constraint. This leads us to our next

Definition The *rank* of a matrix is the number of nonzero rows (i.e., the number of pivots) in its echelon form. It is usually denoted by r.

Then the number of rows of zeroes in the echelon form is $m - r$, and **b** must satisfy $m - r$ constraint equations. We recall that even though a matrix may have lots of different echelon forms, it follows from Theorem 1.2 that they all must have the same number of nonzero rows.

Given a system of m linear equations in n variables, let A denote its coefficient matrix and r the rank of A. Let's now summarize the state of our knowledge:

Proposition 1.3 *The linear system $A\mathbf{x} = \mathbf{b}$ is consistent if and only if the rank of the augmented matrix $[A \mid \mathbf{b}]$ equals the rank of A. In particular, if $r = m$, then the system $A\mathbf{x} = \mathbf{b}$ will be consistent for all vectors $\mathbf{b} \in \mathbb{R}^m$.*

Proof $A\mathbf{x} = \mathbf{b}$ is consistent if and only if the rank of the augmented matrix $[A \mid \mathbf{b}]$, which is the number of nonzero rows in the augmented matrix $[U \mid \mathbf{c}]$, equals the number of nonzero rows in U, i.e., the rank of A. When $r = m$, there is no row of zeroes in U, hence no possibility of inconsistency. ∎

We now turn our attention to the question of how many solutions a given *consistent* system of equations has. Our experience with solving systems of equations suggests that the solutions of a consistent linear system $A\mathbf{x} = \mathbf{b}$ are intimately related to the solutions of the system $A\mathbf{x} = \mathbf{0}$.

Definition A system $A\mathbf{x} = \mathbf{b}$ of linear equations is called *inhomogeneous* when $\mathbf{b} \neq \mathbf{0}$; the corresponding equation $A\mathbf{x} = \mathbf{0}$ is called the associated *homogeneous system*.

1 Gaussian Elimination and the Theory of Linear Systems

The solutions of the inhomogeneous system $A\mathbf{x} = \mathbf{b}$ and those of the associated homogeneous system $A\mathbf{x} = \mathbf{0}$ are related by the following

Proposition 1.4 *Assume the system $A\mathbf{x} = \mathbf{b}$ is consistent, and let $\mathbf{u}_1$ be a "particular solution." Then all the solutions are of the form*

$$\mathbf{u} = \mathbf{u}_1 + \mathbf{v}$$

for some solution $\mathbf{v}$ of the associated homogeneous system $A\mathbf{x} = \mathbf{0}$.

Proof First we observe that any such vector $\mathbf{u}$ is a solution of $A\mathbf{x} = \mathbf{b}$. By linearity, we have

$$A\mathbf{u} = A(\mathbf{u}_1 + \mathbf{v}) = A\mathbf{u}_1 + A\mathbf{v} = \mathbf{b} + \mathbf{0} = \mathbf{b}.$$

Conversely, *every* solution of $A\mathbf{x} = \mathbf{b}$ can be written in this form, for if $\mathbf{u}$ is an arbitrary solution of $A\mathbf{x} = \mathbf{b}$, then, by linearity again,

$$A(\mathbf{u} - \mathbf{u}_1) = A\mathbf{u} - A\mathbf{u}_1 = \mathbf{b} - \mathbf{b} = \mathbf{0},$$

so $\mathbf{v} = \mathbf{u} - \mathbf{u}_1$ is a solution of the associated homogeneous system; now we just solve for $\mathbf{u}$, obtaining $\mathbf{u} = \mathbf{u}_1 + \mathbf{v}$, as required. ∎

Remark As Figure 1.1 suggests, when the inhomogeneous system $A\mathbf{x} = \mathbf{b}$ is consistent, its solutions are obtained by *translating* the set of solutions of the associated homogeneous system by a particular solution $\mathbf{u}_1$.

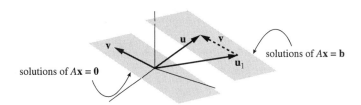

Figure 1.1

Of course, a homogeneous system is always consistent since the *trivial solution*, $\mathbf{x} = \mathbf{0}$, is always a solution of $A\mathbf{x} = \mathbf{0}$. Now, if the rank of A is r, then there will be r pivot variables and $n - r$ free variables in the general solution of $A\mathbf{x} = \mathbf{0}$. In particular, if $r = n$, then $\mathbf{x} = \mathbf{0}$ is the only solution of $A\mathbf{x} = \mathbf{0}$.

Definition If the system of equations $A\mathbf{x} = \mathbf{b}$ has precisely one solution, then we say that the system has a *unique* solution.

Thus, a homogeneous system $A\mathbf{x} = \mathbf{0}$ has a *unique* solution when $r = n$ and *infinitely many* solutions when $r < n$. Note that it is impossible to have $r > n$ since there cannot be more pivots than columns. Similarly, there cannot be more pivots than rows in the matrix, so it follows that whenever $n > m$ (i.e., there are more variables than equations), the homogeneous system $A\mathbf{x} = \mathbf{0}$ must have infinitely many solutions.

From Proposition 1.4 we know that if the inhomogeneous system $A\mathbf{x} = \mathbf{b}$ is consistent, then its solutions are obtained by translating the solutions of the associated homogeneous system $A\mathbf{x} = \mathbf{0}$ by a particular solution. So we have

Proposition 1.5 *Suppose the system $A\mathbf{x} = \mathbf{b}$ is consistent. Then it has a unique solution if and only if the associated homogeneous system $A\mathbf{x} = \mathbf{0}$ has only the trivial solution. This happens exactly when $r = n$.*

We conclude this discussion with an important special case. It is natural to ask when the inhomogeneous system $A\mathbf{x} = \mathbf{b}$ has a *unique* solution for *every* $\mathbf{b} \in \mathbb{R}^m$. From Proposition 1.3 we infer that for the system always to be consistent, we must have $r = m$; from Proposition 1.5 we infer that for solutions to be unique, we must have $r = n$. And so we see that we can only have both conditions when $r = m = n$.

Definition An $n \times n$ matrix of rank $r = n$ is called *nonsingular*. An $n \times n$ matrix of rank $r < n$ is called *singular*.

It is easy—but important—to observe that an $n \times n$ matrix is nonsingular if and only if there is a pivot in each row, hence in each column, of its echelon form.

Proposition 1.6 *Let A be an $n \times n$ matrix. The following are equivalent:*

1. *A is nonsingular.*
2. *$A\mathbf{x} = \mathbf{0}$ has only the trivial solution.*
3. *For every $\mathbf{b} \in \mathbb{R}^n$, the equation $A\mathbf{x} = \mathbf{b}$ has a unique solution.*

EXERCISES 4.1

1. Prove Proposition 1.1.

*2. Decide which of the following matrices are in echelon form, which are in reduced echelon form, and which are neither. Justify your answers.

(a) $\begin{bmatrix} 0 & 1 \\ 2 & 3 \end{bmatrix}$

(b) $\begin{bmatrix} 2 & 1 & 3 \\ 0 & 1 & -1 \end{bmatrix}$

(c) $\begin{bmatrix} 1 & 0 & 2 \\ 0 & 1 & -1 \end{bmatrix}$

(d) $\begin{bmatrix} 1 & 1 & 0 \\ 0 & 0 & 2 \end{bmatrix}$

(e) $\begin{bmatrix} 1 & 1 & 0 \\ 0 & 0 & 0 \\ 0 & 0 & 1 \end{bmatrix}$

(f) $\begin{bmatrix} 1 & 1 & 0 & -1 \\ 0 & 2 & 1 & 0 \\ 0 & 0 & 0 & 1 \end{bmatrix}$

(g) $\begin{bmatrix} 1 & 0 & -2 & 0 & 1 \\ 0 & 1 & 1 & 0 & 1 \\ 0 & 0 & 0 & 1 & 4 \end{bmatrix}$

3. For each of the following matrices A, determine its reduced echelon form and give the general solution of $A\mathbf{x} = \mathbf{0}$ in standard form.

(a) $A = \begin{bmatrix} 1 & 0 & -1 \\ -2 & 3 & -1 \\ 3 & -3 & 0 \end{bmatrix}$

*(b) $A = \begin{bmatrix} 2 & -2 & 4 \\ -1 & 1 & -2 \\ 3 & -3 & 6 \end{bmatrix}$

(c) $A = \begin{bmatrix} 1 & 2 & -1 \\ 1 & 3 & 1 \\ 2 & 4 & 3 \\ -1 & 1 & 6 \end{bmatrix}$

(d) $A = \begin{bmatrix} 1 & -2 & 1 & 0 \\ 2 & -4 & 3 & -1 \end{bmatrix}$

*(e) $A = \begin{bmatrix} 1 & 1 & 1 & 1 \\ 1 & 2 & 1 & 2 \\ 1 & 3 & 2 & 4 \\ 1 & 2 & 2 & 3 \end{bmatrix}$

*(f) $A = \begin{bmatrix} 1 & 2 & 0 & -1 & -1 \\ -1 & -3 & 1 & 2 & 3 \\ 1 & -1 & 3 & 1 & 1 \\ 2 & -3 & 7 & 3 & 4 \end{bmatrix}$

(g) $A = \begin{bmatrix} 1 & -1 & 1 & 1 & 0 \\ 1 & 0 & 2 & 1 & 1 \\ 0 & 2 & 2 & 2 & 0 \\ -1 & 1 & -1 & 0 & -1 \end{bmatrix}$

(h) $A = \begin{bmatrix} 1 & 1 & 0 & 5 & 0 & -1 \\ 0 & 1 & 1 & 3 & -2 & 0 \\ -1 & 2 & 3 & 4 & 1 & -6 \\ 0 & 4 & 4 & 12 & -1 & -7 \end{bmatrix}$

4. Give the general solution of the equation $A\mathbf{x} = \mathbf{b}$ in standard form.

*(a) $A = \begin{bmatrix} 2 & 1 & -1 \\ 1 & 2 & 1 \\ -1 & 1 & 2 \end{bmatrix}, \mathbf{b} = \begin{bmatrix} 3 \\ 0 \\ -3 \end{bmatrix}$

(b) $A = \begin{bmatrix} 1 & 1 & 1 & 1 \\ 3 & 3 & 2 & 0 \end{bmatrix}, \mathbf{b} = \begin{bmatrix} 6 \\ 17 \end{bmatrix}$

(c) $A = \begin{bmatrix} 1 & 1 & 1 & -1 & 0 \\ 2 & 0 & 4 & 1 & -1 \\ 1 & 2 & 0 & -2 & 2 \\ 0 & 1 & -1 & 2 & 4 \end{bmatrix}, \mathbf{b} = \begin{bmatrix} -2 \\ 10 \\ -3 \\ 7 \end{bmatrix}$

*5. Find all the unit vectors $\mathbf{x} \in \mathbb{R}^3$ that make an angle of $\pi/3$ with the vectors $\begin{bmatrix} 1 \\ 0 \\ -1 \end{bmatrix}$ and $\begin{bmatrix} 0 \\ 1 \\ 1 \end{bmatrix}$.

6. Find the normal vector to the hyperplane in $\mathbb{R}^4$ spanned by

*(a) $\begin{bmatrix} 1 \\ 1 \\ 1 \\ 1 \end{bmatrix}, \begin{bmatrix} 1 \\ 2 \\ 1 \\ 2 \end{bmatrix}, \begin{bmatrix} 1 \\ 3 \\ 2 \\ 4 \end{bmatrix}$

(b) $\begin{bmatrix} 1 \\ 1 \\ 1 \\ 1 \end{bmatrix}, \begin{bmatrix} 2 \\ 2 \\ 1 \\ 2 \end{bmatrix}, \begin{bmatrix} 1 \\ 3 \\ 2 \\ 3 \end{bmatrix}$.

*7. A circle C passes through the points $\begin{bmatrix} 2 \\ 6 \end{bmatrix}, \begin{bmatrix} -1 \\ 7 \end{bmatrix}$, and $\begin{bmatrix} -4 \\ -2 \end{bmatrix}$. Find the center and radius of C.
(Hint: The equation of a circle can be written in the form $x^2 + y^2 + ax + by + c = 0$. Why?)

*8. By solving a system of equations, find the linear combination of the vectors
$$\mathbf{v}_1 = \begin{bmatrix} 1 \\ 0 \\ -1 \end{bmatrix}, \quad \mathbf{v}_2 = \begin{bmatrix} 0 \\ 1 \\ 2 \end{bmatrix}, \quad \mathbf{v}_3 = \begin{bmatrix} 2 \\ 1 \\ 1 \end{bmatrix}$$
that gives $\mathbf{b} = \begin{bmatrix} 3 \\ 0 \\ -2 \end{bmatrix}$.

*9. For each of the following vectors $\mathbf{b} \in \mathbb{R}^4$, decide whether $\mathbf{b}$ is a linear combination of
$$\mathbf{v}_1 = \begin{bmatrix} 1 \\ 0 \\ 1 \\ -2 \end{bmatrix}, \quad \mathbf{v}_2 = \begin{bmatrix} 0 \\ -1 \\ 0 \\ 1 \end{bmatrix}, \quad \text{and} \quad \mathbf{v}_3 = \begin{bmatrix} 1 \\ -2 \\ 1 \\ 0 \end{bmatrix}.$$

(a) $\mathbf{b} = \begin{bmatrix} 1 \\ 1 \\ 1 \\ 1 \end{bmatrix}$ (b) $\mathbf{b} = \begin{bmatrix} 1 \\ -1 \\ 1 \\ -1 \end{bmatrix}$ (c) $\mathbf{b} = \begin{bmatrix} 1 \\ 1 \\ 0 \\ -2 \end{bmatrix}$

*10. Decide whether each of the following collections of vectors spans $\mathbb{R}^3$.

(a) $\begin{bmatrix} 1 \\ 1 \\ 1 \end{bmatrix}, \begin{bmatrix} 1 \\ 2 \\ 2 \end{bmatrix}$

(c) $\begin{bmatrix} 1 \\ 0 \\ 1 \end{bmatrix}, \begin{bmatrix} 1 \\ -1 \\ 1 \end{bmatrix}, \begin{bmatrix} 3 \\ 5 \\ 3 \end{bmatrix}, \begin{bmatrix} 2 \\ 3 \\ 2 \end{bmatrix}$

(b) $\begin{bmatrix} 1 \\ 1 \\ 1 \end{bmatrix}, \begin{bmatrix} 1 \\ 2 \\ 2 \end{bmatrix}, \begin{bmatrix} 1 \\ 3 \\ 3 \end{bmatrix}$

(d) $\begin{bmatrix} 1 \\ 0 \\ -1 \end{bmatrix}, \begin{bmatrix} 2 \\ 1 \\ 1 \end{bmatrix}, \begin{bmatrix} 0 \\ 1 \\ 5 \end{bmatrix}$

11. Find the constraint equations that $\mathbf{b}$ must satisfy in order for $A\mathbf{x} = \mathbf{b}$ to be consistent.

(a) $A = \begin{bmatrix} 3 & -1 \\ 6 & -2 \\ -9 & 3 \end{bmatrix}$ *(b) $A = \begin{bmatrix} 1 & 1 & 1 \\ -1 & 1 & 2 \\ 1 & 3 & 4 \end{bmatrix}$ (c) $A = \begin{bmatrix} 1 & -2 & 1 \\ 0 & 1 & 4 \\ -1 & 3 & 0 \\ 1 & 0 & -1 \end{bmatrix}$

12. Find the constraint equations that $\mathbf{b}$ must satisfy in order to be an element of

(a) $V = \text{Span}\left(\begin{bmatrix} 1 \\ 0 \\ 1 \\ 1 \end{bmatrix}, \begin{bmatrix} 0 \\ 1 \\ 1 \\ 2 \end{bmatrix}, \begin{bmatrix} 1 \\ 1 \\ 1 \\ 0 \end{bmatrix}\right)$ (b) $V = \text{Span}\left(\begin{bmatrix} 1 \\ 0 \\ 1 \\ 1 \end{bmatrix}, \begin{bmatrix} 0 \\ 1 \\ 1 \\ 2 \end{bmatrix}, \begin{bmatrix} 2 \\ -1 \\ 1 \\ 0 \end{bmatrix}\right).$

13. Find a matrix A with the given property or explain why none can exist.

*(a) One of the rows of A is $\begin{bmatrix} 1 \\ 0 \\ 1 \end{bmatrix}$ and for some $\mathbf{b} \in \mathbb{R}^2$ both the vectors $\begin{bmatrix} 1 \\ 0 \\ 1 \end{bmatrix}$ and $\begin{bmatrix} 2 \\ 1 \\ 1 \end{bmatrix}$ are solutions of the equation $A\mathbf{x} = \mathbf{b}$;

(b) the rows of A are linear combinations of $\begin{bmatrix} 0 \\ 1 \\ 0 \\ 1 \end{bmatrix}$ and $\begin{bmatrix} 0 \\ 0 \\ 1 \\ 1 \end{bmatrix}$ and for some $\mathbf{b} \in \mathbb{R}^2$ both the vectors $\begin{bmatrix} 1 \\ 2 \\ 1 \\ 2 \end{bmatrix}$ and $\begin{bmatrix} 4 \\ 1 \\ 0 \\ 3 \end{bmatrix}$ are solutions of the equation $A\mathbf{x} = \mathbf{b}$;

(c) the rows of A are orthogonal to $\begin{bmatrix} 1 \\ 0 \\ 1 \\ 0 \end{bmatrix}$ and for some nonzero vector $\mathbf{b} \in \mathbb{R}^2$ both the vectors $\begin{bmatrix} 1 \\ 0 \\ 1 \\ 0 \end{bmatrix}$ and $\begin{bmatrix} 1 \\ 1 \\ 1 \\ 1 \end{bmatrix}$ are solutions of the equation $A\mathbf{x} = \mathbf{b}$;

(d) for some vectors $\mathbf{b}_1, \mathbf{b}_2 \in \mathbb{R}^2$ both the vectors $\begin{bmatrix} 1 \\ 0 \\ 1 \end{bmatrix}$ and $\begin{bmatrix} 2 \\ 1 \\ 1 \end{bmatrix}$ are solutions of the equation $A\mathbf{x} = \mathbf{b}_1$ and both the vectors $\begin{bmatrix} 1 \\ 0 \\ 0 \end{bmatrix}$ and $\begin{bmatrix} 1 \\ 1 \\ 1 \end{bmatrix}$ are solutions of the equation $A\mathbf{x} = \mathbf{b}_2$.

*14. Let $A = \begin{bmatrix} 1 & \alpha \\ \alpha & 3\alpha \end{bmatrix}$.

(a) For which numbers α will A be singular?

(b) For all numbers α *not* on your list in part a, we can solve $A\mathbf{x} = \mathbf{b}$ for every vector $\mathbf{b} \in \mathbb{R}^2$. For each of the numbers α on your list, give the vectors $\mathbf{b}$ for which we can solve $A\mathbf{x} = \mathbf{b}$.

15. Let $A = \begin{bmatrix} 1 & 1 & \alpha \\ \alpha & 2 & \alpha \\ \alpha & \alpha & 1 \end{bmatrix}$.

(a) For which numbers α will A be singular?

(b) For all numbers α *not* on your list in part a, we can solve $A\mathbf{x} = \mathbf{b}$ for every vector $\mathbf{b} \in \mathbb{R}^3$. For each of the numbers α on your list, give the vectors $\mathbf{b}$ for which we can solve $A\mathbf{x} = \mathbf{b}$.

16. Prove or give a counterexample:

(a) If $A\mathbf{x} = \mathbf{0}$ has only the trivial solution $\mathbf{x} = \mathbf{0}$, then $A\mathbf{x} = \mathbf{b}$ always has a unique solution.

(b) Prove or give a counterexample: If $A\mathbf{x} = \mathbf{0}$ and $B\mathbf{x} = \mathbf{0}$ have the same solutions, then the set of vectors $\mathbf{b}$ so that $A\mathbf{x} = \mathbf{b}$ is consistent is the same as the set of the vectors $\mathbf{b}$ so that $B\mathbf{x} = \mathbf{b}$ is consistent.

♯17. (a) Suppose A and B are nonsingular $n \times n$ matrices. Prove that AB is nonsingular. (Hint: Solve $(AB)\mathbf{x} = \mathbf{0}$.)

(b) Suppose A and B are $n \times n$ matrices. Prove that if either A or B is singular, then AB is singular.

18. In each case, give positive integers m and n and an example of an $m \times n$ matrix A with the stated property, or explain why none can exist.

*(a) $A\mathbf{x} = \mathbf{b}$ is inconsistent for every $\mathbf{b} \in \mathbb{R}^m$.

*(b) $A\mathbf{x} = \mathbf{b}$ has one solution for every $\mathbf{b} \in \mathbb{R}^m$.

(c) $A\mathbf{x} = \mathbf{b}$ has either zero or one solution for every $\mathbf{b} \in \mathbb{R}^m$.

(d) $A\mathbf{x} = \mathbf{b}$ has infinitely many solutions for every $\mathbf{b} \in \mathbb{R}^m$.

*(e) $A\mathbf{x} = \mathbf{b}$ has infinitely many solutions whenever it is consistent.

(f) There are vectors $\mathbf{b}_1, \mathbf{b}_2, \mathbf{b}_3$ so that $A\mathbf{x} = \mathbf{b}_1$ has no solution, $A\mathbf{x} = \mathbf{b}_2$ has exactly one solution, and $A\mathbf{x} = \mathbf{b}_3$ has infinitely many solutions.

19. ♯(a) Suppose $A \in \mathcal{M}_{m \times n}, B \in \mathcal{M}_{n \times m}$, and $BA = I_n$. Prove that if for some $\mathbf{b} \in \mathbb{R}^m$ the equation $A\mathbf{x} = \mathbf{b}$ has a solution, then that solution is unique.

(b) Suppose $A \in \mathcal{M}_{m \times n}, C \in \mathcal{M}_{n \times m}$, and $AC = I_m$. Prove that the system $A\mathbf{x} = \mathbf{b}$ is consistent for every $\mathbf{b} \in \mathbb{R}^m$.

♯(c) Suppose $A \in \mathcal{M}_{m \times n}$ and $B, C \in \mathcal{M}_{n \times m}$ are matrices that satisfy $BA = I_n$ and $AC = I_m$. Prove that $B = C$.

20. Let A be an $m \times n$ matrix with row vectors $\mathbf{A}_1, \ldots, \mathbf{A}_m \in \mathbb{R}^n$.

(a) Suppose $\mathbf{A}_1 + \cdots + \mathbf{A}_m = \mathbf{0}$. Prove that $\text{rank}(A) < m$. (Hint: Why must there be a row of zeroes in the echelon form of A?)

(b) More generally, suppose there is some nontrivial linear combination $c_1 \mathbf{A}_1 + \cdots + c_m \mathbf{A}_m = \mathbf{0}$. Prove $\text{rank}(A) < m$.

21. Let A be an $m \times n$ matrix with column vectors $\mathbf{a}_1, \ldots, \mathbf{a}_n \in \mathbb{R}^m$.

(a) Suppose $\mathbf{a}_1 + \cdots + \mathbf{a}_n = \mathbf{0}$. Prove that $\text{rank}(A) < n$. (Hint: Consider solutions of $A\mathbf{x} = \mathbf{0}$.)

(b) More generally, suppose there is some nontrivial linear combination $c_1 \mathbf{a}_1 + \cdots + c_n \mathbf{a}_n = \mathbf{0}$. Prove $\text{rank}(A) < n$.

22. Let $P_i = \begin{bmatrix} x_i \\ y_i \end{bmatrix} \in \mathbb{R}^2, i = 1, 2, 3$. Assume $x_1, x_2,$ and x_3 are distinct.

(a) Show that the matrix
$$\begin{bmatrix} 1 & x_1 & x_1^2 \\ 1 & x_2 & x_2^2 \\ 1 & x_3 & x_3^2 \end{bmatrix}$$
is nonsingular.

(b) Show that the system of equations
$$\begin{bmatrix} x_1^2 & x_1 & 1 \\ x_2^2 & x_2 & 1 \\ x_3^2 & x_3 & 1 \end{bmatrix} \begin{bmatrix} a \\ b \\ c \end{bmatrix} = \begin{bmatrix} y_1 \\ y_2 \\ y_3 \end{bmatrix}$$
always has a unique solution. Deduce that if $P_1, P_2,$ and P_3 are not collinear, then they lie on a unique parabola $y = ax^2 + bx + c$.

23. Let $P_i = \begin{bmatrix} x_i \\ y_i \end{bmatrix} \in \mathbb{R}^2, i = 1, 2, 3$. Let
$$A = \begin{bmatrix} x_1 & y_1 & 1 \\ x_2 & y_2 & 1 \\ x_3 & y_3 & 1 \end{bmatrix}.$$

(a) Prove that the three points $P_1, P_2,$ and P_3 are collinear if and only if the equation $A\mathbf{x} = \mathbf{0}$ has a nontrivial solution. (Hint: A general line in $\mathbb{R}^2$ is of the form $ax + by + c = 0$, where a and b are not both 0.)

(b) Prove that if the three given points are not collinear, then there is a unique circle passing through them. (Hint: If you set up a system of linear equations as suggested by the hint for Exercise 7, you should use part a to deduce that the appropriate coefficient matrix is nonsingular.)

2 ELEMENTARY MATRICES AND CALCULATING INVERSE MATRICES

So far we have focused on the interpretation of matrix multiplication in terms of columns, namely, the fact that the jth column of AB is the product of A with the jth column vector of B. But equally à propos is the observation that

> the ith row of AB is the product of the ith row vector of A with B.

Just as multiplying the matrix A by a column vector **x** on the right,

$$\begin{bmatrix} | & | & & | \\ \mathbf{a}_1 & \mathbf{a}_2 & \cdots & \mathbf{a}_n \\ | & | & & | \end{bmatrix} \begin{bmatrix} x_1 \\ x_2 \\ \vdots \\ x_n \end{bmatrix},$$

gives us the linear combination $x_1\mathbf{a}_1 + x_2\mathbf{a}_2 + \cdots + x_n\mathbf{a}_n$ of the columns of A, the reader can easily check that multiplying A on the *left* by the row vector $[x_1 \ x_2 \ \cdots \ x_m]$,

$$\begin{bmatrix} x_1 & x_2 & \cdots & x_m \end{bmatrix} \begin{bmatrix} \text{---} & \mathbf{A}_1 & \text{---} \\ \text{---} & \mathbf{A}_2 & \text{---} \\ & \vdots & \\ \text{---} & \mathbf{A}_m & \text{---} \end{bmatrix},$$

yields the linear combination $x_1\mathbf{A}_1 + x_2\mathbf{A}_2 + \cdots + x_m\mathbf{A}_m$ of the *rows* of A.

It should come as no surprise, then, that we can perform row operations on a matrix A by multiplying on the *left* by appropriately chosen matrices. For example, if

$$A = \begin{bmatrix} 1 & 2 \\ 3 & 4 \\ 5 & 6 \end{bmatrix},$$

$$E_1 = \begin{bmatrix} 0 & 1 & 0 \\ 1 & 0 & 0 \\ 0 & 0 & 1 \end{bmatrix}, \quad E_2 = \begin{bmatrix} 1 & 0 & 0 \\ 0 & 1 & 0 \\ 0 & 0 & 4 \end{bmatrix}, \quad \text{and} \quad E_3 = \begin{bmatrix} 1 & 0 & 0 \\ -2 & 1 & 0 \\ 0 & 0 & 1 \end{bmatrix},$$

then

$$E_1 A = \begin{bmatrix} 3 & 4 \\ 1 & 2 \\ 5 & 6 \end{bmatrix}, \quad E_2 A = \begin{bmatrix} 1 & 2 \\ 3 & 4 \\ 20 & 24 \end{bmatrix}, \quad \text{and} \quad E_3 A = \begin{bmatrix} 1 & 2 \\ 1 & 0 \\ 5 & 6 \end{bmatrix}.$$

Such matrices that give corresponding elementary row operations are called *elementary matrices*. Note that each *elementary matrix* differs from the identity matrix only in a small way. (N.B. Here we establish the custom that blank spaces in a matrix represent 0's.)

i. To interchange rows i and j, we should multiply by an elementary matrix of the form

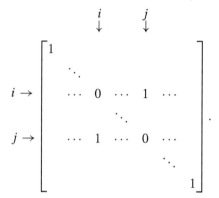

ii. To multiply row i by a scalar c, we should multiply by an elementary matrix of the form

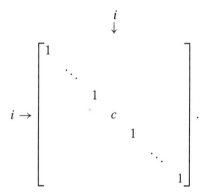

iii. To add c times row i to row j, we should multiply by an elementary matrix of the form

$$\begin{array}{c} i j \\ \downarrow \downarrow \\ \begin{array}{c} \\ \\ i \to \\ \\ j \to \\ \\ \\ \end{array} \left[\begin{array}{ccccccc} 1 & & & & & & \\ & \ddots & & & & & \\ & & 1 & & & & \\ & & & \ddots & & & \\ & & \cdots & c & \cdots & 1 & \\ & & & & & & \ddots \\ & & & & & & & 1 \end{array} \right] \end{array}.$$

Here's an easy way to remember the form of these matrices: Each elementary matrix is obtained by performing the corresponding elementary row operation on the identity matrix.

▶ **EXAMPLE 1**

Let $A = \begin{bmatrix} 4 & 3 & 5 \\ 1 & 2 & 5 \end{bmatrix}$. We put A in reduced echelon form by the following sequence of row operations:

$$\begin{bmatrix} 4 & 3 & 5 \\ 1 & 2 & 5 \end{bmatrix} \rightsquigarrow \begin{bmatrix} 1 & 2 & 5 \\ 4 & 3 & 5 \end{bmatrix} \rightsquigarrow \begin{bmatrix} 1 & 2 & 5 \\ 0 & -5 & -15 \end{bmatrix} \rightsquigarrow \begin{bmatrix} 1 & 2 & 5 \\ 0 & 1 & 3 \end{bmatrix} \rightsquigarrow \begin{bmatrix} 1 & 0 & -1 \\ 0 & 1 & 3 \end{bmatrix}.$$

These steps correspond to multiplying, in sequence from right to left, by the elementary matrices

$$E_1 = \begin{bmatrix} & 1 \\ 1 & \end{bmatrix}, \quad E_2 = \begin{bmatrix} 1 & \\ -4 & 1 \end{bmatrix}, \quad E_3 = \begin{bmatrix} 1 & \\ & -\frac{1}{5} \end{bmatrix}, \quad E_4 = \begin{bmatrix} 1 & -2 \\ & 1 \end{bmatrix};$$

now the reader can check that

$$E = E_4 E_3 E_2 E_1 = \begin{bmatrix} 1 & -2 \\ & 1 \end{bmatrix} \begin{bmatrix} 1 & \\ & -\frac{1}{5} \end{bmatrix} \begin{bmatrix} 1 & \\ -4 & 1 \end{bmatrix} \begin{bmatrix} & 1 \\ 1 & \end{bmatrix} = \begin{bmatrix} \frac{2}{5} & -\frac{3}{5} \\ -\frac{1}{5} & \frac{4}{5} \end{bmatrix}$$

and, indeed,

$$EA = \begin{bmatrix} \frac{2}{5} & -\frac{3}{5} \\ -\frac{1}{5} & \frac{4}{5} \end{bmatrix} \begin{bmatrix} 4 & 3 & 5 \\ 1 & 2 & 5 \end{bmatrix} = \begin{bmatrix} 1 & 0 & -1 \\ 0 & 1 & 3 \end{bmatrix},$$

as it should. ◀

▶ **EXAMPLE 2**

Let's revisit Example 4 on p. 133. Let

$$A = \begin{bmatrix} 1 & 1 & 3 & -1 & 0 \\ -1 & 1 & 1 & 1 & 2 \\ 0 & 1 & 2 & 2 & -1 \\ 2 & -1 & 0 & 1 & -6 \end{bmatrix}.$$

To clear out the entries below the first pivot, we must multiply by the product of the two elementary matrices E_1 and E_2:

$$E_2 E_1 = \begin{bmatrix} 1 & & & \\ & 1 & & \\ & & 1 & \\ -2 & & & 1 \end{bmatrix} \begin{bmatrix} 1 & & & \\ 1 & 1 & & \\ & & 1 & \\ & & & 1 \end{bmatrix} = \begin{bmatrix} 1 & & & \\ 1 & 1 & & \\ & & 1 & \\ -2 & & & 1 \end{bmatrix};$$

to change the pivot in the second row to 1 and then clear out below, we multiply first by

$$E_3 = \begin{bmatrix} 1 & & & \\ & \frac{1}{2} & & \\ & & 1 & \\ & & & 1 \end{bmatrix}$$

and then by the product

$$E_5 E_4 = \begin{bmatrix} 1 & & & \\ & 1 & & \\ & & 1 & \\ & 3 & & 1 \end{bmatrix} \begin{bmatrix} 1 & & & \\ & 1 & & \\ & -1 & 1 & \\ & & & 1 \end{bmatrix} = \begin{bmatrix} 1 & & & \\ & 1 & & \\ & -1 & 1 & \\ & 3 & & 1 \end{bmatrix}.$$

We then change the pivot in the third row to 1 and clear out below, multiplying by

$$E_6 = \begin{bmatrix} 1 & & & \\ & 1 & & \\ & & \frac{1}{2} & \\ & & 1 & \end{bmatrix} \quad \text{and} \quad E_7 = \begin{bmatrix} 1 & & & \\ & 1 & & \\ & & 1 & \\ & & -3 & 1 \end{bmatrix}.$$

Now we clear out above the pivots by multiplying by

$$E_8 = \begin{bmatrix} 1 & & 1 & \\ & 1 & & \\ & & 1 & \\ & & & 1 \end{bmatrix} \quad \text{and} \quad E_9 = \begin{bmatrix} 1 & -1 & & \\ & 1 & & \\ & & 1 & \\ & & & 1 \end{bmatrix}.$$

The net result is this: When we multiply the product

$$E_9 E_8 E_7 E_6 (E_5 E_4) E_3 (E_2 E_1) = \begin{bmatrix} \frac{1}{4} & -\frac{3}{4} & \frac{1}{2} & 0 \\ \frac{1}{2} & \frac{1}{2} & 0 & 0 \\ -\frac{1}{4} & -\frac{1}{4} & \frac{1}{2} & 0 \\ \frac{1}{4} & \frac{9}{4} & -\frac{3}{2} & 1 \end{bmatrix}$$

by the original matrix, we do in fact get the reduced echelon form. ◀

Recall from Section 1 that if we want to find the constraint equations that a vector $\mathbf{b}$ must satisfy in order for $A\mathbf{x} = \mathbf{b}$ to be consistent, we reduce the augmented matrix $[A \mid \mathbf{b}]$ to echelon form $[U \mid \mathbf{c}]$ and set equal to 0 those entries of $\mathbf{c}$ corresponding to the rows of zeroes in U. That is, when A is an $m \times n$ matrix of rank r, the constraint equations are merely the equations $c_{r+1} = \cdots = c_m = 0$. Letting E be the product of the elementary matrices corresponding to the elementary row operations required to put A in echelon form, we have $U = EA$ and so

(†) $$[U \mid \mathbf{c}] = [EA \mid E\mathbf{b}].$$

That is, the constraint equations are the equations

$$\mathbf{E}_{r+1} \cdot \mathbf{b} = 0, \quad \ldots, \quad \mathbf{E}_m \cdot \mathbf{b} = 0.$$

Interestingly, we can use the equation (†) to find a simple way to compute E: When we reduce the augmented matrix $[A \mid \mathbf{b}]$ to echelon form $[U \mid \mathbf{c}]$, E is the matrix so that $E\mathbf{b} = \mathbf{c}$.

► EXAMPLE 3

Taking the matrix A from Example 2, let's find the constraint equations for $A\mathbf{x} = \mathbf{b}$ to be consistent. We start with the augmented matrix

$$[A \mid \mathbf{b}] = \begin{bmatrix} 1 & 1 & 3 & -1 & 0 & b_1 \\ -1 & 1 & 1 & 1 & 2 & b_2 \\ 0 & 1 & 2 & 2 & -1 & b_3 \\ 2 & -1 & 0 & 1 & -6 & b_4 \end{bmatrix}$$

and reduce to echelon form

$$[U \mid \mathbf{c}] = \begin{bmatrix} 1 & 1 & 3 & -1 & 0 & b_1 \\ 0 & 2 & 4 & 0 & 2 & b_1 + b_2 \\ 0 & 0 & 0 & 4 & -4 & -b_1 - b_2 + 2b_3 \\ 0 & 0 & 0 & 0 & 0 & b_1 + 9b_2 - 6b_3 + 4b_4 \end{bmatrix}.$$

Now it is easy to see that if

$$E\mathbf{b} = \begin{bmatrix} b_1 \\ b_1 + b_2 \\ -b_1 - b_2 + 2b_3 \\ b_1 + 9b_2 - 6b_3 + 4b_4 \end{bmatrix}, \quad \text{then} \quad E = \begin{bmatrix} 1 & & & \\ 1 & 1 & & \\ -1 & -1 & 2 & \\ 1 & 9 & -6 & 4 \end{bmatrix}.$$

The reader should check that, in fact, $EA = U$.

We could continue our Gaussian elimination to reach reduced echelon form:

$$[R \mid \mathbf{d}] = \begin{bmatrix} 1 & 0 & 1 & 0 & -2 & \frac{1}{4}b_1 - \frac{3}{4}b_2 + \frac{1}{2}b_3 \\ 0 & 1 & 2 & 0 & 1 & \frac{1}{2}b_1 + \frac{1}{2}b_2 \\ 0 & 0 & 0 & 1 & -1 & -\frac{1}{4}b_1 - \frac{1}{4}b_2 + \frac{1}{2}b_3 \\ 0 & 0 & 0 & 0 & 0 & b_1 + 9b_2 - 6b_3 + 4b_4 \end{bmatrix}.$$

From this we see that $R = E'A$, where

$$E' = \begin{bmatrix} \frac{1}{4} & -\frac{3}{4} & \frac{1}{2} & 0 \\ \frac{1}{2} & \frac{1}{2} & 0 & 0 \\ -\frac{1}{4} & -\frac{1}{4} & \frac{1}{2} & 0 \\ 1 & 9 & -6 & 4 \end{bmatrix},$$

which is very close to—but not the same as—the product of elementary matrices we obtained at the end of Example 2. Can you explain why the first three rows must agree here, but not the last? ◄

We now concentrate on square ($n \times n$) matrices. Recall that the inverse of the $n \times n$ matrix A is the matrix A^{-1} satisfying $AA^{-1} = A^{-1}A = I_n$. It is convenient to have an inverse matrix if we wish to solve the system $A\mathbf{x} = \mathbf{b}$ for numerous vectors $\mathbf{b}$. If A is invertible, we can solve as follows[3]:

---

[3] We will write the "implies" symbol "⟹" vertically so that we can indicate the reasoning in each step.

$$A\mathbf{x} = \mathbf{b}$$
$$\Downarrow \text{ multiplying both sides of the equation by } A^{-1} \text{ on the left}$$
$$A^{-1}(A\mathbf{x}) = A^{-1}\mathbf{b}$$
$$\Downarrow \text{ using the associative property}$$
$$(A^{-1}A)\mathbf{x} = A^{-1}\mathbf{b}$$
$$\Downarrow \text{ using the definition of } A^{-1}$$
$$\mathbf{x} = I_n\mathbf{x} = A^{-1}\mathbf{b}.$$

We aren't done! We've shown that *if* $\mathbf{x}$ is a solution, *then* it must satisfy $\mathbf{x} = A^{-1}\mathbf{b}$. That is, we've shown that the vector $A^{-1}\mathbf{b}$ is a candidate for a solution. But now we check that it truly is a solution by straightforward calculation:

$$A\mathbf{x} = A(A^{-1}\mathbf{b}) = (AA^{-1})\mathbf{b} = I_n\mathbf{b} = \mathbf{b},$$

as required; but note that we have used *both* pieces of the definition of the inverse matrix to prove that the system has a unique solution (which we "discovered" along the way).

It is a consequence of this computation that if A is an invertible $n \times n$ matrix, then $A\mathbf{x} = \mathbf{c}$ has a unique solution for every $\mathbf{c} \in \mathbb{R}^n$, and so it follows from Proposition 1.6 that A must be nonsingular. What about the converse? If A is nonsingular, must A be invertible? Well, if A is nonsingular, we know that every equation $A\mathbf{x} = \mathbf{c}$ has a unique solution. In particular, for $j = 1, \ldots, n$, there is a unique vector $\mathbf{b}_j$ that solves $A\mathbf{b}_j = \mathbf{e}_j$, the jth standard basis vector. If we let B be the $n \times n$ matrix whose column vectors are $\mathbf{b}_1, \ldots, \mathbf{b}_n$, then we have

$$AB = A \begin{bmatrix} | & | & & | \\ \mathbf{b}_1 & \mathbf{b}_2 & \cdots & \mathbf{b}_n \\ | & | & & | \end{bmatrix} = \begin{bmatrix} | & | & & | \\ \mathbf{e}_1 & \mathbf{e}_2 & \cdots & \mathbf{e}_n \\ | & | & & | \end{bmatrix} = I_n.$$

This suggests that the matrix we've constructed should be the inverse matrix of A. But we need to know that $BA = I_n$ as well. Here is a very elegant way to understand why this is so. We can find the matrix B by forming the giant augmented matrix

$$\left[\begin{array}{c|ccc} & | & & | \\ A & \mathbf{e}_1 & \cdots & \mathbf{e}_n \\ & | & & | \end{array} \right] = \left[\begin{array}{c|c} A & I_n \end{array} \right]$$

and using Gaussian elimination to obtain the reduced echelon form

$$\left[\begin{array}{c|c} I_n & B \end{array} \right].$$

(Note that the reduced echelon form of A must be I_n because A is nonsingular.) But this tells us that if E is the product of the elementary matrices required to put A in reduced echelon form, then we have

$$E\,[A \mid I] = [I \mid B],$$

and so $B = E$ and $BA = I_n$, which is what we needed to check. In conclusion, we have proved the following

Theorem 2.1 *An $n \times n$ matrix is nonsingular if and only if it is invertible.*

Note that Gaussian elimination will also let us know when A is not invertible: If we come to a row of zeroes while reducing A to echelon form, then, of course, A is singular and so it cannot be invertible. The following observation is often very useful.

Corollary 2.2 *If A and B are $n \times n$ matrices satisfying $BA = I_n$, then $B = A^{-1}$ and $A = B^{-1}$.*

Proof By Exercise 4.1.19a, the equation $A\mathbf{x} = \mathbf{0}$ has only the trivial solution. Hence, by Proposition 1.6, A is nonsingular; according to Theorem 2.1, A is therefore invertible. Since A has an inverse matrix, A^{-1}, we deduce that

$$BA = I_n$$

⇓ multiplying both sides of the equation by A^{-1} on the right

$$(BA)A^{-1} = I_n A^{-1}$$

⇓ using the associative property

$$B(AA^{-1}) = A^{-1}$$

⇓ using the definition of A^{-1}

$$B = A^{-1},$$

as desired. Since $AB = I_n$ and $BA = I_n$, it now follows that $A = B^{-1}$, as well. ∎

▶ **EXAMPLE 4**

We wish to determine the inverse of the matrix

$$A = \begin{bmatrix} 1 & -1 & 1 \\ 2 & -1 & 0 \\ 1 & -2 & 2 \end{bmatrix}$$

(if it exists). We apply Gaussian elimination to the augmented matrix:

$$\left[\begin{array}{ccc|ccc} 1 & -1 & 1 & 1 & 0 & 0 \\ 2 & -1 & 0 & 0 & 1 & 0 \\ 1 & -2 & 2 & 0 & 0 & 1 \end{array}\right] \rightsquigarrow \left[\begin{array}{ccc|ccc} 1 & -1 & 1 & 1 & 0 & 0 \\ 0 & 1 & -2 & -2 & 1 & 0 \\ 0 & -1 & 1 & -1 & 0 & 1 \end{array}\right]$$

$$\rightsquigarrow \left[\begin{array}{ccc|ccc} 1 & -1 & 1 & 1 & 0 & 0 \\ 0 & 1 & -2 & -2 & 1 & 0 \\ 0 & 0 & -1 & -3 & 1 & 1 \end{array}\right] \rightsquigarrow \left[\begin{array}{ccc|ccc} 1 & -1 & 1 & 1 & 0 & 0 \\ 0 & 1 & -2 & -2 & 1 & 0 \\ 0 & 0 & 1 & 3 & -1 & -1 \end{array}\right]$$

$$\rightsquigarrow \begin{bmatrix} 1 & -1 & 0 & | & -2 & 1 & 1 \\ 0 & 1 & 0 & | & 4 & -1 & -2 \\ 0 & 0 & 1 & | & 3 & -1 & -1 \end{bmatrix} \rightsquigarrow \begin{bmatrix} 1 & 0 & 0 & | & 2 & 0 & -1 \\ 0 & 1 & 0 & | & 4 & -1 & -2 \\ 0 & 0 & 1 & | & 3 & -1 & -1 \end{bmatrix}.$$

It follows that

$$A^{-1} = \begin{bmatrix} 2 & 0 & -1 \\ 4 & -1 & -2 \\ 3 & -1 & -1 \end{bmatrix}.$$

(The reader should check our arithmetic by multiplying AA^{-1} or $A^{-1}A$.) ◀

▶ **EXAMPLE 5**

It is convenient to derive the formula for the inverse of a general 2×2 matrix first given in Example 9 of Chapter 1, Section 4. Let

$$A = \begin{bmatrix} a & b \\ c & d \end{bmatrix}.$$

We assume $a \neq 0$ to start with.

$$\begin{bmatrix} a & b & | & 1 & 0 \\ c & d & | & 0 & 1 \end{bmatrix} \rightsquigarrow \begin{bmatrix} 1 & \frac{b}{a} & | & \frac{1}{a} & 0 \\ c & d & | & 0 & 1 \end{bmatrix} \rightsquigarrow \begin{bmatrix} 1 & \frac{b}{a} & | & \frac{1}{a} & 0 \\ 0 & d - \frac{bc}{a} & | & -\frac{c}{a} & 1 \end{bmatrix} \quad \text{(assuming } ad - bc \neq 0\text{)}$$

$$\rightsquigarrow \begin{bmatrix} 1 & \frac{b}{a} & | & \frac{1}{a} & 0 \\ 0 & 1 & | & -\frac{c}{ad-bc} & \frac{a}{ad-bc} \end{bmatrix} \rightsquigarrow \begin{bmatrix} 1 & 0 & | & \frac{1}{a} - \frac{b}{a}(-\frac{c}{ad-bc}) & -\frac{b}{a}\frac{a}{ad-bc} \\ 0 & 1 & | & -\frac{c}{ad-bc} & \frac{a}{ad-bc} \end{bmatrix}$$

$$= \begin{bmatrix} 1 & 0 & | & \frac{d}{ad-bc} & -\frac{b}{ad-bc} \\ 0 & 1 & | & -\frac{c}{ad-bc} & \frac{a}{ad-bc} \end{bmatrix},$$

and so we see that, provided $ad - bc \neq 0$,

$$A^{-1} = \frac{1}{ad-bc}\begin{bmatrix} d & -b \\ -c & a \end{bmatrix}.$$

As a check, we have

$$\begin{bmatrix} a & b \\ c & d \end{bmatrix}\frac{1}{ad-bc}\begin{bmatrix} d & -b \\ -c & a \end{bmatrix} = I_2 = \frac{1}{ad-bc}\begin{bmatrix} d & -b \\ -c & a \end{bmatrix}\begin{bmatrix} a & b \\ c & d \end{bmatrix}.$$

Of course, we have derived this by assuming $a \neq 0$, but the reader can check easily that the formula works fine even when $a = 0$. We do see, however, from the row reduction that

$$\boxed{\begin{bmatrix} a & b \\ c & d \end{bmatrix} \text{ is nonsingular} \iff ad - bc \neq 0.}$$ ◀

We have shown in the course of proving Theorem 2.1 that when A is square, any B that satisfies $AB = I$ (a so-called *right inverse* of A) must also satisfy $BA = I$ (and thus is a *left inverse* of A). Likewise, we have established in Corollary 2.2 that when A is square, any left inverse of A is a bona fide inverse of A. Indeed, it will never happen that a nonsquare matrix has *both* a left and a right inverse (see Exercise 9).

2 Elementary Matrices and Calculating Inverse Matrices

Remark Even when A is square, the left and right inverses have rather different interpretations. As we saw in the proof of Theorem 2.1, the columns of the right inverse arise as the solutions of $A\mathbf{x} = \mathbf{e}_j$. On the other hand, the left inverse of A is the product of the elementary matrices by which we reduce A to its reduced echelon form, I. (See Exercise 8.)

▶ EXERCISES 4.2

*1. For each of the matrices A in Exercise 4.1.3, find a product of elementary matrices $E = \cdots E_2 E_1$ so that EA is the reduced echelon form of A. Use the matrix E you've found to give constraint equations for $A\mathbf{x} = \mathbf{b}$ to be consistent.

2. Use Gaussian elimination to find A^{-1} (if it exists):

(a) $A = \begin{bmatrix} 1 & 2 \\ -1 & 3 \end{bmatrix}$

(b) $A = \begin{bmatrix} 1 & 2 & 3 \\ 1 & 1 & 2 \\ 0 & 1 & 2 \end{bmatrix}$

*(c) $A = \begin{bmatrix} 1 & 0 & 1 \\ 0 & 2 & 1 \\ -1 & 3 & 1 \end{bmatrix}$

(d) $A = \begin{bmatrix} 1 & 2 & 3 \\ 4 & 5 & 6 \\ 7 & 8 & 9 \end{bmatrix}$

*(e) $A = \begin{bmatrix} 2 & 3 & 4 \\ 2 & 1 & 1 \\ -1 & 1 & 2 \end{bmatrix}$

3. In each case, given A and $\mathbf{b}$,

    i. Find A^{-1}.

    ii. Use your answer to (i) to solve $A\mathbf{x} = \mathbf{b}$.

    iii. Use your answer to (ii) to express $\mathbf{b}$ as a linear combination of the columns of A.

(a) $A = \begin{bmatrix} 2 & 3 \\ 3 & 5 \end{bmatrix}, \mathbf{b} = \begin{bmatrix} 3 \\ 4 \end{bmatrix}$

*(b) $A = \begin{bmatrix} 1 & 1 & 1 \\ 0 & 2 & 3 \\ 3 & 2 & 2 \end{bmatrix}, \mathbf{b} = \begin{bmatrix} 1 \\ 1 \\ 2 \end{bmatrix}$

(c) $A = \begin{bmatrix} 1 & 1 & 1 \\ 0 & 1 & 1 \\ 1 & 2 & 1 \end{bmatrix}, \mathbf{b} = \begin{bmatrix} 3 \\ 0 \\ 1 \end{bmatrix}$

*(d) $A = \begin{bmatrix} 1 & 1 & 1 & 1 \\ 0 & 1 & 1 & 1 \\ 0 & 0 & 1 & 3 \\ 0 & 0 & 1 & 4 \end{bmatrix}, \mathbf{b} = \begin{bmatrix} 2 \\ 0 \\ 1 \\ 1 \end{bmatrix}$

4. (a) Find two different right inverses of the matrix $A = \begin{bmatrix} 1 & -1 & 1 \\ 2 & -1 & 0 \end{bmatrix}$.

(b) Give a nonzero matrix that has no right inverse.

(c) Find two left inverses of the matrix $A = \begin{bmatrix} 1 & 2 \\ 0 & -1 \\ 1 & 1 \end{bmatrix}$.

(d) Give a nonzero matrix that has no left inverse.

5. Prove that the inverse of every elementary matrix is again an elementary matrix. Indeed, give a simple prescription for determining the inverse of each type of elementary matrix.

6. Using Theorem 2.1 and Proposition 4.3 of Chapter 1, prove that if AB and B are nonsingular, then A is nonsingular. (See Exercise 4.1.17.)

♯7. Suppose A is an invertible $m \times m$ matrix and B is an invertible $n \times n$ matrix.
(a) Prove that the matrix
$$\left[\begin{array}{c|c} A & O \\ \hline O & B \end{array}\right]$$
is invertible and give a formula for its inverse.
(b) Suppose C is an arbitrary $m \times n$ matrix. Is the matrix
$$\left[\begin{array}{c|c} A & C \\ \hline O & B \end{array}\right]$$
invertible?
(See Exercise 1.4.12 for the notion of block multiplication.)

8. Complete the following alternative argument that the matrix obtained by Gaussian elimination must be the inverse matrix of A. Suppose A is nonsingular.
(a) Show there are finitely many elementary matrices $E_1, E_2, \ldots, E_k$ so that $E_k E_{k-1} \cdots E_2 E_1 A = I$.
(b) Let $B = E_k E_{k-1} \cdots E_2 E_1$. Prove that $AB = I$. (Hint: Use Proposition 4.3 of Chapter 1.)

9. Let A be an $m \times n$ matrix. Recall that the $n \times m$ matrix B is a left inverse of A if $BA = I_n$ and a right inverse if $AB = I_m$.
(a) Show that A has a right inverse if and only if we can solve $A\mathbf{x} = \mathbf{b}$ for every $\mathbf{b} \in \mathbb{R}^m$ if and only if $\text{rank}(A) = m$.
(b) Show that A has a left inverse if and only if $A\mathbf{x} = \mathbf{0}$ has the unique solution $\mathbf{x} = \mathbf{0}$ if and only if $\text{rank}(A) = n$. (Hint for $\Longleftarrow$: If $\text{rank}(A) = n$, what is the reduced echelon form of A?)
(c) Show that A has both a left inverse and a right inverse if and only if A is invertible if and only if $m = n = \text{rank}(A)$.

▶ 3 LINEAR INDEPENDENCE, BASIS, AND DIMENSION

Given vectors $\mathbf{v}_1, \ldots, \mathbf{v}_k \in \mathbb{R}^n$ and $\mathbf{v} \in \mathbb{R}^n$, it is natural to ask whether $\mathbf{v} \in \text{Span}(\mathbf{v}_1, \ldots, \mathbf{v}_k)$. That is, do there *exist* scalars $c_1, \ldots, c_k$ so that $\mathbf{v} = c_1\mathbf{v}_1 + c_2\mathbf{v}_2 + \cdots + c_k\mathbf{v}_k$? This is in turn a question of whether a certain (inhomogeneous) system of linear equations has a solution. As we saw in Section 1, one is often interested in the allied question: Is that solution *unique*?

▶ EXAMPLE 1

Let
$$\mathbf{v}_1 = \begin{bmatrix} 1 \\ 1 \\ 2 \end{bmatrix}, \quad \mathbf{v}_2 = \begin{bmatrix} 1 \\ -1 \\ 0 \end{bmatrix}, \quad \mathbf{v}_3 = \begin{bmatrix} 1 \\ 0 \\ 1 \end{bmatrix}, \quad \text{and} \quad \mathbf{v} = \begin{bmatrix} 1 \\ 1 \\ 0 \end{bmatrix}.$$

We ask first of all whether $\mathbf{v} \in \text{Span}(\mathbf{v}_1, \mathbf{v}_2, \mathbf{v}_3)$. This is a familiar question when we recast it in matrix notation: Let
$$A = \begin{bmatrix} 1 & 1 & 1 \\ 1 & -1 & 0 \\ 2 & 0 & 1 \end{bmatrix} \quad \text{and} \quad \mathbf{b} = \begin{bmatrix} 1 \\ 1 \\ 0 \end{bmatrix}.$$

Is the system $A\mathbf{x} = \mathbf{b}$ consistent? Immediately we write down the appropriate augmented matrix and reduce to echelon form:

$$\begin{bmatrix} 1 & 1 & 1 & | & 1 \\ 1 & -1 & 0 & | & 1 \\ 2 & 0 & 1 & | & 0 \end{bmatrix} \rightsquigarrow \begin{bmatrix} 1 & 1 & 1 & | & 1 \\ 0 & 2 & 1 & | & 0 \\ 0 & 0 & 0 & | & -2 \end{bmatrix},$$

so the system is obviously inconsistent. The answer: No, $\mathbf{v}$ is not in $\operatorname{Span}(\mathbf{v}_1, \mathbf{v}_2, \mathbf{v}_3)$.

What about

$$\mathbf{w} = \begin{bmatrix} 2 \\ 3 \\ 5 \end{bmatrix}?$$

As the reader can easily check, $\mathbf{w} = 3\mathbf{v}_1 - \mathbf{v}_3$, so $\mathbf{w} \in \operatorname{Span}(\mathbf{v}_1, \mathbf{v}_2, \mathbf{v}_3)$. What's more, $\mathbf{w} = 2\mathbf{v}_1 - \mathbf{v}_2 + \mathbf{v}_3$, as well. So, obviously, there is no unique expression for $\mathbf{w}$ as a linear combination of $\mathbf{v}_1$, $\mathbf{v}_2$, and $\mathbf{v}_3$. But we can conclude more: Setting the two expressions for $\mathbf{w}$ equal, we obtain

$$3\mathbf{v}_1 - \mathbf{v}_3 = 2\mathbf{v}_1 - \mathbf{v}_2 + \mathbf{v}_3, \quad \text{i.e.,} \quad \mathbf{v}_1 + \mathbf{v}_2 - 2\mathbf{v}_3 = \mathbf{0}.$$

That is, there is a nontrivial relation among the vectors $\mathbf{v}_1$, $\mathbf{v}_2$, and $\mathbf{v}_3$, and this is the reason we have different ways of expressing $\mathbf{w}$ as a linear combination of the three of them. Indeed, since $\mathbf{v}_1 = -\mathbf{v}_2 + 2\mathbf{v}_3$, we can see easily that any linear combination of $\mathbf{v}_1$, $\mathbf{v}_2$, and $\mathbf{v}_3$ is a linear combination just of $\mathbf{v}_2$ and $\mathbf{v}_3$:

$$c_1\mathbf{v}_1 + c_2\mathbf{v}_2 + c_3\mathbf{v}_3 = c_1(-\mathbf{v}_2 + 2\mathbf{v}_3) + c_2\mathbf{v}_2 + c_3\mathbf{v}_3 = (c_2 - c_1)\mathbf{v}_2 + (c_3 + 2c_1)\mathbf{v}_3.$$

The vector $\mathbf{v}_1$ was redundant because

$$\operatorname{Span}(\mathbf{v}_1, \mathbf{v}_2, \mathbf{v}_3) = \operatorname{Span}(\mathbf{v}_2, \mathbf{v}_3).$$

We might surmise that the vector $\mathbf{w}$ can now be written *uniquely* as a linear combination of $\mathbf{v}_2$ and $\mathbf{v}_3$, and this is easy to check:

$$[A' | \mathbf{w}] = \begin{bmatrix} 1 & 1 & | & 2 \\ -1 & 0 & | & 3 \\ 0 & 1 & | & 5 \end{bmatrix} \rightsquigarrow \begin{bmatrix} 1 & 1 & | & 2 \\ 0 & 1 & | & 5 \\ 0 & 0 & | & 0 \end{bmatrix},$$

and from the fact that the matrix A' has rank 2 we infer that the system of equations has a unique solution. ◂

Remark In the language of functions, if A is the standard matrix of a linear map $T: \mathbb{R}^n \to \mathbb{R}^m$, we are interested in the *image* of T (i.e., the set of $\mathbf{w} \in \mathbb{R}^m$ so that $\mathbf{w} = T(\mathbf{v})$ for some $\mathbf{v} \in \mathbb{R}^n$) and the issue of whether T is one-to-one (i.e., given $\mathbf{w}$ in the image, is there exactly *one* $\mathbf{v} \in \mathbb{R}^n$ so that $T(\mathbf{v}) = \mathbf{w}$?).

Generalizing the preceding example, we now recast Proposition 1.5:

Proposition 3.1 *Let $\mathbf{v}_1, \ldots, \mathbf{v}_k \in \mathbb{R}^n$ and let $V = \operatorname{Span}(\mathbf{v}_1, \ldots, \mathbf{v}_k)$. An arbitrary vector $\mathbf{v} \in \operatorname{Span}(\mathbf{v}_1, \ldots, \mathbf{v}_k)$ has a unique expression as a linear combination of $\mathbf{v}_1, \ldots, \mathbf{v}_k$ if and only if the zero vector has a unique expression as a linear combination of $\mathbf{v}_1, \ldots, \mathbf{v}_k$; i.e.,*

$$c_1\mathbf{v}_1 + c_2\mathbf{v}_2 + \cdots + c_k\mathbf{v}_k = \mathbf{0} \implies c_1 = c_2 = \cdots = c_k = 0.$$

Proof Suppose for some $\mathbf{v} \in V$ there are two different expressions
$$\mathbf{v} = c_1 \mathbf{v}_1 + c_2 \mathbf{v}_2 + \cdots + c_k \mathbf{v}_k \quad \text{and}$$
$$\mathbf{v} = d_1 \mathbf{v}_1 + d_2 \mathbf{v}_2 + \cdots + d_k \mathbf{v}_k.$$

Then, subtracting, we obtain
$$\mathbf{0} = (c_1 - d_1)\mathbf{v}_1 + \cdots + (c_k - d_k)\mathbf{v}_k,$$
and so the zero vector has a nontrivial representation as a linear combination of $\mathbf{v}_1, \ldots, \mathbf{v}_k$ (by which we mean that not all the coefficients are 0).

Conversely, suppose there is a nontrivial linear combination
$$\mathbf{0} = s_1 \mathbf{v}_1 + \cdots + s_k \mathbf{v}_k.$$

Then, given any vector $\mathbf{v} \in V$, we can express $\mathbf{v}$ as a linear combination of $\mathbf{v}_1, \ldots, \mathbf{v}_k$ in several ways: for instance, adding
$$\mathbf{v} = c_1 \mathbf{v}_1 + c_2 \mathbf{v}_2 + \cdots + c_k \mathbf{v}_k \quad \text{and}$$
$$\mathbf{0} = s_1 \mathbf{v}_1 + s_2 \mathbf{v}_2 + \cdots + s_k \mathbf{v}_k,$$
we obtain another formula for $\mathbf{v}$, namely,
$$\mathbf{v} = (c_1 + s_1)\mathbf{v}_1 + \cdots + (c_k + s_k)\mathbf{v}_k.$$

This completes the proof. ∎

This discussion leads us to make the following

Definition The (indexed) set of vectors $\{\mathbf{v}_1, \ldots, \mathbf{v}_k\}$ is called *linearly independent* if
$$c_1 \mathbf{v}_1 + c_2 \mathbf{v}_2 + \cdots + c_k \mathbf{v}_k = \mathbf{0} \implies c_1 = c_2 = \cdots = c_k = 0,$$
i.e., if the *only* way of expressing the zero vector as a linear combination of $\mathbf{v}_1, \ldots, \mathbf{v}_k$ is the *trivial* linear combination $0\mathbf{v}_1 + \cdots + 0\mathbf{v}_k$.

The set of vectors $\{\mathbf{v}_1, \ldots, \mathbf{v}_k\}$ is called *linearly dependent* if it is not linearly independent, i.e., if there is some expression
$$c_1 \mathbf{v}_1 + c_2 \mathbf{v}_2 + \cdots + c_k \mathbf{v}_k = \mathbf{0}, \quad \text{where not all the } c_i\text{'s are 0.}$$

Remark The language is problematic here. Many mathematicians—often including the author of this text—tend to say things like "the vectors $\mathbf{v}_1, \ldots, \mathbf{v}_k$ are linearly independent." But linear independence (or dependence) is a property of the whole *collection* of vectors, not of the individual vectors. What's worse, we really should refer to an *ordered list* of vectors rather than to a set of vectors: For example, any list in which some vector, $\mathbf{v}$, appears twice is obviously giving a linearly dependent collection; but the set $\{\mathbf{v}, \mathbf{v}\}$ is indistinguishable from the set $\{\mathbf{v}\}$. There seems to be no ideal route out of this morass. Having said all this, we warn the gentle reader that we may occasionally say "the vectors $\mathbf{v}_1, \ldots, \mathbf{v}_k$ are linearly (in)dependent" where it would be too clumsy to be more pedantic. Just stay alert!

3 Linear Independence, Basis, and Dimension

Remark Here is a piece of advice: It is virtually always the case that when you are presented with a set of vectors $\{\mathbf{v}_1, \ldots, \mathbf{v}_k\}$ that you are to prove linearly independent, you should write

"Suppose $c_1\mathbf{v}_1 + c_2\mathbf{v}_2 + \cdots + c_k\mathbf{v}_k = \mathbf{0}$. I must show that $c_1 = \cdots = c_k = 0$."

You then use whatever hypotheses you're given to arrive at that conclusion.

► EXAMPLE 2

We wish to decide whether the vectors

$$\mathbf{v}_1 = \begin{bmatrix} 1 \\ 0 \\ 1 \\ 2 \end{bmatrix}, \quad \mathbf{v}_2 = \begin{bmatrix} 2 \\ 1 \\ 1 \\ 1 \end{bmatrix}, \quad \text{and} \quad \mathbf{v}_3 = \begin{bmatrix} 1 \\ 1 \\ 0 \\ -1 \end{bmatrix} \in \mathbb{R}^4$$

form a linearly independent set. Suppose $c_1\mathbf{v}_1 + c_2\mathbf{v}_2 + c_3\mathbf{v}_3 = \mathbf{0}$; i.e.,

$$c_1 \begin{bmatrix} 1 \\ 0 \\ 1 \\ 2 \end{bmatrix} + c_2 \begin{bmatrix} 2 \\ 1 \\ 1 \\ 1 \end{bmatrix} + c_3 \begin{bmatrix} 1 \\ 1 \\ 0 \\ -1 \end{bmatrix} = \mathbf{0}.$$

Can we conclude that $c_1 = c_2 = c_3 = 0$? We recognize this as a homogeneous system of linear equations:

$$\begin{bmatrix} 1 & 2 & 1 \\ 0 & 1 & 1 \\ 1 & 1 & 0 \\ 2 & 1 & -1 \end{bmatrix} \begin{bmatrix} c_1 \\ c_2 \\ c_3 \end{bmatrix} = \mathbf{0}.$$

By now we are old hands at solving such systems. We find that the echelon form of the coefficient matrix is

$$\begin{bmatrix} 1 & 2 & 1 \\ 0 & 1 & 1 \\ 0 & 0 & 0 \\ 0 & 0 & 0 \end{bmatrix},$$

and so our system of equations in fact has infinitely many solutions. For example, we can take $c_1 = 1$, $c_2 = -1$, and $c_3 = 1$. The vectors therefore form a linearly dependent set. ◄

► EXAMPLE 3

Suppose $\mathbf{u}, \mathbf{v}, \mathbf{w} \in \mathbb{R}^n$. If $\{\mathbf{u}, \mathbf{v}, \mathbf{w}\}$ is linearly independent, then we wish to show next that $\{\mathbf{u} + \mathbf{v}, \mathbf{v} + \mathbf{w}, \mathbf{u} + \mathbf{w}\}$ is likewise linearly independent. Suppose

$$c_1(\mathbf{u} + \mathbf{v}) + c_2(\mathbf{v} + \mathbf{w}) + c_3(\mathbf{u} + \mathbf{w}) = \mathbf{0}.$$

We must show that $c_1 = c_2 = c_3 = 0$. We use the distributive property to rewrite our equation as

$$(c_1 + c_3)\mathbf{u} + (c_1 + c_2)\mathbf{v} + (c_2 + c_3)\mathbf{w} = \mathbf{0}.$$

Since $\{u, v, w\}$ is linearly independent, we may infer that

$$c_1 + c_3 = 0$$
$$c_1 + c_2 = 0$$
$$ c_2 + c_3 = 0,$$

and we leave it to the reader to check that the only solution of this system of equations is, in fact, $c_1 = c_2 = c_3 = 0$, as desired. ◀

▶ **EXAMPLE 4**

Any time one has a list of vectors $v_1, \ldots, v_k$ in which one of the vectors is the zero vector, say $v_1 = \mathbf{0}$, then the set of vectors must be linearly dependent because the equation

$$1v_1 = \mathbf{0}$$

is a nontrivial linear combination of the vectors yielding the zero vector. ◀

▶ **EXAMPLE 5**

How can two nonzero vectors u and v give rise to a linearly dependent set? By definition, this means that there is a linear combination

$$au + bv = \mathbf{0},$$

where either $a \neq 0$ or $b \neq 0$. Suppose $a \neq 0$. Then we may write $u = -\frac{b}{a}v$, so u is a scalar multiple of v. (Similarly, you may show that if $b \neq 0$, v must be a scalar multiple of u.) So two linearly dependent vectors are parallel (and vice versa).

How can a collection of three nonzero vectors be linearly dependent? As before, there must be a linear combination

$$au + bv + cw = \mathbf{0},$$

where (at least) one of a, b, and c is nonzero. Say $a \neq 0$. This means that we can solve:

$$u = -\frac{1}{a}(bv + cw) = (-\frac{b}{a})v + (-\frac{c}{a})w,$$

so $u \in \text{Span}(v, w)$. In particular, $\text{Span}(u, v, w)$ is either a line (if all three vectors u, v, w are parallel) or a plane. ◀

The appropriate generalization of the last example is the following useful

Proposition 3.2 *Suppose $v_1, \ldots, v_k \in \mathbb{R}^n$ form a linearly independent set, and suppose $x \in \mathbb{R}^n$. Then $\{v_1, \ldots, v_k, x\}$ is linearly independent if and only if $x \notin \text{Span}(v_1, \ldots, v_k)$.*

Proof Although Figure 3.1 suggests the result is quite plausible, we will prove the contrapositive:

$\{v_1, \ldots, v_k, x\}$ is linearly dependent if and only if $x \in \text{Span}(v_1, \ldots, v_k)$.

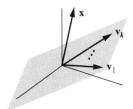

Figure 3.1

Suppose $\mathbf{x} \in \text{Span}(\mathbf{v}_1, \ldots, \mathbf{v}_k)$. Then $\mathbf{x} = c_1\mathbf{v}_1 + c_2\mathbf{v}_2 + \cdots + c_k\mathbf{v}_k$ for some scalars $c_1, \ldots, c_k$, so

$$c_1\mathbf{v}_1 + c_2\mathbf{v}_2 + \cdots + c_k\mathbf{v}_k + (-1)\mathbf{x} = \mathbf{0},$$

from which we conclude that $\{\mathbf{v}_1, \ldots, \mathbf{v}_k, \mathbf{x}\}$ is linearly dependent (since at least one of the coefficients is nonzero).

Now suppose $\{\mathbf{v}_1, \ldots, \mathbf{v}_k, \mathbf{x}\}$ is linearly dependent. This means that there are scalars $c_1, \ldots, c_k$, and c, not all 0, so that

$$c_1\mathbf{v}_1 + c_2\mathbf{v}_2 + \cdots + c_k\mathbf{v}_k + c\mathbf{x} = \mathbf{0}.$$

Note that we cannot have $c = 0$, for if c were 0, we'd have $c_1\mathbf{v}_1 + c_2\mathbf{v}_2 + \cdots + c_k\mathbf{v}_k = \mathbf{0}$, and linear independence of $\{\mathbf{v}_1, \ldots, \mathbf{v}_k\}$ implies $c_1 = \cdots = c_k = 0$, which contradicts our assumption that $\{\mathbf{v}_1, \ldots, \mathbf{v}_k, \mathbf{x}\}$ is linearly dependent. Therefore, $c \neq 0$, and so

$$\mathbf{x} = -\frac{1}{c}(c_1\mathbf{v}_1 + c_2\mathbf{v}_2 + \cdots + c_k\mathbf{v}_k) = (-\frac{c_1}{c})\mathbf{v}_1 + (-\frac{c_2}{c})\mathbf{v}_2 + \cdots + (-\frac{c_k}{c})\mathbf{v}_k,$$

which tells us that $\mathbf{x} \in \text{Span}(\mathbf{v}_1, \ldots, \mathbf{v}_k)$, as required. ∎

Proposition 3.2 has the following consequence: If $\{\mathbf{v}_1, \ldots, \mathbf{v}_k\}$ is linearly independent, then

$$\text{Span}(\mathbf{v}_1) \subsetneq \text{Span}(\mathbf{v}_1, \mathbf{v}_2) \subsetneq \cdots \subsetneq \text{Span}(\mathbf{v}_1, \ldots, \mathbf{v}_k).$$

That is, with each additional vector, the subspace spanned gets larger. We now formalize the notion of "size" of a subspace. But we now understand that when we have a set of linearly independent vectors, no proper subset will yield the same span. In other words, we will have an "efficient" set of spanning vectors (i.e., there is no redundancy in the vectors we've chosen: No proper subset will do). This motivates the following

Definition Let $V \subset \mathbb{R}^n$ be a subspace. The set of vectors $\{\mathbf{v}_1, \ldots, \mathbf{v}_k\}$ is called a *basis* for V if

i. $\mathbf{v}_1, \ldots, \mathbf{v}_k$ span V; i.e., $V = \text{Span}(\mathbf{v}_1, \ldots, \mathbf{v}_k)$, and

ii. $\{\mathbf{v}_1, \ldots, \mathbf{v}_k\}$ is linearly independent.

We comment that the plural of *basis* is *bases*.

EXAMPLE 6

Recall that the vectors

$$\mathbf{e}_1 = \begin{bmatrix} 1 \\ 0 \\ \vdots \\ 0 \\ 0 \end{bmatrix}, \quad \mathbf{e}_2 = \begin{bmatrix} 0 \\ 1 \\ \vdots \\ 0 \\ 0 \end{bmatrix}, \quad \ldots, \quad \mathbf{e}_n = \begin{bmatrix} 0 \\ 0 \\ \vdots \\ 0 \\ 1 \end{bmatrix} \in \mathbb{R}^n$$

are called the *standard basis vectors* for $\mathbb{R}^n$. To check that they make up a basis, we must establish that properties (i) and (ii) above hold for $V = \mathbb{R}^n$. The first is obvious: If $\mathbf{x} \in \mathbb{R}^n$, then $\mathbf{x} = x_1 \mathbf{e}_1 + x_2 \mathbf{e}_2 + \cdots + x_n \mathbf{e}_n$. The second is not much harder. Suppose $c_1 \mathbf{e}_1 + c_2 \mathbf{e}_2 + \cdots + c_n \mathbf{e}_n = \mathbf{0}$. Then this means that

$$\mathbf{c} = \begin{bmatrix} c_1 \\ c_2 \\ \vdots \\ c_n \end{bmatrix} = \begin{bmatrix} 0 \\ 0 \\ \vdots \\ 0 \end{bmatrix},$$

and so $c_1 = c_2 = \cdots = c_n = 0$. ◀

EXAMPLE 7

Consider the plane given by $V = \{\mathbf{x} \in \mathbb{R}^3 : x_1 - x_2 + 2x_3 = 0\} \subset \mathbb{R}^3$. Our algorithms of Section 1 tell us that the vectors

$$\mathbf{v}_1 = \begin{bmatrix} 1 \\ 1 \\ 0 \end{bmatrix} \quad \text{and} \quad \mathbf{v}_2 = \begin{bmatrix} -2 \\ 0 \\ 1 \end{bmatrix}$$

span V. Since these vectors are not parallel, we can deduce (see Example 5) that they must be linearly independent.

For the practice, however, we give a direct argument. Suppose

$$c_1 \mathbf{v}_1 + c_2 \mathbf{v}_2 = c_1 \begin{bmatrix} 1 \\ 1 \\ 0 \end{bmatrix} + c_2 \begin{bmatrix} -2 \\ 0 \\ 1 \end{bmatrix} = \mathbf{0}.$$

Writing out the entries explicitly, we obtain

$$\begin{bmatrix} c_1 - 2c_2 \\ c_1 \\ c_2 \end{bmatrix} = \begin{bmatrix} 0 \\ 0 \\ 0 \end{bmatrix},$$

from which we conclude that $c_1 = c_2 = 0$, as required. (For future reference, we note that this information came from the free variable "slots.") Therefore, $\{\mathbf{v}_1, \mathbf{v}_2\}$ is linearly independent and gives a basis for V, as required. ◀

The following observation may prove useful.

Corollary 3.3 *Let $V \subset \mathbb{R}^n$ be a subspace, and let $\mathbf{v}_1, \ldots, \mathbf{v}_k \in V$. Then $\{\mathbf{v}_1, \ldots, \mathbf{v}_k\}$ is a basis for V if and only if every vector of V can be written uniquely as a linear combination of $\mathbf{v}_1, \ldots, \mathbf{v}_k$.*

Proof This is immediate from Proposition 3.1. ∎

Definition When we write $\mathbf{v} = c_1 \mathbf{v}_1 + c_2 \mathbf{v}_2 + \cdots + c_k \mathbf{v}_k$, we refer to $c_1, \ldots, c_k$ as the *coordinates* of $\mathbf{v}$ with respect to the (ordered) basis $\{\mathbf{v}_1, \ldots, \mathbf{v}_k\}$.

▶ **EXAMPLE 8**

Consider the three vectors

$$\mathbf{v}_1 = \begin{bmatrix} 1 \\ 2 \\ 1 \end{bmatrix}, \quad \mathbf{v}_2 = \begin{bmatrix} 1 \\ 1 \\ 2 \end{bmatrix}, \quad \text{and} \quad \mathbf{v}_3 = \begin{bmatrix} 1 \\ 0 \\ 2 \end{bmatrix}.$$

Let's take a general vector $\mathbf{b} \in \mathbb{R}^3$ and ask first of all whether it has a unique expression as a linear combination of $\mathbf{v}_1, \mathbf{v}_2$, and $\mathbf{v}_3$. Forming the augmented matrix and row reducing, we find

$$\begin{bmatrix} 1 & 1 & 1 & | & b_1 \\ 2 & 1 & 0 & | & b_2 \\ 1 & 2 & 2 & | & b_3 \end{bmatrix} \rightsquigarrow \begin{bmatrix} 1 & 0 & 0 & | & 2b_1 - b_3 \\ 0 & 1 & 0 & | & -4b_1 + b_2 + 2b_3 \\ 0 & 0 & 1 & | & 3b_1 - b_2 - b_3 \end{bmatrix}.$$

It follows from Corollary 3.3 that $\{\mathbf{v}_1, \mathbf{v}_2, \mathbf{v}_3\}$ is a basis for $\mathbb{R}^3$, for an arbitrary vector $\mathbf{b} \in \mathbb{R}^3$ can be written in the form

$$\mathbf{b} = \underbrace{(2b_1 - b_3)}_{c_1} \mathbf{v}_1 + \underbrace{(-4b_1 + b_2 + 2b_3)}_{c_2} \mathbf{v}_2 + \underbrace{(3b_1 - b_2 - b_3)}_{c_3} \mathbf{v}_3.$$

And, what's more,

$$c_1 = 2b_1 - b_3,$$
$$c_2 = -4b_1 + b_2 + 2b_3, \quad \text{and}$$
$$c_3 = 3b_1 - b_2 - b_3$$

give the coordinates of $\mathbf{b}$ with respect to the basis $\{\mathbf{v}_1, \mathbf{v}_2, \mathbf{v}_3\}$. ◀

Another example, which will be quite important to us in the future, is

Proposition 3.4 *Let A be an $n \times n$ matrix. Then A is nonsingular if and only if its column vectors form a basis for $\mathbb{R}^n$.*

Proof As usual, let's denote the column vectors of A by $\mathbf{a}_1, \mathbf{a}_2, \ldots, \mathbf{a}_n$. Using Corollary 3.3, we are to prove that A is nonsingular if and only if every vector in $\mathbb{R}^n$ can be written uniquely as a linear combination of $\mathbf{a}_1, \mathbf{a}_2, \ldots, \mathbf{a}_n$. But this is exactly what Proposition 1.6 tells us. ∎

Given a subspace $V \subset \mathbb{R}^n$, how do we know there is some basis for it? This is a consequence of Proposition 3.2 as well.

Theorem 3.5 *Any subspace $V \subset \mathbb{R}^n$ other than the trivial subspace has a basis.*

Proof Since $V \neq \{\mathbf{0}\}$, we choose a nonzero vector $\mathbf{v}_1 \in V$. If $\mathbf{v}_1$ spans V, then we know $\{\mathbf{v}_1\}$ will constitute a basis for V. If not, choose $\mathbf{v}_2 \notin \mathrm{Span}(\mathbf{v}_1)$. From Proposition 3.2 we infer that $\{\mathbf{v}_1, \mathbf{v}_2\}$ is linearly independent. If $\mathbf{v}_1, \mathbf{v}_2$ span V, then $\{\mathbf{v}_1, \mathbf{v}_2\}$ will be a basis for V. If not, choose $\mathbf{v}_3 \notin \mathrm{Span}(\mathbf{v}_1, \mathbf{v}_2)$. Once again, we know that $\{\mathbf{v}_1, \mathbf{v}_2, \mathbf{v}_3\}$ will be linearly independent and hence will form a basis for V if the three vectors span V. We continue in this fashion, and we are guaranteed that the process will terminate in at most n steps because, according to Exercise 6, once we have $n+1$ vectors in $\mathbb{R}^n$, they must form a linearly dependent set. ∎

Once we realize that every subspace $V \subset \mathbb{R}^n$ has *some* basis, we are confronted with the problem that it has *many* of them. For example, Proposition 3.4 gives us a way of finding zillions of bases for $\mathbb{R}^n$. As we shall now show, all bases for a given subspace have one thing in common: They all consist of the same number of elements.

Proposition 3.6 *Let $V \subset \mathbb{R}^n$ be a subspace, let $\{\mathbf{v}_1, \ldots, \mathbf{v}_k\}$ be a basis for V, and let $\mathbf{w}_1, \ldots, \mathbf{w}_\ell \in V$. If $\ell > k$, then $\{\mathbf{w}_1, \ldots, \mathbf{w}_\ell\}$ must be linearly dependent.*

Proof Each vector in V can be written uniquely as a linear combination of $\mathbf{v}_1, \ldots, \mathbf{v}_k$. So let's write each vector $\mathbf{w}_1, \ldots, \mathbf{w}_\ell$ as such:

$$\mathbf{w}_1 = a_{11}\mathbf{v}_1 + a_{21}\mathbf{v}_2 + \cdots + a_{k1}\mathbf{v}_k$$
$$\mathbf{w}_2 = a_{12}\mathbf{v}_1 + a_{22}\mathbf{v}_2 + \cdots + a_{k2}\mathbf{v}_k$$
$$\vdots$$
$$\mathbf{w}_\ell = a_{1\ell}\mathbf{v}_1 + a_{2\ell}\mathbf{v}_2 + \cdots + a_{k\ell}\mathbf{v}_k.$$

We now form the $k \times \ell$ matrix $A = [a_{ij}]$. This gives the matrix equation

$$(*) \qquad \begin{bmatrix} | & | & & | \\ \mathbf{v}_1 & \mathbf{v}_2 & \cdots & \mathbf{v}_k \\ | & | & & | \end{bmatrix} A = \begin{bmatrix} | & | & & | \\ \mathbf{w}_1 & \mathbf{w}_2 & \cdots & \mathbf{w}_\ell \\ | & | & & | \end{bmatrix}.$$

Since $\ell > k$, there cannot be a pivot in every column of A, and so there is a *nonzero* vector $\mathbf{c}$ satisfying

$$A \begin{bmatrix} c_1 \\ c_2 \\ \vdots \\ c_\ell \end{bmatrix} = \mathbf{0}.$$

Using (∗) and associativity, we have

$$\begin{bmatrix} | & | & & | \\ \mathbf{w}_1 & \mathbf{w}_2 & \cdots & \mathbf{w}_\ell \\ | & | & & | \end{bmatrix} \begin{bmatrix} c_1 \\ c_2 \\ \vdots \\ c_\ell \end{bmatrix} = \begin{bmatrix} | & | & & | \\ \mathbf{v}_1 & \mathbf{v}_2 & \cdots & \mathbf{v}_k \\ | & | & & | \end{bmatrix} \left(A \begin{bmatrix} c_1 \\ c_2 \\ \vdots \\ c_\ell \end{bmatrix} \right) = \mathbf{0}.$$

That is, we have found a nontrivial linear combination

$$c_1 \mathbf{w}_1 + \cdots + c_\ell \mathbf{w}_\ell = \mathbf{0},$$

which means that $\{\mathbf{w}_1, \ldots, \mathbf{w}_\ell\}$ is linearly dependent, as was claimed. ∎

Remark We can easily avoid equation (∗) in its matrix form. Since

$$\mathbf{w}_j = \sum_{i=1}^{k} a_{ij} \mathbf{v}_i,$$

we have

(∗∗) $$\sum_{j=1}^{\ell} c_j \mathbf{w}_j = \sum_{j=1}^{\ell} c_j \left(\sum_{i=1}^{k} a_{ij} \mathbf{v}_i \right) = \sum_{i=1}^{k} \left(\sum_{j=1}^{\ell} a_{ij} c_j \right) \mathbf{v}_i.$$

As before, since $\ell > k$, there is a nonzero vector $\mathbf{c}$ so that $A\mathbf{c} = \mathbf{0}$; this choice of $\mathbf{c}$ makes the right-hand side of (∗∗) the zero vector. Consequently, there is a nontrivial relation among $\mathbf{w}_1, \ldots, \mathbf{w}_\ell$.

This proposition leads directly to our main result.

Theorem 3.7 *Let $V \subset \mathbb{R}^n$ be a subspace, and let $\{\mathbf{v}_1, \ldots, \mathbf{v}_k\}$ and $\{\mathbf{w}_1, \ldots, \mathbf{w}_\ell\}$ be two bases for V. Then we have $k = \ell$.*

Proof Since $\{\mathbf{v}_1, \ldots, \mathbf{v}_k\}$ forms a basis for V and $\{\mathbf{w}_1, \ldots, \mathbf{w}_\ell\}$ is known to be linearly independent, we use Proposition 3.6 to conclude that $\ell \leq k$. Now here's the trick: $\{\mathbf{w}_1, \ldots, \mathbf{w}_\ell\}$ is likewise a basis for V and $\{\mathbf{v}_1, \ldots, \mathbf{v}_k\}$ is known to be linearly independent, so we infer from Proposition 3.6 that $k \leq \ell$. The only way both inequalities can hold is for k and ℓ to be equal, as we wished to show. ∎

We now make the official

Definition The *dimension* of a subspace $V \subset \mathbb{R}^n$ is the number of vectors in any basis for V. We denote the dimension of V by dim V. By convention, dim$\{\mathbf{0}\} = 0$.

As we shall see in our applications, dimension is a powerful tool. Here is the first instance.

Lemma 3.8 *Suppose V and W are subspaces of $\mathbb{R}^n$ with the property that $W \subset V$. If dim $V = $ dim W, then $V = W$.*

Proof Let dim $W = k$ and let $\{\mathbf{v}_1, \ldots, \mathbf{v}_k\}$ be a basis for W. If $W \subsetneq V$, then there must be a vector $\mathbf{v} \in V$ with $\mathbf{v} \notin W$. By virtue of Proposition 3.2, we know that $\{\mathbf{v}_1, \ldots, \mathbf{v}_k, \mathbf{v}\}$ is linearly independent, so dim $V \geq k + 1$. This is a contradiction. Therefore, $V = W$. ∎

The next result is quite useful.

Proposition 3.9 *Let $V \subset \mathbb{R}^n$ be a k-dimensional subspace. Then any k vectors that span V must be linearly independent and any k linearly independent vectors in V must span V.*

Proof Left to the reader in Exercise 17. ∎

▶ **EXAMPLE 9**

Let $V = \text{Span}(\mathbf{v}_1, \mathbf{v}_2, \mathbf{v}_3, \mathbf{v}_4) \subset \mathbb{R}^3$, where

$$\mathbf{v}_1 = \begin{bmatrix} 1 \\ 1 \\ 2 \end{bmatrix}, \quad \mathbf{v}_2 = \begin{bmatrix} 2 \\ 2 \\ 4 \end{bmatrix}, \quad \mathbf{v}_3 = \begin{bmatrix} 0 \\ 1 \\ 1 \end{bmatrix}, \quad \text{and} \quad \mathbf{v}_4 = \begin{bmatrix} 3 \\ 4 \\ 7 \end{bmatrix}.$$

We want a subset of $\{\mathbf{v}_1, \mathbf{v}_2, \mathbf{v}_3, \mathbf{v}_4\}$ that will give us a basis for V. Of course, this set of four vectors must be linearly dependent since $V \subset \mathbb{R}^3$ and $\mathbb{R}^3$ is only 3-dimensional. But let's examine the solutions of

$$c_1\mathbf{v}_1 + c_2\mathbf{v}_2 + c_3\mathbf{v}_3 + c_4\mathbf{v}_4 = \mathbf{0},$$

or, in matrix form,

$$\begin{bmatrix} 1 & 2 & 0 & 3 \\ 1 & 2 & 1 & 4 \\ 2 & 4 & 1 & 7 \end{bmatrix} \begin{bmatrix} c_1 \\ c_2 \\ c_3 \\ c_4 \end{bmatrix} = \mathbf{0}.$$

As usual, we proceed to reduced echelon form:

$$R = \begin{bmatrix} 1 & 2 & 0 & 3 \\ 0 & 0 & 1 & 1 \\ 0 & 0 & 0 & 0 \end{bmatrix},$$

from which we find that the vectors

$$\begin{bmatrix} -2 \\ 1 \\ 0 \\ 0 \end{bmatrix} \quad \text{and} \quad \begin{bmatrix} -3 \\ 0 \\ -1 \\ 1 \end{bmatrix}$$

span the space of solutions. In particular, this tells us that

$$-2\mathbf{v}_1 + \mathbf{v}_2 = \mathbf{0} \quad \text{and} \quad -3\mathbf{v}_1 - \mathbf{v}_3 + \mathbf{v}_4 = \mathbf{0},$$

and so the vectors $\mathbf{v}_2$ and $\mathbf{v}_4$ can be expressed as linear combinations of the vectors $\mathbf{v}_1$ and $\mathbf{v}_3$. On the other hand, $\{\mathbf{v}_1, \mathbf{v}_3\}$ is linearly independent (why?), so this gives a basis for V. ◀

3.1 Abstract Vector Spaces

We have not yet dealt with vector spaces other than Euclidean spaces. In general, a *vector space* is a set endowed with the operations of addition and scalar multiplication, subject to the properties listed in Exercise 1.1.12. Notions of linear independence and basis proceed analogously; the Remark on p. 165 shows that dimension is well defined in the general setting.

▶ **EXAMPLE 10**

Here are a few examples of so-called "abstract" vector spaces. Others appear in the exercises.

a. Let $\mathcal{M}_{m \times n}$ denote the set of all $m \times n$ matrices. As we've seen in Proposition 4.1 of Chapter 1, $\mathcal{M}_{m \times n}$ is a vector space, using the operations of matrix addition and scalar multiplication we've already defined. The zero "vector" is the zero matrix O. This space can naturally be identified with $\mathbb{R}^{mn}$ (see Exercise 24).

b. Let $\mathcal{F}(U)$ denote the collection of all real valued functions defined on some subset $U \subset \mathbb{R}^n$. If $f \in \mathcal{F}(U)$ and $c \in \mathbb{R}$, then we can define a new function $cf \in \mathcal{F}(U)$ by multiplying the *value* of f at each point by the scalar c; i.e.,

$$(cf)(t) = cf(t) \quad \text{for each } t \in U.$$

Similarly, if $f, g \in \mathcal{F}(U)$, then we can define the new function $f + g \in \mathcal{F}(U)$ by adding the *values* of f and g at each point; i.e.,

$$(f+g)(t) = f(t) + g(t) \quad \text{for each } t \in U.$$

By these formulas we define scalar multiplication and vector addition in $\mathcal{F}(U)$. The zero "vector" in $\mathcal{F}(U)$ is the zero function. The various properties of a vector space follow from the corresponding properties of the real numbers (as everything is defined in terms of the *values* of the function at every point t). Since an element of $\mathcal{F}(U)$ is a function, $\mathcal{F}(U)$ is often called a *function space*.

c. Let $\mathbb{R}^\omega$ denote the collection of all infinite sequences of real numbers. That is, an element of $\mathbb{R}^\omega$ looks like $\mathbf{x} = \begin{bmatrix} x_1 \\ x_2 \\ x_3 \\ \vdots \end{bmatrix}$, where $x_i \in \mathbb{R}, i = 1, 2, 3, \ldots$. Operations are defined in the obvious way: If $c \in \mathbb{R}$ and $\mathbf{y} = \begin{bmatrix} y_1 \\ y_2 \\ y_3 \\ \vdots \end{bmatrix}$, then we set $c\mathbf{x} = \begin{bmatrix} cx_1 \\ cx_2 \\ cx_3 \\ \vdots \end{bmatrix}$ and $\mathbf{x} + \mathbf{y} = \begin{bmatrix} x_1 + y_1 \\ x_2 + y_2 \\ x_3 + y_3 \\ \vdots \end{bmatrix}$. ◀

The vector space of functions on an open subset $U \subset \mathbb{R}^n$ has various subspaces that will be of particular interest to us. For any $k \geq 0$ we have $\mathcal{C}^k(U)$, the space of $\mathcal{C}^k$ functions on U; indeed, we have the hierarchy

$$\mathcal{C}^\infty(U) \subset \cdots \subset \mathcal{C}^{k+1}(U) \subset \mathcal{C}^k(U) \subset \cdots \subset \mathcal{C}^2(U) \subset \mathcal{C}^1(U) \subset \mathcal{C}^0(U).$$

(That these are all subspaces follows from the standard fact that sums and scalar multiples of $\mathcal{C}^k$ functions are again $\mathcal{C}^k$.) We can also consider the subspaces of polynomial functions. We denote by $\mathcal{P}_k$ the vector space of polynomials of degree $\leq k$ in one variable.

As we ask the reader to check in Exercise 26, the vector space $\mathcal{P}_k$ has dimension $k+1$. In general, we say a vector space is *finite-dimensional* if it has dimension n for some $n \in \mathbb{N}$ and *infinite-dimensional* if not. The vector space $\mathcal{C}^\infty(\mathbb{R})$ is infinite-dimensional, as it contains polynomials of arbitrarily high degree.

EXERCISES 4.3

1. Let $\mathbf{v}_1 = \begin{bmatrix} 1 \\ 2 \\ 3 \end{bmatrix}$, $\mathbf{v}_2 = \begin{bmatrix} 2 \\ 4 \\ 5 \end{bmatrix}$, and $\mathbf{v}_3 = \begin{bmatrix} 2 \\ 4 \\ 6 \end{bmatrix} \in \mathbb{R}^3$. Is each of the following statements correct or incorrect? Explain.
   (a) The set $\{\mathbf{v}_1, \mathbf{v}_2, \mathbf{v}_3\}$ is linearly dependent.
   (b) Each of the vectors $\mathbf{v}_1$, $\mathbf{v}_2$, and $\mathbf{v}_3$ can be written as a linear combination of the others.

*2. Decide whether each of the following sets of vectors is linearly independent.

(a) $\left\{ \begin{bmatrix} 1 \\ 4 \end{bmatrix}, \begin{bmatrix} 2 \\ 9 \end{bmatrix} \right\} \subset \mathbb{R}^2$

(b) $\left\{ \begin{bmatrix} 1 \\ 4 \\ 0 \end{bmatrix}, \begin{bmatrix} 2 \\ 9 \\ 0 \end{bmatrix} \right\} \subset \mathbb{R}^3$

(c) $\left\{ \begin{bmatrix} 1 \\ 4 \\ 0 \end{bmatrix}, \begin{bmatrix} 2 \\ 9 \\ 0 \end{bmatrix}, \begin{bmatrix} 3 \\ -2 \\ 0 \end{bmatrix} \right\} \subset \mathbb{R}^3$

(d) $\left\{ \begin{bmatrix} 1 \\ 1 \\ 1 \end{bmatrix}, \begin{bmatrix} 2 \\ 3 \\ 3 \end{bmatrix}, \begin{bmatrix} 0 \\ 1 \\ 2 \end{bmatrix} \right\} \subset \mathbb{R}^3$

(e) $\left\{ \begin{bmatrix} 1 \\ 1 \\ 1 \\ 3 \end{bmatrix}, \begin{bmatrix} 1 \\ 1 \\ 3 \\ 1 \end{bmatrix}, \begin{bmatrix} 1 \\ 3 \\ 1 \\ 1 \end{bmatrix}, \begin{bmatrix} 3 \\ 1 \\ 1 \\ 1 \end{bmatrix} \right\} \subset \mathbb{R}^4$

(f) $\left\{ \begin{bmatrix} 1 \\ 1 \\ 1 \\ -3 \end{bmatrix}, \begin{bmatrix} 1 \\ 1 \\ -3 \\ 1 \end{bmatrix}, \begin{bmatrix} 1 \\ -3 \\ 1 \\ 1 \end{bmatrix}, \begin{bmatrix} -3 \\ 1 \\ 1 \\ 1 \end{bmatrix} \right\} \subset \mathbb{R}^4$

3. Suppose $\mathbf{v}, \mathbf{w} \in \mathbb{R}^n$ and $\{\mathbf{v}, \mathbf{w}\}$ is linearly independent. Prove that $\{\mathbf{v} - \mathbf{w}, 2\mathbf{v} + \mathbf{w}\}$ is linearly independent as well.

4. Suppose $\{\mathbf{u}, \mathbf{v}, \mathbf{w}\} \subset \mathbb{R}^3$ is linearly independent.
   (a) Prove that $\mathbf{u} \cdot (\mathbf{v} \times \mathbf{w}) \neq 0$.
   (b) Prove that $\{\mathbf{u} \times \mathbf{v}, \mathbf{v} \times \mathbf{w}, \mathbf{w} \times \mathbf{u}\}$ is linearly independent as well.

♯5. Suppose $\mathbf{v}_1, \ldots, \mathbf{v}_k$ are nonzero vectors with the property that $\mathbf{v}_i \cdot \mathbf{v}_j = 0$ whenever $i \neq j$. Prove that $\{\mathbf{v}_1, \ldots, \mathbf{v}_k\}$ is linearly independent. (Hint: Suppose $c_1 \mathbf{v}_1 + c_2 \mathbf{v}_2 + \cdots + c_k \mathbf{v}_k = \mathbf{0}$. Start by showing $c_1 = 0$.)

♯6. Suppose $k > n$. Prove that any k vectors in $\mathbb{R}^n$ must form a linearly dependent set. (So what can you conclude if you have k linearly independent vectors in $\mathbb{R}^n$?)

7. Suppose $v_1, \ldots, v_k \in \mathbb{R}^n$ form a linearly dependent set. Prove that for some $1 \le j \le k$ we have $v_j \in \text{Span}(v_1, \ldots, v_{j-1}, v_{j+1}, \ldots, v_k)$. That is, one of the vectors $v_1, \ldots, v_k$ can be written as a linear combination of the remaining vectors.

8. Suppose $v_1, \ldots, v_k \in \mathbb{R}^n$ form a linearly dependent set. Prove that either $v_1 = 0$ or $v_{i+1} \in \text{Span}(v_1, \ldots, v_i)$ for some $i = 1, 2, \ldots, k-1$.

9. Let A be an $m \times n$ matrix and $b_1, \ldots, b_k \in \mathbb{R}^m$. Suppose $\{b_1, \ldots, b_k\}$ is linearly independent. Suppose that $v_1, \ldots, v_k \in \mathbb{R}^n$ are chosen so that $Av_1 = b_1, \ldots, Av_k = b_k$. Prove that $\{v_1, \ldots, v_k\}$ must be linearly independent.

10. Suppose $T: \mathbb{R}^n \to \mathbb{R}^n$ is a linear map. Prove that if $[T]$ is nonsingular and $\{v_1, \ldots, v_k\}$ is linearly independent, then $\{T(v_1), T(v_2), \ldots, T(v_k)\}$ is likewise linearly independent.

♯11. Suppose $T: \mathbb{R}^n \to \mathbb{R}^m$ is a linear map and $[T]$ has rank n. Suppose $v_1, \ldots, v_k \in \mathbb{R}^n$ and $\{v_1, \ldots, v_k\}$ is linearly independent. Prove that $\{T(v_1), \ldots, T(v_k)\} \subset \mathbb{R}^m$ is likewise linearly independent. (N.B.: If you did not explicitly make use of the assumption that $\text{rank}([T]) = n$, your proof cannot be correct. Why?)

*12. Decide whether the following sets of vectors give a basis for the indicated space.

(a) $\left\{ \begin{bmatrix} 1 \\ 2 \\ 1 \end{bmatrix}, \begin{bmatrix} 2 \\ 4 \\ 5 \end{bmatrix}, \begin{bmatrix} 1 \\ 2 \\ 3 \end{bmatrix} \right\}; \mathbb{R}^3$

(b) $\left\{ \begin{bmatrix} 1 \\ 0 \\ 1 \end{bmatrix}, \begin{bmatrix} 1 \\ 2 \\ 4 \end{bmatrix}, \begin{bmatrix} 2 \\ 2 \\ 5 \end{bmatrix}, \begin{bmatrix} 2 \\ 2 \\ -1 \end{bmatrix} \right\}; \mathbb{R}^3$

(c) $\left\{ \begin{bmatrix} 1 \\ 0 \\ 2 \\ 3 \end{bmatrix}, \begin{bmatrix} 0 \\ 1 \\ 1 \\ 1 \end{bmatrix}, \begin{bmatrix} 1 \\ 1 \\ 4 \\ 4 \end{bmatrix} \right\}; \mathbb{R}^4$

(d) $\left\{ \begin{bmatrix} 1 \\ 0 \\ 2 \\ 3 \end{bmatrix}, \begin{bmatrix} 0 \\ 1 \\ 1 \\ 1 \end{bmatrix}, \begin{bmatrix} 1 \\ 1 \\ 4 \\ 4 \end{bmatrix}, \begin{bmatrix} 2 \\ -2 \\ 1 \\ 2 \end{bmatrix} \right\}; \mathbb{R}^4$

13. Find a basis for each of the given subspaces and determine its dimension.

*(a) $V = \text{Span}\left(\begin{bmatrix} 1 \\ 2 \\ 3 \end{bmatrix}, \begin{bmatrix} 3 \\ 4 \\ 7 \end{bmatrix}, \begin{bmatrix} 5 \\ -2 \\ 3 \end{bmatrix} \right) \subset \mathbb{R}^3$

(b) $V = \{x \in \mathbb{R}^4 : x_1 + x_2 + x_3 + x_4 = 0, \ x_2 + x_4 = 0\} \subset \mathbb{R}^4$

(c) $V = \left(\text{Span}\left(\begin{bmatrix} 1 \\ 2 \\ 3 \end{bmatrix} \right) \right)^\perp \subset \mathbb{R}^3$

(d) $V = \{x \in \mathbb{R}^5 : x_1 = x_2, \ x_3 = x_4\} \subset \mathbb{R}^5$

14. In each case, check that $\{v_1, \ldots, v_n\}$ is a basis for $\mathbb{R}^n$ and give the coordinates of the given vector $b \in \mathbb{R}^n$ with respect to that basis.

(a) $v_1 = \begin{bmatrix} 2 \\ 3 \end{bmatrix}, v_2 = \begin{bmatrix} 3 \\ 5 \end{bmatrix}; b = \begin{bmatrix} 3 \\ 4 \end{bmatrix}$

*(b) $v_1 = \begin{bmatrix} 1 \\ 0 \\ 3 \end{bmatrix}, v_2 = \begin{bmatrix} 1 \\ 2 \\ 2 \end{bmatrix}, v_3 = \begin{bmatrix} 1 \\ 3 \\ 2 \end{bmatrix}; b = \begin{bmatrix} 1 \\ 1 \\ 2 \end{bmatrix}$

(c) $v_1 = \begin{bmatrix} 1 \\ 0 \\ 1 \end{bmatrix}, v_2 = \begin{bmatrix} 1 \\ 1 \\ 2 \end{bmatrix}, v_3 = \begin{bmatrix} 1 \\ 1 \\ 1 \end{bmatrix}; b = \begin{bmatrix} 3 \\ 0 \\ 1 \end{bmatrix}$

*(d) $\mathbf{v}_1 = \begin{bmatrix} 1 \\ 0 \\ 0 \\ 0 \end{bmatrix}$, $\mathbf{v}_2 = \begin{bmatrix} 1 \\ 1 \\ 0 \\ 0 \end{bmatrix}$, $\mathbf{v}_3 = \begin{bmatrix} 1 \\ 1 \\ 1 \\ 1 \end{bmatrix}$, $\mathbf{v}_4 = \begin{bmatrix} 1 \\ 1 \\ 3 \\ 4 \end{bmatrix}$; $\mathbf{b} = \begin{bmatrix} 2 \\ 0 \\ 1 \\ 1 \end{bmatrix}$

15. Find a basis for the intersection of the subspaces

$$V = \text{Span}\left(\begin{bmatrix} 1 \\ 0 \\ 1 \\ 1 \end{bmatrix}, \begin{bmatrix} 2 \\ 1 \\ 1 \\ 2 \end{bmatrix} \right) \quad \text{and} \quad W = \text{Span}\left(\begin{bmatrix} 0 \\ 1 \\ 1 \\ 0 \end{bmatrix}, \begin{bmatrix} 2 \\ 0 \\ 1 \\ 2 \end{bmatrix} \right) \subset \mathbb{R}^4.$$

♯16. Suppose $\mathbf{v}_1, \ldots, \mathbf{v}_n$ are nonzero, mutually orthogonal vectors in $\mathbb{R}^n$.
 (a) Prove that they form a basis for $\mathbb{R}^n$.
 (b) Given any $\mathbf{x} \in \mathbb{R}^n$, give an explicit formula for the coordinates of $\mathbf{x}$ with respect to the basis $\{\mathbf{v}_1, \ldots, \mathbf{v}_n\}$.
 (c) Deduce from your answer to part b that $\mathbf{x} = \sum_{i=1}^{n} \text{proj}_{\mathbf{v}_i} \mathbf{x}$.

17. Prove Proposition 3.9. (Hint: Exercise 7 and Lemma 3.8 may be of help.)

♯18. Let $V \subset \mathbb{R}^n$ be a subspace, and suppose you are given a linearly independent set of vectors $\{\mathbf{v}_1, \ldots, \mathbf{v}_k\} \subset V$. Prove that there are vectors $\mathbf{v}_{k+1}, \ldots, \mathbf{v}_\ell \in V$ so that $\{\mathbf{v}_1, \ldots, \mathbf{v}_\ell\}$ forms a basis for V.

19. Suppose V and W are subspaces of $\mathbb{R}^n$ and $W \subset V$. Prove that $\dim W \leq \dim V$. (Hint: Start with a basis for W and apply Exercise 18.)

20. Suppose A is an $n \times n$ matrix, and let $\mathbf{v}_1, \ldots, \mathbf{v}_n \in \mathbb{R}^n$. Suppose $\{A\mathbf{v}_1, \ldots, A\mathbf{v}_n\}$ is linearly independent. Prove that A is nonsingular.

21.*(a) Suppose U and V are subspaces of $\mathbb{R}^n$ with $U \cap V = \{\mathbf{0}\}$. If $\{\mathbf{u}_1, \ldots, \mathbf{u}_k\}$ is a basis for U and $\{\mathbf{v}_1, \ldots, \mathbf{v}_\ell\}$ is a basis for V, prove that $\{\mathbf{u}_1, \ldots, \mathbf{u}_k, \mathbf{v}_1, \ldots, \mathbf{v}_\ell\}$ is a basis for $U + V$.
 (b) Let U and V be subspaces of $\mathbb{R}^n$. Prove that if $U \cap V = \{\mathbf{0}\}$, then $\dim(U + V) = \dim U + \dim V$.
 (c) Let U and V be subspaces of $\mathbb{R}^n$. Prove that $\dim(U + V) = \dim U + \dim V - \dim(U \cap V)$. (Hint: Start with a basis for $U \cap V$, and use Exercise 18.)

22. Let $T: \mathbb{R}^n \to \mathbb{R}^m$ be a linear map. Define
$$\ker(T) = \{\mathbf{x} \in \mathbb{R}^n : T(\mathbf{x}) = \mathbf{0}\} \quad \text{and}$$
$$\text{image}(T) = \{\mathbf{y} \in \mathbb{R}^m : \mathbf{y} = T(\mathbf{x}) \text{ for some } \mathbf{x} \in \mathbb{R}^n\}.$$
 (a) Check that $\ker(T)$ and $\text{image}(T)$ are subspaces of $\mathbb{R}^n$ and $\mathbb{R}^m$, respectively.
 (b) Let $\{\mathbf{v}_1, \ldots, \mathbf{v}_k\}$ be a basis for $\ker(T)$ and, using Exercise 18, extend to a basis $\{\mathbf{v}_1, \ldots, \mathbf{v}_k, \mathbf{v}_{k+1}, \ldots, \mathbf{v}_n\}$ for $\mathbb{R}^n$. Prove that $\{T(\mathbf{v}_{k+1}), \ldots, T(\mathbf{v}_n)\}$ gives a basis for $\text{image}(T)$.
 (c) Deduce that $\dim \ker(T) + \dim \text{image}(T) = n$.

*23. Decide whether the following sets of vectors are linearly independent.
 (a) $\left\{ \begin{bmatrix} 1 & 0 \\ 0 & 1 \end{bmatrix}, \begin{bmatrix} 0 & 1 \\ 1 & 0 \end{bmatrix}, \begin{bmatrix} 1 & 1 \\ 1 & -1 \end{bmatrix} \right\} \subset \mathcal{M}_{2\times 2}$
 (b) $\{f_1, f_2, f_3\} \subset \mathcal{P}_1$, where $f_1(t) = t$, $f_2(t) = t+1$, $f_3(t) = t+2$
 (c) $\{f_1, f_2, f_3\} \subset \mathcal{C}^\infty(\mathbb{R})$, where $f_1(t) = 1$, $f_2(t) = \cos t$, $f_3(t) = \sin t$
 (d) $\{f_1, f_2, f_3\} \subset \mathcal{C}^0(\mathbb{R})$, where $f_1(t) = 1$, $f_2(t) = \sin^2 t$, $f_3(t) = \cos^2 t$

(e) $\{f_1, f_2, f_3\} \subset \mathcal{C}^\infty(\mathbb{R})$, where $f_1(t) = 1$, $f_2(t) = \cos t$, $f_3(t) = \cos 2t$
(f) $\{f_1, f_2, f_3\} \subset \mathcal{C}^\infty(\mathbb{R})$, where $f_1(t) = 1$, $f_2(t) = \cos 2t$, $f_3(t) = \cos^2 t$

24. Recall that $\mathcal{M}_{m \times n}$ denotes the vector space of $m \times n$ matrices.
   (a) Give a basis for and determine the dimension of $\mathcal{M}_{m \times n}$.
   (b) Show that the set of diagonal matrices, the set of upper triangular matrices, and the set of lower triangular matrices are all subspaces of $\mathcal{M}_{n \times n}$ and determine their dimensions.
   (c) Show that the set of symmetric matrices, $\mathcal{S}$, and the set of skew-symmetric matrices, $\mathcal{K}$, are subspaces of $\mathcal{M}_{n \times n}$. What are their dimensions? Show that $\mathcal{S} + \mathcal{K} = \mathcal{M}_{n \times n}$. (See Exercise 1.4.36.)

♯**25.** Let V be a vector space.
   (a) Let V^* denote the set of all linear transformations from V to $\mathbb{R}$. Show that V^* is a vector space.
   (b) Suppose $\{\mathbf{v}_1, \ldots, \mathbf{v}_n\}$ is a basis for V. For $i = 1, \ldots, n$, define $\mathbf{f}_i \in V^*$ by
   $$\mathbf{f}_i(a_1 \mathbf{v}_1 + a_2 \mathbf{v}_2 + \cdots + a_n \mathbf{v}_n) = a_i.$$
   Prove that $\{\mathbf{f}_1, \ldots, \mathbf{f}_n\}$ gives a basis for V^*.
   (c) Deduce that whenever V is finite-dimensional, $\dim V^* = \dim V$.

26. Show that the set $\mathcal{P}_k$ of polynomials in one variable of degree $\leq k$ is a vector space of dimension $k + 1$. (Hint: Suppose $c_0 + c_1 x + \cdots + c_k x^k = 0$ for all x. Differentiate.)

27. Recall that $f: \mathbb{R}^n - \{\mathbf{0}\} \to \mathbb{R}$ is homogeneous of degree k if $f(t\mathbf{x}) = t^k f(\mathbf{x})$ for all $t > 0$.
   (a) Show that the set $\mathcal{P}_{k,n}$ of homogeneous polynomials of degree k in n variables is a vector space.
   (b) Fix $k \in \mathbb{N}$. Show that the monomials $x_1^{i_1} x_2^{i_2} \cdots x_n^{i_n}$, where $i_1 + i_2 + \cdots + i_n = k$ and $0 \leq i_j \leq k$ for $j = 1, \ldots, n$, form a basis for $\mathcal{P}_{k,n}$.
   (c) Show that $\dim \mathcal{P}_{k,n} = \binom{n-1+k}{k}$.[4] (Hint: It may help to remember that $\binom{j}{k} = \binom{j}{j-k}$.)
   (d) Using the interpretation in part c, prove that $\sum_{i=0}^{k} \binom{n+i}{i} = \binom{n+k+1}{k}$.

▶ 4 THE FOUR FUNDAMENTAL SUBSPACES

Given an $m \times n$ matrix A (or, more conceptually, a linear map $T: \mathbb{R}^n \to \mathbb{R}^m$), there are four natural subspaces to consider. It is one of our goals to understand the relations among them. We begin with the *column space* and *row space*.

Definition Let A be an $m \times n$ matrix with row vectors $\mathbf{A}_1, \ldots, \mathbf{A}_m \in \mathbb{R}^n$ and column vectors $\mathbf{a}_1, \ldots, \mathbf{a}_n \in \mathbb{R}^m$. We define the *column space* of A to be the subspace of $\mathbb{R}^m$ spanned by $\mathbf{a}_1, \ldots, \mathbf{a}_n$:
$$\mathbf{C}(A) = \mathrm{Span}(\mathbf{a}_1, \ldots, \mathbf{a}_n) \subset \mathbb{R}^m.$$
We define the *row space* of A to be the subspace of $\mathbb{R}^n$ spanned by $\mathbf{A}_1, \ldots, \mathbf{A}_m$:
$$\mathbf{R}(A) = \mathrm{Span}(\mathbf{A}_1, \ldots, \mathbf{A}_m) \subset \mathbb{R}^n.$$

Our work in Section 1 gives an important alternative interpretation of the column space.

---

[4] Recall that the *binomial coefficient* $\binom{n}{k} = n!/k!(n-k)!$ gives the number of k-element subsets of a given n-element set.

Proposition 4.1 Let A be an $m \times n$ matrix. Let $\mathbf{b} \in \mathbb{R}^m$. Then $\mathbf{b} \in \mathbf{C}(A)$ if and only if $\mathbf{b} = A\mathbf{x}$ for some $\mathbf{x} \in \mathbb{R}^n$. That is,

$$\mathbf{C}(A) = \{\mathbf{b} \in \mathbb{R}^m : A\mathbf{x} = \mathbf{b} \text{ is consistent}\}.$$

Proof By definition, $\mathbf{C}(A) = \text{Span}(\mathbf{a}_1, \ldots, \mathbf{a}_n)$, and so $\mathbf{b} \in \mathbf{C}(A)$ if and only if $\mathbf{b}$ is a linear combination of the vectors $\mathbf{a}_1, \ldots, \mathbf{a}_n$; i.e., $\mathbf{b} = x_1\mathbf{a}_1 + \cdots + x_n\mathbf{a}_n$ for some scalars $x_1, \ldots, x_n$. Recalling our crucial observation $(*)$ on p. 135, we conclude that $\mathbf{b} \in \mathbf{C}(A)$ if and only if $\mathbf{b} = A\mathbf{x}$ for some $\mathbf{x} \in \mathbb{R}^n$. The final reformulation is straightforward as long as we remember that the system $A\mathbf{x} = \mathbf{b}$ is consistent if it has a solution. ∎

Remark If we think of A as the standard matrix of a linear map $T: \mathbb{R}^n \to \mathbb{R}^m$, then $\mathbf{C}(A) \subset \mathbb{R}^m$ is the set of all the values of T, i.e., its *image*, denoted image(T).

Perhaps the most natural subspace of all comes from solving a *homogeneous* system of linear equations.

Definition Let A be an $m \times n$ matrix. The *nullspace* of A is the set of solutions of the system $A\mathbf{x} = \mathbf{0}$:

$$\mathbf{N}(A) = \{\mathbf{x} \in \mathbb{R}^n : A\mathbf{x} = \mathbf{0}\}.$$

Recall (see Exercise 1.4.3) that $\mathbf{N}(A)$ is in fact a subspace. If we think of A as the standard matrix of a linear map $T: \mathbb{R}^n \to \mathbb{R}^m$, then $\mathbf{N}(A) \subset \mathbb{R}^n$ is often called the *kernel* of T, denoted ker(T).

We might surmise that our algorithm in Section 1 for finding the general solution of the homogeneous linear system $A\mathbf{x} = \mathbf{0}$ produces a basis for $\mathbf{N}(A)$.

▶ **EXAMPLE 1**

Let's find a basis for the nullspace of the matrix

$$A = \begin{bmatrix} 1 & 2 & 1 & -1 \\ 1 & 0 & 1 & 1 \end{bmatrix}.$$

Of course, we bring A to its reduced echelon form

$$R = \begin{bmatrix} 1 & 0 & 1 & 1 \\ 0 & 1 & 0 & -1 \end{bmatrix}$$

and read off the general solution

$$\begin{aligned} x_1 &= -x_3 - x_4 \\ x_2 &= x_4 \\ x_3 &= x_3 \\ x_4 &= x_4\,; \end{aligned}$$

i.e.,

$$\mathbf{x} = \begin{bmatrix} x_1 \\ x_2 \\ x_3 \\ x_4 \end{bmatrix} = \begin{bmatrix} -x_3 - x_4 \\ x_4 \\ x_3 \\ x_4 \end{bmatrix} = x_3 \begin{bmatrix} -1 \\ 0 \\ 1 \\ 0 \end{bmatrix} + x_4 \begin{bmatrix} -1 \\ 1 \\ 0 \\ 1 \end{bmatrix}.$$

From this we see that the vectors

$$\mathbf{v}_1 = \begin{bmatrix} -1 \\ 0 \\ 1 \\ 0 \end{bmatrix} \quad \text{and} \quad \mathbf{v}_2 = \begin{bmatrix} -1 \\ 1 \\ 0 \\ 1 \end{bmatrix}$$

span $\mathbf{N}(A)$. On the other hand, they are clearly linearly independent, for if

$$c_1 \begin{bmatrix} -1 \\ 0 \\ 1 \\ 0 \end{bmatrix} + c_2 \begin{bmatrix} -1 \\ 1 \\ 0 \\ 1 \end{bmatrix} = \begin{bmatrix} -c_1 - c_2 \\ c_2 \\ c_1 \\ c_2 \end{bmatrix} = \begin{bmatrix} 0 \\ 0 \\ 0 \\ 0 \end{bmatrix},$$

then $c_1 = c_2 = 0$. Thus, $\{\mathbf{v}_1, \mathbf{v}_2\}$ gives a basis for $\mathbf{N}(A)$. ◀

One of the most beautiful and powerful relations among these subspaces is the following:

Proposition 4.2 *Let A be an m × n matrix. Then* $\mathbf{N}(A) = \mathbf{R}(A)^\perp$.

Proof If $\mathbf{x} \in \mathbf{N}(A)$, then, by definition, $\mathbf{A}_i \cdot \mathbf{x} = 0$ for all $i = 1, 2, \ldots, m$. (Remember that $\mathbf{A}_1, \ldots, \mathbf{A}_m$ denote the row vectors of the matrix A.) So it follows (see Exercise 1.3.3) that $\mathbf{x}$ is orthogonal to any linear combination of $\mathbf{A}_1, \ldots, \mathbf{A}_m$, hence to any vector in $\mathbf{R}(A)$. That is, $\mathbf{x} \in \mathbf{R}(A)^\perp$, so $\mathbf{N}(A) \subset \mathbf{R}(A)^\perp$. Now we need only show that $\mathbf{R}(A)^\perp \subset \mathbf{N}(A)$. If $\mathbf{x} \in \mathbf{R}(A)^\perp$, this means that $\mathbf{x}$ is orthogonal to every vector in $\mathbf{R}(A)$, so, in particular, $\mathbf{x}$ is orthogonal to each of the row vectors $\mathbf{A}_1, \ldots, \mathbf{A}_m$. But this means that $A\mathbf{x} = \mathbf{0}$, so $\mathbf{x} \in \mathbf{N}(A)$, as required. ∎

It is also the case that $\mathbf{R}(A) = \mathbf{N}(A)^\perp$, but we are not quite yet in a position to establish this.

Since $\mathbf{C}(A) = \mathbf{R}(A^\mathsf{T})$, the following is immediate:

Corollary 4.3 *Let A be an m × n matrix. Then* $\mathbf{N}(A^\mathsf{T}) = \mathbf{C}(A)^\perp$.

In fact, we really came across this earlier, when we found constraint equations for $A\mathbf{x} = \mathbf{b}$ to be consistent. Just as multiplying A by $\mathbf{x}$ takes linear combinations of the columns of A, so then does multiplying A^T by $\mathbf{x}$ take linear combinations of the *rows* of A (perhaps it helps to think of $A^\mathsf{T}\mathbf{x}$ as $(\mathbf{x}^\mathsf{T} A)^\mathsf{T}$). Corollary 4.3 is the statement that any linear combination of the rows of A that gives $\mathbf{0}$ corresponds to a constraint on $\mathbf{C}(A)$ and vice

EXAMPLE 2

Let
$$A = \begin{bmatrix} 1 & 2 \\ 1 & 1 \\ 0 & 1 \\ 1 & 2 \end{bmatrix}.$$

We wish to find a homogeneous system of linear equations describing $\mathbf{C}(A)$. That is, we seek the equations $\mathbf{b} \in \mathbb{R}^4$ must satisfy in order for $A\mathbf{x} = \mathbf{b}$ to be consistent. By row reduction, we find:

$$\begin{bmatrix} 1 & 2 & | & b_1 \\ 1 & 1 & | & b_2 \\ 0 & 1 & | & b_3 \\ 1 & 2 & | & b_4 \end{bmatrix} \rightsquigarrow \begin{bmatrix} 1 & 2 & | & b_1 \\ 0 & -1 & | & b_2 - b_1 \\ 0 & 1 & | & b_3 \\ 0 & 0 & | & b_4 - b_1 \end{bmatrix} \rightsquigarrow \begin{bmatrix} 1 & 2 & | & b_1 \\ 0 & 1 & | & b_1 - b_2 \\ 0 & 0 & | & -b_1 + b_2 + b_3 \\ 0 & 0 & | & -b_1 + b_4 \end{bmatrix},$$

and so the constraint equations are

$$\begin{aligned} -b_1 + b_2 + b_3 &= 0 \\ -b_1 + b_4 &= 0. \end{aligned}$$

Now, if we keep track of the row operations involved in reducing A to echelon form, we find that

$$\begin{bmatrix} 1 & 0 & 0 & 0 \\ -1 & 1 & 0 & 0 \\ -1 & 1 & 1 & 0 \\ -1 & 0 & 0 & 1 \end{bmatrix} A = \begin{bmatrix} 1 & 2 \\ 0 & -1 \\ 0 & 0 \\ 0 & 0 \end{bmatrix},$$

from which we see that

$$-\mathbf{A}_1 + \mathbf{A}_2 + \mathbf{A}_3 = -\mathbf{A}_1 + \mathbf{A}_4 = \mathbf{0}.$$

Thus, we infer that

$$\begin{bmatrix} -1 \\ 1 \\ 1 \\ 0 \end{bmatrix} \text{ and } \begin{bmatrix} -1 \\ 0 \\ 0 \\ 1 \end{bmatrix}$$

span $\mathbf{N}(A^\mathsf{T})$. On the other hand, in this instance, it is easy to see they are linearly independent and hence give a basis for $\mathbf{N}(A^\mathsf{T})$. ◀

▶ **EXAMPLE 3**

Let
$$A = \begin{bmatrix} 1 & 1 & 0 & 1 & 4 \\ 1 & 2 & 1 & 1 & 6 \\ 0 & 1 & 1 & 1 & 3 \\ 2 & 2 & 0 & 1 & 7 \end{bmatrix}.$$

Gaussian elimination gives us the reduced echelon form R:

$$R = \begin{bmatrix} 1 & 0 & -1 & 0 \\ -1 & 1 & 0 & 0 \\ 1 & -1 & 1 & 0 \\ -1 & -1 & 1 & 1 \end{bmatrix} \begin{bmatrix} 1 & 1 & 0 & 1 & 4 \\ 1 & 2 & 1 & 1 & 6 \\ 0 & 1 & 1 & 1 & 3 \\ 2 & 2 & 0 & 1 & 7 \end{bmatrix} = \begin{bmatrix} 1 & 0 & -1 & 0 & 1 \\ 0 & 1 & 1 & 0 & 2 \\ 0 & 0 & 0 & 1 & 1 \\ 0 & 0 & 0 & 0 & 0 \end{bmatrix}.$$

From this information, we wish to read off bases for each of the subspaces $\mathbf{R}(A)$, $\mathbf{N}(A)$, $\mathbf{C}(A)$, and $\mathbf{N}(A^T)$.

Using the result of Exercise 1, $\mathbf{R}(A) = \mathbf{R}(R)$, so the nonzero rows of R span $\mathbf{R}(A)$; now we need only check that they form a linearly independent set. We keep an eye on the pivot "slots": Suppose

$$c_1 \begin{bmatrix} 1 \\ 0 \\ -1 \\ 0 \\ 1 \end{bmatrix} + c_2 \begin{bmatrix} 0 \\ 1 \\ 1 \\ 0 \\ 2 \end{bmatrix} + c_3 \begin{bmatrix} 0 \\ 0 \\ 0 \\ 1 \\ 1 \end{bmatrix} = \mathbf{0}.$$

This means that

$$\begin{bmatrix} c_1 \\ c_2 \\ -c_1 + c_2 \\ c_3 \\ c_1 + 2c_2 + c_3 \end{bmatrix} = \begin{bmatrix} 0 \\ 0 \\ 0 \\ 0 \\ 0 \end{bmatrix},$$

and so $c_1 = c_2 = c_3 = 0$, as promised.

From the reduced echelon form R, we read off the vectors that span $\mathbf{N}(A)$: The general solution of $A\mathbf{x} = \mathbf{0}$ is

$$\mathbf{x} = \begin{bmatrix} x_3 - x_5 \\ -x_3 - 2x_5 \\ x_3 \\ -x_5 \\ x_5 \end{bmatrix} = x_3 \begin{bmatrix} 1 \\ -1 \\ 1 \\ 0 \\ 0 \end{bmatrix} + x_5 \begin{bmatrix} -1 \\ -2 \\ 0 \\ -1 \\ 1 \end{bmatrix},$$

so

$$\begin{bmatrix} 1 \\ -1 \\ 1 \\ 0 \\ 0 \end{bmatrix} \quad \text{and} \quad \begin{bmatrix} -1 \\ -2 \\ 0 \\ -1 \\ 1 \end{bmatrix}$$

span $\mathbf{N}(A)$. On the other hand, these vectors are linearly independent, for if we take a linear combination

$$x_3 \begin{bmatrix} 1 \\ -1 \\ 1 \\ 0 \\ 0 \end{bmatrix} + x_5 \begin{bmatrix} -1 \\ -2 \\ 0 \\ -1 \\ 1 \end{bmatrix} = \mathbf{0},$$

we infer (from the free variable slots) that $x_3 = x_5 = 0$. Thus, these two vectors form a basis for $\mathbf{N}(A)$.

Obviously, $\mathbf{C}(A)$ is spanned by the five column vectors of A. But these vectors cannot be linearly independent—that's what vectors in the nullspace of A tell us. From our vectors spanning $\mathbf{N}(A)$, we know that

(*) $\qquad \mathbf{a}_1 - \mathbf{a}_2 + \mathbf{a}_3 = \mathbf{0} \qquad$ and $\qquad -\mathbf{a}_1 - 2\mathbf{a}_2 - \mathbf{a}_4 + \mathbf{a}_5 = \mathbf{0}$.

These equations tell us that $\mathbf{a}_3$ and $\mathbf{a}_5$ can be written as linear combinations of $\mathbf{a}_1$, $\mathbf{a}_2$, and $\mathbf{a}_4$, and so these latter three vectors span $\mathbf{C}(A)$. If we can check that they form a linearly independent set, we'll know they give a basis for $\mathbf{C}(A)$. We form a matrix A' with these columns (easier: cross out the third and fifth columns of A) and reduce it to echelon form (easier: cross out the third and fifth columns of R). Well, we have

$$A' = \begin{bmatrix} 1 & 1 & 1 \\ 1 & 2 & 1 \\ 0 & 1 & 1 \\ 2 & 2 & 1 \end{bmatrix} \rightsquigarrow \begin{bmatrix} 1 & 0 & 0 \\ 0 & 1 & 0 \\ 0 & 0 & 1 \\ 0 & 0 & 0 \end{bmatrix} = R',$$

and so only the trivial linear combination of the columns of A' will yield the zero vector. In conclusion, the vectors

$$\mathbf{a}_1 = \begin{bmatrix} 1 \\ 1 \\ 0 \\ 2 \end{bmatrix}, \quad \mathbf{a}_2 = \begin{bmatrix} 1 \\ 2 \\ 1 \\ 2 \end{bmatrix}, \quad \text{and} \quad \mathbf{a}_4 = \begin{bmatrix} 1 \\ 1 \\ 1 \\ 1 \end{bmatrix}$$

are linearly independent and span $\mathbf{C}(A)$.

What about $\mathbf{N}(A^T)$? The only row of zeroes in R arises as the linear combination

$$-\mathbf{A}_1 - \mathbf{A}_2 + \mathbf{A}_3 + \mathbf{A}_4 = \mathbf{0}$$

of the rows of A, so we expect the vector

$$\mathbf{v} = \begin{bmatrix} -1 \\ -1 \\ 1 \\ 1 \end{bmatrix}$$

to give a basis for $\mathbf{N}(A^T)$. ◀

We now state the formal results regarding the four fundamental subspaces.

Theorem 4.4 *Let A be an $m \times n$ matrix. Let U and R, respectively, denote the echelon and reduced echelon form, respectively, of A, and write $EA = U$ (so E is the product of the elementary matrices by which we reduce A to echelon form).*

1. *The nonzero rows of U (or of R) give a basis for $\mathbf{R}(A)$.*
2. *The vectors obtained by setting each free variable equal to 1 and the remaining free variables equal to 0 in the general solution of $A\mathbf{x} = \mathbf{0}$ (which we read off from $R\mathbf{x} = \mathbf{0}$) give a basis for $\mathbf{N}(A)$.*
3. *The pivot columns of A (i.e., the columns of the original matrix A corresponding to the pivots in U) give a basis for $\mathbf{C}(A)$.*
4. *The (transposes of the) rows of E that correspond to the zero rows of U give a basis for $\mathbf{N}(A^T)$. (The same works with E' if we write $E'A = R$.)*

Proof For simplicity of exposition, let's assume that the reduced echelon form takes the shape

$$R = \begin{matrix} r \left\{ \vphantom{\begin{bmatrix}1\\ \\1\\ \hline 0\end{bmatrix}}\right. \\ m-r \left\{ \vphantom{\begin{bmatrix}0\end{bmatrix}}\right. \end{matrix} \left[\begin{array}{ccc|cccc} 1 & & & b_{1,r+1} & b_{1,r+2} & \cdots & b_{1n} \\ & \ddots & & \vdots & \vdots & \ddots & \vdots \\ & & 1 & b_{r,r+1} & b_{r,r+2} & \cdots & b_{rn} \\ \hline & \mathbf{0} & & & \mathbf{0} & & \end{array} \right].$$

1. Since row operations are invertible, $\mathbf{R}(A) = \mathbf{R}(U)$ (see Exercise 1). Clearly the nonzero rows of U span $\mathbf{R}(U)$. Moreover, they are linearly independent because of the pivots. Let $\mathbf{U}_1, \ldots, \mathbf{U}_r$ denote the nonzero rows of U; because of our simplifying assumption on R, we know that the pivots of U occur in the first r columns as well. Suppose now that

$$c_1\mathbf{U}_1 + \cdots + c_r\mathbf{U}_r = \mathbf{0}.$$

The first entry of the left-hand side is $c_1 u_{11}$ (since the first entry of the vectors $\mathbf{U}_2, \ldots, \mathbf{U}_r$ is 0 by definition of echelon form). Since $u_{11} \neq 0$ by definition of pivot, we must have $c_1 = 0$. Continuing in this fashion, we find that $c_1 = c_2 = \cdots = c_r = 0$. In conclusion, $\{\mathbf{U}_1, \ldots, \mathbf{U}_r\}$ forms a basis for $\mathbf{R}(U)$, hence for $\mathbf{R}(A)$.

2. $A\mathbf{x} = \mathbf{0}$ if and only if $R\mathbf{x} = \mathbf{0}$, which means that

$$\begin{aligned} x_1 & + b_{1,r+1}x_{r+1} + b_{1,r+2}x_{r+2} + \cdots + b_{1n}x_n = 0 \\ x_2 & + b_{2,r+1}x_{r+1} + b_{2,r+2}x_{r+2} + \cdots + b_{2n}x_n = 0 \\ & \vdots \phantom{+ b_{1,r+1}x_{r+1}} \vdots \phantom{+ b_{1,r+2}x_{r+2}} \vdots \vdots \\ x_r &+ b_{r,r+1}x_{r+1} + b_{r,r+2}x_{r+2} + \cdots + b_{rn}x_n = 0 \ . \end{aligned}$$

Thus, an arbitrary element of $\mathbf{N}(A)$ can be written in the form

$$\mathbf{x} = \begin{bmatrix} x_1 \\ \vdots \\ x_r \\ x_{r+1} \\ x_{r+2} \\ \vdots \\ x_n \end{bmatrix} = x_{r+1} \begin{bmatrix} -b_{1,r+1} \\ \vdots \\ -b_{r,r+1} \\ 1 \\ 0 \\ \vdots \\ 0 \end{bmatrix} + x_{r+2} \begin{bmatrix} -b_{1,r+2} \\ \vdots \\ -b_{r,r+2} \\ 0 \\ 1 \\ \vdots \\ 0 \end{bmatrix} + \cdots + x_n \begin{bmatrix} -b_{1n} \\ \vdots \\ -b_{rn} \\ 0 \\ 0 \\ \vdots \\ 1 \end{bmatrix}.$$

The assertion is then that the vectors

$$\begin{bmatrix} -b_{1,r+1} \\ \vdots \\ -b_{r,r+1} \\ 1 \\ 0 \\ \vdots \\ 0 \end{bmatrix}, \begin{bmatrix} -b_{1,r+2} \\ \vdots \\ -b_{r,r+2} \\ 0 \\ 1 \\ \vdots \\ 0 \end{bmatrix}, \ldots, \begin{bmatrix} -b_{1n} \\ \vdots \\ -b_{rn} \\ 0 \\ 0 \\ \vdots \\ 1 \end{bmatrix}$$

give a basis for $\mathbf{N}(A)$. They obviously span (since every vector in $\mathbf{N}(A)$ can be expressed as a linear combination of them). We need to check linear independence: The key is the pattern of 1's and 0's in the free variable "slots." Suppose

$$\mathbf{0} = \begin{bmatrix} 0 \\ \vdots \\ 0 \\ 0 \\ 0 \\ \vdots \\ 0 \end{bmatrix} = x_{r+1} \begin{bmatrix} -b_{1,r+1} \\ \vdots \\ -b_{r,r+1} \\ 1 \\ 0 \\ \vdots \\ 0 \end{bmatrix} + x_{r+2} \begin{bmatrix} -b_{1,r+2} \\ \vdots \\ -b_{r,r+2} \\ 0 \\ 1 \\ \vdots \\ 0 \end{bmatrix} + \cdots + x_n \begin{bmatrix} -b_{1n} \\ \vdots \\ -b_{rn} \\ 0 \\ 0 \\ \vdots \\ 1 \end{bmatrix}.$$

Then we get $x_{r+1} = x_{r+2} = \cdots = x_n = 0$, as required.

3. Let's continue with the notational simplification that the pivots occur in the first r columns. Then we need to establish the fact that the first r column vectors of the

original matrix A give a basis for $\mathbf{C}(A)$. These vectors form a linearly independent set since the only solution of

$$c_1\mathbf{a}_1 + \cdots + c_r\mathbf{a}_r = \mathbf{0}$$

is $c_1 = c_2 = \cdots = c_r = 0$ (look only at the first r columns of A and the first r columns of R). It is more interesting to understand why $\mathbf{a}_1, \ldots, \mathbf{a}_r$ span $\mathbf{C}(A)$. Consider each of the basis vectors for $\mathbf{N}(A)$ given above: Each one gives us a linear combination of the column vectors of A that results in the zero vector. In particular, we find that

$$\begin{aligned}
-b_{1,r+1}\mathbf{a}_1 - \cdots - b_{r,r+1}\mathbf{a}_r + \mathbf{a}_{r+1} &= \mathbf{0} \\
-b_{1,r+2}\mathbf{a}_1 - \cdots - b_{r,r+2}\mathbf{a}_r \quad\quad + \mathbf{a}_{r+2} &= \mathbf{0} \\
\vdots \quad\quad\quad\quad\quad\quad\quad\quad \vdots \quad\quad\quad\quad\quad\quad\quad \ddots \quad\quad\quad \vdots \\
-b_{1n}\mathbf{a}_1 - \cdots - b_{rn}\mathbf{a}_r \quad\quad\quad\quad\quad\quad + \mathbf{a}_n &= \mathbf{0},
\end{aligned}$$

from which we conclude that the vectors $\mathbf{a}_{r+1}, \ldots, \mathbf{a}_n$ are all linear combinations of $\mathbf{a}_1, \ldots, \mathbf{a}_r$. It follows that $\mathbf{C}(A)$ is spanned by $\mathbf{a}_1, \ldots, \mathbf{a}_r$, as required.

4. We are interested in the linear relations among the *rows* of A. The key point here is that the first r rows of the echelon matrix U form a linearly independent set, whereas the last $m - r$ rows of U consist just of $\mathbf{0}$. Thus, $\mathbf{N}(U^\mathsf{T})$ is spanned by the last $m - r$ standard basis vectors for $\mathbb{R}^m$. Using $EA = U$, we see that

$$A^\mathsf{T} = (E^{-1}U)^\mathsf{T} = U^\mathsf{T}(E^{-1})^\mathsf{T} = U^\mathsf{T}(E^\mathsf{T})^{-1},$$

and so

$$\begin{aligned}
\mathbf{x} \in \mathbf{N}(A^\mathsf{T}) &\iff \mathbf{x} \in \mathbf{N}(U^\mathsf{T}(E^\mathsf{T})^{-1}) \iff (E^\mathsf{T})^{-1}\mathbf{x} \in \mathbf{N}(U^\mathsf{T}) \\
&\iff \mathbf{x} = E^\mathsf{T}\mathbf{y} \text{ for some } \mathbf{y} \in \mathbf{N}(U^\mathsf{T}).
\end{aligned}$$

This tells us that the last $m - r$ rows of E span $\mathbf{N}(A^\mathsf{T})$. But these vectors are linearly independent since E is nonsingular. ∎

Remark Referring to our earlier discussion of (†) on p. 150 and our discussion in Sections 1 and 2 of this chapter, we finally know that finding the constraint equations for $\mathbf{C}(A)$ will give a *basis* for $\mathbf{N}(A^\mathsf{T})$. It is also worth noting that to find bases for the four fundamental subspaces of the matrix A, we need only find the echelon form of A to deal with $\mathbf{R}(A)$ and $\mathbf{C}(A)$, the reduced echelon form of A to deal with $\mathbf{N}(A)$, and the echelon form of the augmented matrix $[A \mid \mathbf{b}]$ to deal with $\mathbf{N}(A^\mathsf{T})$.

▶ **EXAMPLE 4**

We want bases for $\mathbf{R}(A)$, $\mathbf{N}(A)$, $\mathbf{C}(A)$, and $\mathbf{N}(A^\mathsf{T})$, given the matrix

$$A = \begin{bmatrix} 1 & 1 & 2 & 0 & 0 \\ 0 & 1 & 1 & -1 & -1 \\ 1 & 1 & 2 & 1 & 2 \\ 2 & 1 & 3 & -1 & -3 \end{bmatrix}.$$

We leave it to the reader to check that the reduced echelon form of A is

$$R = \begin{bmatrix} 1 & 0 & 1 & 0 & -1 \\ 0 & 1 & 1 & 0 & 1 \\ 0 & 0 & 0 & 1 & 2 \\ 0 & 0 & 0 & 0 & 0 \end{bmatrix}$$

and that $EA = U$, where

$$E = \begin{bmatrix} 1 & 0 & 0 & 0 \\ 0 & 1 & 0 & 0 \\ -1 & 0 & 1 & 0 \\ -4 & 1 & 2 & 1 \end{bmatrix} \quad \text{and} \quad U = \begin{bmatrix} 1 & 1 & 2 & 0 & 0 \\ 0 & 1 & 1 & -1 & -1 \\ 0 & 0 & 0 & 1 & 2 \\ 0 & 0 & 0 & 0 & 0 \end{bmatrix}.$$

Alternatively, the echelon form of the augmented matrix $[A \mid \mathbf{b}]$ is

$$[EA \mid E\mathbf{b}] = \begin{bmatrix} 1 & 1 & 2 & 0 & 0 & \bigm| & b_1 \\ 0 & 1 & 1 & -1 & -1 & \bigm| & b_2 \\ 0 & 0 & 0 & 1 & 2 & \bigm| & -b_1 + b_3 \\ 0 & 0 & 0 & 0 & 0 & \bigm| & -4b_1 + b_2 + 2b_3 + b_4 \end{bmatrix}.$$

Then we have the following bases for the respective subspaces:

$$\mathbf{R}(A): \left\{ \begin{bmatrix} 1 \\ 1 \\ 2 \\ 0 \\ 0 \end{bmatrix}, \begin{bmatrix} 0 \\ 1 \\ 1 \\ -1 \\ -1 \end{bmatrix}, \begin{bmatrix} 0 \\ 0 \\ 0 \\ 1 \\ 2 \end{bmatrix} \right\}$$

$$\mathbf{N}(A): \left\{ \begin{bmatrix} -1 \\ -1 \\ 1 \\ 0 \\ 0 \end{bmatrix}, \begin{bmatrix} 1 \\ -1 \\ 0 \\ -2 \\ 1 \end{bmatrix} \right\}$$

$$\mathbf{C}(A): \left\{ \begin{bmatrix} 1 \\ 0 \\ 1 \\ 2 \end{bmatrix}, \begin{bmatrix} 1 \\ 1 \\ 1 \\ 1 \end{bmatrix}, \begin{bmatrix} 0 \\ -1 \\ 1 \\ -1 \end{bmatrix} \right\}$$

$$\mathbf{N}(A^T): \left\{ \begin{bmatrix} -4 \\ 1 \\ 2 \\ 1 \end{bmatrix} \right\}.$$

The reader should check these all carefully. Note that $\dim \mathbf{R}(A) = \dim \mathbf{C}(A) = 3$, $\dim \mathbf{N}(A) = 2$, and $\dim \mathbf{N}(A^T) = 1$. ◀

We now deduce the following results on dimension. Recall that the rank of a matrix is the number of pivots in its echelon form.

Theorem 4.5 *Let A be an m × n matrix of rank r. Then*

1. $\dim \mathbf{R}(A) = \dim \mathbf{C}(A) = r$.
2. $\dim \mathbf{N}(A) = n - r$.
3. $\dim \mathbf{N}(A^\mathsf{T}) = m - r$.

Proof There are r pivots and a pivot in each nonzero row of U, so $\dim \mathbf{R}(A) = r$. Similarly, we have a basis vector for $\mathbf{C}(A)$ for every pivot, so $\dim \mathbf{C}(A) = r$, as well. We see that $\dim \mathbf{N}(A)$ is equal to the number of free variables, and this is the difference between the total number of variables (n) and the number of pivot variables (r). Last, the number of zero rows in U is the difference between the total number of rows (m) and the number of nonzero rows (r), so $\dim \mathbf{N}(A^\mathsf{T}) = m - r$. ∎

An immediate corollary of Theorem 4.5 is the following. The dimension of the nullspace of A is often called the *nullity* of A, denoted $\mathrm{null}(A)$. (Cf. also Exercise 4.3.22.)

Corollary 4.6 (Nullity-Rank Theorem) *Let A be an m × n matrix. Then*

$$\mathrm{null}(A) + \mathrm{rank}(A) = n.$$

Now we are in a position to complete our discussion of orthogonal complements.

Proposition 4.7 *Let $V \subset \mathbb{R}^n$ be a k-dimensional subspace. Then $\dim V^\perp = n - k$.*

Proof Choose a basis $\{\mathbf{v}_1, \ldots, \mathbf{v}_k\}$ for V, and let these be the *rows* of a $k \times n$ matrix A. By construction, we have $\mathbf{R}(A) = V$. Notice also that $\mathrm{rank}(A) = \dim \mathbf{R}(A) = \dim V = k$. By Proposition 4.2, we have $V^\perp = \mathbf{N}(A)$, so $\dim V^\perp = \dim \mathbf{N}(A) = n - k$. ∎

Now we arrive at our desired conclusion:

Proposition 4.8 *Let $V \subset \mathbb{R}^n$ be a subspace. Then $(V^\perp)^\perp = V$.*

Proof By Exercise 1.3.10, we have $V \subset (V^\perp)^\perp$. Now we calculate dimensions: If $\dim V = k$, then $\dim V^\perp = n - k$, and $\dim(V^\perp)^\perp = n - (n-k) = k$. Applying Lemma 3.8, we deduce that $V = (V^\perp)^\perp$. ∎

We can finally bring this discussion to a close with the geometric characterization of the relations among the four fundamental subspaces. Note that this result completes the story of Theorem 4.5.

Theorem 4.9 *Let A be an m × n matrix. Then*

1. $\mathbf{R}(A)^\perp = \mathbf{N}(A)$;
2. $\mathbf{N}(A)^\perp = \mathbf{R}(A)$;
3. $\mathbf{C}(A)^\perp = \mathbf{N}(A^\mathsf{T})$;
4. $\mathbf{N}(A^\mathsf{T})^\perp = \mathbf{C}(A)$.

Proof These are immediate from Proposition 4.2, Corollary 4.3, and Proposition 4.8. ∎

Now, using Theorem 4.9, we have an alternative way of expressing a subspace V spanned by a given set of vectors $\mathbf{v}_1, \ldots, \mathbf{v}_k$ as the solution set of a homogeneous system of linear equations. We use the vectors as *rows* of a matrix A; let $\{\mathbf{w}_1, \ldots, \mathbf{w}_\ell\}$ give a basis for $\mathbf{N}(A)$. Since $V = \mathbf{R}(A) = \mathbf{N}(A)^\perp$, we see that V is defined by the equations

$$\mathbf{w}_1 \cdot \mathbf{x} = 0, \quad \ldots, \quad \mathbf{w}_\ell \cdot \mathbf{x} = 0.$$

▶ **EXAMPLE 5**

Let

$$\mathbf{v}_1 = \begin{bmatrix} 1 \\ 1 \\ 0 \\ 1 \end{bmatrix} \quad \text{and} \quad \mathbf{v}_2 = \begin{bmatrix} 2 \\ 1 \\ 1 \\ 2 \end{bmatrix}.$$

We wish to write $V = \mathrm{Span}(\mathbf{v}_1, \mathbf{v}_2)$ as the solution set of a homogeneous system of linear equations. We introduce the matrix

$$A = \begin{bmatrix} 1 & 1 & 0 & 1 \\ 2 & 1 & 1 & 2 \end{bmatrix}$$

and find that

$$\mathbf{w}_1 = \begin{bmatrix} -1 \\ 1 \\ 1 \\ 0 \end{bmatrix} \quad \text{and} \quad \mathbf{w}_2 = \begin{bmatrix} -1 \\ 0 \\ 0 \\ 1 \end{bmatrix}$$

give a basis for $\mathbf{N}(A)$. By our earlier comments,

$$V = \mathbf{R}(A) = \mathbf{N}(A)^\perp$$
$$= \{\mathbf{x} \in \mathbb{R}^4 : \mathbf{w}_1 \cdot \mathbf{x} = 0, \; \mathbf{w}_2 \cdot \mathbf{x} = 0\}$$
$$= \{\mathbf{x} \in \mathbb{R}^4 : -x_1 + x_2 + x_3 = 0, \; -x_1 + x_4 = 0\}. \quad \blacktriangleleft$$

Earlier, e.g., in Example 2, we determined the constraint equations for the column space. The column space, as we've seen, is the intersection of hyperplanes whose normal vectors are the basis vectors for $\mathbf{N}(A^\mathsf{T})$. This is an application of the result that $\mathbf{C}(A) = \mathbf{N}(A^\mathsf{T})^\perp$. As we interchange A and A^T, we turn one method of solving the problem into the other.

To close our discussion now, we introduce in Figure 4.1 a schematic diagram summarizing the geometric relation among our four fundamental subspaces. We know that $\mathbf{N}(A)$ and $\mathbf{R}(A)$ are orthogonal complements of one another in $\mathbb{R}^n$ and that, similarly, $\mathbf{N}(A^\mathsf{T})$ and $\mathbf{C}(A)$ are orthogonal complements of one another in $\mathbb{R}^m$. But there is more to be said.

Recall that, given an $m \times n$ matrix A, we have linear maps $T: \mathbb{R}^n \to \mathbb{R}^m$ and $S: \mathbb{R}^m \to \mathbb{R}^n$ whose standard matrices are A and A^T, respectively. T sends all of $\mathbf{N}(A)$ to $\mathbf{0} \in \mathbb{R}^m$ and S sends all of $\mathbf{N}(A^\mathsf{T})$ to $\mathbf{0} \in \mathbb{R}^n$. Now, the column space of A consists of all vectors of

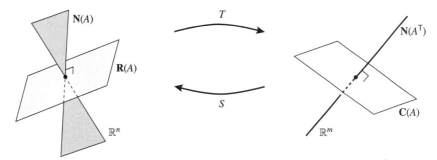

Figure 4.1

the form $A\mathbf{x}$ for some $\mathbf{x} \in \mathbb{R}^n$; that is, it is the image of the function T. Since $\dim \mathbf{R}(A) = \dim \mathbf{C}(A)$, this suggests that T maps the subspace $\mathbf{R}(A)$ one-to-one and onto $\mathbf{C}(A)$. (And, symmetrically, S maps $\mathbf{C}(A)$ one-to-one and onto $\mathbf{R}(A)$. These are, however, generally **not** inverse functions. Why? See Exercise 18.)

Proposition 4.10 *For each $\mathbf{b} \in \mathbf{C}(A)$, there is a unique vector $\mathbf{x} \in \mathbf{R}(A)$ so that $A\mathbf{x} = \mathbf{b}$.*

(See Figure 4.2.)

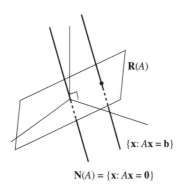

Figure 4.2

Proof Let $\{\mathbf{v}_1, \ldots, \mathbf{v}_r\}$ be a basis for $\mathbf{R}(A)$. Then $A\mathbf{v}_1, \ldots, A\mathbf{v}_r$ are r vectors in $\mathbf{C}(A)$. They are linearly independent (by a modification of the proof of Exercise 4.3.11 that we leave to the reader). Therefore, by Proposition 3.9, these vectors must span $\mathbf{C}(A)$. This tells us that every vector $\mathbf{b} \in \mathbf{C}(A)$ is of the form $\mathbf{b} = A\mathbf{x}$ for some $\mathbf{x} \in \mathbf{R}(A)$ (why?). And there can be only one such vector $\mathbf{x}$ because $\mathbf{R}(A) \cap \mathbf{N}(A) = \{\mathbf{0}\}$. ∎

Remark There is a further geometric interpretation of the vector $\mathbf{x} \in \mathbf{R}(A)$ that arises in the preceding proposition. Of all the solutions of $A\mathbf{x} = \mathbf{b}$, it is the one of least length. Why?

EXERCISES 4.4

*1. Show that if B is obtained from A by performing one or more row operations, then $\mathbf{R}(B) = \mathbf{R}(A)$.

2. Let $A = \begin{bmatrix} 1 & 2 & 1 & 1 \\ -1 & 0 & 3 & 4 \\ 2 & 2 & -2 & -3 \end{bmatrix}$.

(a) Give constraint equations for $\mathbf{C}(A)$. (b) Find a basis for $\mathbf{N}(A^T)$.

3. For each of the following matrices A, give bases for $\mathbf{R}(A)$, $\mathbf{N}(A)$, $\mathbf{C}(A)$, and $\mathbf{N}(A^T)$. Check dimensions and orthogonality.

(a) $A = \begin{bmatrix} 1 & 2 & 3 \\ 2 & 4 & 6 \end{bmatrix}$

(b) $A = \begin{bmatrix} 2 & 1 & 3 \\ 4 & 3 & 5 \\ 3 & 3 & 3 \end{bmatrix}$

(c) $A = \begin{bmatrix} 1 & -2 & 1 & 0 \\ 2 & -4 & 3 & -1 \end{bmatrix}$

(d) $A = \begin{bmatrix} 1 & -1 & 1 & 1 & 0 \\ 1 & 0 & 2 & 1 & 1 \\ 0 & 2 & 2 & 2 & 0 \\ -1 & 1 & -1 & 0 & -1 \end{bmatrix}$

(e) $A = \begin{bmatrix} 1 & 1 & 0 & 1 & -1 \\ 1 & 1 & 2 & -1 & 1 \\ 2 & 2 & 2 & 0 & 0 \\ -1 & -1 & 2 & -3 & 3 \end{bmatrix}$

*(f) $A = \begin{bmatrix} 1 & 1 & 0 & 5 & 0 & -1 \\ 0 & 1 & 1 & 3 & -2 & 0 \\ -1 & 2 & 3 & 4 & 1 & -6 \\ 0 & 4 & 4 & 12 & -1 & -7 \end{bmatrix}$

4. Given each matrix A, find matrices X and Y so that $\mathbf{C}(A) = \mathbf{N}(X)$ and $\mathbf{N}(A) = \mathbf{C}(Y)$.

*(a) $A = \begin{bmatrix} 3 & -1 \\ 6 & -2 \\ -9 & 3 \end{bmatrix}$

(b) $A = \begin{bmatrix} 1 & 1 & 0 \\ 2 & 1 & 1 \\ 1 & -1 & 2 \end{bmatrix}$

(c) $A = \begin{bmatrix} 1 & 1 & 1 \\ 1 & 2 & 0 \\ 1 & 1 & 1 \\ 1 & 0 & 2 \end{bmatrix}$

5. In each case, construct a matrix with the requisite properties or explain why no such matrix exists.

(a) The column space contains $\begin{bmatrix} 1 \\ 1 \\ 1 \end{bmatrix}$ and $\begin{bmatrix} 0 \\ 1 \\ 1 \end{bmatrix}$ and the nullspace contains $\begin{bmatrix} 1 \\ 0 \\ 1 \end{bmatrix}$ and $\begin{bmatrix} 0 \\ 1 \\ 0 \end{bmatrix}$.

*(b) The column space contains $\begin{bmatrix} 1 \\ 1 \\ 1 \end{bmatrix}$ and $\begin{bmatrix} 0 \\ 1 \\ 1 \end{bmatrix}$ and the nullspace contains $\begin{bmatrix} 1 \\ 0 \\ 1 \\ 0 \end{bmatrix}$ and $\begin{bmatrix} 1 \\ 0 \\ 0 \\ 1 \end{bmatrix}$.

*(c) The column space has basis $\begin{bmatrix} 1 \\ 0 \\ 1 \end{bmatrix}$ and the nullspace contains $\begin{bmatrix} 1 \\ 2 \\ 0 \end{bmatrix}$.

(d) The nullspace contains $\begin{bmatrix} 1 \\ 0 \\ 1 \end{bmatrix}, \begin{bmatrix} -1 \\ 2 \\ 1 \end{bmatrix}$, and the row space contains $\begin{bmatrix} 1 \\ 1 \\ -1 \end{bmatrix}$.

*(e) The column space has basis $\begin{bmatrix} 1 \\ 0 \\ 1 \end{bmatrix}, \begin{bmatrix} 0 \\ 1 \\ 1 \end{bmatrix}$, and the row space has basis $\begin{bmatrix} 1 \\ 1 \\ 1 \end{bmatrix}, \begin{bmatrix} 2 \\ 0 \\ 1 \end{bmatrix}$.

(f) The column space and the nullspace both have basis $\begin{bmatrix} 1 \\ 0 \end{bmatrix}$.

(g) The column space and the nullspace both have basis $\begin{bmatrix} 1 \\ 0 \\ 0 \end{bmatrix}$.

6. (a) Construct a 3×3 matrix A with $\mathbf{C}(A) \subset \mathbf{N}(A)$.
   (b) Construct a 3×3 matrix A with $\mathbf{N}(A) \subset \mathbf{C}(A)$.
   (c) Can there be a 3×3 matrix A with $\mathbf{N}(A) = \mathbf{C}(A)$? Why or why not?
   (d) Can there be a 4×4 matrix A with $\mathbf{N}(A) = \mathbf{C}(A)$? Why or why not?

7. Let $V \subset \mathbb{R}^5$ be spanned by $\begin{bmatrix} 1 \\ 0 \\ 1 \\ 1 \\ 1 \end{bmatrix}$ and $\begin{bmatrix} 0 \\ 1 \\ -1 \\ 0 \\ 2 \end{bmatrix}$. Give a homogeneous system of equations having V as its solution set.

*8. Give a basis for the orthogonal complement of each of the following subspaces of $\mathbb{R}^4$:

(a) $V = \mathrm{Span}\left(\begin{bmatrix} 1 \\ 0 \\ 3 \\ 4 \end{bmatrix}, \begin{bmatrix} 0 \\ 1 \\ 2 \\ -5 \end{bmatrix}\right)$.

(b) $W = \{\mathbf{x} \in \mathbb{R}^4 : x_1 + 3x_3 + 4x_4 = 0, \ x_2 + 2x_3 - 5x_4 = 0\}$.

9. (a) Give a basis for the orthogonal complement of the subspace $V \subset \mathbb{R}^4$ given by
$$V = \{\mathbf{x} \in \mathbb{R}^4 : x_1 + x_2 - 2x_4 = 0, \ x_1 - x_2 - x_3 + 6x_4 = 0, \ x_2 + x_3 - 4x_4 = 0\}.$$

(b) Give a basis for the orthogonal complement of the subspace
$$W = \mathrm{Span}\left(\begin{bmatrix} 1 \\ 1 \\ 0 \\ -2 \end{bmatrix}, \begin{bmatrix} 1 \\ -1 \\ -1 \\ 6 \end{bmatrix}, \begin{bmatrix} 0 \\ 1 \\ 1 \\ -4 \end{bmatrix}\right) \subset \mathbb{R}^4.$$

(c) Give a matrix B so that the subspace W defined in part b can be written in the form $W = \mathbf{N}(B)$.

*10. Let A be an $m \times n$ matrix with rank r. Suppose $A = BU$, where U is in echelon form. Prove that the first r columns of B give a basis for $\mathbf{C}(A)$. (In particular, if $EA = U$, where U is the echelon form of A and E is the product of elementary matrices by which we reduce A to U, then the first r columns of E^{-1} give a basis for $\mathbf{C}(A)$.)

11. According to Proposition 4.10, if A is an $m \times n$ matrix, then for each $\mathbf{b} \in \mathbf{C}(A)$, there is a unique $\mathbf{x} \in \mathbf{R}(A)$ with $A\mathbf{x} = \mathbf{b}$. In each case, give a formula for that $\mathbf{x}$.

(a) $A = \begin{bmatrix} 1 & 2 & 3 \\ 1 & 2 & 3 \end{bmatrix}$ *(b) $A = \begin{bmatrix} 1 & 1 & 1 \\ 0 & 1 & -1 \end{bmatrix}$

♯12. Let A be an $m \times n$ matrix and B be an $n \times p$ matrix. Prove that
(a) $\mathbf{N}(B) \subset \mathbf{N}(AB)$.
(b) $\mathbf{C}(AB) \subset \mathbf{C}(A)$. (Hint: Use Proposition 4.1.)
(c) $\mathbf{N}(B) = \mathbf{N}(AB)$ when A is $n \times n$ and nonsingular.
(d) $\mathbf{C}(AB) = \mathbf{C}(A)$ when B is $n \times n$ and nonsingular.

13. Continuing Exercise 12: Let A be an $m \times n$ matrix and B be an $n \times p$ matrix.
(a) Prove that $\text{rank}(AB) \le \text{rank}(A)$. (Hint: See part b of Exercise 12.)
(b) Prove that if $n = p$ and B is nonsingular, then $\text{rank}(AB) = \text{rank}(A)$.
(c) Prove that $\text{rank}(AB) \le \text{rank}(B)$. (Hint: Use part a of Exercise 12 and Theorem 4.5.)
(d) Prove that if $m = n$ and A is nonsingular, then $\text{rank}(AB) = \text{rank}(B)$.
(e) Prove that if $\text{rank}(AB) = n$, then $\text{rank}(A) = \text{rank}(B) = n$.

♯14. Let A be an $m \times n$ matrix. Prove that $\mathbf{N}(A^\mathsf{T} A) = \mathbf{N}(A)$. (Hint: Use Exercise 12 and Exercise 1.4.32.)

15. Let A be an $m \times n$ matrix.
(a) Use Theorem 4.9 to prove that $\mathbf{N}(A^\mathsf{T} A) = \mathbf{N}(A)$. (Hint: You've already proved $\supset$ in Exercise 12. Now, if $\mathbf{x} \in \mathbf{N}(A^\mathsf{T} A)$, then $A\mathbf{x} \in \mathbf{C}(A) \cap \mathbf{N}(A^\mathsf{T})$.)
(b) Prove that $\text{rank}(A) = \text{rank}(A^\mathsf{T} A)$.
(c) Prove that $\mathbf{C}(A^\mathsf{T} A) = \mathbf{C}(A^\mathsf{T})$. (Hint: You've already proved $\subset$ in Exercise 12. Use part b to see that the two spaces have the same dimension.)

16. Suppose A is an $n \times n$ matrix with the property that $A^2 = A$.
(a) Prove that $\mathbf{C}(A) = \{\mathbf{x} \in \mathbb{R}^n : \mathbf{x} = A\mathbf{x}\}$.
(b) Prove that $\mathbf{N}(A) = \{\mathbf{x} \in \mathbb{R}^n : \mathbf{x} = \mathbf{u} - A\mathbf{u} \text{ for some } \mathbf{u} \in \mathbb{R}^n\}$.
(c) Prove that $\mathbf{C}(A) \cap \mathbf{N}(A) = \{\mathbf{0}\}$.
(d) Prove that $\mathbf{C}(A) + \mathbf{N}(A) = \mathbb{R}^n$.

17. Suppose U and V are subspaces of $\mathbb{R}^n$. Prove that $(U \cap V)^\perp = U^\perp + V^\perp$. (Hint: Use Exercise 1.3.12 and Proposition 4.8.)

18. (a) Show that if the $m \times n$ matrix A has rank 1, then there are nonzero vectors $\mathbf{u} \in \mathbb{R}^m$ and $\mathbf{v} \in \mathbb{R}^n$ so that $A = \mathbf{u}\mathbf{v}^\mathsf{T}$. Describe the geometry of the four fundamental subspaces in terms of $\mathbf{u}$ and $\mathbf{v}$.
Pursuing the discussion on p. 183,
(b) Suppose A is an $m \times n$ matrix of rank n. Show that $A^\mathsf{T} A = I_n$ if and only if the column vectors $\mathbf{a}_1, \ldots, \mathbf{a}_n \in \mathbb{R}^m$ are mutually orthogonal unit vectors.
(c) Suppose A is an $m \times n$ matrix of rank 1. Using the notation of part a, show that $(S \circ T)(\mathbf{x}) = \mathbf{x}$ for each $\mathbf{x} \in \mathbf{R}(A)$ if and only if $\|\mathbf{u}\|\|\mathbf{v}\| = 1$. Interpret T geometrically.
(d) Can you generalize? (See Exercise 9.1.15.)

5 THE NONLINEAR CASE: INTRODUCTION TO MANIFOLDS

We have seen that given a *linear* subspace V of $\mathbb{R}^n$, we can represent it either explicitly (parametrically) as the span of its basis vectors or implicitly as the solution set of a homogeneous system of linear equations (i.e., the nullspace of an appropriate matrix A). Proposition 4.2 gives a geometric interpretation of that matrix: Its row vectors must span the orthogonal complement of V.

In the nonlinear case, sometimes we are just as fortunate. Given the hyperbola with equation $xy = 1$, it is easy to solve (everywhere) explicitly for either x or y as a function of the other. In the case of the circle $x^2 + y^2 = 1$, we can solve for y as a function of x *locally* near any point not on the x-axis (viz., $y = \pm\sqrt{1-x^2}$), and for x as a function of y near any point not on the y-axis (analogously).

But it is important to understand that going back and forth between these two approaches can be far more difficult—if not impossible—in the nonlinear case. For example, with a bit of luck, we can see that the parametric curve

$$\mathbf{g}(t) = \begin{bmatrix} t^2 - 1 \\ t(t^2 - 1) \end{bmatrix}, \quad t \in \mathbb{R},$$

is given by the algebraic equation $y^2 = x^2(x+1)$ (the curve pictured in Figure 1.4(b) on p. 55). On the other hand, the cycloid, presented parametrically as the image of the function

$$\mathbf{g}(t) = \begin{bmatrix} t - \sin t \\ 1 - \cos t \end{bmatrix}, \quad t \in \mathbb{R},$$

(see Figure 1.6 on p. 57) is obviously the graph $y = f(x)$ for some function f, but I believe no one can find f explicitly. Nor is there a function on $\mathbb{R}^2$ whose zero-set is the cycloid. Nevertheless, it is easy to see that *locally* we can write x as a function of y away from the cusps. On the other hand, given the *hypocycloid* $x^{2/3} + y^{2/3} = 1$, we can find the parametrization

$$\mathbf{g}(t) = \begin{bmatrix} \cos^3 t \\ \sin^3 t \end{bmatrix}, \quad t \in [0, 2\pi],$$

but giving an explicit (global) parametrization of the curve $y^2 - x^3 + x = 0$ in terms of elementary functions is impossible. However, as Figure 5.1 suggests, away from the points

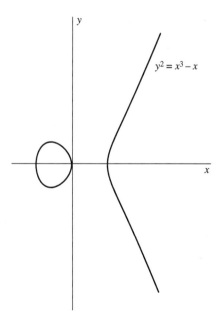

Figure 5.1

lying on the x-axis, we can write y as a function of x (explicitly in this case: $y = \pm\sqrt{x^3 - x}$), and near each of those three points we can write x as a function of y (explicitly only if you know how to solve the cubic equation $x^3 - x = y^2$ explicitly).

Given the hyperplane $\mathbf{a} \cdot \mathbf{x} = 0$ in $\mathbb{R}^n$, we can solve for x_n as a function of $x_1, \ldots, x_{n-1}$—i.e., we can represent the hyperplane as a *graph* over the $x_1 \cdots x_{n-1}$-plane—if and only if $a_n \neq 0$ (and, likewise, we can solve for x_k in terms of the remaining variables if and only if $a_k \neq 0$). More generally, given a system of linear equations, we apply Gaussian elimination and solve for the *pivot* variables as functions of the *free* variables. In particular, as Theorem 4.4 shows, if rank$(A) = r$, then we solve for the r pivot variables as functions of the $n - r$ free variables.

Now, since the derivative gives us the best linear approximation of a function, we expect that if the tangent plane to a surface at a point is a graph, then so locally should be the surface, as depicted in Figure 5.2. We suggested in Section 4 of Chapter 3 that, given a level surface $f = c$ of a differentiable function $f : \mathbb{R}^n \to \mathbb{R}$, the vector $\nabla f(\mathbf{a})$—provided it is nonzero—should be the normal vector to the tangent plane at $\mathbf{a}$; equivalently, the subspace of $\mathbb{R}^n$ parallel to the tangent plane should be the nullspace of the matrix $[Df(\mathbf{a})]$. To establish these facts we need the Implicit Function Theorem, whose proof we delay to Chapter 6.

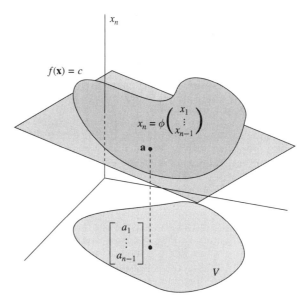

Figure 5.2

Theorem 5.1 (Implicit Function Theorem, Simple Case) *Suppose $U \subset \mathbb{R}^n$ is open, $\mathbf{a} \in U$, and $f : U \to \mathbb{R}$ is $\mathcal{C}^1$. Suppose that $f(\mathbf{a}) = 0$ and $\dfrac{\partial f}{\partial x_n}(\mathbf{a}) \neq 0$. Then there are neighborhoods V of $\begin{bmatrix} a_1 \\ \vdots \\ a_{n-1} \end{bmatrix}$ and W of a_n and a $\mathcal{C}^1$ function $\phi : V \to W$ so that*

5 The Nonlinear Case: Introduction to Manifolds ◀ 189

$$f(\mathbf{x}) = \mathbf{0}, \quad \begin{bmatrix} x_1 \\ \vdots \\ x_{n-1} \end{bmatrix} \in V, \quad \text{and} \quad x_n \in W \iff x_n = \phi \begin{pmatrix} x_1 \\ \vdots \\ x_{n-1} \end{pmatrix}.$$

That is, near **a**, *the level surface* $f = 0$ *can be expressed as a graph over the* $x_1 \cdots x_{n-1}$-*plane; i.e., near* **a**, *the equation* $f = 0$ *defines* x_n *implicitly as a function of the remaining variables.*

More generally, provided $Df(\mathbf{a}) \neq \mathbf{0}$, we know that *some* partial derivative $\dfrac{\partial f}{\partial x_k}(\mathbf{a}) \neq 0$, and so locally the equation $f = 0$ expresses x_k *implicitly* as a function of $x_1, \ldots, x_{k-1}, x_{k+1}, \ldots, x_n$.

▶ **EXAMPLE 1**

Consider the curve

$$f\begin{pmatrix} x \\ y \end{pmatrix} = y^3 - 3y - x = 0,$$

as shown in Figure 5.3. Although it is globally a graph of x as a function of y, we see that $\dfrac{\partial f}{\partial y} = 3(y^2 - 1) = 0$ at the points $\pm \begin{bmatrix} -2 \\ 1 \end{bmatrix}$. Away from these points, y is given (implicitly) locally as a function of x. We recognize these as the three ($\mathcal{C}^1$) local inverse functions ϕ_1, ϕ_2, and ϕ_3 of $g(x) = x^3 - 3x$, defined, respectively, on the intervals $(-2, \infty)$, $(-2, 2)$, and $(-\infty, 2)$. ◀

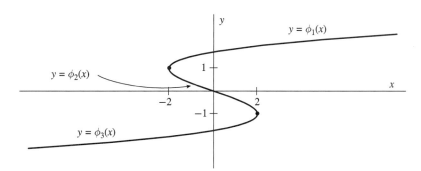

Figure 5.3

▶ **EXAMPLE 2**

Consider the surface

$$f\begin{pmatrix} x \\ y \\ z \end{pmatrix} = z^2 + xz + y = 0,$$

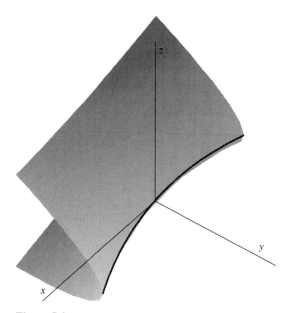

Figure 5.4

pictured in Figure 5.4. Note first of all that it is globally a graph: $y = -(z^2 + xz)$. On the other hand, $\dfrac{\partial f}{\partial z} = 2z + x = 0$ on $f = 0$ precisely when $x = -2z$ and $y = z^2$. That is, away from points of the form $\begin{bmatrix} -2t \\ t^2 \\ t \end{bmatrix}$ for some $t \in \mathbb{R}$, we can locally write $z = \phi\begin{pmatrix} x \\ y \end{pmatrix}$. Of course, it doesn't take a wizard to do so: We have

$$z = \frac{-x \pm \sqrt{x^2 - 4y}}{2},$$

and away from points of the designated form we can choose either the positive or negative square root. It is along the curve $4y = x^2$ (in the xy-plane) that the two roots of this quadratic equation in z coalesce. (Note that this curve is the projection of the locus of points on the surface where $\dfrac{\partial f}{\partial z} = 0$.) ◀

Now we can legitimize (finally) the process of *implicit differentiation* introduced in beginning calculus classes. Suppose $U \subset \mathbb{R}^n$ is open, $\mathbf{a} \in U$, $f \colon U \to \mathbb{R}$ is $\mathcal{C}^1$, and $\dfrac{\partial f}{\partial x_n}(\mathbf{a}) \neq 0$. For convenience here, let's write

$$\overline{\mathbf{x}} = \begin{bmatrix} x_1 \\ \vdots \\ x_{n-1} \end{bmatrix}.$$

Then, by Theorem 5.1, $f = 0$ defines x_n implicitly as a $\mathcal{C}^1$ function $\phi(\overline{\mathbf{x}})$ near $\mathbf{a}$. Then we have

Lemma 5.2 *For $j = 1, \ldots, n-1$, we have*
$$\frac{\partial \phi}{\partial x_j}(\bar{\mathbf{a}}) = -\frac{\frac{\partial f}{\partial x_j}(\mathbf{a})}{\frac{\partial f}{\partial x_n}(\mathbf{a})}.$$

Proof Define $g \colon V \to \mathbb{R}^n$ by $g(\bar{\mathbf{x}}) = \begin{bmatrix} \bar{\mathbf{x}} \\ \phi(\bar{\mathbf{x}}) \end{bmatrix}$. Then $(f \circ g)(\bar{\mathbf{x}}) = 0$ for all $\bar{\mathbf{x}} \in V$. Thus, by the Chain Rule, $D(f \circ g)(\bar{\mathbf{a}}) = Df(g(\bar{\mathbf{a}}))Dg(\bar{\mathbf{a}}) = \mathbf{0}$, so that

$$\begin{bmatrix} \frac{\partial f}{\partial x_1} & \cdots & \frac{\partial f}{\partial x_{n-1}} & \frac{\partial f}{\partial x_n} \end{bmatrix} \begin{bmatrix} 1 & \cdots & 0 \\ \vdots & \ddots & \vdots \\ 0 & \cdots & 1 \\ \frac{\partial \phi}{\partial x_1} & \cdots & \frac{\partial \phi}{\partial x_{n-1}} \end{bmatrix} = \begin{bmatrix} 0 & \cdots & 0 \end{bmatrix}.$$

(Here all the derivatives of ϕ are evaluated at $\bar{\mathbf{a}}$, and all the derivatives of f are evaluated at $g(\bar{\mathbf{a}}) = \mathbf{a}$.) In particular, for any $j = 1, \ldots, n-1$, we have

$$\frac{\partial f}{\partial x_j}(\mathbf{a}) + \frac{\partial f}{\partial x_n}(\mathbf{a})\frac{\partial \phi}{\partial x_j}(\bar{\mathbf{a}}) = 0,$$

from which the result is immediate. ∎

Now we can officially prove our assertion from Chapter 3.

Proposition 5.3 *Suppose $U \subset \mathbb{R}^n$ is open, $\mathbf{a} \in U$, $f \colon U \to \mathbb{R}$ is $\mathcal{C}^1$, and $Df(\mathbf{a}) \neq \mathbf{0}$. Suppose $f(\mathbf{a}) = c$. Then the tangent hyperplane at $\mathbf{a}$ of the level surface $M = f^{-1}(\{c\})$ is given by*
$$T_{\mathbf{a}}M = \{\mathbf{x} \in \mathbb{R}^n : Df(\mathbf{a})(\mathbf{x} - \mathbf{a}) = 0\};$$
that is, $\nabla f(\mathbf{a})$ is normal to the tangent hyperplane.

Proof Since $Df(\mathbf{a}) \neq \mathbf{0}$, we may assume without loss of generality that $\frac{\partial f}{\partial x_n}(\mathbf{a}) \neq 0$. Applying Theorem 5.1 to the function $f - c$, we know that M can be expressed near $\mathbf{a}$ as the graph $x_n = \phi(\bar{\mathbf{x}})$ for some $\mathcal{C}^1$ function ϕ. Now, the tangent plane to the graph $x_n = \phi(\bar{\mathbf{x}})$ at $\mathbf{a} = \begin{bmatrix} \bar{\mathbf{a}} \\ \phi(\bar{\mathbf{a}}) \end{bmatrix}$ is the graph of $D\phi(\bar{\mathbf{a}})$, translated so that it passes through $\mathbf{a}$:

$$x_n - a_n = D\phi(\bar{\mathbf{a}})(\bar{\mathbf{x}} - \bar{\mathbf{a}}) = \sum_{j=1}^{n-1} \frac{\partial \phi}{\partial x_j}(\bar{\mathbf{a}})(x_j - a_j)$$

$$= \sum_{j=1}^{n-1} \left(-\frac{\frac{\partial f}{\partial x_j}(\mathbf{a})}{\frac{\partial f}{\partial x_n}(\mathbf{a})}\right)(x_j - a_j) \quad \text{(by Lemma 5.2)},$$

192 ▶ Chapter 4. Implicit and Explicit Solutions of Linear Systems

and so, by simple algebra, we obtain

$$\sum_{j=1}^{n-1} \frac{\partial f}{\partial x_j}(\mathbf{a})(x_j - a_j) + \frac{\partial f}{\partial x_n}(\mathbf{a})(x_n - a_n) = Df(\mathbf{a})(\mathbf{x} - \mathbf{a}) = 0,$$

as required. ∎

From Theorem 5.1 we infer that if $f: \mathbb{R}^n \to \mathbb{R}$ is $\mathcal{C}^1$ and $\nabla f \neq \mathbf{0}$ on the level surface $M = f^{-1}(\{c\})$, then at each point $\mathbf{a} \in M$, we can locally represent M as a graph over (at least) one of the n coordinate hyperplanes. We call such a set M a *smooth hypersurface* or $(n-1)$-dimensional *manifold*. More generally, a subset $M \subset \mathbb{R}^n$ is an $(n-m)$-dimensional manifold if each point has a neighborhood that is a $\mathcal{C}^1$ graph over some $(n-m)$-dimensional coordinate plane. The general version of the Implicit Function Theorem, which we shall prove in Chapter 6, tells us that this is true whenever M is the level set of a $\mathcal{C}^1$ function $\mathbf{F}: \mathbb{R}^n \to \mathbb{R}^m$ with the property that rank$(D\mathbf{F}(\mathbf{x})) = m$ at every point $\mathbf{x} \in M$. Moreover, if we generalize the result of Proposition 5.3, the $(n-m)$-dimensional tangent plane of M at a point $\mathbf{a}$ is then obtained by translating the $(n-m)$-dimensional subspace $\mathbf{N}([D\mathbf{F}(\mathbf{a})])$ so that it passes through $\mathbf{a}$.

▶ **EXAMPLE 3**

Suppose $a, b > 0$. Consider the intersection M of the cylinders $x^2 + y^2 = a^2$ and $x^2 + z^2 = b^2$. We claim that as long as $a \neq b$, M is a smooth curve (1-dimensional manifold), as pictured in Figure 5.5. If we define $\mathbf{F}: \mathbb{R}^3 \to \mathbb{R}^2$ by

$$\mathbf{F}\begin{pmatrix} x \\ y \\ z \end{pmatrix} = \begin{bmatrix} x^2 + y^2 - a^2 \\ x^2 + z^2 - b^2 \end{bmatrix},$$

then $M = \mathbf{F}^{-1}(\{\mathbf{0}\})$. To see that M is a 1-dimensional manifold, we check that rank$(D\mathbf{F}(\mathbf{x})) = 2$ for every $\mathbf{x} \in M$. We have

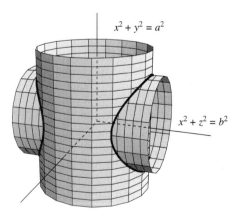

Figure 5.5

$$DF\begin{pmatrix}x\\y\\z\end{pmatrix} = \begin{bmatrix} 2x & 2y & 0 \\ 2x & 0 & 2z \end{bmatrix} = 2\begin{bmatrix} x & y & 0 \\ x & 0 & z \end{bmatrix}.$$

If $x \neq 0$, this matrix will have two pivots, since y and z can't be simultaneously 0. If $x = 0$, then both y and z are nonzero, and once again the matrix has two pivots. Thus, as claimed, the rank of $DF(\mathbf{x})$ is 2 for every $\mathbf{x} \in M$, and so M is a smooth curve. ◀

▶ EXERCISES 4.5

1. Can one solve for one of the variables in terms of the other to express each of the following as a graph? What about locally?
   (a) $xy = 0$
   (b) $2\sin(xy) = 1$

2. Decide whether each of the following is a smooth curve (1-dimensional manifold). If not, what are the trouble points?
   (a) $y^2 - x^3 + x = 0$
   (b) $y^2 - x^3 - x^2 = 0$
   (c) $z - xy = y - x^2 = 0$
   (d) $x^2 + y^2 + z^2 - 1 = x^2 - x + y^2 = 0$
   (e) $x^2 + y^2 + z^2 - 1 = z^2 - xy = 0$

*3. Let
$$f\begin{pmatrix}x\\y\\z\end{pmatrix} = xy^2 + \sin(xz) + e^z \quad \text{and} \quad \mathbf{a} = \begin{bmatrix} 1 \\ -1 \\ 0 \end{bmatrix}.$$

   (a) Show that the equation $f = 2$ defines z as a $\mathcal{C}^1$ function $z = \phi\begin{pmatrix}x\\y\end{pmatrix}$ near $\mathbf{a}$.
   (b) Find $\dfrac{\partial \phi}{\partial x}\begin{pmatrix}1\\-1\end{pmatrix}$ and $\dfrac{\partial \phi}{\partial y}\begin{pmatrix}1\\-1\end{pmatrix}$.
   (c) Find the equation of the tangent plane of the surface $f^{-1}(\{2\})$ at $\mathbf{a}$ in two ways.

4. Suppose $h \colon \mathbb{R}^2 \to \mathbb{R}$ is $\mathcal{C}^1$ and $\dfrac{\partial h}{\partial x_2} \neq 0$. Show that the equation $h\begin{pmatrix}y/x\\z/x\end{pmatrix} = 0$ defines z (locally) implicitly as a $\mathcal{C}^1$ function $z = \phi\begin{pmatrix}x\\y\end{pmatrix}$ and show that
$$x\frac{\partial \phi}{\partial x} + y\frac{\partial \phi}{\partial y} = \phi\begin{pmatrix}x\\y\end{pmatrix}.$$

5. Prove that $S^{n-1} = \{\mathbf{x} \in \mathbb{R}^n : \|\mathbf{x}\| = 1\}$ is an $(n-1)$-dimensional manifold. (Hint: Note that $\|\mathbf{x}\| = 1 \iff \|\mathbf{x}\|^2 = 1$.)

*6. Let $f \colon \mathbb{R}^3 \to \mathbb{R}$ be given by
$$f\begin{pmatrix}x\\y\\z\end{pmatrix} = z^2 + 4x^3z - 6xyz + 4y^3 - 3x^2y^2.$$

Is $M = f^{-1}(\{0\})$ a smooth surface (2-dimensional manifold)? If not, at what points does it fail to be so?

7. Show that the intersection of the surfaces $x^2 + 2y^2 + 3z^2 = 9$ and $x^2 + y^2 = z^2$ is a smooth curve.
Find its tangent line at the point $\mathbf{a} = \begin{bmatrix} 1 \\ 1 \\ \sqrt{2} \end{bmatrix}$.

8. Investigate what happens in Example 3 when $a = b$.

9. Show that the set of nonzero singular 2×2 matrices is a 3-dimensional manifold in $\mathcal{M}_{2 \times 2} = \mathbb{R}^4$.

10. Consider the curve $f\begin{pmatrix} x \\ y \end{pmatrix} = 4y^3 - 3y - x = 0$.

   (a) Sketch the curve.
   (b) Check that y is given (locally) by the following $\mathcal{C}^1$ functions of x on the given intervals:
   $$\phi_1(x) = \tfrac{1}{2}\big((x + \sqrt{x^2 - 1})^{1/3} + (x + \sqrt{x^2 - 1})^{-1/3}\big), \ x \in (1, \infty);$$
   $$\phi_2(x) = \cos(\tfrac{1}{3} \arccos x), \ x \in (-1, 1).$$
   Give the remaining functions (two defined on $(-1, 1)$, one on $(-\infty, -1)$).
   (c) Show that the function $\phi \colon (-1, \infty) \to \mathbb{R}$ defined by
   $$\phi(x) = \begin{cases} \phi_2(x), & x \in (-1, 1) \\ 1, & x = 1 \\ \phi_1(x), & x \in (1, \infty) \end{cases}$$
   is $\mathcal{C}^1$ and that the value of $\phi'(1)$ agrees with that given by Lemma 5.2.

*11. Let $M = \{\mathbf{x} \in \mathbb{R}^4 : x_1^2 + x_2^2 + x_3^2 + x_4^2 = 1, x_1 x_2 = x_3 x_4\}$.
   (a) Show that M is a smooth surface (2-dimensional manifold).
   (b) Find the tangent plane of M at $\mathbf{a} = \begin{bmatrix} 1 \\ 0 \\ 0 \\ 0 \end{bmatrix}$ and at $\mathbf{a} = \begin{bmatrix} 1/2 \\ -1/2 \\ -1/2 \\ 1/2 \end{bmatrix}$.

12. Suppose $f \colon \mathbb{R}^3 \to \mathbb{R}$ is $\mathcal{C}^2$ and $\dfrac{\partial f}{\partial z}(\mathbf{a}) \ne 0$, so that $f = 0$ defines z implicitly as a $\mathcal{C}^2$ function ϕ of x and y near $\bar{\mathbf{a}}$. Show that
$$\frac{\partial^2 \phi}{\partial x^2}(\bar{\mathbf{a}}) = -\frac{\dfrac{\partial^2 f}{\partial z^2}\left(\dfrac{\partial f}{\partial x}\right)^2 - 2\dfrac{\partial^2 f}{\partial x \partial z}\dfrac{\partial f}{\partial x}\dfrac{\partial f}{\partial z} + \dfrac{\partial^2 f}{\partial x^2}\left(\dfrac{\partial f}{\partial z}\right)^2}{\left(\dfrac{\partial f}{\partial z}\right)^3},$$
where all the partial derivatives on the right-hand side are evaluated at $\mathbf{a}$.

13. Consider the three (pairwise) skew lines
$$\ell_1 : \quad \mathbf{x} = s\begin{bmatrix} 1 \\ 0 \\ 0 \end{bmatrix};$$
$$\ell_2 : \quad \mathbf{x} = \begin{bmatrix} 0 \\ 1 \\ 0 \end{bmatrix} + t\begin{bmatrix} 1 \\ 0 \\ 1 \end{bmatrix};$$

$$\ell_3: \quad \mathbf{x} = \begin{bmatrix} 1 \\ 2 \\ 2 \end{bmatrix} + u \begin{bmatrix} 1 \\ 0 \\ 2 \end{bmatrix}.$$

Show that through each point of ℓ_3 there is a single line that intersects both ℓ_1 and ℓ_2. Now, find the equation of the surface formed by all the lines intersecting the three lines ℓ_1, ℓ_2, and ℓ_3. Is it everywhere smooth? Sketch it.

14. Suppose $X \subset \mathbb{R}^n$ is a k-dimensional manifold and $Y \subset \mathbb{R}^p$ is an ℓ-dimensional manifold. Prove that

$$X \times Y = \left\{ \begin{bmatrix} \mathbf{x} \\ \mathbf{y} \end{bmatrix} \in \mathbb{R}^n \times \mathbb{R}^p : \mathbf{x} \in X \text{ and } \mathbf{y} \in Y \right\}$$

is a $(k + \ell)$-dimensional manifold in $\mathbb{R}^{n+p}$. (Hint: Recall that X is locally a graph over a k-dimensional coordinate plane in $\mathbb{R}^n$ and Y is locally a graph over an ℓ-dimensional coordinate plane in $\mathbb{R}^p$.)

15. (a) Suppose A is an $n \times (n + 1)$ matrix of rank n. Show that the 1-dimensional solution space of $A\mathbf{x} = \mathbf{b}$ varies continuously with $\mathbf{b} \in \mathbb{R}^n$. (First you must decide what this means!)
(b) Generalize.

CHAPTER

5

EXTREMUM PROBLEMS

In this chapter we turn to one of the standard topics in differential calculus, solving maximum/minimum problems. In single-variable calculus, the strategy is to invoke the Maximum Value Theorem (which guarantees that a continuous function on a *closed interval* achieves its maximum and minimum) and then to examine all critical points and the endpoints of the interval. In problems that are posed on open intervals, one must work harder to understand the global behavior of the function. For example, it is not too hard to prove that if a differentiable function has precisely one critical point on an interval and that critical point is a *local* maximum point, then it must indeed be the *global* maximum point. As we shall see, all of these issues are—not surprisingly—rather more subtle in higher dimensions. But just to stimulate the reader's geometric intuition, we pose a direct question here.

Query: Suppose $f\colon \mathbb{R}^2 \to \mathbb{R}$ is $\mathcal{C}^1$ and there is exactly one point **a** at which the tangent plane of the graph of f is horizontal. Suppose **a** is a local minimum point. Must it be a global minimum point?

We close the chapter with a discussion of projections and inconsistent linear systems, along with a brief treatment of inner product spaces.

▶ 1 COMPACTNESS AND THE MAXIMUM VALUE THEOREM

In Section 2 of Chapter 2 we introduced the basic topological notions of open and closed sets and sequences. Here we return to a few more questions of the topology of $\mathbb{R}^n$ in order to frame the higher-dimensional version of the Maximum Value Theorem. Let's begin by reminding ourselves why a closed interval is needed in the case of a continuous function of one variable: As Figure 1.1 illustrates, when an endpoint is missing or the interval extends to

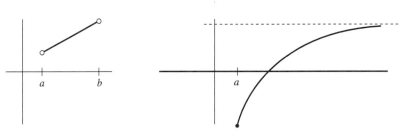

Figure 1.1

infinity, the function may have no maximum value. We now make the "obvious" definition in higher dimensions:

Definition We say $S \subset \mathbb{R}^n$ is *bounded* if all the points of S lie in some ball centered at the origin, i.e., if there is a constant M so that $\|\mathbf{x}\| \leq M$ for all $\mathbf{x} \in S$. We say $S \subset \mathbb{R}^n$ is *compact* if it is a bounded, closed subset. That is, all the points of S lie in some ball centered at the origin, and any convergent sequence of points in S converges to a point in S.

▶ **EXAMPLE 1**

We saw in Example 6 of Chapter 2, Section 2, that a closed interval in $\mathbb{R}$ is a closed subset, and it is obviously bounded, so it is in fact compact. Here are a few more examples.

  a. The unit sphere $S^{n-1} = \{\mathbf{x} \in \mathbb{R}^n : \|\mathbf{x}\| = 1\}$ is compact. Indeed, by Corollary 3.7 of Chapter 2, any level set of a continuous function is closed, so provided we have a bounded set, it will also be compact. (Note that we write S^{n-1} because the sphere is an $(n-1)$-dimensional manifold, as Exercise 4.5.5 shows.)

  b. Any rectangle $[a_1, b_1] \times \cdots \times [a_n, b_n] \subset \mathbb{R}^n$ is compact. This set is obviously bounded, and it is closed because of Exercise 2.2.4.

  c. The set of 2×2 matrices of determinant 1 is a closed subset of $\mathbb{R}^4$ (because the determinant is a polynomial expression in the entries of the matrix) but is not compact. The set is unbounded, as we can take matrices of the form $\begin{bmatrix} k & 0 \\ 0 & 1/k \end{bmatrix}$ for arbitrarily large k. ◀

One of the most important features of a compact set is the following

Theorem 1.1 *If $A \subset \mathbb{R}^n$ is compact, and $\{\mathbf{a}_k\}$ is a sequence of points in A, then there is a convergent subsequence $\{\mathbf{a}_{k_j}\}$ (which a fortiori converges to a point in A).*

Proof We first prove that any sequence of points in a rectangle $[a_1, b_1] \times \cdots \times [a_n, b_n] \subset \mathbb{R}^n$ has a convergent subsequence. (This was the result of Exercise 2.2.15, but the argument is sufficiently subtle that we include the proof here.) We proceed by induction on n.

Step (i): Suppose $n = 1$. Given a sequence $\{x_k\}$ of real numbers with $a \leq x_k \leq b$ for all k, we claim that there is a convergent subsequence. If there are only finitely many *distinct* numbers x_k, this is easy: At least one value must be taken on infinitely often, and we choose $k_1 < k_2 < \ldots$ so that $x_{k_1} = x_{k_2} = \ldots$.

If there are infinitely many distinct numbers among the x_k, then we use the famous "successive bisection" argument. Let $I_0 = [a, b]$. There must be infinitely many distinct elements of our sequence either to the left of the midpoint of I_0 or to the right; let $I_1 = [a_1, b_1]$ be the half that contains infinitely many (if both do, let's agree to choose the left half). Choose $x_{k_1} \in I_1$. At the next step, there must be infinitely many distinct elements of our sequence either to the left or to the right of the midpoint of I_1. Let $I_2 = [a_2, b_2]$ be the half that contains infinitely many (and choose the left half if both do), and choose $x_{k_2} \in I_2$ with $k_1 < k_2$. Continue this process inductively. Suppose we have the interval $I_j = [a_j, b_j]$

containing infinitely many distinct elements of our sequence, as well as $k_1 < k_2 < \cdots < k_j$ with $x_{k_\ell} \in I_\ell$ for $\ell = 1, 2, \ldots, j$. Then there must be infinitely many distinct elements of our sequence either to the left or to the right of the midpoint of the interval I_j, and we let $I_{j+1} = [a_{j+1}, b_{j+1}]$ be the half that contains infinitely many (once again choosing the left half if both do). We also choose $x_{k_{j+1}} \in I_{j+1}$ with $k_j < k_{j+1}$.

At the end of all this, why does the subsequence $\{x_{k_j}\}$ converge? Well, in fact, we know what its limit must be. The set of left endpoints a_j is nonempty and bounded above by b, hence has a least upper bound, α. First of all, the left endpoints a_j must converge to α, because (see Figure 1.2)

$$a_1 \leq a_2 \leq \cdots \leq a_j \leq \cdots \leq \alpha \leq \cdots \leq b_j \leq \cdots \leq b_2 \leq b_1,$$

and so $\alpha - a_j \leq b_j - a_j = (b-a)/2^j \to 0$ as $j \to \infty$. But since α and x_{k_j} both lie in the interval $[a_j, b_j]$, it follows that $|\alpha - x_{k_j}| \leq b_j - a_j \to 0$ as $j \to \infty$.

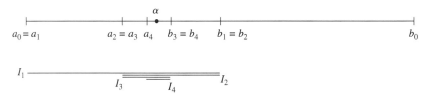

Figure 1.2

Step (ii): Suppose now $n \geq 2$ and we know the result to be true in $\mathbb{R}^{n-1}$. We introduce some notation: Given $\mathbf{x} = \begin{bmatrix} x_1 \\ \vdots \\ x_n \end{bmatrix} \in \mathbb{R}^n$, we write $\bar{\mathbf{x}} = \begin{bmatrix} x_1 \\ \vdots \\ x_{n-1} \end{bmatrix} \in \mathbb{R}^{n-1}$. Given a sequence $\{\mathbf{x}_k\}$ of points in the rectangle $[a_1, b_1] \times \cdots \times [a_n, b_n] \subset \mathbb{R}^n$, consider the sequence $\{\bar{\mathbf{x}}_k\}$ of points in the rectangle $[a_1, b_1] \times \cdots \times [a_{n-1}, b_{n-1}] \subset \mathbb{R}^{n-1}$. By our induction hypothesis, there is a convergent subsequence $\{\bar{\mathbf{x}}_{k_j}\}$. Now the sequence of nth coordinates of the corresponding vectors $\mathbf{x}_{k_j}$, lying in the closed interval $[a_n, b_n]$, has in turn a convergent subsequence, indexed by $k_{j_1} < k_{j_2} < \cdots < k_{j_\ell} < \ldots$. But then, by Exercises 2.2.6 and 2.2.2, it now follows that the subsequence $\{\mathbf{x}_{k_{j_\ell}}\}$ converges, as required.

Step (iii): Now we turn to the case of our general compact subset A. Since it is bounded, it is contained in some ball $B(\mathbf{0}, R)$ centered at the origin, hence in some cube $[-R, R] \times \cdots \times [-R, R]$. Thus, given a sequence $\{\mathbf{x}_k\}$ of points in A, it lies in this cube, and hence by what we've already proved has a convergent subsequence. The limit of that subsequence is, of course, a point of the cube but must in fact lie in A since A is also *closed*. This completes the proof. ■

The result that is the cornerstone of our work in this chapter is the following

Theorem 1.2 (Maximum Value Theorem) *Let $X \subset \mathbb{R}^n$ be compact, and let $f: X \to \mathbb{R}$ be a continuous function.*[1] *Then f takes on its maximum and minimum values; that is, there are points $\mathbf{y}$ and $\mathbf{z} \in X$ so that*

$$f(\mathbf{y}) \leq f(\mathbf{x}) \leq f(\mathbf{z}) \quad \text{for all } \mathbf{x} \in X.$$

Proof First we show that f is bounded (by which we mean that the set of its *values* is a bounded subset of $\mathbb{R}$). Assume to the contrary that the values of f are arbitrarily large. Then for each $k \in \mathbb{N}$ there is a point $\mathbf{x}_k \in X$ so that $f(\mathbf{x}_k) > k$. By Theorem 1.1, since X is compact, the sequence $\{\mathbf{x}_k\}$ has a convergent subsequence, say, $\mathbf{x}_{k_j} \to \mathbf{a}$. Since f is continuous, by Proposition 3.6 of Chapter 2, $f(\mathbf{a}) = \lim_{j \to \infty} f(\mathbf{x}_{k_j})$, but this is impossible since $f(\mathbf{x}_{k_j}) \to \infty$ as $j \to \infty$. An identical argument shows that the values of f are bounded below as well.

Since the set of values of f is bounded above, it has a least upper bound, M. By the definition of *least* upper bound, for each $k \in \mathbb{N}$ there is $\mathbf{x}_k \in X$ so that $M - f(\mathbf{x}_k) < 1/k$. As before, since X is compact, the sequence $\{\mathbf{x}_k\}$ has a convergent subsequence, say, $\mathbf{x}_{k_j} \to \mathbf{z}$. Then, by continuity, $f(\mathbf{z}) = \lim_{j \to \infty} f(\mathbf{x}_{k_j}) = M$, so f takes on its maximum value at $\mathbf{z}$. An identical argument shows that f takes on its minimum value as well. ∎

We infer from Theorem 1.2 that, given any linear map $T: \mathbb{R}^n \to \mathbb{R}^m$, the function

$$f: S^{n-1} \to \mathbb{R}$$
$$f(\mathbf{x}) = \|T(\mathbf{x})\|$$

is continuous (see Exercises 2.3.2 and 2.3.7 and Proposition 3.5 of Chapter 2). Therefore, f takes on its maximum value, which we denote by $\|T\|$, called the *norm* of T:

$$\|T\| = \max_{\|\mathbf{x}\|=1} \|T(\mathbf{x})\|.$$

Since T is linear, the following formula follows immediately:

Proposition 1.3 *Let $T: \mathbb{R}^n \to \mathbb{R}^m$ be a linear map. Then for any $\mathbf{x} \in \mathbb{R}^n$, we have*

$$\|T(\mathbf{x})\| \leq \|T\| \|\mathbf{x}\|.$$

Moreover, for any scalar c we have $\|cT\| = |c| \|T\|$; and if $S: \mathbb{R}^n \to \mathbb{R}^m$ is another linear map, we have $\|S + T\| \leq \|S\| + \|T\|$.

---

[1] Although we have not heretofore defined continuity of a function defined on an arbitrary subset of $\mathbb{R}^n$, there is no serious problem. We say $f: X \to \mathbb{R}$ is continuous at $\mathbf{a} \in X$ if, given any $\varepsilon > 0$, there is $\delta > 0$ so that

$$|f(\mathbf{x}) - f(\mathbf{a})| < \varepsilon \quad \text{whenever} \quad \|\mathbf{x} - \mathbf{a}\| < \delta \text{ and } \mathbf{x} \in X.$$

Proof There is nothing to prove when $\mathbf{x} = \mathbf{0}$. When $\mathbf{x} \neq \mathbf{0}$, we have
$$\left\| T\left(\frac{\mathbf{x}}{\|\mathbf{x}\|}\right) \right\| \leq \|T\|,$$
by definition of the norm, and so, using the linearity of T, we have
$$\|T(\mathbf{x})\| = \left\| T\left(\|\mathbf{x}\|\frac{\mathbf{x}}{\|\mathbf{x}\|}\right) \right\| = \|\mathbf{x}\| \left\| T\left(\frac{\mathbf{x}}{\|\mathbf{x}\|}\right) \right\| \leq \|T\|\|\mathbf{x}\|,$$
as required.

That $\max_{\|\mathbf{x}\|=1} \|cT(\mathbf{x})\| = |c| \max_{\|\mathbf{x}\|=1} \|T(\mathbf{x})\| = |c|\|T\|$ is evident. Now, last, since
$$\|(S+T)(\mathbf{x})\| \leq \|S(\mathbf{x})\| + \|T(\mathbf{x})\|,$$
we have
$$\max_{\|\mathbf{x}\|=1} \|(S+T)(\mathbf{x})\| \leq \max_{\|\mathbf{x}\|=1} \left(\|S(\mathbf{x})\| + \|T(\mathbf{x})\|\right)$$
$$\leq \max_{\|\mathbf{x}\|=1} \|S(\mathbf{x})\| + \max_{\|\mathbf{x}\|=1} \|T(\mathbf{x})\| = \|S\| + \|T\|. \blacksquare$$

We will compute a few nontrivial examples of the norm of a linear map in the Exercises of Section 4, but in the meantime we have the following.

▶ **EXAMPLE 2**

Let A be an $n \times n$ *diagonal* matrix, with diagonal entries $d_1, \ldots, d_n$. Then for any $\mathbf{x} \in S^{n-1}$ we have
$$\|A\mathbf{x}\|^2 = (d_1 x_1)^2 + (d_2 x_2)^2 + \cdots + (d_n x_n)^2$$
$$\leq \max(d_1^2, d_2^2, \ldots, d_n^2)(x_1^2 + \cdots + x_n^2) = \max(d_1^2, d_2^2, \ldots, d_n^2).$$
Note, moreover, that this maximum value is achieved, for if $\max(|d_1|, |d_2|, \ldots, |d_n|) = |d_i|$, then $A\mathbf{e}_i = d_i \mathbf{e}_i$ and $\|A\mathbf{e}_i\| = |d_i|$. Thus, we conclude that
$$\|A\| = \max(|d_1|, |d_2|, \ldots, |d_n|). \triangleleft$$

For future reference, we include the following important and surprising result.

Theorem 1.4 (Uniform Continuity Theorem) *Let $X \subset \mathbb{R}^n$ be compact and let $f: X \to \mathbb{R}$ be continuous. Then f is uniformly continuous; i.e., given $\varepsilon > 0$, there is $\delta > 0$ so that whenever $\|\mathbf{x} - \mathbf{y}\| < \delta$, $\mathbf{x}, \mathbf{y} \in X$, we have $|f(\mathbf{x}) - f(\mathbf{y})| < \varepsilon$.*

Proof We argue by contradiction. Suppose that for some $\varepsilon_0 > 0$ there were no such $\delta > 0$. Then for every $m \in \mathbb{N}$, we could find $\mathbf{x}_m, \mathbf{y}_m \in X$ with $\|\mathbf{x}_m - \mathbf{y}_m\| < 1/m$ and $|f(\mathbf{x}_m) - f(\mathbf{y}_m)| \geq \varepsilon_0$. Since X is compact, we may choose a convergent subsequence $\mathbf{x}_{m_k} \to \mathbf{a}$. Now since $\|\mathbf{x}_m - \mathbf{y}_m\| \to 0$ as $m \to \infty$, it must be the case that $\mathbf{y}_{m_k} \to \mathbf{a}$ as well. Since f is continuous at $\mathbf{a}$, given $\varepsilon_0 > 0$, there is $\delta_0 > 0$ so that whenever $\|\mathbf{x} - \mathbf{a}\| < \delta_0$, we have $|f(\mathbf{x}) - f(\mathbf{a})| < \varepsilon_0/2$. By the triangle inequality, whenever k is sufficiently large that $\|\mathbf{x}_{m_k} - \mathbf{a}\| < \delta_0$ and $\|\mathbf{y}_{m_k} - \mathbf{a}\| < \delta_0$, we have

$$|f(\mathbf{x}_{m_k}) - f(\mathbf{y}_{m_k})| \le |f(\mathbf{x}_{m_k}) - f(\mathbf{a})| + |f(\mathbf{y}_{m_k}) - f(\mathbf{a})| < \varepsilon_0,$$

contradicting our hypothesis that $|f(\mathbf{x}_m) - f(\mathbf{y}_m)| \ge \varepsilon_0$ for all m. ∎

EXERCISES 5.1

*1. Which of the following are compact subsets of the given $\mathbb{R}^n$? Give your reasoning. (Identify the space of all $n \times n$ matrices with $\mathbb{R}^{n^2}$.)

(a) $\left\{ \begin{bmatrix} x \\ y \end{bmatrix} \in \mathbb{R}^2 : x^2 + y^2 = 1 \right\}$

(b) $\left\{ \begin{bmatrix} x \\ y \end{bmatrix} \in \mathbb{R}^2 : x^2 + y^2 \le 1 \right\}$

(c) $\left\{ \begin{bmatrix} x \\ y \end{bmatrix} \in \mathbb{R}^2 : x^2 - y^2 = 1 \right\}$

(d) $\left\{ \begin{bmatrix} x \\ y \end{bmatrix} \in \mathbb{R}^2 : x^2 - y^2 \le 1 \right\}$

(e) $\left\{ \begin{bmatrix} x \\ y \end{bmatrix} \in \mathbb{R}^2 : y = \sin \frac{1}{x} \text{ for some } 0 < x \le 1 \right\}$

(f) $\left\{ \begin{bmatrix} e^t \cos t \\ e^t \sin t \end{bmatrix} \in \mathbb{R}^2 : t \in \mathbb{R} \right\}$

(g) $\left\{ \begin{bmatrix} e^t \cos t \\ e^t \sin t \end{bmatrix} \in \mathbb{R}^2 : t \le 0 \right\}$

(h) $\left\{ \begin{bmatrix} x \\ y \\ z \end{bmatrix} \in \mathbb{R}^3 : x^2 + y^2 + z^2 \le 1 \right\}$

(i) $\left\{ \begin{bmatrix} x \\ y \\ z \end{bmatrix} \in \mathbb{R}^3 : x^3 + y^3 + z^3 \le 1 \right\}$

(j) $\{3 \times 3 \text{ matrices } A : \det A = 1\}$

(k) $\{2 \times 2 \text{ matrices } A : A^\mathsf{T} A = I\}$

(l) $\{3 \times 3 \text{ matrices } A : A^\mathsf{T} A = I\}$

2. If $X \subset \mathbb{R}^n$ is not compact, then show that there is an unbounded continuous function $f \colon X \to \mathbb{R}$.

3. Let $T \colon \mathbb{R}^n \to \mathbb{R}$ be a linear map. Prove that there is a vector $\mathbf{a} \in \mathbb{R}^n$ so that $T(\mathbf{x}) = \mathbf{a} \cdot \mathbf{x}$, and deduce that $\|T\| = \|\mathbf{a}\|$.

4. Find $\|A\|$ if

*(a) $A = \begin{bmatrix} 1 & 1 \\ 1 & 1 \end{bmatrix}$;

(b) $A = \begin{bmatrix} 3 & 4 \\ 3 & 4 \end{bmatrix}$.

♯5. Suppose A is an $m \times n$ matrix. Prove that $\|A\| \le \sqrt{\sum_{i,j} a_{ij}^2} \le \sqrt{n}\|A\|$.

♯6. Suppose $T \colon \mathbb{R}^n \to \mathbb{R}^m$ and $S \colon \mathbb{R}^m \to \mathbb{R}^\ell$ are linear maps. Show that $\|S \circ T\| \le \|S\|\|T\|$. (In particular, when A is an $\ell \times m$ matrix and B is an $m \times n$ matrix, we have $\|AB\| \le \|A\|\|B\|$.)

7. Let A be an $m \times n$ matrix. Show that $\|A\| = \|A^\mathsf{T}\|$. (Hint: Start by showing that $\|A\| \le \|A^\mathsf{T}\|$ by using Proposition 4.5 of Chapter 1.)

8. Suppose $S \subset \mathbb{R}^n$ is compact and $\mathbf{a} \in \mathbb{R}^n$ is fixed. Show that there is a point of S closest to $\mathbf{a}$. (Hint: Use Exercise 2.3.2.)

*9. Suppose $S \subset \mathbb{R}^n$ has the property that any sequence of points in S has a subsequence converging to a point in S. Prove that S is compact.

10. Suppose $\mathbf{f}\colon X \to \mathbb{R}^m$ is continuous and X is compact. Prove that the set $\mathbf{f}(X) = \{\mathbf{y} \in \mathbb{R}^m : \mathbf{y} = \mathbf{f}(\mathbf{x}) \text{ for some } \mathbf{x} \in X\}$ is compact. (Hint: Use Exercise 9.)

11. Suppose $S_1 \supset S_2 \supset S_3 \supset \ldots$ are nonempty compact subsets of $\mathbb{R}^n$. Prove that there is $\mathbf{x} \in \mathbb{R}^n$ so that $\mathbf{x} \in S_k$ for all $k \in \mathbb{N}$. (Cf. Exercise 2.2.10.)

♯12. Suppose $X \subset \mathbb{R}^n$ is a compact set. Suppose $U_1, U_2, U_3, \ldots \subset \mathbb{R}^n$ are open sets whose union contains X. Prove that for some $N \in \mathbb{N}$ we have $X \subset U_1 \cup \cdots \cup U_N$. (Hint: If not, for each k, choose $\mathbf{x}_k \in X$ so that $\mathbf{x}_k \notin U_1 \cup \cdots \cup U_k$.)

13. Suppose $X \subset \mathbb{R}^n$ is compact, and $U_1, U_2, U_3, \ldots \subset \mathbb{R}^n$ are open sets whose union contains X. Prove that there is a number $\delta > 0$ so that for every $\mathbf{x} \in X$, there is some $j \in \mathbb{N}$ so that $B(\mathbf{x}, \delta) \subset U_j$. (Hint: If not, for each $k \in \mathbb{N}$, what happens with $\delta = 1/k$?)

2 MAXIMUM/MINIMUM PROBLEMS

Definition Let $X \subset \mathbb{R}^n$, and let $\mathbf{a} \in X$. The function $f\colon X \to \mathbb{R}$ has a *global maximum* at $\mathbf{a}$ if $f(\mathbf{x}) \leq f(\mathbf{a})$ for all $\mathbf{x} \in X$; the function f has a *local maximum* at $\mathbf{a}$ if, for some $\delta > 0$, we have $f(\mathbf{x}) \leq f(\mathbf{a})$ for all $\mathbf{x} \in B(\mathbf{a}, \delta) \cap X$. We say $\mathbf{a}$ is a (local or global) *maximum point* of f.

Analogously, f has a *global minimum* at $\mathbf{a}$ if $f(\mathbf{x}) \geq f(\mathbf{a})$ for all $\mathbf{x} \in X$; the function f has a *local minimum* at $\mathbf{a}$ if, for some $\delta > 0$, we have $f(\mathbf{x}) \geq f(\mathbf{a})$ for all $\mathbf{x} \in B(\mathbf{a}, \delta) \cap X$. We say $\mathbf{a}$ is a (local or global) *minimum point* of f.

If $\mathbf{a}$ is either a local maximum or local minimum point, we say it is an *extremum*.

We begin with a somewhat silly example:

▶ EXAMPLE 1

If $f(x) = \begin{cases} 1, & x \in \mathbb{Q} \\ 0, & x \notin \mathbb{Q} \end{cases}$, then every point $\mathbf{a} \in \mathbb{Q}$ is a global maximum point and every point $\mathbf{a} \notin \mathbb{Q}$ is a global minimum point. ◀

Now for something a bit more substantial.

▶ EXAMPLE 2

Let $f\colon \mathbb{R}^2 \to \mathbb{R}$ be defined by $f\begin{pmatrix} x \\ y \end{pmatrix} = x^2 + 2xy + 3y^2$. Then

$$f\begin{pmatrix} x \\ y \end{pmatrix} = (x+y)^2 + 2y^2 \geq 0 \quad \text{for all } x, y.$$

From this we infer that $\mathbf{0}$ is a global minimum point. Indeed, $(x+y)^2 + 2y^2 = 0$ if and only if $x + y = y = 0$ if and only if $x = y = 0$, so $\mathbf{0}$ is the only global minimum point of f. But is $\mathbf{0}$ the only extremum? ◀

Lemma 2.1 *Suppose f is defined on some neighborhood of the extremum $\mathbf{a}$ and f is differentiable at $\mathbf{a}$. Then $Df(\mathbf{a}) = \mathbf{0}$ (or, equivalently, $\nabla f(\mathbf{a}) = \mathbf{0}$).*

Proof Suppose that **a** is a local minimum (the case of a local maximum is left to the reader). Then for any $\mathbf{v} \in \mathbb{R}^n$, there is $\delta > 0$ so that we have

$$f(\mathbf{a} + t\mathbf{v}) - f(\mathbf{a}) \geq 0 \quad \text{for all real numbers } t \text{ with } |t| < \delta.$$

This means that

$$\lim_{t \to 0^+} \frac{f(\mathbf{a} + t\mathbf{v}) - f(\mathbf{a})}{t} \geq 0 \quad \text{and} \quad \lim_{t \to 0^-} \frac{f(\mathbf{a} + t\mathbf{v}) - f(\mathbf{a})}{t} \leq 0.$$

Since f is differentiable at **a**, the directional derivative $D_\mathbf{v} f(\mathbf{a})$ exists, and so we must have

$$Df(\mathbf{a})\mathbf{v} = D_\mathbf{v} f(\mathbf{a}) = \lim_{t \to 0} \frac{f(\mathbf{a} + t\mathbf{v}) - f(\mathbf{a})}{t} = 0.$$

Since **v** is arbitrary, we infer that $Df(\mathbf{a}) = \mathbf{0}$. ∎

Remark Geometrically, if we consider f as a function of x_i only, fixing all the other variables, we get a curve with a local minimum at a_i, which must therefore have a flat tangent line. That is, all partial derivatives of f at **a** must be 0, and so the tangent plane must be horizontal.

Definition Suppose f is differentiable at **a**. We say **a** is a *critical point* if $Df(\mathbf{a}) = \mathbf{0}$. A critical point **a** with the property that $f(\mathbf{x}) < f(\mathbf{a})$ for some **x** near **a** and $f(\mathbf{x}) > f(\mathbf{a})$ for other **x** near **a** is called a *saddle point*.

In Section 3 we will devise a second-derivative test to attempt to distinguish among local maxima, local minima, and saddle points, typical ones of which are shown in Figure 2.1. In the sketch in Figure 2.2(a), we cannot tell whether we are at a local maximum or a local minimum; however, in (b) and (c) we strongly suspect a saddle point.

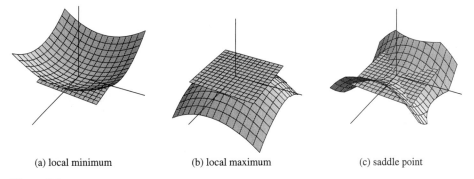

(a) local minimum (b) local maximum (c) saddle point

Figure 2.1

▶ **EXAMPLE 3**

The prototypical example of a saddle point is provided by the function $f\begin{pmatrix} x \\ y \end{pmatrix} = x^2 - y^2$. The origin is a critical point, and clearly $f\begin{pmatrix} x \\ 0 \end{pmatrix} > 0$ for $x \neq 0$ and $f\begin{pmatrix} 0 \\ y \end{pmatrix} < 0$ for $y \neq 0$. In the graph we see

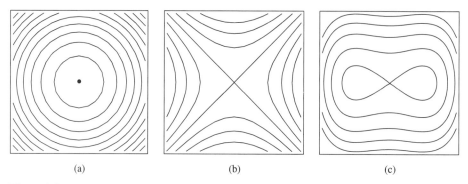

(a) (b) (c)

Figure 2.2

parabolas opening upward in the x-direction and those opening downward in the y-direction (see Figure 2.3(a)).

A somewhat more interesting example is provided by the so-called monkey saddle, pictured in Figure 2.3(b), which is the graph of $f\begin{pmatrix} x \\ y \end{pmatrix} = 3xy^2 - x^3$. Note that whereas the usual saddle surface allows room for the legs, in the case of the monkey saddle there is also room for the monkey's tail. ◀

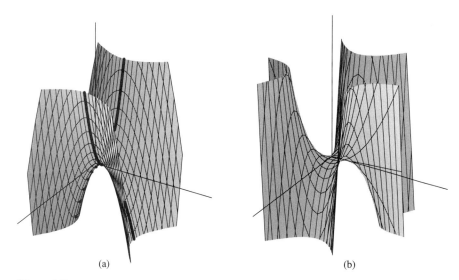

(a) (b)

Figure 2.3

Now we turn to the standard fare in differential calculus, the typical "applied extremum problems." If we are fortunate enough to have a differentiable function on a compact region X, then the Maximum Value Theorem guarantees both a global maximum and a global minimum, and we can test for critical points on the interior of X (points having a neighborhood wholly contained in X). It still remains to examine the function on the boundary of X, as well.

2 Maximum/Minimum Problems ◁ 205

▶ **EXAMPLE 4**

We want to find the hottest and coldest points on the metal plate $R = [0, \pi] \times [0, \pi]$, whose temperature is given by $f\begin{pmatrix} x \\ y \end{pmatrix} = \sin x + \cos 2y$. Since f is continuous and R is compact, we know the global maximum and minimum exist. We find that

$$Df\begin{pmatrix} x \\ y \end{pmatrix} = \begin{bmatrix} \cos x & -2\sin 2y \end{bmatrix},$$

and so the only critical point in the interior of R is $\begin{bmatrix} \pi/2 \\ \pi/2 \end{bmatrix}$. The boundary of R consists of four line segments, as indicated in Figure 2.4. On C_1 and C_3 we have $f\begin{pmatrix} x \\ 0 \end{pmatrix} = f\begin{pmatrix} x \\ \pi \end{pmatrix} = \sin x + 1$, $x \in [0, \pi]$, which achieves a maximum at $\pi/2$ and minima at 0 and π. Similarly, on C_2 and C_4 we have $f\begin{pmatrix} 0 \\ y \end{pmatrix} = f\begin{pmatrix} \pi \\ y \end{pmatrix} = \cos 2y$, $y \in [0, \pi]$, which achieves its maximum at 0 and π and its minimum at $\pi/2$. We now mark the values of f at the nine points we've unearthed. We see that the hottest points are $\begin{bmatrix} \pi/2 \\ 0 \end{bmatrix}$ and $\begin{bmatrix} \pi/2 \\ \pi \end{bmatrix}$ and the coldest points are $\begin{bmatrix} 0 \\ \pi/2 \end{bmatrix}$ and $\begin{bmatrix} \pi \\ \pi/2 \end{bmatrix}$. On the other hand, the critical point at the center of the square is a saddle point (why?). ◁

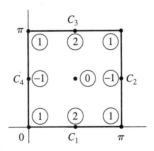

Figure 2.4

Somewhat more challenging are extremum problems where the domain is not naturally compact. Consider the following

▶ **EXAMPLE 5**

Of all rectangular boxes with no lid and having a volume of 4 m^3, we wish to determine the dimensions of the one with least total surface area. Let x, y, and z represent the length, width, and height of the box, respectively, measured in meters (see Figure 2.5). Given that $xyz = 4$, we wish to minimize the surface area $xy + 2z(x + y)$. Substituting $z = 4/xy$, we then define the surface area as a function of the independent variables x and y:

$$f\begin{pmatrix} x \\ y \end{pmatrix} = xy + \frac{8}{xy}(x + y) = xy + 8\left(\frac{1}{x} + \frac{1}{y}\right).$$

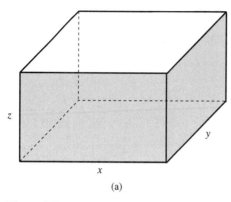
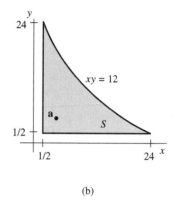

(a)   (b)

Figure 2.5

Note that the domain of f is the open first quadrant, i.e., $X = \left\{ \begin{bmatrix} x \\ y \end{bmatrix} : x > 0 \text{ and } y > 0 \right\}$, which is definitely not compact. What guarantees that our function f achieves a minimum value on X? (Note, for example, that f has no *maximum* value on X.) The heuristic answer is this: If either x or y gets either very small or very large, the value of f gets very large. We shall make this precise soon.

Let's first of all find the critical points of f. We have

$$Df\begin{pmatrix} x \\ y \end{pmatrix} = \left[y - \frac{8}{x^2} \quad x - \frac{8}{y^2} \right],$$

so at a critical point we must have

$$y - \frac{8}{x^2} = x - \frac{8}{y^2} = 0,$$

whence $x = y = 2$. The sole critical point is $\mathbf{a} = \begin{bmatrix} 2 \\ 2 \end{bmatrix}$ and $f\begin{pmatrix} 2 \\ 2 \end{pmatrix} = 12$. Now it is not difficult to establish the fact that $\mathbf{a}$ is the *global* minimum point of f. Let

$$S = \left\{ \begin{bmatrix} x \\ y \end{bmatrix} : \frac{1}{2} \le x \le 24, \ \frac{1}{2} \le y \le \frac{12}{x} \right\},$$

as in Figure 2.5(b). Then S is compact, so the restriction of f to the set S attains its global minimum value. Here is the crucial point: Whenever $\begin{bmatrix} x \\ y \end{bmatrix}$ is on the boundary of or outside S, we have $f\begin{pmatrix} x \\ y \end{pmatrix} > 12$. (For if either $0 < x \le \frac{1}{2}$ or $0 < y \le \frac{1}{2}$, then we have $f\begin{pmatrix} x \\ y \end{pmatrix} > 8(\frac{1}{x} + \frac{1}{y}) > 16$; and if $xy \ge 12$, then we have $f\begin{pmatrix} x \\ y \end{pmatrix} > 12$.) Since $f(\mathbf{a}) = 12$, it follows that the global minimum of f on S cannot occur on the boundary of S, hence must occur at an interior point, and therefore at a critical point of f. It follows that $\mathbf{a}$ is the global minimum point of f on S, hence on all of X since $f(\mathbf{x}) > f(\mathbf{a})$ whenever $\mathbf{x} \notin S$.

In summary, the box of the least surface area has the dimensions 2 m × 2 m × 1 m. ◀

EXERCISES 5.2

1. Find all the critical points of the following scalar functions:

   *(a) $f\begin{pmatrix}x\\y\end{pmatrix} = x^2 + 3x - 2y^2 + 4y$

   (b) $f\begin{pmatrix}x\\y\end{pmatrix} = xy + x - y$

   (c) $f\begin{pmatrix}x\\y\end{pmatrix} = \sin x + \sin y$

   (d) $f\begin{pmatrix}x\\y\end{pmatrix} = x^2 - 3x^2y + y^3$

   (e) $f\begin{pmatrix}x\\y\end{pmatrix} = x^2y + x^3 - x^2 + y^2$

   (f) $f\begin{pmatrix}x\\y\end{pmatrix} = (x^2 + y^2)e^{-y}$

   *(g) $f\begin{pmatrix}x\\y\end{pmatrix} = (x-y)e^{-(x^2+y^2)/4}$

   (h) $f\begin{pmatrix}x\\y\end{pmatrix} = x^2y - 4xy - y^2$

   *(i) $f\begin{pmatrix}x\\y\\z\end{pmatrix} = xyz - x^2 - y^2 + z^2$

   (j) $f\begin{pmatrix}x\\y\\z\end{pmatrix} = x^3 + xz^2 - 3x^2 + y^2 + 2z^2$

   (k) $f\begin{pmatrix}x\\y\\z\end{pmatrix} = e^{-(x^2+y^2+z^2)/6}(x - y + z)$

   (l) $f\begin{pmatrix}x\\y\\z\end{pmatrix} = xyz - x^2 - y^2 - z^2$

2. A rectangular box with edges parallel to the coordinate axes has one corner at the origin and the opposite corner on the plane $x + 2y + 3z = 6$. What is the maximum possible volume of the box?

*3. A rectangular box is inscribed in a *hemisphere* of radius r. Find the dimensions of the box of maximum volume.

*4. The temperature of the circular plate $D = \{\mathbf{x} : \|\mathbf{x}\| \leq \sqrt{2}\} \subset \mathbb{R}^2$ is given by the function $f\begin{pmatrix}x\\y\end{pmatrix} = x^2 + 2y^2 - 2x$. Find the maximum and minimum values of the temperature on D.

5. Two non-overlapping rectangles with their sides parallel to the coordinate axes are inscribed in the triangle with vertices at $\begin{bmatrix}0\\0\end{bmatrix}$, $\begin{bmatrix}1\\0\end{bmatrix}$, and $\begin{bmatrix}0\\1\end{bmatrix}$. What configuration will maximize the sum of their areas?

*6. A post office employee has 12 ft² of cardboard from which to construct a rectangular box with no lid. Find the dimensions of the box with the largest possible volume.

7. Show that the rectangular box of maximum volume with a given surface area is a cube.

8. The material for the sides of a rectangular box cost twice as much per ft² as that for the top and bottom. Find the relative dimensions of the box with greatest volume that can be constructed for a given cost.

9. Find the equation of the plane through the point $\begin{bmatrix}1\\2\\2\end{bmatrix}$ that cuts off the smallest possible volume in the first octant.

*10. A long, flat piece of sheet metal, 12″ wide, is to be bent to form a long trough with cross sections an isosceles trapezoid. Find the shape of the trough with maximum cross-sectional area. (Hint: It will help to use an angle as one of your variables.)

11. A pentagon is formed by placing an isosceles triangle atop a rectangle. If the perimeter P of the pentagon is fixed, find the dimensions of the rectangle and the height of the triangle that give the pentagon of maximum area.

12. An ellipse is formed by intersecting the cylinder $x^2 + y^2 = 1$ and the plane $x + 2y + z = 0$. Find the highest and lowest points on the ellipse. (As usual, the z-axis is vertical.)

13. Suppose x, y, and z are positive numbers with $xy^2z^3 = 108$. Find (with proof) the minimum value of their sum.

14. Let $\mathbf{a}_1, \ldots, \mathbf{a}_k \in \mathbb{R}^n$ be fixed points. Show that the function
$$f(\mathbf{x}) = \sum_{j=1}^{k} \|\mathbf{x} - \mathbf{a}_j\|^2$$
has a global minimum and find the global minimum point.

15. (Cf. Exercise 14.) Let $\mathbf{a}_1, \mathbf{a}_2, \mathbf{a}_3 \in \mathbb{R}^2$ be three noncollinear points. Show that the function
$$f(\mathbf{x}) = \sum_{j=1}^{3} \|\mathbf{x} - \mathbf{a}_j\|$$
has a global minimum and characterize the global minimum point. (Hint: Your answer will be geometric in nature. Can you give an explicit geometric construction?)

3 QUADRATIC FORMS AND THE SECOND DERIVATIVE TEST

Just as the second derivative test in single-variable calculus often allows us to differentiate between local minima and local maxima, there is something quite analogous in the multivariable case, to which we now turn. Of course, even with just one variable, if $f'(a) = f''(a) = 0$, we do not have enough information, and we need higher derivatives to infer the local behavior of f at a; lying behind this is the theory of the *Taylor polynomial*, which works analogously in the multivariable case. In the interest of time, however, we shall content ourselves here with just the second derivative.

First, we need a one-variable generalization of the Mean Value Theorem. (In truth, it is Taylor's Theorem with Remainder for the first-degree Taylor polynomial. See Chapter 20 of Spivak.)

Lemma 3.1 *Suppose $g \colon [0, 1] \to \mathbb{R}$ is twice differentiable. Then*
$$g(1) = g(0) + g'(0) + \tfrac{1}{2} g''(\xi) \qquad \text{for some } 0 < \xi < 1.$$

Proof Define the polynomial P by $P(t) = g(0) + g'(0)t + Ct^2$, where $C = g(1) - g(0) - g'(0)$. This choice of C makes $P(1) = g(1)$, and it is easy to see that $P(0) = g(0)$ and $P'(0) = g'(0)$ as well, as shown in Figure 3.1. Then the function $h = g - P$ satisfies $h(0) = h'(0) = h(1) = 0$. By Rolle's Theorem, since $h(0) = h(1) = 0$, there is $c \in (0, 1)$

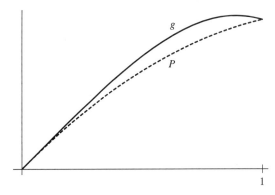

Figure 3.1

so that $h'(c) = 0$. By Rolle's Theorem applied to h', since $h'(0) = h'(c) = 0$, there is $\xi \in (0, c)$ so that $h''(\xi) = 0$. This means that $g''(\xi) = P''(\xi) = 2C$, and so

$$g(1) = P(1) = g(0) + g'(0) + \tfrac{1}{2}g''(\xi),$$

as required. ∎

The derivative in the multivariable setting becomes a linear map (or vector); as we shall soon see, the second derivative should become a *quadratic form*, i.e., a quadratic function of a vector variable.

Definition Assume f is a $\mathcal{C}^2$ function in a neighborhood of $\mathbf{a}$. Define the symmetric matrix

$$\mathrm{Hess}(f)(\mathbf{a}) = \left[\frac{\partial^2 f}{\partial x_i \partial x_j}(\mathbf{a})\right] = \begin{bmatrix} \frac{\partial^2 f}{\partial x_1^2}(\mathbf{a}) & \frac{\partial^2 f}{\partial x_1 \partial x_2}(\mathbf{a}) & \cdots & \frac{\partial^2 f}{\partial x_1 \partial x_n}(\mathbf{a}) \\ \frac{\partial^2 f}{\partial x_2 \partial x_1}(\mathbf{a}) & \frac{\partial^2 f}{\partial x_2^2}(\mathbf{a}) & \cdots & \frac{\partial^2 f}{\partial x_2 \partial x_n}(\mathbf{a}) \\ \vdots & \vdots & \ddots & \vdots \\ \frac{\partial^2 f}{\partial x_n \partial x_1}(\mathbf{a}) & \frac{\partial^2 f}{\partial x_n \partial x_2}(\mathbf{a}) & \cdots & \frac{\partial^2 f}{\partial x_n^2}(\mathbf{a}) \end{bmatrix}.$$

$\mathrm{Hess}(f)(\mathbf{a})$ is called the *Hessian* matrix of f at $\mathbf{a}$. Define the associated quadratic form

$$\mathcal{H}_{f,\mathbf{a}} \colon \mathbb{R}^n \to \mathbb{R} \quad \text{by}$$

$$\mathcal{H}_{f,\mathbf{a}}(\mathbf{h}) = \mathbf{h}^\mathsf{T}\bigl(\mathrm{Hess}(f)(\mathbf{a})\bigr)\mathbf{h} = \sum_{i,j=1}^n \frac{\partial^2 f}{\partial x_i \partial x_j}(\mathbf{a}) h_i h_j.$$

Now we are in a position to state the generalization of Lemma 3.1 to functions of several variables. This will enable us to deduce the appropriate second derivative test for extrema.

Proposition 3.2 *Suppose $f \colon B(\mathbf{a}, r) \to \mathbb{R}$ is $\mathcal{C}^2$. Then for all $\mathbf{h}$ with $\|\mathbf{h}\| < r$ we have*

$$f(\mathbf{a} + \mathbf{h}) = f(\mathbf{a}) + Df(\mathbf{a})\mathbf{h} + \tfrac{1}{2}\mathcal{H}_{f,\mathbf{a}+\xi\mathbf{h}}(\mathbf{h}) \quad \text{for some } 0 < \xi < 1.$$

Consequently,

$$f(\mathbf{a}+\mathbf{h}) = f(\mathbf{a}) + Df(\mathbf{a})\mathbf{h} + \tfrac{1}{2}\mathcal{H}_{f,\mathbf{a}}(\mathbf{h}) + \epsilon(\mathbf{h}), \quad \text{where } \epsilon(\mathbf{h})/\|\mathbf{h}\|^2 \to 0 \text{ as } \mathbf{h}\to \mathbf{0}.$$

Remark Just as the derivative gives the best *linear* approximation to f at $\mathbf{a}$, so adding the quadratic term $\tfrac{1}{2}\mathcal{H}_{f,\mathbf{a}}(\mathbf{h})$ gives the best possible *quadratic* approximation to f at $\mathbf{a}$. This is the second-degree *Taylor polynomial* of f at $\mathbf{a}$. For further reading on multivariable Taylor polynomials, consult, e.g., Edwards's *Advanced Calculus of Several Variables* or Hubbard and Hubbard's, *Vector Calculus, Linear Algebra, and Differential Forms: A Unified Approach*.

Proof We apply Lemma 3.1 to the function $g(t) = f(\mathbf{a}+t\mathbf{h})$. Using the chain rule twice (and applying Theorem 6.1 of Chapter 3 as well), we have

$$g'(t) = Df(\mathbf{a}+t\mathbf{h})\mathbf{h} = \sum_{i=1}^n \frac{\partial f}{\partial x_i}(\mathbf{a}+t\mathbf{h})h_i$$

$$g''(t) = \sum_{i=1}^n \Big(\sum_{j=1}^n \frac{\partial^2 f}{\partial x_j \partial x_i}(\mathbf{a}+t\mathbf{h})h_j\Big)h_i = \sum_{i,j=1}^n \frac{\partial^2 f}{\partial x_j \partial x_i}(\mathbf{a}+t\mathbf{h})h_i h_j$$

$$= \mathcal{H}_{f,\mathbf{a}+t\mathbf{h}}(\mathbf{h}).$$

Now substitution yields the first result.

Since f is $\mathcal{C}^2$, given any $\varepsilon > 0$, there is $\delta > 0$ so that whenever $\|\mathbf{v}\| < \delta$ we have

$$\|\mathrm{Hess}(f)(\mathbf{a}+\mathbf{v}) - \mathrm{Hess}(f)(\mathbf{a})\| < \varepsilon.$$

Using the Cauchy-Schwarz inequality, Proposition 2.3 of Chapter 1, and Proposition 1.3, we find that $|\mathbf{h}^\mathsf{T} A\mathbf{h}| \le \|A\|\|\mathbf{h}\|^2$. So whenever $\|\mathbf{h}\| < \delta$, we have, for any $0 < \xi < 1$,

$$|\mathcal{H}_{f,\mathbf{a}+\xi\mathbf{h}}(\mathbf{h}) - \mathcal{H}_{f,\mathbf{a}}(\mathbf{h})| < \varepsilon\|\mathbf{h}\|^2.$$

By definition, $\epsilon(\mathbf{h}) = \tfrac{1}{2}\big(\mathcal{H}_{f,\mathbf{a}+\xi\mathbf{h}}(\mathbf{h}) - \mathcal{H}_{f,\mathbf{a}}(\mathbf{h})\big)$, so

$$\frac{|\epsilon(\mathbf{h})|}{\|\mathbf{h}\|^2} = \frac{|\mathcal{H}_{f,\mathbf{a}+\xi\mathbf{h}}(\mathbf{h}) - \mathcal{H}_{f,\mathbf{a}}(\mathbf{h})|}{2\|\mathbf{h}\|^2} < \frac{\varepsilon}{2}$$

whenever $\|\mathbf{h}\| < \delta$. Since $\varepsilon > 0$ was arbitrary, this proves the result. ∎

Definition Given a symmetric $n\times n$ matrix A, we say the associated quadratic form $\mathcal{Q}\colon \mathbb{R}^n \to \mathbb{R}$, $\mathcal{Q}(\mathbf{x}) = \mathbf{x}^\mathsf{T} A\mathbf{x}$, is

- *positive definite* if $\mathcal{Q}(\mathbf{x}) > 0$ for all $\mathbf{x}\ne \mathbf{0}$,
- *negative definite* if $\mathcal{Q}(\mathbf{x}) < 0$ for all $\mathbf{x}\ne \mathbf{0}$,
- *positive semidefinite* if $\mathcal{Q}(\mathbf{x}) \ge 0$ for all $\mathbf{x}$ and $=0$ for some $\mathbf{x}\ne \mathbf{0}$,
- *negative semidefinite* if $\mathcal{Q}(\mathbf{x}) \le 0$ for all $\mathbf{x}$ and $=0$ for some $\mathbf{x}\ne \mathbf{0}$, and
- *indefinite* if $\mathcal{Q}(\mathbf{x}) > 0$ for some $\mathbf{x}$ and $\mathcal{Q}(\mathbf{x}) < 0$ for other $\mathbf{x}$.

3 Quadratic Forms and the Second Derivative Test

EXAMPLE 1

a. The quadratic form $Q(\mathbf{x}) = x_1^2 + 4x_1x_2 + 5x_2^2 = \mathbf{x}^T \begin{bmatrix} 1 & 2 \\ 2 & 5 \end{bmatrix} \mathbf{x}$ is positive definite, as we see by completing the square:

$$x_1^2 + 4x_1x_2 + 5x_2^2 = (x_1 + 2x_2)^2 + x_2^2,$$

being the sum of two squares (with positive coefficients), is nonnegative and can vanish only if $x_2 = x_1 + 2x_2 = 0$, i.e., only if $\mathbf{x} = \mathbf{0}$.

b. The quadratic form $Q(\mathbf{x}) = x_1^2 + 2x_1x_2 - x_2^2 = \mathbf{x}^T \begin{bmatrix} 1 & 1 \\ 1 & -1 \end{bmatrix} \mathbf{x}$ is indefinite, as we can see either by completing the square or merely by observing that $Q\begin{pmatrix} t \\ 0 \end{pmatrix} = t^2 > 0$ and $Q\begin{pmatrix} 0 \\ t \end{pmatrix} = -t^2 < 0$ for $t \neq 0$.

c. The quadratic form $Q(\mathbf{x}) = x_1^2 + 2x_1x_2 + 2x_2^2 + 2x_1x_3 + 2x_3^2 = \mathbf{x}^T \begin{bmatrix} 1 & 1 & 1 \\ 1 & 2 & 0 \\ 1 & 0 & 2 \end{bmatrix} \mathbf{x}$ is, however, positive semidefinite, for

$$x_1^2 + 2x_1x_2 + 2x_2^2 + 2x_1x_3 + 2x_3^2 = (x_1 + x_2 + x_3)^2 + x_2^2 - 2x_2x_3 + x_3^2$$
$$= (x_1 + x_2 + x_3)^2 + (x_2 - x_3)^2 \geq 0,$$

but note that $Q\begin{pmatrix} -2 \\ 1 \\ 1 \end{pmatrix} = 0.$ ◄

Theorem 3.3 *Suppose $f: B(\mathbf{a}, r) \to \mathbb{R}$ is C^2 and $\mathbf{a}$ is a critical point. If $\mathcal{H}_{f,\mathbf{a}}$ is positive (resp., negative) definite, then $\mathbf{a}$ is a local minimum (resp., maximum) point; if $\mathcal{H}_{f,\mathbf{a}}$ is indefinite, then $\mathbf{a}$ is a saddle point. If $\mathcal{H}_{f,\mathbf{a}}$ is semidefinite, we can draw no conclusions.*

Proof By Proposition 3.2, given $\varepsilon > 0$, there is $\delta > 0$ so that

$$f(\mathbf{a} + \mathbf{h}) - f(\mathbf{a}) = \tfrac{1}{2}\mathcal{H}_{f,\mathbf{a}}(\mathbf{h}) + \epsilon(\mathbf{h}) \quad \text{where} \quad \frac{|\epsilon(\mathbf{h})|}{\|\mathbf{h}\|^2} < \varepsilon \quad \text{whenever } \|\mathbf{h}\| < \delta.$$

Suppose now that $\mathcal{H}_{f,\mathbf{a}}$ is positive definite. By the Maximum Value Theorem, Theorem 1.2, there is a number $m > 0$ so that $\mathcal{H}_{f,\mathbf{a}}(\mathbf{x}) \geq m$ for all unit vectors $\mathbf{x}$. This means that $\mathcal{H}_{f,\mathbf{a}}(\mathbf{h}) \geq m\|\mathbf{h}\|^2$ for all $\mathbf{h}$. So now, choosing $\varepsilon = m/4$, we have

$$f(\mathbf{a} + \mathbf{h}) - f(\mathbf{a}) = \tfrac{1}{2}\mathcal{H}_{f,\mathbf{a}}(\mathbf{h}) + \epsilon(\mathbf{h}) > \tfrac{1}{4}m\|\mathbf{h}\|^2 > 0$$

for all $\mathbf{h}$ with $\|\mathbf{h}\| < \delta$. This means that $\mathbf{a}$ is a local minimum, as desired. The negative definite case is analogous.

Now suppose $\mathcal{H}_{f,\mathbf{a}}$ is indefinite. Then there are unit vectors $\mathbf{x}$ and $\mathbf{y}$ so that $\mathcal{H}_{f,\mathbf{a}}(\mathbf{x}) = m_1 > 0$ and $\mathcal{H}_{f,\mathbf{a}}(\mathbf{y}) = m_2 < 0$. Choose $\varepsilon = \tfrac{1}{4}\min(m_1, -m_2)$. Now, letting $\mathbf{h} = t\mathbf{x}$ (resp., $t\mathbf{y}$) with $|t| < \delta$, we see that

$$f(\mathbf{a} + t\mathbf{x}) - f(\mathbf{a}) > \tfrac{1}{4}m_1 t^2 > 0 \quad \text{and} \quad f(\mathbf{a} + t\mathbf{y}) - f(\mathbf{a}) < \tfrac{1}{4}m_2 t^2 < 0.$$

This means that $\mathbf{a}$ is a saddle point of f.

Last, note that if $\mathcal{H}_{f,\mathbf{a}}$ is positive semidefinite, then **a** may be either a local minimum, a local maximum (!), or a saddle. Consider, respectively, the functions $f\begin{pmatrix}x\\y\end{pmatrix} = x^2 + y^4$, $-x^4 - y^4$, and $x^2 + y^3$, all at the origin. ■

Corollary 3.4 When $n = 2$, assume f is $\mathcal{C}^2$ near the critical point **a** and $\text{Hess}(f)(\mathbf{a}) = \begin{bmatrix} A & B \\ B & C \end{bmatrix}$. Then

$AC - B^2 > 0$	and $A > 0$	**a** is a local minimum
	and $A < 0$	**a** is a local maximum
$AC - B^2 < 0$		**a** is a saddle point
$AC - B^2 = 0$		the test is inconclusive

Proof This is just the usual process of completing the square: When $A \neq 0$,

$$Ax^2 + 2Bxy + Cy^2 = A\left(x + \frac{B}{A}y\right)^2 + \left(C - \frac{B^2}{A}\right)y^2 = A\left(x + \frac{B}{A}y\right)^2 + \left(\frac{AC - B^2}{A}\right)y^2,$$

so the quadratic form is positive definite when $A > 0$ and $AC - B^2 > 0$, negative definite when $A < 0$ and $AC - B^2 > 0$, and indefinite when $AC - B^2 < 0$. When $A = 0$, we have $2Bxy + Cy^2 = y(2Bx + Cy)$, and so the quadratic form is indefinite provided $B \neq 0$, i.e., provided $AC - B^2 < 0$. ■

▶ **EXAMPLE 2**

Let's find and classify the critical points of the function $f : \mathbb{R}^2 \to \mathbb{R}$, $f\begin{pmatrix}x\\y\end{pmatrix} = x^3 + y^2 - 6xy$. Then

$$Df\begin{pmatrix}x\\y\end{pmatrix} = \begin{bmatrix} 3x^2 - 6y & 2y - 6x \end{bmatrix},$$

and so at a critical point we must have $2y = x^2 = 6x$. Thus, the critical points are $\mathbf{a} = \begin{bmatrix} 0 \\ 0 \end{bmatrix}$ and $\mathbf{b} = \begin{bmatrix} 6 \\ 18 \end{bmatrix}$.

Now, we calculate the Hessian:

$$\text{Hess}(f)\begin{pmatrix}x\\y\end{pmatrix} = \begin{bmatrix} 6x & -6 \\ -6 & 2 \end{bmatrix},$$

and so

$$\text{Hess}(f)\begin{pmatrix}0\\0\end{pmatrix} = \begin{bmatrix} 0 & -6 \\ -6 & 2 \end{bmatrix} \quad \text{and} \quad \text{Hess}(f)\begin{pmatrix}6\\18\end{pmatrix} = \begin{bmatrix} 36 & -6 \\ -6 & 2 \end{bmatrix}.$$

We see that $\mathcal{H}_{f,\mathbf{a}}$ is indefinite, so **a** is a saddle point, and $\mathcal{H}_{f,\mathbf{b}}$ is positive definite, so **b** is a local minimum point. ◀

3 Quadratic Forms and the Second Derivative Test ◀ 213

The process of completing the square as we've done in Example 1 can be couched in matrix language; indeed, it is intimately related to the reduction to echelon form, as we shall now see.

▶ **EXAMPLE 3**

Suppose we begin to reduce the *symmetric* matrix

$$A = \begin{bmatrix} 1 & 3 & 2 \\ 3 & 4 & -4 \\ 2 & -4 & -10 \end{bmatrix}$$

to echelon form. The first step is

$$A = \begin{bmatrix} 1 & 3 & 2 \\ 3 & 4 & -4 \\ 2 & -4 & -10 \end{bmatrix} \rightsquigarrow \begin{bmatrix} 1 & 3 & 2 \\ 0 & -5 & -10 \\ 0 & -10 & -14 \end{bmatrix} = A',$$

where

$$A' = \underbrace{\begin{bmatrix} 1 & & \\ -3 & 1 & \\ -2 & 0 & 1 \end{bmatrix}}_{E_1} A, \quad \text{and so} \quad A = \underbrace{\begin{bmatrix} 1 & & \\ 3 & 1 & \\ 2 & 0 & 1 \end{bmatrix}}_{E_1^{-1}} A'.$$

There are already two interesting observations to make: The first column of E_1^{-1} is the transpose of the first row of A (hence of A'); and if we remove the first row and column from A', what's left is also symmetric. Indeed, we can write

$$A = \begin{bmatrix} 1 \\ 3 \\ 2 \end{bmatrix} \begin{bmatrix} 1 & 3 & 2 \end{bmatrix} + \begin{bmatrix} 0 & 0 & 0 \\ 0 & -5 & -10 \\ 0 & -10 & -14 \end{bmatrix};$$

since the first term is symmetric (why?), the latter term must be as well. Now we just continue:

$$A' = \begin{bmatrix} 1 & 3 & 2 \\ 0 & -5 & -10 \\ 0 & -10 & -14 \end{bmatrix} \rightsquigarrow \begin{bmatrix} 1 & 3 & 2 \\ 0 & -5 & -10 \\ 0 & 0 & 6 \end{bmatrix} = U,$$

and so, as before,

$$U = \underbrace{\begin{bmatrix} 1 & & \\ & 1 & \\ & -2 & 1 \end{bmatrix}}_{E_2} A', \quad \text{and so} \quad A' = \underbrace{\begin{bmatrix} 1 & & \\ & 1 & \\ & 2 & 1 \end{bmatrix}}_{E_2^{-1}} U.$$

Summarizing, we have $A = LU$, where

$$L = E_1^{-1} E_2^{-1} = \begin{bmatrix} 1 & & \\ 3 & 1 & \\ 2 & 2 & 1 \end{bmatrix}$$

is a lower triangular matrix with 1's on the diagonal. Now here comes the amazing thing: If we factor out the diagonal entries of the echelon matrix U, we are left with L^T:

$$U = \underbrace{\begin{bmatrix} 1 & & \\ & -5 & \\ & & 6 \end{bmatrix}}_{D} \underbrace{\begin{bmatrix} 1 & 3 & 2 \\ & 1 & 2 \\ & & 1 \end{bmatrix}}_{L^T}.$$

Because A is symmetric, we arrive at the formula

$$A = LDL^T,$$

corresponding to the formula we get by completing the square:

$$\begin{aligned} x_1^2 + 6x_1x_2 + 4x_2^2 + 4x_1x_3 - 8x_2x_3 - 10x_3^2 &= (x_1 + 3x_2 + 2x_3)^2 - 5x_2^2 - 20x_2x_3 - 14x_3^2 \\ &= (x_1 + 3x_2 + 2x_3)^2 - 5(x_2^2 + 4x_2x_3) - 14x_3^2 \\ &= (x_1 + 3x_2 + 2x_3)^2 - 5(x_2 + 2x_3)^2 + 6x_3^2. \end{aligned}$$

(∗)

To complete the circle of ideas, note that

$$\mathcal{Q}(\mathbf{x}) = \mathbf{x}^T A \mathbf{x} = \mathbf{x}^T (LDL^T) \mathbf{x} = (L^T \mathbf{x})^T D (L^T \mathbf{x})$$

recaptures the form of (∗). ◀

Remark Of course, not every symmetric matrix can be written in the form LDL^T; e.g., take $A = \begin{bmatrix} 0 & 1 \\ 1 & 0 \end{bmatrix}$. The problem arises when we have to switch rows to get pivots in the appropriate places. Nevertheless, by doing appropriate row operations *together* with the companion column operations (to maintain symmetry), one can show that every symmetric matrix can be written in the form EDE^T, where E is the product of elementary matrices with only 1's on the diagonal (i.e., elementary matrices of type (iii)). See Exercise 8b for the example of the matrix A given just above.

Proposition 3.5 *Suppose A is a symmetric matrix with associated quadratic form $\mathcal{Q}$. Suppose $A = LDL^T$, where L is lower triangular with 1's on the diagonal and D is diagonal. If all the entries of D are positive (resp., negative), then $\mathcal{Q}$ is positive (resp., negative) definite; if all the entries of D are nonnegative (resp., nonpositive) and at least one is 0, then $\mathcal{Q}$ is positive (resp., negative) semidefinite; and if entries of D have opposite sign, then $\mathcal{Q}$ is indefinite.*

Conversely, if $\mathcal{Q}$ is positive (resp., negative) definite, then there are a lower triangular matrix L with 1's on the diagonal and a diagonal matrix D with positive (resp., negative) entries so that $A = LDL^T$. If $\mathcal{Q}$ is semidefinite (resp., indefinite), the matrix EAE^T (where E is a suitable product of elementary matrices of type (iii)) can be written in the form LDL^T, where now there is at least one 0 (resp., real numbers of opposite sign) on the diagonal of D.

Sketch of proof Suppose $A = LDL^T$, where L is lower triangular with 1's on the diagonal (or, more generally, $A = EDE^T$, where E is invertible). Let $d_1, \ldots, d_n$ be the diagonal entries of the diagonal matrix D. Letting $\mathbf{y} = L^T\mathbf{x}$, we have

$$\mathcal{Q}(\mathbf{x}) = \mathbf{x}^T A \mathbf{x} = \mathbf{x}^T (LDL^T) \mathbf{x} = (L^T\mathbf{x})^T D (L^T\mathbf{x}) = \mathbf{y}^T D \mathbf{y} = \sum_{i=1}^{n} d_i y_i^2.$$

Realizing that $\mathbf{y} = \mathbf{0} \iff \mathbf{x} = \mathbf{0}$, the conclusions of the first part of the proposition are now evident.

Suppose $\mathcal{Q}$ is positive definite. Then, in particular, $\mathcal{Q}(\mathbf{e}_1) = a_{11} > 0$, so we can write

$$A = \begin{bmatrix} 1 \\ \frac{a_{12}}{a_{11}} \\ \vdots \\ \frac{a_{1n}}{a_{11}} \end{bmatrix} \begin{bmatrix} a_{11} \end{bmatrix} \begin{bmatrix} 1 & \frac{a_{12}}{a_{11}} & \cdots & \frac{a_{1n}}{a_{11}} \end{bmatrix} + \begin{bmatrix} 0 & 0 & \cdots & 0 \\ 0 & & & \\ \vdots & & B & \\ 0 & & & \end{bmatrix},$$

where B is also symmetric and the quadratic form on $\mathbb{R}^{n-1}$ associated to B is likewise positive definite. We now continue by induction. (For example, if the upper left entry of B were 0, this would mean that $\mathcal{Q}(a_{12}\mathbf{e}_1 - a_{11}\mathbf{e}_2) = 0$, contradicting the hypothesis that $\mathcal{Q}$ is positive definite.)

An analogous argument works when $\mathcal{Q}$ is negative definite. If $A = O$, there is nothing to prove. If not, in the semidefinite or indefinite case, if $a_{11} = 0$, we first find an appropriate elementary matrix, E_1, so that the first entry of the symmetric matrix $B = E_1 A E_1^T$ is nonzero, and then we continue as above. ∎

Remark We will see another way, introduced in the next section and developed fully in Chapter 9, of analyzing the nature of the quadratic form $\mathcal{Q}$ associated to a symmetric matrix A. The signs of the *eigenvalues* of A will tell the whole story.

▶ EXERCISES 5.3

*1. Classify the critical points of the functions in Exercise 5.2.1.

2. Consider the function $f\begin{pmatrix} x \\ y \end{pmatrix} = 2x^4 - 3x^2 y + y^2$.

  (a) Show that the origin is a critical point of f and that, restricting f to any line through the origin, the origin is a local minimum point.
  (b) Is the origin a local minimum point of f?[2]

---

[2]We've seen several textbooks that purportedly prove Theorem 3.3 by showing, for example, that if $\mathcal{H}_{f,\mathbf{a}}$ is positive definite, then the restriction of f to any line through $\mathbf{a}$ has a local minimum at $\mathbf{a}$, and then concluding that $\mathbf{a}$ must be a local minimum point of f. We hope that this exercise will convince you that such a proof must be flawed.

*3. Describe the graph of $f\begin{pmatrix}x\\y\end{pmatrix} = (2x^2 + y^2)e^{-(x^2+y^2)}$.

4. Let $f\begin{pmatrix}x\\y\end{pmatrix} = x^3 + e^{3y} - 3xe^y$.

(a) Show that f has exactly one critical point $\mathbf{a}$, which is a *local* minimum point.

(b) Show that $\mathbf{a}$ is not a *global* minimum point.

5. Suppose $f: \mathbb{R}^2 \to \mathbb{R}$ is $\mathcal{C}^2$ and harmonic (see Example 2 on p. 122). Assume $\dfrac{\partial^2 f}{\partial x^2}(\mathbf{a}) \neq 0$. Prove that $\mathbf{a}$ cannot be an extremum of f.

6. For each of the following symmetric matrices A, write $A = LDL^\mathsf{T}$, as in Example 3. Use your answer to determine whether the associated quadratic form $\mathcal{Q}$ given by $\mathcal{Q}(\mathbf{x}) = \mathbf{x}^\mathsf{T} A\mathbf{x}$ is positive definite, negative definite, indefinite, etc.

(a) $A = \begin{bmatrix} 1 & 3 \\ 3 & 13 \end{bmatrix}$

(b) $A = \begin{bmatrix} 2 & 3 \\ 3 & 4 \end{bmatrix}$

*(c) $A = \begin{bmatrix} 2 & 2 & -2 \\ 2 & -1 & 4 \\ -2 & 4 & 1 \end{bmatrix}$

(d) $A = \begin{bmatrix} 1 & -2 & 2 \\ -2 & 6 & -6 \\ 2 & -6 & 9 \end{bmatrix}$

(e) $A = \begin{bmatrix} 1 & 1 & -3 & 1 \\ 1 & 0 & -3 & 0 \\ -3 & -3 & 11 & -1 \\ 1 & 0 & -1 & 2 \end{bmatrix}$

7. Suppose $A = LDU$, where L is lower triangular with 1's on the diagonal, D is diagonal, and U is upper triangular with 1's on the diagonal. Prove that this decomposition is unique; i.e., if $A = LDU = L'D'U'$, where L', D', and U' have the same defining properties as L, D, and U, respectively, then $L = L'$, $D = D'$, and $U = U'$. (Hint: The product of two lower triangular matrices is lower triangular, and likewise for upper.)

8. (a) Let $A = \begin{bmatrix} 0 & 2 \\ 2 & 1 \end{bmatrix}$. After making a row exchange (and corresponding column exchange to preserve symmetry), we get $B = E_1 A E_1^\mathsf{T} = \begin{bmatrix} 1 & 2 \\ 2 & 0 \end{bmatrix}$. Now write $B = LDL^\mathsf{T}$ and get a corresponding equation for A. How, then, have we expressed the associated quadratic form $\mathcal{Q}(\mathbf{x}) = 4x_1 x_2 + x_2^2$ as a sum (or difference) of squares?

*(b) Let $A = \begin{bmatrix} 0 & 1 \\ 1 & 0 \end{bmatrix}$. By considering $B = E_1 A E_1^\mathsf{T} = \begin{bmatrix} 1 & 1 \\ 1 & 0 \end{bmatrix}$, where E_1 is the elementary matrix corresponding to adding $1/2$ of the second row to the first, show that

$$A = EDE^\mathsf{T} \quad \text{where} \quad E = \begin{bmatrix} \frac{1}{2} & -\frac{1}{2} \\ 1 & 1 \end{bmatrix} \quad \text{and} \quad D = \begin{bmatrix} 1 & \\ & -1 \end{bmatrix}.$$

What is the corresponding expression for the quadratic form $\mathcal{Q}(\mathbf{x}) = 2x_1 x_2$ as a sum (or difference) of squares?

▶ 4 LAGRANGE MULTIPLIERS

Most extremum problems, including those encountered in single-variable calculus, involve functions of several variables with some constraints. Consider, for example, the box of prescribed volume, a cylinder inscribed in a sphere of given radius, or the desire to maximize profit with only a certain amount of working capital. There is an elegant and powerful way to approach all these problems by using multivariable calculus, the method of Lagrange multipliers. A generalization to infinite dimensions, which we shall not study here, is central in the calculus of variations, which is a powerful tool in mechanics, thermodynamics, and differential geometry.

▶ **EXAMPLE 1**

Your boat has sprung a leak in the middle of the lake and you are trying to find the closest point on the shoreline. As suggested by Figure 4.1, we imagine dropping a rock in the water at the location of the boat and watching the circular waves radiate outward. The moment the first wave touches the shoreline, we know that the point **a** at which it touches must be closest to us. And at that point, the circle must be *tangent* to the shoreline.

Let's place the origin at the point at which we drop the rock. Then the circles emanating from this point are level curves of $f(\mathbf{x}) = \|\mathbf{x}\|$. Suppose, moreover, that the shoreline is a level curve of a differentiable function g. By Proposition 5.3 of Chapter 4, the gradient is normal to level sets, so if the tangent line of the circle at **a** and the tangent line of the shoreline at **a** are the same, this means that we should have

$$\nabla f(\mathbf{a}) = \lambda \nabla g(\mathbf{a}) \qquad \text{for some scalar } \lambda. \blacktriangleleft$$

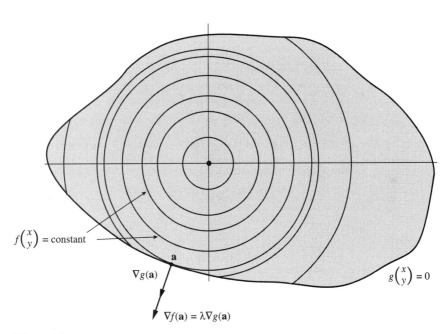

Figure 4.1

We now want to study the calculus of constrained extrema a bit more carefully.

Definition Suppose $U \subset \mathbb{R}^n$ is open and $f: U \to \mathbb{R}$ and $\mathbf{g}: U \to \mathbb{R}^m$ are differentiable. Suppose $\mathbf{g}(\mathbf{a}) = \mathbf{0}$. We say **a** is a *local maximum* (resp., *minimum*) *point of f subject to the constraint* $\mathbf{g}(\mathbf{x}) = \mathbf{0}$ if for some $\delta > 0$, $f(\mathbf{x}) \leq f(\mathbf{a})$ (resp., $f(\mathbf{x}) \geq f(\mathbf{a})$)

for all $\mathbf{x} \in B(\mathbf{a}, \delta)$ satisfying $\mathbf{g}(\mathbf{x}) = \mathbf{0}$. More succinctly, letting $M = \mathbf{g}^{-1}(\{\mathbf{0}\})$, $\mathbf{a}$ is an extremum of the restriction of f to the set M.

Theorem 4.1 *Suppose $U \subset \mathbb{R}^n$ is open, $f: U \to \mathbb{R}$ is differentiable, and $\mathbf{g}: U \to \mathbb{R}^m$ is $\mathcal{C}^1$. Suppose $\mathbf{g}(\mathbf{a}) = \mathbf{0}$ and $\operatorname{rank}(D\mathbf{g}(\mathbf{a})) = m$. If $\mathbf{a}$ is a local extremum of f subject to the constraint $\mathbf{g} = \mathbf{0}$, then there are scalars $\lambda_1, \ldots, \lambda_m$ so that*
$$Df(\mathbf{a}) = \lambda_1 Dg_1(\mathbf{a}) + \cdots + \lambda_m Dg_m(\mathbf{a}).$$
The scalars $\lambda_1, \ldots, \lambda_m$ are called Lagrange multipliers.

Remark As usual, this is a *necessary* condition for a constrained extremum but not a sufficient one. There may be (constrained) saddle points as well.

Proof By the Implicit Function Theorem, we can represent $M = \mathbf{g}^{-1}(\{\mathbf{0}\})$ locally near $\mathbf{a}$ as a graph over some coordinate $(n - m)$-plane. For concreteness, let's say that locally
$$M = \left\{ \begin{bmatrix} \overline{\mathbf{x}} \\ \boldsymbol{\phi}(\overline{\mathbf{x}}) \end{bmatrix} : \overline{\mathbf{x}} \in V \subset \mathbb{R}^{n-m} \right\},$$
where $\boldsymbol{\phi}: V \to \mathbb{R}^m$ is $\mathcal{C}^1$. Thus, we can define a local parametrization of M by
$$\boldsymbol{\Phi}: V \to \mathbb{R}^n, \qquad \boldsymbol{\Phi}(\mathbf{x}) = \begin{bmatrix} \overline{\mathbf{x}} \\ \boldsymbol{\phi}(\overline{\mathbf{x}}) \end{bmatrix},$$
as shown in Figure 4.2, with $\boldsymbol{\Phi}(\overline{\mathbf{a}}) = \mathbf{a}$. Now we have two crucial pieces of information:
$$\mathbf{g} \circ \boldsymbol{\Phi} = \mathbf{0} \qquad \text{and} \qquad f \circ \boldsymbol{\Phi} \text{ has a local extremum at } \overline{\mathbf{a}}.$$
Differentiating by the chain rule, and applying Lemma 2.1, we have
(†) $\qquad D\mathbf{g}(\mathbf{a}) \circ D\boldsymbol{\Phi}(\overline{\mathbf{a}}) = O \qquad \text{and} \qquad Df(\mathbf{a}) \circ D\boldsymbol{\Phi}(\overline{\mathbf{a}}) = \mathbf{0}.$

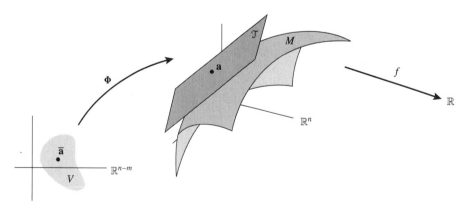

Figure 4.2

The first equation in (†) tells us that $\mathcal{T}$, the $(n-m)$-dimensional image of the linear map $D\Phi(\bar{\mathbf{a}})$, satisfies $\mathcal{T} \subset \ker D\mathbf{g}(\mathbf{a})$ (or $\mathbf{C}([D\Phi(\bar{\mathbf{a}})]) \subset \mathbf{N}([D\mathbf{g}(\mathbf{a})])$). But, by the Nullity-Rank Theorem, Corollary 4.6 of Chapter 4, we have

$$\dim \mathbf{N}([D\mathbf{g}(\mathbf{a})]) = n - \operatorname{rank}([D\mathbf{g}(\mathbf{a})]) = n - m,$$

by hypothesis. Since $\dim \mathcal{T} = n - m = \dim \mathbf{N}([D\mathbf{g}(\mathbf{a})])$, we must have $\mathcal{T} = \mathbf{N}([D\mathbf{g}(\mathbf{a})])$. Moreover,

$$\mathcal{T} = \mathbf{N}([D\mathbf{g}(\mathbf{a})]) = \big(\mathbf{R}([D\mathbf{g}(\mathbf{a})])\big)^{\perp}.$$

On the other hand, the latter equation in (†) tells us that

$$\mathcal{T} \subset \mathbf{N}([Df(\mathbf{a})]) = \mathbf{R}([Df(\mathbf{a})])^{\perp}.$$

Thus,

$$\big(\mathbf{R}([D\mathbf{g}(\mathbf{a})])\big)^{\perp} \subset \mathbf{R}([Df(\mathbf{a})])^{\perp},$$

so, taking orthogonal complements and using Exercise 1.3.9 and Proposition 4.8 of Chapter 4, we have

$$\mathbf{R}([Df(\mathbf{a})]) \subset \mathbf{R}([D\mathbf{g}(\mathbf{a})]),$$

so $Df(\mathbf{a})$ is a linear combination of the linear maps $Dg_1(\mathbf{a}), \ldots, Dg_m(\mathbf{a})$—or, more geometrically, $\nabla f(\mathbf{a})$ is a linear combination of the vectors $\nabla g_1(\mathbf{a}), \ldots, \nabla g_m(\mathbf{a})$—as we needed to show. ∎

Remark The subspace $\mathcal{T} = \operatorname{image}(D\Phi(\bar{\mathbf{a}})) = \big(\mathbf{R}([D\mathbf{g}(\mathbf{a})])\big)^{\perp}$ is called the *tangent space* of M at $\mathbf{a}$. We shall return to such matters in Chapter 6.

▶ **EXAMPLE 2**

The temperature at the point $\begin{bmatrix} x \\ y \\ z \end{bmatrix}$ in space is given by $f\begin{pmatrix} x \\ y \\ z \end{pmatrix} = xy + z^2$. We wish to find the hottest and coldest points on the sphere $x^2 + y^2 + z^2 = 2z$ (the sphere of radius 1 centered at $\begin{bmatrix} 0 \\ 0 \\ 1 \end{bmatrix}$). That is, we must find the extrema of f subject to the constraint $g\begin{pmatrix} x \\ y \\ z \end{pmatrix} = x^2 + y^2 + z^2 - 2z = 0$. By Theorem 4.1, we must find points $\mathbf{x}$ satisfying $g(\mathbf{x}) = 0$ at which $Df(\mathbf{x}) = \lambda Dg(\mathbf{x})$ for some scalar λ. That is, we seek points $\mathbf{x}$ so that

(∗) $$\begin{bmatrix} y & x & 2z \end{bmatrix} = \lambda \begin{bmatrix} x & y & z-1 \end{bmatrix} \quad \text{for some scalar } \lambda.$$

(Notice that we removed the factor of 2 from Dg.)

Eliminating λ, we see that, provided none of our denominators is 0,

$$\frac{y}{x} = \frac{x}{y} = \frac{2z}{z-1}.$$

So either

$$y = x \quad \text{and} \quad 2z = z - 1 \quad \text{or} \quad y = -x \quad \text{and} \quad 2z = 1 - z;$$

the former leads to $z = -1$, which is impossible, and the latter leads to

$$z = \frac{1}{3}, \quad y = -x, \quad x = \pm \frac{1}{3}\sqrt{\frac{5}{2}}.$$

Now, we infer from (∗) that if $x = 0$, then $y = 0$ as well (and vice versa), and then z can be arbitrary, so we also find that the north and south poles of the sphere are constrained critical points. On the other hand, we cannot have the denominator $z - 1 = 0$, for, by (∗), that would require $z = 0$, and these equations cannot hold simultaneously.

Calculating the values of f at our various constrained critical points, we have

$$f\begin{pmatrix} \sqrt{5/2}/3 \\ -\sqrt{5/2}/3 \\ 1/3 \end{pmatrix} = f\begin{pmatrix} -\sqrt{5/2}/3 \\ \sqrt{5/2}/3 \\ 1/3 \end{pmatrix} = -\frac{1}{6}, \quad f\begin{pmatrix} 0 \\ 0 \\ 0 \end{pmatrix} = 0, \quad \text{and} \quad f\begin{pmatrix} 0 \\ 0 \\ 2 \end{pmatrix} = 4.$$

Thus, the topmost point $\begin{bmatrix} 0 \\ 0 \\ 2 \end{bmatrix}$ is the hottest and the two points $\begin{bmatrix} \sqrt{5/2}/3 \\ -\sqrt{5/2}/3 \\ 1/3 \end{bmatrix}$ and $\begin{bmatrix} -\sqrt{5/2}/3 \\ \sqrt{5/2}/3 \\ 1/3 \end{bmatrix}$ are the coldest. ◂

Remark We surmise that the origin is a saddle point. Indeed, representing the sphere locally as a graph near the origin, we have $z = 1 - \sqrt{1 - (x^2 + y^2)}$ and

$$f\begin{pmatrix} x \\ y \\ z \end{pmatrix} = xy + \left(1 - \sqrt{1 - (x^2 + y^2)}\right)^2 = xy + \text{higher order terms}.$$

(This is easiest to see by using $\sqrt{1 + u} = 1 + u/2 + $ higher order terms.) Even easier, the origin is a *nonconstrained* critical point of f. Since f is a quadratic polynomial, $\mathcal{H}_{f,0} = f$, and on the tangent plane of the sphere at $\mathbf{0}$ we just get xy. (Also see Exercise 34.)

▸ **EXAMPLE 3**

Find the shortest possible distance from the ellipse $x^2 + 2y^2 = 2$ to the line $x + y = 2$. We need to consider the (square of the) distance between *pairs* of points, one on the ellipse, the other on the line. This means that we need to work in $\mathbb{R}^2 \times \mathbb{R}^2$, with coordinates $\begin{bmatrix} x \\ y \end{bmatrix}$ and $\begin{bmatrix} u \\ v \end{bmatrix}$, respectively. Let's try to minimize

$$f\begin{pmatrix} x \\ y \\ u \\ v \end{pmatrix} = (x - u)^2 + (y - v)^2$$

subject to the constraints

$$\mathbf{g}\begin{pmatrix}x\\y\\u\\v\end{pmatrix}=\begin{bmatrix}x^2+2y^2-2\\u+v-2\end{bmatrix}=\begin{bmatrix}0\\0\end{bmatrix}.$$

(The rank condition on **g** is easily checked in this case.) So we need to find points at which, for some scalars λ and μ, we have

$$Df = \lambda Dg_1 + \mu Dg_2; \quad \text{i.e.,}$$

$$\begin{bmatrix}x-u & y-v & -(x-u) & -(y-v)\end{bmatrix} = \lambda \begin{bmatrix}x & 2y & 0 & 0\end{bmatrix} + \mu \begin{bmatrix}0 & 0 & 1 & 1\end{bmatrix}.$$

We see that we must have $x - u = y - v$ and so $x = 2y$, as well. Now substituting into the constraint equations yields two critical points:

$$\begin{bmatrix}2/\sqrt{3}\\1/\sqrt{3}\\1+1/(2\sqrt{3})\\1-1/(2\sqrt{3})\end{bmatrix} \quad \text{and} \quad \begin{bmatrix}-2/\sqrt{3}\\-1/\sqrt{3}\\1-1/(2\sqrt{3})\\1+1/(2\sqrt{3})\end{bmatrix}.$$

As a check, note that the vector from $\begin{bmatrix}u\\v\end{bmatrix}$ to $\begin{bmatrix}x\\y\end{bmatrix}$ in each case is normal to both the ellipse and the line, as Figure 4.3 corroborates. Evidently, the first point gives the shortest possible distance, and we leave it to the reader to establish this rigorously. ◂

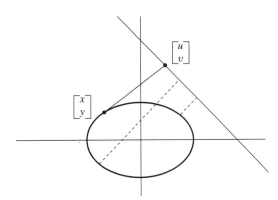

Figure 4.3

We close this section with an application of the method of Lagrange multipliers to linear algebra. Suppose A is a symmetric $n \times n$ matrix. Let's find the extrema of the quadratic form $\mathcal{Q}(\mathbf{x}) = \mathbf{x}^\mathsf{T} A \mathbf{x}$ subject to the constraint $g(\mathbf{x}) = \|\mathbf{x}\|^2 = 1$. By Theorem 4.1, we seek $\mathbf{x} \in \mathbb{R}^n$ so that for some scalar λ we have $D\mathcal{Q}(\mathbf{x}) = \lambda Dg(\mathbf{x})$. Applying the result of Exercise 3.2.14 (and canceling a pair of 2's), this means that at any constrained extremum we must have

$$A\mathbf{x} = \lambda \mathbf{x} \quad \text{for some scalar } \lambda.$$

Such a vector **x** is called an *eigenvector* of A, and the Lagrange multiplier λ is called an *eigenvalue*. Note that by compactness of the unit sphere, Q must have at least a global minimum and a global maximum; hence A must have at least two eigenvalues and corresponding eigenvectors.

EXAMPLE 4

Consider $A = \begin{bmatrix} 6 & 2 \\ 2 & 9 \end{bmatrix}$. Proceeding as above, we arrive at the system of equations

$$6x + 2y = \lambda x$$
$$2x + 9y = \lambda y.$$

Eliminating λ, we obtain

$$\frac{6x + 2y}{x} = \frac{2x + 9y}{y},$$

from which we find the equation

$$2\left(\frac{y}{x}\right)^2 - 3\left(\frac{y}{x}\right) - 2 = \left(2\frac{y}{x} + 1\right)\left(\frac{y}{x} - 2\right) = 0,$$

so either $y = 2x$ or $y = -\frac{1}{2}x$. Substituting into the constraint equation, we obtain the critical points (eigenvectors) $\begin{bmatrix} 1/\sqrt{5} \\ 2/\sqrt{5} \end{bmatrix}$ and $\begin{bmatrix} -2/\sqrt{5} \\ 1/\sqrt{5} \end{bmatrix}$, with respective Lagrange multipliers (eigenvalues) 10 and 5.

EXERCISES 5.4

1. (a) Find the minimum value of $f\begin{pmatrix} x \\ y \end{pmatrix} = x^2 + y^2$ on the curve $x + y = 2$. Why is there no maximum?

   (b) Find the maximum value of $g\begin{pmatrix} x \\ y \end{pmatrix} = x + y$ on the curve $x^2 + y^2 = 2$. Is there a minimum?

   (c) How are the questions (and answers) in parts a and b related?

*2. A wire has the shape of the circle $x^2 + y^2 - 2y = 0$. Its temperature at the point $\begin{bmatrix} x \\ y \end{bmatrix}$ is given by $T\begin{pmatrix} x \\ y \end{pmatrix} = 2x^2 + 3y$. Find the maximum and minimum temperatures of the wire. (Be sure you've found all potential critical points.)

3. Find the maximum value of $f\begin{pmatrix} x \\ y \\ z \end{pmatrix} = 2x + 2y - z$ on the sphere of radius 2 centered at the origin.

4. Find the maximum and minimum values of the function $f\begin{pmatrix} x \\ y \end{pmatrix} = x^2 + xy + y^2$ on the unit disk $D = \{\mathbf{x} \in \mathbb{R}^2 : \|\mathbf{x}\| \leq 1\}$.

5. Find the point(s) on the ellipse $x^2 + 4y^2 = 4$ closest to the point $\begin{bmatrix} 1 \\ 0 \end{bmatrix}$.

6. The temperature at point **x** is given by $f\begin{pmatrix} x \\ y \\ z \end{pmatrix} = x^2 + 2y + 2z$. Find the hottest and coldest points on the sphere $x^2 + y^2 + z^2 = 3$.

7. Find the volume of the largest rectangular box (with all its edges parallel to the coordinate axes) that can be inscribed in the ellipsoid
$$x^2 + \frac{y^2}{2} + \frac{z^2}{3} = 1.$$

8. A space probe in the shape of the ellipsoid $4x^2 + y^2 + 4z^2 = 16$ enters the earth's atmosphere and its surface begins to heat. After 1 hour, the temperature in °C on its surface is given by $f\begin{pmatrix} x \\ y \\ z \end{pmatrix} = 2x^2 + yz - 4z + 600$. Find the hottest and coldest points on the probe's surface.

9. The temperature in space is given by $f\begin{pmatrix} x \\ y \\ z \end{pmatrix} = 3xy + z^3 - 3z$. Prove that there are hottest and coldest points on the sphere $x^2 + y^2 + z^2 - 2z = 0$, and find them.

10. Let $f\begin{pmatrix} x \\ y \\ z \end{pmatrix} = xy + z^3$ and $S = \left\{ \begin{bmatrix} x \\ y \\ z \end{bmatrix} : x^2 + y^2 + z^2 = 1, \ z \geq 0 \right\}$. Prove that f attains its global maximum and minimum on S and determine its global maximum and minimum points.

11. Among all triangles inscribed in the unit circle, which have the greatest area? (Hint: Consider the three small triangles formed by joining the vertices to the center of the circle.)

12. Among all triangles inscribed in the unit circle, which have the greatest perimeter?

*13. Find the ellipse $x^2/a^2 + y^2/b^2 = 1$ that passes through the point $\begin{bmatrix} 3 \\ 1 \end{bmatrix}$ and has the least area. (Recall that the area of the ellipse is πab.)

14. If α, β, and γ are the angles of a triangle, show that
$$\sin\frac{\alpha}{2} \sin\frac{\beta}{2} \sin\frac{\gamma}{2} \leq \frac{1}{8}.$$
For what triangles is the maximum attained?

♯*15. Find the points closest to and farthest from the origin on the ellipse $2x^2 + 4xy + 5y^2 = 1$.

16. Solve Exercise 5.2.8 anew, using Lagrange multipliers.

17. Solve Exercise 5.2.9 anew, using Lagrange multipliers.

18. Find the maximum and minimum values of the function $f(\mathbf{x}) = x_1 + \cdots + x_n$ subject to the constraint $\|\mathbf{x}\| = 1$.

*19. Find the maximum volume of an n-dimensional rectangular parallelepiped of diameter δ.

20. Suppose $x_1, \ldots, x_n$ are positive numbers. Prove that
$$\sqrt[n]{x_1 x_2 \cdots x_n} \leq \frac{x_1 + \cdots + x_n}{n}.$$

Chapter 5. Extremum Problems

21. Suppose $p, q > 0$ and $\frac{1}{p} + \frac{1}{q} = 1$. Suppose $x, y > 0$. Use Lagrange multipliers to prove that
$$\frac{x^p}{p} + \frac{y^q}{q} \geq xy.$$
(Hint: Minimize the left-hand side subject to the constraint $xy =$ constant.)

22. Solve Exercise 5.2.11 anew, using Lagrange multipliers.

23. A silo is built by putting a right circular cone atop a right circular cylinder (both having the same radius). What dimensions will give the silo of maximum volume for a given surface area?

24. Solve Exercise 5.2.12 anew, using Lagrange multipliers.

***25.** Use Lagrange multipliers to find the point closest to the origin on the intersection of the planes $x + 2y + z = 5$ and $2x + y - z = 1$.

26. In each case, find the point in the given subspace V closest to $\mathbf{b}$.

*(a) $V = \{\mathbf{x} \in \mathbb{R}^3 : x_1 - x_2 + 3x_3 = 2x_1 + x_2 = 0\}$, $\mathbf{b} = \begin{bmatrix} 3 \\ 7 \\ 1 \end{bmatrix}$

(b) $V = \{\mathbf{x} \in \mathbb{R}^4 : x_1 + x_2 + x_3 + x_4 = x_1 + 2x_3 + x_4 = 0\}$, $\mathbf{b} = \begin{bmatrix} 3 \\ 1 \\ 1 \\ -1 \end{bmatrix}$

***27.** Find the points on the curve of intersection of the two surfaces $x^2 - xy + y^2 - z^2 = 1$ and $x^2 + y^2 = 1$ that are closest to the origin.

28. Show that of all quadrilaterals with fixed side lengths, the one of maximum area can be inscribed in a circle. (Hint: Use as variables a pair of opposite angles. See also Exercise 1.2.14.)

29. For each of the following symmetric matrices A, find all the extrema of $\mathcal{Q}(\mathbf{x}) = \mathbf{x}^T A \mathbf{x}$ subject to the constraint $\|\mathbf{x}\|^2 = 1$. Also determine the Lagrange multiplier each time.

*(a) $A = \begin{bmatrix} 1 & 2 \\ 2 & -2 \end{bmatrix}$ (b) $A = \begin{bmatrix} 0 & 3 \\ 3 & -8 \end{bmatrix}$

30. Find the norm of each of the following matrices. Note: A calculator will be helpful.

*(a) $\begin{bmatrix} 1 & 1 \\ 0 & 1 \end{bmatrix}$ (b) $\begin{bmatrix} 2 & 1 \\ 0 & 3 \end{bmatrix}$ (c) $\begin{bmatrix} 2 & 1 \\ 1 & 3 \end{bmatrix}$

31. A (frictionless) lasso is thrown around two pegs, as pictured in Figure 4.4, and a large weight hung from the free end. Treating the mass of the rope as insignificant, and supposing the weight hangs freely, what is the equilibrium position of the system?

32. (Interpreting the Lagrange Multiplier)
(a) Suppose $\mathbf{a} = \boldsymbol{\psi}(c)$ is a local extreme point of the function f relative to the constraint $g(\mathbf{x}) = c$; suppose, moreover, that $\boldsymbol{\psi}$ is a differentiable function. Show that $\lambda = (f \circ \boldsymbol{\psi})'(c)$.

(b) Assume that f and g are $\mathcal{C}^2$. Use the Implicit Function Theorem (see Theorem 2.2 of Chapter 6 for the general version) to show that the extreme point $\mathbf{a}$ is given locally as a differentiable function of c whenever the "bordered Hessian"

$$H_{f,g}\begin{pmatrix} \mathbf{a} \\ \lambda \end{pmatrix} = \left[\begin{array}{c|c} \text{Hess}(f)(\mathbf{a}) - \lambda \text{Hess}(g)(\mathbf{a}) & \nabla g(\mathbf{a}) \\ \hline Dg(\mathbf{a}) & 0 \end{array} \right]$$

is invertible.

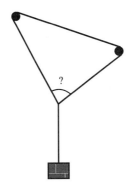

Figure 4.4

33. (An Application of Exercise 32 to Economics) Let $\mathbf{x} \in \mathbb{R}^n$ be the commodity vector, $\mathbf{p} \in \mathbb{R}^n$ the price vector, and $f: \mathbb{R}^n \to \mathbb{R}$ the production function, so that $f(\mathbf{x})$ tells us how many widgets are produced, using x_i units of item i, $i = 1, \ldots, n$. Prove that to produce the greatest number of widgets with a given budget, we must have

$$\lambda = \frac{1}{p_1} \frac{\partial f}{\partial x_1} = \cdots = \frac{1}{p_n} \frac{\partial f}{\partial x_n}.$$

The quantity $\dfrac{1}{p_i} \dfrac{\partial f}{\partial x_i}$ is called the marginal productivity for item i. Explain why this result is believable. What does the result of Exercise 32a tell us in this case?

34. (A Second Derivative Test for Constrained Extrema) Suppose $\mathbf{a}$ is a critical point of f subject to the constraint $g(\mathbf{x}) = c$, $Df(\mathbf{a}) = \lambda Dg(\mathbf{a})$, and $Dg(\mathbf{a}) \neq \mathbf{0}$. Show that $\mathbf{a}$ is a constrained local maximum (resp., minimum) of f on $M = \{\mathbf{x} : g(\mathbf{x}) = c\}$ if the restriction of the Hessian of $f - \lambda g$ to the tangent space $T_{\mathbf{a}}M$ is negative (resp., positive) definite. (Hint: Parametrize the constraint surface M locally by Φ with $\Phi(\bar{\mathbf{a}}) = \mathbf{a}$ and apply Theorem 3.3 to $f \circ \Phi$.) There is an interpretation in terms of the bordered Hessian (see Exercise 32b), which is indicated in Exercise 9.4.21.

5 PROJECTIONS, LEAST SQUARES, AND INNER PRODUCT SPACES

EXAMPLE 1

Suppose we're given the system $A\mathbf{x} = \mathbf{b}$ to solve, where

$$A = \begin{bmatrix} 1 & 2 \\ 0 & 1 \\ 1 & 1 \end{bmatrix} \quad \text{and} \quad \mathbf{b} = \begin{bmatrix} 1 \\ 1 \\ 1 \end{bmatrix}.$$

It is easy to check that $\mathbf{b} \notin \mathbf{C}(A)$, and so this system is inconsistent. The best we can do is to solve $A\mathbf{x} = \mathbf{p}$, where $\mathbf{p}$ is the vector in $\mathbf{C}(A)$ that is *closest* to $\mathbf{b}$. Clearly that point is $\mathbf{p} = \mathbf{b} - \text{proj}_{\mathbf{a}}\mathbf{b}$, where $\mathbf{a}$ is the normal vector to $\mathbf{C}(A) \subset \mathbb{R}^3$, as shown in Figure 5.1. Now we see how to solve our problem. $\mathbf{C}(A)$ is the plane in $\mathbb{R}^3$ with normal vector

$$\mathbf{a} = \begin{bmatrix} -1 \\ 1 \\ 1 \end{bmatrix},$$

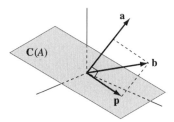

Figure 5.1

and if we compute $\text{proj}_{\mathbf{a}}\mathbf{b}$, then we will have

$$\mathbf{p} = \mathbf{b} - \text{proj}_{\mathbf{a}}\mathbf{b} \in \text{Span}(\mathbf{a})^{\perp} = \mathbf{C}(A) \quad \text{and} \quad \mathbf{b} - \mathbf{p} = \text{proj}_{\mathbf{a}}\mathbf{b} \in \mathbf{C}(A)^{\perp}.$$

In our case, we have

$$\text{proj}_{\mathbf{a}}\mathbf{b} = \frac{\mathbf{b} \cdot \mathbf{a}}{\|\mathbf{a}\|^2}\mathbf{a} = \frac{1}{3}\begin{bmatrix} -1 \\ 1 \\ 1 \end{bmatrix},$$

and so

$$\mathbf{p} = \begin{bmatrix} 1 \\ 1 \\ 1 \end{bmatrix} - \frac{1}{3}\begin{bmatrix} -1 \\ 1 \\ 1 \end{bmatrix} = \begin{bmatrix} 4/3 \\ 2/3 \\ 2/3 \end{bmatrix}.$$

Now it is an easy matter to solve $A\mathbf{x} = \mathbf{p}$; indeed, the solution is

$$\bar{\mathbf{x}} = \begin{bmatrix} 0 \\ 2/3 \end{bmatrix}.$$

This is called the *least squares solution* of the original problem, inasmuch as $A\bar{\mathbf{x}}$ is the vector in $\mathbf{C}(A)$ closest to $\mathbf{b}$. ◀

In general, given $\mathbf{b} \in \mathbb{R}^n$ and an m-dimensional subspace $V \subset \mathbb{R}^n$, we can ask for the *projection* of $\mathbf{b}$ onto V, i.e., the point in V closest to $\mathbf{b}$, which we denote by $\text{proj}_V \mathbf{b}$. We first make the official

Definition Let $V \subset \mathbb{R}^n$ be a subspace, and let $\mathbf{b} \in \mathbb{R}^n$. We define the *projection of* $\mathbf{b}$ *onto* V to be the unique vector $\mathbf{p} \in V$ with the property that $\mathbf{b} - \mathbf{p} \in V^{\perp}$. We write $\mathbf{p} = \text{proj}_V \mathbf{b}$.

We ask the reader to show in Exercise 10 that projection onto a subspace V gives a linear map. As we know from Chapter 4, we can be given V either explicitly (say, $V = \mathbf{C}(A)$ for some $n \times m$ matrix A) or implicitly (say, $V = \mathbf{N}(B)$ for some $m \times n$ matrix B). We will start by applying the methods of this chapter to obtain a simple solution of the problem (and then we will indicate that we could have omitted the calculus completely).

Suppose A is an $n \times m$ matrix of rank m (so that the column vectors $\mathbf{a}_1, \ldots, \mathbf{a}_m$ give a basis for our subspace V). Define
$$f : \mathbb{R}^m \to \mathbb{R} \quad \text{by} \quad f(\mathbf{x}) = \|A\mathbf{x} - \mathbf{b}\|^2.$$
We seek critical points of f. Write $h(\mathbf{x}) = \|\mathbf{x}\|^2$ and $\mathbf{g}(\mathbf{x}) = A\mathbf{x} - \mathbf{b}$, so that $f = h \circ \mathbf{g}$. Then $Dh(\mathbf{y}) = 2\mathbf{y}^\mathsf{T}$ and $D\mathbf{g}(\mathbf{x}) = A$, so, differentiating f by the chain rule, we have $Df(\mathbf{x}) = Dh(\mathbf{g}(\mathbf{x}))D\mathbf{g}(\mathbf{x}) = 2(A\mathbf{x} - \mathbf{b})^\mathsf{T} A$. Thus, $Df(\mathbf{x}) = \mathbf{0} \iff (A\mathbf{x} - \mathbf{b})^\mathsf{T} A = \mathbf{0}$. Transposing for convenience, we deduce that $\mathbf{x}$ is a critical point if and only if

$$(*) \qquad (A^\mathsf{T} A)\mathbf{x} = A^\mathsf{T} \mathbf{b}.$$

Now, $A^\mathsf{T} A$ is an $m \times m$ matrix, and by Exercise 4.4.14 this matrix is nonsingular, so the equation $(*)$ has a unique solution $\bar{\mathbf{x}}$. We claim that $\bar{\mathbf{x}}$ is the global minimum point. This is just the Pythagorean Theorem again: Since $A^\mathsf{T}(A\bar{\mathbf{x}} - \mathbf{b}) = \mathbf{0}$, $A\bar{\mathbf{x}} - \mathbf{b} \in \mathbf{N}(A^\mathsf{T}) = \mathbf{C}(A)^\perp$, so, for any $\mathbf{x} \in \mathbb{R}^m$, $\mathbf{x} \neq \bar{\mathbf{x}}$, as Figure 5.2 shows:

$$f(\mathbf{x}) = \|A\mathbf{x} - \mathbf{b}\|^2 = \|A(\mathbf{x} - \bar{\mathbf{x}}) + (A\bar{\mathbf{x}} - \mathbf{b})\|^2 = \|A(\mathbf{x} - \bar{\mathbf{x}})\|^2 + \|A\bar{\mathbf{x}} - \mathbf{b}\|^2 > f(\bar{\mathbf{x}}).$$

The vector $\bar{\mathbf{x}}$ is called the *least squares solution* of the (inconsistent) linear system $A\mathbf{x} = \mathbf{b}$, and $(*)$ gives the associated *normal equations*.

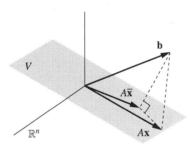

Figure 5.2

Remark When A has rank less than m, the linear system $(*)$ is still consistent (see Exercise 4.4.15) and has infinitely many solutions. We define *the* least squares solution to be the one of smallest length, i.e., the unique vector $\bar{\mathbf{x}} \in \mathbf{R}(A)$ that satisfies the equation. See Proposition 4.10 of Chapter 4. This leads to the *pseudoinverse* that is important in numerical analysis (cf. Strang).

▶ **EXAMPLE 2**

We wish to find the least squares solution of the system $A\mathbf{x} = \mathbf{b}$, where

$$A = \begin{bmatrix} 2 & 1 \\ 1 & 1 \\ 0 & 1 \\ 1 & -1 \end{bmatrix} \quad \text{and} \quad \mathbf{b} = \begin{bmatrix} 2 \\ 1 \\ 1 \\ -1 \end{bmatrix}.$$

We need only solve the normal equations $A^\mathsf{T} A \bar{\mathbf{x}} = A^\mathsf{T} \mathbf{b}$. Now,

$$A^\mathsf{T} A = \begin{bmatrix} 6 & 2 \\ 2 & 4 \end{bmatrix} \quad \text{and} \quad A^\mathsf{T} \mathbf{b} = \begin{bmatrix} 4 \\ 5 \end{bmatrix},$$

and so, using the formula for the inverse of a 2×2 matrix in Example 5 on p. 154,

$$\bar{\mathbf{x}} = (A^\mathsf{T} A)^{-1} A^\mathsf{T} \mathbf{b} = \frac{1}{20} \begin{bmatrix} 4 & -2 \\ -2 & 6 \end{bmatrix} \begin{bmatrix} 4 \\ 5 \end{bmatrix} = \frac{1}{10} \begin{bmatrix} 3 \\ 11 \end{bmatrix}$$

is the least squares solution. ◀

This is all it takes to give an explicit formula for projection onto a subspace $V \subset \mathbb{R}^n$. In particular, denote by

$$\text{proj}_V : \mathbb{R}^n \to \mathbb{R}^n$$

the function that assigns to each vector $\mathbf{b} \in \mathbb{R}^n$ the vector $\mathbf{p} \in V$ closest to $\mathbf{b}$. Start by choosing a basis $\{\mathbf{v}_1, \ldots, \mathbf{v}_m\}$ for V, and let

$$A = \begin{bmatrix} | & | & & | \\ \mathbf{v}_1 & \mathbf{v}_2 & \cdots & \mathbf{v}_m \\ | & | & & | \end{bmatrix}$$

be the $n \times m$ matrix whose column vectors are these basis vectors. Then, given $\mathbf{b} \in \mathbb{R}^n$, we know that if we take $\bar{\mathbf{x}} = (A^\mathsf{T} A)^{-1} A^\mathsf{T} \mathbf{b}$, then $A\bar{\mathbf{x}} = \mathbf{p} = \text{proj}_V \mathbf{b}$. That is,

$$\mathbf{p} = \text{proj}_V \mathbf{b} = \left(A(A^\mathsf{T} A)^{-1} A^\mathsf{T}\right) \mathbf{b},$$

from which we deduce that the matrix

(†) $$P = A(A^\mathsf{T} A)^{-1} A^\mathsf{T}$$

is the appropriate projection matrix: i.e.,

$$[\text{proj}_V] = A(A^\mathsf{T} A)^{-1} A^\mathsf{T}.$$

In Section 5.2, we'll see a bit more of the geometry underlying the formula for the projection matrix.

▶ **EXAMPLE 3**

If $\mathbf{b} \in \mathbf{C}(A)$ to begin with, then $\mathbf{b} = A\mathbf{x}$ for some $\mathbf{x} \in \mathbb{R}^m$, and

$$P\mathbf{b} = \left(A(A^\mathsf{T} A)^{-1} A^\mathsf{T}\right) \mathbf{b} = A(A^\mathsf{T} A)^{-1} (A^\mathsf{T} A) \mathbf{x} = A\mathbf{x} = \mathbf{b},$$

as it should be. And if $\mathbf{b} \in \mathbf{C}(A)^\perp$, then $\mathbf{b} \in \mathbf{N}(A^\mathsf{T})$, so

$$P\mathbf{b} = \left(A(A^\mathsf{T} A)^{-1} A^\mathsf{T}\right) \mathbf{b} = A(A^\mathsf{T} A)^{-1} (A^\mathsf{T} \mathbf{b}) = \mathbf{0},$$

as it should be. ◀

EXAMPLE 4

Note that when dim $V = 1$, we recover our formula for projection onto a line from Section 2 of Chapter 1. If $\mathbf{a} \in \mathbb{R}^n$ is a nonzero vector, we consider it as an $n \times 1$ matrix and the projection formula becomes

$$P = \frac{1}{\|\mathbf{a}\|^2} \mathbf{a}\mathbf{a}^\mathsf{T};$$

that is,

$$P\mathbf{b} = \frac{1}{\|\mathbf{a}\|^2}(\mathbf{a}\mathbf{a}^\mathsf{T})\mathbf{b} = \frac{1}{\|\mathbf{a}\|^2}\mathbf{a}(\mathbf{a}^\mathsf{T}\mathbf{b}) = \frac{\mathbf{a}\cdot\mathbf{b}}{\|\mathbf{a}\|^2}\mathbf{a},$$

as before. ◂

EXAMPLE 5

Let $V \subset \mathbb{R}^3$ be the plane defined by the equation $x_1 - 2x_2 + x_3 = 0$. Then

$$\mathbf{v}_1 = \begin{bmatrix} 2 \\ 1 \\ 0 \end{bmatrix} \quad \text{and} \quad \mathbf{v}_2 = \begin{bmatrix} -1 \\ 0 \\ 1 \end{bmatrix}$$

form a basis for V, and we take

$$A = \begin{bmatrix} 2 & -1 \\ 1 & 0 \\ 0 & 1 \end{bmatrix}.$$

Then, since

$$A^\mathsf{T} A = \begin{bmatrix} 5 & -2 \\ -2 & 2 \end{bmatrix}, \quad \text{we have} \quad (A^\mathsf{T} A)^{-1} = \frac{1}{6}\begin{bmatrix} 2 & 2 \\ 2 & 5 \end{bmatrix},$$

and so

$$P = A(A^\mathsf{T} A)^{-1}A^\mathsf{T} = \frac{1}{6}\begin{bmatrix} 2 & -1 \\ 1 & 0 \\ 0 & 1 \end{bmatrix}\begin{bmatrix} 2 & 2 \\ 2 & 5 \end{bmatrix}\begin{bmatrix} 2 & 1 & 0 \\ -1 & 0 & 1 \end{bmatrix} = \frac{1}{6}\begin{bmatrix} 5 & 2 & -1 \\ 2 & 2 & 2 \\ -1 & 2 & 5 \end{bmatrix}. \;◂$$

Now, what happens if we are given the subspace implicitly? This sounds like the perfect setup for Lagrange multipliers. Suppose the m-dimensional subspace $V \subset \mathbb{R}^n$ is given as the nullspace of an $(n-m) \times n$ matrix B of rank $n-m$. To find the point in V closest to $\mathbf{b} \in \mathbb{R}^n$, we want to minimize the function

$$f: \mathbb{R}^n \to \mathbb{R}, \quad f(\mathbf{x}) = \|\mathbf{x} - \mathbf{b}\|^2, \quad \text{subject to the constraint} \quad \mathbf{g}(\mathbf{x}) = B\mathbf{x} = \mathbf{0}.$$

The method of Lagrange multipliers, Theorem 4.1, tells us that we must have (dropping the factor of 2)

$$(\mathbf{x} - \mathbf{b})^\mathsf{T} = \sum_{i=1}^{n-m} \lambda_i \mathbf{B}_i, \quad \text{for some scalars } \lambda_1, \ldots, \lambda_{n-m},$$

where, as usual, $\mathbf{B}_i$ are the rows of B. Transposing this equation, we have

$$\mathbf{x} - \mathbf{b} = B^\mathsf{T} \boldsymbol{\lambda}, \quad \text{where} \quad \boldsymbol{\lambda} = \begin{bmatrix} \lambda_1 \\ \vdots \\ \lambda_{n-m} \end{bmatrix}.$$

Multiplying this equation by B and using the constraint equation, we get

$$(BB^\mathsf{T})\boldsymbol{\lambda} = -B\mathbf{b}.$$

By analogy with our treatment of the equation (∗), the matrix BB^T has rank $n - m$, and so we can solve for $\boldsymbol{\lambda}$, hence for the constrained extremum $\mathbf{x}_0$:

(‡) $$\mathbf{x}_0 = \mathbf{b} + B^\mathsf{T}\left(-(BB^\mathsf{T})^{-1}B\mathbf{b}\right) = \mathbf{b} - B^\mathsf{T}(BB^\mathsf{T})^{-1}B\mathbf{b}.$$

Note that, according to our projection formula (†), we can interpret this answer as

$$\mathbf{x}_0 = \mathbf{b} - \text{proj}_{\mathbf{C}(B^\mathsf{T})}\mathbf{b} = \mathbf{b} - \text{proj}_{\mathbf{R}(B)}\mathbf{b} = \text{proj}_{\mathbf{R}(B)^\perp}\mathbf{b} = \text{proj}_{\mathbf{N}(B)}\mathbf{b},$$

as it should be.

5.1 Data Fitting

Perhaps the most natural setting in which inconsistent systems of equations arise is that of fitting data to a curve when they won't quite fit. For example, in our laboratory work many of us have tried to find the right constants a and k so that the data points $\begin{bmatrix} x_1 \\ y_1 \end{bmatrix}, \ldots, \begin{bmatrix} x_m \\ y_m \end{bmatrix}$ lie on the curve $y = ax^k$. Taking natural logarithms, we see that this is equivalent to fitting the points $\begin{bmatrix} u_i \\ v_i \end{bmatrix} = \begin{bmatrix} \log x_i \\ \log y_i \end{bmatrix}$, $i = 1, \ldots, m$, to a line $v = ku + \log a$—whence the convenience of log-log paper. The least squares solution of such problems is called the *least squares line* fitting the points (or the *line of regression* in statistics).

▶ **EXAMPLE 6**

Find the least squares line $y = ax + b$ for the data points $\begin{bmatrix} -1 \\ 0 \end{bmatrix}, \begin{bmatrix} 1 \\ 1 \end{bmatrix}$, and $\begin{bmatrix} 2 \\ 3 \end{bmatrix}$. (See Figure 5.3.)

We get the system of equations

$$\begin{aligned} -1a + b &= 0 \\ 1a + b &= 1 \\ 2a + b &= 3, \end{aligned}$$

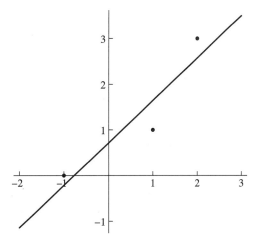

Figure 5.3

which in matrix form becomes

$$A\begin{bmatrix} a \\ b \end{bmatrix} = \begin{bmatrix} -1 & 1 \\ 1 & 1 \\ 2 & 1 \end{bmatrix} \begin{bmatrix} a \\ b \end{bmatrix} = \begin{bmatrix} 0 \\ 1 \\ 3 \end{bmatrix}.$$

The least squares solution is

$$\begin{bmatrix} \bar{a} \\ \bar{b} \end{bmatrix} = (A^T A)^{-1} A^T \begin{bmatrix} 0 \\ 1 \\ 3 \end{bmatrix} = \frac{1}{14} \begin{bmatrix} 3 & -2 \\ -2 & 6 \end{bmatrix} \begin{bmatrix} 7 \\ 4 \end{bmatrix} = \frac{1}{14} \begin{bmatrix} 13 \\ 10 \end{bmatrix}.$$

That is, the least squares line is

$$y = \frac{13}{14}x + \frac{5}{7}. \blacktriangleleft$$

When we find the least squares line $y = \bar{a}x + \bar{b}$ fitting the data points $\begin{bmatrix} x_1 \\ y_1 \end{bmatrix}, \ldots, \begin{bmatrix} x_m \\ y_m \end{bmatrix}$, we are finding the least squares solution of the (inconsistent) system $A\begin{bmatrix} a \\ b \end{bmatrix} = \mathbf{y}$, where

$$A = \begin{bmatrix} x_1 & 1 \\ x_2 & 1 \\ \vdots & \vdots \\ x_m & 1 \end{bmatrix} \quad \text{and} \quad \mathbf{y} = \begin{bmatrix} y_1 \\ y_2 \\ \vdots \\ y_m \end{bmatrix}.$$

Let's denote by $\bar{\mathbf{y}} = A\begin{bmatrix} \bar{a} \\ \bar{b} \end{bmatrix}$ the projection of $\mathbf{y}$ onto $\mathbf{C}(A)$. The least squares solution $\begin{bmatrix} \bar{a} \\ \bar{b} \end{bmatrix}$ has the property that $\|\mathbf{y} - \bar{\mathbf{y}}\|$ is as small as possible. If we define the *error* vector $\boldsymbol{\epsilon} = \mathbf{y} - \bar{\mathbf{y}}$, then we have

$$\epsilon = \begin{bmatrix} \epsilon_1 \\ \epsilon_2 \\ \vdots \\ \epsilon_m \end{bmatrix} = \begin{bmatrix} y_1 - \bar{y}_1 \\ y_2 - \bar{y}_2 \\ \vdots \\ y_m - \bar{y}_m \end{bmatrix} = \begin{bmatrix} y_1 - (\bar{a}x_1 + \bar{b}) \\ y_2 - (\bar{a}x_2 + \bar{b}) \\ \vdots \\ y_m - (\bar{a}x_m + \bar{b}) \end{bmatrix}.$$

The least squares process chooses $\bar{a}$ and $\bar{b}$ so that $\|\epsilon\|^2 = \epsilon_1^2 + \cdots + \epsilon_m^2$ is as small as possible. But something interesting happens. Recall that

$$\epsilon = \mathbf{y} - \bar{\mathbf{y}} \in \mathbf{C}(A)^\perp.$$

Thus, ϵ is orthogonal to each of the column vectors of A, and so, in particular,

$$\begin{bmatrix} \epsilon_1 \\ \vdots \\ \epsilon_m \end{bmatrix} \cdot \begin{bmatrix} 1 \\ \vdots \\ 1 \end{bmatrix} = \epsilon_1 + \cdots + \epsilon_m = 0.$$

That is, in the process of minimizing the sum of the squares of the errors ϵ_i, we have in fact made their (algebraic) sum equal to 0.

5.2 Orthogonal Bases

We have seen how to find the projection of a vector onto a subspace $V \subset \mathbb{R}^n$ by using the so-called normal equations. But the inner workings of the formula (†) on p. 228 escape us. Since we have known since Chapter 1 how to project a vector $\mathbf{x}$ onto a line, it might seem more natural to start with a basis $\{\mathbf{v}_1, \ldots, \mathbf{v}_k\}$ for V and sum up the projections of $\mathbf{x}$ onto the $\mathbf{v}_j$'s. However, as we see in Figure 5.4(a), when we start with $\mathbf{x} \in V$ and add the projections of $\mathbf{x}$ onto the vectors of an arbitrary basis for V, the resulting vector needn't have much to do with $\mathbf{x}$. Nevertheless, the diagram on the right suggests that when we start with a basis consisting of mutually orthogonal vectors, the process may work. We begin by proving this as a lemma.

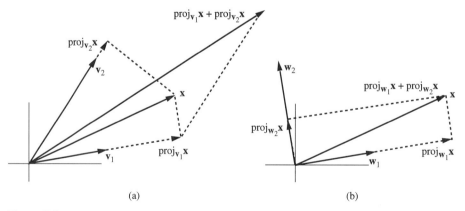

Figure 5.4

Definition Let $\mathbf{v}_1, \ldots, \mathbf{v}_k \in \mathbb{R}^n$. We say $\{\mathbf{v}_1, \ldots, \mathbf{v}_k\}$ is an *orthogonal set* of vectors provided $\mathbf{v}_i \cdot \mathbf{v}_j = 0$ whenever $i \neq j$. We say $\{\mathbf{v}_1, \ldots, \mathbf{v}_k\}$ is an *orthogonal basis* for a subspace V if $\{\mathbf{v}_1, \ldots, \mathbf{v}_k\}$ is both a basis for V and an orthogonal set.

Lemma 5.1 *Suppose $\{\mathbf{v}_1, \ldots, \mathbf{v}_k\}$ is a basis for V and $\mathbf{x} \in V$. Then*

$$\mathbf{x} = \sum_{i=1}^{k} \mathrm{proj}_{\mathbf{v}_i} \mathbf{x} = \sum_{i=1}^{k} \frac{\mathbf{x} \cdot \mathbf{v}_i}{\|\mathbf{v}_i\|^2} \mathbf{v}_i$$

if and only if $\{\mathbf{v}_1, \ldots, \mathbf{v}_k\}$ is an orthogonal basis for V.

Proof Suppose $\{\mathbf{v}_1, \ldots, \mathbf{v}_k\}$ is an orthogonal basis for V. Then there are scalars $c_1, \ldots, c_k$ so that

$$\mathbf{x} = c_1 \mathbf{v}_1 + \cdots + c_i \mathbf{v}_i + \cdots + c_k \mathbf{v}_k.$$

Taking advantage of the orthogonality of the $\mathbf{v}_j$'s, we take the dot product of this equation with $\mathbf{v}_i$:

$$\mathbf{x} \cdot \mathbf{v}_i = c_1(\mathbf{v}_1 \cdot \mathbf{v}_i) + \cdots + c_i(\mathbf{v}_i \cdot \mathbf{v}_i) + \cdots + c_k(\mathbf{v}_k \cdot \mathbf{v}_i)$$
$$= c_i \|\mathbf{v}_i\|^2,$$

and so

$$c_i = \frac{\mathbf{x} \cdot \mathbf{v}_i}{\|\mathbf{v}_i\|^2}.$$

(Note that $\mathbf{v}_i \neq \mathbf{0}$ since $\{\mathbf{v}_1, \ldots, \mathbf{v}_k\}$ forms a basis for V.)

Conversely, suppose that every vector $\mathbf{x} \in V$ is the sum of its projections on $\mathbf{v}_1, \ldots, \mathbf{v}_k$. Let's just examine what this means when $\mathbf{x} = \mathbf{v}_1$: We are given that

$$\mathbf{v}_1 = \sum_{i=1}^{k} \mathrm{proj}_{\mathbf{v}_i} \mathbf{v}_1 = \sum_{i=1}^{k} \frac{\mathbf{v}_1 \cdot \mathbf{v}_i}{\|\mathbf{v}_i\|^2} \mathbf{v}_i.$$

Recall from Proposition 3.1 of Chapter 4 that every vector has a *unique* expansion as a linear combination of basis vectors, so comparing coefficients of $\mathbf{v}_2, \ldots, \mathbf{v}_k$ on either side of this equation, we conclude that

$$\mathbf{v}_1 \cdot \mathbf{v}_i = 0 \quad \text{for all } i = 2, \ldots, k.$$

A similar argument shows that $\mathbf{v}_i \cdot \mathbf{v}_j = 0$ for all $i \neq j$, and the proof is complete. ∎

As we mentioned above, if $\{\mathbf{v}_1, \ldots, \mathbf{v}_k\}$ is a basis for V, then every vector $\mathbf{x} \in V$ can be written uniquely as a linear combination

$$\mathbf{x} = c_1 \mathbf{v}_1 + c_2 \mathbf{v}_2 + \cdots + c_k \mathbf{v}_k.$$

We recall that the coefficients $c_1, c_2, \ldots, c_k$ that appear here are called the *coordinates* of $\mathbf{x}$ with respect to the basis $\{\mathbf{v}_1, \ldots, \mathbf{v}_k\}$. It is worth emphasizing that when $\{\mathbf{v}_1, \ldots, \mathbf{v}_k\}$ forms an *orthogonal basis* for V, it is quite easy to compute the coordinates of $\mathbf{x}$ by using the dot product; that is, $c_i = \mathbf{x} \cdot \mathbf{v}_i / \|\mathbf{v}_i\|^2$. As we saw in Example 8 of Section 3 of Chapter 4 (see also Section 1 of Chapter 9), when the basis is not orthogonal, it is far more tedious to compute these coordinates.

Not only do orthogonal bases make it easy to calculate coordinates, they also make projections quite easy to compute, as we now see.

Proposition 5.2 *Let $V \subset \mathbb{R}^n$ be a k-dimensional subspace. For any vector $\mathbf{b} \in \mathbb{R}^n$,*

(**) $$\operatorname{proj}_V \mathbf{b} = \sum_{i=1}^{k} \operatorname{proj}_{\mathbf{v}_i} \mathbf{b} = \sum_{i=1}^{k} \frac{\mathbf{b} \cdot \mathbf{v}_i}{\|\mathbf{v}_i\|^2} \mathbf{v}_i$$

if and only if $\{\mathbf{v}_1, \ldots, \mathbf{v}_k\}$ is an orthogonal basis for V.

Proof Assume $\{\mathbf{v}_1, \ldots, \mathbf{v}_k\}$ is an orthogonal basis for V and write $\mathbf{b} = \mathbf{p} + (\mathbf{b} - \mathbf{p})$, where $\mathbf{p} = \operatorname{proj}_V \mathbf{b}$ (and so $\mathbf{b} - \mathbf{p} \in V^\perp$). Then, since $\mathbf{p} \in V$, by Lemma 5.1, we know $\mathbf{p} = \sum_{i=1}^{k} \frac{\mathbf{p} \cdot \mathbf{v}_i}{\|\mathbf{v}_i\|^2} \mathbf{v}_i$. Moreover, for $i = 1, \ldots, k$, we have $\mathbf{b} \cdot \mathbf{v}_i = \mathbf{p} \cdot \mathbf{v}_i$ since $\mathbf{b} - \mathbf{p} \in V^\perp$. Thus,

$$\operatorname{proj}_V \mathbf{b} = \mathbf{p} = \sum_{i=1}^{k} \operatorname{proj}_{\mathbf{v}_i} \mathbf{p} = \sum_{i=1}^{k} \frac{\mathbf{p} \cdot \mathbf{v}_i}{\|\mathbf{v}_i\|^2} \mathbf{v}_i = \sum_{i=1}^{k} \frac{\mathbf{b} \cdot \mathbf{v}_i}{\|\mathbf{v}_i\|^2} \mathbf{v}_i = \sum_{i=1}^{k} \operatorname{proj}_{\mathbf{v}_i} \mathbf{b}.$$

Conversely, suppose $\operatorname{proj}_V \mathbf{b} = \sum_{i=1}^{k} \operatorname{proj}_{\mathbf{v}_i} \mathbf{b}$ for all $\mathbf{b} \in \mathbb{R}^n$. In particular, when $\mathbf{b} \in V$, we deduce that $\mathbf{b} = \operatorname{proj}_V \mathbf{b}$ can be written as a linear combination of $\mathbf{v}_1, \ldots, \mathbf{v}_k$, so these vectors span V; since V is k-dimensional, $\{\mathbf{v}_1, \ldots, \mathbf{v}_k\}$ gives a basis for V. By Lemma 5.1, it must be an orthogonal basis. ∎

We now have another way to calculate the projection of a vector on a subspace V, provided we can come up with an orthogonal basis for V.

▶ **EXAMPLE 7**

We return to Example 5 on p. 229. The basis $\{\mathbf{v}_1, \mathbf{v}_2\}$ we used there was certainly not an orthogonal basis, but it is not hard to find one that is. Instead, we take

$$\mathbf{w}_1 = \begin{bmatrix} -1 \\ 0 \\ 1 \end{bmatrix} \quad \text{and} \quad \mathbf{w}_2 = \begin{bmatrix} 1 \\ 1 \\ 1 \end{bmatrix}.$$

(It is immediate that $\mathbf{w}_1 \cdot \mathbf{w}_2 = 0$ and that $\mathbf{w}_1, \mathbf{w}_2$ lie in the plane $x_1 - 2x_2 + x_3 = 0$.) Now, we calculate

$$\operatorname{proj}_V \mathbf{b} = \operatorname{proj}_{\mathbf{w}_1} \mathbf{b} + \operatorname{proj}_{\mathbf{w}_2} \mathbf{b} = \frac{\mathbf{b} \cdot \mathbf{w}_1}{\|\mathbf{w}_1\|^2} \mathbf{w}_1 + \frac{\mathbf{b} \cdot \mathbf{w}_2}{\|\mathbf{w}_2\|^2} \mathbf{w}_2$$

$$= \left(\frac{1}{\|\mathbf{w}_1\|^2} \mathbf{w}_1 \mathbf{w}_1^\mathsf{T} + \frac{1}{\|\mathbf{w}_2\|^2} \mathbf{w}_2 \mathbf{w}_2^\mathsf{T} \right) \mathbf{b}$$

$$= \left(\frac{1}{2} \begin{bmatrix} 1 & 0 & -1 \\ 0 & 0 & 0 \\ -1 & 0 & 1 \end{bmatrix} + \frac{1}{3} \begin{bmatrix} 1 & 1 & 1 \\ 1 & 1 & 1 \\ 1 & 1 & 1 \end{bmatrix} \right) \mathbf{b}$$

$$= \begin{bmatrix} \frac{5}{6} & \frac{1}{3} & -\frac{1}{6} \\ \frac{1}{3} & \frac{1}{3} & \frac{1}{3} \\ -\frac{1}{6} & \frac{1}{3} & \frac{5}{6} \end{bmatrix} \mathbf{b},$$

as we found earlier. ◀

Remark This is exactly what we get from formula (†) on p. 228 when $\{\mathbf{v}_1, \ldots, \mathbf{v}_k\}$ is an orthogonal set. In particular,

$$P = A(A^\mathsf{T} A)^{-1} A^\mathsf{T}$$

$$= \begin{bmatrix} | & | & & | \\ \mathbf{v}_1 & \mathbf{v}_2 & \cdots & \mathbf{v}_k \\ | & | & & | \end{bmatrix} \begin{bmatrix} \frac{1}{\|\mathbf{v}_1\|^2} & & & \\ & \frac{1}{\|\mathbf{v}_2\|^2} & & \\ & & \ddots & \\ & & & \frac{1}{\|\mathbf{v}_k\|^2} \end{bmatrix} \begin{bmatrix} \text{---} & \mathbf{v}_1^\mathsf{T} & \text{---} \\ \text{---} & \mathbf{v}_2^\mathsf{T} & \text{---} \\ & \vdots & \\ \text{---} & \mathbf{v}_k^\mathsf{T} & \text{---} \end{bmatrix}$$

$$= \sum_{i=1}^k \frac{1}{\|\mathbf{v}_i\|^2} \mathbf{v}_i \mathbf{v}_i^\mathsf{T}.$$

Now it is time to develop an algorithm for transforming a given (ordered) basis $\{\mathbf{v}_1, \ldots, \mathbf{v}_k\}$ for a subspace into an orthogonal basis $\{\mathbf{w}_1, \ldots, \mathbf{w}_k\}$, as shown in Figure 5.5. The idea is quite simple. We set

$$\mathbf{w}_1 = \mathbf{v}_1.$$

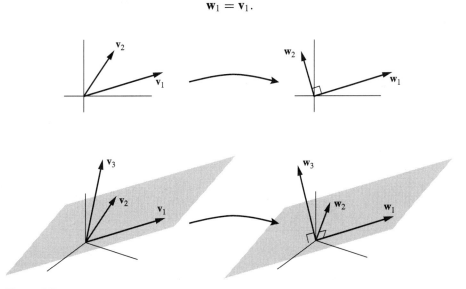

Figure 5.5

If $\mathbf{v}_2$ is orthogonal to $\mathbf{w}_1$, then we set $\mathbf{w}_2 = \mathbf{v}_2$. Of course, in general, it will not be, and we want $\mathbf{w}_2$ to be the part of $\mathbf{v}_2$ that is orthogonal to $\mathbf{w}_1$; i.e., we set

$$\mathbf{w}_2 = \mathbf{v}_2 - \text{proj}_{\mathbf{w}_1} \mathbf{v}_2 = \mathbf{v}_2 - \frac{\mathbf{v}_2 \cdot \mathbf{w}_1}{\|\mathbf{w}_1\|^2} \mathbf{w}_1.$$

Then, by construction, $\mathbf{w}_1$ and $\mathbf{w}_2$ are orthogonal and $\text{Span}(\mathbf{w}_1, \mathbf{w}_2) \subset \text{Span}(\mathbf{v}_1, \mathbf{v}_2)$. Since $\mathbf{w}_2 \neq \mathbf{0}$ (why?), $\{\mathbf{w}_1, \mathbf{w}_2\}$ must be linearly independent and therefore give a basis for $\text{Span}(\mathbf{v}_1, \mathbf{v}_2)$ by Lemma 3.8. We continue, replacing $\mathbf{v}_3$ by its part orthogonal to the plane spanned by $\mathbf{w}_1$ and $\mathbf{w}_2$:

$$\mathbf{w}_3 = \mathbf{v}_3 - \text{proj}_{\text{Span}(\mathbf{w}_1, \mathbf{w}_2)} \mathbf{v}_3 = \mathbf{v}_3 - \text{proj}_{\mathbf{w}_1} \mathbf{v}_3 - \text{proj}_{\mathbf{w}_2} \mathbf{v}_3 = \mathbf{v}_3 - \frac{\mathbf{v}_3 \cdot \mathbf{w}_1}{\|\mathbf{w}_1\|^2} \mathbf{w}_1 - \frac{\mathbf{v}_3 \cdot \mathbf{w}_2}{\|\mathbf{w}_2\|^2} \mathbf{w}_2.$$

Note that we are making definite use of Proposition 5.2 here: We must use $\mathbf{w}_1$ and $\mathbf{w}_2$ in the formula here, rather than $\mathbf{v}_1$ and $\mathbf{v}_2$, because the formula (∗∗) requires an orthogonal basis. Once again, we find that $\mathbf{w}_3 \neq \mathbf{0}$ (why?), and so $\{\mathbf{w}_1, \mathbf{w}_2, \mathbf{w}_3\}$ must be linearly independent and, consequently, an orthogonal basis for $\text{Span}(\mathbf{v}_1, \mathbf{v}_2, \mathbf{v}_3)$. The process continues until we have arrived at $\mathbf{v}_k$ and replaced it by

$$\mathbf{w}_k = \mathbf{v}_k - \text{proj}_{\text{Span}(\mathbf{w}_1, \ldots, \mathbf{w}_{k-1})} \mathbf{v}_k = \mathbf{v}_k - \frac{\mathbf{v}_k \cdot \mathbf{w}_1}{\|\mathbf{w}_1\|^2} \mathbf{w}_1 - \frac{\mathbf{v}_k \cdot \mathbf{w}_2}{\|\mathbf{w}_2\|^2} \mathbf{w}_2 - \cdots - \frac{\mathbf{v}_k \cdot \mathbf{w}_{k-1}}{\|\mathbf{w}_{k-1}\|^2} \mathbf{w}_{k-1}.$$

Summarizing, we have the algorithm that goes by the name of the Gram-Schmidt process.

Theorem 5.3 (Gram-Schmidt Process) *Given a basis $\{\mathbf{v}_1, \ldots, \mathbf{v}_k\}$ for a subspace $V \subset \mathbb{R}^n$, we obtain an orthogonal basis $\{\mathbf{w}_1, \ldots, \mathbf{w}_k\}$ for V as follows:*

$$\mathbf{w}_1 = \mathbf{v}_1$$
$$\mathbf{w}_2 = \mathbf{v}_2 - \frac{\mathbf{v}_2 \cdot \mathbf{w}_1}{\|\mathbf{w}_1\|^2} \mathbf{w}_1$$
$$\vdots$$

and, assuming $\mathbf{w}_1, \ldots, \mathbf{w}_j$ have been defined,

$$\mathbf{w}_{j+1} = \mathbf{v}_{j+1} - \frac{\mathbf{v}_{j+1} \cdot \mathbf{w}_1}{\|\mathbf{w}_1\|^2} \mathbf{w}_1 - \frac{\mathbf{v}_{j+1} \cdot \mathbf{w}_2}{\|\mathbf{w}_2\|^2} \mathbf{w}_2 - \cdots - \frac{\mathbf{v}_{j+1} \cdot \mathbf{w}_j}{\|\mathbf{w}_j\|^2} \mathbf{w}_j$$

$$\vdots$$

$$\mathbf{w}_k = \mathbf{v}_k - \frac{\mathbf{v}_k \cdot \mathbf{w}_1}{\|\mathbf{w}_1\|^2} \mathbf{w}_1 - \frac{\mathbf{v}_k \cdot \mathbf{w}_2}{\|\mathbf{w}_2\|^2} \mathbf{w}_2 - \cdots - \frac{\mathbf{v}_k \cdot \mathbf{w}_{k-1}}{\|\mathbf{w}_{k-1}\|^2} \mathbf{w}_{k-1}.$$

If we so desire, we can arrange for an orthogonal basis consisting of unit *vectors by dividing each of $\mathbf{w}_1, \ldots, \mathbf{w}_k$ by its respective length:*

$$\mathbf{q}_1 = \frac{\mathbf{w}_1}{\|\mathbf{w}_1\|}, \quad \mathbf{q}_2 = \frac{\mathbf{w}_2}{\|\mathbf{w}_2\|}, \quad \ldots, \quad \mathbf{q}_k = \frac{\mathbf{w}_k}{\|\mathbf{w}_k\|}.$$

The set $\{\mathbf{q}_1, \ldots, \mathbf{q}_k\}$ is called an orthonormal basis *for V.*

EXAMPLE 8

Let $\mathbf{v}_1 = \begin{bmatrix} 1 \\ 1 \\ 1 \\ 1 \end{bmatrix}$, $\mathbf{v}_2 = \begin{bmatrix} 3 \\ 1 \\ -1 \\ 1 \end{bmatrix}$, and $\mathbf{v}_3 = \begin{bmatrix} 1 \\ 1 \\ 3 \\ 3 \end{bmatrix}$. We want to use the Gram-Schmidt process to give an orthogonal basis for $V = \text{Span}(\mathbf{v}_1, \mathbf{v}_2, \mathbf{v}_3) \subset \mathbb{R}^4$. We take

$$\mathbf{w}_1 = \mathbf{v}_1 = \begin{bmatrix} 1 \\ 1 \\ 1 \\ 1 \end{bmatrix};$$

$$\mathbf{w}_2 = \mathbf{v}_2 - \frac{\mathbf{v}_2 \cdot \mathbf{w}_1}{\|\mathbf{w}_1\|^2}\mathbf{w}_1 = \begin{bmatrix} 3 \\ 1 \\ -1 \\ 1 \end{bmatrix} - \frac{\begin{bmatrix} 3 \\ 1 \\ -1 \\ 1 \end{bmatrix} \cdot \begin{bmatrix} 1 \\ 1 \\ 1 \\ 1 \end{bmatrix}}{\left\|\begin{bmatrix} 1 \\ 1 \\ 1 \\ 1 \end{bmatrix}\right\|^2} \begin{bmatrix} 1 \\ 1 \\ 1 \\ 1 \end{bmatrix}$$

$$= \begin{bmatrix} 3 \\ 1 \\ -1 \\ 1 \end{bmatrix} - \frac{4}{4}\begin{bmatrix} 1 \\ 1 \\ 1 \\ 1 \end{bmatrix} = \begin{bmatrix} 2 \\ 0 \\ -2 \\ 0 \end{bmatrix};$$

$$\mathbf{w}_3 = \mathbf{v}_3 - \frac{\mathbf{v}_3 \cdot \mathbf{w}_1}{\|\mathbf{w}_1\|^2}\mathbf{w}_1 - \frac{\mathbf{v}_3 \cdot \mathbf{w}_2}{\|\mathbf{w}_2\|^2}\mathbf{w}_2$$

$$= \begin{bmatrix} 1 \\ 1 \\ 3 \\ 3 \end{bmatrix} - \frac{\begin{bmatrix} 1 \\ 1 \\ 3 \\ 3 \end{bmatrix} \cdot \begin{bmatrix} 1 \\ 1 \\ 1 \\ 1 \end{bmatrix}}{\left\|\begin{bmatrix} 1 \\ 1 \\ 1 \\ 1 \end{bmatrix}\right\|^2} \begin{bmatrix} 1 \\ 1 \\ 1 \\ 1 \end{bmatrix} - \frac{\begin{bmatrix} 1 \\ 1 \\ 3 \\ 3 \end{bmatrix} \cdot \begin{bmatrix} 2 \\ 0 \\ -2 \\ 0 \end{bmatrix}}{\left\|\begin{bmatrix} 2 \\ 0 \\ -2 \\ 0 \end{bmatrix}\right\|^2} \begin{bmatrix} 2 \\ 0 \\ -2 \\ 0 \end{bmatrix}$$

$$= \begin{bmatrix} 1 \\ 1 \\ 3 \\ 3 \end{bmatrix} - \frac{8}{4}\begin{bmatrix} 1 \\ 1 \\ 1 \\ 1 \end{bmatrix} - \frac{-4}{8}\begin{bmatrix} 2 \\ 0 \\ -2 \\ 0 \end{bmatrix} = \begin{bmatrix} 0 \\ -1 \\ 0 \\ 1 \end{bmatrix}.$$

And if we desire an *orthonormal* basis, then we take

$$\mathbf{q}_1 = \frac{1}{2}\begin{bmatrix} 1 \\ 1 \\ 1 \\ 1 \end{bmatrix}, \quad \mathbf{q}_2 = \frac{1}{\sqrt{2}}\begin{bmatrix} 1 \\ 0 \\ -1 \\ 0 \end{bmatrix}, \quad \mathbf{q}_3 = \frac{1}{\sqrt{2}}\begin{bmatrix} 0 \\ -1 \\ 0 \\ 1 \end{bmatrix}.$$

It's always a good idea to check that the vectors form an orthogonal (or orthonormal) set, and it's easy—with these numbers—to do so. ◀

5.3 Inner Product Spaces

In certain abstract vector spaces we may define a notion of dot product.

Definition Let V be a real vector space. We say V is an *inner product space* if for every pair of elements $\mathbf{u}, \mathbf{v} \in V$ there is a real number $\langle \mathbf{u}, \mathbf{v} \rangle$, called the *inner product of* $\mathbf{u}$ *and* $\mathbf{v}$, such that

1. $\langle \mathbf{u}, \mathbf{v} \rangle = \langle \mathbf{v}, \mathbf{u} \rangle$ for all $\mathbf{u}, \mathbf{v} \in V$;
2. $\langle c\mathbf{u}, \mathbf{v} \rangle = c \langle \mathbf{u}, \mathbf{v} \rangle$ for all $\mathbf{u}, \mathbf{v} \in V$ and scalars c;
3. $\langle \mathbf{u} + \mathbf{v}, \mathbf{w} \rangle = \langle \mathbf{u}, \mathbf{w} \rangle + \langle \mathbf{v}, \mathbf{w} \rangle$ for all $\mathbf{u}, \mathbf{v}, \mathbf{w} \in V$;
4. $\langle \mathbf{v}, \mathbf{v} \rangle \geq 0$ for all $\mathbf{v} \in V$ and $\langle \mathbf{v}, \mathbf{v} \rangle = 0$ only if $\mathbf{v} = \mathbf{0}$.

▶ **EXAMPLE 9**

a. Fix $k+1$ distinct real numbers $t_1, t_2, \ldots, t_{k+1}$ and define an inner product on $\mathcal{P}_k$, the vector space of polynomials of degree $\leq k$, by the formula

$$\langle p, q \rangle = \sum_{i=1}^{k+1} p(t_i) q(t_i), \quad p, q \in \mathcal{P}_k.$$

All the properties of an inner product are obvious except for the very last. If $\langle p, p \rangle = 0$, then $\sum_{i=1}^{k+1} p(t_i)^2 = 0$, and so we must have $p(t_1) = p(t_2) = \cdots = p(t_{k+1}) = 0$. But if a polynomial of degree $\leq k$ has (at least) $k+1$ roots, then it must be the zero polynomial.

b. Let $\mathcal{C}^0([a, b])$ denote the vector space of continuous functions on the interval $[a, b]$. If $f, g \in \mathcal{C}^0([a, b])$, define

$$\langle f, g \rangle = \int_a^b f(t) g(t) dt.$$

We verify that the defining properties hold.

1. $\langle f, g \rangle = \int_a^b f(t)g(t)dt = \int_a^b g(t)f(t)dt = \langle g, f \rangle$.
2. $\langle cf, g \rangle = \int_a^b (cf)(t)g(t)dt = \int_a^b cf(t)g(t)dt = c\int_a^b f(t)g(t)dt = c\langle f, g \rangle$.
3. $\langle f + g, h \rangle = \int_a^b (f + g)(t)h(t)dt = \int_a^b \big(f(t) + g(t)\big)h(t)dt = \int_a^b \big(f(t)h(t) + g(t)h(t)\big)dt = \int_a^b f(t)h(t)dt + \int_a^b g(t)h(t)dt = \langle f, h \rangle + \langle g, h \rangle$.

4. $\langle f, f \rangle = \int_a^b f(t)^2 dt \geq 0$ since $f(t)^2 \geq 0$ for all t. On the other hand, if $\langle f, f \rangle = \int_a^b (f(t))^2 dt = 0$, then since f is continuous and $f^2 \geq 0$, it must be the case that $f = 0$. (If not, we would have $f(t_0) \neq 0$ for some t_0, and then $f(t)^2$ would be positive on some small interval containing t_0; it would then follow that $\int_a^b f(t)^2 dt > 0$.)

The same inner product can be defined on subspaces of $\mathcal{C}^0([a, b])$, e.g., $\mathcal{P}_k$.

c. We define an inner product on $\mathcal{M}_{n \times n}$ in Exercise 18. ◂

If V is an inner product space, we define length, orthogonality, and the angle between vectors just as we did in $\mathbb{R}^n$. If $\mathbf{v} \in V$, we define its length to be $\|\mathbf{v}\| = \sqrt{\langle \mathbf{v}, \mathbf{v} \rangle}$. We say $\mathbf{v}$ and $\mathbf{w}$ are orthogonal if $\langle \mathbf{v}, \mathbf{w} \rangle = 0$. Since the Cauchy-Schwarz inequality can be established in general by following the proof of Proposition 2.3 of Chapter 1 *verbatim*, we can define the angle θ between $\mathbf{v}$ and $\mathbf{w}$ by the equation

$$\cos \theta = \frac{\langle \mathbf{v}, \mathbf{w} \rangle}{\|\mathbf{v}\| \|\mathbf{w}\|}.$$

We can define orthogonal subspaces, orthogonal complements, and the Gram-Schmidt process analogously.

We can use the inner product defined in Example 9a to prove the following important result about curve fitting.

Theorem 5.4 (Lagrange Interpolation Formula) *Given $k + 1$ points*

$$\begin{bmatrix} t_1 \\ b_1 \end{bmatrix}, \begin{bmatrix} t_2 \\ b_2 \end{bmatrix}, \ldots, \begin{bmatrix} t_{k+1} \\ b_{k+1} \end{bmatrix}$$

in the plane with $t_1, t_2, \ldots, t_{k+1}$ distinct, there is exactly one polynomial $p \in \mathcal{P}_k$ whose graph passes through the points.

Proof We begin by explicitly constructing a basis for $\mathcal{P}_k$ consisting of mutually orthogonal vectors of length 1 with respect to the inner product defined in Example 9a. That is, to start, we seek a polynomial $p_1 \in \mathcal{P}_k$ so that

$$p_1(t_1) = 1, \quad p_1(t_2) = 0, \quad \ldots, \quad p_1(t_{k+1}) = 0.$$

The polynomial $q_1(t) = (t - t_2)(t - t_3) \cdots (t - t_{k+1})$ has the property that $q_1(t_j) = 0$ for $j = 2, 3, \ldots, k + 1$, and $q_1(t_1) = (t_1 - t_2)(t_1 - t_3) \cdots (t_1 - t_{k+1}) \neq 0$ (why?). So now we set

$$p_1(t) = \frac{(t - t_2)(t - t_3) \cdots (t - t_{k+1})}{(t_1 - t_2)(t_1 - t_3) \cdots (t_1 - t_{k+1})};$$

then, as desired, $p_1(t_1) = 1$ and $p_1(t_j) = 0$ for $j = 2, 3, \ldots, k + 1$. Similarly, we can define

$$p_2(t) = \frac{(t - t_1)(t - t_3) \cdots (t - t_{k+1})}{(t_2 - t_1)(t_2 - t_3) \cdots (t_2 - t_{k+1})}$$

and polynomials $p_3, \ldots, p_{k+1}$ so that

$$p_i(t_j) = \begin{cases} 1, & \text{when } i = j \\ 0, & \text{when } i \neq j \end{cases}.$$

Like the standard basis vectors in Euclidean space, $p_1, p_2, \ldots, p_{k+1}$ are unit vectors in $\mathcal{P}_k$ that are orthogonal to one another. It follows from Exercise 4.3.5 that these vectors form a linearly independent set, hence a basis for $\mathcal{P}_k$ (why?). In Figure 5.6 we give the graphs of the Lagrange basis polynomials p_1, p_2, p_3 for $\mathcal{P}_2$ when $t_1 = -1$, $t_2 = 0$, and $t_3 = 2$.

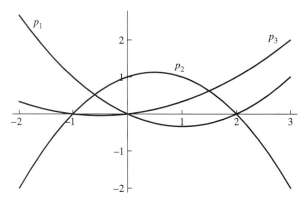

Figure 5.6

Now it is easy to see that the appropriate linear combination

$$p = b_1 p_1 + b_2 p_2 + \cdots + b_{k+1} p_{k+1}$$

has the desired properties: viz., $p(t_j) = b_j$ for $j = 1, 2, \ldots, k+1$. On the other hand, two polynomials of degree $\leq k$ with the same values at $k+1$ points must be equal since their difference is a polynomial of degree $\leq k$ with at least $k+1$ roots. This establishes uniqueness. (More elegantly, any polynomial q with $q(t_j) = b_j$, $j = 1, \ldots, k+1$, must satisfy $\langle q, p_j \rangle = b_j$, $j = 1, \ldots, k+1$.) ∎

▶ EXERCISES 5.5

1. Find the projection of the given vector $\mathbf{b} \in \mathbb{R}^n$ onto the given hyperplane $V \subset \mathbb{R}^n$.

   (a) $V = \{x_1 + x_2 + x_3 = 0\} \subset \mathbb{R}^3$, $\mathbf{b} = \begin{bmatrix} 2 \\ 1 \\ 1 \end{bmatrix}$

   *(b) $V = \{x_1 + x_2 + x_3 = 0\} \subset \mathbb{R}^4$, $\mathbf{b} = \begin{bmatrix} 0 \\ 1 \\ 2 \\ 3 \end{bmatrix}$

(c) $V = \{x_1 - x_2 + x_3 + 2x_4 = 0\} \subset \mathbb{R}^4$, $\mathbf{b} = \begin{bmatrix} 1 \\ 1 \\ 1 \\ 1 \end{bmatrix}$

2. Check from the formula $P = A(A^\mathsf{T} A)^{-1} A^\mathsf{T}$ for the projection matrix that $P = P^\mathsf{T}$ and $P^2 = P$. Show that $I - P$ has the same properties; explain.

3. Let $V = \text{Span}\left(\begin{bmatrix} 1 \\ 0 \\ 1 \end{bmatrix}, \begin{bmatrix} 0 \\ 1 \\ -2 \end{bmatrix} \right) \subset \mathbb{R}^3$. Construct the matrix $[\text{proj}_V]$

(a) by finding $[\text{proj}_{(V^\perp)}]$;
(b) by using the projection matrix P given in formula (†) on p. 228;
(c) by finding an orthogonal basis for V.

*4. (a) Find the least squares solution of
$$\begin{aligned} x_1 + x_2 &= 4 \\ 2x_1 + x_2 &= -2 \\ x_1 - x_2 &= 1. \end{aligned}$$

(b) Find the point on the plane spanned by $\begin{bmatrix} 1 \\ 2 \\ 1 \end{bmatrix}$ and $\begin{bmatrix} 1 \\ 1 \\ -1 \end{bmatrix}$ that is closest to $\begin{bmatrix} 4 \\ -2 \\ 1 \end{bmatrix}$.

5. (a) Find the least squares solution of
$$\begin{aligned} x_1 + x_2 &= 1 \\ x_1 - 3x_2 &= 4 \\ 2x_1 + x_2 &= 3. \end{aligned}$$

(b) Find the point on the plane spanned by $\begin{bmatrix} 1 \\ 1 \\ 2 \end{bmatrix}$ and $\begin{bmatrix} 1 \\ -3 \\ 1 \end{bmatrix}$ that is closest to $\begin{bmatrix} 1 \\ 4 \\ 3 \end{bmatrix}$.

6. Solve Exercise 5.4.26 anew, using (‡) on p. 230.

7. Consider the four data points $\begin{bmatrix} -1 \\ 0 \end{bmatrix}, \begin{bmatrix} 0 \\ 1 \end{bmatrix}, \begin{bmatrix} 1 \\ 3 \end{bmatrix}, \begin{bmatrix} 2 \\ 5 \end{bmatrix}$.

*(a) Find the least squares horizontal line $y = a$ fitting the data points. Check that the sum of the errors is 0.

(b) Find the least squares line $y = ax + b$ fitting the data points. Check that the sum of the errors is 0.

*(c) Find the least squares parabola $y = ax^2 + bx + c$ fitting the data points. (Calculator recommended.) What is true of the sum of the errors in this case?

8. Consider the four data points $\begin{bmatrix} 1 \\ 1 \end{bmatrix}, \begin{bmatrix} 2 \\ 2 \end{bmatrix}, \begin{bmatrix} 3 \\ 1 \end{bmatrix}, \begin{bmatrix} 4 \\ 3 \end{bmatrix}$.

(a) Find the least squares horizontal line $y = a$ fitting the data points. Check that the sum of the errors is 0.

(b) Find the least squares line $y = ax + b$ fitting the data points. Check that the sum of the errors is 0.

(c) Find the least squares parabola $y = ax^2 + bx + c$ fitting the data points. (Calculator recommended.) What is true of the sum of the errors in this case?

9. Derive the equation (∗) on p. 227 by starting with the equation $A\bar{\mathbf{x}} = \mathbf{p}$ and using the result of Theorem 4.9 of Chapter 4.

10. Let $V \subset \mathbb{R}^n$ be a subspace. Prove from the definition of proj_V on p. 226 that
   (a) $\text{proj}_V(\mathbf{x}+\mathbf{y}) = \text{proj}_V\mathbf{x} + \text{proj}_V\mathbf{y}$ for all vectors $\mathbf{x}$ and $\mathbf{y}$;
   (b) $\text{proj}_V(c\mathbf{x}) = c\,\text{proj}_V\mathbf{x}$ for all vectors $\mathbf{x}$ and scalars c.
   (c) for any $\mathbf{b} \in \mathbb{R}^n$ we have $\mathbf{b} = \text{proj}_V\mathbf{b} + \text{proj}_{V^\perp}\mathbf{b}$.
   Parts a and b tell us that proj_V is a linear map.

11. Using the definition of projection on p. 226, prove that
   (a) if $[\text{proj}_V] = A$, then $A = A^2$ and $A = A^\mathsf{T}$. (Hint: For the latter, show that $A\mathbf{x} \cdot \mathbf{y} = \mathbf{x} \cdot A\mathbf{y}$ for all $\mathbf{x}, \mathbf{y}$. It may be helpful to write $\mathbf{x}$ and $\mathbf{y}$ as the sum of vectors in V and $V^\perp$.)
   (b) if $A^2 = A$ and $A = A^\mathsf{T}$, then A is a projection matrix. (Hints: First decide onto which subspace it should be projecting. Then show that for all $\mathbf{x}$, the vector $A\mathbf{x}$ lies in that subspace and $\mathbf{x} - A\mathbf{x}$ is orthogonal to that subspace.)

12. Execute the Gram-Schmidt process in each case to give an orthonormal basis for the subspace spanned by the given vectors.

(a) $\begin{bmatrix} 1 \\ 0 \\ 0 \end{bmatrix}, \begin{bmatrix} 2 \\ 1 \\ 0 \end{bmatrix}, \begin{bmatrix} 3 \\ 2 \\ 1 \end{bmatrix}$

*(c) $\begin{bmatrix} 1 \\ 0 \\ 1 \\ 0 \end{bmatrix}, \begin{bmatrix} 2 \\ 1 \\ 0 \\ 1 \end{bmatrix}, \begin{bmatrix} 0 \\ 1 \\ 2 \\ -3 \end{bmatrix}$

(b) $\begin{bmatrix} 1 \\ 1 \\ 1 \end{bmatrix}, \begin{bmatrix} 0 \\ 1 \\ 1 \end{bmatrix}, \begin{bmatrix} 0 \\ 0 \\ 1 \end{bmatrix}$

(d) $\begin{bmatrix} -1 \\ 2 \\ 0 \\ 2 \end{bmatrix}, \begin{bmatrix} 2 \\ -4 \\ 1 \\ -4 \end{bmatrix}, \begin{bmatrix} -1 \\ 3 \\ 1 \\ 1 \end{bmatrix}$

*13. Let $V = \text{Span}\left(\begin{bmatrix} 1 \\ -1 \\ 0 \\ 2 \end{bmatrix}, \begin{bmatrix} 1 \\ 0 \\ 1 \\ 1 \end{bmatrix}\right) \subset \mathbb{R}^4$, and let $\mathbf{b} = \begin{bmatrix} 1 \\ -3 \\ 1 \\ 1 \end{bmatrix}$.

(a) Find an orthogonal basis for V.
(b) Use your answer to part a to find $\mathbf{p} = \text{proj}_V \mathbf{b}$.
(c) Letting
$$A = \begin{bmatrix} 1 & 1 \\ -1 & 0 \\ 0 & 1 \\ 2 & 1 \end{bmatrix},$$
use your answer to part b to give the least squares solution of $A\mathbf{x} = \mathbf{b}$.

14. Let $V = \text{Span}\left(\begin{bmatrix} 1 \\ 0 \\ 1 \\ 1 \end{bmatrix}, \begin{bmatrix} 0 \\ 1 \\ -1 \\ 1 \end{bmatrix}\right) \subset \mathbb{R}^4$.

(a) Give an orthogonal basis for V.

(b) Give an orthogonal basis for $V^\perp$.

(c) Given a general vector $\mathbf{x} \in \mathbb{R}^4$, find $\mathbf{v} \in V$ and $\mathbf{w} \in V^\perp$ so that $\mathbf{x} = \mathbf{v} + \mathbf{w}$.

15. According to Proposition 4.10 of Chapter 4, if A is an $m \times n$ matrix, then for each $\mathbf{b} \in \mathbf{C}(A)$, there is a unique $\mathbf{x} \in \mathbf{R}(A)$ with $A\mathbf{x} = \mathbf{b}$. In each case, give a formula for that $\mathbf{x}$.

(a) $A = \begin{bmatrix} 1 & 2 & 3 \\ 1 & 2 & 3 \end{bmatrix}$

(c) $A = \begin{bmatrix} 1 & 1 & 1 & 1 \\ 1 & 1 & 3 & -5 \end{bmatrix}$

*(b) $A = \begin{bmatrix} 1 & 1 & 1 \\ 0 & 1 & -1 \end{bmatrix}$

(d) $A = \begin{bmatrix} 1 & 1 & 1 & 1 \\ 1 & 1 & 3 & -5 \\ 2 & 2 & 4 & -4 \end{bmatrix}$

♯**16.** Let A be an $n \times n$ matrix and, as usual, let $\mathbf{a}_1, \ldots, \mathbf{a}_n$ denote its column vectors.

(a) Suppose $\mathbf{a}_1, \ldots, \mathbf{a}_n$ form an *orthonormal* set. Prove that $A^{-1} = A^\mathsf{T}$.

*(b) Suppose $\mathbf{a}_1, \ldots, \mathbf{a}_n$ form an *orthogonal* set and each is nonzero. Find the appropriate formula for A^{-1}.

17. Let $V = \mathcal{C}^0([-a, a])$ with the inner product $\langle f, g \rangle = \int_{-a}^{a} f(t)g(t)\,dt$. Let $U^+ \subset V$ be the subset of even functions, and let $U^- \subset V$ be the subset of odd functions. That is, $U^+ = \{f \in V : f(-t) = f(t) \text{ for all } t \in [-a, a]\}$ and $U^- = \{f \in V : f(-t) = -f(t) \text{ for all } t \in [-a, a]\}$.

(a) Prove that U^+ and U^- are orthogonal subspaces of V.

(b) Use the fact that every function can be written as the sum of an even and an odd function, viz.,

$$f(t) = \underbrace{\tfrac{1}{2}\big(f(t) + f(-t)\big)}_{\text{even}} + \underbrace{\tfrac{1}{2}\big(f(t) - f(-t)\big)}_{\text{odd}},$$

to prove that $U^- = (U^+)^\perp$ and $U^+ = (U^-)^\perp$.

18. (See Exercise 1.4.22 for the definition and basic properties of trace.)

(a) If $A, B \in \mathcal{M}_{n \times n}$, define $\langle A, B \rangle = \mathrm{tr}(A^\mathsf{T} B)$. Check that this is an inner product on $\mathcal{M}_{n \times n}$.

(b) Check that if A is symmetric and B is skew-symmetric, then $\langle A, B \rangle = 0$. (Hint: Show that $\langle A, B \rangle = -\langle B, A \rangle$.)

(c) Deduce that the subspaces of symmetric and skew-symmetric matrices (cf. Exercise 4.3.24) are orthogonal complements in $\mathcal{M}_{n \times n}$.

19. Let $g_1(t) = 1$ and $g_2(t) = t$. Using the inner product defined in Example 9b, find the orthogonal complement of $\mathrm{Span}(g_1, g_2)$ in

(a) $\mathcal{P}_2 \subset \mathcal{C}^0([-1, 1])$; *(b) $\mathcal{P}_2 \subset \mathcal{C}^0([0, 1])$; (c) $\mathcal{P}_3 \subset \mathcal{C}^0([-1, 1])$.

*20. Show that for any positive integer n, the functions $1, \cos t, \sin t, \cos 2t, \sin 2t, \ldots, \cos nt, \sin nt$ are orthogonal in $\mathcal{C}^\infty([-\pi, \pi]) \subset \mathcal{C}^0([-\pi, \pi])$ (using the inner product defined in Example 9b).

CHAPTER 6

SOLVING NONLINEAR PROBLEMS

In this brief chapter we introduce some important techniques for dealing with nonlinear problems (and in the infinite-dimensional setting as well, although that is too far off-track for us here). As we've said all along, we expect the derivative of a nonlinear function to dictate *locally* how the function behaves. In this chapter we come to the rigorous treatment of the inverse and implicit function theorems, to which we alluded at the end of Chapter 4, and to a few equivalent descriptions of a k-dimensional manifold, which will play a prominent role in Chapter 8.

▶ 1 THE CONTRACTION MAPPING PRINCIPLE

We begin with a useful result about summing series of vectors. It will be important not just in our immediate work but also in our treatment of matrix exponentials in Chapter 9.

Proposition 1.1 *Suppose* $\{\mathbf{a}_k\}$ *is a sequence of vectors in* $\mathbb{R}^n$ *and the series*

$$\sum_{k=1}^{\infty} \|\mathbf{a}_k\|$$

converges (i.e., the sequence of partial sums $t_k = \|\mathbf{a}_1\| + \cdots + \|\mathbf{a}_k\|$ *is a convergent sequence of real numbers). Then the series*

$$\sum_{k=1}^{\infty} \mathbf{a}_k$$

of vectors converges (i.e., the sequence of partial sums $\mathbf{s}_k = \mathbf{a}_1 + \cdots + \mathbf{a}_k$ *is a convergent sequence of vectors in* $\mathbb{R}^n$*).*

Proof We first prove the result in the case $n = 1$. Given a sequence $\{a_k\}$ of real numbers, define $b_k = a_k + |a_k|$. Note that

$$b_k = \begin{cases} 2a_k, & \text{if } a_k \geq 0 \\ 0, & \text{otherwise} \end{cases}.$$

Now, the series $\sum_{k=1}^{\infty} b_k$ converges by comparison with $\sum 2|a_k|$. (Directly: Since $b_k \geq 0$, the partial sums form a nondecreasing sequence that is bounded above by $2\sum |a_k|$. That nondecreasing sequence must converge to its least upper bound. See Example 4c of Chapter 2, Section 2.) Since $a_k = b_k - |a_k|$, the series $\sum a_k$ converges, being the sum of the two convergent series $\sum b_k$ and $-\sum |a_k|$.

We use this case to derive the general result. Denote by $a_{k,j}$, $j = 1, \ldots, n$, the jth component of the vector $\mathbf{a}_k$. Obviously, we have $|a_{k,j}| \leq \|\mathbf{a}_k\|$. By comparison with the convergent series $\sum \|\mathbf{a}_k\|$, for any $j = 1, \ldots, n$, the series $\sum_k |a_{k,j}|$ converges, and hence, by what we've just proved, so does the series $\sum_k a_{k,j}$. Since this is true for each $j = 1, \ldots, n$, the series

$$\sum_k \mathbf{a}_k = \begin{bmatrix} \sum_k a_{k,1} \\ \vdots \\ \sum_k a_{k,n} \end{bmatrix}$$

converges as well, as we wished to establish. ∎

Remark The result holds even if we use something other than the Euclidean length in $\mathbb{R}^n$. For example, we can apply the result by using the norm defined on the vector space of $m \times n$ matrices in Section 1 of Chapter 5, since the triangle inequality $\|A + B\| \leq \|A\| + \|B\|$ holds (see Proposition 1.3 of Chapter 5) and $|a_{ij}| \leq \|A\|$ for any matrix $A = [a_{ij}]$ (why?).

The following result is crucial in both pure and applied mathematics, and applies in infinite-dimensional settings as well.

Definition Let X be a subset of $\mathbb{R}^n$. A function $\mathbf{f} \colon X \to X$ is called a *contraction mapping* if there is a constant c, $0 < c < 1$, so that

$$\|\mathbf{f}(\mathbf{x}) - \mathbf{f}(\mathbf{y})\| \leq c\|\mathbf{x} - \mathbf{y}\| \quad \text{for all } \mathbf{x}, \mathbf{y} \in X.$$

(It is crucial that c be strictly less than 1, as Exercise 2 illustrates.)

▶ **EXAMPLE 1**

Consider $f \colon [0, \pi/3] \to [0, 1] \subset [0, \pi/3]$ given by $f(x) = \cos x$. Then by the mean value theorem, for any $x, y \in [0, \pi/3]$,

$$|f(x) - f(y)| = |\sin z||x - y| \quad \text{for some } z \text{ between } x \text{ and } y$$

$$\leq \frac{\sqrt{3}}{2}|x - y| \quad \text{since } 0 < z < \pi/3.$$

Since $\sqrt{3}/2 < 1$, f is a contraction mapping. ◀

Theorem 1.2 (Contraction Mapping Principle) *Let $X \subset \mathbb{R}^n$ be closed. Let $\mathbf{f} \colon X \to X$ be a contraction mapping. Then there is a unique point $\mathbf{x} \in X$ such that $\mathbf{f}(\mathbf{x}) = \mathbf{x}$. (Not surprisingly, $\mathbf{x}$ is called a* fixed point *of $\mathbf{f}$.)*

Proof Let $\mathbf{x}_0 \in X$ be arbitrary, and define a sequence recursively by

$$\mathbf{x}_{k+1} = \mathbf{f}(\mathbf{x}_k).$$

Our goal is to show that, inasmuch as $\mathbf{f}$ is a contraction mapping, this sequence converges to some point $\mathbf{x} \in X$. Then, by continuity of $\mathbf{f}$ (see Exercise 1), we will have

$$\mathbf{f}(\mathbf{x}) = \lim_{k \to \infty} \mathbf{f}(\mathbf{x}_k) = \lim_{k \to \infty} \mathbf{x}_{k+1} = \mathbf{x}.$$

Consider the equation

$$\mathbf{x}_k = \mathbf{x}_0 + (\mathbf{x}_1 - \mathbf{x}_0) + (\mathbf{x}_2 - \mathbf{x}_1) + \cdots + (\mathbf{x}_k - \mathbf{x}_{k-1}) = \mathbf{x}_0 + \sum_{j=1}^{k} (\mathbf{x}_j - \mathbf{x}_{j-1}).$$

This suggests that we set

$$\mathbf{a}_k = \mathbf{x}_k - \mathbf{x}_{k-1}$$

and try to determine whether the series $\sum \mathbf{a}_k$ converges. To this end, we wish to apply Proposition 1.1, and so we begin by estimating $\|\mathbf{a}_k\|$: By the definition of the sequence $\{\mathbf{x}_k\}$ and the definition of a contraction mapping, we have

$$\|\mathbf{a}_k\| = \|\mathbf{x}_k - \mathbf{x}_{k-1}\| = \|\mathbf{f}(\mathbf{x}_{k-1}) - \mathbf{f}(\mathbf{x}_{k-2})\| \leq c\|\mathbf{x}_{k-1} - \mathbf{x}_{k-2}\| = c\|\mathbf{a}_{k-1}\|$$

for some constant $0 < c < 1$, so that

$$\|\mathbf{a}_k\| \leq c\|\mathbf{a}_{k-1}\| \leq c^2 \|\mathbf{a}_{k-2}\| \leq \cdots \leq c^{k-1} \|\mathbf{a}_1\|.$$

Therefore,

$$\sum_{k=1}^{K} \|\mathbf{a}_k\| \leq \left(\sum_{k=1}^{K} c^{k-1} \right) \|\mathbf{a}_1\| = \frac{1 - c^K}{1 - c} \|\mathbf{a}_1\|.$$

Since $0 < c < 1$,

$$\lim_{K \to \infty} \frac{1 - c^K}{1 - c} = \frac{1}{1 - c},$$

and so the series $\sum \|\mathbf{a}_k\|$ converges. By Proposition 1.1, we infer that the series $\sum \mathbf{a}_k$ converges to some vector $\mathbf{a} \in \mathbb{R}^n$. It follows, then, that $\mathbf{x}_k \to \mathbf{x}_0 + \mathbf{a} = \mathbf{x}$, as required.

Two issues remain. Since $\mathbf{x}_k \to \mathbf{x}$, all the $\mathbf{x}_k$'s are elements of X, and X is closed, then we know that $\mathbf{x} \in X$ as well. The uniqueness of the fixed point is left to the reader in Exercise 1. ∎

▶ EXAMPLE 2

According to Theorem 1.2, the function f introduced in Example 1 must have a unique fixed point in the interval $[0, \pi/3]$. Following the proof with $x_0 = 0$, we obtain the following values:

k	x_k	k	x_k
1	1.	11	0.744237
2	0.540302	12	0.735604
3	0.857553	13	0.741425
4	0.654289	14	0.737506
5	0.793480	15	0.740147
6	0.701368	16	0.738369
7	0.763959	17	0.739567
8	0.722102	18	0.738760
9	0.750417	19	0.739303
10	0.731404	20	0.738937

Indeed, as Figure 1.1 illustrates, the values x_k are converging to the x-coordinate of the intersection of the graph of $f(x) = \cos x$ with the diagonal $y = x$. ◄

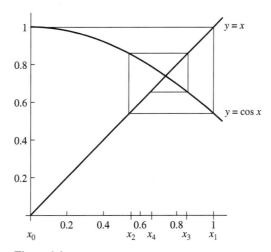

Figure 1.1

Example 2 shows that this is a very slow method to obtain the solution of $\cos x = x$. Far better is Newton's method, familiar to every student of calculus. Given a differentiable function $g: \mathbb{R} \to \mathbb{R}$, we start at x_k, draw the tangent line to the graph of g at x_k, and let x_{k+1} be the x-intercept of that tangent line, as shown in Figure 1.2. We obtain in this way a sequence, and one hopes that if x_0 is sufficiently close to a root a, then the sequence will converge to a. It is easy to see that the recursion formula for this sequence is

$$x_{k+1} = x_k - \frac{g(x_k)}{g'(x_k)},$$

so, in fact, we are looking for a fixed point of the mapping $f(x) = x - \frac{g(x)}{g'(x)}$. If we assume g is twice differentiable, then we find that $f' = gg''/(g')^2$, so f will be a contraction mapping whenever $|gg''/(g')^2| \leq c < 1$. In particular, if $|g''| \leq M$ and $|g'| \geq m$, then

248 Chapter 6. Solving Nonlinear Problems

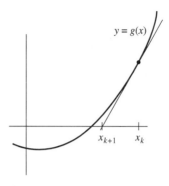

Figure 1.2

iterating f will converge to a root a of g if we start in any closed interval containing a on which $|g| < m^2/M$ (provided f maps that interval back to itself). For the strongest result, see Exercise 8.

▶ **EXAMPLE 3**

Reconsidering the problem of Example 2, let's use Newton's method to approximate the root of $\cos x = x$ by taking $g(x) = x - \cos x$ and iterating the map $f(x) = x - \dfrac{x - \cos x}{1 + \sin x}$.

k	x_k
0	1.
1	0.750364
2	0.739113
3	0.739085
4	0.739085

k	x_k
0	0.523599
1	0.751883
2	0.739121
3	0.739085
4	0.739085

Here we see that, whether we start at either $x_0 = 1$ or at $x_0 = \pi/6$, Newton's method converges to the root quite rapidly. Indeed, on the interval $[\pi/6, \pi/3]$, we have $m = 1.5$, $M = .87$, and $|g| \le .55$, which is far smaller than $m^2/M \approx 2.6$. ◀

To move to higher dimensions, we need a multivariable Mean Value Theorem. The Mean Value Theorem, although often misinterpreted in beginning calculus courses, tells us that if we have bounds on the size of the derivative of a differentiable function, then we have bounds on how much the function itself can change from one point to another. A crucial tool here will be the norm of a linear map, introduced in Chapters 3 and 5.

Proposition 1.3 (The Mean Value Inequality) *Suppose $U \subset \mathbb{R}^n$ is open, $\mathbf{f} \colon U \to \mathbb{R}^m$ is $\mathcal{C}^1$, and $\mathbf{a}$ and $\mathbf{b}$ are points in U so that the line segment between them is contained in U.*[1] *Then*

$$\|\mathbf{f}(\mathbf{b}) - \mathbf{f}(\mathbf{a})\| \le \left(\max_{\mathbf{x} \in [\mathbf{a}, \mathbf{b}]} \|D\mathbf{f}(\mathbf{x})\| \right) \|\mathbf{b} - \mathbf{a}\|.$$

---

[1] More generally, all we need is a $\mathcal{C}^1$ path in U joining $\mathbf{a}$ and $\mathbf{b}$.

Proof Define $\mathbf{g}\colon [0,1] \to \mathbb{R}^m$ by $\mathbf{g}(t) = \mathbf{f}(\mathbf{a} + t(\mathbf{b}-\mathbf{a}))$. Note that

$$\mathbf{f}(\mathbf{b}) - \mathbf{f}(\mathbf{a}) = \mathbf{g}(1) - \mathbf{g}(0).$$

By the chain rule, $\mathbf{g}$ is differentiable and

(∗) $\qquad\qquad\qquad \mathbf{g}'(t) = D\mathbf{f}(\mathbf{a} + t(\mathbf{b}-\mathbf{a}))(\mathbf{b}-\mathbf{a}).$

Applying Lemma 5.3 of Chapter 3, we have

$$\|\mathbf{f}(\mathbf{b}) - \mathbf{f}(\mathbf{a})\| = \left\|\int_0^1 \mathbf{g}'(t)\,dt\right\| \le \int_0^1 \|\mathbf{g}'(t)\|\,dt \le \max_{t \in [0,1]} \|\mathbf{g}'(t)\|.$$

By (∗), we have $\|\mathbf{g}'(t)\| \le \|D\mathbf{f}(\mathbf{a} + t(\mathbf{b}-\mathbf{a}))\|\,\|\mathbf{b}-\mathbf{a}\|$, and so

$$\max_{t \in [0,1]} \|\mathbf{g}'(t)\| \le \left(\max_{\mathbf{x} \in [\mathbf{a},\mathbf{b}]} \|D\mathbf{f}(\mathbf{x})\|\right) \|\mathbf{b}-\mathbf{a}\|.$$

This completes the proof. ∎

▶ EXERCISES 6.1

1. Prove that any contraction mapping is continuous and has at most one fixed point.

2. Let $f\colon \mathbb{R} \to \mathbb{R}$ be given by $f(x) = \sqrt{x^2+1}$. Show that f has no fixed point and that $|f'(x)| < 1$ for all $x \in \mathbb{R}$. Why does this not contradict Theorem 1.2?

*3. For the sequence $\{\mathbf{x}_k\}$ defined in the proof of Theorem 1.2, prove that $\|\mathbf{x}_k - \mathbf{x}\| \le \dfrac{c^k}{1-c}\|\mathbf{x}_1 - \mathbf{x}_0\|$. This gives an a priori estimate on how fast the sequence converges to the fixed point.

4. A sequence $\{\mathbf{x}_k\}$ of points in $\mathbb{R}^n$ is called a *Cauchy sequence* if for all $\varepsilon > 0$ there is K so that whenever $k,\ell > K$, we have $\|\mathbf{x}_k - \mathbf{x}_\ell\| < \varepsilon$. It is a fact that any Cauchy sequence in $\mathbb{R}^n$ is convergent. (See Exercise 2.2.14.) Suppose $0 < c < 1$ and $\{\mathbf{x}_k\}$ is a sequence of points in $\mathbb{R}^n$ so that $\|\mathbf{x}_{k+1} - \mathbf{x}_k\| < c\|\mathbf{x}_k - \mathbf{x}_{k-1}\|$ for all $k \in \mathbb{N}$. Prove that $\{\mathbf{x}_k\}$ is a Cauchy sequence, hence convergent. (Hint: Show that whenever $k,\ell > K$, we have $\|\mathbf{x}_k - \mathbf{x}_\ell\| < \dfrac{c^K}{1-c}\|\mathbf{x}_1 - \mathbf{x}_0\|$.)

5. Use the result of Exercise 2.2.14 to give a different proof of Proposition 1.1.

♯6. (a) Show that if H is any square matrix with $\|H\| < 1$, then $I - H$ is invertible. (Hint: Consider the geometric series $\sum_{k=0}^\infty H^k$. You will need to use the result of Exercise 5.1.6.)

(b) Suppose, more generally, that A is an invertible $n \times n$ matrix. Show that when $\|H\| < 1/\|A^{-1}\|$, the matrix $A + H$ is invertible as well. (Hint: Write $A + H = A(I + A^{-1}H)$.)

(c) Prove that the set of invertible $n \times n$ matrices is an open subset of $\mathcal{M}_{n \times n} = \mathbb{R}^{n^2}$. This set is denoted $GL(n)$, the *general linear group*. (Hint: By Exercise 5.1.5, if $\left(\sum h_{ij}^2\right)^{1/2} < \delta$, then $\|H\| < \delta$.)

♯7. Continuing Exercise 6:
(a) Show that if $\|H\| < \varepsilon < 1$, then $\|(I+H)^{-1} - I\| < \dfrac{\varepsilon}{1-\varepsilon}$.

(b) More generally, if A is invertible and $\|A^{-1}\|\,\|H\| < \varepsilon < 1$, then estimate $\|(A+H)^{-1} - A^{-1}\|$.

(c) Let $X \subset \mathcal{M}_{n \times n}$ be the set of invertible $n \times n$ matrices (by Exercise 6, this is an open subset). Prove that the function $\mathbf{f}\colon X \to X$, $\mathbf{f}(A) = A^{-1}$, is continuous.

8. Suppose $x_0 \in U \subset \mathbb{R}$, $g: U \to \mathbb{R}$ is $\mathcal{C}^2$, and $g'(x_0) \neq 0$. Set $h_0 = -g(x_0)/g'(x_0)$ and $x_1 = x_0 + h_0$. Prove that if $|g''| \leq M$ on the interval $[x_1 - |h_0|, x_1 + |h_0|]$ and $|g(x_0)|M \leq \frac{1}{2}(g'(x_0))^2$, then Newton's method, starting at x_0, to the unique root of g in that interval,[2] as follows.

   (a) According to Proposition 3.2 of Chapter 5, we have $g(x_1) = g(x_0) + g'(x_0)h_0 + \frac{1}{2}g''(\xi)h_0^2$ for some ξ between x_0 and x_1. Prove that $|g(x_1)| \leq \frac{1}{4}|g(x_0)|$.

   (b) Using the fact that $g'(x_1) = g'(x_0) + g''(c)h_0$ for some c between x_0 and x_1, show that
   $$\frac{1}{|g'(x_1)|} \leq \frac{2}{|g'(x_0)|}.$$
   Now deduce that
   $$\frac{|g(x_1)|}{g'(x_1)^2} \leq \frac{|g(x_0)|}{g'(x_0)^2} \quad \text{and hence that} \quad |g(x_1)|M \leq \frac{1}{2}(g'(x_1))^2.$$

   (c) Deduce that $|g(x_1)/g'(x_1)| \leq |h_0|/2$.

   (d) Prove analogously that if, when we apply Newton's method, we set $x_{k+1} = x_k + h_k$, then $|h_k| \leq |h_0|/2^k$. Deduce that iterating Newton's method converges to a point in the given interval.

9. Using the result of Exercise 8:

   *(a) Let $g(x) = x^2 - 2$. Carry out two steps of Newton's method starting at $x_0 = 1$. Give an interval that is guaranteed to contain a nearby root of g.

   (b) Let $g(x) = x^3 - 2$. Carry out two steps of Newton's method starting at $x_0 = 5/4$. Give an interval that is guaranteed to contain a nearby root of g.

   *(c) Let $g(x) = x - \cos 2x$. Carry out two steps of Newton's method starting at $x_0 = \pi/4$. Give an interval that is guaranteed to contain a nearby root of g.

10. Suppose $\mathbf{x}_0 \in U \subset \mathbb{R}^n$, $\mathbf{g}: U \to \mathbb{R}^n$ is $\mathcal{C}^2$, and $D\mathbf{g}(\mathbf{x}_0)$ is invertible. Newton's method in n dimensions is given by iterating the map
    $$\mathbf{f}(\mathbf{x}) = \mathbf{x} - D\mathbf{g}(\mathbf{x})^{-1}\mathbf{g}(\mathbf{x}),$$
    starting at $\mathbf{x}_0$. Set $\mathbf{h}_0 = -D\mathbf{g}(\mathbf{x}_0)^{-1}\mathbf{g}(\mathbf{x}_0)$ and $\mathbf{x}_1 = \mathbf{x}_0 + \mathbf{h}_0$. Let $B = \overline{B}(\mathbf{x}_1, \|\mathbf{h}_0\|)$; suppose $\|\text{Hess}(g_i)\| \leq M_i$ on B, and set $M = \sqrt{\sum_{i=1}^n M_i^2}$. Suppose, moreover, that
    $$\|D\mathbf{g}(\mathbf{x}_0)^{-1}\|^2 \|\mathbf{g}(\mathbf{x}_0)\| M \leq 1/2.$$
    Prove that Newton's method converges, starting at $\mathbf{x}_0$, to a point of B, as follows.

    (a) Using Proposition 3.2 of Chapter 5, prove that $\|\mathbf{g}(\mathbf{x}_1)\| \leq \frac{1}{2}M\|\mathbf{h}_0\|^2 \leq \frac{1}{4}\|\mathbf{g}(\mathbf{x}_0)\|$.

    (b) Show that $\|D\mathbf{g}(\mathbf{x}_1) - D\mathbf{g}(\mathbf{x}_0)\| \leq M\|\mathbf{h}_0\|$. Using Exercise 7, deduce that $\|D\mathbf{g}(\mathbf{x}_1)^{-1}\| \leq 2\|D\mathbf{g}(\mathbf{x}_0)^{-1}\|$. (Hint: Let $H = D\mathbf{g}(\mathbf{x}_0)^{-1}(D\mathbf{g}(\mathbf{x}_1) - D\mathbf{g}(\mathbf{x}_0))$; show that $\|H\| \leq 1/2$.)

    (c) Now show that $\|D\mathbf{g}(\mathbf{x}_1)^{-1}\|^2 \|\mathbf{g}(\mathbf{x}_1)\| \leq \|D\mathbf{g}(\mathbf{x}_0)^{-1}\|^2 \|\mathbf{g}(\mathbf{x}_0)\|$, and conclude that $\|D\mathbf{g}(\mathbf{x}_1)^{-1}\|^2 \|\mathbf{g}(\mathbf{x}_1)\| M \leq 1/2$.

    (d) Prove that $\|D\mathbf{g}(\mathbf{x}_1)^{-1}\mathbf{g}(\mathbf{x}_1)\| \leq \|\mathbf{h}_0\|/2$.

    (e) Letting $\mathbf{h}_k = -D\mathbf{g}(\mathbf{x}_k)^{-1}\mathbf{g}(\mathbf{x}_k)$, prove analogously that $\|\mathbf{h}_k\| \leq \|\mathbf{h}_0\|/2^k$. Deduce that iterating Newton's method converges to a point in the given ball.

11. Using the result of Exercise 10:

    *(a) Let $\mathbf{g}: \mathbb{R}^2 \to \mathbb{R}^2$ be defined by $\mathbf{g}\begin{pmatrix} x_1 \\ x_2 \end{pmatrix} = \begin{bmatrix} 4x_1 + x_2^2 - 4 \\ 4x_1 x_2 - 1 \end{bmatrix}$. Do one step of Newton's method to solve $\mathbf{g}(\mathbf{x}) = 0$, starting at $\mathbf{x}_0 = \begin{bmatrix} 1 \\ 0 \end{bmatrix}$, and find a ball in $\mathbb{R}^2$ that is guaranteed to contain a root of $\mathbf{g}$.

---

[2] We learned of the n-dimensional version of this result, which we give in Exercise 10, called Kantarovich's Theorem, in Hubbard and Hubbard's *Vector Calculus, Linear Algebra, and Differential Forms*.

(b) Let $\mathbf{g}\colon \mathbb{R}^2 \to \mathbb{R}^2$ be defined by $\mathbf{g}\begin{pmatrix} x_1 \\ x_2 \end{pmatrix} = \begin{bmatrix} x_1^2 + x_2^2 - 5 \\ \frac{3}{2}x_1 x_2 - 1 \end{bmatrix}$. Do one step of Newton's method to solve $\mathbf{g}(\mathbf{x}) = \mathbf{0}$, starting at $\mathbf{x}_0 = \begin{bmatrix} 2 \\ 0 \end{bmatrix}$, and find a ball in $\mathbb{R}^2$ that is guaranteed to contain a root of $\mathbf{g}$.

(c) Let $\mathbf{g}\colon \mathbb{R}^2 \to \mathbb{R}^2$ be defined by $\mathbf{g}\begin{pmatrix} x_1 \\ x_2 \end{pmatrix} = \begin{bmatrix} 4 \sin x_1 + x_2^2 \\ 2 x_1 x_2 - 1 \end{bmatrix}$. Do one step of Newton's method to solve $\mathbf{g}(\mathbf{x}) = \mathbf{0}$, starting at $\mathbf{x}_0 = \begin{bmatrix} \pi \\ 0 \end{bmatrix}$, and find a ball in $\mathbb{R}^2$ that is guaranteed to contain a root of $\mathbf{g}$.

(d) Let $\mathbf{g}\colon \mathbb{R}^2 \to \mathbb{R}^2$ be defined by $\mathbf{g}\begin{pmatrix} x_1 \\ x_2 \end{pmatrix} = \begin{bmatrix} x_1 - \frac{1}{4}x_2^2 \\ x_2 - \cos x_1 \end{bmatrix}$. Do one step of Newton's method to solve $\mathbf{g}(\mathbf{x}) = \mathbf{0}$, starting at $\mathbf{x}_0 = \begin{bmatrix} 0 \\ 1 \end{bmatrix}$, and find a ball in $\mathbb{R}^2$ that is guaranteed to contain a root of $\mathbf{g}$.

12. Prove the following, slightly stronger version of Proposition 1.3. Suppose $U \subset \mathbb{R}^n$ is open, $\mathbf{f}\colon U \to \mathbb{R}^m$ is differentiable, and $\mathbf{a}$ and $\mathbf{b}$ are points in U so that the line segment between them is contained in U. Then prove that there is a point $\boldsymbol{\xi}$ on that line segment so that $\|\mathbf{f}(\mathbf{b}) - \mathbf{f}(\mathbf{a})\| \le \|D\mathbf{f}(\boldsymbol{\xi})\|\|\mathbf{b} - \mathbf{a}\|$. (Hints: Define $\mathbf{g}$ as before, let $\mathbf{v} = \mathbf{g}(1) - \mathbf{g}(0)$, and define $\phi\colon [0,1] \to \mathbb{R}$ by $\phi(t) = \mathbf{g}(t) \cdot \mathbf{v}$. Apply the usual mean value theorem and the Cauchy-Schwarz inequality, Proposition 2.3 of Chapter 1, to show that $\|\mathbf{v}\|^2 = \phi(1) - \phi(0) \le \|\mathbf{g}'(c)\|\|\mathbf{v}\|$ for some $c \in (0,1)$.)

2 THE INVERSE AND IMPLICIT FUNCTION THEOREMS

When we study functions $f\colon \mathbb{R} \to \mathbb{R}$ in single-variable calculus, it is usually quite simple to decide when a function has an inverse function. Any increasing (or decreasing) function certainly has an inverse, even if we are unable to give it explicitly (e.g., what is the inverse of the function $f(x) = x^5 + x + 1$?). Sometimes we make up names for inverse functions, such as log, the inverse of exp, and arcsin, the inverse of sin (restricted to the interval $[-\pi/2, \pi/2]$).

Since a differentiable function on an interval in $\mathbb{R}$ with nowhere zero derivative has a differentiable inverse, it is tempting to think that if the derivative $f'(a) \ne 0$, then f should have a *local* inverse at a.

EXAMPLE 1

Let $f(x) = \begin{cases} \frac{x}{2} + x^2 \sin \frac{1}{x}, & x \ne 0 \\ 0, & x = 0 \end{cases}$. Then, calculating from the definition, we find

$$f'(0) = \tfrac{1}{2} + \lim_{h \to 0} h \sin \tfrac{1}{h} = \tfrac{1}{2} > 0.$$

On the other hand, if $x \ne 0$,

$$f'(x) = \tfrac{1}{2} + 2x \sin \tfrac{1}{x} - \cos \tfrac{1}{x},$$

so there are points (e.g., $x = 1/2\pi n$ for any nonzero integer n) arbitrarily close to 0 where $f'(x) < 0$. That is, despite the fact that $f'(0) > 0$, there is no interval around 0 on which f is increasing, as Figure 2.1 suggests. Thus, f has no inverse on any neighborhood of 0. ◀

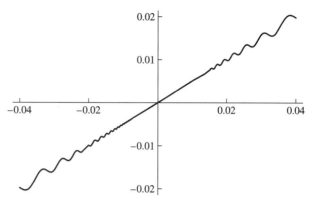

Figure 2.1

All right, so we need a stronger hypothesis. If we assume f is $\mathcal{C}^1$, then it will follow that if $f'(a) > 0$, then $f' > 0$ on an interval around a, and so f will be increasing—hence invertible—on that interval. That is the result that generalizes nicely to higher dimensions.

Theorem 2.1 (Inverse Function Theorem) *Suppose $U \subset \mathbb{R}^n$ is open, $\mathbf{x}_0 \in U$, $\mathbf{f}\colon U \to \mathbb{R}^n$ is $\mathcal{C}^1$, and $D\mathbf{f}(\mathbf{x}_0)$ is invertible. Then there is a neighborhood $V \subset U$ of $\mathbf{x}_0$ on which $\mathbf{f}$ has a $\mathcal{C}^1$ inverse function. That is, there are neighborhoods V of $\mathbf{x}_0$ and W of $\mathbf{f}(\mathbf{x}_0) = \mathbf{y}_0$ and a $\mathcal{C}^1$ function $\mathbf{g}\colon W \to V$ so that*

$$\mathbf{f}(\mathbf{g}(\mathbf{y})) = \mathbf{y} \quad \text{for all } \mathbf{y} \in W \quad \text{and} \quad \mathbf{g}(\mathbf{f}(\mathbf{x})) = \mathbf{x} \quad \text{for all } \mathbf{x} \in V.$$

Moreover, if $\mathbf{f}(\mathbf{x}) = \mathbf{y}$, we have

$$D\mathbf{g}(\mathbf{y}) = \big(D\mathbf{f}(\mathbf{x})\big)^{-1}.$$

Proof Without loss of generality, we assume that $\mathbf{x}_0 = \mathbf{y}_0 = \mathbf{0}$ and that $D\mathbf{f}(\mathbf{0}) = I$. (We make appropriate translations and then replace $\mathbf{f}(\mathbf{x})$ by $D\mathbf{f}(\mathbf{0})^{-1}\mathbf{f}(\mathbf{x})$.) Since $\mathbf{f}$ is $\mathcal{C}^1$, there is $r > 0$ so that

$$\|D\mathbf{f}(\mathbf{x}) - I\| \le \tfrac{1}{2} \quad \text{whenever } \|\mathbf{x}\| \le r.$$

Now, fix $\mathbf{y}$ with $\|\mathbf{y}\| < r/2$, and define the function $\boldsymbol{\phi}$ by

$$\boldsymbol{\phi}(\mathbf{x}) = \mathbf{x} - \mathbf{f}(\mathbf{x}) + \mathbf{y}.$$

Note that $\|D\boldsymbol{\phi}(\mathbf{x})\| = \|D\mathbf{f}(\mathbf{x}) - I\|$. Whenever $\|\mathbf{x}\| \le r$, we have (by Proposition 1.3)

$$\|\boldsymbol{\phi}(\mathbf{x})\| \le \|\mathbf{x} - \mathbf{f}(\mathbf{x})\| + \|\mathbf{y}\| < \frac{r}{2} + \frac{r}{2} = r,$$

and so $\boldsymbol{\phi}$ maps the closed ball $\overline{B}(\mathbf{0}, r)$ to itself. Moreover, if $\mathbf{x}, \mathbf{y} \in B(\mathbf{0}, r)$, by Proposition 1.3 we have

$$\|\boldsymbol{\phi}(\mathbf{x}) - \boldsymbol{\phi}(\mathbf{y})\| \le \tfrac{1}{2}\|\mathbf{x} - \mathbf{y}\|,$$

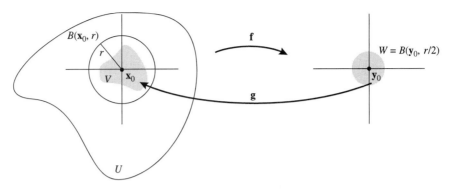

Figure 2.2

so ϕ is a contraction mapping on $\overline{B}(0, r)$. By Theorem 1.2, ϕ has a unique fixed point $\mathbf{x_y} \in \overline{B}(0, r)$. That is, there is a unique point $\mathbf{x_y} \in \overline{B}(0, r)$ so that $\mathbf{f}(\mathbf{x_y}) = \mathbf{y}$. We leave it to the reader to check in Exercise 10 that in fact $\mathbf{x_y} \in B(0, r)$.

As pictured in Figure 2.2, take $W = B(0, r/2)$ and $V = \mathbf{f}^{-1}(W) \cap B(0, r)$ (note that V is open because $\mathbf{f}$ is continuous; see also Exercise 2.2.7). Define $\mathbf{g} \colon W \to V$ by $\mathbf{g}(\mathbf{y}) = \mathbf{x_y}$. We claim first of all that $\mathbf{g}$ is continuous. Indeed, define $\boldsymbol{\psi} \colon B(0, r) \to \mathbb{R}^n$ by $\boldsymbol{\psi}(\mathbf{x}) = \mathbf{f}(\mathbf{x}) - \mathbf{x}$. Then, by Proposition 1.3 we have

$$\left\|\bigl(\mathbf{f}(\mathbf{u}) - \mathbf{u}\bigr) - \bigl(\mathbf{f}(\mathbf{v}) - \mathbf{v}\bigr)\right\| = \|\boldsymbol{\psi}(\mathbf{u}) - \boldsymbol{\psi}(\mathbf{v})\| \le \tfrac{1}{2}\|\mathbf{u} - \mathbf{v}\|.$$

Thus, we have

$$\left\|\bigl(\mathbf{f}(\mathbf{u}) - \mathbf{f}(\mathbf{v})\bigr) - \bigl(\mathbf{u} - \mathbf{v}\bigr)\right\| \le \tfrac{1}{2}\|\mathbf{u} - \mathbf{v}\|.$$

It follows from the triangle inequality (see Exercise 1.2.17) that

$$\|\mathbf{u} - \mathbf{v}\| - \|\mathbf{f}(\mathbf{u}) - \mathbf{f}(\mathbf{v})\| \le \tfrac{1}{2}\|\mathbf{u} - \mathbf{v}\|,$$

and so

$$\tfrac{1}{2}\|\mathbf{u} - \mathbf{v}\| \le \|\mathbf{f}(\mathbf{u}) - \mathbf{f}(\mathbf{v})\|.$$

Writing $\mathbf{f}(\mathbf{u}) = \mathbf{y}$ and $\mathbf{f}(\mathbf{v}) = \mathbf{z}$, we have

(∗) $$\|\mathbf{g}(\mathbf{y}) - \mathbf{g}(\mathbf{z})\| \le 2\|\mathbf{y} - \mathbf{z}\|.$$

It follows that $\mathbf{g}$ is continuous (e.g., given $\varepsilon > 0$, take $\delta = \varepsilon/2$).

Next, we check that $\mathbf{g}$ is differentiable. Fix $\mathbf{y} \in W$ and write $\mathbf{g}(\mathbf{y}) = \mathbf{x}$; and we wish to prove that $D\mathbf{g}(\mathbf{y}) = \bigl(D\mathbf{f}(\mathbf{x})\bigr)^{-1}$. Choose $\mathbf{k}$ sufficiently small that $\mathbf{y} + \mathbf{k} \in W$. Set $\mathbf{g}(\mathbf{y} + \mathbf{k}) = \mathbf{x} + \mathbf{h}$, so that $\mathbf{h} = \mathbf{g}(\mathbf{y} + \mathbf{k}) - \mathbf{g}(\mathbf{y})$. For ease of notation, write $A = D\mathbf{f}(\mathbf{x})$. We are to prove that

$$\frac{\mathbf{g}(\mathbf{y} + \mathbf{k}) - \mathbf{g}(\mathbf{y}) - A^{-1}\mathbf{k}}{\|\mathbf{k}\|} \to \mathbf{0} \quad \text{as } \mathbf{k} \to \mathbf{0}.$$

We consider instead the result of multiplying this quantity by (the fixed matrix) A:

$$\frac{A\big(\mathbf{g}(\mathbf{y}+\mathbf{k})-\mathbf{g}(\mathbf{y})\big)-\mathbf{k}}{\|\mathbf{k}\|} = \frac{A\mathbf{h}-\mathbf{k}}{\|\mathbf{k}\|} = -\frac{\mathbf{f}(\mathbf{x}+\mathbf{h})-\mathbf{f}(\mathbf{x})-D\mathbf{f}(\mathbf{x})\mathbf{h}}{\|\mathbf{h}\|} \cdot \frac{\|\mathbf{h}\|}{\|\mathbf{k}\|}.$$

We infer from (∗) that $\|\mathbf{h}\| \le 2\|\mathbf{k}\|$, so as $\mathbf{k} \to \mathbf{0}$, it follows that $\mathbf{h} \to \mathbf{0}$ as well. Note, moreover, that $\mathbf{h} \ne \mathbf{0}$ when $\mathbf{k} \ne \mathbf{0}$ (why?). Now we analyze the final product above: The first term approaches $\mathbf{0}$ by the differentiability of $\mathbf{f}$; the second is bounded above by 2. Thus, the product approaches $\mathbf{0}$, as desired.

The last order of business is to see that $\mathbf{g}$ is $\mathcal{C}^1$. We have

$$D\mathbf{g}(\mathbf{y}) = \big(D\mathbf{f}(\mathbf{g}(\mathbf{y}))\big)^{-1},$$

so we see that $D\mathbf{g}$ is the composition of the function $\mathbf{y} \rightsquigarrow D\mathbf{f}(\mathbf{g}(\mathbf{y}))$ and the function $A \rightsquigarrow A^{-1}$ on the space of invertible matrices. Since $\mathbf{g}$ is continuous and $\mathbf{f}$ is $\mathcal{C}^1$, the former is continuous. By Exercise 6.1.7, the latter is continuous (indeed, we will prove much more in Corollary 5.19 of Chapter 7 when we study determinants in detail). Since the composition of continuous functions is continuous, the function $\mathbf{y} \rightsquigarrow D\mathbf{g}(\mathbf{y})$ is continuous, as required. ∎

Remark More generally, with a bit more work, one can show that if $\mathbf{f}$ is $\mathcal{C}^k$ (or smooth), then the local inverse $\mathbf{g}$ is likewise $\mathcal{C}^k$ (or smooth).

It is important to remember that this theorem guarantees only a *local* inverse function. It may be rather difficult to determine whether $\mathbf{f}$ is globally one-to-one. Indeed, as the following example shows, even if $D\mathbf{f}$ is *everywhere* invertible, the function $\mathbf{f}$ may be very much not one-to-one.

▶ **EXAMPLE 2**

Define $\mathbf{f}\colon \mathbb{R}^2 \to \mathbb{R}^2$ by

$$\mathbf{f}\begin{pmatrix} u \\ v \end{pmatrix} = \begin{bmatrix} e^u \cos v \\ e^u \sin v \end{bmatrix}.$$

Then $\mathbf{f}$ is $\mathcal{C}^1$, and

$$D\mathbf{f}\begin{pmatrix} u \\ v \end{pmatrix} = \begin{bmatrix} e^u \cos v & -e^u \sin v \\ e^u \sin v & e^u \cos v \end{bmatrix}$$

is everywhere nonsingular since its determinant is $e^{2u} \ne 0$. Nevertheless, since sine and cosine are periodic, it is clear that $\mathbf{f}$ is not one-to-one: We have $\mathbf{f}\begin{pmatrix} u \\ v \end{pmatrix} = \mathbf{f}\begin{pmatrix} u \\ v + 2\pi k \end{pmatrix}$ for any integer k.

On the other hand, if $\mathbf{f}\begin{pmatrix} u \\ v \end{pmatrix} = \begin{bmatrix} x \\ y \end{bmatrix}$, then we apparently can solve for u and v:

(†)
$$\mathbf{g}\begin{pmatrix} x \\ y \end{pmatrix} = \begin{bmatrix} \tfrac{1}{2}\log(x^2+y^2) \\ \arctan(y/x) \end{bmatrix}.$$

2 The Inverse and Implicit Function Theorems ◀ 255

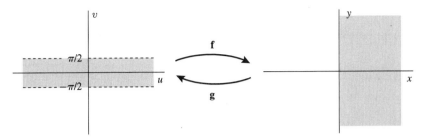

Figure 2.3

certainly satisfies $\mathbf{f} \circ \mathbf{g} \begin{pmatrix} x \\ y \end{pmatrix} = \begin{bmatrix} x \\ y \end{bmatrix}$. So, why is $\mathbf{g}$ not the inverse function of $\mathbf{f}$? Recall that arctan: $\mathbb{R} \to (-\pi/2, \pi/2)$. So, as shown in Figure 2.3, if we consider the domain of $\mathbf{f}$ to be $\left\{ \begin{bmatrix} u \\ v \end{bmatrix} : -\pi/2 < v < \pi/2 \right\}$ and the domain of $\mathbf{g}$ to be $\left\{ \begin{bmatrix} x \\ y \end{bmatrix} : x > 0 \right\}$, then $\mathbf{f}$ and $\mathbf{g}$ will be inverse functions.

Let's calculate the derivative of *any* local inverse $\mathbf{g}$ according to Theorem 2.1. If $\mathbf{f} \begin{pmatrix} u \\ v \end{pmatrix} = \begin{bmatrix} x \\ y \end{bmatrix}$, then

$$D\mathbf{g}\begin{pmatrix} x \\ y \end{pmatrix} = \left(D\mathbf{f}\begin{pmatrix} u \\ v \end{pmatrix}\right)^{-1} = \begin{bmatrix} e^{-u}\cos v & e^{-u}\sin v \\ -e^{-u}\sin v & e^{-u}\cos v \end{bmatrix} = \begin{bmatrix} \dfrac{x}{x^2+y^2} & \dfrac{y}{x^2+y^2} \\ -\dfrac{y}{x^2+y^2} & \dfrac{x}{x^2+y^2} \end{bmatrix}.$$

Note that we get the same formula by differentiating our specific inverse function (†). It is a bit surprising that the derivative of any other inverse function, with different domain and range, must be given by the identical formula. ◀

Now we are finally in a position to prove the Implicit Function Theorem, which first arose in our informal discussion of manifolds in Section 5 of Chapter 4. It is without question one of the most important theorems in higher mathematics.

Theorem 2.2 (Implicit Function Theorem) *Suppose $U \subset \mathbb{R}^n$ is open and $\mathbf{F}: U \to \mathbb{R}^m$ is $\mathcal{C}^1$. Writing a vector in $\mathbb{R}^n$ as $\begin{bmatrix} \mathbf{x} \\ \mathbf{y} \end{bmatrix}$, with $\mathbf{x} \in \mathbb{R}^{n-m}$ and $\mathbf{y} \in \mathbb{R}^m$, suppose that $\mathbf{F}\begin{pmatrix} \mathbf{x}_0 \\ \mathbf{y}_0 \end{pmatrix} = \mathbf{0}$ and the $m \times m$ matrix $\dfrac{\partial \mathbf{F}}{\partial \mathbf{y}}\begin{pmatrix} \mathbf{x}_0 \\ \mathbf{y}_0 \end{pmatrix}$ is invertible. Then there are neighborhoods V of $\mathbf{x}_0$ and W of $\mathbf{y}_0$ and a $\mathcal{C}^1$ function $\boldsymbol{\phi}: V \to W$ so that*

$$\mathbf{F}\begin{pmatrix} \mathbf{x} \\ \mathbf{y} \end{pmatrix} = \mathbf{0}, \ \mathbf{x} \in V, \ \text{and} \ \mathbf{y} \in W \iff \mathbf{y} = \boldsymbol{\phi}(\mathbf{x}).$$

Moreover,

$$D\boldsymbol{\phi}(\mathbf{x}) = -\left(\dfrac{\partial \mathbf{F}}{\partial \mathbf{y}}\begin{pmatrix} \mathbf{x} \\ \boldsymbol{\phi}(\mathbf{x}) \end{pmatrix}\right)^{-1} \dfrac{\partial \mathbf{F}}{\partial \mathbf{x}}\begin{pmatrix} \mathbf{x} \\ \boldsymbol{\phi}(\mathbf{x}) \end{pmatrix}.$$

256 ▶ Chapter 6. Solving Nonlinear Problems

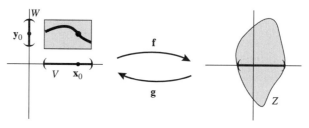

Figure 2.4

Proof Define $\mathbf{f} \colon U \to \mathbb{R}^n = \mathbb{R}^{n-m} \times \mathbb{R}^m$ by

$$\mathbf{f}\begin{pmatrix} \mathbf{x} \\ \mathbf{y} \end{pmatrix} = \begin{bmatrix} \mathbf{x} \\ \mathbf{F}\begin{pmatrix} \mathbf{x} \\ \mathbf{y} \end{pmatrix} \end{bmatrix}.$$

Note that the linear map

$$D\mathbf{f}\begin{pmatrix} \mathbf{x}_0 \\ \mathbf{y}_0 \end{pmatrix} = \begin{bmatrix} I & O \\ \dfrac{\partial \mathbf{F}}{\partial \mathbf{x}}\begin{pmatrix} \mathbf{x}_0 \\ \mathbf{y}_0 \end{pmatrix} & \dfrac{\partial \mathbf{F}}{\partial \mathbf{y}}\begin{pmatrix} \mathbf{x}_0 \\ \mathbf{y}_0 \end{pmatrix} \end{bmatrix}$$

is invertible (see Exercise 4.2.7). This means that—as illustrated in Figure 2.4—there are neighborhoods $V \subset \mathbb{R}^{n-m}$ of $\mathbf{x}_0$, $W \subset \mathbb{R}^m$ of $\mathbf{y}_0$, and $Z \subset \mathbb{R}^n$ of $\mathbf{0}$ and a $\mathcal{C}^1$ function $\mathbf{g} \colon Z \to V \times W$ so that $\mathbf{g}$ is an inverse of $\mathbf{f}$ on $V \times W$. Now define $\boldsymbol{\phi} \colon V \to W$ by

$$\begin{bmatrix} \mathbf{x} \\ \boldsymbol{\phi}(\mathbf{x}) \end{bmatrix} = \mathbf{g}\begin{pmatrix} \mathbf{x} \\ \mathbf{0} \end{pmatrix};$$

$\boldsymbol{\phi}$ is obviously $\mathcal{C}^1$ since $\mathbf{g}$ is. And

$$\begin{bmatrix} \mathbf{x} \\ \mathbf{F}\begin{pmatrix} \mathbf{x} \\ \boldsymbol{\phi}(\mathbf{x}) \end{pmatrix} \end{bmatrix} = \mathbf{f}\begin{pmatrix} \mathbf{x} \\ \boldsymbol{\phi}(\mathbf{x}) \end{pmatrix} = \mathbf{f}\left(\mathbf{g}\begin{pmatrix} \mathbf{x} \\ \mathbf{0} \end{pmatrix}\right) = \begin{bmatrix} \mathbf{x} \\ \mathbf{0} \end{bmatrix},$$

so $\mathbf{F}\begin{pmatrix} \mathbf{x} \\ \boldsymbol{\phi}(\mathbf{x}) \end{pmatrix} = \mathbf{0}$, as desired. On the other hand, if $\mathbf{F}\begin{pmatrix} \mathbf{x} \\ \mathbf{y} \end{pmatrix} = \mathbf{0}$, $\mathbf{x} \in V$, and $\mathbf{y} \in W$, then $\begin{bmatrix} \mathbf{x} \\ \mathbf{y} \end{bmatrix} = \mathbf{g}\begin{pmatrix} \mathbf{x} \\ \mathbf{0} \end{pmatrix}$, so $\mathbf{y}$ must be equal to $\boldsymbol{\phi}(\mathbf{x})$.

Since we know that $\boldsymbol{\phi}$ is $\mathcal{C}^1$, we can calculate the derivative by implicit differentiation: Define $\mathbf{h} \colon V \to \mathbb{R}^m$ by $\mathbf{h}(\mathbf{x}) = \mathbf{F}\begin{pmatrix} \mathbf{x} \\ \boldsymbol{\phi}(\mathbf{x}) \end{pmatrix}$. Then $\mathbf{h}$ is $\mathcal{C}^1$, and since $\mathbf{h}(\mathbf{x}) = \mathbf{0}$ for all $\mathbf{x} \in V$, we have

$$O = D\mathbf{h}(\mathbf{x}) = \frac{\partial \mathbf{F}}{\partial \mathbf{x}}\binom{\mathbf{x}}{\phi(\mathbf{x})} + \frac{\partial \mathbf{F}}{\partial \mathbf{y}}\binom{\mathbf{x}}{\phi(\mathbf{x})} D\phi(\mathbf{x}).$$

Since by hypothesis $\dfrac{\partial \mathbf{F}}{\partial \mathbf{y}}\binom{\mathbf{x}}{\phi(\mathbf{x})}$ is invertible, the desired result is immediate. ∎

Remark With not much more work, one can prove analogously that if $\mathbf{F}$ is $\mathcal{C}^k$ (or smooth), then $\mathbf{y}$ is given locally as a $\mathcal{C}^k$ (or smooth) function of $\mathbf{x}$. We may take this for granted in our later work.

▶ **EXAMPLE 3**

Consider the function $F\colon \mathbb{R}^2 \to \mathbb{R}$, $F\binom{x}{y} = x^3 e^y + 2x\cos(xy)$. We assert that the equation $F\binom{x}{y} = 3$ defines y locally as a function of x near the point $\begin{bmatrix} x_0 \\ y_0 \end{bmatrix} = \begin{bmatrix} 1 \\ 0 \end{bmatrix}$. By the Implicit Function Theorem, Theorem 2.2, we need only check that $\dfrac{\partial F}{\partial y}\binom{1}{0} \neq 0$. Well,

$$\frac{\partial F}{\partial y} = x^3 e^y - 2x^2 \sin(xy) \qquad \text{and so} \qquad \frac{\partial F}{\partial y}\binom{1}{0} = 1,$$

so we know that in a neighborhood of $x_0 = 1$ there is a $\mathcal{C}^1$ function ϕ with $\phi(1) = 0$ whose graph is (the appropriate piece of) the level curve $F = 3$, as shown in Figure 2.5. Of course, farther away, the curve apparently gets quite crazy.

If we're interested in the best linear approximation to the curve at the point $\begin{bmatrix} 1 \\ 0 \end{bmatrix}$, then we also know from Theorem 2.2 that

$$\phi'(1) = -\frac{\frac{\partial F}{\partial x}}{\frac{\partial F}{\partial y}}\binom{1}{0} = -\frac{5}{1} = -5,$$

so the line $5x + y = 5$ is the desired tangent line of the curve at that point. ◀

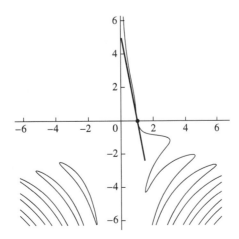

Figure 2.5

EXAMPLE 4

Consider $\mathbf{F}\colon \mathbb{R}^5 \to \mathbb{R}^3$ given by

$$\mathbf{F}\begin{pmatrix} x_1 \\ x_2 \\ y_1 \\ y_2 \\ y_3 \end{pmatrix} = \begin{bmatrix} 2x_1 + x_2 + y_1 + y_3 - 1 \\ x_1 x_2^3 + x_1 y_1 + x_2^2 y_2^2 - y_2 y_3 \\ x_2 y_1 y_3 + x_1 y_1^2 + y_2 y_3^2 \end{bmatrix}, \quad \text{and let} \quad \mathbf{a} = \begin{bmatrix} 0 \\ 1 \\ -1 \\ 1 \\ 1 \end{bmatrix}.$$

Does the equation $\mathbf{F} = \mathbf{0}$ define $\mathbf{y} = \begin{bmatrix} y_1 \\ y_2 \\ y_3 \end{bmatrix}$ as implicitly as a function of $\mathbf{x} = \begin{bmatrix} x_1 \\ x_2 \end{bmatrix}$ near $\mathbf{a}$? Note first of all that $\mathbf{F}$ is $\mathcal{C}^1$. We begin by calculating the derivative of $\mathbf{F}$:

$$D\mathbf{F}\begin{pmatrix} x_1 \\ x_2 \\ y_1 \\ y_2 \\ y_3 \end{pmatrix} = \begin{bmatrix} 2 & 1 & 1 & 0 & 1 \\ x_2^3 + y_1 & 3x_1 x_2^2 + 2x_2 y_2^2 & x_1 & 2x_2^2 y_2 - y_3 & -y_2 \\ y_1^2 & y_1 y_3 & x_2 y_3 + 2x_1 y_1 & y_3^2 & x_2 y_1 + 2y_2 y_3 \end{bmatrix},$$

and so

$$D\mathbf{F}\begin{pmatrix} 0 \\ 1 \\ -1 \\ 1 \\ 1 \end{pmatrix} = \begin{bmatrix} 2 & 1 & 1 & 0 & 1 \\ 0 & 2 & 0 & 1 & -1 \\ 1 & -1 & 1 & 1 & 1 \end{bmatrix}.$$

In particular, we see that

$$\frac{\partial \mathbf{F}}{\partial \mathbf{y}}(\mathbf{a}) = \begin{bmatrix} 1 & 0 & 1 \\ 0 & 1 & -1 \\ 1 & 1 & 1 \end{bmatrix},$$

which is easily checked to be nonsingular, and so the hypotheses of the Implicit Function Theorem, Theorem 2.2, are fulfilled. There is a neighborhood of $\mathbf{a}$ in which we have $\mathbf{y} = \boldsymbol{\phi}(\mathbf{x})$. Moreover, we have

$$D\boldsymbol{\phi}\begin{pmatrix} 0 \\ 1 \end{pmatrix} = -\left(\frac{\partial \mathbf{F}}{\partial \mathbf{y}}(\mathbf{a})\right)^{-1}\left(\frac{\partial \mathbf{F}}{\partial \mathbf{x}}(\mathbf{a})\right) = -\begin{bmatrix} 2 & 1 & -1 \\ -1 & 0 & 1 \\ -1 & -1 & 1 \end{bmatrix}\begin{bmatrix} 2 & 1 \\ 0 & 2 \\ 1 & -1 \end{bmatrix} = \begin{bmatrix} -3 & -5 \\ 1 & 2 \\ 1 & 4 \end{bmatrix}.$$

With this information, we can easily give the tangent plane at $\mathbf{a}$ of the surface $\mathbf{F} = \mathbf{0}$.

Remark In general, we shall not always be so chivalrous (nor shall life) as to set up the notation precisely as in the statement of Theorem 2.2. Just as in the case of linear equations where the first r variables needn't always be the pivot variables, here the last m variables needn't always be (locally) the dependent variables. In general, it is a matter of finding m pivots in *some* m columns of the $m \times n$ derivative matrix.

EXERCISES 6.2

1. By applying the Inverse Function Theorem, Theorem 2.1, determine at which points $\mathbf{x}_0$ the given function $\mathbf{f}$ has a local $\mathcal{C}^1$ inverse $\mathbf{g}$, and calculate $D\mathbf{g}(\mathbf{f}(\mathbf{x}_0))$.

*(a) $\mathbf{f}\begin{pmatrix}x\\y\end{pmatrix} = \begin{bmatrix}x^2 - y^2\\2xy\end{bmatrix}$

(b) $\mathbf{f}\begin{pmatrix}x\\y\end{pmatrix} = \begin{bmatrix}x/(x^2+y^2)\\y/(x^2+y^2)\end{bmatrix}$

(c) $\mathbf{f}\begin{pmatrix}x\\y\end{pmatrix} = \begin{bmatrix}x+h(y)\\y\end{bmatrix}$ for any $\mathcal{C}^1$ function $h\colon \mathbb{R} \to \mathbb{R}$

(d) $\mathbf{f}\begin{pmatrix}x\\y\end{pmatrix} = \begin{bmatrix}x+e^y\\y+e^x\end{bmatrix}$

(e) $\mathbf{f}\begin{pmatrix}x\\y\\z\end{pmatrix} = \begin{bmatrix}x+y+z\\xy+xz+yz\\xyz\end{bmatrix}$ (cf. also Exercise 2)

2. Let $U = \left\{\begin{bmatrix}u\\v\end{bmatrix} : 0 < v < u\right\}$, and define $\mathbf{f}\colon U \to \mathbb{R}^2$ by $\mathbf{f}\begin{pmatrix}u\\v\end{pmatrix} = \begin{bmatrix}u+v\\uv\end{bmatrix}$.

(a) Show that $\mathbf{f}$ has a global inverse function $\mathbf{g}$. Determine the domain of and an explicit formula for $\mathbf{g}$.

(b) Calculate $D\mathbf{g}$ both directly and by the formula given in the Inverse Function Theorem. Compare your answers.

(c) What does this exercise have to do with Example 2 in Chapter 4, Section 5? In particular, give a concrete interpretation of your answer to part b.

3. Check that in each of the following cases, the equation $\mathbf{F} = \mathbf{0}$ defines $\mathbf{y}$ locally as a $\mathcal{C}^1$ function $\boldsymbol{\phi}(\mathbf{x})$ near $\mathbf{a} = \begin{bmatrix}\mathbf{x}_0\\ \mathbf{y}_0\end{bmatrix}$, and calculate $D\boldsymbol{\phi}(\mathbf{x}_0)$.

(a) $F\begin{pmatrix}x\\y\end{pmatrix} = y^2 - x^3 - 2\sin(\pi(x-y))$, $x_0 = 1$, $y_0 = -1$

*(b) $F\begin{pmatrix}x_1\\x_2\\y\end{pmatrix} = e^{x_1 y} + y^2 \cos x_1 x_2 - 1$, $\mathbf{x}_0 = \begin{bmatrix}1\\2\end{bmatrix}$, $y_0 = 0$

(c) $F\begin{pmatrix}x_1\\x_2\\y\end{pmatrix} = e^{x_1 y} + y^2 \arctan x_2 - (1 + \pi/4)$, $\mathbf{x}_0 = \begin{bmatrix}0\\1\end{bmatrix}$, $y_0 = 1$

(d) $\mathbf{F}\begin{pmatrix}x\\y_1\\y_2\end{pmatrix} = \begin{bmatrix}x^2 - y_1^2 - y_2^2 - 2\\x - y_1 + y_2 - 2\end{bmatrix}$, $x_0 = 2$, $\mathbf{y}_0 = \begin{bmatrix}1\\1\end{bmatrix}$

(e) $\mathbf{F}\begin{pmatrix}x_1\\x_2\\y_1\\y_2\end{pmatrix} = \begin{bmatrix}x_1^2 - x_2^2 - y_1^3 + y_2^2 + 4\\2x_1 x_2 + x_2^2 - 2y_1^2 + 3y_2^4 + 8\end{bmatrix}$, $\mathbf{x}_0 = \begin{bmatrix}2\\-1\end{bmatrix}$, $\mathbf{y}_0 = \begin{bmatrix}2\\1\end{bmatrix}$

*4. Show that the equations $x^2 y + xy^2 + t^2 - 1 = 0$ and $x^2 + y^2 - 2yt = 0$ define x and y implicitly as $\mathcal{C}^1$ functions of t near $\begin{bmatrix}x\\y\\t\end{bmatrix} = \begin{bmatrix}-1\\1\\1\end{bmatrix}$. Find the tangent line at this point to the curve so defined.

5. Let $F\begin{pmatrix}x\\y\\z\end{pmatrix} = x^2 + 2y^2 - 2xz - z^2 = 0$. Show that near the point $\mathbf{a} = \begin{bmatrix}1\\1\\1\end{bmatrix}$, z is given implicitly as a $\mathcal{C}^1$ function of x and y. Find the largest neighborhood of $\mathbf{a}$ on which this is true.

*6. Using the law of cosines (see Exercise 1.2.12) and Theorem 2.2, show that the angles of a triangle are $\mathcal{C}^1$ functions of the sides. To a small change in which one of the sides (keeping the other two fixed) is an angle most sensitive?

7. Define $\mathbf{f}: \mathcal{M}_{n\times n} \to \mathcal{M}_{n\times n}$ by $\mathbf{f}(A) = A^2$.
   (a) By applying the Inverse Function Theorem, Theorem 2.1, show that every matrix B in a neighborhood of I has (at least) two square roots A (i.e., $A^2 = B$), each varying as a $\mathcal{C}^1$ function of B. (See Exercise 3.1.13.)
   (b) Can you decide if there are precisely two or more? (Hint: In the 2×2 case, what is $D\mathbf{f}\left(\begin{bmatrix}1 & 0\\0 & -1\end{bmatrix}\right)$?)

8. Suppose $U \subset \mathbb{R}^3$ is an open set, $F: U \to \mathbb{R}$ is a $\mathcal{C}^1$ function, and on $F = 0$ we have $\dfrac{\partial F}{\partial p} \neq 0$, $\dfrac{\partial F}{\partial V} \neq 0$, and $\dfrac{\partial F}{\partial T} \neq 0$. (You might use, as an example, the equation $F\begin{pmatrix}p\\V\\T\end{pmatrix} = pV - RT = 0$ for one mole of ideal gas; here R is the so-called gas constant.) Then it is a consequence of the Implicit Function Theorem that in some neighborhood of $\begin{bmatrix}p_0\\V_0\\T_0\end{bmatrix}$, each of p, V, and T can be written as a differentiable function of the remaining two variables. Physical chemists denote by $\left(\dfrac{\partial p}{\partial V}\right)_T$ the partial derivative of the function $p = p\begin{pmatrix}V\\T\end{pmatrix}$ with respect to V, holding T constant, etc. Prove the thermodynamicist's magic formula
$$\left(\frac{\partial p}{\partial V}\right)_T \left(\frac{\partial V}{\partial T}\right)_p \left(\frac{\partial T}{\partial p}\right)_V = -1.$$

9. Using the notation of Exercise 8, physical chemists define the expansion coefficient α and isothermal compressibility β to be, respectively,
$$\alpha = \frac{1}{V}\left(\frac{\partial V}{\partial T}\right)_p \quad \text{and} \quad \beta = -\frac{1}{V}\left(\frac{\partial V}{\partial p}\right)_T.$$

*(a) Calculate α and β for an ideal gas.
(b) Show that in general we have $\left(\dfrac{\partial p}{\partial T}\right)_V = \dfrac{\alpha}{\beta}$.

10. Check that, under the hypotheses in place in the proof of Theorem 2.1, if $\|\mathbf{x}\| = r$, then $\|\mathbf{f}(\mathbf{x})\| \geq r/2$. (Hint: Use Exercise 1.2.17.)

♯11. Let $B = B(\mathbf{0}, r) \subset \mathbb{R}^n$. Suppose $U \subset \mathbb{R}^n$ is an open subset containing the closed ball $\overline{B}$, $\mathbf{f}: U \to \mathbb{R}^n$ is $\mathcal{C}^1$, $\mathbf{f}(\mathbf{0}) = \mathbf{0}$, and $\|D\mathbf{f}(\mathbf{x}) - I\| \leq s < 1$ for all $\mathbf{x} \in B$. Prove that if $\|\mathbf{y}\| < r(1-s)$, then there is $\mathbf{x} \in B$ such that $\mathbf{f}(\mathbf{x}) = \mathbf{y}$.

12. Suppose $U \subset \mathbb{R}^n$ is open and $\mathbf{f}: U \to \mathbb{R}^m$ is $\mathcal{C}^1$ with $\mathbf{f}(\mathbf{a}) = \mathbf{0}$ and $\text{rank}(D\mathbf{f}(\mathbf{a})) = m$. Prove that for every $\mathbf{c}$ sufficiently close to $\mathbf{0} \in \mathbb{R}^m$ the equation $\mathbf{f}(\mathbf{x}) = \mathbf{c}$ has a solution near $\mathbf{a}$.

13. **(The Envelope of a Family of Curves)** Suppose $f\colon \mathbb{R}^2 \times (a,b) \to \mathbb{R}$ is $\mathcal{C}^2$ and for each $t \in (a,b)$, $\nabla f\begin{pmatrix}\mathbf{x}\\t\end{pmatrix} \neq \mathbf{0}$ on the level curve $C_t = \left\{f\begin{pmatrix}\mathbf{x}\\t\end{pmatrix} = 0\right\}$. (Here the gradient denotes differentiation with respect only to $\mathbf{x}$.) The curve C is called the *envelope* of the family of curves $\{C_t : t \in (a,b)\}$ if each member of the family is tangent to C at some point (depending on t).

(a) Suppose the matrix

$$\begin{bmatrix} \dfrac{\partial f}{\partial x}\begin{pmatrix}\mathbf{x}_0\\t_0\end{pmatrix} & \dfrac{\partial f}{\partial y}\begin{pmatrix}\mathbf{x}_0\\t_0\end{pmatrix} \\ \dfrac{\partial^2 f}{\partial x \partial t}\begin{pmatrix}\mathbf{x}_0\\t_0\end{pmatrix} & \dfrac{\partial^2 f}{\partial y \partial t}\begin{pmatrix}\mathbf{x}_0\\t_0\end{pmatrix} \end{bmatrix}$$

is nonsingular. Show that for some $\delta > 0$, there is a $\mathcal{C}^1$ curve $\mathbf{g}\colon (t_0 - \delta, t_0 + \delta) \to \mathbb{R}^2$ so that

$$f\begin{pmatrix}\mathbf{g}(t)\\t\end{pmatrix} = \frac{\partial f}{\partial t}\begin{pmatrix}\mathbf{g}(t)\\t\end{pmatrix} = 0.$$

Conclude that $\mathbf{g}$ is a parametrization of the envelope C near $\mathbf{x}_0$.

(b) Find the envelopes of the following families of curves (portions of which are sketched in Figure 2.6).

  i. $f\begin{pmatrix}\mathbf{x}\\t\end{pmatrix} = (\cos t)x + (\sin t)y - 1 = 0$

  ii. $f\begin{pmatrix}\mathbf{x}\\t\end{pmatrix} = y + t^2 x - t = 0$

  iii. $f\begin{pmatrix}\mathbf{x}\\t\end{pmatrix} = \left(\dfrac{x}{t}\right)^2 + \left(\dfrac{y}{1-t}\right)^2 - 1 = 0,\ t \in (0,1)$

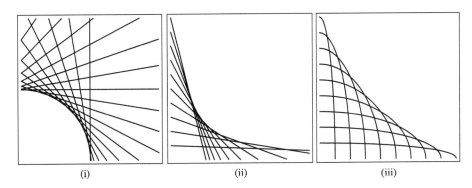

(i)   (ii)   (iii)

Figure 2.6

3 MANIFOLDS REVISITED

In Chapter 4, we introduced k-dimensional manifolds in $\mathbb{R}^n$ informally as being locally the graph of $\mathcal{C}^1$ function over an open subset of a k-dimensional coordinate plane. We suggested that, because of the Implicit Function Theorem, under the appropriate hypotheses, a level set of a $\mathcal{C}^1$ function is a prototypical example. Indeed, as we now wish to make clear, there are three equivalent formulations, roughly these:

Explicit: Near each point, M is a graph over some k-dimensional coordinate plane.

Implicit: Near each point, M is the level set of some function whose derivative has maximum rank.

Parametric: Near each point, M is parametrized by some one-to-one function whose derivative has maximum rank (e.g., a parametrized curve with nonzero velocity).

We've seen that the implicit formulation arises in working with Lagrange multipliers, and the parametric formulation will be crucial for our work with integration in Chapter 8. In this brief section, we are going to make the three definitions quite precisely and then prove their equivalence in Theorem 3.1. To make our life easier in Chapter 8, we will replace the $\mathcal{C}^1$ condition with "smooth."

Definition We say $M \subset \mathbb{R}^n$ is a *k-dimensional manifold* if any one of the following three criteria holds:

1. For any $\mathbf{p} \in M$, there is a neighborhood $W \subset \mathbb{R}^n$ of $\mathbf{p}$ so that $M \cap W$ is the graph of a smooth function $\mathbf{f}\colon V \to \mathbb{R}^{n-k}$, where $V \subset \mathbb{R}^k$ is an open set. Here we are allowed to choose any k integers $1 \le i_1 < \cdots < i_k \le n$; then $\mathbb{R}^k$ is the $x_{i_1}\cdots x_{i_k}$-plane, and $\mathbb{R}^{n-k}$ is the plane of the complementary coordinates.

2. For any $\mathbf{p} \in M$, there are a neighborhood $W \subset \mathbb{R}^n$ of $\mathbf{p}$ and a smooth function $\mathbf{F}\colon W \to \mathbb{R}^{n-k}$ so that $\mathbf{F}^{-1}(\mathbf{0}) = M \cap W$ and $\mathrm{rank}(D\mathbf{F}(\mathbf{x})) = n - k$ for every $\mathbf{x} \in M \cap W$.

3. For any $\mathbf{p} \in M$, there is a neighborhood $W \subset \mathbb{R}^n$ of $\mathbf{p}$ so that $M \cap W$ is the image of a smooth function $\mathbf{g}\colon U \to \mathbb{R}^n$ for some open set $U \subset \mathbb{R}^k$, with the properties that $\mathbf{g}$ is one-to-one, $\mathrm{rank}(D\mathbf{g}(\mathbf{u})) = k$ for all $\mathbf{u} \in U$, and $\mathbf{g}^{-1}\colon M \cap W \to U$ is continuous. (See Figure 3.1.)

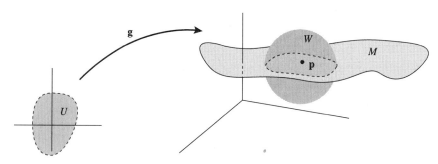

Figure 3.1

If the curious reader wonders why the last (and obviously technical) condition is included in the third definition, see Exercises 2 and 3.

Theorem 3.1 *The three criteria given in this definition are all equivalent.*

Proof The Implicit Function Theorem, Theorem 2.2, tells us precisely that (2)$\Longrightarrow$(1). And (1)$\Longrightarrow$(3) is obvious since we can set $\mathbf{g}(\mathbf{u}) = \begin{bmatrix} \mathbf{u} \\ \mathbf{f}(\mathbf{u}) \end{bmatrix}$ (where, for ease of notation, we assume here that $\mathbb{R}^k$ is the $x_1 \cdots x_k$-plane). So it remains only to check that (3)$\Longrightarrow$(2).

Suppose, as in the third definition, that we are given a neighborhood $\widetilde{W} \subset \mathbb{R}^n$ of $\mathbf{p} \in M$ so that $M \cap \widetilde{W}$ is the image of a smooth function $\mathbf{g}\colon U \to \mathbb{R}^n$ for some open set $U \subset \mathbb{R}^k$, with the properties that $\mathbf{g}$ is one-to-one, rank$(D\mathbf{g}(\mathbf{u})) = k$ for all $\mathbf{u} \in U$, and $\mathbf{g}^{-1}\colon M \cap \widetilde{W} \to U$ is continuous. The last condition tells us that that if $\mathbf{g}(\mathbf{u}_0) = \mathbf{p}$, then points sufficiently close to $\mathbf{p}$ in M must map by $\mathbf{g}^{-1}$ close to $\mathbf{u}_0$; that is, all points of $M \cap \widetilde{W}$ are the image under $\mathbf{g}$ of a neighborhood of $\mathbf{u}_0$.

We may assume that $\mathbf{g}(\mathbf{0}) = \mathbf{p}$ and (renumbering coordinates in $\mathbb{R}^n$ as necessary) $D\mathbf{g}(\mathbf{0}) = \begin{bmatrix} A \\ B \end{bmatrix}$, where A is an invertible $k \times k$ matrix. We define $\mathbf{G}\colon U \times \mathbb{R}^{n-k} \to \mathbb{R}^n$ by $\mathbf{G}\begin{pmatrix} \mathbf{u} \\ \mathbf{v} \end{pmatrix} = \mathbf{g}(\mathbf{u}) + \begin{bmatrix} \mathbf{0} \\ \mathbf{v} \end{bmatrix}$. Since

$$D\mathbf{G}\begin{pmatrix} \mathbf{0} \\ \mathbf{0} \end{pmatrix} = \left[\begin{array}{c|c} A & O \\ \hline B & I_{n-k} \end{array}\right]$$

is invertible (see Exercise 4.2.7), it follows from the Inverse Function Theorem, Theorem 2.1, that there are neighborhoods $V = V_1 \times V_2 \subset \mathbb{R}^k \times \mathbb{R}^{n-k}$ of $\begin{bmatrix} \mathbf{0} \\ \mathbf{0} \end{bmatrix}$ and $W \subset \mathbb{R}^n$ of $\mathbf{p}$ and a local (smooth) inverse $\mathbf{H}\colon W \to V$ of $\mathbf{G}$. (Shrinking W if necessary, we assume $W \subset \widetilde{W}$.) Writing $\mathbf{H}(\mathbf{x}) = \begin{bmatrix} \mathbf{H}_1(\mathbf{x}) \\ \mathbf{H}_2(\mathbf{x}) \end{bmatrix} \in \mathbb{R}^k \times \mathbb{R}^{n-k}$, we define $\mathbf{F}\colon W \to \mathbb{R}^{n-k}$ by $\mathbf{F} = \mathbf{H}_2$. Now suppose $\mathbf{F}(\mathbf{x}) = \mathbf{0}$. Since $\mathbf{x} \in W$, $\mathbf{x} = \mathbf{G}\begin{pmatrix} \mathbf{u} \\ \mathbf{v} \end{pmatrix}$ for a unique vector $\begin{bmatrix} \mathbf{u} \\ \mathbf{v} \end{bmatrix} \in V$. Then

$$\mathbf{F}(\mathbf{x}) = \mathbf{F}\left(\mathbf{G}\begin{pmatrix} \mathbf{u} \\ \mathbf{v} \end{pmatrix}\right) = \mathbf{H}_2\left(\mathbf{G}\begin{pmatrix} \mathbf{u} \\ \mathbf{v} \end{pmatrix}\right) = \mathbf{v},$$

so $\mathbf{F}(\mathbf{x}) = \mathbf{0}$ if and only if $\mathbf{v} = \mathbf{0}$, which means that $\mathbf{x} = \mathbf{g}(\mathbf{u})$. This proves that the equation $\mathbf{F} = \mathbf{0}$ defines that portion of M given by $\mathbf{g}(\mathbf{u})$ for all $\mathbf{u} \in V_1$. But because $W \subset \widetilde{W}$, we know that such points comprise all of $M \cap W$. ∎

▶ **EXAMPLE 1**

Perhaps an explicit example will make this proof a bit more understandable. Suppose $\mathbf{g}\colon \mathbb{R} \to \mathbb{R}^3$ is given by $\mathbf{g}(u) = \begin{bmatrix} u \\ u^2 \\ u^3 \end{bmatrix}$ and M is the image of $\mathbf{g}$. We wish to write M (perhaps locally) as the level set of a function near $\mathbf{p} = \mathbf{0}$. As in the proof, we define

$$\mathbf{G}\begin{pmatrix} u \\ v_1 \\ v_2 \end{pmatrix} = \begin{bmatrix} u \\ u^2 \\ u^3 \end{bmatrix} + \begin{bmatrix} 0 \\ v_1 \\ v_2 \end{bmatrix} = \begin{bmatrix} u \\ u^2 + v_1 \\ u^3 + v_2 \end{bmatrix}.$$

We can explicitly construct the inverse function

$$\mathbf{G}^{-1}\begin{pmatrix} x \\ y \\ z \end{pmatrix} = \mathbf{H}\begin{pmatrix} x \\ y \\ z \end{pmatrix} = \begin{bmatrix} x \\ y - x^2 \\ z - x^3 \end{bmatrix}.$$

The proof tells us to define $\mathbf{F} = \mathbf{H}_2$, and, indeed, this works. M is the zero-set of the function

$$\mathbf{F} \colon \mathbb{R}^3 \to \mathbb{R}^2 \quad \text{given by} \quad \mathbf{F}\begin{pmatrix} x \\ y \\ z \end{pmatrix} = \begin{bmatrix} y - x^2 \\ z - x^3 \end{bmatrix}.$$

We ask the reader to carry this procedure out in Exercise 6 in a situation where it will only work locally. ◀

There are corresponding notions of the *tangent space* of the manifold M at $\mathbf{p}$. (Recall that we shall attempt to refer to the tangent *space* as a subspace, whereas the tangent *plane* is obtained by translating it to pass through the point $\mathbf{p}$.)

Definition If the manifold M is presented in the three respective forms above, then its *tangent space* at $\mathbf{p}$, denoted $T_\mathbf{p} M$, is defined as follows.

1. Assuming M is locally the graph of $\mathbf{f}$ with $\mathbf{p} = \begin{bmatrix} \mathbf{a} \\ \mathbf{f}(\mathbf{a}) \end{bmatrix}$, then $T_\mathbf{p} M$ is the graph of $D\mathbf{f}(\mathbf{a})$.
2. Assuming M is locally a level set of $\mathbf{F}$, then $T_\mathbf{p} M = \mathbf{N}([D\mathbf{F}(\mathbf{p})])$.
3. Assuming M is locally parametrized by $\mathbf{g}$ with $\mathbf{p} = \mathbf{g}(\mathbf{a})$, then $T_\mathbf{p} M$ is the image of the linear map $D\mathbf{g}(\mathbf{a}) \colon \mathbb{R}^k \to \mathbb{R}^n$.

Once again, we need to check that these three recipes all give the same k-dimensional subspace of $\mathbb{R}^n$. The ideas involved in this check have all emerged already in the preceding chapters. Since (1) is a special case of (3) (why?), we need only check that $\mathbf{N}([D\mathbf{F}(\mathbf{p})]) = \text{image}(D\mathbf{g}(\mathbf{a}))$. Note that both of these are k-dimensional subspaces because of our rank conditions on $\mathbf{F}$ and $\mathbf{g}$. So it suffices to show that $\text{image}(D\mathbf{g}(\mathbf{a})) \subset \mathbf{N}([D\mathbf{F}(\mathbf{p})])$. But this is easy: The function $\mathbf{F} \circ \mathbf{g} \colon U \to \mathbb{R}^{n-k}$ is identically $\mathbf{0}$, so, by the chain rule, $D\mathbf{F}(\mathbf{p}) \circ D\mathbf{g}(\mathbf{a}) = O$, which says precisely that any vector in the image of $D\mathbf{g}(\mathbf{a})$ is in the kernel of $D\mathbf{F}(\mathbf{p})$.

▶ EXERCISES 6.3

*1. Show that the set $X = \left\{ \begin{bmatrix} x \\ y \end{bmatrix} : y = |x| \right\}$ is not a 1-dimensional manifold, even though the function $\mathbf{g}(t) = \begin{bmatrix} t^3 \\ |t^3| \end{bmatrix}$ gives a $\mathcal{C}^1$ "parametrization" of it. What's going on?

2. Show that the parametric curve $\mathbf{g}(t) = \begin{bmatrix} \cos 2t \cos t \\ \cos 2t \sin t \end{bmatrix}$, $t \in (-\pi/2, \pi/4)$, is not a 1-dimensional manifold. (Hint: Stare at Figure 3.2.)

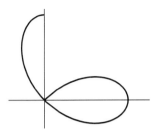

Figure 3.2

3. Consider the following union of parallel lines:

$$X = \left\{ \begin{bmatrix} x \\ y \end{bmatrix} : y = q \text{ for some } q \in \mathbb{Q} \right\} \subset \mathbb{R}^2.$$

Is X a 1-dimensional manifold? (Here $\mathbb{Q}$ denotes the set of rational numbers.)

4. Is the union of the hyperbola $xy = 1$ and its asymptote $y = 0$ a 1-dimensional manifold? Give your reasoning.

5. Show the equivalence of the three definitions for each of the following 1-dimensional manifolds:

*(a) parametric curve $\begin{bmatrix} t^2 \\ t \\ t^4 \end{bmatrix}$

(b) parametric curve $\begin{bmatrix} \cos t \\ 3 \sin t \end{bmatrix}$

(c) implicit curve $x^2 + y^2 = 1$, $x^2 + y^2 + z^2 = 2x$

(d) implicit curve $x^2 + y^2 = 1$, $z^2 + w^2 = 1$, $xz + yw = 0$

6. Suppose $\mathbf{g}: \mathbb{R} \to \mathbb{R}^3$ is given by $\mathbf{g}(u) = \begin{bmatrix} u + u^2 \\ u^2 \\ u^3 \end{bmatrix}$. Let M be the image of $\mathbf{g}$.

(a) Show that $\mathbf{g}$ is globally one-to-one.

(b) Following the proof given of Theorem 3.1, find a neighborhood W of $\mathbf{0} \in \mathbb{R}^3$ and $\mathbf{F}: W \to \mathbb{R}^2$ so that $M \cap W = \mathbf{F}^{-1}(\mathbf{0})$.

7. Show the equivalence of the three definitions for each of the following 2-dimensional manifolds:

(a) implicit surface $x^2 + y^2 = 1$ (in $\mathbb{R}^3$)

(b) implicit surface $x^2 + y^2 = z^2$ (in $\mathbb{R}^3 - \{\mathbf{0}\}$)

*(c) parametric surface $\begin{bmatrix} u \cos v \\ u \sin v \\ v \end{bmatrix}$, $u > 0$, $v \in \mathbb{R}$

(d) parametric surface $\begin{bmatrix} \sin u \cos v \\ \sin u \sin v \\ \cos u \end{bmatrix}$, $0 < u < \pi$, $0 < v < 2\pi$

(e) parametric surface $\begin{bmatrix} \sin u \cos v \\ \sin u \sin v \\ 2\cos u \end{bmatrix}$, $0 < u < \pi, 0 < v < 2\pi$

(f) parametric surface $\begin{bmatrix} (3+2\cos u)\cos v \\ (3+2\cos u)\sin v \\ 2\sin u \end{bmatrix}$, $0 \le u, v \le 2\pi$

8. (a) Show that
$$X = \left\{ \begin{bmatrix} x \\ y \\ z \end{bmatrix} : (x^2 + y^2 + z^2)^2 - 10(x^2 + y^2) + 6z^2 + 9 = 0 \right\}$$
is a 2-manifold.

(b) Check that $\begin{bmatrix} x \\ y \\ z \end{bmatrix} \in X \iff (\sqrt{x^2+y^2} - 2)^2 + z^2 = 1$. Use this to sketch X.

9. At what points is
$$X = \left\{ \begin{pmatrix} x \\ y \\ z \end{pmatrix} : (x^2 + y^2 + z^2)^2 - 4(x^2 + y^2) = 0 \right\}$$
a smooth surface? Proof? Give the equation of its tangent space at such a point.

10. Prove that the equations
$$x_1^2 + x_2^2 + x_3^2 + x_4^2 = 4 \quad \text{and} \quad x_1 x_2 + x_3 x_4 = 0$$
define a smooth surface in $\mathbb{R}^4$. Give a basis for its tangent space at $\begin{bmatrix} 1 \\ 1 \\ -1 \\ 1 \end{bmatrix}$.

11. Prove (1)$\Longrightarrow$(2) in Theorem 3.1 directly.

12. Writing $\begin{bmatrix} \mathbf{x} \\ \mathbf{y} \end{bmatrix} \in \mathbb{R}^3 \times \mathbb{R}^3$, show that the equations
$$\|\mathbf{x}\|^2 = \|\mathbf{y}\|^2 = 1 \quad \text{and} \quad \mathbf{x} \cdot \mathbf{y} = 0$$
define a 3-dimensional manifold in $\mathbb{R}^6$. Give a geometric interpretation of this manifold.

13. Recall from Exercise 1.4.34 that an $n \times n$ matrix A is called orthogonal if $A^\mathsf{T} A = I$.
(a) Prove that the set $O(n)$ of $n \times n$ orthogonal matrices forms a $\frac{n(n-1)}{2}$-dimensional manifold in $\mathcal{M}_{n \times n} = \mathbb{R}^{n^2}$. (Hint: Consider $\mathbf{F}: \mathcal{M}_{n \times n} \to \{\text{symmetric } n \times n \text{ matrices}\} = \mathbb{R}^{n(n+1)/2}$ defined by $\mathbf{F}(A) = A^\mathsf{T} A - I$. Use Exercise 3.1.13.)
(b) Show that the tangent space of $O(n)$ at I is the set of skew-symmetric $n \times n$ matrices.

14. Prove (3)$\Longrightarrow$(1) in Theorem 3.1 directly. (Hint: Suppose $\mathbf{g}(\mathbf{0}) = \mathbf{p}$ and $D\mathbf{g}(\mathbf{0}) = \begin{bmatrix} A \\ B \end{bmatrix}$, where A is an invertible $k \times k$ matrix. Write $\mathbf{g} = \begin{bmatrix} \mathbf{g}_1 \\ \mathbf{g}_2 \end{bmatrix}: U \to \mathbb{R}^k \times \mathbb{R}^{n-k}$ and observe that $\mathbf{g}_1$ has a local inverse. What about the general case?)

CHAPTER

7

INTEGRATION

We turn now to the integral, with which, intuitively, we chop a large problem into small, understandable bits and add them up, then proceeding to a limit in some fashion. We start with the definition and then proceed to the computation, which is, once again, based on reducing the problem to several one-variable calculus problems. We then learn how to exploit symmetry by using different coordinate systems and tackle various standard physical applications (e.g., center of mass, moment of inertia, and gravitational attraction). The discussion of determinants, initiated in Chapter 1, culminates here with a complete treatment and their role in integration and the change of variables theorem.

▶ 1 MULTIPLE INTEGRALS

In single-variable calculus the integral is motivated by the problem of finding the area under a curve $y = f(x)$ over an interval $[a, b]$. Now we want to find the volume of the region in $\mathbb{R}^3$ lying under the graph $z = f\begin{pmatrix} x \\ y \end{pmatrix}$ and over the rectangle $R = [a, b] \times [c, d]$ in the xy-plane. Once we see how partitions, upper and lower sums, and the integral are defined for rectangles in $\mathbb{R}^2$, then it is simple (although notationally discomforting) to generalize to higher dimensions.

Definition Let $R = [a, b] \times [c, d]$ be a *rectangle* in $\mathbb{R}^2$. Let $f: R \to \mathbb{R}$ be a bounded function. Given partitions $\mathcal{P}_1 = \{a = x_0 < x_1 < \cdots < x_k = b\}$ and $\mathcal{P}_2 = \{c = y_0 < y_1 < \cdots < y_\ell = d\}$ of $[a, b]$ and $[c, d]$, respectively, denote by $\mathcal{P} = \mathcal{P}_1 \times \mathcal{P}_2$ the *partition* of the rectangle R into subrectangles

$$R_{ij} = [x_{i-1}, x_i] \times [y_{j-1}, y_j], \quad 1 \le i \le k, \ 1 \le j \le \ell.$$

Let $M_{ij} = \sup_{\mathbf{x} \in R_{ij}} f(\mathbf{x})$ and $m_{ij} = \inf_{\mathbf{x} \in R_{ij}} f(\mathbf{x})$, as indicated in Figure 1.1. Define the *upper sum* of f with respect to the partition $\mathcal{P}$,

$$U(f, \mathcal{P}) = \sum_{i,j} M_{ij} \operatorname{area}(R_{ij}),$$

and the analogous *lower sum*

$$L(f, \mathcal{P}) = \sum_{i,j} m_{ij} \operatorname{area}(R_{ij}).$$

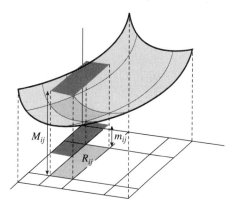

Figure 1.1

We say f is *integrable* on R if there is a *unique* number I satisfying
$$L(f, \mathcal{P}) \leq I \leq U(f, \mathcal{P}) \quad \text{for all partitions } \mathcal{P}.$$
In that event, we denote $I = \int_R f\, dA$, called the *integral of f over R*.

(Note that the inequality $L(f, \mathcal{P}) \leq U(f, \mathcal{P})$ is obvious, as $m_{ij} \leq M_{ij}$ for all i and j.)

▶ **EXAMPLE 1**

Let f be a constant function, viz., $f(\mathbf{x}) = \alpha$ for all $\mathbf{x} \in R$. Then for any partition $\mathcal{P}$ of R we have $L(f, \mathcal{P}) = \alpha\,\text{area}(R) = U(f, \mathcal{P})$, so f is integrable on R and $\int_R f\, dA = \alpha\,\text{area}(R)$. ◀

In higher dimensions, we proceed analogously, but the notation is horrendous. Let $R = [a_1, b_1] \times [a_2, b_2] \times \cdots \times [a_n, b_n] \subset \mathbb{R}^n$ be a *rectangle* in $\mathbb{R}^n$. We obtain a *partition* of R by dividing each of the intervals into subintervals,

$$a_1 = x_{1,0} < x_{1,1} < \cdots < x_{1,k_1} = b_1,$$
$$a_2 = x_{2,0} < x_{2,1} < \cdots < x_{2,k_2} = b_2,$$
$$\vdots$$
$$a_n = x_{n,0} < x_{n,1} < \cdots < x_{n,k_n} = b_n,$$

and in such a way forming a "paving" of R by subrectangles

$$R_{j_1 j_2 \ldots j_n} = [x_{1, j_1-1}, x_{1, j_1}] \times [x_{2, j_2-1}, x_{2, j_2}] \times \cdots \times [x_{n, j_n-1}, x_{n, j_n}]$$
$$\text{for some } 1 \leq j_s \leq k_s, s = 1, \ldots, n.$$

We will usually suppress all the subscripts and just refer to the partition as $\{R_i\}$. We define the *volume* of a rectangle $R = [a_1, b_1] \times [a_2, b_2] \times \cdots \times [a_n, b_n] \subset \mathbb{R}^n$ to be

$$\text{vol}(R) = (b_1 - a_1)(b_2 - a_2) \cdots (b_n - a_n).$$

Then upper sums, lower sums, and the integral are defined as before, substituting volume (of a rectangle in $\mathbb{R}^n$) for area (of a rectangle in $\mathbb{R}^2$). In dimensions $n \geq 3$, we denote the integral by $\int_R f\,dV$.

We need some criteria to detect integrability of functions. Then we will find soon that we can evaluate integrals by reverting to our techniques from one-variable calculus.

Definition Let $\mathcal{P}$ and $\mathcal{P}'$ be partitions of a given rectangle R. We say $\mathcal{P}'$ is a *refinement* of $\mathcal{P}$ if for every rectangle $Q' \in \mathcal{P}'$ there is a rectangle $Q \in \mathcal{P}$ so that $Q' \subset Q$. (See Figure 1.2.)

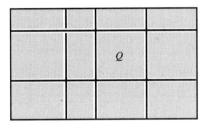
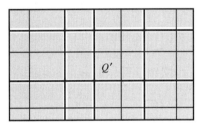

partition $\mathcal{P}$ of the rectangle R refinement $\mathcal{P}'$ of the partition $\mathcal{P}$

Figure 1.2

Lemma 1.1 *Let $\mathcal{P}$ and $\mathcal{P}'$ be partitions of a given rectangle R, and suppose $\mathcal{P}$ is a refinement of $\mathcal{P}'$. Suppose f is a bounded function on R. Then we have*

$$L(f, \mathcal{P}') \leq L(f, \mathcal{P}) \leq U(f, \mathcal{P}) \leq U(f, \mathcal{P}').$$

Proof It suffices to check the following: Let Q be a single rectangle, and let $\mathcal{Q} = \{Q_1, \ldots, Q_r\}$ be a partition of Q. Let $m = \inf_{\mathbf{x} \in Q} f(\mathbf{x})$, $m_i = \inf_{\mathbf{x} \in Q_i} f(\mathbf{x})$, $M = \sup_{\mathbf{x} \in Q} f(\mathbf{x})$, and $M_i = \sup_{\mathbf{x} \in Q_i} f(\mathbf{x})$. Then we claim that

$$m\,\text{area}(Q) \leq \sum_{i=1}^{r} m_i\,\text{area}(Q_i) \leq \sum_{i=1}^{r} M_i\,\text{area}(Q_i) \leq M\,\text{area}(Q).$$

This is immediate from the fact that $m \leq m_i \leq M_i \leq M$ for all $i = 1, \ldots, r$. ∎

Corollary 1.2 *If $\mathcal{P}'$ and $\mathcal{P}''$ are two partitions of R, we have $L(f, \mathcal{P}') \leq U(f, \mathcal{P}'')$.*

Proof Let $\mathcal{P}$ be the partition of R formed by taking the *union* of the respective partitions in each coordinate, as indicated in Figure 1.3. $\mathcal{P}$ is called the *common refinement* of $\mathcal{P}'$ and $\mathcal{P}''$. Then by Lemma 1.1, we have

$$L(f, \mathcal{P}') \leq L(f, \mathcal{P}) \leq U(f, \mathcal{P}) \leq U(f, \mathcal{P}''),$$

as required. ∎

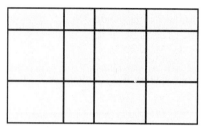
partition $\mathcal{P}'$ of the rectangle R

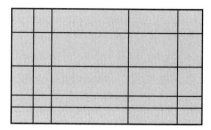
partition $\mathcal{P}''$ of the rectangle R

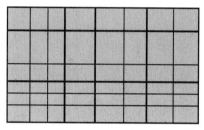
common refinement of partitions $\mathcal{P}'$ and $\mathcal{P}''$

Figure 1.3

Proposition 1.3 (Convenient Criterion) *Given a bounded function on a rectangle R, f is integrable on R if and only if, for any $\varepsilon > 0$, there is a partition $\mathcal{P}$ of R so that $U(f, \mathcal{P}) - L(f, \mathcal{P}) < \varepsilon$.*

Proof $\Longleftarrow$: Suppose there were two different numbers I_1 and I_2 satisfying $L(f, \mathcal{P}) \leq I_j \leq U(f, \mathcal{P})$ for all partitions $\mathcal{P}$. Choosing $\varepsilon = |I_2 - I_1|$ yields a contradiction.

$\Longrightarrow$: Now suppose f is integrable, so that there is a unique number I satisfying $L(f, \mathcal{P}) \leq I \leq U(f, \mathcal{P})$ for all partitions $\mathcal{P}$. Given $\varepsilon > 0$, we can find partitions $\mathcal{P}'$ and $\mathcal{P}''$ so that

$$I - L(f, \mathcal{P}') < \varepsilon/2 \quad \text{and} \quad U(f, \mathcal{P}'') - I < \varepsilon/2.$$

(If we could not get as close as desired to I with upper and lower sums, we would violate uniqueness of I.) Let $\mathcal{P}$ be the common refinement of $\mathcal{P}'$ and $\mathcal{P}''$. Then

$$L(f, \mathcal{P}') \leq L(f, \mathcal{P}) \leq U(f, \mathcal{P}) \leq U(f, \mathcal{P}''),$$

so

$$U(f, \mathcal{P}) - L(f, \mathcal{P}) \leq U(f, \mathcal{P}'') - L(f, \mathcal{P}') < \varepsilon,$$

as required. ∎

We need to be aware of the basic properties of the integral (which we leave to the reader as exercises).

Proposition 1.4 *Suppose f and g are integrable functions on R. Then $f + g$ is integrable on R and we have*
$$\int_R (f+g)\, dV = \int_R f\, dV + \int_R g\, dV.$$

Proof See Exercise 9. ■

Proposition 1.5 *Suppose f is an integrable function on R and α is a scalar. Then αf is integrable on R and we have*
$$\int_R (\alpha f)\, dV = \alpha \int_R f\, dV.$$

Proof See Exercise 9. ■

Proposition 1.6 *Suppose $R = R' \cup R''$ is the union of two subrectangles. Then f is integrable on R if and only if f is integrable on both R' and R'', in which case we have*
$$\int_R f\, dV = \int_{R'} f\, dV + \int_{R''} f\, dV.$$

Proof See Exercise 9. ■

Proposition 1.7 *Let $R \subset \mathbb{R}^n$ be a rectangle. Suppose $f: R \to \mathbb{R}$ is continuous. Then f is integrable.*

Proof Given $\varepsilon > 0$, we must find a partition $\mathcal{P}$ of R so that $U(f, \mathcal{P}) - L(f, \mathcal{P}) < \varepsilon$. Since f is continuous on the compact set R, it follows from Theorem 1.4 of Chapter 5 that f is uniformly continuous. That means that given any $\varepsilon > 0$, there is $\delta > 0$ so that whenever $\|\mathbf{x} - \mathbf{y}\| < \delta$, $\mathbf{x}, \mathbf{y} \in R$, we have $|f(\mathbf{x}) - f(\mathbf{y})| < \dfrac{\varepsilon}{\mathrm{vol}(R)}$. Partition R into subrectangles R_i, $i = 1, \ldots, k$, of diameter less than δ (e.g., whose sidelengths are less than $\delta/\sqrt{n}$). Then on any such subrectangle R_i, we will have $M_i - m_i < \dfrac{\varepsilon}{\mathrm{vol}(R)}$, and so
$$U(f, \mathcal{P}) - L(f, \mathcal{P}) = \sum_{i=1}^{k} (M_i - m_i)\mathrm{vol}(R_i) < \frac{\varepsilon}{\mathrm{vol}(R)} \mathrm{vol}(R) = \varepsilon,$$
as needed. ■

Definition We say $X \subset \mathbb{R}^n$ has (n-dimensional) *volume zero* if for every $\varepsilon > 0$, there are finitely many rectangles $R_1, \ldots, R_s$ so that $X \subset R_1 \cup \cdots \cup R_s$ and $\sum_{i=1}^{s} \mathrm{vol}(R_i) < \varepsilon$.

Proposition 1.8 *Suppose $f: R \to \mathbb{R}$ is a bounded function and the set $X = \{\mathbf{x} \in R : f$ is not continuous at $\mathbf{x}\}$ has volume zero. Then f is integrable on R.*

Proof Let $\varepsilon > 0$ be given. We must find a partition $\mathcal{P}$ of R so that $U(f, \mathcal{P}) - L(f, \mathcal{P}) < \varepsilon$. Since f is bounded, there is a real number M so that $|f| \leq M$. Because X has volume zero, we can find finitely many rectangles $R'_1, \ldots, R'_s$, as shown in Figure 1.4, that cover X and satisfy $\sum_{j=1}^{s} \text{vol}(R'_j) < \varepsilon/4M$. We can also ensure that no point of X is a frontier point of the union of these rectangles (see Exercise 2.2.8). Now create a partition of R in such a way that each of R'_j, $j = 1, \ldots, s$, will be a union of subrectangles of this partition, as shown in Figure 1.5. Consider the closure Y of $R - \bigcup_{j=1}^{s} R'_j$; it, too, is compact, and f is continuous on Y, hence uniformly continuous. Proceeding as in the proof of Proposition 1.7, we can refine the partition to obtain a partition $\mathcal{P} = \{R_1, \ldots, R_t\}$ of R with the property that

$$\sum_{R_i \subset Y} (M_i - m_i) \text{vol}(R_i) < \frac{\varepsilon}{2}.$$

But we already know that

$$\sum_{R_i \subset \bigcup_{j=1}^{s} R'_j} (M_i - m_i) \text{vol}(R_i) < (2M) \frac{\varepsilon}{4M} = \frac{\varepsilon}{2}.$$

Therefore, $U(f, \mathcal{P}) - L(f, \mathcal{P}) < \varepsilon$, as required. ∎

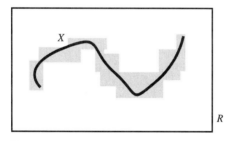

Figure 1.4   **Figure 1.5**

If we want to integrate over a nonrectangular bounded set Ω, we pick a rectangle R with $\Omega \subset R$. Given a bounded function $f : \Omega \to \mathbb{R}$, define

$$\tilde{f} : R \to \mathbb{R}$$

$$\tilde{f}(\mathbf{x}) = \begin{cases} f(\mathbf{x}), & \mathbf{x} \in \Omega \\ 0, & \text{otherwise} \end{cases}.$$

We then define f to be integrable when $\tilde{f}$ is, and set

$$\int_\Omega f \, dV = \int_R \tilde{f} \, dV.$$

(We leave it to the reader to check in Exercise 8 that this is well defined.)

Definition We say a subset $\Omega \subset \mathbb{R}^n$ is a *region* if it is the closure of a bounded open subset of $\mathbb{R}^n$ and its frontier, i.e., the set of its frontier points, has volume 0.

Remark Note, first of all, that any region is compact.

As we ask the reader to check in Exercise 12, if $m < n$, $X \subset \mathbb{R}^m$ is compact, and $\boldsymbol{\phi}: X \to \mathbb{R}^n$ is $\mathcal{C}^1$, then $\boldsymbol{\phi}(X)$ has volume 0 in $\mathbb{R}^n$. So, any time the frontier of Ω is a finite union of such sets, it has volume 0 and Ω is a region.

Corollary 1.9 *If $\Omega \subset \mathbb{R}^n$ is a region and $f: \Omega \to \mathbb{R}$ is continuous, then f is integrable on Ω.*

Proof Recall that to integrate f over Ω we must integrate $\tilde{f}$, as defined above, over some rectangle R containing Ω. The function $\tilde{f}$ is continuous on all of R except for the frontier of Ω, which is a set of volume zero. ∎

Corollary 1.10 *If $\Omega \subset \mathbb{R}^n$ is a region, then $\mathrm{vol}(\Omega) = \int_\Omega 1\, dV$ is well defined.*

Proof The constant function 1 is continuous on Ω. ∎

The following result is often quite useful.

Proposition 1.11 *Suppose f and g are integrable functions on the region Ω and $f \le g$. Then*

$$\int_\Omega f\, dV \le \int_\Omega g\, dV.$$

Proof Let R be a rectangle containing Ω and let $\tilde{f}$ and $\tilde{g}$ be the functions as defined above. Then we have $\tilde{f} \le \tilde{g}$ everywhere on R. Then, applying Propositions 1.4 and 1.5, the function $h = \tilde{g} - \tilde{f}$ is integrable and $\int_R h\, dV = \int_\Omega g\, dV - \int_\Omega f\, dV$. On the other hand, since $h \ge 0$, for any partition $\mathcal{P}$ of R, the lower sum $L(h, \mathcal{P}) \ge 0$, and therefore $\int_R h\, dV \ge 0$. The desired result now follows immediately. ∎

EXERCISES 7.1

*1. Suppose $f\begin{pmatrix} x \\ y \end{pmatrix} = \begin{cases} 0, & 0 \le y \le \frac{1}{2} \\ 1, & \frac{1}{2} < y \le 1 \end{cases}$. Prove that f is integrable on $R = [0, 1] \times [0, 1]$ and find $\int_R f\, dA$.

2. Show directly that the function
$$f\begin{pmatrix}x\\y\end{pmatrix} = \begin{cases} 1, & x = y \\ 0, & \text{otherwise} \end{cases}$$
is integrable on $R = [0, 1] \times [0, 1]$ and find $\int_R f\,dA$. (Hint: Partition R into $1/N$ by $1/N$ squares.)

3. Show directly that the function
$$f\begin{pmatrix}x\\y\end{pmatrix} = \begin{cases} 1, & y < x \\ 0, & \text{otherwise} \end{cases}$$
is integrable on $R = [0, 1] \times [0, 1]$ and find $\int_R f\,dA$.

4. Show directly that the function
$$f\begin{pmatrix}x\\y\end{pmatrix} = \begin{cases} 1, & x = \frac{1}{2}, \frac{1}{3}, \frac{1}{4}, \frac{1}{5}, \ldots \\ 0, & \text{otherwise} \end{cases}$$
is integrable on $R = [0, 1] \times [0, 1]$ and find $\int_R f\,dA$.

♯5. (a) Suppose $f: R \to \mathbb{R}$ is nonnegative, continuous, and positive at some point of R. Prove that $\int_R f\,dV > 0$.

(b) Give an example to show the result of part a is false if we remove the hypothesis of continuity.

6. Let $\Omega \subset \mathbb{R}^n$ be a region and suppose $f: \Omega \to \mathbb{R}$ is continuous.
(a) If m and M are, respectively, the minimum and maximum values of f, prove that
$$m\operatorname{vol}(\Omega) \le \int_\Omega f\,dV \le M\operatorname{vol}(\Omega).$$
(Hint: Use Proposition 1.11.)

(b) (**Mean Value Theorem for Integrals**) Suppose Ω is connected (this means that any pair of points in Ω can be joined by a path in Ω). Prove that there is a point $\mathbf{c} \in \Omega$ so that
$$\int_\Omega f\,dV = f(\mathbf{c})\operatorname{vol}(\Omega).$$
(Hint: Apply the Intermediate Value Theorem.)

♯7. Suppose f is continuous at $\mathbf{a}$ and integrable on a neighborhood of $\mathbf{a}$. Prove that
$$\lim_{\varepsilon \to 0^+} \frac{1}{\operatorname{vol} B(\mathbf{a}, \varepsilon)} \int_{B(\mathbf{a}, \varepsilon)} f\,dV = f(\mathbf{a}).$$

*8. Check that $\int_\Omega f\,dV$ is well defined. That is, if R and R' are two rectangles containing Ω and $\tilde{f}$ and $\tilde{f}'$ are the corresponding functions, check that $\tilde{f}$ is integrable over R if and only if $\tilde{f}'$ is integrable over R' and that $\int_R \tilde{f}\,dV = \int_{R'} \tilde{f}'\,dV$.

9. (a) Prove Proposition 1.4. (Hint: If $\mathcal{P} = \{R_i\}$ is a partition and m_i^f, m_i^g, m_i^{f+g}, M_i^f, M_i^g, M_i^{f+g} denote the obvious, show that
$$m_i^f + m_i^g \le m_i^{f+g} \le M_i^{f+g} \le M_i^f + M_i^g.$$

It will also be helpful to see that $\int_R f\,dV + \int_R g\,dV$ is the unique number between $L(f,\mathcal{P}) + L(g,\mathcal{P})$ and $U(f,\mathcal{P}) + U(g,\mathcal{P})$ for all partitions $\mathcal{P}$.)

(b) Prove Proposition 1.5.

(c) Prove Proposition 1.6.

♯10. Suppose f is integrable on R. Given $\varepsilon > 0$, prove there is $\delta > 0$ so that whenever all the rectangles of a partition $\mathcal{P}$ have diameter less than δ, we have $U(f,\mathcal{P}) - L(f,\mathcal{P}) < \varepsilon$. (Hint: By Proposition 1.3, there is a partition $\mathcal{P}'$ (as indicated by the darker lines in Figure 1.6) so that $U(f,\mathcal{P}') - L(f,\mathcal{P}') < \varepsilon/2$. Show that covering the dividing hyperplanes (of total area A) of the partition by rectangles of diameter $< \delta$ requires at most volume $A\delta/\sqrt{n}$. If $|f| \le M$, then we can pick δ so that that total volume is at most $\varepsilon/4M$. Show that this δ works.)

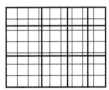

Figure 1.6

♯11. Let $X \subset \mathbb{R}^n$ be a set of volume 0.

(a) Show that for every $\varepsilon > 0$, there are finitely many *cubes* $C_1, \ldots, C_r$ so that $X \subset C_1 \cup \cdots \cup C_r$ and $\sum_{i=1}^{r} \text{vol}(C_i) < \varepsilon$. (Hint: If R is a rectangle with $\text{vol}(R) < \delta$, show that there is a rectangle R' containing R with $\text{vol}(R') < \delta$ and whose sidelengths are *rational* numbers.)

(b) Let $T: \mathbb{R}^n \to \mathbb{R}^n$ be a linear map. Prove that $T(X)$ has volume 0 as well. (Hint: Show that there is a constant k so that for any cube C, the image $T(C)$ is contained in a cube whose volume is at most k times the volume of C.) Query: What goes wrong with this if $T: \mathbb{R}^n \to \mathbb{R}^m$ and $m < n$?

♯12. Let $m < n$, let $X \subset \mathbb{R}^m$ be compact, and let $U \subset \mathbb{R}^m$ be an open set containing X. Suppose $\boldsymbol{\phi}: U \to \mathbb{R}^n$ is $\mathcal{C}^1$. Prove $\boldsymbol{\phi}(X)$ has volume 0 in $\mathbb{R}^n$. (Hints: Take $X \subset C$, where C is a cube. Show that if N is sufficiently large and we divide C into N^m subcubes, then X is covered by such cubes all contained in U,[1] and $\boldsymbol{\phi}(X)$ will be contained in at most N^m cubes in $\mathbb{R}^n$. Argue by continuity of $D\boldsymbol{\phi}$ that there is a constant k (not depending on N) so that each of these will have volume less than $(k/N)^n$.)

13. We've seen in Proposition 1.8 a sufficient condition for f to be integrable. Show that it isn't necessary by considering the famous function $f: [0,1] \to \mathbb{R}$ given by

$$f(x) = \begin{cases} \frac{1}{q}, & x = \frac{p}{q} \text{ in lowest terms} \\ 0, & \text{otherwise} \end{cases}.$$

(Hint: Why is $\mathbb{Q} \cap [0,1]$ not a set of length zero?)

14. A subset $X \subset \mathbb{R}^n$ has *measure zero* if, given any $\varepsilon > 0$, there is a sequence of rectangles $R_1, R_2, R_3, \ldots, R_k, \ldots$, so that

$$X \subset \bigcup_{i=1}^{\infty} R_i \quad \text{and} \quad \sum_{i=1}^{\infty} \text{vol}(R_i) < \varepsilon.$$

(a) Prove that any set of volume 0 has measure 0.

(b) Give an example of a set of measure 0 that does not have volume 0.

---

[1] This follows from Exercise 5.1.13.

(c) Prove that if X is compact and has measure 0, then X has volume 0. (Hint: See Exercise 5.1.12.)

(d) Suppose $X_1, X_2, \ldots$ is a sequence of sets of measure 0. Prove that $\bigcup_{i=1}^{\infty} X_i$ has measure 0.

15. In this (somewhat challenging) exercise, we discover precisely which bounded functions are integrable. Let $f: R \to \mathbb{R}$ be a bounded function.

(a) Let $\mathbf{a} \in R$ and $\delta > 0$. Define
$$M(f, \mathbf{a}, \delta) = \sup_{\mathbf{x} \in B(\mathbf{a}, \delta) \cap R} f(\mathbf{x})$$
$$m(f, \mathbf{a}, \delta) = \inf_{\mathbf{x} \in B(\mathbf{a}, \delta) \cap R} f(\mathbf{x})$$
$$o(f, \mathbf{a}) = \lim_{\delta \to 0^+} M(f, \mathbf{a}, \delta) - m(f, \mathbf{a}, \delta).$$

Prove that $o(f, \mathbf{a})$ makes sense (i.e., the limit exists) and is nonnegative; it is called the *oscillation* of f at $\mathbf{a}$. Prove that f is continuous at $\mathbf{a}$ if and only if $o(f, \mathbf{a}) = 0$.

(b) For any $\varepsilon > 0$, set $D_\varepsilon = \{\mathbf{x} \in R : o(f, \mathbf{x}) \ge \varepsilon\}$, and let $D = \{\mathbf{x} \in R : f \text{ is discontinuous at } \mathbf{a}\}$. Show that $D = D_1 \cup D_{1/2} \cup D_{1/3} \cup \cdots$ and that D_ε is a closed set.

(c) Suppose that f is integrable on R. Prove that for any $k \in \mathbb{N}$, $D_{1/k}$ has volume 0. Deduce that if f is integrable on R, then D has measure 0. (Hint: Use Exercise 14.)

(d) Conversely, prove that if D has measure 0, then f is integrable. (Hints: Choose $\varepsilon > 0$ and apply the convenient criterion. If D has measure 0, then so has D_ε, and so it has volume 0 (why?). Create a partition consisting of rectangles disjoint from D_ε and of rectangles of small total volume that cover D_ε.)

▶ 2 ITERATED INTEGRALS AND FUBINI'S THEOREM

In one-variable integral calculus, we learned that we could compute the volume of a solid region by slicing it by parallel planes and integrating the cross-sectional area. In particular, given a rectangle $R = [a, b] \times [c, d]$, if we are interested in finding the volume over R and under the graph $z = f\begin{pmatrix} x \\ y \end{pmatrix}$, we could slice by planes perpendicular to the x-axis, as shown in Figure 2.1, obtaining

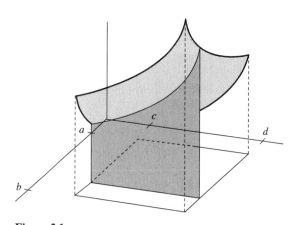

Figure 2.1

$$\text{volume} = \int_a^b (\text{cross-sectional area at } x)\,dx$$

$$= \int_a^b \underbrace{\left(\int_c^d f\left(\frac{x}{y}\right)dy\right)}_{x \text{ fixed}} dx = \int_a^b \int_c^d f\left(\frac{x}{y}\right)dy\,dx.$$

This expression is called an *iterated integral*. Perhaps it would be more suggestive to call it a nested integral. Calculating iterated integrals reverts to one-variable calculus skills (finding antiderivatives and applying the Fundamental Theorem of Calculus) along with a healthy dose of neat bookkeeping.

▶ **EXAMPLE 1**

$$\int_0^1 \int_1^2 (1 + x^2 + xy)\,dy\,dx = \int_0^1 \underbrace{(1 + x^2)y + \tfrac{1}{2}xy^2 \Big]_{y=1}^{2}}_{x \text{ fixed}} dx$$

$$= \int_0^1 \left(1 + x^2 + \tfrac{3}{2}x\right)dx = x + \tfrac{1}{3}x^3 + \tfrac{3}{4}x^2 \Big]_{x=0}^{1}$$

$$= 1 + \frac{1}{3} + \frac{3}{4} = \frac{25}{12}. \quad \blacktriangleleft$$

▶ **EXAMPLE 2**

Let's investigate an obvious question.

a. We wish to evaluate $\int_{-1}^{1} \int_0^2 xye^{x+y^2}\,dx\,dy$.

$$\int_{-1}^{1}\left(\int_0^2 xye^{x+y^2}dx\right)dy = \int_{-1}^{1}\left(ye^{y^2}(x-1)e^x\Big]_{x=0}^{2}\right)dy \quad (\text{recalling that } \int xe^x\,dx = xe^x - e^x)$$

$$= \int_{-1}^{1} ye^{y^2}(e^2+1)dy = \tfrac{1}{2}(e^2+1)e^{y^2}\Big]_{y=-1}^{1} = 0.$$

b. Now let's consider $\int_0^2 \int_{-1}^{1} xye^{x+y^2}\,dy\,dx$.

$$\int_0^2 \int_{-1}^{1} xye^{x+y^2}\,dy\,dx = \int_0^2\left(\int_{-1}^{1}(xe^x)(ye^{y^2})dy\right)dx$$

$$= \int_0^2 \left(\tfrac{1}{2}(xe^x)e^{y^2}\Big]_{y=-1}^{1}\right)dx$$

$$= 0.$$

More to the point, we should observe that for fixed x, the function $(xe^x)(ye^{y^2})$ is an *odd* function of y, and hence the integral as y varies from -1 to 1 must be 0.

We shall prove in a moment that for reasonable functions the iterated integrals in either order are equal, and so it behooves us to think a minute about symmetry (or about the difficulty of finding an antiderivative) and choose the more convenient order of integration. ◀

EXAMPLE 3

Suppose we wish to find the volume of the region lying over the triangle $\Omega \subset \mathbb{R}^2$ with vertices at $\begin{bmatrix} 0 \\ 0 \end{bmatrix}$, $\begin{bmatrix} 1 \\ 0 \end{bmatrix}$, and $\begin{bmatrix} 1 \\ 1 \end{bmatrix}$ and bounded above by $z = f\begin{pmatrix} x \\ y \end{pmatrix} = xy$. Then we wish to find the integral of f over the region Ω. By definition, we consider Ω as a subset of, say, the square $R = [0, 1] \times [0, 1]$ and define $\tilde{f}: R \to \mathbb{R}$ by

$$\tilde{f}\begin{pmatrix} x \\ y \end{pmatrix} = \begin{cases} xy, & \begin{bmatrix} x \\ y \end{bmatrix} \in \Omega \\ 0, & \text{otherwise} \end{cases},$$

whose graph is sketched in Figure 2.2. Note that for x fixed, $\tilde{f}\begin{pmatrix} x \\ y \end{pmatrix} = xy$ when $0 \leq y \leq x$ and is 0 otherwise. So

$$\int_0^1 \tilde{f}\begin{pmatrix} x \\ y \end{pmatrix} dy = \int_0^x xy\,dy + \int_x^1 0\,dy = \int_0^x xy\,dy.$$

Thus, we have

$$\int_0^1 \int_0^1 \tilde{f}\begin{pmatrix} x \\ y \end{pmatrix} dy\,dx = \int_0^1 \left(\int_0^x xy\,dy \right) dx$$
$$= \int_0^1 \left(\tfrac{1}{2}xy^2 \right]_{y=0}^x \right) dx$$
$$= \int_0^1 \tfrac{1}{2}x^3\,dx = \frac{1}{8}.$$

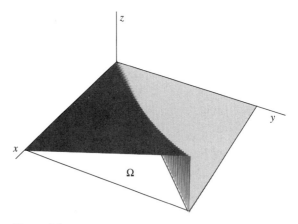

Figure 2.2

EXAMPLE 4

Suppose we slice into a cylindrical tree trunk, $x^2 + y^2 \leq a^2$, and remove the wedge bounded below by $z = 0$ and above by $z = y$, as depicted in Figure 2.3. What is the volume of the chunk we remove?

2 Iterated Integrals and Fubini's Theorem ◀ 279

We see that the plane $z = y$ lies above the plane $z = 0$ when $y \geq 0$, so we let $\Omega = \left\{ \begin{bmatrix} x \\ y \end{bmatrix} : x^2 + y^2 \leq a^2, \ y \geq 0 \right\}$, as indicated in Figure 2.4, and to obtain the volume we calculate:

$$\int_\Omega y \, dA = \int_{-a}^{a} \int_0^{\sqrt{a^2-x^2}} y \, dy \, dx = \int_{-a}^{a} \left(\tfrac{1}{2} y^2 \right]_{y=0}^{\sqrt{a^2-x^2}} dx$$

$$= \frac{1}{2} \int_{-a}^{a} (a^2 - x^2) \, dx = \int_0^a (a^2 - x^2) \, dx = \frac{2}{3} a^3. \quad \blacktriangleleft$$

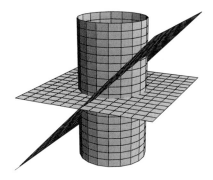

Figure 2.3 **Figure 2.4**

The fact that we can compute volume by using either a multiple integral or an iterated integral suggests that, at least for "reasonable" functions, we should in general be able to calculate multiple integrals by computing iterated integrals. The crucial theorem that allows us to calculate multiple integrals with relative ease is the following

Theorem 2.1 (Fubini's Theorem, 2-Dimensional Case) *Suppose f is integrable on a rectangle $R = [a, b] \times [c, d] \subset \mathbb{R}^2$. Suppose that for each $x \in [a, b]$, the function $f\begin{pmatrix} x \\ y \end{pmatrix}$ is integrable on $[c, d]$; i.e., $F(x) = \int_c^d f\begin{pmatrix} x \\ y \end{pmatrix} dy$ exists. Suppose next that the function F is integrable on $[a, b]$; i.e.,*

$$\int_a^b F(x) \, dx = \int_a^b \left(\int_c^d f\begin{pmatrix} x \\ y \end{pmatrix} dy \right) dx$$

exists. Then we have

$$\int_R f \, dA = \int_a^b \left(\int_c^d f\begin{pmatrix} x \\ y \end{pmatrix} dy \right) dx.$$

Proof Let $\mathcal{P}$ be an arbitrary partition of R into rectangles $R_{ij} = [x_{i-1}, x_i] \times [y_{j-1}, y_j]$, $i = 1, \ldots, k$, $j = 1, \ldots, \ell$. When $\begin{bmatrix} x \\ y \end{bmatrix} \in R_{ij}$, we have

$$m_{ij} \leq f\begin{pmatrix} x \\ y \end{pmatrix} \leq M_{ij}, \quad \text{and so}$$

$$m_{ij}(y_j - y_{j-1}) \leq \int_{y_{j-1}}^{y_j} f\begin{pmatrix} x \\ y \end{pmatrix} dy \leq M_{ij}(y_j - y_{j-1}).$$

So now when $x \in [x_{i-1}, x_i]$, we have

$$\sum_{j=1}^{\ell} m_{ij}(y_j - y_{j-1}) \leq \int_c^d f\begin{pmatrix} x \\ y \end{pmatrix} dy \leq \sum_{j=1}^{\ell} M_{ij}(y_j - y_{j-1}), \quad \text{whence}$$

$$\left(\sum_{j=1}^{\ell} m_{ij}(y_j - y_{j-1})\right)(x_i - x_{i-1}) \leq \int_{x_{i-1}}^{x_i} \left(\int_c^d f\begin{pmatrix} x \\ y \end{pmatrix} dy\right) dx$$

$$\leq \left(\sum_{j=1}^{\ell} M_{ij}(y_j - y_{j-1})\right)(x_i - x_{i-1}).$$

Finally, summing over i, we have

$$\sum_{i=1}^{k}\left(\sum_{j=1}^{\ell} m_{ij}(y_j - y_{j-1})\right)(x_i - x_{i-1}) \leq \int_a^b \left(\int_c^d f\begin{pmatrix} x \\ y \end{pmatrix} dy\right) dx$$

$$\leq \sum_{i=1}^{k}\left(\sum_{j=1}^{\ell} M_{ij}(y_j - y_{j-1})\right)(x_i - x_{i-1}).$$

But this can be rewritten as

$$\sum_{i=1}^{k}\sum_{j=1}^{\ell} m_{ij}\,\text{area}(R_{ij}) \leq \int_a^b \left(\int_c^d f\begin{pmatrix} x \\ y \end{pmatrix} dy\right) dx \leq \sum_{i=1}^{k}\sum_{j=1}^{\ell} M_{ij}\,\text{area}(R_{ij});$$

i.e., $\quad L(f, \mathcal{P}) \leq \int_a^b \left(\int_c^d f\begin{pmatrix} x \\ y \end{pmatrix} dy\right) dx \leq U(f, \mathcal{P}).$

Since f is integrable on R, if a number I satisfies

$$L(f, \mathcal{P}) \leq I \leq U(f, \mathcal{P}) \quad \text{for all partitions } \mathcal{P} \text{ of } [a, b],$$

then $I = \int_R f\,dA$. This completes the proof. ∎

Corollary 2.2 *Suppose f is integrable on the rectangle $R = [a, b] \times [c, d]$ and the iterated integrals*

$$\int_a^b \int_c^d f\begin{pmatrix} x \\ y \end{pmatrix} dy\,dx \quad \text{and} \quad \int_c^d \int_a^b f\begin{pmatrix} x \\ y \end{pmatrix} dx\,dy$$

both exist. (That is, for each x, the integral $\int_c^d f\begin{pmatrix}x\\y\end{pmatrix} dy$ exists and defines a function of x that is integrable on $[a, b]$. And, likewise, for each y, the integral $\int_a^b f\begin{pmatrix}x\\y\end{pmatrix} dx$ exists and defines a function of y that is integrable on $[c, d]$.) Then

$$\int_a^b \int_c^d f\begin{pmatrix}x\\y\end{pmatrix} dy\,dx = \int_R f\,dA = \int_c^d \int_a^b f\begin{pmatrix}x\\y\end{pmatrix} dx\,dy.$$

In general, in n dimensions, we have

Theorem 2.3 (Fubini's Theorem, General Case) *Let $R \subset \mathbb{R}^n$ be a rectangle, say,*

$$R = [a_1, b_1] \times \cdots \times [a_n, b_n].$$

Suppose $f : R \to \mathbb{R}$ is integrable and that, moreover, the integrals

$$\int_{a_n}^{b_n} f(\mathbf{x})dx_n, \quad \int_{a_{n-1}}^{b_{n-1}} \left(\int_{a_n}^{b_n} f(\mathbf{x})dx_n \right) dx_{n-1}, \quad \ldots,$$

$$\int_{a_1}^{b_1} \cdots \left(\int_{a_{n-1}}^{b_{n-1}} \left(\int_{a_n}^{b_n} f(\mathbf{x})dx_n \right) dx_{n-1} \right) \cdots dx_1$$

all exist. Then the multiple integral and the iterated integral are equal:

$$\int_R f(\mathbf{x})dV = \int_{a_1}^{b_1} \cdots \int_{a_n}^{b_n} f(\mathbf{x})dx_n \cdots dx_1.$$

(The same is true for the iterated integral in any order, provided all the intermediate integrals exist.) In particular, whenever f is continuous on R, then the multiple integral equals any of the $n!$ possible iterated integrals.

▶ **EXAMPLE 5**

It is easy to find a function f on the rectangle $R = [0, 1] \times [0, 1]$ that is integrable but whose iterated integral doesn't exist. Take

$$f\begin{pmatrix}x\\y\end{pmatrix} = \begin{cases} 1, & x = 0,\ y \in \mathbb{Q} \\ 0, & \text{otherwise} \end{cases}.$$

The integral $\int_0^1 f\begin{pmatrix}0\\y\end{pmatrix} dy$ does not exist, but it is easy to see that f is integrable and $\int_R f\,dA = 0$. ◀

▶ **EXAMPLE 6**

It is somewhat harder to find a function whose iterated integral exists but that is not integrable. Let

$$f\begin{pmatrix}x\\y\end{pmatrix} = \begin{cases} 1, & y \in \mathbb{Q} \\ 2x, & y \notin \mathbb{Q} \end{cases}.$$

Then $\int_0^1 f\begin{pmatrix}x\\y\end{pmatrix} dx = 1$ for every $y \in [0, 1]$, so the iterated integral $\int_0^1 \int_0^1 f\begin{pmatrix}x\\y\end{pmatrix} dx\,dy$ exists and equals 1. Whether f is integrable on $R = [0, 1] \times [0, 1]$ is more subtle. Probably the easiest way to see that it is not is this: If it were, by Proposition 1.6, then it would also be integrable on $R' = [0, \frac{1}{2}] \times [0, 1]$. For any partition $\mathcal{P}$ of R', we have $U(f, \mathcal{P}) = \frac{1}{2}$, whereas we can make $L(f, \mathcal{P})$ as close to $\int_0^1 \int_0^{1/2} 2x\,dx\,dy = \frac{1}{4}$ as we wish.

We ask the reader to decide in Exercise 4 whether the other iterated integral, $\int_0^1 \int_0^1 f\begin{pmatrix}x\\y\end{pmatrix} dy\,dx$, exists. ◀

▶ **EXAMPLE 7**

More subtle yet is a nonintegrable function on $R = [0, 1] \times [0, 1]$ both of whose iterated integrals exist. Define

$$f\begin{pmatrix}x\\y\end{pmatrix} = \begin{cases} 1, & x = \frac{m}{q} \text{ and } y = \frac{n}{q} \text{ for some } m, n, q \in \mathbb{N} \text{ with } q \text{ prime} \\ 0, & \text{otherwise} \end{cases}.$$

First of all, f is not integrable on R since $L(f, \mathcal{P}) = 0$ and $U(f, \mathcal{P}) = 1$ for every partition $\mathcal{P}$ of R (see Exercise 5). Next, we claim that for any x, $\int_0^1 f\begin{pmatrix}x\\y\end{pmatrix} dy$ exists and equals 0. When $x \notin \mathbb{Q}$, this is obvious. When $x = \frac{m}{q}$, only for *finitely* many $y \in [0, 1]$ is $f\begin{pmatrix}x\\y\end{pmatrix}$ not equal to 0, and so the integral exists. Obviously, then, the iterated integral $\int_0^1 \int_0^1 f\begin{pmatrix}x\\y\end{pmatrix} dy\,dx$ exists. The same argument applies when we reverse the order. ◀

▶ **EXAMPLE 8**

(Changing the Order of Integration) You are asked to evaluate the iterated integral

$$\int_0^1 \int_y^1 \frac{\sin x}{x} dx\,dy.$$

It is a classical fact that $\int \frac{\sin x}{x} dx$ cannot be evaluated in elementary terms, and so (other than resorting to numerical integration) we are stymied. To be careful, we define

$$f\begin{pmatrix}x\\y\end{pmatrix} = \begin{cases} \frac{\sin x}{x}, & x \neq 0 \\ 1, & x = 0 \end{cases}.$$

Then f is continuous and we recognize (applying Theorem 2.1) that the iterated integral is equal to the double integral $\int_\Omega f\,dA$, where

$$\Omega = \left\{ \begin{bmatrix} x \\ y \end{bmatrix} : 0 \leq y \leq 1, \ y \leq x \leq 1 \right\},$$

which is the triangle pictured in Figure 2.5. Once we have a picture of Ω, we see that we can equally well represent it in the form

$$\Omega = \left\{ \begin{bmatrix} x \\ y \end{bmatrix} : 0 \le x \le 1,\ 0 \le y \le x \right\},$$

and so, writing the iterated integral in the other order,

$$\int_\Omega f\,dA = \int_0^1 \left(\underbrace{\int_0^x \frac{\sin x}{x} dy}_{x\text{ fixed}} \right) dx$$

$$= \int_0^1 \left(\left(\frac{\sin x}{x} \right) y \Big]_{y=0}^{x} \right) dx$$

$$= \int_0^1 \left(\frac{\sin x}{x} \cdot x \right) dx = \int_0^1 \sin x\,dx = 1 - \cos 1.$$

The moral of this story is that, when confronted by an iterated integral that cannot be evaluated in elementary terms, it doesn't hurt to change the order of integration and see what happens. ◀

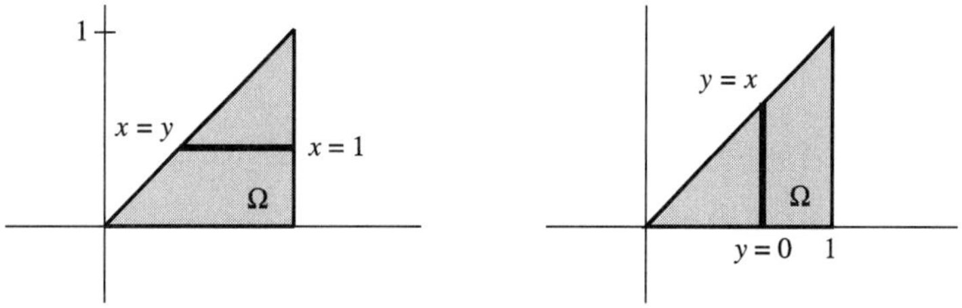

Figure 2.5

▶ EXAMPLE 9

Let $\Omega \subset \mathbb{R}^3$ be the region in the first octant bounded below by the paraboloid $z = x^2 + y^2$ and above by the plane $z = 4$, shown in Figure 2.6. Evaluate $\int_\Omega x\,dV$. It is most natural to integrate first with respect to z; notice that the projection of Ω onto the xy-plane is the quarter of the disk of radius 2 centered at the origin lying in the first quadrant. For each point $\begin{bmatrix} x \\ y \end{bmatrix}$ in that quarter-disk, z varies from $x^2 + y^2$ to 4. Thus, we have

$$\int_\Omega x\,dV = \int_0^2 \int_0^{\sqrt{4-x^2}} \int_{x^2+y^2}^4 x\,dz\,dy\,dx$$

$$= \int_0^2 \int_0^{\sqrt{4-x^2}} x\left(4 - (x^2 + y^2)\right) dy\,dx$$

$$= \int_0^2 x\left((4-x^2)^{3/2} - \tfrac{1}{3}(4-x^2)^{3/2} \right) dx$$

$$= -\tfrac{2}{15}(4-x^2)^{5/2} \Big]_0^2 = \frac{64}{15}.$$

We will revisit this example in Section 3. ◀

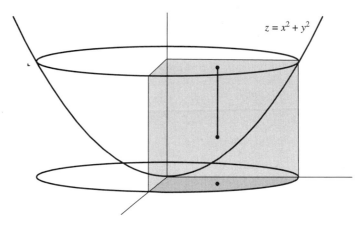

Figure 2.6

▶ **EXAMPLE 10**

Let $\Omega = \{\mathbf{x} \in \mathbb{R}^n : 0 \leq x_n \leq x_{n-1} \leq \cdots \leq x_2 \leq x_1 \leq 1\}$. This region is pictured in the case $n = 3$ in Figure 2.7. Then

$$\text{vol}(\Omega) = \int_0^1 \int_0^{x_1} \cdots \int_0^{x_{n-1}} dx_n \cdots dx_2 dx_1$$

$$= \int_0^1 \int_0^{x_1} \cdots \int_0^{x_{n-2}} x_{n-1} dx_{n-1} \cdots dx_2 dx_1$$

$$= \int_0^1 \int_0^{x_1} \cdots \int_0^{x_{n-3}} \tfrac{1}{2} x_{n-2}^2 dx_{n-2} \cdots dx_2 dx_1$$

$$= \cdots = \int_0^1 \frac{1}{(n-1)!} x_1^{n-1} dx_1 = \frac{1}{n!} \quad \blacktriangleleft$$

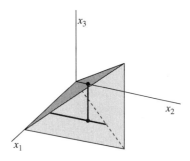

Figure 2.7

▶ **EXERCISES 7.2**

1. Evaluate the integrals $\int_R f\,dV$ for the given function f and rectangle R.

   *(a) $f\begin{pmatrix}x\\y\end{pmatrix} = e^x \cos y$, $R = [0, 1] \times [0, \tfrac{\pi}{2}]$

(b) $f\begin{pmatrix}x\\y\end{pmatrix} = \frac{y}{x}$, $R = [1, 3] \times [2, 4]$

*(c) $f\begin{pmatrix}x\\y\end{pmatrix} = \frac{x}{x^2 + y}$, $R = [0, 1] \times [1, 3]$

(d) $f\begin{pmatrix}x\\y\\z\end{pmatrix} = (x + y)z$, $R = [-1, 1] \times [1, 2] \times [2, 3]$

2. Interpret each of the following iterated integrals as a double integral $\int_\Omega f\,dA$ for the appropriate region Ω, sketch Ω, and change the order of integration. (You may assume f is continuous.)

(a) $\int_0^1 \int_x^1 f\begin{pmatrix}x\\y\end{pmatrix} dy\,dx$

(d) $\int_{-1}^1 \int_0^{\sqrt{1-y^2}} f\begin{pmatrix}x\\y\end{pmatrix} dx\,dy$

*(b) $\int_0^1 \int_0^{2x} f\begin{pmatrix}x\\y\end{pmatrix} dy\,dx$

(e) $\int_0^1 \int_{x^2}^x f\begin{pmatrix}x\\y\end{pmatrix} dy\,dx$

(c) $\int_1^2 \int_{y^2}^4 f\begin{pmatrix}x\\y\end{pmatrix} dx\,dy$

*(f) $\int_{-1}^2 \int_{x^2}^{x+2} f\begin{pmatrix}x\\y\end{pmatrix} dy\,dx$

3. Evaluate each of the following iterated integrals. In addition, interpret each as a double integral $\int_\Omega f\,dA$, sketch the region Ω, change the order of integration, and evaluate the alternative iterated integral.

(a) $\int_0^1 \int_0^x (x+y)\,dy\,dx$

(b) $\int_0^1 \int_{-\sqrt{1-y^2}}^{\sqrt{1-y^2}} y\,dx\,dy$

*(c) $\int_0^1 \int_{x^2}^x \frac{x}{1+y^2}\,dy\,dx$

4. Given the function f in Example 6, does the iterated integral $\int_0^1 \int_0^1 f\begin{pmatrix}x\\y\end{pmatrix} dy\,dx$ exist?

5. Check that for the function f defined in Example 7, for every partition $\mathcal{P}$ of R, $U(f, \mathcal{P}) = 1$ and $L(f, \mathcal{P}) = 0$. (Hint: Show that for every $\delta > 0$, if $1/q < \delta$, then every interval of length δ in $[0, 1]$ contains a point of the form k/q.)

6. Let

$$f\begin{pmatrix}x\\y\end{pmatrix} = \begin{cases} \frac{1}{q}, & x = \frac{p}{q} \text{ in lowest terms, } y \in \mathbb{Q} \\ 0, & \text{otherwise} \end{cases}.$$

Decide whether f is integrable on $R = [0, 1] \times [0, 1]$ and whether the iterated integrals

$$\int_0^1 \int_0^1 f\begin{pmatrix}x\\y\end{pmatrix} dx\,dy \quad \text{and} \quad \int_0^1 \int_0^1 f\begin{pmatrix}x\\y\end{pmatrix} dy\,dx$$

exist.

7. Is there an integrable function on a rectangle *neither* of whose iterated integrals exists?

8. Evaluate the following iterated integrals:

*(a) $\int_0^4 \int_{\sqrt{y}}^2 \frac{1}{1+x^3}\,dx\,dy$

(b) $\int_0^1 \int_{\sqrt[3]{x}}^1 e^{y^4}\,dy\,dx$

(c) $\int_0^1 \int_{\sqrt{y}}^1 e^{y/x} dx dy$ (Be careful: Why does the double integral even exist?)

9. Find the volume of the region in the first octant of $\mathbb{R}^3$ bounded below by the xy-plane, on the sides by $x = 0$ and $y = 2x$, and above by $y^2 + z^2 = 16$.

10. Find the volume of the region in the $\mathbb{R}^3$ bounded below by the xy-plane, above by $z = y$, and on the sides by $y = 4 - x^2$.

*11. Find the volume of the region in $\mathbb{R}^3$ bounded by the cylinders $x^2 + y^2 = 1$ and $x^2 + z^2 = 1$.

12. Interpret each of the following iterated integrals as a triple integral $\int_\Omega f dV$ for the appropriate region Ω, sketch Ω, and change the order of integration so that the innermost integral is taken with respect to y. (You may assume f is continuous.)

*(a) $\int_0^1 \int_0^{1-x} \int_0^{1-x-y} f\begin{pmatrix} x \\ y \\ z \end{pmatrix} dz dy dx$

*(d) $\int_0^1 \int_0^{1-x^2} \int_0^{x+y} f\begin{pmatrix} x \\ y \\ z \end{pmatrix} dz dy dx$

(b) $\int_0^1 \int_0^{1-x^2} \int_0^y f\begin{pmatrix} x \\ y \\ z \end{pmatrix} dz dy dx$

(e) $\int_0^1 \int_0^{1-x} \int_0^{x+y} f\begin{pmatrix} x \\ y \\ z \end{pmatrix} dz dy dx$

(c) $\int_{-1}^1 \int_{-\sqrt{1-x^2}}^{\sqrt{1-x^2}} \int_{\sqrt{x^2+y^2}}^1 f\begin{pmatrix} x \\ y \\ z \end{pmatrix} dz dy dx$

*13. Suppose a, b, and c are positive. Find the volume of the tetrahedron bounded by the coordinate planes and the plane $x/a + y/b + z/c = 1$.

14. Find the volume of the region in $\mathbb{R}^3$ bounded by $z = 1 - x^2$, $z = x^2 - 1$, $y + z = 1$, and $y = 0$.

*15. Let $\Omega \subset \mathbb{R}^3$ be the portion of the cube $0 \le x, y, z \le 1$ lying above the plane $y + z = 1$ and below the plane $x + y + z = 2$. Evaluate $\int_\Omega x dV$.

16. Let

$$f\begin{pmatrix} x \\ y \end{pmatrix} = \frac{x - y}{(x + y)^3}.$$

Calculate the iterated integrals $\int_0^1 \int_0^1 f\begin{pmatrix} x \\ y \end{pmatrix} dx dy$ and $\int_0^1 \int_0^1 f\begin{pmatrix} x \\ y \end{pmatrix} dy dx$. Explain your results.

17. Let $R = [0, 1] \times [0, 1]$. Define $f: R \to \mathbb{R}$ by

$$f\begin{pmatrix} x \\ y \end{pmatrix} = \begin{cases} \frac{k\ell(k+1)(\ell+1)}{2^{\ell-k}}, & \frac{1}{k+1} < x \le \frac{1}{k}, \frac{1}{\ell+1} < y \le \frac{1}{\ell}, k < \ell \\ -k^2(k+1)^2, & \frac{1}{k+1} < x, y \le \frac{1}{k} \\ 0, & \text{otherwise} \end{cases}$$

Decide if both iterated integrals exist and if they are equal. Is f integrable on R? (Hint: To see where this function came from, calculate $\int_{[\frac{1}{k+1}, \frac{1}{k}] \times [\frac{1}{\ell+1}, \frac{1}{\ell}]} f dA$.)

18. **(Exploiting Symmetry)** Let $R \subset \mathbb{R}^n$. Suppose $f: R \to \mathbb{R}$ is integrable.
(a) Suppose R is a rectangle that is symmetric about the hyperplane $x_1 = 0$; i.e.,

$$\begin{bmatrix} x_1 \\ x_2 \\ \vdots \\ x_n \end{bmatrix} \in R \iff \begin{bmatrix} -x_1 \\ x_2 \\ \vdots \\ x_n \end{bmatrix} \in R,$$

and $f\begin{pmatrix} -x_1 \\ x_2 \\ \vdots \\ x_n \end{pmatrix} = -f\begin{pmatrix} x_1 \\ x_2 \\ \vdots \\ x_n \end{pmatrix}$. Prove that $\int_R f\,dV = 0$.

(b) Suppose R is a rectangle that is symmetric about the origin; i.e., $\mathbf{x} \in R \iff -\mathbf{x} \in R$, and suppose f is an odd function, so that $f(-\mathbf{x}) = -f(\mathbf{x})$. Prove that $\int_R f\,dV = 0$.

(c) Generalize the results of parts a and b to allow regions other than rectangles.

19. Assume f is $\mathcal{C}^2$. Prove Theorem 6.1 of Chapter 3 by applying Fubini's Theorem. (Hint: Proceed by contradiction: If the mixed partials are not equal at some point, apply Exercise 2.3.5 to show we can find a rectangle on which, say, $\dfrac{\partial^2 f}{\partial x \partial y} > \dfrac{\partial^2 f}{\partial y \partial x}$. Exercise 7.1.5 may also be useful.)

♯20. **(Differentiating Under the Integral Sign)** Suppose $f: [a,b] \times [c,d] \to \mathbb{R}$ is continuous and $\dfrac{\partial f}{\partial x}$ is continuous. Define $F(x) = \int_c^d f\begin{pmatrix} x \\ y \end{pmatrix} dy$.

(a) Prove that F is continuous. (Hint: You will need to use *uniform* continuity of f.)

(b) Prove that F is differentiable and that $F'(x) = \int_c^d \dfrac{\partial f}{\partial x}\begin{pmatrix} x \\ y \end{pmatrix} dy$. (Hint: Let $\phi(t) = \int_c^d \dfrac{\partial f}{\partial x}\begin{pmatrix} t \\ y \end{pmatrix} dy$, and let $\Phi(x) = \int_a^x \phi(t)\,dt$. Show that ϕ is continuous and that $F(x) = \Phi(x) + \text{const.}$)

21. Let $F(x) = \int_0^1 \dfrac{y^x - 1}{\log y}\,dy$. Use Exercise 20 to calculate $F'(x)$ and prove that $\int_0^1 \dfrac{y-1}{\log y}\,dy = F(1) = \log 2$.

22. Let $f(x) = \left(\int_0^x e^{-t^2}\,dt\right)^2$ and $g(x) = \int_0^1 \dfrac{e^{-x^2(t^2+1)}}{t^2+1}\,dt$.

(a) Using Exercise 20 as necessary, prove that $f'(x) + g'(x) = 0$ for all x.

(b) Prove that $f(x) + g(x) = \pi/4$ for all x. Deduce that $\int_0^\infty e^{-t^2}\,dt = \lim_{N \to \infty} \int_0^N e^{-t^2}\,dt = \sqrt{\pi}/2$.

23. Suppose $f: [a,b] \times [c,d] \to \mathbb{R}$ is continuous and $\dfrac{\partial f}{\partial x}$ is continuous. Suppose $g: [a,b] \to (c,d)$ is differentiable. Let $h(x) = \int_c^{g(x)} f\begin{pmatrix} x \\ y \end{pmatrix} dy$. Use the chain rule and Exercise 20 to show that

$$h'(x) = \int_c^{g(x)} \frac{\partial f}{\partial x}\begin{pmatrix} x \\ y \end{pmatrix} dy + f\begin{pmatrix} x \\ g(x) \end{pmatrix} g'(x).$$

(Hint: Consider $F\begin{pmatrix} x \\ z \end{pmatrix} = \int_c^z f\begin{pmatrix} x \\ y \end{pmatrix} dy$.)

24. Evaluate $\int_0^x \frac{dy}{x^2 + y^2}$. Use Exercise 23 to evaluate

*(a) $\int_0^x \frac{dy}{(x^2 + y^2)^2}$

(b) $\int_0^x \frac{dy}{(x^2 + y^2)^3}$

25. Suppose f is continuous. Let $h(x) = \int_0^x \sin(x - y) f(y) dy$. Show that $h(0) = h'(0) = 0$ and $h''(x) + h(x) = f(x)$.

26. Suppose f is continuous. Prove that
$$\int_0^x \int_0^{x_1} \int_0^{x_2} \cdots \int_0^{x_{n-1}} f(x_n) dx_n \cdots dx_3 dx_2 dx_1 = \frac{1}{(n-1)!} \int_0^x (x-t)^{n-1} f(t) dt.$$
(Hint: Start by doing the cases $n = 2$ and $n = 3$.)

▶ 3 POLAR, CYLINDRICAL, AND SPHERICAL COORDINATES

In this section we introduce three extremely useful alternative coordinate systems in two and three dimensions. We treat the question of changes of variables in multiple integrals intuitively here, leaving the official proofs for Section 6.

Suppose one wished to calculate $\int_S f\begin{pmatrix}x\\y\end{pmatrix} dA$, where S is the annular region between two concentric circles, as shown in Figure 3.1. As we quickly realize if we try to write down iterated integrals in xy-coordinates, although it is not impossible to evaluate them, it is far from a pleasant task. It would be much more sensible to work in a coordinate system that is built around the radial symmetry. This is the place of *polar coordinates*.

Polar coordinates on the xy-plane are defined as follows: As shown in Figure 3.2, let $r = \sqrt{x^2 + y^2}$ denote the distance of the point $\begin{bmatrix}x\\y\end{bmatrix}$ from the origin, and let θ denote the angle from the positive x-axis to the vector from the origin to the point. Ordinarily, we adopt the convention that

$$r \geq 0 \quad \text{and} \quad 0 \leq \theta < 2\pi \quad \text{or} \quad -\pi \leq \theta < \pi.$$

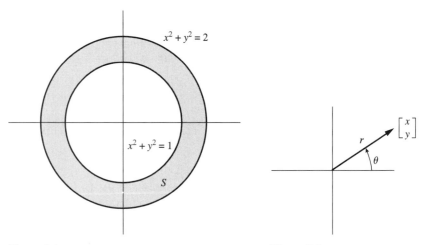

Figure 3.1   Figure 3.2

3 Polar, Cylindrical, and Spherical Coordinates

It is better to express x and y in terms of r and θ, and we do this by means of the mapping

$$\mathbf{g}: [0, \infty) \times [0, 2\pi) \to \mathbb{R}^2$$

$$\mathbf{g}\begin{pmatrix} r \\ \theta \end{pmatrix} = \begin{bmatrix} r\cos\theta \\ r\sin\theta \end{bmatrix} = \begin{bmatrix} x \\ y \end{bmatrix}.$$

To evaluate a double integral $\int_S f\begin{pmatrix} x \\ y \end{pmatrix} dA$ in polar coordinates, we first determine the region Ω in the $r\theta$-plane that maps to S. We substitute $x = r\cos\theta$ and $y = r\sin\theta$, and then realize that a little rectangle Δr by $\Delta\theta$ in the $r\theta$-plane maps to an "annular chunk" whose area is approximately $\Delta r(r\Delta\theta)$ in the xy-plane (see Figure 3.3). That is, partitioning the region Ω into little rectangles corresponds to "partitioning" S into such annular pieces. Summing over all the subrectangles of a partition suggests a formula like

$$\int_S f\begin{pmatrix} x \\ y \end{pmatrix} dA = \int_\Omega f\begin{pmatrix} r\cos\theta \\ r\sin\theta \end{pmatrix} r\,dr\,d\theta.$$

A rigorous justification will come in Section 6.

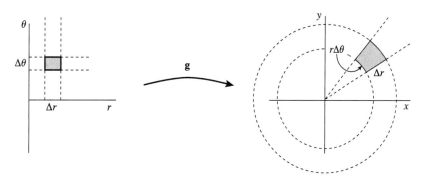

Figure 3.3

▶ EXAMPLE 1

Let S be the annular region $\left\{ \begin{bmatrix} x \\ y \end{bmatrix} : 1 \leq x^2 + y^2 \leq 2 \right\}$ pictured in Figure 3.1. We wish to evaluate $\int_S \sqrt{x^2 + y^2}\,dA$:

$$\int_S \sqrt{x^2 + y^2}\,dA = \int_0^{2\pi} \int_1^{\sqrt{2}} \underbrace{r}_{\sqrt{x^2+y^2}} \underbrace{r\,dr\,d\theta}_{dA}$$

$$= \int_0^{2\pi} \int_1^{\sqrt{2}} r^2\,dr\,d\theta$$

$$= \frac{2\pi(2\sqrt{2} - 1)}{3}.$$

If you are not yet convinced, try doing this in Cartesian coordinates! ◀

EXAMPLE 2

Let $S \subset \mathbb{R}^2$ be the region inside the circle $x^2 + y^2 = 9$, below the line $y = x$, above the x-axis, and lying to the right of $x = 1$, as shown in Figure 3.4. Evaluate $\int_S xy\, dA$. We begin by finding the region Ω in the $r\theta$-plane that maps to S, as shown in Figure 3.5. Clearly θ goes from 0 to $\pi/4$, and for each fixed θ, we see that r starts at $r = \sec\theta$ (as we enter S at the line $x = 1$) and increases to $r = 3$ (as we exit S at the circle). (We think naturally of determining r as a function of θ, so naturally we would place θ on the horizontal axis and r on the vertical; for reasons we'll see in Chapter 8, this is not a good idea.)

Therefore, we have

$$\int_S xy\, dA = \int_0^{\pi/4} \int_{\sec\theta}^3 \underbrace{(r\cos\theta)}_{x}\underbrace{(r\sin\theta)}_{y}\underbrace{r\,dr\,d\theta}_{dA}$$

$$= \int_0^{\pi/4} \int_{\sec\theta}^3 r^3 \cos\theta \sin\theta\, dr\, d\theta$$

$$= \frac{1}{4}\int_0^{\pi/4} (81 - \sec^4\theta)\cos\theta \sin\theta\, d\theta$$

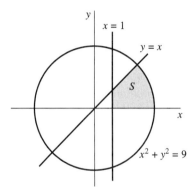

Figure 3.4

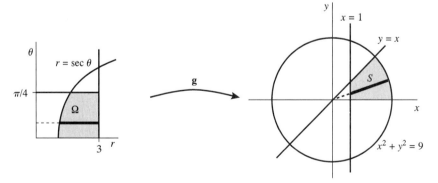

Figure 3.5

$$= \frac{1}{4} \int_0^{\pi/4} \left(81 \cos\theta \sin\theta - \frac{\sin\theta}{\cos^3\theta}\right) d\theta$$

$$= \frac{1}{8} \left(81 \sin^2\theta - \frac{1}{\cos^2\theta}\right)\bigg]_0^{\pi/4} = \frac{79}{16}. \blacktriangleleft$$

▶ **EXAMPLE 3**

We wish to evaluate the improper integral $\int_0^\infty e^{-x^2} dx$. This "Gaussian integral" is ubiquitous in probability, statistics, and statistical mechanics. Although one way of doing so was given in Exercise 7.2.22, the approach we take here is more amenable to generalization.

Taking advantage of the property $e^{a+b} = e^a e^b$, we exploit radial symmetry by calculating instead the double integral

$$\left(\int_0^\infty e^{-x^2} dx\right)^2 = \int_0^\infty \int_0^\infty e^{-x^2} e^{-y^2} dy\, dx = \int_{[0,\infty)\times[0,\infty)} e^{-(x^2+y^2)} dA.$$

Converting to polar coordinates, we have

$$\int_{[0,\infty)\times[0,\infty)} e^{-(x^2+y^2)} dA = \int_0^{\pi/2} \int_0^\infty e^{-r^2} r\, dr\, d\theta$$

$$= \lim_{R \to \infty} \int_0^{\pi/2} \int_0^R e^{-r^2} r\, dr\, d\theta$$

$$= \lim_{R \to \infty} \frac{\pi}{2} \left(-\tfrac{1}{2} e^{-r^2}\right)\bigg]_0^R = \lim_{R \to \infty} \frac{\pi}{4} \left(1 - e^{-R^2}\right)$$

$$= \frac{\pi}{4},$$

and so our original integral is equal to $\frac{\sqrt{\pi}}{2}$. ◀

Remark We should probably stop to worry for a moment about convergence of these improper integrals. First of all,

$$\int_0^\infty e^{-x^2} dx = \lim_{N \to \infty} \int_0^N e^{-x^2} dx$$

exists because, for example, when $x \geq 1$, we have $0 < e^{-x^2} \leq e^{-x}$, and so the integrals $\int_0^N e^{-x^2} dx$ increase as $N \to \infty$ and are all bounded above by $1 + \int_1^\infty e^{-x} dx = 1 + e^{-1}$. Now it is easy to see, as Figure 3.6 suggests, that the integral of $e^{-(x^2+y^2)}$ over the square $[0, N] \times [0, N]$ lies between the integral over the quarter-disk of radius N and the integral over the quarter-disk of radius $N\sqrt{2}$, both of which approach $\pi/4$.

In general, it is good to use polar coordinates when either the form of the integrand or the shape of the region recommends it.

Next we come to three dimensions. *Cylindrical coordinates* r, θ, z are merely polar coordinates (used in the xy-plane) along with the cartesian coordinate z:

$$\mathbf{g} \colon [0, \infty) \times [0, 2\pi) \times \mathbb{R} \to \mathbb{R}^3$$

292 ▶ Chapter 7. Integration

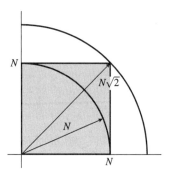

Figure 3.6

$$\mathbf{g}\begin{pmatrix} r \\ \theta \\ z \end{pmatrix} = \begin{bmatrix} r\cos\theta \\ r\sin\theta \\ z \end{bmatrix}.$$

The intuitive argument we gave earlier for polar coordinates suggests now that a little rectangle Δr by $\Delta\theta$ by Δz in $r\theta z$-space corresponds to a "chunk" with approximate volume $\Delta V \approx \Delta r(r\Delta\theta)\Delta z$, as pictured in Figure 3.7. If $\mathbf{g}$ maps the region Ω in $r\theta z$-space to our region $S \subset \mathbb{R}^3$, then we expect

$$\int_S f\begin{pmatrix} x \\ y \\ z \end{pmatrix} dV = \int_\Omega f\begin{pmatrix} r\cos\theta \\ r\sin\theta \\ z \end{pmatrix} r\,dr\,d\theta\,dz = \iint \left(\int f\begin{pmatrix} r\cos\theta \\ r\sin\theta \\ z \end{pmatrix} dz \right) r\,dr\,d\theta.$$

Indeed, as suggested by the last integral above, it is almost always preferable to set up an iterated integral with dz innermost, and then the usual $r\,dr\,d\theta$ outside (integrating over the projection of Ω onto the xy-plane).

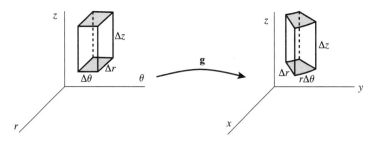

Figure 3.7

▶ **EXAMPLE 4**

Revisiting Example 9 of Section 2, we let $S \subset \mathbb{R}^3$ be the region in the first octant bounded below by the paraboloid $z = x^2 + y^2$ and above by the plane $z = 4$. To evaluate $\int_S x\,dV$ by using cylindrical coordinates, we realize that S is the image under $\mathbf{g}$ of the region

$$\Omega = \left\{ \begin{bmatrix} r \\ \theta \\ z \end{bmatrix} : 0 \leq r \leq 2,\ 0 \leq \theta \leq \pi/2,\ r^2 \leq z \leq 4 \right\}.$$

Thus, we have

$$\int_S x\,dV = \int_0^{\pi/2} \int_0^2 \int_{r^2}^4 \underbrace{r\cos\theta}_{x}\,\underbrace{r\,dz\,dr\,d\theta}_{dV}$$

$$= \int_0^{\pi/2} \int_0^2 \int_{r^2}^4 r^2 \cos\theta\,dz\,dr\,d\theta$$

$$= \int_0^{\pi/2} \int_0^2 r^2 \cos\theta (4 - r^2)\,dr\,d\theta$$

$$= \int_0^{\pi/2} \frac{64}{15} \cos\theta\,d\theta = \frac{64}{15},$$

which, reassuringly, is the same answer we obtained earlier.

▶ **EXAMPLE 5**

Let S be the region bounded above by the paraboloid $z = 6 - x^2 - y^2$ and below by the cone $z = \sqrt{x^2 + y^2}$, as pictured in Figure 3.8. Find $\int_S z\,dV$. The symmetry of S about the z-axis makes cylindrical coordinates a natural. The surfaces $z = 6 - r^2$ and $z = r$ intersect when $r = 2$, so we see that S is the image under $\mathbf{g}$ of the region

$$\Omega = \left\{ \begin{bmatrix} r \\ \theta \\ z \end{bmatrix} : 0 \leq r \leq 2,\ 0 \leq \theta \leq 2\pi,\ r \leq z \leq 6 - r^2 \right\}.$$

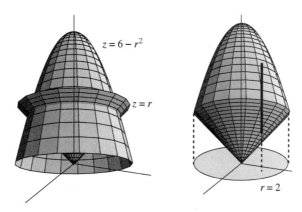

Figure 3.8

Thus, we have

$$\int_S z\,dV = \int_0^{2\pi}\int_0^2\int_r^{6-r^2} z\,\underbrace{r\,dz\,dr\,d\theta}_{dV}$$

$$= \int_0^{2\pi}\int_0^2 \tfrac{1}{2}\big((6-r^2)^2 - r^2\big)r\,dr\,d\theta$$

$$= \pi\int_0^2 (36 - 13r^2 + r^4)r\,dr = \frac{92}{3}\pi. \blacktriangleleft$$

Last, we come to *spherical coordinates*: ρ represents the distance from the origin to the point, ϕ the angle from the positive z-axis to the vector from the origin to the point, and θ the angle from the positive x-axis to the projection of that vector into the xy-plane. That is, in some sense, ϕ specifies the latitude of the point and θ specifies its longitude. (As shown in Figure 3.9, when ρ and ϕ are held constant, we get a circle parallel to the xy-plane; when ρ and θ are held constant, we get a great circle going from the north pole to the south pole.) Notice that we make the convention that

$$\rho \geq 0, \qquad 0 \leq \phi \leq \pi, \qquad \text{and} \qquad 0 \leq \theta < 2\pi.$$

As usual, we use basic trigonometry to express x, y, and z in terms of our new coordinates ρ, ϕ, and θ (see also Figure 3.10):

$$\mathbf{g}\colon [0,\infty) \times [0,\pi] \times [0,2\pi) \to \mathbb{R}^3$$

$$\mathbf{g}\begin{pmatrix}\rho\\ \phi\\ \theta\end{pmatrix} = \begin{bmatrix}\rho\sin\phi\cos\theta\\ \rho\sin\phi\sin\theta\\ \rho\cos\phi\end{bmatrix}.$$

As suggested by Figure 3.11, a rectangle $\Delta\rho$ by $\Delta\phi$ by $\Delta\theta$ in $\rho\phi\theta$-space maps to a spherical chunk of volume approximately

$$\Delta V \approx (\Delta\rho)(\rho\Delta\phi)(\rho\sin\phi\Delta\theta) = \rho^2\sin\phi\Delta\rho\Delta\phi\Delta\theta.$$

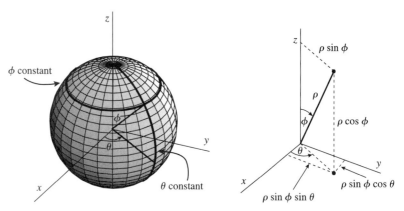

Figure 3.9   **Figure 3.10**

3 Polar, Cylindrical, and Spherical Coordinates ◀ 295

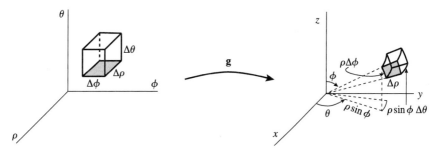

Figure 3.11

So, if **g** maps the region Ω to S, we expect that

$$\int_S f\begin{pmatrix} x \\ y \\ z \end{pmatrix} dV = \int_\Omega f\begin{pmatrix} \rho \sin\phi \cos\theta \\ \rho \sin\phi \sin\theta \\ \rho \cos\phi \end{pmatrix} \rho^2 \sin\phi \, d\rho \, d\phi \, d\theta.$$

▶ **EXAMPLE 6**

Let $S \subset \mathbb{R}^3$ be the "ice-cream cone" bounded above by the sphere $x^2 + y^2 + z^2 = a^2$ and below by the cone $z = c\sqrt{x^2 + y^2}$, where c is a fixed positive constant, as depicted in Figure 3.12. It is easy to see that the region Ω in $\rho\phi\theta$-space mapping to S is given by

$$\Omega = \left\{ \begin{bmatrix} \rho \\ \phi \\ \theta \end{bmatrix} : 0 \leq \rho \leq a, \ 0 \leq \phi \leq \phi_0, \ 0 \leq \theta \leq 2\pi \right\},$$

where $\phi_0 = \arctan(1/c)$.

The volume of S is calculated by using spherical coordinates as follows:

$$\text{vol}(S) = \int_S 1 \, dV = \int_0^{2\pi} \int_0^{\phi_0} \int_0^a \rho^2 \sin\phi \, d\rho \, d\phi \, d\theta$$

$$= \frac{2\pi}{3} a^3 (1 - \cos\phi_0) = \frac{2\pi}{3} a^3 \left(1 - \frac{c}{\sqrt{1+c^2}} \right).$$

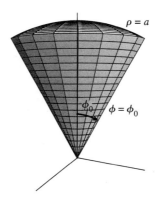

Figure 3.12

We can calculate $\int_S z\,dV$ as well:

$$\int_S z\,dV = \int_0^{2\pi}\int_0^{\phi_0}\int_0^a \underbrace{(\rho\cos\phi)}_{z}\underbrace{\rho^2\sin\phi\,d\rho\,d\phi\,d\theta}_{dV}$$

$$= \int_0^{2\pi}\int_0^{\phi_0}\int_0^a \rho^3\sin\phi\cos\phi\,d\rho\,d\phi\,d\theta$$

$$= \frac{\pi}{4}a^4\sin^2\phi_0 = \frac{\pi}{4}a^4\left(\frac{1}{1+c^2}\right). \triangleleft$$

▶ EXAMPLE 7

Let S be the sphere of radius a centered at $\begin{bmatrix} 0 \\ 0 \\ a \end{bmatrix}$. We wish to evaluate $\int_S z^2\,dV$. We observe first that, by Exercise 1.2.14, the triangle shown in Figure 3.13 is a right triangle, and so the equation of the sphere is $\rho = 2a\cos\phi$, $0 \le \phi \le \pi/2$. So we have

$$\int_S z^2\,dV = \int_0^{2\pi}\int_0^{\pi/2}\int_0^{2a\cos\phi} \underbrace{(\rho^2\cos^2\phi)}_{z^2}\underbrace{\rho^2\sin\phi\,d\rho\,d\phi\,d\theta}_{dV}$$

$$= \int_0^{2\pi}\int_0^{\pi/2}\int_0^{2a\cos\phi} \rho^4\cos^2\phi\sin\phi\,d\rho\,d\phi\,d\theta$$

$$= \frac{64}{5}\pi a^5\int_0^{\pi/2}\cos^7\phi\sin\phi\,d\phi = \frac{8}{5}\pi a^5. \triangleleft$$

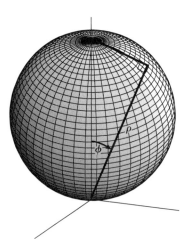

Figure 3.13

EXERCISES 7.3

1. Sketch the curves:
(a) $r = 4\cos\theta$
(b) $r = 3\sec\theta$
(c) $r = 1 - \sin\theta$
(d) $r = 1/(\cos\theta + \sin\theta)$

2. Find the area of the region bounded on the left by $x = 1$ and on the right by $x^2 + y^2 = 4$. Check your answer with simple geometry.

3. Find the area of the cardioid $r = 1 + \cos\theta$, pictured in Figure 3.14.

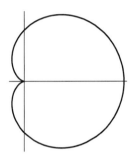

Figure 3.14

4. For $\varepsilon > 0$, let $S_\varepsilon = \{\mathbf{x} : \varepsilon \le \|\mathbf{x}\| \le 1\} \subset \mathbb{R}^2$. Evaluate $\lim_{\varepsilon \to 0^+} \int_{S_\varepsilon} \dfrac{1}{\sqrt{x^2 + y^2}}\, dA$. (This is often expressed as the *improper integral* $\int_{B(0,1)} (x^2 + y^2)^{-1/2}\, dA$.)

*5. Let S be the annular region shown in Figure 3.1. Evaluate $\int_S y^2\, dA$
(a) directly;
(b) by instead calculating $\int_S (x^2 + y^2)\, dA$.

*6. Calculate $\int_S y(x^2 + y^2)^{-5/2}\, dA$, where S is the planar region lying above the x-axis, bounded on the left by $x = 1$ and above by $x^2 + y^2 = 2$.

7. Calculate $\int_S (x^2 + y^2)^{-3/2}\, dA$, where S is the planar region bounded below by $y = 1$ and above by $x^2 + y^2 = 4$.

8. Let $f\begin{pmatrix} x \\ y \end{pmatrix} = \dfrac{y^2}{\sqrt{x^2 + y^2}}$. Let S be the planar region lying inside the circle $x^2 + y^2 = 2x$, above the x-axis, and to the right of $x = 1$. Evaluate $\int_S f\, dA$.

*9. Evaluate $\displaystyle\int_0^1 \int_y^1 \dfrac{xe^x}{x^2 + y^2}\, dx\, dy$.

10. Find the volume of the region bounded above by $z = 2y$ and below by $z = x^2 + y^2$.

11. Find the volume of the "doughnut with no hole," $\rho = \sin\phi$, pictured in Figure 3.15.

*12. Sketch and find the volume of the "pillow" $\rho = \sin\theta$, $0 \le \theta \le \pi$.

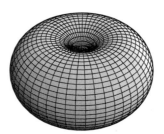

Figure 3.15

13. Find the volume of the region inside both $x^2 + y^2 = 1$ and $x^2 + y^2 + z^2 = 2$.

14. Find the volume of the region inside both $x^2 + y^2 + z^2 = 4a^2$ and $x^2 + y^2 = 2ay$.

15. Find the volume of the region bounded above by $x^2 + y^2 + z^2 = 2$ and below by $z = x^2 + y^2$.

16. Find the volume of the region inside the sphere $x^2 + y^2 + z^2 = a^2$ by integrating in
    (a) cylindrical coordinates; (b) spherical coordinates.

*17. Find the volume of a right circular cone of base radius a and height h by integrating in
    (a) cylindrical coordinates; (b) spherical coordinates.

18. Find the volume of the region lying above the cone $z = \sqrt{x^2 + y^2}$ and inside the sphere $x^2 + y^2 + z^2 = 2$ by integrating in
    (a) cylindrical coordinates; (b) spherical coordinates.

19. Find the volume of the region lying above the plane $z = a$ and inside the sphere $x^2 + y^2 + z^2 = 4a^2$ by integrating in
    (a) cylindrical coordinates; (b) spherical coordinates.

*20. Let $S \subset \mathbb{R}^3$ be the unit ball. Use symmetry principles to compute $\int_S x^2 \, dV$ as easily as possible.

21. (a) Evaluate $\int_{\mathbb{R}^3} e^{-(x^2+y^2+z^2)} dV$. (b) Evaluate $\int_{\mathbb{R}^3} e^{-(x^2+2y^2+3z^2)} dV$.

*22. Find the volume of the region in $\mathbb{R}^3$ bounded above by the plane $z = 3x + 4y$ and below by the paraboloid $z = x^2 + y^2$.

23. Evaluate $\int_S \dfrac{z}{(x^2+y^2+z^2)^{3/2}} dV$, where S is the region bounded below by the sphere $x^2 + y^2 + z^2 = 2z$ and above by the sphere $x^2 + y^2 + z^2 = 1$.

24. Find the volume of the region in $\mathbb{R}^3$ bounded by the cylinders $x^2 + y^2 = 1$, $y^2 + z^2 = 1$, and $x^2 + z^2 = 1$. (Hint: Make full use of symmetry.)

▶ 4 PHYSICAL APPLICATIONS

So far we have focused on area and volume as our interpretation of the multiple integral. Now we discuss average value and mass (which have both physical and probabilistic interpretations), center of mass, moment of inertia, and gravitational attraction.

Recall from one-variable calculus the notion of the *average value* of an integrable function. Given a real-valued function f on an interval $[a, b]$, we may take the uniform partition $\mathcal{P}_k$ of the interval into k equal subintervals, $x_i = a + i\left(\frac{b-a}{k}\right)$, $i = 1, \ldots, k$, and calculate the average of the values $f(x_1), \ldots, f(x_k)$:

$$\bar{y}^{(k)} = \frac{1}{k} \sum_{i=1}^{k} f(x_i).$$

Multiplying and dividing by $b - a$ gives

$$\bar{y}^{(k)} = \frac{1}{b-a} \sum_{i=1}^{k} f(x_i) \frac{b-a}{k}.$$

Now let's suppose that f is bounded. Then, as usual, $m_i \leq f(x_i) \leq M_i$ for each $i = 1, \ldots, k$, and so

$$\frac{1}{b-a} L(f, \mathcal{P}_k) \leq \bar{y}^{(k)} \leq \frac{1}{b-a} U(f, \mathcal{P}_k)$$

for every uniform partition $\mathcal{P}_k$ of the interval $[a, b]$. Now assume that f is integrable. Then it follows from Exercise 7.1.10 that $L(f, \mathcal{P}_k)$ and $U(f, \mathcal{P}_k)$ both approach $\int_a^b f(x)dx$ as $k \to \infty$, and so

$$\bar{y}^{(k)} \to \frac{1}{b-a} \int_a^b f(x)dx \quad \text{as } k \to \infty.$$

This motivates the following

Definition Let f be an integrable function on the interval $[a, b]$. We define the *average value* of f on $[a, b]$ to be

$$\bar{f} = \frac{1}{b-a} \int_a^b f(x)dx.$$

In general, if $\Omega \subset \mathbb{R}^n$ is a region and $f: \Omega \to \mathbb{R}$ is integrable, we define the average value of f on Ω to be

$$\bar{f} = \frac{1}{\text{vol}(\Omega)} \int_\Omega f \, dV.$$

▶ **EXAMPLE 1**

A round hotplate S is given by the disk $r \leq \pi/2$. Its temperature is given by $f\begin{pmatrix} x \\ y \end{pmatrix} = \cos\sqrt{x^2 + y^2}$. We want to determine the average temperature of the plate. We calculate

$$\bar{f} = \frac{1}{\text{area}(S)} \int_S f \, dA$$

by proceeding in polar coordinates:

$$\int_S f \, dA = \int_0^{2\pi} \int_0^{\pi/2} (\cos r) r \, dr \, d\theta = 2\pi (r \sin r + \cos r) \Big|_0^{\pi/2} = 2\pi \left(\tfrac{\pi}{2} - 1\right),$$

and so
$$\bar{f} = \frac{2\pi\left(\frac{\pi}{2}-1\right)}{\pi\left(\frac{\pi}{2}\right)^2} = \frac{4(\pi-2)}{\pi^2} \approx 0.463.$$ ◀

It is useful to define the integral of a vector-valued function $\mathbf{f}\colon \Omega \to \mathbb{R}^m$ component by component (generalizing what we did in Lemma 5.3 of Chapter 3): Assuming each of the component functions $f_1, \ldots, f_m$ is integrable on Ω, we set

$$\int_\Omega \mathbf{f}\,dV = \begin{bmatrix} \int_\Omega f_1\,dV \\ \vdots \\ \int_\Omega f_m\,dV \end{bmatrix}.$$

Then we can define the average value of $\mathbf{f}$ on Ω in the obvious way:

$$\bar{\mathbf{f}} = \frac{1}{\mathrm{vol}(\Omega)}\int_\Omega \mathbf{f}\,dV.$$

In particular, we define the *centroid* or *center of mass* of the region Ω to be

$$\bar{\mathbf{x}} = \frac{1}{\mathrm{vol}(\Omega)}\int_\Omega \mathbf{x}\,dV.$$

▶ **EXAMPLE 2**

We want to find the centroid of the plane region Ω bounded below by $y=0$, above by $y=x^2$, and on the right by $x=1$. Its area is given by

$$\mathrm{area}(\Omega) = \int_0^1 \int_0^{x^2} dy\,dx = \frac{1}{3}.$$

Now, integrating the position vector $\mathbf{x}$ over Ω gives

$$\int_\Omega \mathbf{x}\,dA = \int_0^1 \int_0^{x^2} \begin{bmatrix} x \\ y \end{bmatrix} dy\,dx = \int_0^1 \begin{bmatrix} x^3 \\ \frac{1}{2}x^4 \end{bmatrix} dx = \begin{bmatrix} 1/4 \\ 1/10 \end{bmatrix},$$

so $\bar{\mathbf{x}} = \begin{bmatrix} 3/4 \\ 3/10 \end{bmatrix}$, which makes physical sense (see Figure 4.1). ◀

It is useful to observe that when the region Ω is symmetric about an axis, its centroid will lie on that axis. (See Exercise 7.2.18.)

When a mass distribution Ω is nonuniform, it is important to understand the idea of density. Much like instantaneous velocity (or slope of a curve), which is defined as a limit of average velocities (or slopes of secant lines), we define the density $\delta(\mathbf{x})$ to be the limit as $r \to 0^+$ of the average density (mass/volume) of a cube of sidelength r centered at $\mathbf{x}$.[2] Then it is quite plausible that, with some reasonable assumptions on the behavior of "mass," it should be recaptured by integrating the density function.

---

[2] More precisely, the average density of that portion of the cube lying in Ω.

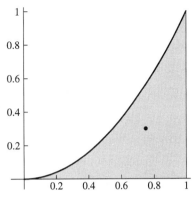

Figure 4.1

Lemma 4.1 *Let $\Omega \subset \mathbb{R}^n$ be a region. Assume the density function $\delta\colon \Omega \to \mathbb{R}$ is integrable. Then* $\operatorname{mass}(\Omega) = \int_\Omega \delta\, dV$.

Proof As usual, it suffices to assume Ω is a rectangle R. For any partition $\mathcal{P} = \{R_i\}$ of R, let $m_i = \inf_{\mathbf{x} \in R_i} \delta(\mathbf{x})$ and $M_i = \sup_{\mathbf{x} \in R_i} \delta(\mathbf{x})$. Then $m_i \operatorname{vol}(R_i) \le \operatorname{mass}(R_i) \le M_i \operatorname{vol}(R_i)$. (Suppose, for example, that
$$M_i < \frac{\operatorname{mass}(R_i)}{\operatorname{vol}(R_i)} = \delta^*.$$
Then, in particular, for all $\mathbf{x} \in R_i$, we have $\delta(\mathbf{x}) < \delta^*$, and so, by the definition of δ, for each $\mathbf{x}$ there is a cube centered at $\mathbf{x}$ whose average density is less than δ^*. By compactness, we can cover R_i by finitely many such cubes, and we see that the average density of R_i itself is less than δ^*, which is a contradiction.) It now follows that $L(\delta, \mathcal{P}) \le \operatorname{mass}(R) \le U(\delta, \mathcal{P})$ for any partition $\mathcal{P}$ of R, and so, since we've assumed δ is integrable, we must have
$$\operatorname{mass}(R) = \int_R \delta\, dV. \quad \blacksquare$$

Remark We should be a little bit careful here. The Fundamental Theorem of Calculus tells us that we can recover f by differentiating its integral $F(x) = \int_a^x f(t)\,dt$ *provided f is continuous*. If we start with an arbitrary integrable function f, e.g., the function in Exercise 7.1.13, this will, of course, not work. A similar situation occurs if we start with an integrable δ, define the mass by integrating, and then try to recapture δ by "differentiating" (taking the limit of average densities). Since we are concerned here with physical applications, we will tacitly assume δ is continuous (see Exercise 7.1.7). In more sophisticated treatments, we really would like to allow point masses and "generalized functions," called *distributions*; this will have to wait for a more advanced course.

Now, generalizing our earlier definition of center of mass, if Ω is a mass distribution with density function δ, then we define the center of mass to be the *weighted average*
$$\bar{\mathbf{x}} = \frac{1}{\operatorname{mass}(\Omega)} \int_\Omega \delta(\mathbf{x})\mathbf{x}\, dV.$$

This is a natural generalization of the weighted average we see with a system of finitely many point masses $m_1, \ldots, m_N$ at positions $\mathbf{x}_1, \ldots, \mathbf{x}_N$, respectively, as shown in Figure 4.2. In this case, the weighted average is

$$\bar{\mathbf{x}} = \frac{\sum_{i=1}^{N} m_i \mathbf{x}_i}{\sum_{i=1}^{N} m_i},$$

and it has the following physical interpretation. If external forces $\mathbf{F}_i$ act on the point masses m_i, they impart accelerations $\mathbf{x}_i''$ according to Newton's second law: $\mathbf{F}_i = m_i \mathbf{x}_i''$. Consider the resultant force $\mathbf{F} = \sum_{i=1}^{N} \mathbf{F}_i$ acting on the total mass $m = \sum_{i=1}^{N} m_i$ (any internal forces cancel ultimately by Newton's third law). Then

$$\mathbf{F} = \sum_{i=1}^{N} \mathbf{F}_i = \sum_{i=1}^{N} m_i \mathbf{x}_i'' = m \bar{\mathbf{x}}''.$$

That is, as the forces act and time passes, the center of mass of the system translates exactly as if we concentrated the total mass m at $\bar{\mathbf{x}}$ and let the resultant force $\mathbf{F}$ act there.

Next, let's consider a rigid body[3] consisting of point masses $m_1, \ldots, m_N$ rotating about an axis ℓ; a typical such mass is pictured in Figure 4.3. The kinetic energy of the system is

$$\text{K.E.} = \sum_{i=1}^{N} \frac{1}{2} m_i \|\mathbf{x}_i'\|^2 = \frac{1}{2} \sum_{i=1}^{N} m_i (r_i \omega)^2,$$

where ω is the angular speed with which the body is rotating about the axis and r_i is the distance from the axis of rotation to the point mass m_i. (Remember that each mass is

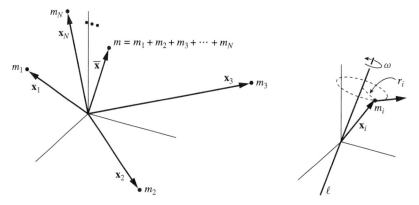

Figure 4.2   **Figure 4.3**

---

[3] A rigid body does not move relative to itself; imagine the masses connected to one another by inflexible rods.

moving in a circle whose center lies on ℓ.) Regrouping, we get

$$\text{K.E.} = \tfrac{1}{2}I\omega^2, \qquad \text{where} \qquad I = \sum_{i=1}^{N} m_i r_i^2.$$

I is called the *moment of inertia* of the rigid body *about ℓ*.

In the case of a mass distribution Ω forming a rigid body, we define by analogy (partitioning it and approximating it by a finite number of masses) its *moment of inertia about an axis ℓ* to be

$$I = \int_{\Omega} \delta r^2 dV,$$

where r is the distance from ℓ.

▶ **EXAMPLE 3**

Let's find the moment of inertia of a uniform solid ball Ω of radius a about an axis through its center. We may as well place the ball with its center at the origin and let the axis be the z-axis. Then, using spherical coordinates, we have (since δ is constant)

$$I = \int_{\Omega} \delta r^2 dV = \delta \int_0^{2\pi} \int_0^{\pi} \int_0^{a} \underbrace{(\rho \sin \phi)^2}_{r^2} \underbrace{\rho^2 \sin \phi \, d\rho \, d\phi \, d\theta}_{dV}$$

$$= 2\pi \delta \int_0^{\pi} \int_0^{a} \rho^4 \sin^3 \phi \, d\rho \, d\phi$$

$$= 2\pi \delta \cdot \frac{a^5}{5} \cdot \frac{4}{3} = \left(\frac{4}{3}\pi a^3 \delta\right) \frac{2}{5} a^2 = \frac{2}{5} m a^2,$$

where $m = \tfrac{4}{3}\pi a^3 \delta$ is the total mass of Ω. ◀

▶ **EXAMPLE 4**

One of the classic applications of the moment of inertia is to decide which rolling object wins the race down a ramp. Given a hula hoop, a wooden nickel, a hollow ball, a solid ball, or something more imaginative like a solid cone, as pictured in Figure 4.4, which one gets to the bottom first?

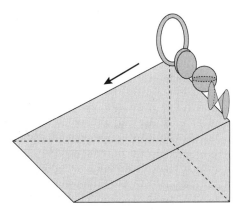

Figure 4.4

We use the basic result from physics (see the remark on p. 352 and Example 6 of Chapter 8, Section 3) that, if we ignore friction, total energy—potential plus kinetic—is conserved.[4] We measure potential energy relative to ground level, so a mass m has potential energy mgh at (relatively small) heights h. If the rolling radius is a, its angular speed is ω, and its *linear* speed is v, then we have $a\omega = v$, so when the mass has descended a vertical height h, we have

original (potential) energy = final (kinetic) energy

$$mgh = \underbrace{\frac{1}{2}mv^2}_{\text{translational K.E.}} + \underbrace{\frac{1}{2}I\left(\frac{v}{a}\right)^2}_{\text{rotational K.E.}} = \frac{1}{2}m\left(1 + \frac{I}{ma^2}\right)v^2.$$

Thus, the object's speed is greatest when the fraction I/ma^2 is smallest. We calculated in Example 3 that this fraction is $2/5$ for a solid ball. For a hula-hoop of radius a or for a hollow cylinder of radius a, it is obviously 1 (why?). So the solid ball beats the hula-hoop or hollow cylinder. What about the other shapes? (See Exercises 16, 17, and 19.) And is there an optimal shape? ◀

Newton's law of gravitation applies to point masses: The force $\mathbf{F}$ exerted by a mass m at position $\mathbf{x}$ on a test mass (which we take to have mass 1 unit) at the origin is given by

$$\mathbf{F} = Gm\frac{\mathbf{x}}{\|\mathbf{x}\|^3}.$$

Thus, the gravitational force exerted by a collection of masses m_i, $i = 1, \ldots, N$, at positions $\mathbf{x}_i$ on the test mass is given by

$$\mathbf{F} = \sum_{i=1}^{N} \mathbf{F}_i = G\sum_{i=1}^{N} m_i \frac{\mathbf{x}_i}{\|\mathbf{x}_i\|^3},$$

and, thus, the gravitational force exerted by a continuous mass distribution Ω with density function δ is

$$\mathbf{F} = G\int_{\Omega} \delta \frac{\mathbf{x}}{\|\mathbf{x}\|^3} dV.$$

▶ **EXAMPLE 5**

Find the gravitational attraction on a unit mass at the origin of the uniform region Ω bounded above by the sphere $x^2 + y^2 + z^2 = 2a^2$ and below by the paraboloid $az = x^2 + y^2$, pictured in Figure 4.5. (Take $\delta = 1$.)

Since Ω is symmetric about the z-axis, the net force will lie entirely in the z-direction, so we calculate only the $\mathbf{e}_3$-component of $\mathbf{F}$. Working in cylindrical coordinates, we see that Ω lies over the disk of radius a centered at the origin in the xy-plane, and so

$$F_3 = G\int_{\Omega} \frac{z}{(x^2 + y^2 + z^2)^{3/2}} dV$$

$$= G\int_0^{2\pi} \int_0^a \int_{r^2/a}^{\sqrt{2a^2-r^2}} \frac{z}{(r^2 + z^2)^{3/2}} r\, dz\, dr\, d\theta$$

---

[4] Of course, for the objects to roll, there must be *some* friction.

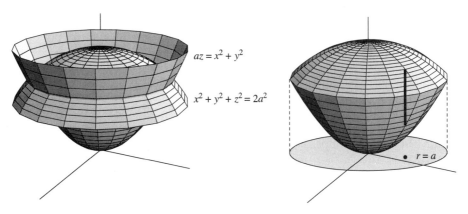

Figure 4.5

$$= 2\pi G \int_0^a r\left(-(r^2+z^2)^{-1/2}\right)\Big]_{r^2/a}^{\sqrt{2a^2-r^2}} dr$$

$$= 2\pi G \int_0^a \left(\frac{a}{\sqrt{a^2+r^2}} - \frac{r}{a\sqrt{2}}\right) dr = 2\pi G a \left(\log(1+\sqrt{2}) - \frac{1}{2\sqrt{2}}\right).$$

We leave it to the reader to set the problem up in spherical coordinates (see Exercise 24).

▶ **EXAMPLE 6**

Newton wanted to understand the gravitational attraction of the earth, which he took to be a uniform ball. Most of us are taught nowadays that the gravitational attraction of the earth on a point mass *outside* the earth is that of a point mass M concentrated at the center of the earth. But what happens if the point mass is *inside* the earth? We put the earth (a ball of radius R) with its center at the origin and the point mass at $\begin{bmatrix} 0 \\ 0 \\ b \end{bmatrix}$, $b > 0$, as shown in Figure 4.6. By symmetry, the net force will lie in the z-direction, so we compute only that component. If the earth has (constant) density δ, we have

$$F_3 = \int_\Omega -G\delta \frac{\cos\alpha}{d^2} dV = -G\delta \int_\Omega \frac{b-\rho\cos\phi}{(b^2+\rho^2-2b\rho\cos\phi)^{3/2}} dV$$

$$= -2\pi G\delta \int_0^R \int_0^\pi \frac{b-\rho\cos\phi}{(b^2+\rho^2-2b\rho\cos\phi)^{3/2}} \rho^2 \sin\phi\, d\phi\, d\rho.$$

(Note that we are going to do the ϕ integral first.)

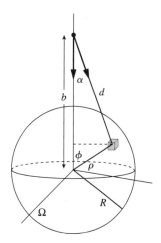

Figure 4.6

Fixing ρ, let $u = b^2 + \rho^2 - 2b\rho \cos\phi$, $du = 2b\rho \sin\phi\, d\phi$, so

$$\int_0^\pi \frac{b - \rho\cos\phi}{(b^2 + \rho^2 - 2b\rho\cos\phi)^{3/2}} \rho^2 \sin\phi\, d\phi = \int_{(b-\rho)^2}^{(b+\rho)^2} \frac{b - \frac{b^2+\rho^2-u}{2b}}{u^{3/2}} \frac{\rho}{2b} du$$

$$= \frac{\rho}{4b^2} \int_{(b-\rho)^2}^{(b+\rho)^2} \left((b^2 - \rho^2)u^{-3/2} + u^{-1/2}\right) du$$

$$= \frac{\rho}{4b^2} \left((b^2 - \rho^2)(-2u^{-1/2}) + 2u^{1/2}\right)\Big|_{(b-\rho)^2}^{(b+\rho)^2}$$

$$= \frac{\rho}{2b^2}\left[(b^2 - \rho^2)\left(\frac{1}{|b-\rho|} - \frac{1}{b+\rho}\right) + (b+\rho) - |b-\rho|\right]$$

$$= \begin{cases} 2\rho^2/b^2, & \rho \le b \\ 0, & \rho > b \end{cases}.$$

Now we do the ρ integral. In the event that $b \ge R$, we have

$$F_3 = -2\pi G\delta \int_0^R \frac{2\rho^2}{b^2} d\rho = -\frac{4\pi G\delta}{b^2} \frac{R^3}{3} = -\frac{GM}{b^2},$$

where $M = 4\pi\delta R^3/3$ is the total mass of the earth. On the other hand, if $b < R$, then, since the integrand vanishes whenever $\rho > b$, we have

$$F_3 = -2\pi G\delta \int_0^b \frac{2\rho^2}{b^2} d\rho = -\frac{4\pi G\delta}{3} b = -G\frac{M}{R^3}b,$$

which, interestingly, is *linear* in b. (When $b = R$, of course, the two answers agree.) Incidentally, we will be able to rederive these results in a matter of seconds in Section 6 of Chapter 8. ◀

▶ EXERCISES 7.4

*1. Find the average distance from the origin to the points in the ball $B(\mathbf{0}, a) \subset \mathbb{R}^2$.

2. Find the average distance from the origin to the points in the ball $B(\mathbf{0}, a) \subset \mathbb{R}^3$.

*3. Find the average distance from a point on the boundary of a ball of radius a in $\mathbb{R}^2$ to the points inside the ball.

*4. Find the average distance from a point on the boundary of a ball of radius a in $\mathbb{R}^3$ to the points inside the ball.

5. Find the average distance from one corner of a square of sidelength a to the points inside the square.

6. Consider the region Ω lying inside the circle $x^2 + y^2 = 2x$, above the x-axis, and to the right of $x = 1$, with density $\delta \begin{pmatrix} x \\ y \end{pmatrix} = \dfrac{y}{x^2 + y^2}$. Find the center of mass of Ω.

*7. Consider the region Ω lying inside the circle $x^2 + y^2 = 2x$ and outside the circle $x^2 + y^2 = 1$. If its density function is given by $\delta \begin{pmatrix} x \\ y \end{pmatrix} = (x^2 + y^2)^{-1/2}$, find its center of mass.

8. Find the center of mass of a uniform semicircular plate of radius a in $\mathbb{R}^2$.

9. Find the center of mass of a uniform solid hemisphere of radius a in $\mathbb{R}^3$.

10. Find the center of mass of the uniform region in Exercise 7.3.19.

*11. Find the center of mass of the uniform tetrahedron bounded by the coordinate planes and the plane $x/a + y/b + z/c = 1$.

*12. Find the mass of a solid cylinder of height h and base radius a if its density at $\mathbf{x}$ is equal to the distance from $\mathbf{x}$ to the axis of the cylinder. Next find its moment of inertia about the axis.

13. Find the moment of inertia about the z-axis of a solid ball of radius a centered at the origin, whose density is given by $\delta(\mathbf{x}) = \|\mathbf{x}\|$.

14. Let Ω be the region bounded above by $x^2 + y^2 + z^2 = 4$ and below by $z = \sqrt{x^2 + y^2}$. Calculate the moment of inertia of Ω about the z-axis by integrating in both cylindrical and spherical coordinates.

15. Find the moment of inertia about the z-axis of the region of constant density $\delta = 1$ bounded above by the sphere $x^2 + y^2 + z^2 = 4$ and below by the cone $z\sqrt{3} = \sqrt{x^2 + y^2}$.

*16. Find the moment of inertia about the z-axis of each of the following uniform objects:
(a) a hollow cylindrical can $x^2 + y^2 = a^2$, $0 \le z \le h$
(b) the solid cylinder $x^2 + y^2 \le a^2$, $0 \le z \le h$
(c) the solid cone of base radius a and height h symmetric about the z-axis
Express each of your answers in the form $I = kma^2$ for the appropriate constant k.

17. (a) Let $0 < b < a$. Find the moment of inertia $I_{a,b}$ about the z-axis of the uniform region $b^2 \le x^2 + y^2 + z^2 \le a^2$.

(b) Find $\lim\limits_{b \to a^-} \dfrac{I_{a,b}}{a^3 - b^3}$.

(c) Use your answer to part b to show that the moment of inertia of a uniform hollow spherical shell $x^2 + y^2 + z^2 = a^2$ about the z-axis is $\frac{2}{3}ma^2$, where m is its total mass.

18. Let $\Omega \subset \mathbb{R}^n$ be a region. For what value of $\mathbf{a} \in \mathbb{R}^n$ is the integral
$$\int_\Omega \|\mathbf{x} - \mathbf{a}\|^2 dV$$
minimized? (Cf. Exercise 5.2.14.)

19. Let Ω be the uniform solid of revolution obtained by rotating the graph of $y = |x|^n$, $|x| \le a^{1/n}$, about the x-axis, as indicated in Figure 4.7. Let I be the moment of inertia about the x-axis. Show that $\dfrac{I}{ma^2} = \dfrac{2n+1}{2(4n+1)}$.

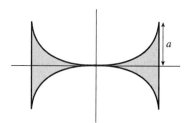

Figure 4.7

20. Let Ω_ε denote the uniform solid region described in spherical coordinates by $0 \leq \rho \leq a$, $0 \leq \phi \leq \varepsilon$.
(a) Find the center of mass of Ω_ε.
(b) Find the limiting position of the center of mass as $\varepsilon \to 0^+$. Explain your answer.

21. (Pappus's Theorem) Suppose $R \subset \mathbb{R}^2$ is a plane region (say, that bounded by the graphs of f and g on the interval $[a, b]$), and let $\Omega \subset \mathbb{R}^3$ be obtained by revolving R about the x-axis. Prove that the volume of Ω is equal to
$$\text{vol}(\Omega) = 2\pi \bar{y} \cdot \text{area}(R).$$

22. Let Ω denote a mass distribution. Denote by I the moment of inertia of Ω about a given axis ℓ, and by I_0 the moment of inertia about the axis ℓ_0 parallel to ℓ and passing through the center of mass of Ω. Then prove the parallel axis theorem:
$$I = I_0 + mh^2,$$
where m is the total mass of Ω and h is the distance between ℓ and ℓ_0.

23. Calculate the gravitational attraction of a solid ball of radius R on a unit mass on its boundary if its density is equal to distance from the center of the ball.

24. Set up Example 5 in spherical coordinates and verify the calculations.

25. Prove or give a counterexample: The gravitational force on a test mass of a body with total mass M is equal to that of a point mass M located at the center of mass of the body.

26. Show that Newton's first result in Example 6 still works for a nonuniform earth, as long as the density δ is radially symmetric (i.e., is a function of ρ only). What happens to the second result?

27. Consider the solid region Ω bounded by $(x^2 + y^2 + z^2)^{3/2} = kz$ ($k > 0$), with k chosen so that the volume of Ω is equal to the volume of the unit ball.
(a) Find k.
(b) Taking $\delta = 1$, find the gravitational attraction of Ω on a unit test mass at the origin.

Remark Your answer to part b should be somewhat larger than $4\pi G/3$, the gravitational attraction of the unit ball (with $\delta = 1$) on a unit mass on its boundary. In fact, Ω is the region of appropriate mass that *maximizes* the gravitational attraction on a point mass at the origin. Can you think of any explanation—physical, geometric, or otherwise?

28. A completely uniform forest is in the shape of a plane region Ω. The forest service will locate a helipad somewhere in the forest and, in the event of fire, will dispatch helicopters to fight it. If a fire is equally likely to start anywhere in the forest, where should the forest service locate the helipad to minimize fire damage? (Let's take the simplest model possible: Assume that fire spreads radially at a constant rate and that the helicopters fly at a constant rate and take off as soon as the fire starts. So what are we trying to minimize here?)

5 DETERMINANTS AND n-DIMENSIONAL VOLUME

We now want to complete the discussion of determinants initiated in Section 5 of Chapter 1. We will see soon the relation between such "multilinear" functions and n-dimensional volume. Indeed, determinants will play a central role in all our remaining work.

Theorem 5.1 *For each $n \geq 1$, there is exactly one function $\mathcal{D}\colon \underbrace{\mathbb{R}^n \times \cdots \times \mathbb{R}^n}_{n \text{ times}} \to \mathbb{R}$ having the following properties:*

1. *If any pair of the vectors $\mathbf{v}_1, \ldots, \mathbf{v}_n$ are exchanged, $\mathcal{D}$ changes sign. That is,*

$$\mathcal{D}(\mathbf{v}_1, \ldots, \mathbf{v}_i, \ldots, \mathbf{v}_j, \ldots, \mathbf{v}_n) = -\mathcal{D}(\mathbf{v}_1, \ldots, \mathbf{v}_j, \ldots, \mathbf{v}_i, \ldots, \mathbf{v}_n)$$

for any $1 \leq i < j \leq n$.

2. *For all $\mathbf{v}_1, \ldots, \mathbf{v}_n \in \mathbb{R}^n$ and $c \in \mathbb{R}$, we have*

$$\mathcal{D}(c\mathbf{v}_1, \mathbf{v}_2, \ldots, \mathbf{v}_n) = \mathcal{D}(\mathbf{v}_1, c\mathbf{v}_2, \ldots, \mathbf{v}_n) = \cdots$$
$$= \mathcal{D}(\mathbf{v}_1, \ldots, \mathbf{v}_{n-1}, c\mathbf{v}_n) = c\mathcal{D}(\mathbf{v}_1, \ldots, \mathbf{v}_n).$$

3. *For any vectors $\mathbf{v}_1, \ldots, \mathbf{v}_n$ and $\mathbf{v}'_i$, we have*

$$\mathcal{D}(\mathbf{v}_1, \ldots, \mathbf{v}_{i-1}, \mathbf{v}_i + \mathbf{v}'_i, \mathbf{v}_{i+1}, \ldots, \mathbf{v}_n) = $$
$$\mathcal{D}(\mathbf{v}_1, \ldots, \mathbf{v}_{i-1}, \mathbf{v}_i, \mathbf{v}_{i+1}, \ldots, \mathbf{v}_n) + \mathcal{D}(\mathbf{v}_1, \ldots, \mathbf{v}_{i-1}, \mathbf{v}'_i, \mathbf{v}_{i+1}, \ldots, \mathbf{v}_n).$$

4. *If $\{\mathbf{e}_1, \ldots, \mathbf{e}_n\}$ is the standard basis for $\mathbb{R}^n$, then we have*

$$\mathcal{D}(\mathbf{e}_1, \ldots, \mathbf{e}_n) = 1.$$

Properties (2) and (3) indicate that $\mathcal{D}$ is linear as a function of each of its variables (whence "*multi*linear"); property (1) indicates that $\mathcal{D}$ is "alternating." Property (4) can be interpreted as saying that the unit cube should have volume 1.

Definition Given an $n \times n$ matrix A with column vectors $\mathbf{a}_1, \ldots, \mathbf{a}_n \in \mathbb{R}^n$, set

$$\det A = \mathcal{D}(\mathbf{a}_1, \ldots, \mathbf{a}_n).$$

This is called the *determinant* of A.

Since most of our work with matrices has centered on row operations, it would perhaps be more convenient to define the determinant in terms of the rows of A. But it really is inconsequential for two reasons: First, everything we proved using row operations (and, correspondingly, left multiplication by elementary matrices) works *verbatim* for column operations (and, correspondingly, right multiplication by elementary matrices); second, we will prove shortly that $\det A^\top = \det A$.

Properties (1)–(3) of $\mathcal{D}$ listed in Theorem 5.1 allow us to see the effect of elementary column operations on the determinant of a matrix. Indeed, Property (1) corresponds to a column interchange; Property (2) corresponds to multiplying a column by a scalar; and

Property (3) tells us—in combination with Property (1)—that adding a multiple of one column to another does not change the determinant.

▶ EXAMPLE 1

We can calculate the determinant of the matrix

$$A = \begin{bmatrix} 0 & 0 & 0 & 4 \\ 0 & 2 & 0 & 0 \\ 0 & 0 & 1 & 0 \\ 3 & 0 & 0 & 0 \end{bmatrix}$$

as follows. First we factor out the 3 from the first column to get

$$\begin{vmatrix} 0 & 0 & 0 & 4 \\ 0 & 2 & 0 & 0 \\ 0 & 0 & 1 & 0 \\ 3 & 0 & 0 & 0 \end{vmatrix} = 3 \begin{vmatrix} 0 & 0 & 0 & 4 \\ 0 & 2 & 0 & 0 \\ 0 & 0 & 1 & 0 \\ 1 & 0 & 0 & 0 \end{vmatrix}$$

by Property (2). Repeating this process with the 4 and the 2, we obtain

$$3 \begin{vmatrix} 0 & 0 & 0 & 4 \\ 0 & 2 & 0 & 0 \\ 0 & 0 & 1 & 0 \\ 1 & 0 & 0 & 0 \end{vmatrix} = 2 \cdot 4 \cdot 3 \begin{vmatrix} 0 & 0 & 0 & 1 \\ 0 & 1 & 0 & 0 \\ 0 & 0 & 1 & 0 \\ 1 & 0 & 0 & 0 \end{vmatrix}.$$

Now interchanging columns 1 and 4 introduces a factor of -1 by Property (1), and we have

$$\det A = 24 \begin{vmatrix} 0 & 0 & 0 & 1 \\ 0 & 1 & 0 & 0 \\ 0 & 0 & 1 & 0 \\ 1 & 0 & 0 & 0 \end{vmatrix} = -24 \begin{vmatrix} 1 & 0 & 0 & 0 \\ 0 & 1 & 0 & 0 \\ 0 & 0 & 1 & 0 \\ 0 & 0 & 0 & 1 \end{vmatrix} = -24$$

since Property (4) tells us that $\det I_4 = 1$. ◀

To calculate the effect of the third type of column operation—adding a multiple of one column to another—we need the following observation.

Lemma 5.2 *If two columns of a matrix A are equal, then $\det A = 0$.*

Proof If $\mathbf{a}_i = \mathbf{a}_j$, then the matrix is unchanged when we switch columns i and j. On the other hand, by Property (1), its determinant changes sign when we do so. That is, we have $\det A = -\det A$. This can happen only when $\det A = 0$. ∎

Now we can easily prove the

Proposition 5.3 *Let A be an $n \times n$ matrix and let A' be the matrix obtained by adding a multiple of one column of A to another. Then $\det A' = \det A$.*

Proof Suppose A' is obtained from A by replacing the ith column by its sum with c times the jth column; i.e., $\mathbf{a}'_i = \mathbf{a}_i + c\mathbf{a}_j$, with $i \neq j$. (As a notational convenience, we assume $i < j$, but that really is inconsequential.) We wish to show that

$$\det A' = \mathcal{D}(\mathbf{a}_1, \ldots, \mathbf{a}_{i-1}, \mathbf{a}_i + c\mathbf{a}_j, \mathbf{a}_{i+1}, \ldots, \mathbf{a}_j, \ldots, \mathbf{a}_n)$$
$$= \mathcal{D}(\mathbf{a}_1, \ldots, \mathbf{a}_{i-1}, \mathbf{a}_i, \mathbf{a}_{i+1}, \ldots, \mathbf{a}_j, \ldots, \mathbf{a}_n) = \det A.$$

By Property (3), we have

$$\det A' = \mathcal{D}(\mathbf{a}_1, \ldots, \mathbf{a}_{i-1}, \mathbf{a}_i + c\mathbf{a}_j, \mathbf{a}_{i+1}, \ldots, \mathbf{a}_j, \ldots, \mathbf{a}_n)$$
$$= \mathcal{D}(\mathbf{a}_1, \ldots, \mathbf{a}_{i-1}, \mathbf{a}_i, \mathbf{a}_{i+1}, \ldots, \mathbf{a}_j, \ldots, \mathbf{a}_n)$$
$$\qquad + \mathcal{D}(\mathbf{a}_1, \ldots, \mathbf{a}_{i-1}, c\mathbf{a}_j, \mathbf{a}_{i+1}, \ldots, \mathbf{a}_j, \ldots, \mathbf{a}_n)$$
$$= \mathcal{D}(\mathbf{a}_1, \ldots, \mathbf{a}_{i-1}, \mathbf{a}_i, \mathbf{a}_{i+1}, \ldots, \mathbf{a}_j, \ldots, \mathbf{a}_n)$$
$$\qquad + c\mathcal{D}(\mathbf{a}_1, \ldots, \mathbf{a}_{i-1}, \mathbf{a}_j, \mathbf{a}_{i+1}, \ldots, \mathbf{a}_j, \ldots, \mathbf{a}_n)$$
$$= \mathcal{D}(\mathbf{a}_1, \ldots, \mathbf{a}_{i-1}, \mathbf{a}_i, \mathbf{a}_{i+1}, \ldots, \mathbf{a}_j, \ldots, \mathbf{a}_n)$$

since $\mathcal{D}(\mathbf{a}_1, \ldots, \mathbf{a}_{i-1}, \mathbf{a}_j, \mathbf{a}_{i+1}, \ldots, \mathbf{a}_j, \ldots, \mathbf{a}_n) = 0$ by the preceding Lemma. ∎

▶ EXAMPLE 2

We now use column operations to calculate the determinant of the matrix

$$A = \begin{bmatrix} 2 & 2 & 1 \\ 4 & 1 & 0 \\ 6 & 0 & 1 \end{bmatrix}.$$

First we exchange columns 1 and 3, and then we proceed to (column) echelon form:

$$\det A = \begin{vmatrix} 2 & 2 & 1 \\ 4 & 1 & 0 \\ 6 & 0 & 1 \end{vmatrix} = -\begin{vmatrix} 1 & 2 & 2 \\ 0 & 1 & 4 \\ 1 & 0 & 6 \end{vmatrix} = -\begin{vmatrix} 1 & 0 & 0 \\ 0 & 1 & 4 \\ 1 & -2 & 4 \end{vmatrix} = -\begin{vmatrix} 1 & 0 & 0 \\ 0 & 1 & 0 \\ 1 & -2 & 12 \end{vmatrix}.$$

But

$$\begin{vmatrix} 1 & 0 & 0 \\ 0 & 1 & 0 \\ 1 & -2 & 12 \end{vmatrix} = 12\begin{vmatrix} 1 & 0 & 0 \\ 0 & 1 & 0 \\ 1 & -2 & 1 \end{vmatrix},$$

and now we can use the pivots to column-reduce to the identity matrix without changing the determinant. Thus,

$$\det A = -12\begin{vmatrix} 1 & 0 & 0 \\ 0 & 1 & 0 \\ 1 & -2 & 1 \end{vmatrix} = -12\begin{vmatrix} 1 & 0 & 0 \\ 0 & 1 & 0 \\ 0 & 0 & 1 \end{vmatrix} = -12. \quad ◀$$

This is altogether too brain-twisting. We will now go back to the theory and soon show that it's perfectly all right to use row operations. First, let's summarize what we've established so far: We have

Proposition 5.4 *Let A be an n × n matrix.*

1. *Let A' be obtained from A by exchanging two columns. Then $\det A' = -\det A$.*
2. *Let A' be obtained from A by multiplying some column by the number c. Then $\det A' = c \det A$.*
3. *Let A' be obtained from A by adding a multiple of one column to another. Then $\det A' = \det A$.*

Generalizing our discovery in Example 5 of Section 2 of Chapter 4, we have the following characterization of nonsingular matrices that will be critical both here and in Chapter 9.

Theorem 5.5 *Let A be a square matrix. Then A is nonsingular if and only if $\det A \neq 0$.*

Proof Suppose A is nonsingular. Then its reduced (column) echelon form is the identity matrix. Turning this upside down, we can start with the identity matrix and perform a sequence of column operations to obtain A. If we keep track of their effects on the determinant, we see that we've started with $\det I = 1$ and multiplied it by a nonzero number to obtain $\det A$. That is, $\det A \neq 0$. Conversely, suppose A is singular. Then its (column) echelon form U has a column of zeroes and therefore (see Exercise 2) $\det U = 0$. It follows as in the previous case that $\det A = 0$. ∎

Reinterpreting Proposition 5.4, we have

Corollary 5.6 *Let E be an elementary matrix and let A be an arbitrary square matrix. Then*

$$\det(AE) = \det E \det A.$$

Proof Left to the reader in Exercise 3. ∎

Of especial interest is the "product rule" for determinants.

Proposition 5.7 *Let A and B be n × n matrices. Then*

$$\det(AB) = \det A \det B.$$

Proof Suppose B is singular, so that there is some nontrivial linear relation among its column vectors:

$$c_1 \mathbf{b}_1 + \cdots + c_n \mathbf{b}_n = \mathbf{0}.$$

Then, multiplying by A on the left, we find that

$$c_1 (A\mathbf{b}_1) + \cdots + c_n (A\mathbf{b}_n) = \mathbf{0},$$

from which we conclude that there is (the same) nontrivial linear relation among the column vectors of AB, and so AB is singular as well. We infer from Theorem 5.5 that both $\det B = 0$ and $\det AB = 0$, and so the result holds in this case.

Now, if B is nonsingular, we know that we can write B as a product of elementary matrices, viz., $B = E_1 E_2 \cdots E_m$. We now apply Corollary 5.6 twice: First, we have

$$\det B = \det(I E_1 E_2 \cdots E_m) = \det E_1 \det E_2 \cdots \det E_m \det I = \det E_1 \det E_2 \cdots \det E_m;$$

but then we have

$$\det AB = \det(A E_1 E_2 \cdots E_m) = \det E_1 \det E_2 \cdots \det E_m \det A$$
$$= \det A (\det E_1 \det E_2 \cdots \det E_m) = \det A \det B,$$

as claimed. ∎

A consequence of this proposition is that $\det(AB) = \det(BA)$, even though matrix multiplication is not commutative. Thus, we have

Corollary 5.8 *Let E be an elementary matrix and let A be an arbitrary square matrix. Then*

$$\det(EA) = \det E \det A.$$

Now we infer that, analogous to Proposition 5.4, we have

Proposition 5.9 *Let A be an $n \times n$ matrix.*

1. *Let A' be obtained from A by exchanging two rows. Then $\det A' = -\det A$.*
2. *Let A' be obtained from A by multiplying some row by the number c. Then $\det A' = c \det A$.*
3. *Let A' be obtained from A by adding a multiple of one row to another. Then $\det A' = \det A$.*

Another useful observation is the following

Corollary 5.10 *If A is nonsingular, then $\det(A^{-1}) = \dfrac{1}{\det A}$.*

Proof From the equation $AA^{-1} = I$ and Proposition 5.7, we deduce that $\det A \det A^{-1} = 1$, so $\det A^{-1} = 1/\det A$. ∎

Since we've seen that row and column operations have the same effect on determinant, it should not come as a surprise that a matrix and its transpose have the same determinant.

Proposition 5.11 *Let A be a square matrix. Then*

$$\det A^\mathsf{T} = \det A.$$

Proof Suppose A is singular. Then so is A^T (why?). Thus, $\det A^\mathsf{T} = 0 = \det A$, and so the result holds in this case. Suppose now that A is nonsingular. As in the preceding proof, we write $A = E_1 E_2 \cdots E_m$. Now we have $A^\mathsf{T} = (E_1 E_2 \cdots E_m)^\mathsf{T} = E_m^\mathsf{T} \cdots E_2^\mathsf{T} E_1^\mathsf{T}$, and so, using the product rule and Exercise 4, we obtain

$$\det A^\mathsf{T} = \det(E_m^\mathsf{T}) \cdots \det(E_2^\mathsf{T}) \det(E_1^\mathsf{T}) = \det E_1 \det E_2 \cdots \det E_m = \det A. \blacksquare$$

The following result can be useful:

Proposition 5.12 *If A is an upper (lower) triangular $n \times n$ matrix, then $\det A = a_{11} a_{22} \cdots a_{nn}$.*

Proof If $a_{ii} = 0$ for some i, then A is singular (why?) and so $\det A = 0$, and the desired equality holds in this case. Now assume all the a_{ii} are nonzero. Let $\mathbf{A}_i$ be the ith row vector of A, as usual, and write $\mathbf{A}_i = a_{ii} \mathbf{B}_i$, where the ith entry of $\mathbf{B}_i$ is 1. Then, using Property (2) repeatedly, we have $\det A = a_{11} \cdots a_{nn} \det B$. Now B is an upper triangular matrix with 1's on the diagonal, so we can use the pivots to clear out the upper (lower) entries without changing the determinant, and thus $\det B = \det I = 1$. So $\det A = a_{11} a_{22} \cdots a_{nn}$, as promised. $\blacksquare$

Remark As we shall prove in Theorem 1.1 of Chapter 9, any two matrices A and A' representing a linear map T are related by the equation $A' = P^{-1} A P$ for some invertible matrix P. As a consequence of Proposition 5.7, we have

$$\det A' = \det(P^{-1} A P) = \det(A P P^{-1}) = \det A \det(P P^{-1}) = \det A,$$

and so it makes sense to define $\det T = \det A$ for *any* matrix representative of T.

We now come to the geometric meaning of $\det T$: It gives the factor by which signed volume is distorted under the mapping by T. (See Exercise 24 for another approach.)

Proposition 5.13 *Let $T \colon \mathbb{R}^n \to \mathbb{R}^n$ be a linear map, and let R be a parallelepiped. Then $\operatorname{vol}(T(R)) = |\det T| \operatorname{vol}(R)$. Indeed, if $\Omega \subset \mathbb{R}^n$ is a general region, then $\operatorname{vol}(T(\Omega)) = |\det T| \operatorname{vol}(\Omega)$.*

Proof When T has rank $< n$, $\det T = 0$ and the image of T lies in a subspace of dimension $< n$; hence, by Exercise 7.1.12, $T(R)$ has volume zero. When T has rank n, we can write $[T]$ as a product of elementary matrices. Because of Proposition 5.7, it now suffices to prove the result when $[T]$ is an elementary matrix itself.

Recall that there are three kinds of elementary matrices (see p. 148). When R is a rectangle, it is clear that the first type does not change volume, and the second multiplies the volume by $|c|$; the third (a shear) does not change the volume, for the following reason. The transformation is the identity in all directions other than the $x_i x_j$-plane, and we've already checked that in two dimensions the determinant gives the signed area. (See also Exercise 24.)

Suppose Ω is a region. Then we can take a rectangle $\overline{R}$ containing Ω and consider the function

$$\chi : \overline{R} \to \mathbb{R}, \qquad \chi(\mathbf{x}) = \begin{cases} 1, & \mathbf{x} \in \Omega \\ 0, & \text{otherwise} \end{cases}.$$

Since by our definition of region, χ is integrable, given $\varepsilon > 0$, we can find a partition $\mathcal{P}$ of $\overline{R}$ so that $U(\chi, \mathcal{P}) - L(\chi, \mathcal{P}) < \varepsilon$. That is, the sum of the volumes of those subrectangles of $\mathcal{P}$ that intersect the frontier of Ω is less than ε. In particular, this means Ω contains a union, S_1, of subrectangles of $\mathcal{P}$ and is contained in a union, S_2, of subrectangles of $\mathcal{P}$, as shown in Figure 5.1, with the property that $\text{vol}(S_2) - \text{vol}(S_1) < \varepsilon$. And, likewise, $T(\Omega)$ contains a union, $T(S_1)$, of parallelepipeds and is contained in a union, $T(S_2)$, of parallelepipeds, with $\text{vol}(T(S_i)) = |c|\text{vol}(S_i)$ or $\text{vol}(T(S_i)) = \text{vol}(S_i)$, depending on the nature of the elementary matrix. In either event, we see that

$$|\det T|\text{vol}(S_1) \le \text{vol}(T(\Omega)) \le |\det T|\text{vol}(S_2),$$

and since $\varepsilon > 0$ was arbitrary, we are done. (Note that, by Exercise 7.1.11 and Corollary 1.10, $T(\Omega)$ has a well-defined volume.) ∎

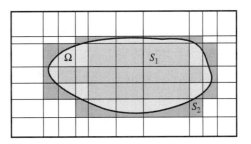

Figure 5.1

5.1 Formulas for the Determinant

In Chapter 1 we had explicit formulas for the determinant of 2×2 and 3×3 matrices. It is sometimes more useful to have a recursive way of calculating the determinant. Given an $n \times n$ matrix A with $n \ge 2$, denote by A_{ij} the $(n-1) \times (n-1)$ matrix obtained by deleting the ith row and the jth column from A. Define the ij^{th} *cofactor* of the matrix to be

$$c_{ij} = (-1)^{i+j} \det A_{ij}.$$

Note that we include the coefficient of ± 1 according to the "checkerboard" pattern as indicated below:

$$\begin{bmatrix} + & - & + & \cdots \\ - & + & - & \cdots \\ + & - & + & \cdots \\ \vdots & \vdots & \vdots & \ddots \end{bmatrix}.$$

Then we have the following formula, called the *expansion in cofactors along the ith row*.

Proposition 5.14 *Let A be an $n \times n$ matrix. Then for any fixed i, we have*

$$\det A = \sum_{j=1}^{n} a_{ij} C_{ij}.$$

Using *rows* here allows us to check that the expression on the right-hand side of this equation satisfies the properties of a determinant as set forth in Theorem 5.1. However, using the fact that $\det A^{\mathsf{T}} = \det A$, we can transpose this result to obtain the *expansion in cofactors along the jth column*.

Proposition 5.15 *Let A be an $n \times n$ matrix. Then for any fixed j, we have*

$$\det A = \sum_{i=1}^{n} a_{ij} C_{ij}.$$

Note that when we define the determinant of a 1×1 matrix by the obvious rule,

$$\det [a] = a,$$

Proposition 5.15 yields the familiar formula for the determinant of a 2×2 matrix and, again, that of a 3×3 matrix.

▶ **EXAMPLE 3**

Let

$$A = \begin{bmatrix} 2 & 1 & 3 \\ 1 & -2 & 3 \\ 0 & 2 & 1 \end{bmatrix}.$$

We calculate $\det A$ by expanding in cofactors along the second row:

$$\det A = (-1)^{(2+1)}(1) \begin{vmatrix} 1 & 3 \\ 2 & 1 \end{vmatrix} + (-1)^{(2+2)}(-2) \begin{vmatrix} 2 & 3 \\ 0 & 1 \end{vmatrix} + (-1)^{(2+3)}(3) \begin{vmatrix} 2 & 1 \\ 0 & 2 \end{vmatrix}$$

$$= -(1)(-5) + (-2)(2) - (3)(4) = -11.$$

Of course, because of the 0 entry in the third row, we'd have been smarter to expand in cofactors along the third row, obtaining

$$\det A = (-1)^{(3+1)}(0) \begin{vmatrix} 1 & 3 \\ -2 & 3 \end{vmatrix} + (-1)^{(3+2)}(2) \begin{vmatrix} 2 & 3 \\ 1 & 3 \end{vmatrix} + (-1)^{(3+3)}(1) \begin{vmatrix} 2 & 1 \\ 1 & -2 \end{vmatrix}$$

$$= -2(3) + 1(-5) = -11. \quad \blacktriangleleft$$

Sketch of proof of Proposition 5.15 As we mentioned earlier, we must check that the expression on the right-hand side has the requisite properties. When we form a new matrix A' by switching two adjacent columns (say, columns k and $k+1$) of A, then whenever $j \neq k$

and $j \neq k+1$, we have $a'_{ij} = a_{ij}$ and $c'_{ij} = -c_{ij}$; on the other hand, when $j = k$, we have $a'_{ik} = a_{i,k+1}$ and $c'_{ik} = -c_{i,k+1}$; when $j = k+1$, we have $a'_{i,k+1} = a_{ik}$ and $c'_{i,k+1} = -c_{ik}$, so

$$\sum_{j=1}^n a'_{ij} c'_{ij} = -\sum_{j=1}^n a_{ij} c_{ij},$$

as required. We can exchange an arbitrary pair of columns by exchanging an *odd number* of adjacent pairs in succession (see Exercise 16), so the general result follows.

The remaining properties are easier to check. If we multiply the kth column by c, then for $j \neq k$, we have $a'_{ij} = a_{ij}$ and $c'_{ij} = cc_{ij}$, whereas for $j = k$, we have $c'_{ik} = c_{ik}$ and $a'_{ik} = ca_{ik}$. Thus,

$$\sum_{j=1}^n a'_{ij} c'_{ij} = c \sum_{j=1}^n a_{ij} c_{ij},$$

as required. Suppose now that we replace the kth column by the sum of two column vectors, viz., $\mathbf{a}'_k = \mathbf{a}_k + \mathbf{a}''_k$. Then for $j \neq k$, we have $c'_{ij} = c_{ij} + c''_{ij}$ and $a'_{ij} = a_{ij} = a''_{ij}$. When $j = k$, we likewise have $c'_{ik} = c_{ik} = c''_{ik}$, but $a'_{ik} = a_{ik} + a''_{ik}$. So

$$\sum_{j=1}^n a'_{ij} c'_{ij} = \sum_{j=1}^n a_{ij} c_{ij} + \sum_{j=1}^n a''_{ij} c''_{ij},$$

as required. ■

Proof of Theorem 5.1 Proposition 5.15 establishes the existence of a function $\mathcal{D}$ satisfying the properties listed in the statement of the theorem. On the other hand, as we saw, calculating determinants by just using the properties, there can only be one such function because, by reducing the matrix to echelon form by column or row operations, we are able to compute the determinant. (See also Exercise 21.) ■

Remark It is worth remarking that expansion in cofactors is an important theoretical tool but a computational nightmare. Even with calculators and computers, to compute an $n \times n$ determinant by expanding in cofactors requires more than $n!$ multiplications[5] (and lots of additions). On the other hand, to compute an $n \times n$ determinant by row reducing the matrix to upper triangular form requires slightly fewer than $\frac{1}{3}n^3$ multiplications (and additions). Now, Stirling's formula tells us that $n!$ grows faster than $(n/e)^n$, which gets large *much* faster than does n^3. Indeed, consider the following table, displaying the number

---

[5] In fact, as n gets larger, the number of multiplications is essentially $(e-1)n!$.

of operations required:

n	cofactors	row or column operations
2	2	2
3	9	8
4	40	20
5	205	40
6	1236	70
7	8659	112
8	69280	168
9	623529	240
10	6235300	330

Thus, we see that once $n > 4$, it is sheer folly to calculate a determinant by the cofactor method (unless almost all the entries of the matrix happen to be 0).

We conclude this section with a few classic formulas. The first is particularly useful for solving 2×2 systems of equations and may be useful even for larger n if you are interested only in a certain component x_i of the solution vector.

Proposition 5.16 (Cramer's Rule) *Let A be a nonsingular $n \times n$ matrix, and let $\mathbf{b} \in \mathbb{R}^n$. Then the ith coordinate of the vector $\mathbf{x}$ solving $A\mathbf{x} = \mathbf{b}$ is*

$$x_i = \frac{\det B_i}{\det A},$$

where B_i is the matrix obtained by replacing the ith column of A by the vector $\mathbf{b}$.

Proof This is amazingly simple. We calculate the determinant of the matrix B_i obtained by replacing the ith column of A by $\mathbf{b} = A\mathbf{x} = x_1\mathbf{a}_1 + \cdots + x_n\mathbf{a}_n$:

$$\det B_i = \begin{vmatrix} | & | & & | & & | \\ \mathbf{a}_1 & \mathbf{a}_2 & \cdots & x_1\mathbf{a}_1 + \cdots + x_n\mathbf{a}_n & \cdots & \mathbf{a}_n \\ | & | & & | & & | \end{vmatrix}$$

$$= \begin{vmatrix} | & | & & | & & | \\ \mathbf{a}_1 & \mathbf{a}_2 & \cdots & x_i\mathbf{a}_i & \cdots & \mathbf{a}_n \\ | & | & & | & & | \end{vmatrix} = x_i \det A$$

since the multiples of columns other than the ith do not contribute to the determinant. ∎

▶ **EXAMPLE 4**

We wish to solve

$$\begin{bmatrix} 2 & 3 \\ 4 & 7 \end{bmatrix} \begin{bmatrix} x_1 \\ x_2 \end{bmatrix} = \begin{bmatrix} 3 \\ -1 \end{bmatrix}.$$

We have

$$B_1 = \begin{bmatrix} 3 & 3 \\ -1 & 7 \end{bmatrix} \quad \text{and} \quad B_2 = \begin{bmatrix} 2 & 3 \\ 4 & -1 \end{bmatrix},$$

so $\det B_1 = 24$, $\det B_2 = -14$, and $\det A = 2$. Therefore, $x_1 = 12$ and $x_2 = -7$. ◀

We now deduce from Cramer's rule an "explicit" formula for the inverse of a non-singular matrix. Students always seem to want an alternative to Gaussian elimination, but what follows is practical only for the 2×2 case (where it gives us our familiar formula from Example 5 on p. 154) and—barely—for the 3×3 case.

Proposition 5.17 *Let A be a nonsingular matrix, and let $C = [c_{ij}]$ be the matrix of its cofactors. Then*

$$A^{-1} = \frac{1}{\det A} C^\mathsf{T}.$$

Proof We recall from p. 152 that the jth column vector of A^{-1} is the solution of $A\mathbf{x} = \mathbf{e}_j$, where $\mathbf{e}_j$ is the jth standard basis vector for $\mathbb{R}^n$. Now, Cramer's rule tells us that the ith coordinate of the jth column of A^{-1} is

$$(A^{-1})_{ij} = \frac{1}{\det A} \det \mathcal{A}_{ji},$$

where $\mathcal{A}_{ji}$ is the matrix obtained by replacing the ith column of A by $\mathbf{e}_j$. Now, we calculate $\det \mathcal{A}_{ji}$ by expanding in cofactors along the ith column of the matrix $\mathcal{A}_{ji}$. Since the only nonzero entry of that column is the jth, and since all its remaining columns are those of the original matrix A, we find that

$$\det \mathcal{A}_{ji} = (-1)^{i+j} \det A_{ji} = c_{ji},$$

and this proves the result. ∎

For 3×3 matrices, this formula isn't bad when $\det A$ would cause troublesome arithmetic in doing Gaussian elimination.

▶ **EXAMPLE 5**

Consider the matrix

$$A = \begin{bmatrix} 1 & 2 & 1 \\ -1 & 1 & 2 \\ 2 & 0 & 3 \end{bmatrix};$$

then

$$\det A = (1) \begin{vmatrix} 1 & 2 \\ 0 & 3 \end{vmatrix} - (2) \begin{vmatrix} -1 & 2 \\ 2 & 3 \end{vmatrix} + (1) \begin{vmatrix} -1 & 1 \\ 2 & 0 \end{vmatrix} = 15,$$

and so we suspect the fractions would not be fun if we implemented Gaussian elimination. Undaunted, we calculate the cofactor matrix:

$$C = \begin{bmatrix} 3 & 7 & -2 \\ -6 & 1 & 4 \\ 3 & -3 & 3 \end{bmatrix},$$

and so

$$A^{-1} = \frac{1}{\det A} C^\mathsf{T} = \frac{1}{15} \begin{bmatrix} 3 & -6 & 3 \\ 7 & 1 & -3 \\ -2 & 4 & 3 \end{bmatrix}. \blacktriangleleft$$

In general, the determinant of an $n \times n$ matrix can be written as the sum of $n!$ terms, each ($\pm$) the product of n entries of the matrix, one from each row and column. This can be deduced either from the recursive formula of Proposition 5.15 or directly from the properties of Theorem 5.1.

Definition A *permutation* σ of the numbers $1, \ldots, n$ is a one-to-one function σ mapping $\{1, \ldots, n\}$ to itself. The *sign* of the permutation σ, denoted $\text{sign}(\sigma)$, is $+1$ if an even number of exchanges is required to change the ordered set $\{1, \ldots, n\}$ to the ordered set $\{\sigma(1), \ldots, \sigma(n)\}$ and -1 if an odd number of exchanges is required.

Remark It is a consequence of the well-definedness of the determinant, which we've already proved, that the sign of a permutation is itself well defined. One can then define $\text{sign}(\sigma)$ to be the determinant of the permutation matrix whose ij-entry is 1 when $j = \sigma(i)$ and 0 otherwise.

Proposition 5.18 *Let A be an $n \times n$ matrix. Then*

$$\det A = \sum_{\text{permutations } \sigma \text{ of } 1,\ldots,n} \text{sign}(\sigma) a_{\sigma(1)1} a_{\sigma(2)2} \cdots a_{\sigma(n)n}$$

$$= \sum_{\text{permutations } \sigma \text{ of } 1,\ldots,n} \text{sign}(\sigma) a_{1\sigma(1)} a_{2\sigma(2)} \cdots a_{n\sigma(n)}.$$

Proof The jth column of A is the vector $\mathbf{a}_j = \sum_{i=1}^{n} a_{ij} \mathbf{e}_i$, and so, by Properties (2) and (3), we have

$$\det A = \mathcal{D}(\mathbf{a}_1, \ldots, \mathbf{a}_n) = \mathcal{D}\left(\sum_{i_1=1}^{n} a_{i_1 1} \mathbf{e}_{i_1}, \sum_{i_2=1}^{n} a_{i_2 2} \mathbf{e}_{i_2}, \ldots, \sum_{i_n=1}^{n} a_{i_n n} \mathbf{e}_{i_n} \right)$$

$$= \sum_{i_1,\ldots,i_n=1}^{n} a_{i_1 1} a_{i_2 2} \cdots a_{i_n n} \mathcal{D}(\mathbf{e}_{i_1}, \mathbf{e}_{i_2}, \ldots, \mathbf{e}_{i_n}),$$

which, by Property (1),

$$= \sum_{\text{permutations } \sigma \text{ of } 1,\ldots,n} \text{sign}(\sigma) a_{\sigma(1)1} a_{\sigma(2)2} \cdots a_{\sigma(n)n} \mathcal{D}(\mathbf{e}_1, \ldots, \mathbf{e}_n)$$

$$= \sum_{\text{permutations } \sigma \text{ of } 1,\ldots,n} \text{sign}(\sigma) a_{\sigma(1)1} a_{\sigma(2)2} \cdots a_{\sigma(n)n},$$

by Property (4). To obtain the second equality, we apply Proposition 5.11. ∎

Recall that $GL(n)$ denotes the set of invertible $n \times n$ matrices (which, by Exercise 6.1.6, is an open subset of $\mathcal{M}_{n \times n}$).

Corollary 5.19 *The function* $f: GL(n) \to GL(n)$, $f(A) = A^{-1}$, *is smooth.*

Proof Proposition 5.18 shows that the determinant of an $n \times n$ matrix is a polynomial expression (of degree n) in its n^2 entries. Thus, we infer from Proposition 5.17 that each entry of A^{-1} is a rational function (quotient of polynomials) of the entries of A. ∎

▶ EXERCISES 7.5

1. Calculate the following determinants:

(a) $\begin{vmatrix} -1 & 6 & -2 \\ 3 & 4 & 5 \\ 5 & 2 & 1 \end{vmatrix}$

(c) $\begin{vmatrix} 1 & 4 & 1 & -3 \\ 2 & 10 & 0 & 1 \\ 0 & 0 & 2 & 2 \\ 0 & 0 & -2 & 1 \end{vmatrix}$

*(b) $\begin{vmatrix} 1 & 0 & 2 & 0 \\ -1 & 2 & -2 & 0 \\ 0 & 1 & 2 & 6 \\ 1 & 1 & 3 & 2 \end{vmatrix}$

*(d) $\begin{vmatrix} 2 & -1 & 0 & 0 & 0 \\ -1 & 2 & -1 & 0 & 0 \\ 0 & -1 & 2 & -1 & 0 \\ 0 & 0 & -1 & 2 & -1 \\ 0 & 0 & 0 & -1 & 2 \end{vmatrix}$

2. Suppose one column of the matrix A consists only of 0 entries; i.e., $\mathbf{a}_i = \mathbf{0}$ for some i. Prove that $\det A = 0$.

3. Prove Corollary 5.6.

♯4. Prove (without using Proposition 5.11) that for any elementary matrix E, we have $\det E^\top = \det E$. (Hint: Consider each of the three types of elementary matrices.)

5. Let A be an $n \times n$ matrix and let c be a scalar. Prove $\det(cA) = c^n \det A$.

6. Prove that if the entries of a matrix A are integers, then $\det A$ is an integer. (Hint: Use Proposition 5.14 and induction or Proposition 5.18.)

7. Given that 1898, 3471, 7215, and 8164 are all divisible by 13, use only the properties of determinants and the result of Exercise 6 to prove that

$$\begin{vmatrix} 1 & 8 & 9 & 8 \\ 3 & 4 & 7 & 1 \\ 7 & 2 & 1 & 5 \\ 8 & 1 & 6 & 4 \end{vmatrix}$$

is divisible by 13.

8. Let $A = \begin{bmatrix} a_1 \\ a_2 \end{bmatrix}$, $B = \begin{bmatrix} b_1 \\ b_2 \end{bmatrix}$, and $C = \begin{bmatrix} c_1 \\ c_2 \end{bmatrix}$ be points in $\mathbb{R}^2$. Show that the signed area of $\triangle ABC$ is given by

$$\frac{1}{2} \begin{vmatrix} a_1 & b_1 & c_1 \\ a_2 & b_2 & c_2 \\ 1 & 1 & 1 \end{vmatrix}.$$

9. Let A be an $n \times n$ matrix. Prove that

$$\det \begin{bmatrix} 1 & 0 & \cdots & 0 \\ \hline 0 & & & \\ \vdots & & A & \\ 0 & & & \end{bmatrix} = \det A.$$

What's the interpretation in terms of (signed) volume?

10. Generalizing Exercise 9, we have the following:
(a) Suppose $A \in \mathcal{M}_{k \times k}$, $B \in \mathcal{M}_{k \times \ell}$, and $D \in \mathcal{M}_{\ell \times \ell}$. Prove that

$$\det \begin{bmatrix} A & B \\ \hline O & D \end{bmatrix} = \det A \det D.$$

(b) Suppose now that A, B, and D are as in part a, and $C \in \mathcal{M}_{\ell \times k}$. Prove that if A is invertible, then

$$\det \begin{bmatrix} A & B \\ \hline C & D \end{bmatrix} = \det A \det(D - CA^{-1}B).$$

(c) If we assume, moreover, that $k = \ell$ and $AC = CA$, then deduce that

$$\det \begin{bmatrix} A & B \\ \hline C & D \end{bmatrix} = \det(AD - CB).$$

(d) Give examples to show that the result of part c needn't hold when A is singular or when A and C do not commute.

*11. Suppose A is an *orthogonal* $n \times n$ matrix. (Recall that this means that $A^\mathsf{T} A = I_n$.) Compute $\det A$.

12. Suppose A is a skew-symmetric $n \times n$ matrix. (Recall that this means that $A^\mathsf{T} = -A$.) Prove that when n is odd, $\det A = 0$. Give an example to show this needn't be true when n is even. (Hint: Use Exercise 5.)

*13. Let $A = \begin{bmatrix} 1 & 2 & 1 \\ 2 & 3 & 0 \\ 1 & 4 & 2 \end{bmatrix}$.

(a) If $A\mathbf{x} = \begin{bmatrix} 1 \\ 2 \\ -1 \end{bmatrix}$, use Cramer's rule to find x_2.

(b) Find A^{-1} by using cofactors.

*14. Using cofactors, find the determinant and the inverse of the matrix
$$A = \begin{bmatrix} -1 & 2 & 3 \\ 2 & 1 & 0 \\ 0 & 2 & 3 \end{bmatrix}.$$

♯15. (a) Suppose A is an $n \times n$ matrix with integer entries and $\det A = \pm 1$. Prove that A^{-1} has all integer entries.

(b) Conversely, suppose A and A^{-1} are both matrices with integer entries. Prove that $\det A = \pm 1$.

16. Prove that the exchange of any pair of rows (or columns) of a matrix can be accomplished by an odd number of exchanges of adjacent pairs.

17. Suppose A is an orthogonal $n \times n$ matrix. Show that the cofactor matrix $C = \pm A$.

18. Generalizing the result of Proposition 5.17, prove that $AC^T = (\det A)I$ even if A happens to be singular. In particular, when A is singular, what can you conclude about the columns of C^T?

19. (a) Show that if $\begin{bmatrix} x_1 \\ y_1 \end{bmatrix}$ and $\begin{bmatrix} x_2 \\ y_2 \end{bmatrix}$ are distinct points in $\mathbb{R}^2$, then the unique line passing through them is given by the equation
$$\begin{vmatrix} 1 & 1 & 1 \\ x & x_1 & x_2 \\ y & y_1 & y_2 \end{vmatrix} = 0.$$

(b) Show that if $\begin{bmatrix} x_1 \\ y_1 \\ z_1 \end{bmatrix}$, $\begin{bmatrix} x_2 \\ y_2 \\ z_2 \end{bmatrix}$, and $\begin{bmatrix} x_3 \\ y_3 \\ z_3 \end{bmatrix}$ are noncollinear points in $\mathbb{R}^3$, then the unique plane passing through them is given by the equation
$$\begin{vmatrix} 1 & 1 & 1 & 1 \\ x & x_1 & x_2 & x_3 \\ y & y_1 & y_2 & y_3 \\ z & z_1 & z_2 & z_3 \end{vmatrix} = 0.$$

20. As we saw in Exercises 4.1.22 and 4.1.23, through any three noncollinear points in $\mathbb{R}^2$ there pass a unique parabola[6] $y = ax^2 + bx + c$ and a unique circle $x^2 + y^2 + ax + by + c = 0$. Given three such points, $\begin{bmatrix} x_1 \\ y_1 \end{bmatrix}$, $\begin{bmatrix} x_2 \\ y_2 \end{bmatrix}$, and $\begin{bmatrix} x_3 \\ y_3 \end{bmatrix}$, show that the equation of the parabola and circle are, respectively,
$$\begin{vmatrix} 1 & 1 & 1 & 1 \\ x & x_1 & x_2 & x_3 \\ x^2 & x_1^2 & x_2^2 & x_3^2 \\ y & y_1 & y_2 & y_3 \end{vmatrix} = 0 \quad \text{and} \quad \begin{vmatrix} 1 & 1 & 1 & 1 \\ x & x_1 & x_2 & x_3 \\ y & y_1 & y_2 & y_3 \\ x^2+y^2 & x_1^2+y_1^2 & x_2^2+y_2^2 & x_3^2+y_3^2 \end{vmatrix} = 0.$$

21. Using Corollary 5.6, prove that the determinant function is uniquely determined by the properties listed in Theorem 5.1. (Hint: Mimic the proof of Proposition 5.7. It might be helpful to consider two functions, det and $\widetilde{\det}$, that have these properties and prove that $\det(A) = \widetilde{\det}(A)$ for every square matrix A.)

---

[6]Here we must also assume that no pair of the points lies on a vertical line.

22. Let $v_1, \ldots, v_k \in \mathbb{R}^n$. Show that

$$\begin{vmatrix} v_1 \cdot v_1 & v_1 \cdot v_2 & \cdots & v_1 \cdot v_k \\ v_2 \cdot v_1 & v_2 \cdot v_2 & & v_2 \cdot v_k \\ \vdots & \vdots & \ddots & \vdots \\ v_k \cdot v_1 & v_k \cdot v_2 & \cdots & v_k \cdot v_k \end{vmatrix}$$

is the square of the (k-dimensional) volume of the k-dimensional parallelepiped spanned by $v_1, \ldots, v_k$. (Hints: First take care of the case that $\{v_1, \ldots, v_k\}$ is linearly dependent. Now, supposing they are linearly independent and therefore span a k-dimensional subspace V, choose an orthonormal basis $\{u_{k+1}, \ldots, u_n\}$ for $V^\perp$. What is the relation between the k-dimensional volume of the parallelepiped spanned by $v_1, \ldots, v_k$ and the n-dimensional volume of the parallelepiped spanned by $v_1, \ldots, v_k, u_{k+1}, \ldots, u_n$?)

23. (a) Using Proposition 5.18, prove that $D(\det)(I)B = \operatorname{tr} B = b_{11} + \cdots + b_{nn}$. (See Exercise 1.4.22.)

(b) More generally, show that for any invertible matrix A, $D(\det)(A)B = \det A \operatorname{tr}(A^{-1}B)$.

24. Give an alternative proof of Proposition 5.13 for general parallelepipeds as follows. Let $R \subset \mathbb{R}^n$ be a parallelepiped. Suppose $T: \mathbb{R}^n \to \mathbb{R}^n$ is a linear map of either of the forms

$$T\begin{pmatrix} x_1 \\ x_2 \\ \vdots \\ x_n \end{pmatrix} = \begin{bmatrix} cx_1 \\ x_2 \\ \vdots \\ x_n \end{bmatrix} \quad \text{or} \quad T\begin{pmatrix} x_1 \\ x_2 \\ \vdots \\ x_n \end{pmatrix} = \begin{bmatrix} x_1 + cx_2 \\ x_2 \\ \vdots \\ x_n \end{bmatrix}.$$

Calculate the volume of R and of $T(R)$ by applying Fubini's Theorem, putting the x_1 integral innermost. (This is in essence a proof of Cavalieri's principle.)

25. **(From the 1994 Putnam Exam)** Find the value of m so that the line $y = mx$ bisects the region

$$\left\{ \begin{bmatrix} x \\ y \end{bmatrix} \in \mathbb{R}^2 : \frac{x^2}{4} + y^2 \leq 1,\ x \geq 0,\ y \geq 0 \right\}.$$

26. Given any ellipse, show that there are infinitely many inscribed triangles of maximal area.

27. **(From the 1994 Putnam Exam)** Let A and B be 2×2 matrices with integer entries such that A, $A + B$, $A + 2B$, $A + 3B$, and $A + 4B$ are all invertible matrices whose inverses have integer entries. Prove that $A + 5B$ is invertible and that its inverse has integer entries. (Hint: Use Exercise 15.)

▶ 6 CHANGE OF VARIABLES THEOREM

We end this chapter with a general theorem justifying our formulas for integration in polar, cylindrical, and spherical coordinates. Since we know that the determinant tells us the factor by which linear maps distort signed volume, and since the derivative gives the best linear approximation, we expect a change of variables formula to involve the determinant of the derivative matrix. Giving a rigorous proof is, however, another matter.

Since integration is based upon rectangles rather than balls, it is most convenient to choose (for this section only) a different norm to measure vectors and linear maps, which, for obvious reasons, we dub the *cubical norm*.

Definition If $\mathbf{x} \in \mathbb{R}^n$, set $\|\mathbf{x}\|_\square = \max(|x_1|, |x_2|, \ldots, |x_n|)$. If $T: \mathbb{R}^n \to \mathbb{R}^m$ is a linear map, set $\|T\|_\square = \max_{\|\mathbf{x}\|_\square = 1} \|T(\mathbf{x})\|_\square$.

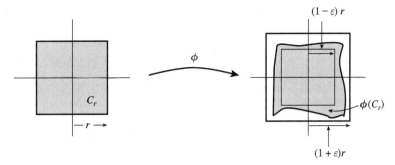

Figure 6.1

We leave it to the reader to check in Exercise 1 that these are indeed norms and, as will be crucial for us, that $\|T(\mathbf{x})\|_\square \le \|T\|_\square \|\mathbf{x}\|_\square$ for all $\mathbf{x} \in \mathbb{R}^n$. Our first result, depicted in Figure 6.1, estimates how much a $\mathcal{C}^1$ map can distort a cube.

Lemma 6.1 *Let C_r denote the cube in $\mathbb{R}^n$ of sidelength $2r$ centered at $\mathbf{0}$. Suppose $U \subset \mathbb{R}^n$ is an open set containing C_r and $\boldsymbol{\phi} \colon U \to \mathbb{R}^n$ is a $\mathcal{C}^1$ function with the property that $\boldsymbol{\phi}(\mathbf{0}) = \mathbf{0}$ and $\|D\boldsymbol{\phi}(\mathbf{x}) - I\|_\square < \varepsilon$ for all $\mathbf{x} \in C_r$ and some $0 < \varepsilon < 1$. Then*

$$C_{(1-\varepsilon)r} \subset \boldsymbol{\phi}(C_r) \subset C_{(1+\varepsilon)r}.$$

Proof One can check that Proposition 1.3 of Chapter 6 holds when we use the $\|\cdot\|_\square$ norm instead of the usual one (see Exercise 1f). Then if $\mathbf{x} \in C_r$, we have

$$\|\boldsymbol{\phi}(\mathbf{x})\|_\square \le \max_{\mathbf{y} \in [\mathbf{0}, \mathbf{x}]} \|D\boldsymbol{\phi}(\mathbf{y})\|_\square \|\mathbf{x}\|_\square < (1+\varepsilon)r,$$

so $\boldsymbol{\phi}(C_r) \subset C_{(1+\varepsilon)r}$. The other inclusion can be proved by applying Exercise 6.2.11 in the $\|\cdot\|_\square$ norm. ∎

The crucial ingredient in the proof of the Change of Variables Theorem is the following result, which says that for sufficiently small cubes C, the image $\mathbf{g}(C)$ is well approximated by the image under the derivative at the center of C.

Proposition 6.2 *Suppose $U \subset \mathbb{R}^n$ is open, $\mathbf{g} \colon U \to \mathbb{R}^n$ is $\mathcal{C}^1$, and $D\mathbf{g}(\mathbf{x})$ is invertible for every $\mathbf{x} \in U$. Let $C \subset U$ be a cube with center $\mathbf{a}$, and suppose*

$$\|D\mathbf{g}(\mathbf{a})^{-1} \circ D\mathbf{g}(\mathbf{x}) - I\|_\square < \varepsilon < 1 \quad \text{for all } \mathbf{x} \in C.$$

Then $\mathbf{g}(C)$ is a region (and hence has volume) and

$$(1-\varepsilon)^n |\det D\mathbf{g}(\mathbf{a})| \operatorname{vol}(C) \le \operatorname{vol}(\mathbf{g}(C)) \le (1+\varepsilon)^n |\det D\mathbf{g}(\mathbf{a})| \operatorname{vol}(C).$$

Proof Since $\mathbf{g}$ is $\mathcal{C}^1$ with invertible derivative at each point of U, $\mathbf{g}$ maps open sets to open sets and the frontier of $\mathbf{g}(C)$ is the image of the frontier of C, hence a set of zero volume (see Exercise 7.1.12). Therefore, $\mathbf{g}(C)$ is a region.

Suppose the sidelength of the cube C is $2r$. We apply Lemma 6.1 to the function $\boldsymbol{\phi}$ defined by
$$\boldsymbol{\phi}(\mathbf{x}) = D\mathbf{g}(\mathbf{a})^{-1}\big(\mathbf{g}(\mathbf{x} + \mathbf{a}) - \mathbf{g}(\mathbf{a})\big), \quad \mathbf{x} \in C_r.$$
Then $\boldsymbol{\phi}(\mathbf{0}) = \mathbf{0}$, $D\boldsymbol{\phi}(\mathbf{0}) = I$, and $D\boldsymbol{\phi}(\mathbf{x}) = D\mathbf{g}(\mathbf{a})^{-1} \circ D\mathbf{g}(\mathbf{x} + \mathbf{a})$, so, by the hypothesis, $\|D\boldsymbol{\phi}(\mathbf{x}) - I\|_\square < \varepsilon$ for all $\mathbf{x} \in C_r$. Therefore, we have
$$C_{(1-\varepsilon)r} \subset \boldsymbol{\phi}(C_r) \subset C_{(1+\varepsilon)r},$$
and so
$$\mathbf{g}(\mathbf{a}) + D\mathbf{g}(\mathbf{a})\big(C_{(1-\varepsilon)r}\big) \subset \mathbf{g}(C) \subset \mathbf{g}(\mathbf{a}) + D\mathbf{g}(\mathbf{a})\big(C_{(1+\varepsilon)r}\big).$$
Applying Proposition 5.5.13, using the fact that $\mathrm{vol}(C_{\alpha r}) = \alpha^n \mathrm{vol}(C_r)$, and remembering that translation preserves volume, we obtain the result. ∎

We begin our onslaught on the Change of Variables Theorem with a very simple case, whose proof is left to the reader in Exercise 2.

Lemma 6.3 *Suppose $T: \mathbb{R}^n \to \mathbb{R}^n$ is a linear map whose standard matrix is diagonal and nonsingular. Let $R \subset \mathbb{R}^n$ be a rectangle, and suppose f is integrable on $T(R)$. Then $f \circ T$ is integrable on R and*
$$\int_{T(R)} f(\mathbf{y}) dV_{\mathbf{y}} = |\det T| \int_R (f \circ T)(\mathbf{x}) dV_{\mathbf{x}}.$$

Theorem 6.4 (Change of Variables Theorem) *Let $\Omega \subset \mathbb{R}^n$ be a region and let U be an open set containing Ω so that $\mathbf{g}: U \to \mathbb{R}^n$ is one-to-one and $\mathcal{C}^1$ with invertible derivative at each point. Suppose $f: \mathbf{g}(\Omega) \to \mathbb{R}$ and $(f \circ \mathbf{g})|\det D\mathbf{g}|: \Omega \to \mathbb{R}$ are both integrable. Then*
$$\int_{\mathbf{g}(\Omega)} f(\mathbf{y}) dV_{\mathbf{y}} = \int_\Omega (f \circ \mathbf{g})(\mathbf{x}) |\det D\mathbf{g}(\mathbf{x})| dV_{\mathbf{x}}.$$

Remark One can strengthen the theorem, in particular by allowing $D\mathbf{g}(\mathbf{x})$ to fail to be invertible on a set of volume 0. This is important for many applications—e.g., polar, cylindrical, and spherical coordinates. But we won't bother justifying it here.

Proof First, we may assume Ω is a rectangle R (as usual, by choosing a rectangle R with $\Omega \subset R$ and working with the function $\tilde{f}$). Next, by applying Lemma 6.3, we may assume R is a cube. That is, choose a cube C and a linear map $T: \mathbb{R}^n \to \mathbb{R}^n$ so that $T(C) = R$. Then, by the chain rule (Theorem 3.2 of Chapter 3) and the product rule for determinants (Proposition 5.7) and recalling that a linear map T is its own derivative, we have
$$\det D(\mathbf{g} \circ T)(\mathbf{u}) = \det \big(D\mathbf{g}(T(\mathbf{u}))DT(\mathbf{u})\big) = \det \big(D\mathbf{g}(T(\mathbf{u}))\big) \det T,$$

and so

$$\int_R (f \circ \mathbf{g})(\mathbf{x})|\det D\mathbf{g}(\mathbf{x})|dV_\mathbf{x} = |\det T| \int_C ((f \circ \mathbf{g}) \circ T)(\mathbf{u})|\det D\mathbf{g}(T(\mathbf{u}))|dV_\mathbf{u}$$

by the lemma

$$= \int_C ((f \circ \mathbf{g}) \circ T)(\mathbf{u})|\det D(\mathbf{g} \circ T)(\mathbf{u})|dV_\mathbf{u}$$

by the previous comment

$$= \int_C (f \circ (\mathbf{g} \circ T))(\mathbf{u})|\det D(\mathbf{g} \circ T)(\mathbf{u})|dV_\mathbf{u}.$$

Thus, to prove the theorem, we substitute $\mathbf{g} \circ T$ for $\mathbf{g}$ and work on the cube C; that is, it suffices to assume R is a cube.

There are positive constants M and N so that $|f| \leq M$ (by integrability) and $\|(D\mathbf{g})^{-1}\|_\square \leq N$ (by continuity and compactness). Choose $0 < \varepsilon < 1$. By uniform continuity, Theorem 1.4 of Chapter 5, there is $\delta_1 > 0$ so that $\|D\mathbf{g}(\mathbf{x}) - D\mathbf{g}(\mathbf{y})\|_\square \leq \varepsilon/N$ whenever $\|\mathbf{x} - \mathbf{y}\| < \delta_1, \mathbf{x}, \mathbf{y} \in R$. Similarly, there is $\delta_2 > 0$ so that $|\det D\mathbf{g}(\mathbf{x}) - \det D\mathbf{g}(\mathbf{y})| < \varepsilon/M$ whenever $\|\mathbf{x} - \mathbf{y}\| < \delta_2, \mathbf{x}, \mathbf{y} \in R$. And by integrability of $(f \circ \mathbf{g})|\det D\mathbf{g}|$, there is $\delta_3 > 0$ so that whenever the diameter of the cubes of a cubical partition $\mathcal{P}$ is less than δ_3, we have $U((f \circ \mathbf{g})|\det D\mathbf{g}|, \mathcal{P}) - L((f \circ \mathbf{g})|\det D\mathbf{g}|, \mathcal{P}) < \varepsilon$ (see Exercise 7.1.10).

Suppose $\mathcal{P} = \{R_1, \ldots, R_s\}$ is a partition of R into cubes of diameter less than $\delta = \min(\delta_1, \delta_2, \delta_3)$. Let

$$M_i = \sup_{\mathbf{x} \in R_i}(f \circ \mathbf{g})(\mathbf{x});$$
$$m_i = \inf_{\mathbf{x} \in R_i}(f \circ \mathbf{g})(\mathbf{x});$$
$$\widetilde{M}_i = \sup_{\mathbf{x} \in R_i}(f \circ \mathbf{g})(\mathbf{x})|\det D\mathbf{g}(\mathbf{x})|;$$
$$\widetilde{m}_i = \inf_{\mathbf{x} \in R_i}(f \circ \mathbf{g})(\mathbf{x})|\det D\mathbf{g}(\mathbf{x})|.$$

We claim that if $\mathbf{a}_i$ is the center of the cube R_i, then

(*) $\qquad \widetilde{m}_i - \varepsilon \leq m_i |\det D\mathbf{g}(\mathbf{a}_i)| \quad$ and $\quad M_i |\det D\mathbf{g}(\mathbf{a}_i)| \leq \widetilde{M}_i + \varepsilon.$

We check the latter: Choose a sequence of points $\mathbf{x}_k \in R_i$ so that $(f \circ \mathbf{g})(\mathbf{x}_k) \to M_i$ (and we assume $M_i > 0$ and all $(f \circ \mathbf{g})(\mathbf{x}_k) > 0$ for convenience). We have $|\det D\mathbf{g}(\mathbf{a}_i)| < |\det D\mathbf{g}(\mathbf{x}_k)| + \varepsilon/M$ and so

$$(f \circ \mathbf{g})(\mathbf{x}_k)|\det D\mathbf{g}(\mathbf{a}_i)| < (f \circ \mathbf{g})(\mathbf{x}_k)|\det D\mathbf{g}(\mathbf{x}_k)| + (f \circ \mathbf{g})(\mathbf{x}_k)\frac{\varepsilon}{M}$$
$$\leq (f \circ \mathbf{g})(\mathbf{x}_k)|\det D\mathbf{g}(\mathbf{x}_k)| + \varepsilon \leq \widetilde{M}_i + \varepsilon.$$

Taking the limit as $k \to \infty$, we conclude that

$$M_i |\det D\mathbf{g}(\mathbf{a}_i)| \leq \widetilde{M}_i + \varepsilon,$$

as required.

On any cube R_i with center $\mathbf{a}_i$, we have

$$\|D\mathbf{g}(\mathbf{a}_i)^{-1} \circ D\mathbf{g}(\mathbf{x}) - I\|_\square \leq \|D\mathbf{g}(\mathbf{a}_i)^{-1}\|_\square \|D\mathbf{g}(\mathbf{x}) - D\mathbf{g}(\mathbf{a})\|_\square < N\frac{\varepsilon}{N} = \varepsilon$$

for all $\mathbf{x} \in R_i$. By Proposition 6.2, we have

$$(1-\varepsilon)^n |\det D\mathbf{g}(\mathbf{a}_i)| \text{vol}(R_i) \le \text{vol}(\mathbf{g}(R_i)) \le (1+\varepsilon)^n |\det D\mathbf{g}(\mathbf{a}_i)| \text{vol}(R_i).$$

Now, $\displaystyle\int_{\mathbf{g}(R)} f\, dV = \sum_{i=1}^{s} \int_{\mathbf{g}(R_i)} f\, dV$, and

$$m_i \text{vol}(\mathbf{g}(R_i)) \le \int_{\mathbf{g}(R_i)} f\, dV \le M_i \text{vol}(\mathbf{g}(R_i)) \quad \text{for } i = 1, \ldots, s.$$

Therefore, we have

$$(1-\varepsilon)^n \sum_{i=1}^{s} m_i |\det D\mathbf{g}(\mathbf{a}_i)| \text{vol}(R_i) \le \int_{\mathbf{g}(R)} f\, dV \le (1+\varepsilon)^n \sum_{i=1}^{s} M_i |\det D\mathbf{g}(\mathbf{a}_i)| \text{vol}(R_i).$$

Substituting the inequalities $(*)$, we find

$$(1-\varepsilon)^n \sum_{i=1}^{s} (\widetilde{m}_i - \varepsilon) \text{vol}(R_i) \le \int_{\mathbf{g}(R)} f\, dV \le (1+\varepsilon)^n \sum_{i=1}^{s} (\widetilde{M}_i + \varepsilon) \text{vol}(R_i).$$

Now, since $0 < \varepsilon < 1$, we have

$$(1+\varepsilon)^n < 2^n, \qquad\qquad (1+\varepsilon)^n - 1 < 2^{n-1} n\varepsilon \quad \text{(by the mean value theorem)};$$
$$(1-\varepsilon)^n < 1, \quad \text{and} \quad 1 - (1-\varepsilon)^n < n\varepsilon \quad \text{(by the mean value theorem)}.$$

Therefore,

$$\sum_{i=1}^{s} \widetilde{m}_i \text{vol}(R_i) - \varepsilon(\text{vol}(R) + Mn) \le \int_{\mathbf{g}(R)} f\, dV$$
$$\le \sum_{i=1}^{s} \widetilde{M}_i \text{vol}(R_i) + \varepsilon(2^n \text{vol}(R) + 2^{n-1} Mn).$$

We've almost arrived at the end. For convenience, let $\beta = 2^n \text{vol}(R) + 2^{n-1} Mn$. Recall that since $(f \circ \mathbf{g})|\det D\mathbf{g}|$ is integrable, its integral is the unique number lying between all its upper and lower sums. Suppose now that $\displaystyle\int_{\mathbf{g}(R)} f\, dV \ne \int_R (f \circ \mathbf{g})|\det D\mathbf{g}|\, dV$. In particular, suppose $\displaystyle\int_{\mathbf{g}(R)} f\, dV = \int_R (f \circ \mathbf{g})|\det D\mathbf{g}|\, dV + \gamma$ for some $\gamma > 0$. Let $\varepsilon > 0$ be chosen small enough so that $(\beta + 1)\varepsilon < \gamma$. We have

$$\int_{\mathbf{g}(R)} f\, dV \le U((f \circ \mathbf{g})|\det D\mathbf{g}|, \mathcal{P}) + \beta\varepsilon < \int_R (f \circ \mathbf{g})|\det D\mathbf{g}|\, dV + (\beta+1)\varepsilon$$
$$< \int_R (f \circ \mathbf{g})|\det D\mathbf{g}|\, dV + \gamma = \int_{\mathbf{g}(R)} f\, dV,$$

which is a contradiction. Similarly, supposing that $\gamma < 0$ leads to a contradiction. Thus,

$$\int_{\mathbf{g}(R)} f\, dV = \int_R (f \circ \mathbf{g})|\det D\mathbf{g}|\, dV,$$

as desired. ∎

EXAMPLE 1

First, to be official, we check that the formulas we derived in a heuristic manner in Section 4 are valid.

a. **Polar coordinates:** Let $\mathbf{g}\begin{pmatrix} r \\ \theta \end{pmatrix} = \begin{bmatrix} r \cos\theta \\ r \sin\theta \end{bmatrix}$. Then

$$D\mathbf{g}\begin{pmatrix} r \\ \theta \end{pmatrix} = \begin{bmatrix} \cos\theta & -r\sin\theta \\ \sin\theta & r\cos\theta \end{bmatrix} \quad \text{and} \quad \det D\mathbf{g}\begin{pmatrix} r \\ \theta \end{pmatrix} = r.$$

b. **Cylindrical coordinates:** $\mathbf{g}\begin{pmatrix} r \\ \theta \\ z \end{pmatrix} = \begin{bmatrix} r\cos\theta \\ r\sin\theta \\ z \end{bmatrix}$. Then

$$D\mathbf{g}\begin{pmatrix} r \\ \theta \\ z \end{pmatrix} = \begin{bmatrix} \cos\theta & -r\sin\theta & 0 \\ \sin\theta & r\cos\theta & 0 \\ 0 & 0 & 1 \end{bmatrix} \quad \text{and} \quad \det D\mathbf{g}\begin{pmatrix} r \\ \theta \\ z \end{pmatrix} = r.$$

c. **Spherical coordinates:** Let $\mathbf{g}\begin{pmatrix} \rho \\ \phi \\ \theta \end{pmatrix} = \begin{bmatrix} \rho\sin\phi\cos\theta \\ \rho\sin\phi\sin\theta \\ \rho\cos\phi \end{bmatrix}$. Then

$$D\mathbf{g}\begin{pmatrix} \rho \\ \phi \\ \theta \end{pmatrix} = \begin{bmatrix} \sin\phi\cos\theta & \rho\cos\phi\cos\theta & -\rho\sin\phi\sin\theta \\ \sin\phi\sin\theta & \rho\cos\phi\sin\theta & \rho\sin\phi\cos\theta \\ \cos\phi & -\rho\sin\phi & 0 \end{bmatrix},$$

and, expanding in cofactors along the third row, we find that

$$\det D\mathbf{g}\begin{pmatrix} \rho \\ \phi \\ \theta \end{pmatrix} = \cos\phi(\rho^2 \sin\phi\cos\phi) + \rho\sin\phi(\rho\sin^2\phi) = \rho^2 \sin\phi.$$

EXAMPLE 2

Let $S \subset \mathbb{R}^2$ be the parallelogram with vertices $\begin{bmatrix} 0 \\ 0 \end{bmatrix}, \begin{bmatrix} 3 \\ 1 \end{bmatrix}, \begin{bmatrix} 4 \\ 3 \end{bmatrix}$, and $\begin{bmatrix} 1 \\ 2 \end{bmatrix}$, as pictured in Figure 6.2. Evaluate $\int_S x\, dA$. Of course, with a bit of patience, we could evaluate this by three different iterated integrals in cartesian coordinates, but it makes sense to take a linear transformation $\mathbf{g}$ that maps the unit square, R, to the region S; e.g.,

$$\mathbf{g}\begin{pmatrix} u \\ v \end{pmatrix} = \begin{bmatrix} 3 & 1 \\ 1 & 2 \end{bmatrix}\begin{bmatrix} u \\ v \end{bmatrix} = \begin{bmatrix} x \\ y \end{bmatrix}.$$

Then, applying the Change of Variables Theorem, we have

$$\int_S x\, dA_{xy} = \int_R \underbrace{(3u+v)}_{x} \underbrace{5}_{|\det D\mathbf{g}|}\, dA_{uv}$$

$$= 5\int_0^1 \int_0^1 (3u+v)\, dv\, du = 5\int_0^1 (3u+\tfrac{1}{2})\, du = 5\cdot 2 = 10.$$

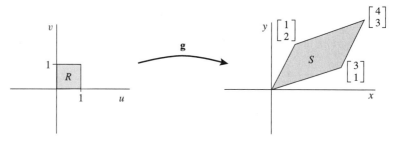

Figure 6.2

EXAMPLE 3

Let $S \subset \mathbb{R}^2$ be the region bounded by the curves $y = x$, $y = 3$, $xy = 1$, and $xy = 4$. We wish to evaluate $\int_S y\, dA$. The equations of the boundary curves suggest a substitution $u = xy$, $v = y/x$. To determine the function $\mathbf{g}$ so that $\mathbf{g}\begin{pmatrix} u \\ v \end{pmatrix} = \begin{bmatrix} x \\ y \end{bmatrix}$, we need the inverse function (note that S lies in the first quadrant):

$$\mathbf{g}\begin{pmatrix} u \\ v \end{pmatrix} = \begin{bmatrix} \sqrt{\frac{u}{v}} \\ \sqrt{uv} \end{bmatrix}.$$

If we look at Figure 6.3, it is easy to check that $\mathbf{g}$ maps the region $\Omega = \left\{ \begin{bmatrix} u \\ v \end{bmatrix} : 1 \leq u \leq 4,\ 1 \leq v \leq \frac{9}{u} \right\}$ to S. Now,

$$D\mathbf{g}\begin{pmatrix} u \\ v \end{pmatrix} = \begin{bmatrix} \frac{1}{2\sqrt{uv}} & -\frac{1}{2}\sqrt{\frac{u}{v^3}} \\ \frac{1}{2}\sqrt{\frac{v}{u}} & \frac{1}{2}\sqrt{\frac{u}{v}} \end{bmatrix}, \quad \text{so} \quad \det D\mathbf{g}\begin{pmatrix} u \\ v \end{pmatrix} = \frac{1}{2v}.$$

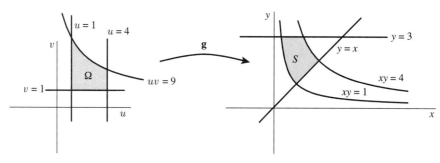

Figure 6.3

Then, by the Change of Variables Theorem, we have

$$\int_S y\, dA_{xy} = \int_\Omega \sqrt{uv}\,\frac{1}{2v}\,dA_{uv}$$
$$= \frac{1}{2}\int_1^4 \int_1^{9/u} \sqrt{\frac{u}{v}}\,dv\,du$$
$$= \int_1^4 \left(\sqrt{u}\,\sqrt{v}\right]_1^{9/u}\,du = \int_1^4 (3 - \sqrt{u})\,du = \frac{13}{3}.$$

EXERCISES 7.6

1. Suppose $\mathbf{x}, \mathbf{y} \in \mathbb{R}^n$, S and T are linear maps from $\mathbb{R}^n$ to $\mathbb{R}^m$, and $c \in \mathbb{R}$.
   (a) Prove that $\|\mathbf{x} + \mathbf{y}\|_\square \leq \|\mathbf{x}\|_\square + \|\mathbf{y}\|_\square$ and $\|c\mathbf{x}\|_\square = |c|\|\mathbf{x}\|_\square$.
   (b) Prove that $\|S + T\|_\square \leq \|S\|_\square + \|T\|_\square$ and $\|cT\|_\square = |c|\|T\|_\square$.
   (c) Prove that $\|T(\mathbf{x})\|_\square \leq \|T\|_\square \|\mathbf{x}\|_\square$.
   (d) Suppose the standard matrix for T is the $m \times n$ matrix A. Prove that
   $$\|T\|_\square = \max_{1 \leq i \leq m} \sum_{j=1}^n |a_{ij}|.$$
   (e) Check that $\|\mathbf{x}\|_\square \leq \|\mathbf{x}\| \leq \sqrt{n}\|\mathbf{x}\|_\square$ and $\frac{1}{\sqrt{n}}\|T\|_\square \leq \|T\| \leq \sqrt{n}\|T\|_\square$.
   (f) Suppose $\mathbf{g}\colon [a, b] \to \mathbb{R}^n$ is continuous. Prove that $\left\|\int_a^b \mathbf{g}(t)\,dt\right\|_\square \leq \int_a^b \|\mathbf{g}(t)\|_\square\,dt.$ (This is needed to prove Proposition 1.3 of Chapter 6 with the $\|\cdot\|_\square$ norm.)

2. Prove Lemma 6.3.

3. Find the area of the ellipse $\dfrac{x^2}{a^2} + \dfrac{y^2}{b^2} \leq 1$ and the volume of the ellipsoid $\dfrac{x^2}{a^2} + \dfrac{y^2}{b^2} + \dfrac{z^2}{c^2} \leq 1$. (Cf. also Exercise 9.4.17.)

4. Let S be the triangle with vertices at $\begin{bmatrix}0\\0\end{bmatrix}, \begin{bmatrix}1\\0\end{bmatrix},$ and $\begin{bmatrix}0\\1\end{bmatrix}$. Let $f\begin{pmatrix}x\\y\end{pmatrix} = e^{(x-y)/(x+y)}$. Evaluate the integral $\int_S f\,dA$
   (a) by changing to polar coordinates,
   (b) by making the change of variables $u = x - y$, $v = x + y$.

*5. Let S be the plane region bounded by $y = 0$, $2x + y = 1$, $2x + y = 5$, and $-x + 3y = 1$. Evaluate $\int_S \dfrac{x - 3y}{2x + y}\,dA.$

6. Rework Example 3 with the substitution $u = xy$, $v = y$.

*7. Let S be the plane region bounded by $x = 0$, $y = 0$, and $x + y = 1$. Evaluate $\int_S \cos\left(\dfrac{x-y}{x+y}\right)dA.$
(Remark: The integrand is undefined at the origin. Does this cause a problem?)

8. Find the volume of the region bounded below by the plane $z = 0$ and above by the elliptical paraboloid $z = 16 - x^2 - 4y^2$.

9. Let S be the plane region in the first quadrant bounded by the curves $y = x$, $y = 2x$, and $xy = 3$. Evaluate $\int_S x\,dA.$

*10. Let S be the plane region in the first quadrant bounded by the curves $y = x$, $y = 2x$, $xy = 3$, and $xy = 1$. Evaluate $\int_S \dfrac{x}{y} dA$.

11. Let S be the region in the first quadrant bounded by $y = 0$, $y = x$, $xy = 1$, and $x^2 - y^2 = 1$. Evaluate $\int_S (x^2 + y^2) dA$. (Hint: The obvious change of variables is $u = xy$, $v = x^2 - y^2$. Here it is too hard to find $\begin{bmatrix} x \\ y \end{bmatrix} = \mathbf{g}\begin{pmatrix} u \\ v \end{pmatrix}$ explicitly, but how can you find det $D\mathbf{g}$ another way?)

12. Let S be the region bounded by $y = -x$, $y = \tfrac{1}{3}$, $y = 2x$, and $y = 2x - 1$. Evaluate $\int_S \dfrac{x+y}{(2x-y+1)^4} dA$.

*13. Let S be the region with $x \geq 0$ bounded by $y + x^2 = 0$, $x - y = 2$, and $x^2 - 2x + 4y = 0$. Evaluate $\int_S (x - y + 1)^{-2} dA$. (Hint: Consider $x = u + v$, $y = v - u^2$.)

*14. Suppose $0 < b < a$. Define $\mathbf{g}: (0, b) \times (0, 2\pi) \times (0, 2\pi) \to \mathbb{R}^3$ by
$$\mathbf{g}\begin{pmatrix} r \\ \theta \\ \phi \end{pmatrix} = \begin{bmatrix} (a + r\cos\phi)\cos\theta \\ (a + r\cos\phi)\sin\theta \\ r\sin\phi \end{bmatrix}.$$
Describe and sketch the image of $\mathbf{g}$, and find its volume.

15. Let
$$A = \begin{bmatrix} 1 & 1 & 1 & \cdots & 1 \\ 1 & 2 & 1 & \cdots & 1 \\ 1 & 2 & 3 & \cdots & 1 \\ \vdots & \vdots & \vdots & \ddots & \vdots \\ 1 & 2 & 3 & \cdots & n \end{bmatrix}.$$
Given that $\int_{\mathbb{R}^n} f dV = 1$, evaluate $\int_{\mathbb{R}^n} f(A^{-1}\mathbf{x}) dV$.

16. Let $S = \{\mathbf{x} \in \mathbb{R}^n : x_i \geq 0 \text{ for all } i, \ x_1 + 2x_2 + 3x_3 + \cdots + nx_n \leq n\}$. Find vol($S$).

*17. Define spherical coordinates in $\mathbb{R}^4$ and calculate $\int_{B(\mathbf{0},a)} \|\mathbf{x}\| dV$.

18. Let $R = [0, 1] \times [0, 1]$, and consider the integral $I = \int_R \dfrac{1}{1-xy} dA$.

(a) By expanding the integrand in a geometric series, show that $I = \sum_{k=1}^{\infty} \dfrac{1}{k^2}$. (To be completely rigorous, you will need to write I as the limit of integrals over $[0, 1] \times [0, 1 - \delta]$ as $\delta \to 0^+$. Why?)
(b) Evaluate I by rotating the plane through $\pi/4$. A reasonable amount of cleverness will be required.[7]

19. Let a_n denote the n-dimensional volume of the n-dimensional unit ball $B(\mathbf{0}, 1) \subset \mathbb{R}^n$. Prove that
$$a_n = \begin{cases} \pi^m/m!, & n = 2m \\ \pi^m 2^{2m+1} m!/(2m+1)!, & n = 2m+1 \end{cases}.$$
(Hint: Proceed by induction with gaps of 2.)

---

[7] We learned of this calculation from Simmons's *Calculus with Analytic Geometry*, First Edition, pp. 751–52.

CHAPTER

8

DIFFERENTIAL FORMS AND INTEGRATION ON MANIFOLDS

In this chapter we come to the culmination of our study of multivariable calculus. Just as in single-variable calculus, we've studied two seemingly unrelated topics—the derivative and the integral. Now the time has come to make the connection between the two, namely, the multivariable version of the Fundamental Theorem of Calculus. After building up to the ultimate theorem, we consider some nontrivial applications to physics and topology.

▶ 1 MOTIVATION

We want to be able to integrate on k-dimensional manifolds, so we begin by introducing the appropriate integrands, which are called (differential) k-forms. These integrals should generalize the ideas of work (done by a force field along a directed curve) and flux (of a vector field outward across a surface). But not only are k-forms invented to be integrated, they can also be differentiated. There is a natural operator d, called the *exterior derivative*, which will turn k-forms into $k+1$-forms. The classical Fundamental Theorem of Calculus, we recall, tells us that

$$\int_a^b f'(t)dt = f(b) - f(a)$$

whenever f is $\mathcal{C}^1$. We should think of this as relating the integral of the derivative over the interval $[a, b]$ to the "integral" of f over the boundary of the interval, which in this case is the signed sum of the values $f(b)$ and $f(a)$. Notice that there is a notion of *direction* or *orientation* built into the integral, inasmuch as $\int_b^a f(t)dt = -\int_a^b f(t)dt$. In this guise, we can write the Fundamental Theorem of Calculus in the form

$$\int_{[a,b]} df = \int_{\partial[a,b]} f = f(b) - f(a).$$

More generally, we will prove Stokes's Theorem, which says that

$$\int_M d\omega = \int_{\partial M} \omega$$

333

for any k-form ω and compact, oriented k-dimensional manifold M with boundary ∂M. The original versions of Stokes's Theorem all arose in the first half of the nineteenth century in connection with physics, particularly potential theory and electrostatics.

Just as the Fundamental Theorem of Calculus tells us that our displacement is the integral of our velocity, so can it tell us the area of a plane region by tracing around its boundary (see Exercises 1.5.3 and 8.3.26). Another instance of the Fundamental Theorem of Calculus is Gauss's Law in physics, which tells us that the total flux of the electric field across a "Gaussian surface" is proportional to the total charge contained inside that surface. And, as we shall see in Section 7, another application is the Hairy Ball Theorem, which tells us we can't comb the hairs on a billiard ball. The elegant modern-day theory of *calibrated geometries*, which grew out of understanding minimal surfaces (the surfaces of least area with a given boundary curve), is based on differential forms and Stokes's Theorem.

As we've seen in Sections 5 and 6 of Chapter 7, determinants play a crucial role in the understanding of n-dimensional volume, and so it is not surprising that k-forms, the objects we wish to integrate over k-dimensional surfaces, will be built out of determinants. We turn to this multilinear algebra in the next section.

▶ EXERCISES 8.1

1. Why does a (plane) mirror reverse left and right but not up and down?

2. Appropriating from Tom and Ray Magliozzi's "Car Talk":

    RAY: Picture this. It's 1936. You're in your second year of high school. Europe is on the brink of yet another war.
    TOM: Second senior year in high school.
    RAY: In a secret location in Germany, German officers are gathered around a table with the designers and builders of its new personnel carrier. They're going over every little detail and leaving no stone unturned. They want everything to be flawless. One of the officers stands up and says, "I have a question about the fan belt, about the longevity of the fan belt." You with me?
    TOM: They spoke English there?
    RAY: Oh, yeah.
    TOM: Just like in all the movies?
    RAY: I'm reading the subtitles.
    TOM: Just like in all the movies. I often wondered how come they all spoke English?
    RAY: Well, it's so close to German, after all.
    TOM: Yeah. You just add an ish or ein to the end of everything.
    RAY: Anyway, this fan belt looks just like the belt around your waist. It's a flat piece of rubber, and it's designed to run around the fan and the generator. So, he asks, "How long do you expect the belt to last?" The engineer says, "30 to 40 thousand kilometers." The officer says, "Not good enough."
    TOM: He said, how many miles is that?
    RAY: The colonel says . . .
    TOM: That's why I never made any money in scriptwriting.
    RAY: Yeah. The colonel says, "Not good enough. We need it to last at least 60K." The engineer says, "Huh. Not a problem. It's just a question of taking off the belt and flipping it over, right?"
    TOM: Sure.
    RAY: Turning it inside-out.

TOM: Yeah.

RAY: The officer says, "That's unacceptable. Our soldiers will be engaged in battle. We can't ask them to change fan belts in the middle of the battlefield."

TOM: Well, it's a good point.

RAY: That's right.

TOM: I mean, come on. You can't tell the guys to stop shooting, your fan belt's got to be replaced.

RAY: Exactly. Hold your fire. So, the engineers huddle together, and they come up with a clever design change. And I think I mentioned they do not change the material of the belt in any way, yet they satisfy the new longevity requirement quite easily. What did they do?

TOM: Whew!

(*Source: Tom and Ray Magliozzi from Car Talk on NPR.*)

▶ 2 DIFFERENTIAL FORMS

We have learned how to calculate multiple integrals over regions in $\mathbb{R}^n$. Our next goal is to be able to integrate over compact manifolds, e.g., curves and surfaces in $\mathbb{R}^3$. In some sense, the most basic question is this: We know that the determinant gives the signed volume of an n-dimensional parallelepiped in $\mathbb{R}^n$; how do we find the signed volume of a k-dimensional parallelepiped in $\mathbb{R}^n$, and what does "signed" mean in this instance?

2.1 The Multilinear Setup

We begin by using the determinant to define various multilinear functions of (ordered) sets of k vectors in $\mathbb{R}^n$. First, we define n different linear maps $dx_i \colon \mathbb{R}^n \to \mathbb{R}$, $i = 1, \ldots, n$, as follows: If

$$\mathbf{v} = \begin{bmatrix} v_1 \\ v_2 \\ \vdots \\ v_n \end{bmatrix} \in \mathbb{R}^n, \quad \text{then set} \quad dx_i(\mathbf{v}) = v_i.$$

(The reason for the bizarre notation will soon become clear.) Note that the set of linear maps from $\mathbb{R}^n$ to $\mathbb{R}$ is an n-dimensional vector space, often denoted $(\mathbb{R}^n)^*$, and $\{dx_1, \ldots, dx_n\}$ is a basis for it. (See Exercise 4.3.25.) For if $\phi \colon \mathbb{R}^n \to \mathbb{R}$ is a linear map, then, letting $\{\mathbf{e}_1, \ldots, \mathbf{e}_n\}$ be the standard basis for $\mathbb{R}^n$, we set $a_i = \phi(\mathbf{e}_i)$, $i = 1, \ldots, n$. Then $\phi = a_1 dx_1 + \cdots + a_n dx_n$, so $dx_1, \ldots, dx_n$ span $(\mathbb{R}^n)^*$. Why do they form a linearly independent set? Well, suppose $\phi = c_1 dx_1 + \cdots + c_n dx_n$ is the zero linear map. Then, in particular, $\phi(\mathbf{e}_i) = c_i = 0$ for all $i = 1, \ldots, n$, as required.

Now, if $I = (i_1, \ldots, i_k)$ is an ordered k-tuple, define

$$d\mathbf{x}_I \colon \underbrace{\mathbb{R}^n \times \cdots \times \mathbb{R}^n}_{k \text{ times}} \to \mathbb{R} \quad \text{by}[1]$$

---

[1] Here we revert to the usual notation for functions, inasmuch as $\mathbf{v}_1, \ldots, \mathbf{v}_k$ are all vectors.

336 ▶ Chapter 8. Differential Forms and Integration on Manifolds

$$d\mathbf{x}_I(\mathbf{v}_1, \ldots, \mathbf{v}_k) = \begin{vmatrix} dx_{i_1}(\mathbf{v}_1) & \cdots & dx_{i_1}(\mathbf{v}_k) \\ \vdots & \ddots & \vdots \\ dx_{i_k}(\mathbf{v}_1) & \cdots & dx_{i_k}(\mathbf{v}_k) \end{vmatrix}.$$

As is the case with the determinant, $d\mathbf{x}_I$ defines an alternating, multilinear function of k vectors in $\mathbb{R}^n$. If we write

$$\mathbf{v}_i = \begin{bmatrix} v_{i,1} \\ v_{i,2} \\ \vdots \\ v_{i,n} \end{bmatrix}, \quad i = 1, \ldots, k,$$

then

$$d\mathbf{x}_I(\mathbf{v}_1, \ldots, \mathbf{v}_k) = \begin{vmatrix} v_{1,i_1} & \cdots & v_{k,i_1} \\ \vdots & \ddots & \vdots \\ v_{1,i_k} & \cdots & v_{k,i_k} \end{vmatrix}.$$

When $i_1 < i_2 < \cdots < i_k$, this is of course the determinant of the $k \times k$ matrix obtained by taking rows $i_1, \ldots, i_k$ of the matrix

$$\begin{bmatrix} | & | & & | \\ \mathbf{v}_1 & \mathbf{v}_2 & \cdots & \mathbf{v}_k \\ | & | & & | \end{bmatrix}.$$

▶ **EXAMPLE 1**

Let $n = 3$, $I = (1, 3)$, $\mathbf{v}_1 = \begin{bmatrix} 2 \\ 4 \\ 5 \end{bmatrix}$, and $\mathbf{v}_2 = \begin{bmatrix} -1 \\ 0 \\ 3 \end{bmatrix}$. Then

$$d\mathbf{x}_{(1,3)}\left(\begin{bmatrix} 2 \\ 4 \\ 5 \end{bmatrix}, \begin{bmatrix} -1 \\ 0 \\ 3 \end{bmatrix}\right) = \begin{vmatrix} dx_1\left(\begin{bmatrix} 2 \\ 4 \\ 5 \end{bmatrix}\right) & dx_1\left(\begin{bmatrix} -1 \\ 0 \\ 3 \end{bmatrix}\right) \\ dx_3\left(\begin{bmatrix} 2 \\ 4 \\ 5 \end{bmatrix}\right) & dx_3\left(\begin{bmatrix} -1 \\ 0 \\ 3 \end{bmatrix}\right) \end{vmatrix} = \begin{vmatrix} 2 & -1 \\ 5 & 3 \end{vmatrix} = 11. \blacktriangleleft$$

▶ **EXAMPLE 2**

Let $n = 4$, $I = (3, 1, 4)$,

$$\mathbf{v}_1 = \begin{bmatrix} 1 \\ -1 \\ 0 \\ 2 \end{bmatrix}, \quad \mathbf{v}_2 = \begin{bmatrix} 0 \\ 3 \\ -2 \\ 1 \end{bmatrix}, \quad \text{and} \quad \mathbf{v}_3 = \begin{bmatrix} 2 \\ 1 \\ 1 \\ 1 \end{bmatrix}.$$

Then

$$d\mathbf{x}_{(3,1,4)}\left(\begin{bmatrix}1\\-1\\0\\2\end{bmatrix},\begin{bmatrix}0\\3\\-2\\1\end{bmatrix},\begin{bmatrix}2\\1\\1\\1\end{bmatrix}\right) = \begin{vmatrix} dx_3\left(\begin{bmatrix}1\\-1\\0\\2\end{bmatrix}\right) & dx_3\left(\begin{bmatrix}0\\3\\-2\\1\end{bmatrix}\right) & dx_3\left(\begin{bmatrix}2\\1\\1\\1\end{bmatrix}\right) \\ dx_1\left(\begin{bmatrix}1\\-1\\0\\2\end{bmatrix}\right) & dx_1\left(\begin{bmatrix}0\\3\\-2\\1\end{bmatrix}\right) & dx_1\left(\begin{bmatrix}2\\1\\1\\1\end{bmatrix}\right) \\ dx_4\left(\begin{bmatrix}1\\-1\\0\\2\end{bmatrix}\right) & dx_4\left(\begin{bmatrix}0\\3\\-2\\1\end{bmatrix}\right) & dx_4\left(\begin{bmatrix}2\\1\\1\\1\end{bmatrix}\right) \end{vmatrix}$$

$$= \begin{vmatrix} 0 & -2 & 1 \\ 1 & 0 & 2 \\ 2 & 1 & 1 \end{vmatrix} = -5. \blacktriangleleft$$

When $i_1 < i_2 < \cdots < i_k$, we say that the ordered k-tuple $I = (i_1, \ldots, i_k)$ is *increasing*. If I is a k-tuple with no repeated index, we denote by $I^<$ the associated increasing k-tuple. For example, if $I = (2, 4, 5, 1)$, then $I^< = (1, 2, 4, 5)$, and we observe that

$$d\mathbf{x}_{(2,4,5,1)} = -d\mathbf{x}_{(2,4,1,5)} = +d\mathbf{x}_{(2,1,4,5)} = -d\mathbf{x}_{(1,2,4,5)}.$$

In general, $d\mathbf{x}_I = (-1)^s d\mathbf{x}_{I^<}$, where s is the number of exchanges required to move from I to $I^<$. Note that if we switch two of the indices in the ordered k-tuple, this amounts to switching two rows in the matrix, and the determinant changes sign. Similarly, if two of the indices are equal, the determinant will always be 0, so $d\mathbf{x}_I = 0$ whenever there is a repeated index in I.

It follows from Theorem 5.1 or Proposition 5.18 of Chapter 7 that the set of $d\mathbf{x}_I$ with I increasing spans the vector space of alternating multilinear functions from $(\mathbb{R}^n)^k$ to $\mathbb{R}$, denoted $\Lambda^k(\mathbb{R}^n)^*$. In particular, if $T \in \Lambda^k(\mathbb{R}^n)^*$, then for any increasing k-tuple I, set $a_I = T(\mathbf{e}_{i_1}, \ldots, \mathbf{e}_{i_k})$. Then we leave it to the reader to check that

$$T = \sum_{I \text{ increasing}} a_I d\mathbf{x}_I$$

and that the set of $d\mathbf{x}_I$ with I increasing forms a linearly independent set (see Exercise 1). Since counting the increasing sequences of k numbers between 1 and n is the same as counting the number of k-element subsets of an n-element set, we have

$$\dim\left(\Lambda^k(\mathbb{R}^n)^*\right) = \binom{n}{k}.$$

Remark Suppose I is an increasing k-tuple. We have the following geometric interpretation: Given vectors $\mathbf{v}_1, \ldots, \mathbf{v}_k \in \mathbb{R}^n$, the number $d\mathbf{x}_I(\mathbf{v}_1, \ldots, \mathbf{v}_k)$ is the signed volume

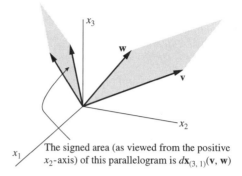

The signed area (as viewed from the positive x_2-axis) of this parallelogram is $d\mathbf{x}_{(3,1)}(\mathbf{v}, \mathbf{w})$

Figure 2.1

of the projection onto the $x_{i_1} x_{i_2} \ldots x_{i_k}$-plane of the parallelepiped spanned by $\mathbf{v}_1, \ldots, \mathbf{v}_k$. See Figure 2.1.

Generalizing the cross product of vectors in $\mathbb{R}^3$ (see Exercise 3), we define the product of these alternating multilinear functions, as follows. If I and J are ordered k- and ℓ-tuples, respectively, we define

$$d\mathbf{x}_I \wedge d\mathbf{x}_J = d\mathbf{x}_{(I,J)},$$

where by (I, J) we mean the ordered $(k+\ell)$-tuple obtained by concatenating I and J.

▶ **EXAMPLE 3**

$$d\mathbf{x}_{(1,2)} \wedge dx_3 = d\mathbf{x}_{(1,2,3)}$$
$$d\mathbf{x}_{(1,5)} \wedge d\mathbf{x}_{(4,2)} = d\mathbf{x}_{(1,5,4,2)} = -d\mathbf{x}_{(1,2,4,5)}$$
$$d\mathbf{x}_{(1,3,2)} \wedge d\mathbf{x}_{(3,4)} = d\mathbf{x}_{(1,3,2,3,4)} = 0 \quad \triangleleft$$

We extend by linearity: If $\omega = \sum a_I d\mathbf{x}_I$ and $\eta = \sum b_J d\mathbf{x}_J$, then we set $\omega \wedge \eta = \sum (a_I b_J) d\mathbf{x}_I \wedge d\mathbf{x}_J = \sum (a_I b_J) d\mathbf{x}_{(I,J)}$. This is called the *wedge product* of ω and η.

▶ **EXAMPLE 4**

Suppose $\omega = a_1 dx_1 + a_2 dx_2$ and $\eta = b_1 dx_1 + b_2 dx_2 \in \Lambda^1(\mathbb{R}^2)^* = (\mathbb{R}^2)^*$. Then let's compute $\omega \wedge \eta \in \Lambda^2(\mathbb{R}^2)^*$:

$$\begin{aligned}\omega \wedge \eta &= (a_1 dx_1 + a_2 dx_2) \wedge (b_1 dx_1 + b_2 dx_2) \\ &= a_1 b_1 dx_1 \wedge dx_1 + a_2 b_1 dx_2 \wedge dx_1 + a_1 b_2 dx_1 \wedge dx_2 + a_2 b_2 dx_2 \wedge dx_2 \\ &= a_1 b_1 d\mathbf{x}_{(1,1)} + a_2 b_1 d\mathbf{x}_{(2,1)} + a_1 b_2 d\mathbf{x}_{(1,2)} + a_2 b_2 d\mathbf{x}_{(2,2)} \\ &= (a_1 b_2 - a_2 b_1) d\mathbf{x}_{(1,2)}.\end{aligned}$$

Of course, it should not be altogether surprising that the determinant of the coefficient matrix $\begin{bmatrix} a_1 & a_2 \\ b_1 & b_2 \end{bmatrix}$ has emerged here. ◀

Proposition 2.1 *The wedge product enjoys the following properties.*

1. *It is bilinear:* $(\omega + \phi) \wedge \eta = \omega \wedge \eta + \phi \wedge \eta$ and $(c\omega) \wedge \eta = c(\omega \wedge \eta)$.
2. *It is* skew-commutative: $\omega \wedge \eta = (-1)^{k\ell} \eta \wedge \omega$, when $\omega \in \Lambda^k(\mathbb{R}^n)^*$ and $\eta \in \Lambda^\ell(\mathbb{R}^n)^*$.
3. *It is associative:* $(\omega \wedge \eta) \wedge \phi = \omega \wedge (\eta \wedge \phi)$.

Proof Properties (1) and (3) are obvious from the definition. For (2), we observe that to change the ordered $(k+\ell)$-tuple $(i_1, \ldots, i_k, j_1, \ldots, j_\ell)$ to the ordered $(k+\ell)$-tuple $(j_1, \ldots, j_\ell, i_1, \ldots, i_k)$ requires $k\ell$ exchanges: To move j_1 past $i_1, \ldots, i_k$ requires k exchanges, to move j_2 past $i_1, \ldots, i_k$ requires k more, and so on. ∎

Now that we've established associativity, we can make the crucial observation that

$$dx_i \wedge dx_j = d\mathbf{x}_{(i,j)} \quad \text{and, moreover,}$$
$$dx_{i_1} \wedge dx_{i_2} \wedge \cdots \wedge dx_{i_k} = d\mathbf{x}_{(i_1, \ldots, i_k)}.$$

As has been our custom throughout this text, when we work in $\mathbb{R}^3$, it is often more convenient to write x, y, z for x_1, x_2, x_3.

2.2 Differential Forms on $\mathbb{R}^n$ and the Exterior Derivative

A (differential) 0-form on $\mathbb{R}^n$ is a smooth function. An n-form on $\mathbb{R}^n$ is an expression of the form[2]

$$\omega = f(\mathbf{x}) dx_1 \wedge \cdots \wedge dx_n$$

for some smooth function f. As we shall soon see, these (rather than functions) are precisely what it makes sense to integrate over regions in $\mathbb{R}^n$. A (differential) k-form on $\mathbb{R}^n$ is an expression

$$\omega = \sum_{\text{increasing } k\text{-tuples } I} f_I(\mathbf{x}) d\mathbf{x}_I = \sum_{i_1 < \cdots < i_k} f_I dx_{i_1} \wedge \cdots \wedge dx_{i_k}$$

for some smooth functions f_I. (Remember that the $d\mathbf{x}_I$ with I increasing gives a basis for $\Lambda^k(\mathbb{R}^n)^*$.) As usual, if $k > n$, the only k-form is 0.

We can perform the obvious algebraic manipulations with forms: We can add two k-forms; we can multiply a k-form by a function; we can form the wedge product of a k-form and an ℓ-form. The set of k-forms on $\mathbb{R}^n$ is naturally a vector space, which we denote by $\mathcal{A}^k(\mathbb{R}^n)$.[3] For reference we list the relevant algebraic properties:

Proposition 2.2 *Let $U \subset \mathbb{R}^n$ be an open set. Let $\omega \in \mathcal{A}^k(U)$, $\eta \in \mathcal{A}^\ell(U)$, and $\phi \in \mathcal{A}^m(U)$.*

---

[2] Sorry about that. You think of a better word!

[3] For those of you who may see such words in the future, it is in fact a *module* over the *ring* of smooth functions. Indeed, because we can multiply by using the wedge product, if we put all the k-forms together, $k = 0, 1, \ldots, n$, we get what is called a *graded algebra*.

1. When $k = \ell = m$, $\omega + \eta = \eta + \omega$ and $(\omega + \eta) + \phi = \omega + (\eta + \phi)$.
2. $\omega \wedge \eta = (-1)^{k\ell} \eta \wedge \omega$.
3. $(\omega \wedge \eta) \wedge \phi = \omega \wedge (\eta \wedge \phi)$.
4. When $k = \ell$, $(\omega + \eta) \wedge \phi = (\omega \wedge \phi) + (\eta \wedge \phi)$.

Determinants (and hence volume) are already built into the structure of k-forms. As the name "differential form" suggests, their substantial power comes, however, from our ability to differentiate them. We begin with the case of a 0-form, i.e., a smooth function $f: U \to \mathbb{R}$. Then for any $\mathbf{x} \in U$ we want $df(\mathbf{x}) = Df(\mathbf{x})$ as a linear map on $\mathbb{R}^n$. In other words, we have

$$df = \sum_{j=1}^n \frac{\partial f}{\partial x_j} dx_j.$$

In particular, note that if we take f to be the ith coordinate function, then $df = dx_i$ and $dx_i(\mathbf{v}) = Dx_i(\mathbf{v}) = v_i$, so this explains (in part) our original choice of notation. If $\omega = \sum_I f_I(\mathbf{x}) d\mathbf{x}_I$ is a k-form, then we define

$$d\omega = \sum_I df_I \wedge d\mathbf{x}_I = \sum_I \sum_{j=1}^n \frac{\partial f_I}{\partial x_j} dx_j \wedge dx_{i_1} \wedge \cdots \wedge dx_{i_k}.$$

(Note that for a fixed k-tuple I, only the terms dx_j where j is different from $i_1, \ldots, i_k$ will appear.)

▶ **EXAMPLE 5**

a. Suppose $f: \mathbb{R} \to \mathbb{R}$ is smooth. Then we have $df = f'(x)dx$.

b. Let $\omega = ydx + xdy \in \mathcal{A}^1(\mathbb{R}^2)$. Then $d\omega = dy \wedge dx + dx \wedge dy = 0$.

c. Let $\omega = -ydx + xdy \in \mathcal{A}^1(\mathbb{R}^2)$. Then $d\omega = -dy \wedge dx + dx \wedge dy = 2dx \wedge dy$.

d. Let $\omega = d\left(\arctan \frac{y}{x}\right) = \frac{-ydx + xdy}{x^2 + y^2} \in \mathcal{A}^1(\mathbb{R}^2 - \{\mathbf{0}\})$. Then

$$d\omega = d\left(-\frac{y}{x^2 + y^2}\right) \wedge dx + d\left(\frac{x}{x^2 + y^2}\right) \wedge dy$$

$$= -\frac{\partial}{\partial y}\left(\frac{y}{x^2 + y^2}\right) dy \wedge dx + \frac{\partial}{\partial x}\left(\frac{x}{x^2 + y^2}\right) dx \wedge dy$$

$$= \frac{(x^2 + y^2) - 2y^2}{(x^2 + y^2)^2} dx \wedge dy + \frac{(x^2 + y^2) - 2x^2}{(x^2 + y^2)^2} dx \wedge dy = 0.$$

e. Let $\omega = x_1 dx_2 + x_3 dx_4 + x_5 dx_6 \in \mathcal{A}^1(\mathbb{R}^6)$. Then $d\omega = dx_1 \wedge dx_2 + dx_3 \wedge dx_4 + dx_5 \wedge dx_6$.

f. Let $\omega = (x^2 + e^{yz})dy \wedge dz + (y^2 + \sin(x^3 z))dz \wedge dx + (z^2 + \arctan(x^2 + y^2))dx \wedge dy \in \mathcal{A}^2(\mathbb{R}^3)$. Then

$$d\omega = 2x dx \wedge dy \wedge dz + 2y dy \wedge dz \wedge dx + 2z dz \wedge dx \wedge dy$$
$$= 2(x + y + z) dx \wedge dy \wedge dz. \quad \triangleleft$$

The operator d, called the *exterior derivative*, enjoys the following properties.

Proposition 2.3 Let $\omega \in \mathcal{A}^k(U)$ and $\eta \in \mathcal{A}^\ell(U)$. Let f be a smooth function.
1. When $k = \ell$, we have $d(\omega + \eta) = d\omega + d\eta$.
2. $d(f\omega) = df \wedge \omega + f d\omega$.
3. $d(\omega \wedge \eta) = d\omega \wedge \eta + (-1)^k \omega \wedge d\eta$.
4. $d(d\omega) = 0$.

Proof Properties (1) and (2) are immediate; indeed, (2) is a consequence of (3). To prove (3), we note that because d commutes with sums, it suffices to consider the case that $\omega = f d\mathbf{x}_I$ and $\eta = g d\mathbf{x}_J$. Then, since the product rule gives $d(fg) = gdf + fdg$, we have

$$d(\omega \wedge \eta) = d(fg d\mathbf{x}_I \wedge d\mathbf{x}_J) = d(fg) \wedge d\mathbf{x}_I \wedge d\mathbf{x}_J$$
$$= (gdf + fdg) \wedge d\mathbf{x}_I \wedge d\mathbf{x}_J = gdf \wedge d\mathbf{x}_I \wedge d\mathbf{x}_J + fdg \wedge d\mathbf{x}_I \wedge d\mathbf{x}_J$$
$$= (df \wedge d\mathbf{x}_I) \wedge (g d\mathbf{x}_J) + (-1)^k (f d\mathbf{x}_I) \wedge (dg \wedge d\mathbf{x}_J)$$

(since we must switch $dg \in \mathcal{A}^1(U)$ and $d\mathbf{x}_I \in \mathcal{A}^k(U)$)

$$= d\omega \wedge \eta + (-1)^k \omega \wedge d\eta.$$

To prove (4), suppose $\omega = f d\mathbf{x}_I$. Then

$$d\omega = \sum_{j=1}^n \frac{\partial f}{\partial x_j} dx_j \wedge d\mathbf{x}_I$$

and

(*) $$d(d\omega) = \sum_{i=1}^n \sum_{j=1}^n \frac{\partial^2 f}{\partial x_i \partial x_j} dx_i \wedge dx_j \wedge d\mathbf{x}_I.$$

Since $dx_i \wedge dx_j = -dx_j \wedge dx_i$, we can rewrite the right-hand side of (*) as

$$\sum_{i<j} \left(\frac{\partial^2 f}{\partial x_i \partial x_j} - \frac{\partial^2 f}{\partial x_j \partial x_i} \right) dx_i \wedge dx_j \wedge d\mathbf{x}_I.$$

But by Theorem 6.1 of Chapter 3, we have

$$\frac{\partial^2 f}{\partial x_i \partial x_j} = \frac{\partial^2 f}{\partial x_j \partial x_i},$$

and so this sum is 0, as required. ∎

2.3 Pullback

All the algebraic and differential structure inherent in differential forms endows them with a very natural behavior under mappings. The main point is to generalize the procedure of "integration by substitution," familiar to all calculus students: When confronted with the integral $\int_a^b f(g(u))g'(u)du$, we substitute $x = g(u)$, formally write $dx = g'(u)du$, and say $\int_a^b f(g(u))g'(u)du = \int_{g(a)}^{g(b)} f(x)dx$. The proof that this works is, of course, the chain rule. Now we put this procedure in the proper setting.

Definition Let $U \subset \mathbb{R}^m$ be open, and let $\mathbf{g}: U \to \mathbb{R}^n$ be smooth. If $\omega \in \mathcal{A}^k(\mathbb{R}^n)$, then we define $\mathbf{g}^*\omega \in \mathcal{A}^k(U)$ (the *pullback* of ω by $\mathbf{g}$) as follows. To pull back a function (0-form) f, we just compose functions:

$$\mathbf{g}^* f = f \circ \mathbf{g}.$$

To pull back the basis 1-forms, if $\mathbf{g}(\mathbf{u}) = \mathbf{x}$, then set

$$\mathbf{g}^* dx_i = dg_i = \sum_{j=1}^{m} \frac{\partial g_i}{\partial u_j} du_j.$$

Note that the coefficients of $\mathbf{g}^* dx_i$, written as a linear combination of $du_1, \ldots, du_m$, are the entries of the ith row of the derivative matrix of $\mathbf{g}$. Now just let the pullback of a wedge product be the wedge product of the pullbacks:

$$\mathbf{g}^*(dx_{i_1} \wedge \cdots \wedge dx_{i_k}) = dg_{i_1} \wedge \cdots \wedge dg_{i_k}, \quad \text{which we can abbreviate as } dg_I.$$

Last, we take the pullback of a sum to be the sum of the pullbacks:

$$\mathbf{g}^*\left(\sum_I f_I d\mathbf{x}_I\right) = \sum_I (f_I \circ \mathbf{g}) d\mathbf{g}_I = \sum_I (f_I \circ \mathbf{g}) dg_{i_1} \wedge \cdots \wedge dg_{i_k}.$$

▶ **EXAMPLE 6**

a. If $g: \mathbb{R} \to \mathbb{R}$, then $g^*(f(x)dx) = f(g(u))g'(u)du$.

b. Let $\mathbf{g}: \mathbb{R} \to \mathbb{R}^2$ be given by

$$\mathbf{g}(t) = \begin{bmatrix} \cos t \\ \sin t \end{bmatrix}.$$

Then $\mathbf{g}^* dx = -\sin t\, dt$ and $\mathbf{g}^* dy = \cos t\, dt$, so $\mathbf{g}^*(-ydx + xdy) = (-\sin t)(-\sin t\, dt) + (\cos t)(\cos t\, dt) = dt$.

c. Let $\mathbf{g}: \mathbb{R}^2 \to \mathbb{R}^2$ be given by

$$\mathbf{g}\begin{pmatrix} u \\ v \end{pmatrix} = \begin{bmatrix} u\cos v \\ u\sin v \end{bmatrix}.$$

If $\omega = xdx + ydy$, then

$$\mathbf{g}^*\omega = (u\cos v)(\cos v du - u\sin v dv) + (u\sin v)(\sin v du + u\cos v dv)$$
$$= u(\cos^2 v + \sin^2 v)du + u^2(-\cos v \sin v + \cos v \sin v)dv = udu.$$

Moreover,

$$\mathbf{g}^*(dx \wedge dy) = \mathbf{g}^*dx \wedge \mathbf{g}^*dy = (\cos v du - u\sin v dv) \wedge (\sin v du + u\cos v dv)$$
$$= u(\cos^2 v + \sin^2 v)du \wedge dv = udu \wedge dv,$$

so $\mathbf{g}^*\left(e^{-(x^2+y^2)}dx \wedge dy\right) = ue^{-u^2}du \wedge dv$.

d. Let $\mathbf{g}\colon \mathbb{R}^2 \to \mathbb{R}^3$ be given by

$$\mathbf{g}\begin{pmatrix} u \\ v \end{pmatrix} = \begin{bmatrix} u\cos v \\ u\sin v \\ v \end{bmatrix}.$$

Then

$$\mathbf{g}^*dx = \cos v du - u\sin v dv$$
$$\mathbf{g}^*dy = \sin v du + u\cos v dv$$
$$\mathbf{g}^*dz = dv$$

and so

$$\mathbf{g}^*(dx \wedge dy) = udu \wedge dv$$
$$\mathbf{g}^*(dx \wedge dz) = \cos v du \wedge dv$$
$$\mathbf{g}^*(dy \wedge dz) = \sin v du \wedge dv.$$

Therefore, if $\omega = (x^2 + y^2)dx \wedge dy + xdx \wedge dz + ydy \wedge dz$, then we have

$$\mathbf{g}^*\omega = u^2(udu \wedge dv) + (u\cos v)(\cos v du \wedge dv) + (u\sin v)(\sin v du \wedge dv)$$
$$= u(u^2 + 1)du \wedge dv. \blacktriangleleft$$

It is impossible to miss the appearance of determinants of the derivative matrix in the calculations we just performed. Indeed, if I is an ordered k-tuple,

$$\mathbf{g}^*d\mathbf{x}_I = \sum_{\text{increasing } k\text{-tuples } J} \det\left[\frac{\partial \mathbf{g}_I}{\partial \mathbf{u}_J}\right]d\mathbf{u}_J; \quad \text{i.e.,}$$

$$\mathbf{g}^*\left(dx_{i_1} \wedge \cdots \wedge dx_{i_k}\right) = \sum_{1 \leq j_1 < \cdots < j_k \leq m} \begin{vmatrix} \frac{\partial g_{i_1}}{\partial u_{j_1}} & \cdots & \frac{\partial g_{i_1}}{\partial u_{j_k}} \\ \vdots & \ddots & \vdots \\ \frac{\partial g_{i_k}}{\partial u_{j_1}} & \cdots & \frac{\partial g_{i_k}}{\partial u_{j_k}} \end{vmatrix} du_{j_1} \wedge \cdots \wedge du_{j_k}.$$

We need one last technical result before we turn to integrating.

Proposition 2.4 Let $U \subset \mathbb{R}^m$ be open, and let $\mathbf{g}: U \to \mathbb{R}^n$ be smooth. If $\omega \in \mathcal{A}^k(\mathbb{R}^n)$, then
$$\mathbf{g}^*(d\omega) = d(\mathbf{g}^*\omega).$$

Proof The statement for $k=0$ is the chain rule (Theorem 3.2 of Chapter 3):

$$d(\mathbf{g}^* f) = d(f \circ \mathbf{g}) = \sum_{j=1}^{m}\left(\sum_{i=1}^{n}\left(\frac{\partial f}{\partial x_i}\circ \mathbf{g}\right)\frac{\partial g_i}{\partial u_j}\right)du_j$$

$$= \sum_{i=1}^{n}\left(\frac{\partial f}{\partial x_i}\circ \mathbf{g}\right)\left(\sum_{j=1}^{m}\frac{\partial g_i}{\partial u_j}du_j\right) = \sum_{i=1}^{n}\mathbf{g}^*\left(\frac{\partial f}{\partial x_i}\right)\mathbf{g}^* dx_i$$

$$= \mathbf{g}^*\left(\sum_{i=1}^{n}\frac{\partial f}{\partial x_i}dx_i\right) = \mathbf{g}^*(df).$$

Since the pullback of a wedge product is the wedge product of the pullbacks, we infer that $\mathbf{g}^*(d\mathbf{x}_I) = d\mathbf{g}_I$. Because d and pullback are linear, it suffices to prove the result for $\omega = f\,d\mathbf{x}_I$. Well,

$$\mathbf{g}^*\big(d(f\,d\mathbf{x}_I)\big) = \mathbf{g}^*\big(df \wedge d\mathbf{x}_I\big) = \mathbf{g}^*(df) \wedge \mathbf{g}^*(d\mathbf{x}_I) = \mathbf{g}^*(df) \wedge d\mathbf{g}_I$$
$$= d(\mathbf{g}^* f) \wedge d\mathbf{g}_I = d\big((\mathbf{g}^* f)d\mathbf{g}_I\big) = d\big(\mathbf{g}^*(f\,d\mathbf{x}_I)\big).$$

(Notice that at the penultimate step we use the rule for differentiating the wedge product and the fact that $d(dg_i) = 0$.) ∎

Now we come to integration. Given an n-form $\omega = f(\mathbf{x})dx_1 \wedge \cdots \wedge dx_n$ on a region $\Omega \subset \mathbb{R}^n$, we define

$$\int_\Omega \omega = \int_\Omega f\,dV.$$

Note that since f is smooth, it is continuous and hence integrable on any region Ω. It is very important to emphasize here that the n-form ω must be written as a functional multiple of the *standard n-form* $dx_1 \wedge \cdots \wedge dx_n$.

In some sense, the whole point of differential forms is the following restatement of the Change of Variables Theorem:

Proposition 2.5 Let $\Omega \subset \mathbb{R}^n$ be a region, and let $\mathbf{g}: \Omega \to \mathbb{R}^n$ be smooth and one-to-one, with $\det(D\mathbf{g}) > 0$. Then for any n-form $\omega = f\,dx_1 \wedge \cdots \wedge dx_n$ on $S = \mathbf{g}(\Omega)$, we have

$$\int_S \omega = \int_\Omega \mathbf{g}^*\omega.$$

Let $\Omega \subset \mathbb{R}^k$ be a region, and let $\mathbf{g}: \Omega \to \mathbb{R}^n$ be a smooth, one-to-one map whose derivative has rank k at every point. (Actually, it is allowed to have lesser rank on a set of volume 0, but we won't bother with this now.) We say that $M = \mathbf{g}(\Omega) \subset \mathbb{R}^n$ is a *parametrized k-dimensional manifold*. If ω is a k-form on $\mathbb{R}^n$, we define

$$\int_M \omega = \int_\Omega \mathbf{g}^* \omega.$$

If $\mathbf{g}_1 \colon \Omega_1 \to \mathbb{R}^n$ and $\mathbf{g}_2 \colon \Omega_2 \to \mathbb{R}^n$ are two parametrizations of the same k-manifold M, then, provided $\det D(\mathbf{g}_2^{-1} \circ \mathbf{g}_1) > 0$ (which, as we shall soon see, means that $\mathbf{g}_1$ and $\mathbf{g}_2$ parametrize M with the same orientation),

$$\int_{\Omega_2} \mathbf{g}_2^* \omega = \int_{\Omega_1} (\mathbf{g}_2^{-1} \circ \mathbf{g}_1)^* (\mathbf{g}_2^* \omega) \quad \text{(by Proposition 2.5)}$$

$$= \int_{\Omega_1} \left(\mathbf{g}_2 \circ (\mathbf{g}_2^{-1} \circ \mathbf{g}_1) \right)^* \omega \quad \text{(see Exercise 16)}$$

$$= \int_{\Omega_1} \left((\mathbf{g}_2 \circ \mathbf{g}_2^{-1}) \circ \mathbf{g}_1 \right)^* \omega \quad \text{by associativity}$$

$$= \int_{\Omega_1} \mathbf{g}_1^* \omega.$$

That is, the integral of ω over the (oriented) parametrized manifold M is well defined.

▶ EXERCISES 8.2

1. Prove that as I ranges over all increasing k-tuples, the $d\mathbf{x}_I$ form a linearly independent set in $\Lambda^k(\mathbb{R}^n)^*$. Also check that for any $T \in \Lambda^k(\mathbb{R}^n)^*$, $T = \sum_{I \text{ increasing}} a_I d\mathbf{x}_I$, where $a_I = T(\mathbf{e}_{i_1}, \ldots, \mathbf{e}_{i_k})$.

2. (a) Suppose $\omega \in \Lambda^k(\mathbb{R}^n)^*$ and k is odd. Prove that $\omega \wedge \omega = 0$.
   (b) Give an example to show that the result of part a need not hold when k is even.

3. Suppose $\mathbf{v}, \mathbf{w} \in \mathbb{R}^3$. Show that $dx(\mathbf{v} \times \mathbf{w}) = dy \wedge dz(\mathbf{v}, \mathbf{w})$, $dy(\mathbf{v} \times \mathbf{w}) = dz \wedge dx(\mathbf{v}, \mathbf{w})$, and $dz(\mathbf{v} \times \mathbf{w}) = dx \wedge dy(\mathbf{v}, \mathbf{w})$.

4. Simplify the following expressions:
   *(a) $(2dx + 3dy + 4dz) \wedge (dx - dy + 2dz)$
   (b) $(dx + dy - dz) \wedge (dx + 2dy + dz) \wedge (dx - 2dy + dz)$
   *(c) $(2dx \wedge dy + dy \wedge dz) \wedge (3dx - dy + 4dz)$
   (d) $(dx_1 \wedge dx_2 + dx_3 \wedge dx_4) \wedge (dx_1 \wedge dx_2 + dx_3 \wedge dx_4)$
   (e) $(dx_1 \wedge dx_2 + dx_3 \wedge dx_4 + dx_5 \wedge dx_6) \wedge (dx_1 \wedge dx_2 + dx_3 \wedge dx_4 + dx_5 \wedge dx_6) \wedge (dx_1 \wedge dx_2 + dx_3 \wedge dx_4 + dx_5 \wedge dx_6)$

♯5. Let $\mathbf{n} \in \mathbb{R}^3$ be a unit vector, and let $\mathbf{v}$ and $\mathbf{w}$ be orthogonal to $\mathbf{n}$. Let

$$\phi = n_1 dy \wedge dz + n_2 dz \wedge dx + n_3 dx \wedge dy.$$

Prove that $\phi(\mathbf{v}, \mathbf{w})$ is equal to the signed area of the parallelogram spanned by $\mathbf{v}$ and $\mathbf{w}$ (the sign being determined by whether $\mathbf{n}, \mathbf{v}, \mathbf{w}$ form a right-handed system for $\mathbb{R}^3$).

*6. Calculate the exterior derivatives of the following differential forms:
   (a) $\omega = e^{xy} dx$
   (b) $\omega = z^2 dx + x^2 dy + y^2 dz$
   (c) $\omega = x^2 dy \wedge dz + y^2 dz \wedge dx + z^2 dx \wedge dy$
   (d) $\omega = x_1 x_2 dx_3 \wedge dx_4$

*7. Can there be a function f so that df is the given 1-form ω (everywhere ω is defined)? If so, can you find f?
(a) $\omega = -y\,dx + x\,dy$
(b) $\omega = 2xy\,dx + x^2\,dy$
(c) $\omega = y\,dx + z\,dy + x\,dz$
(d) $\omega = (x^2 + yz)\,dx + (xz + \cos y)\,dy + (z + xy)\,dz$
(e) $\omega = \frac{x}{x^2+y^2}\,dx + \frac{y}{x^2+y^2}\,dy$
(f) $\omega = -\frac{y}{x^2+y^2}\,dx + \frac{x}{x^2+y^2}\,dy$

8. For each of the following k-forms ω, can there be a $(k-1)$-form η (defined wherever ω is) so that $d\eta = \omega$?
(a) $\omega = dx \wedge dy$
(b) $\omega = x\,dx \wedge dy$
(c) $\omega = z\,dx \wedge dy$
(d) $\omega = z\,dx \wedge dy + y\,dx \wedge dz + z\,dy \wedge dz$
(e) $\omega = x\,dy \wedge dz + y\,dx \wedge dz + z\,dx \wedge dy$
(f) $\omega = (x^2 + y^2 + z^2)^{-1}(x\,dy \wedge dz + y\,dz \wedge dx + z\,dx \wedge dy)$
(g) $\omega = x_5\,dx_1 \wedge dx_2 \wedge dx_3 \wedge dx_4 + x_1\,dx_2 \wedge dx_4 \wedge dx_3 \wedge dx_5$

♯9. **(The Star Operator)**
(a) Define $\star\colon \mathcal{A}^1(\mathbb{R}^2) \to \mathcal{A}^1(\mathbb{R}^2)$ by $\star dx = dy$ and $\star dy = -dx$, extending by linearity. If f is a smooth function, show that
$$d\star(df) = \left(\frac{\partial^2 f}{\partial x^2} + \frac{\partial^2 f}{\partial y^2}\right) dx \wedge dy.$$

(b) Define $\star\colon \mathcal{A}^1(\mathbb{R}^3) \to \mathcal{A}^2(\mathbb{R}^3)$ by $\star dx = dy \wedge dz$, $\star dy = dz \wedge dx$, and $\star dz = dx \wedge dy$, extending by linearity. If f is a smooth function, show that
$$d\star(df) = \left(\frac{\partial^2 f}{\partial x^2} + \frac{\partial^2 f}{\partial y^2} + \frac{\partial^2 f}{\partial z^2}\right) dx \wedge dy \wedge dz.$$

(Note that we can generalize the definition of the star operator by declaring that, in $\mathbb{R}^n$, $\star$ of a basis 1-form $\phi = dx_i$ is the "complementary" $(n-1)$-form, subject to the sign requirement that $\phi \wedge \star\phi = dx_1 \wedge \cdots \wedge dx_n$.)

10. Suppose $\omega \in \mathcal{A}^1(\mathbb{R}^n)$ and there is a nowhere-zero function λ so that $\lambda\omega$ is the exterior derivative of some function f. Prove that $\omega \wedge d\omega = 0$. (This problem gives a useful criterion for deciding whether the differential equation $\omega = 0$ has an *integrating factor* λ.)

11. In each case, calculate the pullback $\mathbf{g}^*\omega$ and simplify your answer as much as possible.
(a) $g\colon (-\pi/2, \pi/2) \to \mathbb{R}$, $g(u) = \sin u$, $\omega = dx/\sqrt{1-x^2}$

*(b) $\mathbf{g}\colon \mathbb{R} \to \mathbb{R}^2$, $\mathbf{g}(v) = \begin{bmatrix} 3\cos 2v \\ 3\sin 2v \end{bmatrix}$, $\omega = -y\,dx + x\,dy$

(c) $\mathbf{g}\colon \mathbb{R}^2 \to \mathbb{R}^2$, $\mathbf{g}\begin{pmatrix} u \\ v \end{pmatrix} = \begin{bmatrix} 3u\cos 2v \\ 3u\sin 2v \end{bmatrix}$, $\omega = -y\,dx + x\,dy$

(d) $\mathbf{g}\colon \mathbb{R}^2 \to \mathbb{R}^3$, $\mathbf{g}\begin{pmatrix} u \\ v \end{pmatrix} = \begin{bmatrix} \cos u \\ \sin u \\ v \end{bmatrix}$, $\omega = z\,dx + x\,dy + y\,dz$

*(e) $\mathbf{g}\colon \mathbb{R}^2 \to \mathbb{R}^3$, $\mathbf{g}\begin{pmatrix} u \\ v \end{pmatrix} = \begin{bmatrix} \cos u \\ \sin u \\ v \end{bmatrix}$, $\omega = z\,dx \wedge dy + y\,dz \wedge dx$

(f) $\mathbf{g}\colon \mathbb{R}^2 \to \mathbb{R}^4$, $\mathbf{g}\begin{pmatrix} u \\ v \end{pmatrix} = \begin{bmatrix} \cos u \\ \sin v \\ \sin u \\ \cos v \end{bmatrix}$, $\omega = x_2\,dx_1 + x_3\,dx_4$

(g) $\mathbf{g}\colon \mathbb{R}^2 \to \mathbb{R}^4$, $\mathbf{g}\begin{pmatrix} u \\ v \end{pmatrix} = \begin{bmatrix} \cos u \\ \sin v \\ \sin u \\ \cos v \end{bmatrix}$, $\omega = x_1 dx_3 - x_2 dx_4$

(h) $\mathbf{g}\colon \mathbb{R}^2 \to \mathbb{R}^4$, $\mathbf{g}\begin{pmatrix} u \\ v \end{pmatrix} = \begin{bmatrix} \cos u \\ \sin v \\ \sin u \\ \cos v \end{bmatrix}$, $\omega = (-x_3 dx_1 + x_1 dx_3) \wedge (-x_2 dx_4 + x_4 dx_2)$

12. For each part of Exercise 11, calculate $\mathbf{g}^*(d\omega)$ and $d(\mathbf{g}^*\omega)$ and compare your answers.

13. Let $\mathbf{g}\colon (0,\infty) \times (0,\pi) \times (0, 2\pi) \to \mathbb{R}^3$ be the usual spherical coordinates mapping, given on p. 294. Compute $\mathbf{g}^*(dx \wedge dy \wedge dz)$.

♯14. We say a k-form ω is *closed* if $d\omega = 0$ and *exact* if $\omega = d\eta$ for some $(k-1)$-form η.
 (a) Prove that an exact form is closed. Is every closed form exact? (Hint: Work with Example 5d.)
 (b) Prove that if ω and ϕ are closed, then $\omega \wedge \phi$ is closed.
 (c) Prove that if ω is exact and ϕ is closed, then $\omega \wedge \phi$ is exact.

15. Suppose $k \leq n$. Let $\omega_1, \ldots, \omega_k \in (\mathbb{R}^n)^*$ and suppose that $\sum_{i=1}^{k} dx_i \wedge \omega_i = 0$. Prove that there are scalars a_{ij} such that $a_{ij} = a_{ji}$ and $\omega_i = \sum_{j=1}^{k} a_{ij} dx_j$.

16. Suppose $\mathbb{R}^\ell \xrightarrow{\mathbf{h}} \mathbb{R}^m \xrightarrow{\mathbf{g}} \mathbb{R}^n$. Prove that $(\mathbf{g}\circ \mathbf{h})^* = \mathbf{h}^* \circ \mathbf{g}^*$. (Hint: It suffices to prove $(\mathbf{g}\circ \mathbf{h})^* dx_i = \mathbf{h}^*(\mathbf{g}^* dx_i)$. Why?)

17. (a) Suppose $I = (i_1, \ldots, i_n)$ is an ordered n-tuple and $I^< = (1, 2, \ldots, n)$. Then we can define a permutation σ of the numbers $1, \ldots, n$ by $\sigma(j) = i_j$, $j = 1, \ldots, n$. Show that
$$d\mathbf{x}_I = \operatorname{sign}(\sigma) dx_1 \wedge \cdots \wedge dx_n.$$
 (b) Suppose $\omega_i = \sum_{j=1}^{n} a_{ij} dx_j$, $i = 1, \ldots, n$, are 1-forms on $\mathbb{R}^n$. Use Proposition 5.18 of Chapter 7 to prove that $\omega_1 \wedge \cdots \wedge \omega_n = (\det A) dx_1 \wedge \cdots \wedge dx_n$.
 (c) Suppose $\mathbf{g}\colon \mathbb{R}^n \to \mathbb{R}^n$ is smooth. Show that $dg_1 \wedge \cdots \wedge dg_n = \det(D\mathbf{g}) dx_1 \wedge \cdots \wedge dx_n$.

18. Suppose $\phi_1, \ldots, \phi_k \in (\mathbb{R}^n)^*$ and $\mathbf{v}_1, \ldots, \mathbf{v}_k \in \mathbb{R}^n$. Prove that
$$\phi_1 \wedge \cdots \wedge \phi_k(\mathbf{v}_1, \ldots, \mathbf{v}_k) = \det\left[\phi_i(\mathbf{v}_j)\right].$$
(Hints: First of all, it suffices to check this holds when the $\mathbf{v}_j$ are standard basis vectors. Why? Write out the ϕ_i as linear combinations of the dx_j, $\phi_i = \sum_{j=1}^{n} a_{ij} dx_j$, and show that both sides of the desired equality are
$$\begin{vmatrix} a_{1j_1} & \cdots & a_{1j_k} \\ \vdots & \ddots & \vdots \\ a_{kj_1} & \cdots & a_{kj_k} \end{vmatrix}$$
when we take $\mathbf{v}_1 = \mathbf{e}_{j_1}, \ldots, \mathbf{v}_k = \mathbf{e}_{j_k}$.)

19. Suppose $U \subset \mathbb{R}^m$ is open and $\mathbf{g}\colon U \to \mathbb{R}^n$ is smooth. Prove that for any $\omega \in \mathcal{A}^k(\mathbb{R}^n)$ and $\mathbf{v}_1, \ldots, \mathbf{v}_k \in \mathbb{R}^m$, we have
$$\mathbf{g}^*\omega(\mathbf{a})(\mathbf{v}_1, \ldots, \mathbf{v}_k) = \omega(\mathbf{g}(\mathbf{a}))\bigl(D\mathbf{g}(\mathbf{a})\mathbf{v}_1, \ldots, D\mathbf{g}(\mathbf{a})\mathbf{v}_k\bigr).$$
(Hint: Consider $\omega = d\mathbf{x}_I$.)

20. Prove that there is a unique linear operator d mapping $\mathcal{A}^k(U) \to \mathcal{A}^{k+1}(U)$ for all k that satisfies the properties in Proposition 2.3 *and* $df = \sum_{j=1}^{n} \frac{\partial f}{\partial x_j} dx_j$. (This tells us that, appearances to the contrary notwithstanding, the exterior derivative d does not depend on our coordinate system.)

3 LINE INTEGRALS AND GREEN'S THEOREM

We begin with a 1-form $\omega = \sum F_i \, dx_i$ on $\mathbb{R}^n$ and a parametrized curve, C, given by a $\mathcal{C}^1$ function $\mathbf{g} \colon [a, b] \to \mathbb{R}^n$ (ordinarily with $\mathbf{g}' \neq \mathbf{0}$). Then we define

$$\int_C \omega = \int_{[a,b]} \mathbf{g}^* \omega = \int_a^b \sum_{i=1}^n F_i(\mathbf{g}(t)) g_i'(t) \, dt.$$

Now we define a *vector field* (vector-valued function) $\mathbf{F} \colon \mathbb{R}^n \to \mathbb{R}^n$ by

$$\mathbf{F} = \begin{bmatrix} F_1 \\ F_2 \\ \vdots \\ F_n \end{bmatrix}.$$

We then recognize that

$$\int_C \omega = \int_a^b \mathbf{F}(\mathbf{g}(t)) \cdot \mathbf{g}'(t) \, dt = \int_a^b \mathbf{F}(\mathbf{g}(t)) \cdot \frac{\mathbf{g}'(t)}{\|\mathbf{g}'(t)\|} \|\mathbf{g}'(t)\| \, dt = \int_C \mathbf{F} \cdot \mathbf{T} \, ds,$$

where ds is classically called the "element of arclength" on C and $\mathbf{T}$ is the unit tangent vector (see Section 5 of Chapter 3). The most general path over which we'll be integrating will be a finite union of $\mathcal{C}^1$ paths, as above. In particular, we say the path C is *piecewise-$\mathcal{C}^1$* if $C = C_1 \cup \cdots \cup C_s$, where C_j is the image of the $\mathcal{C}^1$ function $\mathbf{g}_j \colon [a_j, b_j] \to \mathbb{R}^n$.

Remark Let C^- be the curve given by the parametrization $\mathbf{h} \colon [a, b] \to \mathbb{R}^n$, $\mathbf{h}(u) = \mathbf{g}(a + b - u)$. Then

$$\int_{[a,b]} \mathbf{h}^* \omega = \int_a^b \mathbf{F}(\mathbf{h}(u)) \cdot \mathbf{h}'(u) \, du = \int_a^b \mathbf{F}(\mathbf{g}(a+b-u)) \cdot \big(-\mathbf{g}'(a+b-u)\big) du$$

$$= -\int_a^b \mathbf{F}(\mathbf{g}(t)) \cdot \mathbf{g}'(t) \, dt \quad \text{(substituting } t = a + b - u\text{)}$$

$$= -\int_{[a,b]} \mathbf{g}^* \omega.$$

Note that $\mathbf{h}(a) = \mathbf{g}(b)$ and $\mathbf{h}(b) = \mathbf{g}(a)$: When we go backward on C, the integral of ω changes sign. We can think of obtaining C^- by reversing the orientation (or direction) of C.

In comparing C and C^-, the unit tangent vector $\mathbf{T}$ reverses direction, so that $\mathbf{F} \cdot \mathbf{T}$ changes sign but ds does not. That is, the notation notwithstanding, ds is not a 1-form, as its value on a tangent vector to C is the length of that tangent vector; this, in turn, is not a linear function of tangent vectors. It would probably be better to write $|ds|$.

3 Line Integrals and Green's Theorem

▶ **EXAMPLE 1**

Let C be the line segment from $\begin{bmatrix} 1 \\ -1 \\ 0 \end{bmatrix}$ to $\begin{bmatrix} 2 \\ 2 \\ 2 \end{bmatrix}$, and let $\omega = xy\,dz$. We wish to calculate $\int_C \omega$. The first step is to parametrize C:

$$\mathbf{g}(t) = \begin{bmatrix} 1+t \\ -1+3t \\ 2t \end{bmatrix}, \quad 0 \le t \le 1.$$

Then

$$\int_C \omega = \int_{[0,1]} \mathbf{g}^*\omega = \int_0^1 (1+t)(-1+3t)(2\,dt)$$

$$= 2\int_0^1 (3t^2 + 2t - 1)\,dt = 2(t^3 + t^2 - t)\Big]_0^1 = 2. \blacktriangleleft$$

▶ **EXAMPLE 2**

Let $\omega = -y\,dx + x\,dy$. Consider two parametrized curves C_1 and C_2, as shown in Figure 3.1, starting at $A = \begin{bmatrix} 1 \\ 0 \end{bmatrix}$ and ending at $B = \begin{bmatrix} 0 \\ 1 \end{bmatrix}$, and parametrized, respectively, by

$$\mathbf{g}(t) = \begin{bmatrix} \cos t \\ \sin t \end{bmatrix}, \quad 0 \le t \le \frac{\pi}{2}, \quad \text{and} \quad \mathbf{h}(t) = \begin{bmatrix} 1-t \\ t \end{bmatrix}, \quad 0 \le t \le 1;$$

$$\int_{C_1} \omega = \int_{[0,\pi/2]} \mathbf{g}^*\omega = \int_0^{\pi/2} (-\sin t)(-\sin t\,dt) + (\cos t)(\cos t\,dt) = \int_0^{\pi/2} 1\,dt = \frac{\pi}{2};$$

$$\int_{C_2} \omega = \int_{[0,1]} \mathbf{h}^*\omega = \int_0^1 (-t)(-dt) + (1-t)(dt) = \int_0^1 1\,dt = 1.$$

Thus, we see that $\int_A^B \omega$ depends not just on the endpoints of the path but also on the particular path joining them. ◀

Figure 3.1

Recall from your integral calculus (or introductory physics) class the definition of *work* done by a force in displacing an object. When the force and the displacement are parallel, the definition is

$$\text{work} = \text{force} \times \text{displacement},$$

and in general only the component of the force vector $\mathbf{F}$ in the direction of the displacement vector $\mathbf{d}$ is considered to do work, so

$$\text{work} = \mathbf{F} \cdot \mathbf{d}.$$

If a vector field $\mathbf{F}$ moves a particle along a parametrized curve C, then it is reasonable to suggest that the total work should be $\int_C \mathbf{F} \cdot \mathbf{T} ds$: Instantaneously, the particle moves in the direction of $\mathbf{T}$, and only the component of $\mathbf{F}$ in that direction should contribute. Without providing complete rigor, we see from Figure 3.2 that the amount of work done by the force in moving the particle along C during a very small time interval $[t, t+h]$ is approximately $\mathbf{F}(\mathbf{g}(t)) \cdot (\mathbf{g}(t+h) - \mathbf{g}(t)) \approx \mathbf{F}(\mathbf{g}(t)) \cdot \mathbf{g}'(t)h$, which suggests that the total work should be given by $\int_a^b \mathbf{F}(\mathbf{g}(t)) \cdot \mathbf{g}'(t) dt$.

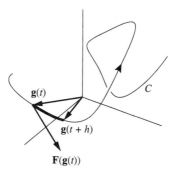

Figure 3.2

▶ EXAMPLE 3

What is the relation between work and energy? As we saw in Section 4 of Chapter 7, the kinetic energy of a particle with mass m and velocity $\mathbf{v}$ is defined to be K.E. $= \frac{1}{2} m \|\mathbf{v}\|^2$. Suppose a particle with mass m moves along a curve C, its position at time t being given by $\mathbf{g}(t)$, $t \in [a, b]$. Then the work done by the force field $\mathbf{F}$ on the particle is given by

$$\begin{aligned}
\text{work} &= \int_C \mathbf{F} \cdot \mathbf{T} ds = \int_a^b \mathbf{F}(\mathbf{g}(t)) \cdot \mathbf{g}'(t) dt \\
&= \int_a^b m \mathbf{g}''(t) \cdot \mathbf{g}'(t) dt \quad \text{by Newton's second law of motion} \\
&= m \int_a^b \tfrac{1}{2} \left(\|\mathbf{g}'\|^2 \right)'(t) dt \\
&= \tfrac{1}{2} m \left(\|\mathbf{g}'(b)\|^2 - \|\mathbf{g}'(a)\|^2 \right) \quad \text{by the Fundamental Theorem of Calculus} \\
&= \Delta(\tfrac{1}{2} m \|\mathbf{v}\|^2) = \Delta(\text{K.E.}).
\end{aligned}$$

That is, assuming **F** is the only force acting on the particle, the work done in moving it along a path is the particle's change in kinetic energy along that path. ◄

3.1 The Fundamental Theorem of Calculus for Line Integrals

Proposition 3.1 *Suppose $\omega = df$ for some $\mathcal{C}^1$ function f. Then for any path (i.e., piecewise-$\mathcal{C}^1$ manifold) C starting at A and ending at B, we have*

$$\int_C \omega = f(B) - f(A).$$

Equivalently, when $\mathbf{F} = \nabla f$, we have

$$\int_C \mathbf{F} \cdot \mathbf{T} ds = f(B) - f(A).$$

Proof It follows from Theorem 3.1 of Chapter 6 that any $\mathcal{C}^1$ segment of C is a finite union of parametrized curves C_j, $j = 1, \ldots, s$, where C_j is the image of a $\mathcal{C}^1$ function $\mathbf{g}_j : [a_j, b_j] \to \mathbb{R}^n$. Let $\mathbf{g}_j(a_j) = A_j$ and $\mathbf{g}_j(b_j) = B_j$. We may arrange that $A_1 = A$, $B_j = A_{j+1}$, $j = 1, \ldots, s - 1$, and $B_s = B$. It suffices to prove the result for C_j, for then we will have

$$\int_C \omega = \sum_{j=1}^{s} \int_{C_j} \omega = \sum_{j=1}^{s} \left(f(B_j) - f(A_j) \right) = f(B) - f(A).$$

Now, we have

$$\int_{C_j} \omega = \int_{a_j}^{b_j} \mathbf{g}_j^* \omega \qquad \text{by definition}$$

$$= \int_{a_j}^{b_j} \mathbf{g}_j^*(df) = \int_{a_j}^{b_j} d(\mathbf{g}_j^* f) \qquad \text{since } d \text{ commutes with pullback}$$

$$= \int_{a_j}^{b_j} d(f \circ \mathbf{g}_j) \qquad \text{by definition of pullback}$$

$$= \int_{a_j}^{b_j} (f \circ \mathbf{g}_j)'(t) dt$$

$$= (f \circ \mathbf{g}_j)(b_j) - (f \circ \mathbf{g}_j)(a_j) \qquad \text{by the Fundamental Theorem of Calculus}$$

$$= f(B_j) - f(A_j),$$

as required. Note that the proof amounts merely to applying the standard Fundamental Theorem of Calculus, along with the definition of line integration by pullback. The fact that d commutes with pullback, in this instance, is simply the chain rule. ∎

Theorem 3.2 *Let $\omega = \sum F_i dx_i$ be a 1-form (or let $\mathbf{F}$ be the corresponding force field) on an open subset $U \subset \mathbb{R}^n$. The following are equivalent:*

1. $\oint_C \omega = 0$ for every closed curve $C \subset U$;
2. $\int_A^B \omega$ is path-independent in U;
3. $\omega = df$ (or $\mathbf{F} = \nabla f$) for some potential function f on U.

Remark In light of Example 3, there is no net work done by $\mathbf{F}$ around closed paths, so that kinetic energy is conserved—which is why such force fields are called *conservative*. Physicists refer to $-f$ as the *potential energy* (P.E.). It then follows from Proposition 3.1 that the total energy, K.E. + P.E., is conserved along all curves, for

$$\Delta(\text{K.E.}) = \text{work} = f(B) - f(A) = -\Delta(\text{P.E.}), \quad \text{and so} \quad \Delta(\text{K.E.} + \text{P.E.}) = 0.$$

Proof (1) $\Longrightarrow$ (2): If C_1 and C_2 are two paths from A to B, then $C = C_1 \cup C_2^-$ is a closed curve, as indicated in Figure 3.3(a). Then

$$0 = \int_C \omega = \int_{C_1} \omega - \int_{C_2} \omega \implies \int_{C_1} \omega = \int_{C_2} \omega.$$

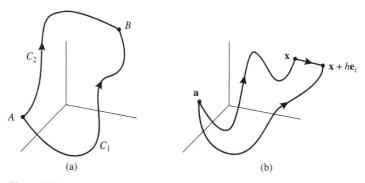

(a) (b)

Figure 3.3

(2) $\Longrightarrow$ (3): (Here we assume any two points of U can be joined by a path. If not, one must repeat the argument on each connected "piece" of U.) Fix $\mathbf{a} \in U$, and define $f : U \to \mathbb{R}$ by

$$f(\mathbf{x}) = \int_\mathbf{a}^\mathbf{x} \omega, \quad \text{where the integral is computed along any path from } \mathbf{a} \text{ to } \mathbf{x}.$$

By path-independence, f is well defined. Now, to show that $df = \omega$, we must evidently establish that $\dfrac{\partial f}{\partial x_i}(\mathbf{x}) = F_i(\mathbf{x})$. As Figure 3.3(b) suggests,

$$\frac{\partial f}{\partial x_i}(\mathbf{x}) = \lim_{h\to 0} \frac{1}{h}\left(f\begin{pmatrix} x_1 \\ \vdots \\ x_i + h \\ \vdots \\ x_n \end{pmatrix} - f\begin{pmatrix} x_1 \\ \vdots \\ x_i \\ \vdots \\ x_n \end{pmatrix} \right)$$

$$= \lim_{h\to 0} \frac{1}{h} \int_{\mathbf{x}}^{\mathbf{x}+h\mathbf{e}_i} \omega = \lim_{h\to 0} \frac{1}{h} \int_0^h F_i(\mathbf{x}+t\mathbf{e}_i)\,dt$$

$$= F_i(\mathbf{x})$$

by the usual Fundamental Theorem of Calculus.

(3) $\implies$ (1): This is immediate from Proposition 3.1. ∎

Remark Given a 1-form ω, by (4) of Proposition 2.3, a necessary condition for $\omega = df$ for some function f (ω exact) is that $d\omega = 0$ (ω closed). As we saw in Example 5d of Section 2, the condition is definitely *not* sufficient. We shall soon see that the topology of the region on which ω is defined is relevant.

3.2 Finding a Potential Function

If we know that $\int \omega$ is path-independent on a region, then we can construct a potential function by choosing a convenient path. We illustrate the general principle with some examples.

▶ **EXAMPLE 4**

Let $\omega = (e^x + 2xy)dx + (x^2 + \cos y)dy$. We show two different ways to calculate a potential function f, i.e., a function f with $df = \omega$.

a. Take the line segment C joining $\mathbf{0} = \begin{bmatrix} 0 \\ 0 \end{bmatrix}$ and $\mathbf{x}_0 = \begin{bmatrix} x_0 \\ y_0 \end{bmatrix}$ as shown in Figure 3.4(a); we take the obvious parametrization:

$$\mathbf{g}(t) = t\mathbf{x}_0 = \begin{bmatrix} tx_0 \\ ty_0 \end{bmatrix}, \quad 0 \le t \le 1.$$

Then

$$f\begin{pmatrix} x_0 \\ y_0 \end{pmatrix} = \int_0^{\mathbf{x}_0} \omega = \int_{[0,1]} \mathbf{g}^*\omega$$

$$= \int_0^1 \left((e^{tx_0} + 2t^2 x_0 y_0)x_0 + (t^2 x_0^2 + \cos(ty_0))y_0 \right) dt$$

$$= \left(e^{tx_0} + \tfrac{2}{3}t^3 x_0^2 y_0 + \tfrac{1}{3}t^3 x_0^2 y_0 + \sin(ty_0) \right) \Big|_0^1$$

$$= \left(e^{x_0} + x_0^2 y_0 + \sin y_0 \right) - 1,$$

and so we set $f\begin{pmatrix} x \\ y \end{pmatrix} = e^x + x^2 y + \sin y - 1$, and it is easy to check that $df = \omega$.

b. Now we take the two-step path, as shown in Figure 3.4(b), first varying x and then varying y, to get from $\mathbf{0}$ to $\mathbf{x}_0$. That is, we have the two parametrizations:

$$C_1 : \mathbf{g}_1(t) = \begin{bmatrix} t \\ 0 \end{bmatrix}, \quad 0 \le t \le x_0, \qquad C_2 : \mathbf{g}_2(t) = \begin{bmatrix} x_0 \\ t \end{bmatrix}, \quad 0 \le t \le y_0.$$

Then we have

$$f\begin{pmatrix} x_0 \\ y_0 \end{pmatrix} = \int_{C_1} \omega + \int_{C_2} \omega$$

$$= \int_0^{x_0} e^t \, dt + \int_0^{y_0} (x_0^2 + \cos t) \, dt$$

$$= (e^{x_0} - 1) + (x_0^2 y_0 + \sin y_0).$$

Once again, we have $f\begin{pmatrix} x \\ y \end{pmatrix} = e^x - 1 + x^2 y + \sin y$.

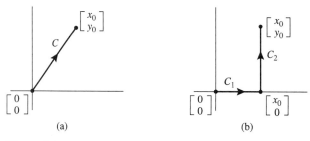

(a)   (b)

Figure 3.4

c. As a variation on the approach of part (b), we proceed purely by antidifferentiating. If we seek a function f with $df = \omega$, then this means that

(∗) $\qquad \dfrac{\partial f}{\partial x} = e^x + 2xy \qquad \text{and} \qquad \dfrac{\partial f}{\partial y} = x^2 + \cos y.$

Integrating the first equation, *holding y fixed*, we obtain

(†) $\qquad f\begin{pmatrix} x \\ y \end{pmatrix} = \int (e^x + 2xy) \, dx = e^x + x^2 y + h(y)$

for some arbitrary function h (this is the "constant of integration"). Differentiating (†) with respect to y and comparing with the latter equation in (∗), we find

$$\frac{\partial f}{\partial y} = x^2 + h'(y) = x^2 + \cos y,$$

whence $h'(y) = \cos y$ and $h(y) = \sin y + C$. Thus, the general potential function is $f\begin{pmatrix} x \\ y \end{pmatrix} = e^x + x^2 y + \sin y + C$ for any constant C.

Note that even though it is computationally more clumsy, the approach in (a) requires only that we be able to draw a line segment from the "base point" (in this case, the origin) to all the other points of our region. The approaches in (b) and (c) require some further sort of convexity: We must be able to start at our base point and reach every other point by a path that is first horizontal and then vertical. ◀

3 Line Integrals and Green's Theorem ◀ 355

We now prove a general result along these lines: Suppose an open subset $U \subset \mathbb{R}^n$ has the property that for some point $\mathbf{a} \in U$, the line segment from $\mathbf{a}$ to each and every point $\mathbf{x} \in U$ lies entirely in U. (Such a region is called *star-shaped* with respect to $\mathbf{a}$, as Figure 3.5 suggests.) Then we have:

Proposition 3.3 *Let ω be a closed 1-form on a star-shaped region. Then ω is exact.*

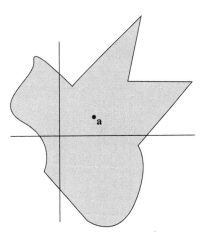

Figure 3.5

Proof Write $\omega = \sum F_i \, dx_i$. For any $\mathbf{x} \in U$, we can parametrize the line segment from $\mathbf{a}$ to $\mathbf{x}$ by

$$\mathbf{g}(t) = \mathbf{a} + t(\mathbf{x} - \mathbf{a}), \quad 0 \le t \le 1.$$

Then we have

$$f(\mathbf{x}) = \int_{\mathbf{a}}^{\mathbf{x}} \omega = \int_0^1 \left(\sum_{j=1}^n F_j(\mathbf{a} + t(\mathbf{x} - \mathbf{a}))(x_j - a_j) \right) dt$$

$$= \sum_{j=1}^n (x_j - a_j) \int_0^1 F_j(\mathbf{a} + t(\mathbf{x} - \mathbf{a})) \, dt.$$

Using Exercise 7.2.20 to calculate the derivative of f, we have

$$\frac{\partial f}{\partial x_i} = \int_0^1 F_i(\mathbf{a} + t(\mathbf{x} - \mathbf{a})) \, dt + \sum_{j=1}^n \int_0^1 t \frac{\partial F_j}{\partial x_i} (\mathbf{a} + t(\mathbf{x} - \mathbf{a}))(x_j - a_j) \, dt$$

$$= \int_0^1 F_i(\mathbf{a} + t(\mathbf{x} - \mathbf{a})) \, dt + \int_0^1 t \left(\sum_{j=1}^n \frac{\partial F_i}{\partial x_j} (\mathbf{a} + t(\mathbf{x} - \mathbf{a}))(x_j - a_j) \right) dt$$

356 ▶ Chapter 8. Differential Forms and Integration on Manifolds

(using the fact that $\dfrac{\partial F_j}{\partial x_i} = \dfrac{\partial F_i}{\partial x_j}$ since $d\omega = 0$)

$$= \int_0^1 F_i(\mathbf{a} + t(\mathbf{x} - \mathbf{a}))\,dt + \int_0^1 t(F_i \circ \mathbf{g})'(t)\,dt \quad \text{(by the chain rule)}$$

$$= \int_0^1 F_i(\mathbf{a} + t(\mathbf{x} - \mathbf{a}))\,dt + t(F_i \circ \mathbf{g})(t)\Big]_0^1 - \int_0^1 F_i(\mathbf{a} + t(\mathbf{x} - \mathbf{a}))\,dt$$

(integrating by parts)

$$= F_i(\mathbf{x}).$$

That is, $df = \omega$, as required. ■

▶ **EXAMPLE 5**

Let C be the parametric curve

$$\mathbf{g}(t) = \begin{bmatrix} e^{t^7} \\ t^6 + 4t^3 - 1 \\ t^4 + (t - t^2)e^{\sin t} \end{bmatrix}, \quad 0 \le t \le 1,$$

and let $\omega = \left(\dfrac{z}{x} + y\right)dx + (x + z)\,dy + \left(\log x + y + 2z\right)dz$. We wish to calculate $\displaystyle\int_C \omega$.

We certainly hope that the 1-form ω is exact (or, equivalently, that the corresponding force field is conservative), for then we can apply the Fundamental Theorem of Calculus for Line Integrals, Proposition 3.1.

If ω is to be equal to df for some function f, we need to solve

$$\frac{\partial f}{\partial x} = \frac{z}{x} + y, \qquad \frac{\partial f}{\partial y} = x + z, \qquad \frac{\partial f}{\partial z} = \log x + y + 2z.$$

Integrating the first equation, we obtain:

$$f\begin{pmatrix}x\\y\\z\end{pmatrix} = \int \left(\frac{z}{x} + y\right)dx = z\log x + xy + g\begin{pmatrix}y\\z\end{pmatrix},$$

where $g\begin{pmatrix}y\\z\end{pmatrix}$ is the "constant of integration." Differentiating with respect to y, we have

$$\frac{\partial f}{\partial y} = x + \frac{\partial g}{\partial y} = x + z,$$

and so we find that $\dfrac{\partial g}{\partial y} = z$. Thus, $g\begin{pmatrix}y\\z\end{pmatrix} = yz + h(z)$ for some appropriate "constant of integration" $h(z)$. So

$$f\begin{pmatrix}x\\y\\z\end{pmatrix} = z\log x + xy + yz + h(z).$$

Now, differentiating with respect to z, we have

$$\frac{\partial f}{\partial z} = \log x + y + h'(z) = \log x + y + 2z,$$

and so—finally—$h(z) = z^2 + c$, whence

$$f\begin{pmatrix} x \\ y \\ z \end{pmatrix} = z \log x + xy + yz + z^2 + c.$$

Now comes the easy part. The curve goes from

$$A = \mathbf{g}(0) = \begin{bmatrix} 1 \\ -1 \\ 0 \end{bmatrix} \quad \text{to} \quad B = \mathbf{g}(1) = \begin{bmatrix} e \\ 4 \\ 1 \end{bmatrix},$$

and so

$$\int_C \omega = f(B) - f(A) = \left(1 + 4e + 4 + 1\right) - \left(-1\right) = 4e + 7. \quad \blacktriangleleft$$

▶ **EXAMPLE 6**

Newton's law of gravitation states that the gravitational force exerted by a point mass M at the origin on a unit test mass is radial and inverse-square in magnitude:

$$\mathbf{F} = -GM \frac{\mathbf{x}}{\|\mathbf{x}\|^3}.$$

The corresponding 1-form is $\omega = -GM(x^2 + y^2 + z^2)^{-3/2}(x\,dx + y\,dy + z\,dz)$. Since

$$d(\|\mathbf{x}\|) = \|\mathbf{x}\|^{-1}(x\,dx + y\,dy + z\,dz)$$

(see Example 1 of Chapter 3, Section 4), it follows immediately that a potential function for the gravitational field is $f(\mathbf{x}) = GM/\|\mathbf{x}\|$. (Physicists ordinarily choose the constant so that the potential goes to 0 as $\mathbf{x}$ goes to infinity.)

Let's now consider the case of the gravitational field of the earth; note that the gravitational acceleration at the surface of the earth is given by $g = GM/R^2$, where R is the radius of the earth. By Proposition 3.1, the work done (against gravity) to lift a unit test mass from a point A on the surface of the earth to a point B height h units above the surface of the earth is therefore

$$-(f(B) - f(A)) = GM\left(\frac{1}{R} - \frac{1}{R+h}\right) = GM\frac{h}{R(R+h)} = \frac{GM}{R^2}\frac{h}{1+\frac{h}{R}} \approx gh,$$

provided h is quite small compared to R. This checks with the standard formula for the potential energy of a mass m at (small) height h above the surface of the earth: P.E. $= mgh$. ◀

3.3 Green's Theorem

We have seen that whenever $\omega = df$ for some function f, it is the case that $\oint_C \omega = 0$ for all closed curves C. So certainly we expect that the size of $d\omega$ on a region will affect the integral of ω around the boundary of that region. The precise statement is the following

Theorem 3.4 (Green's Theorem for a Rectangle) *Let $R \subset \mathbb{R}^2$ be a rectangle, and let ω be a 1-form on R. Then*

$$\int_{\partial R} \omega = \int_R d\omega.$$

(Here the boundary ∂R is traversed counterclockwise.)

Proof Take $R = [a, b] \times [c, d]$, as shown in Figure 3.6, and write $\omega = P\,dx + Q\,dy$. Then

$$d\omega = \left(\frac{\partial Q}{\partial x} - \frac{\partial P}{\partial y}\right) dx \wedge dy.$$

Now we merely calculate, using Fubini's Theorem appropriately:

$$\int_R d\omega = \int_R \left(\frac{\partial Q}{\partial x} - \frac{\partial P}{\partial y}\right) dA$$

$$= \int_c^d \left(\int_a^b \frac{\partial Q}{\partial x} dx\right) dy - \int_a^b \left(\int_c^d \frac{\partial P}{\partial y} dy\right) dx$$

$$= \int_c^d \left(Q\binom{b}{y} - Q\binom{a}{y}\right) dy - \int_a^b \left(P\binom{x}{d} - P\binom{x}{c}\right) dx$$

by the Fundamental Theorem of Calculus

$$= \int_a^b P\binom{x}{c} dx + \int_c^d Q\binom{b}{y} dy - \int_a^b P\binom{x}{d} dx - \int_c^d Q\binom{a}{y} dy$$

$$= \int_{\partial R} \omega,$$

as required. ∎

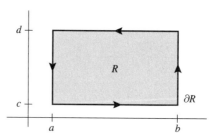

Figure 3.6

For most applications, the following observation is adequate:

Corollary 3.5 *If $S \subset \mathbb{R}^2$ is parametrized by a rectangle, and ω is a 1-form on S, then*

$$\int_{\partial S} \omega = \int_S d\omega.$$

Proof Let $\mathbf{g}\colon R \to S \subset \mathbb{R}^2$ be a parametrization. Then, applying Proposition 2.4, we have

$$\int_{\partial S} \omega = \int_{\partial R} \mathbf{g}^*\omega = \int_R d(\mathbf{g}^*\omega) = \int_R \mathbf{g}^*(d\omega) = \int_S d\omega.$$

(It is important to understand that both S and ∂S inherit an orientation from the parametrization $\mathbf{g}$.) ∎

▶ **EXAMPLE 7**

Suppose ω is a smooth 1-form on the unit disk D in $\mathbb{R}^2$. Can we infer that $\int_{\partial D} \omega = \int_D d\omega$? The naïve answer is "of course," parametrizing by polar coordinates and applying Corollary 3.5. The difficulty that arises is that we only get a bona fide parametrization on $(0, 1] \times (0, 2\pi)$. But we can apply Corollary 3.5 on the rectangle $R_{\delta,\varepsilon} = [\delta, 1] \times [\varepsilon, 2\pi]$ when $\delta, \varepsilon > 0$ are small. Let $D_{\delta,\varepsilon} = \mathbf{g}(R_{\delta,\varepsilon})$, as indicated in Figure 3.7. Because ω is smooth on all of the unit disk, we have

$$\int_D d\omega = \lim_{\delta,\varepsilon\to 0^+} \int_{D_{\delta,\varepsilon}} d\omega = \lim_{\delta,\varepsilon\to 0^+} \int_{R_{\delta,\varepsilon}} \mathbf{g}^* d\omega = \lim_{\delta,\varepsilon\to 0^+} \int_{\partial R_{\delta,\varepsilon}} \mathbf{g}^*\omega = \lim_{\delta,\varepsilon\to 0^+} \int_{\partial D_{\delta,\varepsilon}} \omega = \int_{\partial D} \omega.$$

(We leave it to the reader to justify the first and last equalities.) We shall not belabor such details in the future. ◀

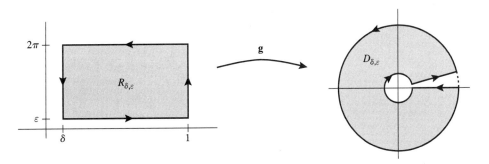

Figure 3.7

More generally, we observe that Green's Theorem holds for any region S that can be decomposed as a finite union of parametrized rectangles overlapping only along their edges.

For, as Figure 3.8 illustrates, if $S = \bigcup_{i=1}^{k} S_i$, because the integrals over interior boundary segments cancel in pairs, we have

$$\int_{\partial S} \omega = \sum_{i=1}^{k} \int_{\partial S_i} \omega = \sum_{i=1}^{k} \int_{S_i} d\omega = \int_S d\omega.$$

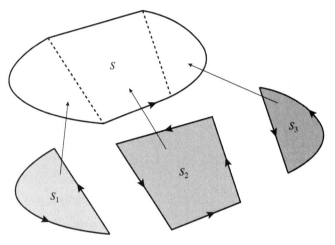

Figure 3.8

Remark We do not usually stop to express every "reasonable" region explicitly as a union of parametrized rectangles. (For most purposes, our work in Section 5 will obviate all such worries.) In Example 7 we already dealt with the case of a disk. To set our minds further at ease, we can easily check that

$$\mathbf{g}_1 : [0,1] \times [0,\pi], \quad \mathbf{g}_1\begin{pmatrix} r \\ \theta \end{pmatrix} = r \begin{bmatrix} \cos\theta \\ \sin\theta \end{bmatrix}$$

maps a rectangle to a half-disk, and that

$$\mathbf{g}_2 : [0,1] \times [0,\pi/2], \quad \mathbf{g}_2\begin{pmatrix} r \\ \theta \end{pmatrix} = \frac{r}{\cos\theta + \sin\theta} \begin{bmatrix} \cos\theta \\ \sin\theta \end{bmatrix}$$

maps a rectangle to the triangle with vertices at $\begin{bmatrix} 0 \\ 0 \end{bmatrix}, \begin{bmatrix} 1 \\ 0 \end{bmatrix},$ and $\begin{bmatrix} 0 \\ 1 \end{bmatrix}$.

▶ **EXAMPLE 8**

We can use Green's Theorem to calculate the area of a planar region S by line integration. Since

$$dx \wedge dy = d(x\,dy) = d(-y\,dx) = d\left(\tfrac{1}{2}(-y\,dx + x\,dy)\right),$$

we have

$$\text{area}(S) = \int_{\partial S} x\,dy = \int_{\partial S} -y\,dx = \frac{1}{2}\int_{\partial S} -y\,dx + x\,dy. \quad \blacktriangleleft$$

Definition A subset $X \subset \mathbb{R}^n$ is called *simply connected* if it is connected and every simple closed curve in X can be continuously shrunk to a point in X.

Corollary 3.6 *Let $\Omega \subset \mathbb{R}^2$ be a simply connected region. If ω is a smooth 1-form on Ω with $d\omega = 0$, then ω is exact; i.e., there is a function f so that $\omega = df$.*

Proof By Green's Theorem, for any rectangle $R \subset \Omega$, we have
$$\int_{\partial R} \omega = \int_R d\omega = 0,$$
and, as the proof of Theorem 3.2 showed, this is sufficient to construct a potential function f. ∎

To emphasize the importance of all the hypotheses, we give an important example.

▶ **EXAMPLE 9**

Let $\omega = -\dfrac{y}{x^2 + y^2}dx + \dfrac{x}{x^2 + y^2}dy$. Then, as we calculated in Example 5d of Section 2, $d\omega = 0$. And yet, letting C be the unit circle, it is easy to check that $\oint_C \omega = 2\pi$. So ω cannot be exact. We shall see further instances of this phenomenon in later sections. ◀

Nevertheless, we can use Green's Theorem to draw a very interesting conclusion.

▶ **EXAMPLE 10**

Suppose C is *any* simple closed curve in the plane that encircles the origin, and let Γ be a circle centered at the origin lying in the interior of C, as shown in Figure 3.9. Let S be the region lying between C and Γ. If we orient C and Γ counterclockwise, then we have $\partial S = C + \Gamma^-$. Once again, let $\omega = -\dfrac{y}{x^2 + y^2}dx + \dfrac{x}{x^2 + y^2}dy$. Then, as in Example 9, we have $d\omega = 0$. But now ω is smooth everywhere on S, and so
$$0 = \int_S d\omega = \int_{\partial S} \omega = \int_C \omega - \int_\Gamma \omega.$$

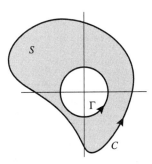

Figure 3.9

That is,
$$\int_C \omega = \int_\Gamma \omega = 2\pi,$$
and this is true for any simple closed curve C with the origin in its interior. More generally, consider the curves shown in Figure 3.10. Then $\int_C \omega = 2\pi, 4\pi,$ and 0, respectively, in parts (a), (b), and (c). For reasons we leave to the reader to surmise, for a closed plane curve not passing through the origin, the integer
$$\frac{1}{2\pi}\int_C -\frac{y}{x^2+y^2}dx + \frac{x}{x^2+y^2}dy$$
is called the *winding number* of C around the origin. ◀

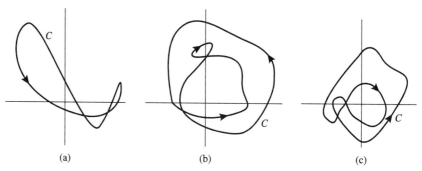

Figure 3.10

▶ EXERCISES 8.3

*1. Let $\omega = ydx + xdy$. Compare and contrast the integrals $\int_C \omega$ for the following parametrized curves C. (Be sure to sketch C.)

(a) $\mathbf{g} \colon [0,1] \to \mathbb{R}^2$, $\mathbf{g}(t) = \begin{bmatrix} t \\ t \end{bmatrix}$

(b) $\mathbf{g} \colon [0,1] \to \mathbb{R}^2$, $\mathbf{g}(t) = \begin{bmatrix} t \\ t^2 \end{bmatrix}$

(c) $\mathbf{g} \colon [0,1] \to \mathbb{R}^2$, $\mathbf{g}(t) = \begin{bmatrix} 1-t \\ 1-t \end{bmatrix}$

(d) $\mathbf{g} \colon [0,\pi/2] \to \mathbb{R}^2$, $\mathbf{g}(t) = \begin{bmatrix} \cos^2 t \\ 1-\sin^2 t \end{bmatrix}$

(e) $\mathbf{g} \colon [0,\pi/4] \to \mathbb{R}^2$, $\mathbf{g}(t) = \begin{bmatrix} \sin 2t \\ 1-\cos 2t \end{bmatrix}$

(f) $\mathbf{g} \colon [0,\pi/2] \to \mathbb{R}^2$, $\mathbf{g}(t) = \begin{bmatrix} \cos t \\ 1-\sin t \end{bmatrix}$

*2. Repeat Exercise 1 with $\omega = y^2 dx + xdy$.

3. Calculate the following line integrals:

(a) $\int_C xy^3 dx$, where C is the unit circle $x^2 + y^2 = 1$, oriented counterclockwise.

(b) $\int_C zdx + xdy + ydz$, where C is the line segment from $\begin{bmatrix} 0 \\ 1 \\ 2 \end{bmatrix}$ to $\begin{bmatrix} 1 \\ -1 \\ 3 \end{bmatrix}$.

(c) $\int_C y^2 dx + zdy - 3xydz$, where C is the line segment from $\begin{bmatrix} 1 \\ 0 \\ 1 \end{bmatrix}$ to $\begin{bmatrix} 2 \\ 3 \\ -1 \end{bmatrix}$.

(d) $\int_C ydx$, where C is the intersection of the unit sphere and the plane $x + y + z = 0$, oriented counterclockwise as viewed from high above the xy-plane. (Hint: Find an orthonormal basis for the plane.)

4. Let C be the curve of intersection of the upper hemisphere $x^2 + y^2 + z^2 = 4$, $z \geq 0$ and the cylinder $x^2 + y^2 = 2x$, oriented counterclockwise as viewed from high above the xy-plane. Evaluate
$$\int_C ydx + zdy + xdz.$$

5. Let $\omega = \dfrac{x}{x^2 + y^2}dx + \dfrac{y}{x^2 + y^2}dy$. If C is an arbitrary path from $\begin{bmatrix} 1 \\ 1 \end{bmatrix}$ to $\begin{bmatrix} 2 \\ 2 \end{bmatrix}$ not passing through the origin, calculate $\int_C \omega$.

6. Determine which of the following 1-forms ω are exact (or, in other words, which of the corresponding vector fields $\mathbf{F}$ are conservative). For those that are, construct (following one of the algorithms in the text) a potential function f. For those that are not, give a closed curve C for which $\oint_C \omega \neq 0$.

(a) $\omega = (x + y)dx + (x + y)dy$
(b) $\omega = y^2 dx + x^2 dy$
(c) $\omega = (e^x + 2xy)dx + (x^2 + y^2)dy$
(d) $\omega = (x^2 + y + z)dx + (x + y^2 + z)dy + (x + y + z^2)dz$
(e) $\omega = y^2 z dx + (2xyz + \sin z)dy + (xy^2 + y \cos z)dz$

7. Let $f: \mathbb{R} \to \mathbb{R}$ and $\omega = f(\|\mathbf{x}\|)\left(\sum_{i=1}^{n} x_i dx_i\right) \in \mathcal{A}^1(\mathbb{R}^n)$.

(a) Assuming f is differentiable, prove that $d\omega = 0$ on $\mathbb{R}^n - \{\mathbf{0}\}$.
(b) Assuming f is continuous, prove that ω is exact.

8. Let C be the parametric curve
$$\mathbf{g}(t) = \begin{bmatrix} e^{t^7} \cos(2\pi t^{21}) \\ t^{17} + 4t^3 - 1 \\ t^4 + (t - t^2)e^{\sin t} \end{bmatrix}, \quad 0 \leq t \leq 1.$$

Calculate $\int_C (3x + y^2 + 2xz)dx + (2xy + ze^{yz} + y)dy + (x^2 + ye^{yz} + ze^{z^2})dz$. (Hint: This problem should involve very little computation.)

9. Let C be any closed curve in the plane. Show that $\oint_C ydx = -\oint_C xdy$. What is the geometric interpretation of these integrals?

10. Calculate each of the following line integrals $\int_C \omega$ directly and by applying Green's Theorem. (In all cases, C is traversed counterclockwise.)

(a) $\omega = (x^2 - y^2)dx + 2xydy$, C is the square with vertices $\begin{bmatrix} 0 \\ 0 \end{bmatrix}, \begin{bmatrix} 1 \\ 0 \end{bmatrix}, \begin{bmatrix} 1 \\ 1 \end{bmatrix}$, and $\begin{bmatrix} 0 \\ 1 \end{bmatrix}$.

(b) $\omega = -y^3 dx + x^3 dy$, C as in part a.
*(c) $\omega = -x^2 ydx + xy^2 dy$, C is the circle of radius a centered at the origin.

*(d) $\omega = \sqrt{x^2+y^2}(-ydx+xdy)$, C is the circle $x^2+y^2 = 2x$.
(e) $\omega = -y^2dx + x^2dy$, C is the boundary of the sector of the circle $r \le a, 0 \le \theta \le \pi/4$.

11. Let C be the circle $x^2 + y^2 = 2x$, oriented counterclockwise. Evaluate $\int_C \omega$, where $\omega = (-y^2 + e^{x^2})dx + (x + \sin(y^3))dy$.

*12. Use Green's Theorem to find the area of the ellipse $\dfrac{x^2}{a^2} + \dfrac{y^2}{b^2} \le 1$.

13. Find the area inside the hypocycloid $x^{2/3} + y^{2/3} = 1$. (Hint: Parametrize by $\mathbf{g}(t) = \begin{bmatrix} \cos^3 t \\ \sin^3 t \end{bmatrix}$.)

*14. Let $0 < b < a$. Find the area beneath one arch of the trochoid (as shown in Figure 3.11)
$$\mathbf{g}(t) = \begin{bmatrix} at - b\sin t \\ a - b\cos t \end{bmatrix}, \quad 0 \le t \le 2\pi.$$

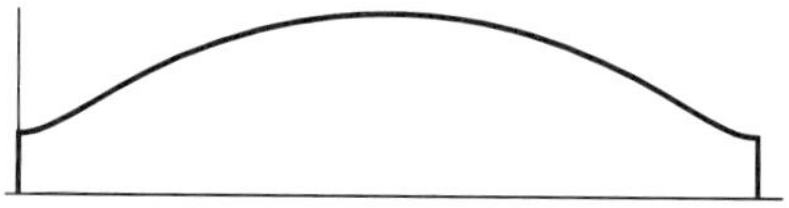

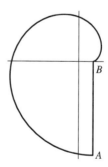

Figure 3.11

15. Find the area of the plane region bounded by the evolute
$$\mathbf{g}(t) = \begin{bmatrix} a(\cos t + t \sin t) \\ a(\sin t - t \cos t) \end{bmatrix}, \quad 0 \le t \le 2\pi,$$
and the line segment $\overline{AB}$, as pictured in Figure 3.12.

Figure 3.12

16. Use symmetry considerations to find the following.
(a) Let C be the polygonal curve shown in Figure 3.13(a). Compute $\oint_C (e^{x^2} - 2xy)dx + (2xy - x^2)dy$.
(b) Let C be the curve pictured in Figure 3.13(b); you might visualize it as a racetrack with two semicircular ends. Compute $\oint_C (4x^3y - 3y^2)dx + (x^4 + e^{\sin y})dy$.

17. Let C be an oriented curve in $\mathbb{R}^2$, and let $\mathbf{n}$ be the unit outward-pointing normal (this means that $\{\mathbf{n}, \mathbf{T}\}$ gives a right-handed basis for $\mathbb{R}^2$). Define the 1-form σ on C by $\sigma = -n_2 dx + n_1 dy$.
(a) Show that $\sigma(\mathbf{T}) = 1$.

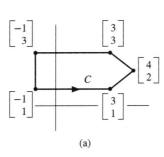
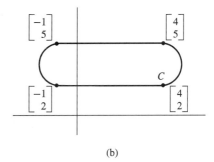

(a)   (b)

Figure 3.13

(b) Show that $\int_C \sigma$ gives the arclength of C.

(c) Can you explain how your answers to parts a and b might be related?

18. Let C be an oriented curve in $\mathbb{R}^2$, and let $\mathbf{n}$ be the unit outward-pointing normal (this means that $\{\mathbf{n}, \mathbf{T}\}$ gives a right-handed basis for $\mathbb{R}^2$). Let $\mathbf{F} = \begin{bmatrix} F_1 \\ F_2 \end{bmatrix}$ be a vector field on the plane, and let $\omega = F_1 dx + F_2 dy$ be the corresponding 1-form. Show that

$$\int_C \mathbf{F} \cdot \mathbf{n}\, ds = \int_C F_1 dy - F_2 dx = \int_C \star\omega.$$

This is called the *flux* of $\mathbf{F}$ across C. (See Exercise 8.2.9.) Conclude that when $C = \partial S$, we have

$$\int_C \mathbf{F} \cdot \mathbf{n}\, ds = \int_S \left(\frac{\partial F_1}{\partial x} + \frac{\partial F_2}{\partial y} \right) dA.$$

19. Prove Green's theorem for the annular region $\Omega = \left\{ \begin{bmatrix} x \\ y \end{bmatrix} : a \leq \sqrt{x^2 + y^2} \leq b \right\}$, pictured in Figure 3.14.

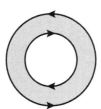

Figure 3.14

20. Give a direct proof of Green's theorem for

(a) a triangle with vertices at $\begin{bmatrix} 0 \\ 0 \end{bmatrix}$, $\begin{bmatrix} a \\ 0 \end{bmatrix}$, and $\begin{bmatrix} 0 \\ b \end{bmatrix}$,

(b) the region $\left\{ \begin{bmatrix} x \\ y \end{bmatrix} : a \leq x \leq b,\ g(x) \leq y \leq h(x) \right\}$. (Hint: Exercise 7.2.23 will be helpful.)

21. Suppose C is a piecewise $\mathcal{C}^1$ closed curve in $\mathbb{R}^2$ that intersects itself finitely many times and does not pass through the origin. Show that the line integral

$$\frac{1}{2\pi} \int_C \frac{-y\, dx + x\, dy}{x^2 + y^2}$$

is always an integer. (See the discussion of Example 10.)

22. Suppose C is a piecewise $\mathcal{C}^1$ closed curve in $\mathbb{R}^2$ that intersects itself finitely many times and does not pass through $\begin{bmatrix} 1 \\ 0 \end{bmatrix}$ or $\begin{bmatrix} -1 \\ 0 \end{bmatrix}$. Show that there are integers m and n so that

$$\frac{1}{2\pi} \int_C \left(A \frac{-y\,dx + (x-1)\,dy}{(x-1)^2 + y^2} + B \frac{-y\,dx + (x+1)\,dy}{(x+1)^2 + y^2} \right) = mA + nB.$$

23. An ant finds himself in the xy-plane in the presence of the force field $\mathbf{F} = \begin{bmatrix} y^3 + x^2 y \\ 2x^2 - 6xy \end{bmatrix}$. Around what simple closed curve beginning and ending at the origin should he travel counterclockwise (once) in order to maximize the work done on him by $\mathbf{F}$?

24. Suppose $\Omega \subset \mathbb{R}^2$ is a region with the property that every simple closed curve in Ω bounds a region contained in Ω that is a finite union of parametrized rectangles. Prove that if ω is 1-form on Ω with $d\omega = 0$, then ω is exact; i.e., there is a potential function f with $\omega = df$.

25. (a) Suppose there is a current c in a river. Show that if we row at a constant ground speed $v > c$ directly downstream a certain distance and then directly back upstream to our beginning point, the time required (ignoring the time to turn around) is always greater than the time it would take with no current. (This is just an elementary algebra problem.)

(b) Show that the same is true no matter what closed path C we take in the river. (Assume we still row with ground velocity $\mathbf{v}$, with $\|\mathbf{v}\| > c$ constant.) (Hint: Express the time of the trip as a line integral over C and do some clever estimates. The diagram in Figure 3.15 may help.)

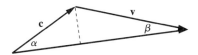

Figure 3.15

26. According to Webster, a *planimeter*, pictured in Figure 3.16, is "an instrument for measuring the area of a regular or irregular plane figure by tracing the perimeter of the figure." As we show a bit more schematically in Figure 3.17, an arm of fixed length b has one fixed end; to the other is attached another arm of length a, which is free to rotate. A wheel (for convenience attached slightly off the near end) turns as the arm rotates about the pivot point. Use Green's Theorem to explain how the amount that the wheel rotates tells us the area of the figure.

Figure 3.16

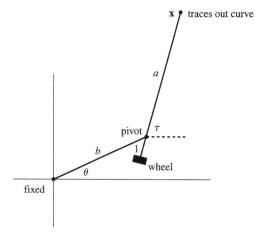

Figure 3.17

4 SURFACE INTEGRALS AND FLUX

Suppose $U \subset \mathbb{R}^2$ is a bounded open set and $\mathbf{g} \colon U \to \mathbb{R}^n$ is a one-to-one smooth map with the property that $D\mathbf{g}(\mathbf{a})$ has rank 2 for all $\mathbf{a} \in U$. Then we call $S = \mathbf{g}(U)$ a parametrized surface.

EXAMPLE 1

a. Consider $\mathbf{g} \colon (0, 2\pi) \times (0, a) \to \mathbb{R}^3$ given by

$$\mathbf{g}\begin{pmatrix} u \\ v \end{pmatrix} = \begin{bmatrix} v \cos u \\ v \sin u \\ v \end{bmatrix}.$$

This is a parametrization of that portion of the cone $z = \sqrt{x^2 + y^2}$ between $z = 0$ and $z = a$, less one ruling, as shown in Figure 4.1.

Figure 4.1

b. Consider $\mathbf{g} \colon (0, 2\pi) \times (0, 2\pi) \to \mathbb{R}^3$ given by

$$\mathbf{g}\begin{pmatrix} u \\ v \end{pmatrix} = \begin{bmatrix} (a + b \cos v) \cos u \\ (a + b \cos v) \sin u \\ b \sin v \end{bmatrix}.$$

If $0 < b < a$, the image of $\mathbf{g}$ is most of a *torus*, as pictured in Figure 4.2, the surface of revolution obtained by rotating a circle of radius b about an axis a units from its center.

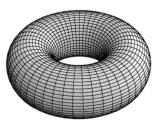

Figure 4.2

c. Consider $\mathbf{g} \colon \mathbb{R} \times (0, \infty) \to \mathbb{R}^3$ given by

$$\mathbf{g}\begin{pmatrix} u \\ v \end{pmatrix} = \begin{bmatrix} v \cos u \\ v \sin u \\ u \end{bmatrix}.$$

This parametrized surface, pictured in part in Figure 4.3, resembles a spiral ramp, and is officially called a *helicoid*. ◀

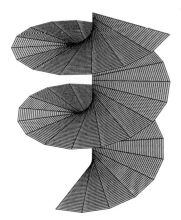

Figure 4.3

As we expect by now, to define the integral of a 2-form over a parametrized surface S, we pull back and integrate: When $\omega \in \mathcal{A}^2(\mathbb{R}^n)$ and $S = \mathbf{g}(U)$, we set

$$\int_S \omega = \int_U \mathbf{g}^*\omega$$

(provided the integral exists).

▶ **EXAMPLE 2**

For these examples, let's fix $\omega = z dx \wedge dy \in \mathcal{A}^2(\mathbb{R}^3)$.

   a. Let $D \subset \mathbb{R}^2$ be the open unit disk, and let $\mathbf{g} \colon D \to \mathbb{R}^3$ be given by

$$\mathbf{g}\begin{pmatrix} r \\ \theta \end{pmatrix} = \begin{bmatrix} r\cos\theta \\ r\sin\theta \\ \sqrt{1-r^2} \end{bmatrix}.$$

Then we recognize that $\mathbf{g}$ is a parametrization of the upper unit hemisphere, S. We then have

$$\mathbf{g}^*(z dx \wedge dy) = \sqrt{1-r^2}\, r dr \wedge d\theta,$$

and so

$$\int_S \omega = \int_D \mathbf{g}^*\omega = \int_0^{2\pi}\int_0^1 \sqrt{1-r^2}\, r dr d\theta = \frac{2\pi}{3}.$$

   b. Now consider $\mathbf{g} \colon (0, \pi/2) \times (0, 2\pi) \to \mathbb{R}^3$ given by

$$g\begin{pmatrix} \phi \\ \theta \end{pmatrix} = \begin{bmatrix} \sin\phi\cos\theta \\ \sin\phi\sin\theta \\ \cos\phi \end{bmatrix}.$$

This is an alternative parametrization of the upper hemisphere, S. Then
$$\mathbf{g}^*(zdx \wedge dy) = \cos\phi(\cos\phi\sin\phi d\phi \wedge d\theta) = \cos^2\phi\sin\phi d\phi \wedge d\theta,$$
and so
$$\int_S \omega = \int_{(0,\pi/2)\times(0,2\pi)} \mathbf{g}^*\omega = \int_0^{2\pi}\int_0^{\pi/2} \cos^2\phi\sin\phi d\phi d\theta = \frac{2\pi}{3}.$$

c. Now let's do the lower hemisphere correspondingly in each of these two ways. Parametrizing by the unit disk, we have
$$\mathbf{h}\begin{pmatrix} r \\ \theta \end{pmatrix} = \begin{bmatrix} r\cos\theta \\ r\sin\theta \\ -\sqrt{1-r^2} \end{bmatrix}.$$

We then have $\mathbf{h}^*(zdx \wedge dy) = -\sqrt{1-r^2}rdr \wedge d\theta$, and so
$$\int_S \omega = \int_D \mathbf{h}^*\omega = -\frac{2\pi}{3}.$$

On the other hand, in spherical coordinates, we have $\mathbf{k}\colon (\pi/2,\pi)\times(0,2\pi) \to \mathbb{R}^3$ given by the same formula as $\mathbf{g}$ in part b above, and so
$$\int_S \omega = \int_{(\pi/2,\pi)\times(0,2\pi)} \mathbf{k}^*\omega = \frac{2\pi}{3}.$$

What gives? ◀

The answer to the query is very simple. Imagine you were walking around on the unit sphere with your feet on the surface (your body pointing radially outward, normal to the sphere). As you look down, you determine that a basis for the tangent plane to the sphere will be "correctly oriented" if you see a positive (counterclockwise) rotation from the first vector ($\mathbf{u}$) to the second ($\mathbf{v}$), as pictured in Figure 4.4. We will say that your body is pointing

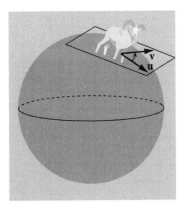

Figure 4.4

in the direction of the *outward-pointing normal* vector to the surface. Note that then **n**, **u**, **v** form a positively-oriented basis for $\mathbb{R}^3$; i.e.,

$$\begin{vmatrix} | & | & | \\ \mathbf{n} & \mathbf{u} & \mathbf{v} \\ | & | & | \end{vmatrix} > 0.$$

More generally, an *orientation* on a surface $S \subset \mathbb{R}^n$ is a continuously varying notion of what a positively oriented basis for the tangent plane at each point should be. In particular, S has an orientation if and only if we can choose various parametrizations $\mathbf{g} \colon U \to \mathbb{R}^n$ of (subsets of) S (the union of whose images covers all of S) so that $\dfrac{\partial \mathbf{g}}{\partial u_1}$ and $\dfrac{\partial \mathbf{g}}{\partial u_2}$ give a positively oriented basis of the tangent plane of S at *every* point of $\mathbf{g}(U)$. We say a surface is *orientable* if there is an orientation on S. (See Exercise 26.)

An alternative characterization of an orientation on a surface S is the following. Recall that $\dim \Lambda^2(\mathbb{R}^2)^* = 1$; i.e., any nonzero element of this vector space is either a positive or a negative multiple of $dx_1 \wedge dx_2$. Given a nonzero element $\phi \in \Lambda^2(\mathbb{R}^2)^*$, it defines an orientation on $\mathbb{R}^2$ in an obvious way: The basis vectors $\mathbf{v}_1, \mathbf{v}_2$ are said to define a positive orientation on $\mathbb{R}^2$ if and only if $\phi(\mathbf{v}_1, \mathbf{v}_2) > 0$. Now, by analogy, a *nowhere-zero* 2-form ω on the surface S defines an orientation on S: For each point $\mathbf{a} \in S$, the tangent vectors $\mathbf{u}$ and $\mathbf{v}$ at $\mathbf{a}$ will form a positively oriented basis for the tangent plane if and only if $\omega(\mathbf{a})(\mathbf{u}, \mathbf{v}) > 0$. (We will abuse notation as follows: Given an orientation on S, i.e., a compatible choice of positively oriented basis for each tangent plane of S, we will say $\omega > 0$ on S if the value of ω on that basis is positive and $\omega < 0$ if the value is negative.) Orienting the sphere as pictured in Figure 4.4, we now see from Figure 4.5 that $dx \wedge dy > 0$ on the upper hemisphere, whereas $dx \wedge dy < 0$ on the lower. This explains the sign disparity in the two calculations in part c of the preceding example.

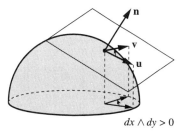
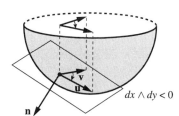

$dx \wedge dy > 0$ $dx \wedge dy < 0$

Figure 4.5

▶ **EXAMPLE 3**

The standard example of a nonorientable surface is the Möbius strip, pictured in Figure 4.6. Observe that if you slide the positive basis $\{\mathbf{u}, \mathbf{v}\}$ once around the strip, it will return with the opposite orientation. Alternatively, if you start with an outward-pointing normal $\mathbf{n}$ and travel once around the Möbius strip, the normal returns pointing in the opposite direction. ◂

Definition If S is an oriented surface, its (oriented) *area 2-form* σ is the 2-form with the property that $\sigma(\mathbf{a})$ assigns to each pair of tangent vectors at $\mathbf{a}$ the signed area of the

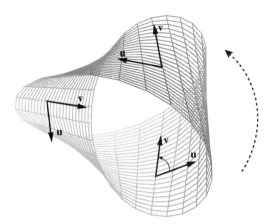

Figure 4.6

parallelogram they span. (By signed area we mean the obvious: The pair of tangent vectors form a positively oriented basis if and only if the signed area is positive.)

4.1 Oriented Surfaces in $\mathbb{R}^3$ and Flux

Let $S \subset \mathbb{R}^3$ be an oriented surface with outward-pointing unit normal $\mathbf{n} = \begin{bmatrix} n_1 \\ n_2 \\ n_3 \end{bmatrix}$. Then we claim that

$$\sigma = n_1 dy \wedge dz + n_2 dz \wedge dx + n_3 dx \wedge dy$$

is its area 2-form. This was the point of Exercise 8.2.5, but we give the argument here. If $\mathbf{u}$ and $\mathbf{v}$ are in the tangent plane to S, then

$$\sigma(\mathbf{u}, \mathbf{v}) = \begin{vmatrix} | & | & | \\ \mathbf{n} & \mathbf{u} & \mathbf{v} \\ | & | & | \end{vmatrix}$$

gives the signed volume of the parallelepiped spanned by $\mathbf{n}$, $\mathbf{u}$, and $\mathbf{v}$. Since $\mathbf{n}$ is a unit vector orthogonal to $\mathbf{u}$ and $\mathbf{v}$, this volume is the area of the parallelogram spanned by $\mathbf{u}$ and $\mathbf{v}$; our definition of orientation dictates that the signs agree.

▶ **EXAMPLE 4**

Consider the surface of revolution S defined by $z = f(r)$, $0 \le r \le a$, oriented so that its outward-pointing normal has a positive $\mathbf{e}_3$-component. We can parametrize S by

$$\mathbf{g} \colon (0, a) \times (0, 2\pi) \to \mathbb{R}^3,$$

$$\mathbf{g}\begin{pmatrix} r \\ \theta \end{pmatrix} = \begin{bmatrix} r \cos \theta \\ r \sin \theta \\ f(r) \end{bmatrix}.$$

Since the vector $\dfrac{\partial \mathbf{g}}{\partial r} \times \dfrac{\partial \mathbf{g}}{\partial \theta}$ has a positive $\mathbf{e}_3$-component, this is an appropriate parametrization. Now, the unit normal is

$$\mathbf{n} = \frac{1}{\sqrt{1+f'(r)^2}} \begin{bmatrix} -\frac{x}{r} f'(r) \\ -\frac{y}{r} f'(r) \\ 1 \end{bmatrix},$$

and so

$$\sigma = \frac{1}{\sqrt{1+f'(r)^2}} \left(-\frac{x}{r} f'(r) dy \wedge dz - \frac{y}{r} f'(r) dz \wedge dx + dx \wedge dy \right).$$

Pulling back, we have

$$\mathbf{g}^* \sigma = r\sqrt{1+f'(r)^2}\, dr \wedge d\theta,$$

and so the surface area of S is given by

$$\int_{(0,a)\times(0,2\pi)} \mathbf{g}^* \sigma = \int_0^{2\pi} \int_0^a r\sqrt{1+f'(r)^2}\, dr\, d\theta,$$

which agrees with the formula usually derived in single variable integral calculus. ◀

▶ **EXAMPLE 5**

Given a plane $\mathbf{n} \cdot \mathbf{x} = c$, with $\|\mathbf{n}\| = 1$, then, assuming $n_3 \neq 0$, we can give a parametrization by thinking of the plane as a graph over the xy-plane:

$$\mathbf{g}\begin{pmatrix} x \\ y \end{pmatrix} = \begin{bmatrix} x \\ y \\ \frac{1}{n_3}(c - n_1 x - n_2 y) \end{bmatrix}.$$

Then

$$\mathbf{g}^* \sigma = n_1 \left(\frac{n_1}{n_3} dx \wedge dy \right) + n_2 \left(\frac{n_2}{n_3} dx \wedge dy \right) + n_3 dx \wedge dy = \frac{1}{n_3} dx \wedge dy.$$

Recall that if $\mathbf{u}$ and $\mathbf{v}$ are two vectors in the plane, then $\sigma(\mathbf{u}, \mathbf{v})$ gives the signed area of the parallelogram they span, whereas $(dx \wedge dy)(\mathbf{u}, \mathbf{v})$ gives the signed area of its projection into the xy-plane. As we see from Figure 4.7, the area of the projection is $|n_3| = |\cos \gamma|$ times the area of the original parallelogram, where γ is the angle between the plane and the xy-plane, so the general theory is compatible with a more intuitive, geometric approach. ◀

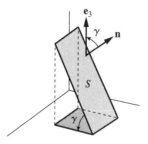

Figure 4.7

Given a vector field $\mathbf{F} = \begin{bmatrix} F_1 \\ F_2 \\ F_3 \end{bmatrix}$ on an open subset of $\mathbb{R}^3$, we saw in Section 3 that integrating the 1-form $\omega = F_1 dx + F_2 dy + F_3 dz$ along an oriented curve computes the work done by $\mathbf{F}$ in moving a test particle along that curve. What is the meaning of integrating the corresponding 2-form $\eta = F_1 dy \wedge dz + F_2 dz \wedge dx + F_3 dx \wedge dy$ over an oriented surface S? (The observant reader who's worked Exercise 8.2.9 will recognize that $\eta = \star\omega$. See also Exercise 8.3.18.) Well, if $\mathbf{u}$ and $\mathbf{v}$ are tangent to S, then

$$\eta(\mathbf{u}, \mathbf{v}) = \begin{vmatrix} | & | & | \\ \mathbf{F} & \mathbf{u} & \mathbf{v} \\ | & | & | \end{vmatrix} = (\mathbf{F} \cdot \mathbf{n}) \times (\text{signed area of the parallelogram spanned by } \mathbf{u} \text{ and } \mathbf{v}).$$

That is, $\int_S \eta$ represents the *flux* of $\mathbf{F}$ outward across S, often written $\int_S \mathbf{F} \cdot \mathbf{n} dS$. Here dS represents an element of (nonoriented) surface area, just as ds represented the element of (nonoriented) arclength on a curve; in neither case should these be interpreted as the exterior derivative of something.

A physical interpretation is the following: Imagine a fluid in motion (not depending on time), and let $\mathbf{F}(\mathbf{x})$ represent the velocity of the fluid at $\mathbf{x}$ multiplied by the density of the fluid at $\mathbf{x}$. (Note that $\mathbf{F}$ points in the direction of the velocity and has units of mass/(area × time).) Then the mass of fluid that flows across a small area ΔS of S in a small amount of time Δt is approximately

$$\Delta m \approx \delta \Delta V \approx \delta(\mathbf{v}\Delta t \cdot \mathbf{n})(\Delta S) \approx (\mathbf{F} \cdot \mathbf{n})\Delta S \Delta t,$$

so that

$$\frac{\Delta m}{\Delta t} \approx \mathbf{F} \cdot \mathbf{n}\Delta S.$$

Taking the limit as $\Delta t \to 0$ and summing over the bits of area ΔS, we infer that $\int_S \eta$ represents the rate at which mass is transferred across S by the fluid flow.

▶ **EXAMPLE 6**

We wish to find the flux of the vector field $\mathbf{F} = \begin{bmatrix} xz^2 \\ yx^2 \\ zy^2 \end{bmatrix}$ outward across the sphere S of radius a centered at the origin. That is, we wish to find the integral over S of the 2-form $\eta = xz^2 dy \wedge dz + yx^2 dz \wedge dx + zy^2 dx \wedge dy$. Calculating the pullback under the spherical coordinate parametrization $\mathbf{g}: (0, \pi) \times (0, 2\pi) \to \mathbb{R}^3$,

$$\mathbf{g}\begin{pmatrix} \phi \\ \theta \end{pmatrix} = a \begin{bmatrix} \sin\phi \cos\theta \\ \sin\phi \sin\theta \\ \cos\phi \end{bmatrix},$$

we have

$$\mathbf{g}^*\eta = a^5\big(\sin\phi\cos\theta\cos^2\phi(\sin^2\phi\cos\theta) + \sin^3\phi\sin\theta\cos^2\theta(\sin^2\phi\sin\theta) \\ + \cos\phi\sin^2\phi\sin^2\theta(\sin\phi\cos\phi)\big)d\phi\wedge d\theta$$
$$= a^5\big(\sin^3\phi\cos^2\phi + \sin^5\phi\cos^2\theta\sin^2\theta\big)d\phi\wedge d\theta,$$

and so

$$\int_S \eta = \int_{(0,\pi)\times(0,2\pi)} a^5\big(\sin^3\phi\cos^2\phi + \sin^5\phi\cos^2\theta\sin^2\theta\big)d\phi\wedge d\theta$$
$$= a^5\int_0^\pi\int_0^{2\pi}\big(\sin^3\phi\cos^2\phi + \sin^5\phi\cos^2\theta\sin^2\theta\big)d\theta\,d\phi$$
$$= 2\pi a^5\int_0^\pi\big(\sin^3\phi\cos^2\phi + \tfrac{1}{8}\sin^5\phi\big)d\phi$$
$$= 2\pi a^5\int_0^\pi\big(\tfrac{1}{8}\sin\phi + \tfrac{3}{4}\cos^2\phi\sin\phi - \tfrac{7}{8}\cos^4\phi\sin\phi\big)d\phi = \frac{4}{5}\pi a^5. \quad\blacktriangleleft$$

4.2 Surface Area

We have pilfered Figure 4.8 from someone who, in turn, plagiarized from the book Математический Анализ на Многообразиях by Михаил Спивак. As this example, discovered by Hermann Schwarz, illustrates, one must be far more careful to define surface area by a limiting process than to define arclength of curves. It seems natural to approximate a surface by inscribed triangles. But, even as the triangles get smaller and smaller, the sum of their areas may go to infinity, even in the case of a surface as simplistic as a cylinder. In particular, by moving the planes of the hexagons closer together, the triangles become more and more orthogonal to the cylinder. The area of the individual triangles approaches $h\ell/2$, and the number of triangles grows without bound.

For an oriented surface $S \subset \mathbb{R}^3$, we can (and did) explicitly write down the 2-form σ that gives the oriented area-form on S. In analogy with our development of arclength of a

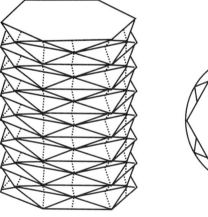
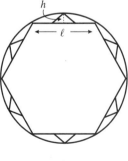

Figure 4.8

curve and our treatment of change of variables in Chapter 7, we next give a definition of surface area that will work for any parametrized surface. We need the result of Exercise 7.5.22: If $\mathbf{u}$ and $\mathbf{v}$ are vectors in $\mathbb{R}^n$, the area of the parallelogram they span is given by

$$\sqrt{\left|\begin{matrix} \mathbf{u}\cdot\mathbf{u} & \mathbf{u}\cdot\mathbf{v} \\ \mathbf{v}\cdot\mathbf{u} & \mathbf{v}\cdot\mathbf{v} \end{matrix}\right|}.$$

(Here is the sketch of a proof. We may assume $\{\mathbf{u}, \mathbf{v}\}$ is linearly independent, and let $\{\mathbf{v}_3, \ldots, \mathbf{v}_n\}$ be an orthonormal basis for $\text{Span}(\mathbf{u}, \mathbf{v})^\perp$. Then we know that the volume of the n-dimensional parallelepiped spanned by $\mathbf{u}, \mathbf{v}, \mathbf{v}_3, \ldots, \mathbf{v}_n$ is the absolute value of the determinant of the matrix

$$A = \begin{bmatrix} | & | & | & & | \\ \mathbf{u} & \mathbf{v} & \mathbf{v}_3 & \cdots & \mathbf{v}_n \\ | & | & | & & | \end{bmatrix}.$$

But by our choice of the vectors $\mathbf{v}_3, \ldots, \mathbf{v}_n$, this volume is evidently the area of the parallelogram spanned by $\mathbf{u}$ and $\mathbf{v}$. But by Propositions 5.11 and 5.7 of Chapter 7, we have

$$(\det A)^2 = \det(A^T A) = \left|\begin{matrix} \mathbf{u}\cdot\mathbf{u} & \mathbf{u}\cdot\mathbf{v} & 0 & \cdots & 0 \\ \mathbf{v}\cdot\mathbf{u} & \mathbf{v}\cdot\mathbf{v} & 0 & \cdots & 0 \\ 0 & 0 & 1 & & \\ \vdots & \vdots & & \ddots & \\ 0 & 0 & & & 1 \end{matrix}\right| = \left|\begin{matrix} \mathbf{u}\cdot\mathbf{u} & \mathbf{u}\cdot\mathbf{v} \\ \mathbf{v}\cdot\mathbf{u} & \mathbf{v}\cdot\mathbf{v} \end{matrix}\right|,$$

as required.) If $\mathbf{g}$ is a parametrization of a smooth surface, then for sufficiently small Δu and Δv, we expect that the area of the image $\mathbf{g}([u, u+\Delta u] \times [v, v+\Delta v])$ should be approximately the area of the parallelogram that is the image of this rectangle under the linear map $D\mathbf{g}\begin{pmatrix} u \\ v \end{pmatrix}$, and that, in turn, is $\Delta u \Delta v$ times the area of the parallelogram spanned by $\dfrac{\partial \mathbf{g}}{\partial u}$ and $\dfrac{\partial \mathbf{g}}{\partial v}$.

With this motivation, we now make the following

Definition Let $S \subset \mathbb{R}^n$ be a parametrized surface, given by $\mathbf{g}: \Omega \to \mathbb{R}^n$, for some region $\Omega \subset \mathbb{R}^2$. Let

$$E = \left\|\frac{\partial \mathbf{g}}{\partial u}\right\|^2, \qquad F = \frac{\partial \mathbf{g}}{\partial u}\cdot\frac{\partial \mathbf{g}}{\partial v}, \qquad G = \left\|\frac{\partial \mathbf{g}}{\partial v}\right\|^2.$$

We define the *surface area* of S to be

$$\text{area}(S) = \int_\Omega \sqrt{EG - F^2}\, dA_{uv}.$$

We leave it to the reader to check in Exercise 20 that for a parametrized, oriented surface in $\mathbb{R}^3$ this gives the same result as integrating the area 2-form σ over the surface.

EXERCISES 8.4

1. Let S be that portion of the plane $x + 2y + 2z = 4$ lying in the first octant, oriented with outward normal pointing upward. Find
   (a) the area of S,
   (b) $\int_S (x - y + 3z)\sigma$,
   (c) $\int_S z\,dx \wedge dy + y\,dz \wedge dx + x\,dy \wedge dz$.

2. Find the area of that portion of the cylinder $x^2 + y^2 = a^2$ lying above the xy-plane and below the plane $z = y$.

3. Find the area of that portion of the cone $z = \sqrt{2(x^2 + y^2)}$ lying beneath the plane $y + z = 1$.

*4. Find the area of that portion of the cylinder $x^2 + y^2 = 2y$ lying inside the sphere $x^2 + y^2 + z^2 = 4$.

♯5. Let S be the sphere of radius a centered at the origin, oriented with normal pointing outward. Evaluate $\int_S x\,dy \wedge dz + y\,dz \wedge dx + z\,dx \wedge dy$ explicitly. What formula do you deduce for the surface area of S?

6. Let S be the surface of the unit sphere, and let its area element be σ.
   (a) Calculate $\int_S x^2 \sigma$ directly.
   (b) Evaluate the integral in part a without doing any calculations. (Hint: Why is $\int_S x^2\sigma = \int_S y^2\sigma = \int_S z^2\sigma$?)

7. Find the surface area of the torus given parametrically in Example 1b.

*8. Find the surface area of that portion of a sphere of radius a lying between two parallel planes (both intersecting the sphere) a distance h apart.

9. Let S be that portion of the helicoid given parametrically by

$$\mathbf{g}\begin{pmatrix} u \\ v \end{pmatrix} = \begin{bmatrix} u \cos v \\ u \sin v \\ v \end{bmatrix}, \quad 0 \le u \le 1,\ 0 \le v \le 2\pi.$$

   (a) With the orientation determined by $\mathbf{g}$, decide whether the outward-pointing normal points upward or downward.
   (b) If we orient S with the normal pointing upward, compute $\int_S x\,dz \wedge dx$.

*10. We can parametrize the unit sphere (except for the north pole) by *stereographic projection* from the north pole, as indicated in Figure 4.9. If $\begin{bmatrix} u \\ v \\ 0 \end{bmatrix}$ is the point where the line through $\begin{bmatrix} 0 \\ 0 \\ 1 \end{bmatrix}$ and $\begin{bmatrix} x \\ y \\ z \end{bmatrix}$ (on the sphere) intersects the plane $z = 0$, solve for u and v. Then solve for $\mathbf{g}\begin{pmatrix} u \\ v \end{pmatrix} = \begin{bmatrix} x \\ y \\ z \end{bmatrix}$.

Explain geometrically why stereographic projection is an orientation-reversing parametrization.

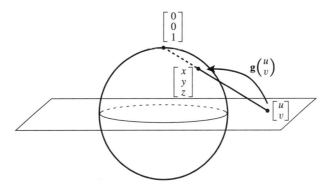

Figure 4.9

11. Let $\omega = xdy \wedge dz$. Let S be the unit sphere, oriented with outward-pointing normal. Calculate $\int_S \omega$ by parametrizing S
   (a) by spherical coordinates,
   (b) as a union of graphs,
   (c) by stereographic projection (see Exercise 10).

12. Let S be the unit upper hemisphere, oriented with outward-pointing normal. Calculate $\int_S z\sigma$ by showing that $z\sigma = dx \wedge dy$ as 2-forms on S.

13. Let S be the cylinder $x^2 + y^2 = a^2$, $0 \le z \le h$, oriented with outward-pointing normal. Calculate $\int_S \omega$ for
   (a) $\omega = zdx \wedge dy$,   **(b)** $\omega = ydx \wedge dz$.

*14. Find the moment of inertia about the z-axis of a uniform spherical shell of radius a centered at the origin.

*15. Find the flux of the vector field $\mathbf{F}(\mathbf{x}) = \mathbf{x}$ outward across the following surfaces (all oriented with outward-pointing normal pointing away from the origin):
   (a) the surface of the sphere of radius a centered at the origin,
   (b) the surface of the cylinder $x^2 + y^2 = a^2$, $-h \le z \le h$,
   (c) the surface of the cylinder $x^2 + y^2 = a^2$, $-h \le z \le h$, together with the two disks, $x^2 + y^2 \le a^2$, $z = \pm h$,
   (d) the surface of the cube with vertices at $\begin{bmatrix} \pm 1 \\ \pm 1 \\ \pm 1 \end{bmatrix}$.

16. Find the flux of the vector field $\mathbf{F} = \begin{bmatrix} x^2 \\ y^2 \\ z^2 \end{bmatrix}$ outward across the given surface S (all oriented with outward-pointing normal pointing away from the origin, unless otherwise specified):
   (a) S is the sphere of radius a centered at the origin.
   (b) S is the upper hemisphere of radius a centered at the origin.
   (c) S is the cone $z = \sqrt{x^2 + y^2}$, $0 < z < 1$, with outward-pointing normal having a negative $\mathbf{e}_3$-component.
   (d) S is the cylinder $x^2 + y^2 = a^2$, $0 \le z \le h$.

(e) S is the cylinder $x^2 + y^2 = a^2$, $0 \le z \le h$, along with the disks $x^2 + y^2 \le a^2$, $z = 0$ and $z = h$.

*17. Calculate the flux of the vector field $\mathbf{F} = \begin{bmatrix} xz \\ yz \\ x^2 + y^2 \end{bmatrix}$ outward across the surface of the paraboloid S given by $z = 4 - x^2 - y^2$, $z \ge 0$ (with outward-pointing normal having positive $\mathbf{e}_3$-component).

*18. Find the flux of the vector field $\mathbf{F}(\mathbf{x}) = \mathbf{x}/\|\mathbf{x}\|^3$ outward across the given surface (oriented with outward-pointing normal pointing away from the origin):
   (a) the surface of the sphere of radius a centered at the origin,
   (b) the surface of the cylinder $x^2 + y^2 = a^2$, $-h \le z \le h$,
   (c) the surface of the cylinder $x^2 + y^2 = a^2$, $-h \le z \le h$, together with the two disks, $x^2 + y^2 \le a^2$, $z = \pm h$,
   (d) the surface of the cube with vertices at $\begin{bmatrix} \pm 1 \\ \pm 1 \\ \pm 1 \end{bmatrix}$.

19. Let S be that portion of the cone $z = \sqrt{x^2 + y^2}$ lying inside the sphere $x^2 + y^2 + z^2 = 2ax$ and oriented with normal pointing downward. Calculate $\int_S \omega$ for
   (a) $\omega = dx \wedge dy$,
   (b) $\omega = \dfrac{x}{z} dy \wedge dz + \dfrac{y}{z} dz \wedge dx - dx \wedge dy$.

20. Suppose $\mathbf{g}\colon \Omega \to \mathbb{R}^3$ gives a parametrized, oriented surface with unit outward normal $\mathbf{n}$. Let $\mathbf{N} = \dfrac{\partial \mathbf{g}}{\partial u} \times \dfrac{\partial \mathbf{g}}{\partial v}$, so that $\mathbf{n} = \mathbf{N}/\|\mathbf{N}\|$. Check that

$$\mathbf{g}^*(n_1 dy \wedge dz + n_2 dz \wedge dx + n_3 dx \wedge dy) = \|\mathbf{N}\| du \wedge dv = \sqrt{EG - F^2}\, du \wedge dv.$$

21. Sketch the parametrized surface $\mathbf{g}\colon [0, 2\pi] \times [-1, 1]$ given by:

$$\mathbf{g}\begin{pmatrix} u \\ v \end{pmatrix} = \begin{bmatrix} (2 + v \sin \tfrac{u}{2}) \cos u \\ (2 + v \sin \tfrac{u}{2}) \sin u \\ v \cos \tfrac{u}{2} \end{bmatrix}.$$

Compare $\mathbf{g}^*(dy \wedge dz)$ at $\begin{bmatrix} 0 \\ 0 \end{bmatrix}$ and $\begin{bmatrix} 2\pi \\ 0 \end{bmatrix}$. Explain.

*22. Consider the "flat torus"

$$X = \left\{ \begin{bmatrix} x_1 \\ x_2 \\ x_3 \\ x_4 \end{bmatrix} : x_1^2 + x_2^2 = 1,\ x_3^2 + x_4^2 = 1 \right\} \subset \mathbb{R}^4.$$

Orient X so that $dx_2 \wedge dx_4 > 0$ at the point $\begin{bmatrix} 1 \\ 0 \\ 1 \\ 0 \end{bmatrix} \in X$. Calculate $\int_X \omega$ for

   (a) $\omega = dx_1 \wedge dx_2 + dx_3 \wedge dx_4$,
   (b) $\omega = dx_1 \wedge dx_3 + dx_2 \wedge dx_4$,
   (c) $\omega = x_2 x_4 dx_1 \wedge dx_3$.

23. Consider the cylinder S with equation $x^2 + y^2 = 1$, $-1 \le z \le 1$, oriented with unit normal pointing outward. Calculate

(a) $\int_S x\,dy \wedge dz - z\,dx \wedge dy$,

(b) $\int_{C_1} xz\,dy$ and $\int_{C_2} xz\,dy$ (See Figure 4.10.)

Compare your answers and explain.

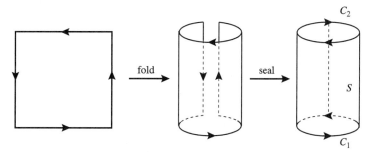

Figure 4.10

24. Let S be the hemisphere $x^2 + y^2 + z^2 = a^2$, $z \ge 0$, oriented with unit normal pointing upward. Let C be the boundary curve, $x^2 + y^2 = a^2$, $z = 0$, oriented counterclockwise. Calculate

(a) $\int_S dx \wedge dy + 2z\,dz \wedge dx$,

(b) $\int_C x\,dy + z^2\,dx$.

Compare your answers and explain.

25. Construct two Möbius strips out of paper: For each, cut out a long rectangle, and attach the short edges with opposite orientations.

(a) Cut along the center circle of the first strip. What happens? Explain. What happens if you repeat the process?

(b) Make parallel cuts in the second strip one-third of the way from either edge. What happens? Explain.

26. Prove or give a counterexample: If S is an orientable surface, then there are exactly two possible orientations on S.

▶ 5 STOKES'S THEOREM

We now come to the generalization of Green's Theorem to higher dimensions. We first stop to make the official definition of the integral of a differential form over a compact, oriented manifold. So far we have dealt only with the integrals of 1- and 2-forms over parametrized curves and surfaces, respectively.

5.1 Integrating over a General Compact, Oriented k-Dimensional Manifold

We know how to integrate a k-form over a parametrized k-dimensional manifold by pulling back. In general, a manifold will be a union of parametrized pieces that overlap, and so summing the integrals will give a meaningless result. To solve this problem, we introduce one of the powerful tools in the study of manifolds, one that allows us to chop a global problem into local ones and then add up the answers.

Chapter 8. Differential Forms and Integration on Manifolds

We start with a

Definition A subset $M \subset \mathbb{R}^n$ is called a k-dimensional *manifold with boundary* if for each point $\mathbf{p} \in M$ there is an open set $W \subset \mathbb{R}^n$ containing $\mathbf{p}$ and a parametrization[4] $\mathbf{g}\colon U \to \mathbb{R}^n$ so that

  i. $\mathbf{g}(U) = V = W \cap M$; and
  ii. U is an open subset either of $\mathbb{R}^k$ or of $\mathbb{R}^k_+ = \{\mathbf{u} \in \mathbb{R}^k : u_k \geq 0\}$.[5]

See Figure 5.1. We say $\mathbf{p}$ is a *boundary point* of M (written $\mathbf{p} \in \partial M$) if $\mathbf{p} = \mathbf{g}(\mathbf{u})$ for some $\mathbf{u} \in \partial \mathbb{R}^k_+ = \{\mathbf{u} \in \mathbb{R}^k : u_k = 0\}$.

$\mathbf{g}(U)$ is sometimes called a *coordinate chart* on M. A *coordinate ball* on M is the image of some ball under some parametrization.

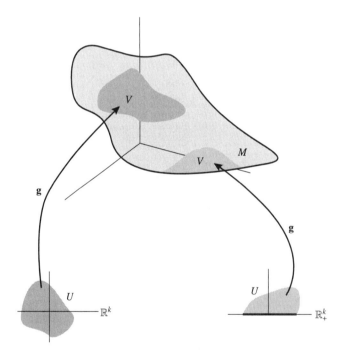

Figure 5.1

As was the case with surfaces, an *orientation* on a manifold with boundary $M \subset \mathbb{R}^n$ is a continuously varying notion of what a positively oriented basis for the tangent space at each point should be. M has an orientation if and only if we can cover M by coordinate

---
[4]Recall from Section 3 of Chapter 6 that this means that $\mathbf{g}$ is a one-to-one smooth map from U to $W \cap M$ so that $D\mathbf{g}(\mathbf{u})$ has rank k for every $\mathbf{u} \in U$ and $\mathbf{g}^{-1}\colon W \cap M \to U$ is continuous.

[5]We say $U \subset \mathbb{R}^k_+$ is an open subset of $\mathbb{R}^k_+$ if it is the intersection of $\mathbb{R}^k_+$ with some open subset of $\mathbb{R}^k$.

charts $\mathbf{g}\colon U \to \mathbb{R}^n$ so that $\left\{\dfrac{\partial \mathbf{g}}{\partial u_1}, \ldots, \dfrac{\partial \mathbf{g}}{\partial u_k}\right\}$ is a positive basis for the tangent space of M at each point. We say M is *orientable* if there is some orientation on M.

We leave it to the reader to prove, using Theorem 5.1, that M is orientable if and only if there is a nowhere-vanishing k-form on M (see Exercise 23). Then we can make the

Definition Let M be an oriented k-dimensional manifold with boundary. Its (oriented) *volume form* is the k-form σ with the property that $\sigma(\mathbf{a})$ assigns to each k-tuple of tangent vectors at $\mathbf{a}$ the signed volume of the parallelepiped they span.

Now we come to the main technical tool that will enable us to define integration on manifolds.

Theorem 5.1 *Let $M \subset \mathbb{R}^n$ be a compact k-dimensional manifold with boundary. Then there are smooth real-valued functions $\rho_1, \ldots, \rho_N$ on M so that*

   i. $0 \le \rho_i \le 1$ *for all i;*
   ii. *each ρ_i is zero outside some coordinate ball;*
   iii. $\sum_{i=1}^{N} \rho_i = 1.$

$\{\rho_i\}$ *is called a* partition of unity *on M.*

Proof

Step 1: Define $h\colon \mathbb{R} \to \mathbb{R}$ by

$$h(x) = \begin{cases} e^{-1/x}, & x > 0 \\ 0, & x \le 0 \end{cases}.$$

Then h is smooth (in particular, all its derivatives at 0 are equal to 0, as we ask the reader to prove in Exercise 25). Set

$$j(x) = \dfrac{\displaystyle\int_0^x h(t)h(1-t)\,dt}{\displaystyle\int_0^1 h(t)h(1-t)\,dt},$$

and define $\psi \colon \mathbb{R}^k \to \mathbb{R}$ by $\psi(\mathbf{x}) = j(3 - 2\|\mathbf{x}\|)$. Then ψ is a smooth function with $\psi(\mathbf{x}) = 1$ whenever $\|\mathbf{x}\| \le 1$ and $\psi(\mathbf{x}) = 0$ whenever $\|\mathbf{x}\| \ge 3/2$; ψ is often called a *bump function*. (See Figure 5.2 for the graph of ψ for $k = 1$.)

Step 2: For each point $\mathbf{p} \in M$, choose a coordinate chart whose domain is a ball of radius 2 in $\mathbb{R}^k$ (why can we do so?).[6] The images of the balls of radius 1 obviously cover all of M; indeed, we can choose a sequence (countable number) of $\mathbf{p}$'s so that this is true. (See Exercise 26.) By Exercise 5.1.12, *finitely* many of these images of balls of radius 1, say, $V_1, \ldots, V_N$, cover all of M. Let $\mathbf{g}_i \colon B(\mathbf{0}, 2) \to V_i$ be the respective coordinate charts,

---

[6] For those $\mathbf{p}$ in the boundary, this will be a half-ball, i.e., the points in the ball with nonnegative k^{th} coordinate.

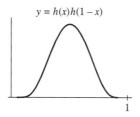 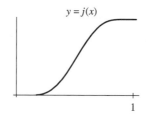 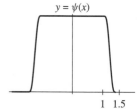

Figure 5.2

and define $\theta_i = \psi \circ \mathbf{g}_i^{-1}$, interpreting θ_i to be defined on all of M by letting it be 0 outside of V_i (note that the fact that ψ is 0 outside the ball of radius $3/2$ means that θ_i will be smooth). Set

$$\rho_i = \frac{\theta_i}{\sum_{j=1}^{N} \theta_j}.$$

Note that for each $\mathbf{p} \in M$, we have $\mathbf{p} = \mathbf{g}_j(\mathbf{u})$ for some j and some $\mathbf{u} \in B(\mathbf{0}, 1)$, and hence $\theta_j(\mathbf{p}) = 1$ for some j. Thus, the sum is everywhere positive. These functions ρ_i fulfill the requirements of the theorem. ∎

Now it is easy to define the integral. Let $M \subset \mathbb{R}^n$ be a compact, oriented k-dimensional manifold (with piecewise-smooth boundary). Let ω be a k-form on M.[7] Let $\{\rho_i\}$ be a partition of unity, and let $\mathbf{g}_i$ be the corresponding parametrizations, which we may take to be orientation-preserving (how?). Now we set

$$\int_M \omega = \int_M \left(\sum_{i=1}^N \rho_i \omega \right) = \sum_{i=1}^N \int_{B(0,2)} \mathbf{g}_i^*(\rho_i \omega) = \sum_{i=1}^N \int_{B(0,2)} \psi \mathbf{g}_i^* \omega.$$

The point is that the form $\rho_i \omega$ is nonzero only inside the image of the parametrization $\mathbf{g}_i$.

One last technical point. Let M be a k-dimensional manifold with boundary, and let $\mathbf{p}$ be a boundary point. The tangent space of ∂M at $\mathbf{p}$ is a $(k-1)$-dimensional subspace of the tangent space of M at $\mathbf{p}$, and its orthogonal complement is 1-dimensional. That 1-dimensional subspace has two possible basis vectors, called the *inward-* and *outward-pointing normal vectors*. By definition, if we follow a curve starting at $\mathbf{p}$ whose tangent vector is the *inward*-pointing normal, we move into M, as shown in Figure 5.3. We endow ∂M with an orientation, called the *boundary orientation*, by saying that the outward normal, $\mathbf{n}$, followed by a positively oriented basis for the tangent space of ∂M should provide a positively oriented basis for the tangent space of M. For examples, see Figure 5.4. We ask

---

[7] We are being a bit casual about what a smooth function or k-form on M ought to mean. We might start with something defined on a neighborhood of M in $\mathbb{R}^n$ or, instead, we might just know the pullbacks under coordinate charts are smooth. Because of Theorem 3.1 of Chapter 6, these notions are equivalent. We leave the technical details to a more advanced course. In practice, except for results such as Theorem 5.1, we will usually start with objects defined on $\mathbb{R}^n$ anyhow.

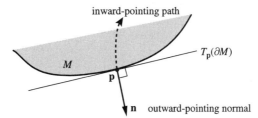

Figure 5.3

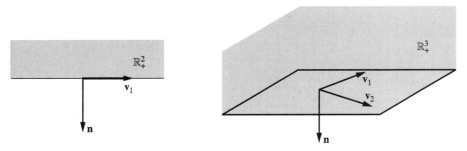

Figure 5.4

the reader to check in Exercise 1 that the boundary orientation on $\partial \mathbb{R}^k_+$ is the usual one on $\mathbb{R}^{k-1}$ precisely when k is even.

5.2 Stokes's Theorem

Now we come to the crowning result. We will give various physical interpretations and applications in the next section, as well as some applications to topology in the last section of the chapter. Here we will give the theorem and some concrete examples.

Theorem 5.2 (Stokes's Theorem) *Let M be a compact, oriented k-dimensional manifold with boundary, and let ω be a smooth $(k-1)$-form on M. Then*

$$\int_{\partial M} \omega = \int_M d\omega.$$

(Here ∂M is endowed with the boundary orientation, as described above.)

Remark Note that the usual Fundamental Theorem of Calculus, the Fundamental Theorem of Calculus for Line Integrals (Proposition 3.1), and Green's Theorem (Corollary 3.5) are all special cases of this theorem. When we're orienting the boundary of an oriented line segment, we assign a + when the outward-pointing normal agrees with the orientation on the segment, and a − when it disagrees. This is compatible with the signs in

$$\int_a^b df = \int_a^b f'(t)dt = f(b) - f(a).$$

Proof Since both sides of the desired equation are linear in ω, we can (by using a partition of unity) reduce to the case that ω is zero outside of a *compact* subset of a *single* coordinate chart, $\mathbf{g}\colon U \to \mathbb{R}^n$ (where U is open in either $\mathbb{R}^k$ or $\mathbb{R}_+^k$). Then we have

$$\int_M d\omega = \int_{\mathbf{g}(U)} d\omega = \int_U \mathbf{g}^*(d\omega) = \int_U d(\mathbf{g}^*\omega).$$

$\mathbf{g}^*\omega$, being a $(k-1)$-form on $U \subset \mathbb{R}^k$, can be written as follows:

$$\mathbf{g}^*\omega = \sum_{i=1}^k f_i(\mathbf{x}) dx_1 \wedge \cdots \wedge \widehat{dx_i} \wedge \cdots \wedge dx_k,$$

where $\widehat{dx_i}$ indicates that the dx_i term is omitted. So we have

$$d(\mathbf{g}^*\omega) = \sum_{i=1}^k \frac{\partial f_i}{\partial x_i} dx_i \wedge dx_1 \wedge \cdots \wedge \widehat{dx_i} \wedge \cdots \wedge dx_k$$

$$= \sum_{i=1}^k \left((-1)^{i-1} \frac{\partial f_i}{\partial x_i}\right) dx_1 \wedge \cdots \wedge dx_i \wedge \cdots \wedge dx_k.$$

Case 1: Suppose U is open in $\mathbb{R}^k$; this means that $\omega = 0$ on ∂M, and so we need only show that $\int_M d\omega = \int_U d(\mathbf{g}^*\omega) = 0$. The crucial point is this: Since $\mathbf{g}^*\omega$ is smooth and 0 outside of a compact subset of U, we may choose a rectangle R containing U, as shown in Figure 5.5, and extend the functions f_i to functions on all of R by setting them equal to 0 outside of U. Finally, we integrate over $R = [a_1, b_1] \times \cdots \times [a_k, b_k]$:

$$\int_U d(\mathbf{g}^*\omega) = \int_R \sum_{i=1}^k \left((-1)^{i-1} \frac{\partial f_i}{\partial x_i}\right) dx_1 \wedge \cdots \wedge dx_i \wedge \cdots \wedge dx_k$$

$$= \sum_{i=1}^k (-1)^{i-1} \int_R \frac{\partial f_i}{\partial x_i} dx_1 dx_2 \cdots dx_k$$

$$= \sum_{i=1}^k (-1)^{i-1} \int_{a_k}^{b_k} \cdots \int_{a_1}^{b_1} \left(\int_{a_i}^{b_i} \frac{\partial f_i}{\partial x_i} dx_i\right) dx_1 \cdots \widehat{dx_i} \cdots dx_k$$

$$= \sum_{i=1}^k (-1)^{i-1} \int_{a_k}^{b_k} \cdots \int_{a_1}^{b_1} \left(f_i\begin{pmatrix}x_1\\\vdots\\b_i\\\vdots\\x_k\end{pmatrix} - f_i\begin{pmatrix}x_1\\\vdots\\a_i\\\vdots\\x_k\end{pmatrix}\right) dx_1 \cdots \widehat{dx_i} \cdots dx_k$$

$$= 0$$

since $f_i = 0$ everywhere on the boundary of R. (Note the applications of Fubini's Theorem and the traditional Fundamental Theorem of Calculus.)

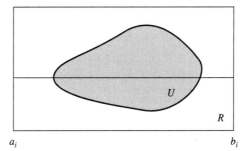

Figure 5.5

Case 2: Now comes the more interesting situation. Suppose U is open in $\mathbb{R}_+^k$, and once again we extend the functions f_i to functions on a rectangle $R \subset \mathbb{R}_+^k$ by letting them be 0 outside of U. In this case, the rectangle is of the form $R = [a_1, b_1] \times \cdots \times [a_{k-1}, b_{k-1}] \times [0, b_k]$, as we see in Figure 5.6. Now we have

$$\int_U d(\mathbf{g}^*\omega) = \int_R \sum_{i=1}^k \left((-1)^{i-1} \frac{\partial f_i}{\partial x_i} \right) dx_1 \wedge \cdots \wedge dx_i \wedge \cdots \wedge dx_k$$

$$= \sum_{i=1}^k (-1)^{i-1} \int_R \frac{\partial f_i}{\partial x_i} dx_1 dx_2 \cdots dx_k$$

$$= \sum_{i=1}^k (-1)^{i-1} \int_{a_k}^{b_k} \cdots \int_{a_1}^{b_1} \left(\int_{a_i}^{b_i} \frac{\partial f_i}{\partial x_i} dx_i \right) dx_1 \cdots \widehat{dx_i} \cdots dx_k$$

$$= \sum_{i=1}^k (-1)^{i-1} \int_{a_k}^{b_k} \cdots \int_{a_1}^{b_1} \left(f_i \begin{pmatrix} x_1 \\ \vdots \\ b_i \\ \vdots \\ x_k \end{pmatrix} - f_i \begin{pmatrix} x_1 \\ \vdots \\ a_i \\ \vdots \\ x_k \end{pmatrix} \right) dx_1 \cdots \widehat{dx_i} \cdots dx_k$$

$$= (-1)^{k-1} \int_{a_{k-1}}^{b_{k-1}} \cdots \int_{a_1}^{b_1} \left(f_k \begin{pmatrix} x_1 \\ \vdots \\ x_{k-1} \\ b_k \end{pmatrix} - f_k \begin{pmatrix} x_1 \\ \vdots \\ x_{k-1} \\ 0 \end{pmatrix} \right) dx_1 \cdots dx_{k-1}$$

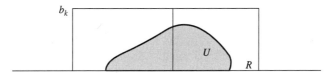

Figure 5.6

(since all the other integrals vanish for the same reason as in Case 1)

$$= (-1)^k \int_{U \cap \partial \mathbb{R}^k_+} f_k \begin{pmatrix} x_1 \\ \vdots \\ x_{k-1} \\ 0 \end{pmatrix} dx_1 \cdots dx_{k-1}$$

$$= \int_{U \cap \partial \mathbb{R}^k_+} \mathbf{g}^* \omega = \int_{\partial M} \omega,$$

as required. Note the crucial sign in the definition of the boundary orientation (see also Exercise 1). ∎

Remark Although we won't take the time to prove it here, Stokes's Theorem is also valid when the boundary, rather than being a manifold itself, is piecewise smooth, e.g., a union of smooth $(k-1)$-dimensional manifolds with boundary intersecting along $(k-2)$-dimensional manifolds. For example, we may take a cube or a solid cylinder, whose boundary is the union of a cylinder and two disks. The theorem also applies to such non-manifolds as a solid cone.

Corollary 5.3 *Let M be a compact, oriented k-dimensional manifold without boundary. Let ω be an exact k-form; i.e., $\omega = d\eta$ for some $(k-1)$-form η. Then $\int_M \omega = 0$.*

Proof This is immediate from Case 1 of the proof of Theorem 5.2. ∎

▶ **EXAMPLE 1**

Let C be the intersection of the unit sphere $x^2 + y^2 + z^2 = 1$ and the plane $x + 2y + z = 0$, oriented counterclockwise as viewed from high above the xy-plane. We wish to evaluate $\int_C (z-x)dx + (x-y)dy + (y-z)dz$.

We let $\omega = (z-x)dx + (x-y)dy + (y-z)dz$ and M be that portion of the plane $x + 2y + z = 0$ lying inside the unit sphere, oriented so that the outward-pointing normal has a positive $\mathbf{e}_3$-component, as shown in Figure 5.7. Then $\partial M = C$, and by Stokes's Theorem we have

$$(*) \qquad \int_C \omega = \int_{\partial M} \omega = \int_M d\omega = \int_M (dy \wedge dz + dz \wedge dx + dx \wedge dy).$$

Parametrizing the plane by projection on the xy-plane, we have $M = \mathbf{g}(D)$, where D is the interior of the ellipse $2x^2 + 4xy + 5y^2 = 1$ (why?), and

$$\mathbf{g}\begin{pmatrix} x \\ y \end{pmatrix} = \begin{bmatrix} x \\ y \\ -x - 2y \end{bmatrix}.$$

(The reader should check that $\mathbf{g}$ is an orientation-preserving parametrization.) Therefore,

$$\int_M d\omega = \int_D \mathbf{g}^* d\omega = \int_D (1 + 2 + 1) dx \wedge dy = 4\operatorname{area}(D).$$

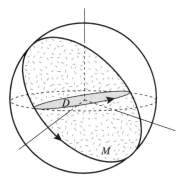

Figure 5.7

Now, by Exercise 5.4.15 or by techniques we shall learn in Chapter 9, this ellipse has semimajor axis 1 and semiminor axis $1/\sqrt{6}$, so, using the result of Exercise 8.3.12, its area is $\pi/\sqrt{6}$, and the integral is $4\pi/\sqrt{6}$.

Alternatively, applying our discussion of flux in Section 4, we recognize the surface integral in $(*)$ as the flux of the constant vector field $\mathbf{F} = \begin{bmatrix} 1 \\ 1 \\ 1 \end{bmatrix}$ outward across M. Since the unit normal of M is $\mathbf{n} = \dfrac{1}{\sqrt{6}} \begin{bmatrix} 1 \\ 2 \\ 1 \end{bmatrix}$, we see that

$$\int_M (dy \wedge dz + dz \wedge dx + dx \wedge dy) = \int_M \left(\begin{bmatrix} 1 \\ 1 \\ 1 \end{bmatrix} \cdot \frac{1}{\sqrt{6}} \begin{bmatrix} 1 \\ 2 \\ 1 \end{bmatrix} \right) dS = \frac{4}{\sqrt{6}} \text{area}(M) = \frac{4}{\sqrt{6}} \pi,$$

since M is, after all, a disk of radius 1. ◀

▶ **EXAMPLE 2**

Let S be the sphere $x^2 + y^2 + (z-1)^2 = 1$, oriented in the customary fashion. We wish to evaluate $\int_S \omega$, where $\omega = xz\,dy \wedge dz + yz\,dz \wedge dx + z^2 dx \wedge dy$. Let M be the compact 3-manifold whose boundary is S; i.e., $M = \{\mathbf{x} \in \mathbb{R}^3 : x^2 + y^2 + (z-1)^2 \le 1\}$, oriented by the standard orientation on $\mathbb{R}^3$. We apply Stokes's Theorem to M:

$$\int_S \omega = \int_{\partial M} \omega = \int_M d\omega = \int_M 4z\,dx \wedge dy \wedge dz$$
$$= \int_M 4z\,dV = 4\bar{z}\,\text{vol}(M) = 4 \cdot 1 \cdot \frac{4\pi}{3} = \frac{16\pi}{3}.$$

(Recall that $\bar{z}$ is the z-component of the center of mass of M.) Of course, we could compute the surface integral directly, parametrizing S by, for example, spherical coordinates centered at $\begin{bmatrix} 0 \\ 0 \\ 1 \end{bmatrix}$. ◀

► EXAMPLE 3

Suppose we wish to calculate the flux of the vector field $\mathbf{F} = \begin{bmatrix} xz \\ yz \\ x^2+y^2 \end{bmatrix}$ outward across the surface of the paraboloid S given by $z = 4 - x^2 - y^2$, $z \geq 0$ (with outward-pointing normal having positive $\mathbf{e}_3$-component). That is, we want to compute the integral of $\omega = xz\,dy \wedge dz + yz\,dz \wedge dx + (x^2+y^2)dx \wedge dy$. How might we do this with Stokes's Theorem? If ω were exact, i.e., if $\omega = d\eta$ for some 1-form η, then we would have $\int_S \omega = \int_{\partial S} \eta$; but since $d\omega = 2z\,dx \wedge dy \wedge dz \neq 0$, we know that ω cannot be exact. What now?

If we attach the disk $D = \{x^2 + y^2 \leq 4,\ z = 0\}$, to S, then we have a (piecewise-smooth) closed surface, which bounds the region $M = \{0 \leq z \leq 4 - x^2 - y^2\} \subset \mathbb{R}^3$, as shown in Figure 5.8. Then we have $\partial M = S \cup D^-$ (where by D^- we mean the disk with outward-pointing normal given by $-\mathbf{e}_3$). Applying Stokes's Theorem, we find

$$\int_{\partial M} \omega = \int_M d\omega = \int_M 2z\,dx \wedge dy \wedge dz = \int_M 2z\,dV = \int_0^{2\pi}\int_0^2\int_0^{4-r^2} 2rz\,dz\,dr\,d\theta = \frac{64}{3}\pi.$$

But we are interested in the integral of ω only over the surface S. Since

$$\int_{\partial M} \omega = \int_S \omega + \int_{D^-} \omega = \int_S \omega - \int_D \omega$$

(where by D we mean the disk with its usual upward orientation), then we have

$$\int_S \omega = \int_{\partial M} \omega + \int_D \omega = \frac{64}{3}\pi + \int_0^{2\pi}\int_0^2 r^2\,r\,dr\,d\theta = \frac{64}{3}\pi + 8\pi = \frac{88}{3}\pi.$$

We leave it to the reader to check this by a direct calculation (see Exercise 8.4.17). ◄

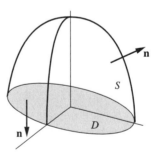

Figure 5.8

► EXAMPLE 4

We come now to the 3-dimensional analogue of Example 9 of Section 3. It will play a major role in physical and topological applications in upcoming sections. Consider the 2-form

$$\omega = \frac{x\,dy \wedge dz + y\,dz \wedge dx + z\,dx \wedge dy}{(x^2 + y^2 + z^2)^{3/2}},$$

which is defined and smooth on $\mathbb{R}^3 - \{\mathbf{0}\}$. The astute reader may recognize that on a sphere of radius a centered at the origin, ω is $1/a^2$ times the area 2-form.

Pulling back by the spherical coordinates parametrization given on p. 329, with a bit of work we see that

$$\mathbf{g}^*\omega = \sin\phi\, d\phi \wedge d\theta,$$

which establishes again the geometric interpretation of ω. It is also clear that $d(\mathbf{g}^*\omega) = 0$; since $\det D\mathbf{g} \neq 0$ whenever $\rho \neq 0$ and $\phi \neq 0, \pi$, it follows that $d\omega = 0$. (Of course, it isn't too hard to calculate this directly.)

So here we have a 2-form whose integral over any sphere centered at the origin (with outward-pointing normal) is 4π, and yet, for any ball B centered at the origin, $\int_B d\omega = 0$. What happened to Stokes's Theorem? The problem is that ω is not defined, let alone smooth, on all of B.

But there is more to be learned here. If $\Omega \subset \mathbb{R}^3$ is a compact 3-manifold with boundary with $\mathbf{0} \notin \partial\Omega$, then we claim that

$$\int_{\partial\Omega} \omega = \begin{cases} 4\pi, & \mathbf{0} \in \Omega \\ 0, & \mathbf{0} \notin \Omega \end{cases},$$

rather like what happened with the winding number in Example 10 of Section 3. When $\mathbf{0} \notin \Omega$, we know that ω is a (smooth) 2-form on all of Ω, and hence Stokes's Theorem applies directly to give

$$\int_{\partial\Omega} \omega = \int_{\Omega} d\omega = 0.$$

When $\mathbf{0} \in \Omega$, however, we choose $\varepsilon > 0$ small enough so that the closed ball $\overline{B}(\mathbf{0}, \varepsilon) \subset \Omega$, and we let $\Omega_\varepsilon = \Omega - B(\mathbf{0}, \varepsilon)$, as pictured in Figure 5.9, recalling that $\partial\Omega_\varepsilon = \partial\Omega + S_\varepsilon^-$. (Here S_ε denotes the sphere of radius ε centered at $\mathbf{0}$, with its usual outward orientation.) Then ω is a smooth form defined on all of Ω_ε and we have

$$0 = \int_{\Omega_\varepsilon} d\omega = \int_{\partial\Omega_\varepsilon} \omega = \int_{\partial\Omega} \omega - \int_{S_\varepsilon} \omega.$$

Therefore, we have

$$\int_{\partial\Omega} \omega = \int_{S_\varepsilon} \omega = 4\pi,$$

as we learned above. ◂

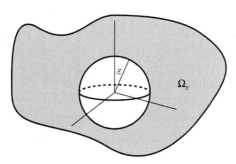

Figure 5.9

EXERCISES 8.5

*1. Check that the boundary orientation on $\partial \mathbb{R}_+^k$ is $(-1)^k$ times the usual orientation on $\mathbb{R}^{k-1}$.

2. Let C be the intersection of the cylinder $x^2 + y^2 = 1$ and the plane $2x + 3y - z = 1$, oriented counterclockwise as viewed from high above the xy-plane. Evaluate
$$\int_C y\,dx - 2z\,dy + x\,dz$$
directly and by applying Stokes's Theorem.

*3. Compute $\int_C (y-z)\,dx + (z-x)\,dy + (x-y)\,dz$, where C is the intersection of the cylinder $x^2 + y^2 = a^2$ and the plane $\dfrac{x}{a} + \dfrac{z}{b} = 1$, oriented clockwise as viewed from high above the xy-plane.

4. Let C be the intersection of the sphere $x^2 + y^2 + z^2 = 2$ and the plane $z = 1$, oriented counterclockwise as viewed from high above the xy-plane. Evaluate
$$\int_C (-y^3 + z)\,dx + (x^3 + 2y)\,dy + (y - x)\,dz.$$

5. Let C be the intersection of the sphere $x^2 + y^2 + z^2 = a^2$ and the plane $x + y + z = 0$, oriented counterclockwise as viewed from high above the xy-plane. Evaluate
$$\int_C 2z\,dx + 3x\,dy - dz.$$

*6. Let $\Omega \subset \mathbb{R}^3$ be the region bounded above by the sphere $x^2 + y^2 + z^2 = a^2$ and below by the plane $z = 0$. Compute
$$\int_{\partial \Omega} xz\,dy \wedge dz + yz\,dz \wedge dx + (x^2 + y^2 + z^2)\,dx \wedge dy$$
directly and by applying Stokes's Theorem.

7. Let $\omega = y^2 dy \wedge dz + x^2 dz \wedge dx + z^2 dx \wedge dy$, and let M be the solid paraboloid $0 \le z \le 1 - x^2 - y^2$. Evaluate $\int_{\partial M} \omega$ directly and by applying Stokes's Theorem.

8. Let M be the surface of the paraboloid $z = 1 - x^2 - y^2$, $z \ge 0$, oriented so that the outward-pointing normal has positive e_3-component. Let $\mathbf{F} = \begin{bmatrix} x^2 z \\ y^2 z \\ x^2 + y^2 \end{bmatrix}$. Compute $\int_M \mathbf{F} \cdot \mathbf{n}\,dS$ directly and by applying Stokes's Theorem. Be careful!

9. Let M be the surface pictured in Figure 5.10, with boundary curve $x^2 + y^2 = 4$, $z = 0$. Calculate
$$\int_M yz\,dy \wedge dz + x^3 dz \wedge dx + y^2 dx \wedge dy.$$

10. Suppose M and M' are two compact, oriented k-dimensional manifolds with boundary, and suppose $\partial M = \partial M'$ (as oriented $(k-1)$-dimensional manifolds). Prove that for any $(k-1)$-form ω, $\int_M d\omega = \int_{M'} d\omega$.

11. Use the result of Exercise 10 to compute $\int_M d\omega$ for the given surface M and 1-form ω:

(a) M is the upper hemisphere $x^2 + y^2 + z^2 = a^2$, $z \ge 0$, oriented with outward-pointing normal having positive e_3-component; $\omega = (x^3 + 3x^2 y - y)\,dx + (y^3 z + x + x^3)\,dy + (x^2 + y^2 + z)\,dz$.

(b) M is that portion of the paraboloid $z = x^2 + y^2$ lying beneath $z = 4$, oriented with outward-pointing normal having negative e_3-component; $\omega = y\,dx + z\,dy + x\,dz$.

Figure 5.10

(c) M is the union of the cylinder $x^2 + y^2 = 1, 0 \leq z \leq 2$, and the disk $x^2 + y^2 \leq 1, z = 0$, oriented so that the normal to the cylindrical portion points radially outward; $\omega = -y^3 z\,dx + x^3 z\,dy + x^2 y^2\,dz$.

12. Let $M = \{\mathbf{x} \in \mathbb{R}^4 : x_1^2 + x_2^2 + x_3^2 \leq x_4 \leq 1\}$, with the standard orientation inherited from $\mathbb{R}^4$. Evaluate $\displaystyle\int_{\partial M} \omega$:

*(a) $\omega = (x_1^3 x_2^4 + x_4)\,dx_1 \wedge dx_2 \wedge dx_3$,

(b) $\omega = \|\mathbf{x}\|^2\,dx_1 \wedge dx_2 \wedge dx_3$.

13. Redo Exercise 8.4.22c by applying Stokes's Theorem.

14. Suppose f is a smooth function on a compact 3-manifold with boundary $M \subset \mathbb{R}^3$. At a point of ∂M, let $D_\mathbf{n} f$ denote the directional derivative of f in the direction of the unit outward normal. Show that

$$\int_{\partial M} D_\mathbf{n} f\,dS = \int_M \nabla^2 f\,dV,$$

where $\nabla^2 f = \dfrac{\partial^2 f}{\partial x^2} + \dfrac{\partial^2 f}{\partial y^2} + \dfrac{\partial^2 f}{\partial z^2}$ is the *Laplacian* of f. (Hint: $\nabla^2 f\,dx \wedge dy \wedge dz = d\star df$. See Exercise 8.2.9.)

15. Let S be that portion of the cylinder $x^2 + y^2 = a^2$ lying above the xy-plane and below the sphere $x^2 + (y-a)^2 + z^2 = 4a^2$. Let C be the intersection of the cylinder and sphere, oriented clockwise as viewed from high above the xy-plane.

(a) Evaluate $\displaystyle\int_S z\,dS$.

(b) Use your answer to part a to evaluate $\displaystyle\int_C y(z^2 - 1)\,dx + x(1 - z^2)\,dy + z^2\,dz$.

16. Let S be that portion of the cylinder $x^2 + y^2 = a^2$ lying above the xy-plane and below the sphere $(x-a)^2 + y^2 + z^2 = 4a^2$. Let C be the intersection of the cylinder and sphere, oriented clockwise as viewed from high above the xy-plane.

(a) Evaluate $\displaystyle\int_S z^2\,dS$.

(b) Use your answer to part a to evaluate $\displaystyle\int_C y(z^3 + 1)\,dx - x(z^3 + 1)\,dy + z\,dz$.

17. Let

$$M = \left\{ \begin{bmatrix} x_1 \\ x_2 \\ y_1 \\ y_2 \end{bmatrix} \in \mathbb{R}^2 \times \mathbb{R}^2 : \|\mathbf{x}\|^2 + \|\mathbf{y}\|^2 = 1, \mathbf{x} \cdot \mathbf{y} = 0 \right\} \subset \mathbb{R}^4,$$

oriented so that $\dfrac{dx_2 \wedge dy_2}{y_1^2 - x_1^2} > 0$ on M. Evaluate $\displaystyle\int_M (y_1^2 - x_1^2)\, dx_2 \wedge dy_2$. (Hint: By applying an appropriate linear transformation, you should be able to recognize M as a torus.)

*18. Let C be the intersection of the sphere $x^2 + y^2 + z^2 = 1$ and the plane $x + y + z = 0$, oriented counterclockwise as viewed from high above the xy-plane. Evaluate
$$\int_C z^3\, dx.$$
(Hint: Give an orthonormal basis for the plane $x + y + z = 0$, and use polar coordinates.)

19. Let C be the intersection of the sphere $x^2 + y^2 + z^2 = 1$ and the plane $x + y + z = 0$, oriented counterclockwise as viewed from high above the xy-plane. Evaluate
$$\int_C xy^2\, dx + yz^2\, dy + zx^2\, dz.$$
(Hint: See Exercise 18.)

20. Suppose $\omega \in \mathcal{A}^{k-2}(\mathbb{R}^k)$. Complete the following proof that $d(d\omega) = 0$. Write $d(d\omega) = f(\mathbf{x})\, dx_1 \wedge \cdots \wedge dx_k$, and suppose $f(\mathbf{a}) > 0$. By considering the integral of $d(d\omega)$ over a small ball centered at $\mathbf{a}$ and applying Corollary 5.3, arrive at a contradiction.

21. We saw in Example 8 of Section 3 that there are 1-forms ω on $\mathbb{R}^2$ with the property that for every region $S \subset \mathbb{R}^2$ we have $\text{area}(S) = \displaystyle\int_{\partial S} \omega$. Can there be such a 1-form on
(a) the unit sphere?
(b) the torus?
(c) the punctured sphere (i.e., the sphere less the north pole)?

22. In this exercise we sketch a proof that the graph of a function f satisfying the minimal surface equation (see p. 124) on a region $\Omega \subset \mathbb{R}^2$ has less area than any other surface with the same boundary curve.[8]
(a) Consider the area 2-form σ of the graph of f:
$$\sigma = \frac{1}{\sqrt{1 + \|\nabla f\|^2}}\left(-\frac{\partial f}{\partial x}\, dy \wedge dz - \frac{\partial f}{\partial y}\, dz \wedge dx + dx \wedge dy\right).$$
Show that $d\sigma = 0$ if and only if f satisfies the minimal surface equation.
(b) Show that for any compact oriented surface $N \subset \mathbb{R}^3$, $\displaystyle\int_N \sigma \le \text{area}(N)$, and equality holds if and only if N is parallel to the graph of f. (Hint: Interpret $\displaystyle\int_N \sigma$ as a flux integral.)
(c) Let M be the graph of f over Ω, and let N be a different oriented surface with $\partial N = \partial M$. Deduce that $\text{area}(M) < \text{area}(N)$.

23. (a) Prove that M is an orientable k-dimensional manifold with boundary if and only if there is a nowhere-zero k-form on M. (Hint: For $\Longrightarrow$, use definition (1) of a manifold on p. 262 and a partition of unity to glue together compatibly chosen forms on coordinate charts. Although we've only proved Theorem 5.1 for a compact manifold M, the proof can easily be adapted to show that for any manifold M and any covering $\{U_j\}$ by coordinate charts, we have a sequence of such functions ρ_i, each of which is zero outside some U_j.)
(b) Conclude that M is orientable if and only if there is a volume form globally defined on M.

---

[8] This is an illustration of the use of *calibrations*, introduced by Reese Harvey and Blaine Lawson in their seminal paper, *Calibrated Geometries*, Acta Math. **148** (1982), pp. 47–157.

24. Let M be a compact, orientable k-dimensional manifold (with no boundary), and let ω be a $(k-1)$-form. Show that $d\omega = 0$ at some point of M. (Hint: Using Exercise 23, write $d\omega = f\sigma$, where σ is the volume form of M. Without loss of generality, you may assume M is connected. Why?)

25. Let $h(x) = \begin{cases} e^{-1/x}, & x > 0 \\ 0, & x \leq 0 \end{cases}$. Because exponential functions grow faster at infinity than any polynomial, it should be plausible that all the derivatives of h at 0 are 0. But give a rigorous proof as follows:
   (a) Let $f(x) = e^{-1/x}$, $x > 0$. Prove by induction that $f^{(k)}$, the kth derivative of f, is given by $f^{(k)}(x) = e^{-1/x} p_k(1/x)$ for some polynomial p_k of degree $2k$.
   (b) Prove by induction that $h^{(k)}(0) = 0$ for all $k \geq 0$.

26. Let $X \subset \mathbb{R}^n$. Prove that given any collection $\{V_\alpha\}$ of open subsets of $\mathbb{R}^n$ whose union contains X, there is a sequence $V_{\alpha_1}, V_{\alpha_2}, \ldots$ of these sets whose union contains X. (Hint: Consider all balls $B(\mathbf{q}, 1/k) \subset \mathbb{R}^n$ (for some $k \in \mathbb{N}$) centered at points $\mathbf{q} \in \mathbb{R}^n$ all of whose coordinates are rational. This collection is countable, i.e., can be arranged in a sequence. Show that we can choose such balls $B(\mathbf{q}_i, 1/k_i)$, $i = 1, 2, \ldots$, covering all of X with the additional property that each is contained in some V_{α_i}.)

6 APPLICATIONS TO PHYSICS

6.1 The Dictionary in $\mathbb{R}^3$

We have already seen that a vector field in $\mathbb{R}^3$ can plausibly be interpreted as either a 1-form or a 2-form, the former when we are calculating work, the latter when we are calculating flux. We have already seen that for any function f, the 1-form df corresponds to the vector field ∇f. We want to give the traditional interpretations of the exterior derivative as it acts on 1- and 2-forms.

Given a 1-form
$$\omega = F_1 dx_1 + F_2 dx_2 + F_3 dx_3 \in \mathcal{A}^1(\mathbb{R}^3), \quad \text{we have}$$

$$d\omega = \left(\frac{\partial F_3}{\partial x_2} - \frac{\partial F_2}{\partial x_3}\right) dx_2 \wedge dx_3 + \left(\frac{\partial F_1}{\partial x_3} - \frac{\partial F_3}{\partial x_1}\right) dx_3 \wedge dx_1 + \left(\frac{\partial F_2}{\partial x_1} - \frac{\partial F_1}{\partial x_2}\right) dx_1 \wedge dx_2.$$

(We stick to the subscript notation here to make the symmetries as clear as possible.) Correspondingly, given the vector field

$$\mathbf{F} = \begin{bmatrix} F_1 \\ F_2 \\ F_3 \end{bmatrix}, \quad \text{we set} \quad \operatorname{curl} \mathbf{F} = \begin{bmatrix} \dfrac{\partial F_3}{\partial x_2} - \dfrac{\partial F_2}{\partial x_3} \\ \dfrac{\partial F_1}{\partial x_3} - \dfrac{\partial F_3}{\partial x_1} \\ \dfrac{\partial F_2}{\partial x_1} - \dfrac{\partial F_1}{\partial x_2} \end{bmatrix}.$$

Note first of all that $d^2 = 0$ tells us that

$$\boxed{\operatorname{curl}(\nabla f) = 0 \quad \text{for all } \mathcal{C}^2 \text{ functions } f.}$$

In somewhat older books one often sees the notation "rot," rather than "curl"; both terms suggest that we think of curl **F** as having something to do with rotation (curling).

Stokes's Theorem can now be phrased in the following classical form:

Theorem 6.1 (Classical Stokes's Theorem) *Let $S \subset \mathbb{R}^3$ be a compact, oriented surface with boundary. Let **F** be a smooth vector field defined on all of S. Then we have*

$$\int_{\partial S} \underbrace{\mathbf{F} \cdot \mathbf{T} ds}_{\omega} = \int_S \underbrace{\text{curl } \mathbf{F} \cdot \mathbf{n} dS}_{d\omega}.$$

If we return to our discussion of flux in Section 4 and visualize **F** as the velocity field of a fluid, then the line integral $\int_C \mathbf{F} \cdot \mathbf{T} ds$ around a closed curve C may be interpreted as the *circulation* of **F** around C, which we might visualize as a measure of the tendency of a piece of wire in the shape of C to turn (or circulate) when dropped in the fluid. Applying the theorem with $S = D_r$, a 2-dimensional disk of radius r centered at **a** with normal vector **n**, and using continuity (see Exercise 7.1.7), we have

$$\text{curl } \mathbf{F}(\mathbf{a}) \cdot \mathbf{n} = \lim_{r \to 0^+} \frac{1}{\pi r^2} \int_{\partial D_r} \mathbf{F} \cdot \mathbf{T} ds.$$

In particular, if, as pictured in Figure 6.1, we stick a very small paddlewheel (of radius r) in the fluid, it will spin the fastest when the axle points in the direction of curl **F** (and—at least in the limit—won't spin at all when the axle is orthogonal to curl **F**). Indeed, if the fluid—and hence the paddlewheel—is spinning about an axis with angular speed v, then $\|\text{curl } \mathbf{F}\| = 2v$ (see Exercise 1).

Now, given the 2-form

$$\omega = F_1 dx_2 \wedge dx_3 + F_2 dx_3 \wedge dx_1 + F_3 dx_1 \wedge dx_2 \in \mathcal{A}^2(\mathbb{R}^3)$$

(which happens to be obtained by applying the star operator, defined in Exercise 8.2.9, to our original 1-form), then

$$d\omega = \left(\frac{\partial F_1}{\partial x_1} + \frac{\partial F_2}{\partial x_2} + \frac{\partial F_3}{\partial x_3} \right) dx_1 \wedge dx_2 \wedge dx_3.$$

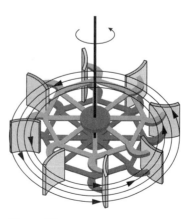

Figure 6.1

Correspondingly, given the vector field

$$\mathbf{F} = \begin{bmatrix} F_1 \\ F_2 \\ F_3 \end{bmatrix}, \quad \text{we set} \quad \text{div } \mathbf{F} = \frac{\partial F_1}{\partial x_1} + \frac{\partial F_2}{\partial x_2} + \frac{\partial F_3}{\partial x_3};$$

"div" is short for *divergence*, a term that is à propos, as we shall soon see. In this case, $d^2 = 0$ can be restated as

$$\boxed{\text{div (curl } \mathbf{F}) = 0 \quad \text{for all } \mathcal{C}^2 \text{ vector fields } \mathbf{F}.}$$

Stokes's Theorem now takes the following form, sometimes called Gauss's Theorem:

Theorem 6.2 (Classical Divergence Theorem) *Suppose $\mathbf{F}$ is a smooth vector field on a compact 3-manifold with boundary, $\Omega \subset \mathbb{R}^3$. Then*

$$\int_{\partial\Omega} \underbrace{\mathbf{F} \cdot \mathbf{n} dS}_{\omega} = \int_{\Omega} \underbrace{\text{div } \mathbf{F} dV}_{d\omega}.$$

Once again, we get from this a limiting interpretation of the divergence: Applying Exercise 7.1.7, we find

$$(*) \qquad \text{div } \mathbf{F}(\mathbf{a}) = \lim_{r \to 0^+} \frac{1}{\frac{4}{3}\pi r^3} \int_{\partial \overline{B}(\mathbf{a},r)} \mathbf{F} \cdot \mathbf{n} dS.$$

That is, div $\mathbf{F}(\mathbf{a})$ is a measure of the flux (per unit volume) outward across very small spheres centered at $\mathbf{a}$. If that flux is positive, we can visualize $\mathbf{a}$ as a *source* of the field, with a net divergence of the fluid flow; if the flux is negative, we can visualize $\mathbf{a}$ as a *sink*, with a net confluence of the fluid. We shall see a beautiful alternative interpretation of the divergence in Chapter 9.

Given a vector field $\mathbf{F}$ (in the context of work) and the corresponding 1-form ω, applying the star operator introduced in Exercise 8.2.9 gives the 2-form $\star\omega$ corresponding to the same vector field $\mathbf{F}$ (in the context of flux)—and vice versa. That is, when we have an oriented surface S, the 2-form $\star\omega$ gives the normal component of $\mathbf{F}$ times area 2-form σ of S. In particular, if we start with a function f, then on S, $\star df = (D_\mathbf{n} f)\sigma$, where $D_\mathbf{n} f = \nabla f \cdot \mathbf{n}$ is the directional derivative of f in the normal direction.

We summarize the relation among forms and vector fields, the d operator and gradient, curl, and divergence in the following table:

	Differential Forms	Fields	
	0-forms	functions (scalar fields)	
	↓ d	↓ grad	
$d^2 = 0$	1-forms	vector fields (work)	curl (grad) = **0**
	↓ d	↓ curl	
$d^2 = 0$	2-forms	vector fields (flux)	div (curl) = 0
	↓ d	↓ div	
	3-forms	functions (scalar fields)	

6.2 Gauss's Law

In this passage we concentrate on inverse square forces, either gravitation (according to Newton's law of gravitation) or electrostatic attraction (according to Coulomb's law). We will stick with the notation of Newton's law of gravitation, as we discussed in Section 4 of Chapter 7: The gravitational attraction of a mass M at the origin on a unit test mass at position $\mathbf{x}$ is given by

$$\mathbf{F} = -GM \frac{\mathbf{x}}{\|\mathbf{x}\|^3}.$$

(Here G is the universal gravitation constant.) As we saw in Example 4 of Section 5, div $\mathbf{F} = 0$ (except at the origin) and for any compact surface $S \subset \mathbb{R}^3$ bounding a region Ω, we have

$$\int_S \mathbf{F} \cdot \mathbf{n} dS = \begin{cases} -4\pi GM, & \mathbf{0} \in \Omega \\ 0, & \text{otherwise} \end{cases}.$$

(We must also stipulate that $\mathbf{0} \notin S$ for the integral to make sense.) More generally, if $\mathbf{F_a}$ is the gravitational force field due to a point mass at point $\mathbf{a} \notin S$, then

$$\int_S \mathbf{F_a} \cdot \mathbf{n} dS = \begin{cases} -4\pi GM, & \mathbf{a} \in \Omega \\ 0, & \text{otherwise} \end{cases}.$$

If we have point masses $M_1, \ldots, M_k$ at points $\mathbf{a}_1, \ldots, \mathbf{a}_k$, then the flux of the resultant gravitation force $\mathbf{F} = \sum_{j=1}^{k} \mathbf{F}_{\mathbf{a}_j}$ outward across the surface S (on which, once again, none of the point masses lies) is given by

$$\int_S \mathbf{F} \cdot \mathbf{n} dS = -4\pi G \sum_{\mathbf{a}_j \in \Omega} M_j.$$

Indeed, given a mass distribution with (integrable) density function δ on a region D, we can, in fact, write an explicit formula for the gravitational field (see Section 4 of Chapter 7):

(†) $$\mathbf{F}(\mathbf{x}) = G \int_D \frac{\mathbf{y} - \mathbf{x}}{\|\mathbf{y} - \mathbf{x}\|^3} \delta(\mathbf{y}) dV_\mathbf{y}.$$

(When $\mathbf{x} \in D$, this integral is improper, yet convergent, as can be verified by using spherical coordinates centered at the point $\mathbf{x}$.) It should come as no surprise, approximating the mass distribution by a finite set of point masses, that the flux of the resulting gravitational force $\mathbf{F}$ is given by

$$\int_S \mathbf{F} \cdot \mathbf{n} dS = -4\pi G \int_\Omega \delta dV = -4\pi GM,$$

where M is the mass inside $S = \partial \Omega$. This is Gauss's law.

Using the limiting formula for divergence given in (∗) on p. 395, we see that, even if $\mathbf{F}$ isn't apparently smooth, it is plausible to define

$$\text{div } \mathbf{F}(\mathbf{x}) = -4\pi G \delta(\mathbf{x})$$

when δ is continuous on D (and div $\mathbf{F}(\mathbf{x}) = 0$ when $\mathbf{x} \notin D$).

Now we can determine, as did Newton (following the lines of Example 6 of Chapter 7, Section 4), the gravitational field **F** inside the earth, assuming—albeit incorrectly—that the earth is a ball of uniform density. Take the earth to be a ball of radius R centered at the origin and to have constant density and total mass M. Fix **x** with $\|\mathbf{x}\| = b < R$. First of all, we have

$$\int_{\partial \overline{B}(\mathbf{0},b)} \mathbf{F} \cdot \mathbf{n}\, dS = -4\pi G(\text{mass of the earth inside } B(\mathbf{0}, b)) = -4\pi G \left(\frac{b}{R}\right)^3 M.$$

Now, by symmetry, **F** points radially inward, and so

$$\int_{\partial \overline{B}(\mathbf{0},b)} \mathbf{F} \cdot \mathbf{n}\, dS = -\|\mathbf{F}\| \text{area}\big(\partial \overline{B}(\mathbf{0}, b)\big) = -\|\mathbf{F}\|(4\pi b^2).$$

Thus, we have $\|\mathbf{F}(\mathbf{x})\| = \dfrac{GM}{R^3}\|\mathbf{x}\|$. Since **F** is radial, we have

$$\mathbf{F}(\mathbf{x}) = -\frac{GM}{R^3}\mathbf{x}.$$

It is often surprising to find that the gravitational force *inside* the earth is linear in the distance from the center. Notice that at the earth's surface, this analysis is in accord with the inverse-square nature of the field. (See Exercise 2.)

As an amusing application, we calculate the time required to travel in a perfectly frictionless tunnel inside the earth from one point on the surface to another. We suppose that we start the trip with zero speed. When the mass is at position **x**, the component of the gravitational force acting in the direction of the tunnel is

$$-\|\mathbf{F}\| \sin \theta = -\frac{GM}{R^3}u,$$

where u is the displacement of the mass from the center of the tunnel (see Figure 6.2). By Newton's second law, we have

$$mu''(t) = -\frac{GM}{R^3}u(t).$$

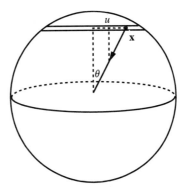

Figure 6.2

The general solution is

$$u(t) = a\cos\left(\sqrt{\tfrac{GM}{R^3}}t\right) + b\sin\left(\sqrt{\tfrac{GM}{R^3}}t\right).$$

If we start with the initial conditions $u(0) = u_0$ and $u'(0) = 0$, then we have

$$u(t) = u_0 \cos\left(\sqrt{\tfrac{GM}{R^3}}t\right),$$

and we see that the mass reaches the opposite end of the tunnel after time

$$T = \frac{\pi}{\sqrt{\tfrac{GM}{R^3}}} = \pi\sqrt{\frac{R}{g}} \approx 42 \text{ min}.$$

As was pointed out to me my freshman year of college, this is rather less time than many of our commutes.

6.3 Maxwell's Equations

Let $\mathbf{E}$ denote the electric field, $\mathbf{B}$ the magnetic field, ρ the charge density, and $\mathbf{J}$ the current density. All of these are functions on (some region in) $\mathbb{R}^3 \times \mathbb{R}$ (space-time), on which we use coordinates x, y, z, and t. The classical presentation of Maxwell's equations is the following system of four partial differential equations (ignoring various constants such as 4π and c, the speed of light):

Gauss's law: $\quad \operatorname{div} \mathbf{E} = \rho$

No magnetic monopoles: $\quad \operatorname{div} \mathbf{B} = 0$

Faraday's law: $\quad \operatorname{curl} \mathbf{E} = -\dfrac{\partial \mathbf{B}}{\partial t}$

Ampère's law: $\quad \operatorname{curl} \mathbf{B} = \dfrac{\partial \mathbf{E}}{\partial t} + \mathbf{J}$

These are all "differential" versions of equivalent "integral" statements obtained by applying Stokes's Theorem, as we already encountered Gauss's Law in the previous subsection. Briefly, suppose S is an oriented surface (perhaps imagined) and ∂S represents a wire. Then Faraday's law states that

$$\int_{\partial S} \mathbf{E} \cdot \mathbf{T}\, ds = -\int_S \frac{\partial \mathbf{B}}{\partial t} \cdot \mathbf{n}\, dS = -\frac{d}{dt}\int_S \mathbf{B} \cdot \mathbf{n}\, dS$$

(using the result of Exercise 7.2.20 to differentiate under the integral sign); i.e., the voltage around the loops ∂S equals the negative of the rate of change of magnetic flux across the loop. (More colloquially, a moving magnetic field induces an electric field that in turn does work, namely, creates a voltage drop across the loop.) On the other hand, Ampère's law states that (in steady state, with no time variation)

$$\int_{\partial S} \mathbf{B} \cdot \mathbf{T}\, ds = \int_S \mathbf{J} \cdot \mathbf{n}\, dS;$$

i.e., the circulation of the magnetic field around the wire is the flux of the current density across the loop.

Let
$$\omega = (E_1 dx + E_2 dy + E_3 dz) \wedge dt + (B_1 dy \wedge dz + B_2 dz \wedge dx + B_3 dx \wedge dy).$$
Then
$$d\omega = \left(\left(\frac{\partial E_3}{\partial y} - \frac{\partial E_2}{\partial z} + \frac{\partial B_1}{\partial t}\right) dy \wedge dz + \left(\frac{\partial E_1}{\partial z} - \frac{\partial E_3}{\partial x} + \frac{\partial B_2}{\partial t}\right) dz \wedge dx \right.$$
$$\left. + \left(\frac{\partial E_2}{\partial x} - \frac{\partial E_1}{\partial y} + \frac{\partial B_3}{\partial t}\right) dx \wedge dy \right) \wedge dt + \left(\frac{\partial B_1}{\partial x} + \frac{\partial B_2}{\partial y} + \frac{\partial B_3}{\partial z}\right) dx \wedge dy \wedge dz,$$
and so we see that
$$d\omega = 0 \iff \text{div } \mathbf{B} = 0 \quad \text{and} \quad \text{curl } \mathbf{E} + \frac{\partial \mathbf{B}}{\partial t} = \mathbf{0}.$$

Next, let
$$\theta = -(E_1 dy \wedge dz + E_2 dz \wedge dx + E_3 dx \wedge dy) + (B_1 dx + B_2 dy + B_3 dz) \wedge dt.$$
(Using the star operator defined in Exercise 8.2.9, one can check that $\theta = \star\omega$. The subtlety is that we're working in space-time, endowed with a Lorentz metric in which the standard orthonormal basis $\{\mathbf{e}_1, \ldots, \mathbf{e}_4\}$ has the property that $\mathbf{e}_4 \cdot \mathbf{e}_4 = -1$; this introduces a minus sign so that $\star(dx \wedge dt) = -dy \wedge dz$, etc.) Then an analogous calculation shows that
$$d\theta = 0 \iff \text{div } \mathbf{E} = 0 \quad \text{and} \quad \text{curl } \mathbf{B} - \frac{\partial \mathbf{E}}{\partial t} = \mathbf{0}.$$
This would hold, for example, in a vacuum, where $\rho = 0$ and $\mathbf{J} = \mathbf{0}$. But, in general, the first and last of Maxwell's equations are equivalent to the equation
$$d\theta = (J_1 dy \wedge dz + J_2 dz \wedge dx + J_3 dx \wedge dy) \wedge dt - \rho dx \wedge dy \wedge dz.$$

Since $d\omega = 0$ on $\mathbb{R}^4$, there is a 1-form
$$\alpha = a_1 dx + a_2 dy + a_3 dz - \varphi dt$$
so that $d\alpha = \omega$ (see Exercise 8.7.12). Of course, α is far from unique; for any function f, we will have $d(\alpha + df) = \omega$ as well. Let $\beta = \alpha + df$, where f is a solution of the inhomogeneous wave equation
$$\nabla^2 f - \frac{\partial^2 f}{\partial t^2} = -\left(\frac{\partial a_1}{\partial x} + \frac{\partial a_2}{\partial y} + \frac{\partial a_3}{\partial z} + \frac{\partial \varphi}{\partial t}\right).$$
This means that $d\star df = -d\star\alpha$, and so $d\star\beta = 0$. If we write
$$\beta = A_1 dx + A_2 dy + A_3 dz - \phi dt,$$
the condition that $d\star\beta = 0$ is equivalent to

(*) $$\frac{\partial A_1}{\partial x} + \frac{\partial A_2}{\partial y} + \frac{\partial A_3}{\partial z} + \frac{\partial \phi}{\partial t} = 0.$$

Since $d\beta = \omega$, $\star d\beta = \star\omega = \theta$, we calculate
$$d\star d\beta = d\theta = (J_1 dy \wedge dz + J_2 dz \wedge dx + J_3 dx \wedge dy) \wedge dt - \rho dx \wedge dy \wedge dz.$$

Using (∗) to substitute

$$\frac{\partial^2 A_1}{\partial x \partial t} + \frac{\partial^2 A_2}{\partial y \partial t} + \frac{\partial^2 A_3}{\partial z \partial t} = -\frac{\partial^2 \phi}{\partial t^2},$$

we can check that solving Maxwell's equations is equivalent to finding $\mathbf{A} = \begin{bmatrix} A_1 \\ A_2 \\ A_3 \end{bmatrix}$ and ϕ satisfying the inhomogeneous wave equations

$$\nabla^2 \mathbf{A} - \frac{\partial^2 \mathbf{A}}{\partial t^2} = -\mathbf{J} \quad \text{and} \quad \nabla^2 \phi - \frac{\partial^2 \phi}{\partial t^2} = -\rho.$$

Solving such equations is a standard topic in an upper-division course in partial differential equations.

▶ EXERCISES 8.6

1. Write down the vector field $\mathbf{F}$ corresponding to a rotation counterclockwise about an axis in the direction of the unit vector $\mathbf{a}$ with angular speed v, and check that curl $\mathbf{F} = 2v\mathbf{a}$.

2. Using Gauss's law, show that the gravitational field of a uniform ball *outside* the ball is that of a point mass at its center.

3. (Green's Formulas) Let $f, g: \Omega \to \mathbb{R}$ be smooth functions on a region $\Omega \subset \mathbb{R}^3$. Recall that $D_n g$ denotes the directional derivative of g in the normal direction.
(a) Prove that

$$\int_{\partial \Omega} (D_n g)\, dS = \int_{\Omega} \nabla^2 g\, dV$$

$$\int_{\partial \Omega} f(D_n g)\, dS = \int_{\Omega} \left(f \nabla^2 g + \nabla f \cdot \nabla g \right) dV$$

$$\int_{\partial \Omega} (f D_n g - g D_n f)\, dS = \int_{\Omega} \left(f \nabla^2 g - g \nabla^2 f \right) dV.$$

(Hint: $\nabla^2 g\, dx \wedge dy \wedge dz = d \star dg$.)
(b) We say f is *harmonic* on Ω if $\nabla^2 f = 0$ on Ω. Prove that if f and g are harmonic on Ω, then

$$\int_{\partial \Omega} (D_n f)\, dS = 0$$

$$\int_{\partial \Omega} f(D_n f)\, dS = \int_{\Omega} \|\nabla f\|^2\, dV$$

$$\int_{\partial \Omega} f(D_n g)\, dS = \int_{\partial \Omega} g(D_n f)\, dS.$$

4. (See Exercise 3.) Prove that if f and g are harmonic on a region Ω and $f = g$ on $\partial \Omega$, then $f = g$ everywhere on Ω. (Hint: Consider $f - g$.)

5. (a) Prove that $g: \mathbb{R}^3 - \{0\} \to \mathbb{R}$, $g(\mathbf{x}) = 1/\|\mathbf{x}\|$, is harmonic. (See Exercise 3.)
(b) Prove that if f is harmonic on $B(\mathbf{0}, R) \subset \mathbb{R}^3$, then f has the *mean value property*: $f(\mathbf{0})$ is the average of the values of f on the sphere of any radius $r < R$ centered at $\mathbf{0}$. (Hint: Apply the appropriate results of Exercise 3 with $\Omega_\varepsilon = \{\mathbf{x} : \varepsilon \leq \|\mathbf{x}\| \leq r\}$ and g as in part a; then let $\varepsilon \to 0^+$.)

(c) Deduce the *maximum principle* for harmonic functions: If f is harmonic on a region Ω, then f takes on its maximum value on $\partial\Omega$.

6. Let $S \subset \mathbb{R}^3$ be a closed, oriented surface. Using the formula (†) for the gravitational field $\mathbf{F}$, show that

(a) the flux of $\mathbf{F}$ outward across S is 0 when no points of D lie on or inside S;

(b) the flux of $\mathbf{F}$ outward across S is $-4\pi G \int_D \delta dV$ when all of D lies inside S.

(Hint: Change the order of integration.)

*__7.__ Try to determine which of the vector fields pictured in Figure 6.3 have zero divergence and which have zero curl. Justify your answers.

8. Let $\mathbf{F}$ be a smooth vector field on an open set $U \subset \mathbb{R}^n$. A parametrized curve $\mathbf{g}$ is a *flow line* for a vector field $\mathbf{F}$ if $\mathbf{g}'(t) = \mathbf{F}(\mathbf{g}(t))$ for all t.

(a) Give a vector field with a closed flow line.

(b) Prove that if $\mathbf{F}$ is conservative, then it can have no closed flow line (other than a single point).

(c) Prove that if $n = 2$ and $\mathbf{F}$ has a closed flow line C, then div $\mathbf{F}$ must equal 0 at some point inside C. (Hint: See Exercise 8.3.18.)

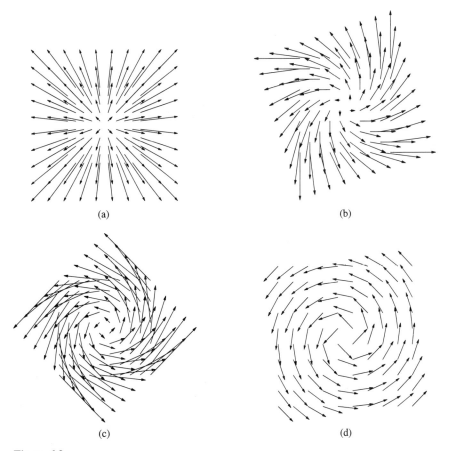

Figure 6.3

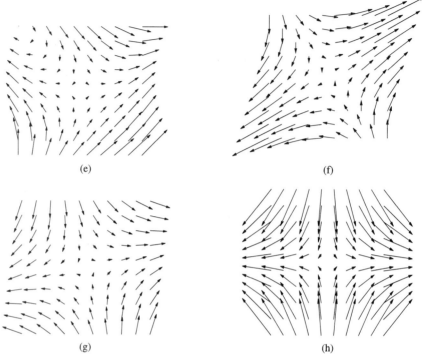

(e) (f) (g) (h)

Figure 6.3 (continued)

9. Let $\Omega \subset \mathbb{R}^3$ be a compact 3-manifold with boundary.

(a) Prove that $\int_{\partial\Omega} f\mathbf{n}\,dS = \int_{\Omega} \nabla f\,dV$. (Hint: Apply Stokes's Theorem to each component.)

(b) Deduce that $\int_{\partial\Omega} \mathbf{n}\,dS = \mathbf{0}$. Give a (geometric) plausibility argument for this result.

10. (Archimedes's Law of Buoyancy) Prove that when a floating body in a uniform liquid is at equilibrium, it displaces its own weight, as follows. Let Ω denote the portion of the body that is submerged.

(a) The force exerted by the pressure of the liquid on a planar piece of surface is directed inward normal to the surface, and pressure is force per unit area. Deduce that the buoyancy force is given by $\mathbf{B} = \int_{\partial\Omega} -p\mathbf{n}\,dS$, where p is the pressure.

(b) Assuming that $\nabla p = \delta \mathbf{g}$, where δ is the (constant) density of the liquid and $\mathbf{g}$ is the acceleration of gravity, deduce that $\mathbf{B} = -M\mathbf{g}$, where M is the mass of the displaced liquid. (Hint: Apply Exercise 9.)

(c) Deduce the result.

11. Let $\mathbf{v}$ be the velocity field of a fluid flow, and let δ be the density of the fluid. (These are both $\mathcal{C}^1$ functions of position and time.) Let $\mathbf{F} = \delta\mathbf{v}$. The *law of conservation of mass* states that

$$\frac{d}{dt}\int_{\Omega} \delta\,dV = -\int_{\partial\Omega} \mathbf{F}\cdot\mathbf{n}\,dS.$$

Show that the validity of this equation for all regions Ω is equivalent to the *equation of continuity*:

$$\text{div } \mathbf{F} + \frac{\partial \delta}{\partial t} = 0.$$

(Hint: Use Exercise 7.2.20.)

12. Suppose a body $\Omega \subset \mathbb{R}^3$ has ($\mathcal{C}^2$) temperature $u\begin{pmatrix}\mathbf{x}\\t\end{pmatrix}$ at position $\mathbf{x} \in \Omega$ at time t. Assume that the heat flow vector $\mathbf{q} = -K\nabla u$, where K is a constant (called the heat conductivity of the body); the flux of $\mathbf{q}$ outward across an oriented surface S represents the rate of heat flow across S.

    (a) Show that the rate of heat flow across $\partial\Omega$ into Ω is $\mathcal{F} = \int_\Omega K\nabla^2 u \, dV$.

    (b) Let c denote the heat capacity of the body; the amount of heat required to raise the temperature of the volume ΔV by ΔT degrees is approximately $(c\Delta T)\Delta V$; thus, the rate at which the volume ΔV absorbs heat is $c\frac{\partial u}{\partial t}\Delta V$. Conclude that the rate of heat flow into Ω is $\mathcal{F} = \int_\Omega c\frac{\partial u}{\partial t} dV$.

    (c) Deduce that the heat flow within Ω is governed by the partial differential equation $c\frac{\partial u}{\partial t} = K\nabla^2 u$.

13. Suppose $\Omega \subset \mathbb{R}^3$ is a region and $u\colon \Omega \times [0, \infty) \to \mathbb{R}$ is a $\mathcal{C}^2$ solution of the heat equation $\nabla^2 u = \frac{\partial u}{\partial t}$. Suppose $u\begin{pmatrix}\mathbf{x}\\0\end{pmatrix} = 0$ for all $\mathbf{x} \in \Omega$ and $D_\mathbf{n} u = 0$ on $\partial\Omega$ (this means the region is insulated along the boundary).

    (a) Consider the "energy" $E(t) = \frac{1}{2}\int_\Omega u^2 dV$. Note that $E(0) = 0$. Prove that $E'(t) \leq 0$ (this means that heat dissipates) and show that $E(t) = 0$ for all $t \geq 0$. (Hint: Use Exercise 7.2.20.)

    (b) Prove that $u\begin{pmatrix}\mathbf{x}\\t\end{pmatrix} = 0$ for all $\mathbf{x} \in \Omega$ and all $t \geq 0$.

    (c) Prove that if u_1 and u_2 are two solutions of the heat equation that agree at $t = 0$ and agree on $\partial\Omega$ for all time $t \geq 0$, then they must agree for all time $t \geq 0$.

14. Suppose $\Omega \subset \mathbb{R}^3$ is a region and $u\colon \Omega \times \mathbb{R} \to \mathbb{R}$ is a $\mathcal{C}^2$ solution of the wave equation $\nabla^2 u = \frac{\partial^2 u}{\partial t^2}$. Suppose that $u\begin{pmatrix}\mathbf{x}\\t\end{pmatrix} = f(\mathbf{x})$ for all $\mathbf{x} \in \partial\Omega$ and all t (e.g., in two dimensions, the drumhead is clamped along the boundary of Ω). Prove that the total energy

$$E(t) = \frac{1}{2}\int_\Omega \left(\left(\frac{\partial u}{\partial t}\right)^2 + \|\nabla u\|^2\right) dV$$

is constant. Here by ∇u we mean the vector of derivatives with respect only to the space variables.

7 APPLICATIONS TO TOPOLOGY

We are going to give a brief introduction to the field of topology by using the techniques of differential forms and Stokes's Theorem to prove three rather deep theorems. The basic ingredient of several of our proofs is the following. Let S^n denote the n-dimensional unit sphere, $S^n = \{\mathbf{x} \in \mathbb{R}^{n+1} : \|\mathbf{x}\| = 1\}$, and D^n the closed unit ball, $D^n = \{\mathbf{x} \in \mathbb{R}^n : \|\mathbf{x}\| \leq 1\}$. (Then $\partial D^{n+1} = S^n$.)

Lemma 7.1 *There is an n-form ω on S^n whose integral is nonzero.*

Proof It is easy to check directly that the volume form

$$\omega = \sum_{i=1}^{n+1}(-1)^{i-1}x_i\,dx_1 \wedge \cdots \wedge \widehat{dx_i} \wedge \cdots \wedge dx_{n+1}$$

is such a form. ∎

Theorem 7.2 *There is no smooth function* $\mathbf{r}\colon D^{n+1} \to S^n$ *with the property that* $\mathbf{r}(\mathbf{x}) = \mathbf{x}$ *for all* $\mathbf{x} \in S^n$.

Proof Suppose there were such an $\mathbf{r}$. Letting ω be an n-form on S^n, as in Lemma 7.1, we have

$$\int_{S^n} \omega = \int_{S^n} \mathbf{r}^*\omega = \int_{D^{n+1}} d(\mathbf{r}^*\omega) = \int_{D^{n+1}} \mathbf{r}^*(d\omega) = 0,$$

inasmuch as the only $(n+1)$-form on an n-dimensional manifold is 0 (and hence $d\omega = 0$). But this is a contradiction since we chose ω with a nonzero integral. ∎

Corollary 7.3 (Brouwer Fixed Point Theorem) *Let* $\mathbf{f}\colon D^n \to D^n$ *be smooth. Then there must be a point* $\mathbf{x} \in D^n$ *so that* $\mathbf{f}(\mathbf{x}) = \mathbf{x}$; *i.e.,* $\mathbf{f}$ *must have a fixed point.*

Proof Suppose it does not. Then for all $\mathbf{x} \in D^n$, the points $\mathbf{x}$ and $\mathbf{f}(\mathbf{x})$ are distinct. Define $\mathbf{r}\colon D^n \to S^{n-1}$ by setting $\mathbf{r}(\mathbf{x})$ to be the point where the ray *starting* at $\mathbf{f}(\mathbf{x})$ and passing through $\mathbf{x}$ intersects the unit sphere, as shown in Figure 7.1. We leave it to the reader to check in Exercise 1 that $\mathbf{r}$ is in fact smooth. By construction, whenever $\mathbf{x} \in S^{n-1}$, we have $\mathbf{r}(\mathbf{x}) = \mathbf{x}$. By Theorem 7.2, no such function can exist, and hence $\mathbf{f}$ must have a fixed point. ∎

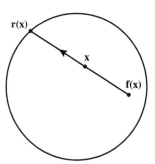

Figure 7.1

Topology is in some sense the study of continuous (or, in our case, smooth) deformations of objects. An old saw is that a topologist is one who cannot tell the difference between a doughnut and a coffee cup. This occurs because we can continuously deform one to the other, assuming we have flexible, plastic objects: The "hole" in the doughnut becomes the "hole" in the handle of the cup. The crucial notion here is the following:

Definition Suppose $X \subset \mathbb{R}^n$ and $Y \subset \mathbb{R}^m$. Let $\mathbf{f}\colon X \to Y$ and $\mathbf{g}\colon X \to Y$ be (smooth) functions. We say they are (smoothly) *homotopic* if there is a (smooth) map $\mathbf{H}\colon X \times [0, 1] \to Y$ so that $\mathbf{H}\begin{pmatrix}\mathbf{x}\\0\end{pmatrix} = \mathbf{f}(\mathbf{x})$ and $\mathbf{H}\begin{pmatrix}\mathbf{x}\\1\end{pmatrix} = \mathbf{g}(\mathbf{x})$ for all $\mathbf{x} \in X$.

▶ **EXAMPLE 1**

The identity function $\mathbf{f}\colon D^n \to D^n$, $\mathbf{f}(\mathbf{x}) = \mathbf{x}$, is homotopic to the constant map $\mathbf{g}(\mathbf{x}) = \mathbf{0}$. We merely set

$$\mathbf{H}\begin{pmatrix}\mathbf{x}\\t\end{pmatrix} = (1-t)\mathbf{x}.$$

The homotopy shrinks the unit ball gradually to its center. ◀

▶ **EXAMPLE 2**

Are the maps $\mathbf{f}, \mathbf{g}\colon S^1 \to S^1$ given by

$$\mathbf{f}\begin{pmatrix}\cos t\\ \sin t\end{pmatrix} = \begin{bmatrix}\cos t\\ \sin t\end{bmatrix} \quad\text{and}\quad \mathbf{g}\begin{pmatrix}\cos t\\ \sin t\end{pmatrix} = \begin{bmatrix}\cos 2t\\ \sin 2t\end{bmatrix}$$

homotopic? These parametrized curves wrap once and twice, respectively, around the unit circle, so the winding numbers of these curves about the origin are 1 and 2, respectively. If we surmise that the winding number should vary continuously as we continuously deform the curve, then we guess that the curves cannot be homotopic. Let's make this precise: Suppose there were a homotopy $\mathbf{H}\colon S^1 \times [0, 1] \to S^1$ between $\mathbf{f}$ and $\mathbf{g}$. Let $\omega = -y\,dx + x\,dy \in \mathcal{A}^1(S^1)$. Then

$$\int_{S^1} \mathbf{f}^*\omega = 2\pi \quad\text{and}\quad \int_{S^1} \mathbf{g}^*\omega = 4\pi.$$

We observe that,

$$\int_{\partial(S^1 \times [0,1])} \mathbf{H}^*\omega = \int_{S^1 \times [0,1]} d(\mathbf{H}^*\omega) = \int_{S^1 \times [0,1]} \mathbf{H}^*(d\omega) = 0$$

since any 2-form on S^1 must be 0. On the other hand, as we see from Figure 7.2,

$$\partial(S^1 \times [0, 1]) = (S^1 \times \{1\})^- \cup (S^1 \times \{0\}),$$

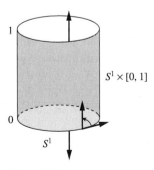

Figure 7.2

so

$$\int_{\partial(S^1\times[0,1])} \mathbf{H}^*\omega = \int_{S^1} \mathbf{f}^*\omega - \int_{S^1} \mathbf{g}^*\omega.$$

In conclusion, if $\mathbf{f}$ and $\mathbf{g}$ are homotopic, then we must have

$$\int_{S^1} \mathbf{f}^*\omega = \int_{S^1} \mathbf{g}^*\omega;$$

since $2\pi \neq 4\pi$, we infer that $\mathbf{f}$ and $\mathbf{g}$ cannot be homotopic. ◀

In general, we have the following important result:

Proposition 7.4 *Suppose X is a compact, oriented k-dimensional manifold and $\mathbf{f}, \mathbf{g}\colon X \to Y$ are homotopic maps. Then for any closed k-form ω on Y, we have*

$$\int_X \mathbf{f}^*\omega = \int_X \mathbf{g}^*\omega.$$

Proof We leave this to the reader in Exercise 3. ∎

By the way, it is time to give a more precise definition of the term "simply connected." A closed curve in $\mathbb{R}^n$ is nothing other than the image of a map $S^1 \to X$.

Definition We say $X \subset \mathbb{R}^n$ is *simply connected* if every pair of points in X can be joined by a path and every map $\mathbf{f}\colon S^1 \to X$ is homotopic to a constant map.

Recall that a k-form ω is *closed* if $d\omega = 0$ and *exact* if $\omega = d\eta$ for some $(k-1)$-form η. As a consequence of Proposition 7.4, we have

Corollary 7.5 *Suppose X is a simply connected manifold. Then every closed 1-form ω on X is exact.*

Proof Let $\mathbf{f}\colon S^1 \to X$ be a closed curve; $\mathbf{f}$ is homotopic to a constant map $\mathbf{g}$. Since $\mathbf{g}^*\omega = 0$, we infer that $\int_{S^1} \mathbf{f}^*\omega = 0$. The result now follows from Theorem 3.2. ∎

Note that this is the generalization of the local result we obtained earlier, Proposition 3.3.

Before moving on to our last topic, we stop to state and prove one of the cornerstones of classical mathematics. We assume a modest familiarity with the complex numbers.

Theorem 7.6 (Fundamental Theorem of Algebra) *Let $n \geq 1$ and $a_0, a_1, \ldots, a_{n-1} \in \mathbb{C}$; consider a polynomial $p(z) = z^n + a_{n-1}z^{n-1} + \cdots + a_1z + a_0$. Then p has n roots in $\mathbb{C}$ (counting multiplicities).*

Proof (We identify $\mathbb{C}$ with $\mathbb{R}^2$ for purposes of the vector calculus.[9]) Since

$$\lim_{z \to \infty} \frac{a_{n-1}z^{n-1} + \cdots + a_1 z + a_0}{z^n} = 0,$$

there is $R > 0$ so that whenever $|z| \geq R$ we have

$$\left| \frac{a_{n-1}z^{n-1} + \cdots + a_1 z + a_0}{z^n} \right| \leq \frac{1}{2}.$$

On $\partial \overline{B}(\mathbf{0}, R)$ we have a homotopy $\mathbf{H} \colon \partial \overline{B}(\mathbf{0}, R) \times [0, 1] \to \mathbb{C} - \{0\}$ between p and $g(z) = z^n$ given by

$$\mathbf{H}\begin{pmatrix} z \\ t \end{pmatrix} = z^n + (1-t)(a_{n-1}z^{n-1} + \cdots + a_1 z + a_0) = tg(z) + (1-t)p(z).$$

The crucial issue is that, by the triangle inequality,

$$|z^n + (1-t)(a_{n-1}z^{n-1} + \cdots + a_1 z + a_0)| \geq |z^n| - (1-t)|a_{n-1}z^{n-1} + \cdots + a_1 z + a_0|$$

$$\geq |z^n|\left(1 - \frac{1}{2}\right) = \frac{R^n}{2},$$

so the function $\mathbf{H}$ indeed takes values in $\mathbb{C} - \{0\}$.

Recall that the 1-form $\omega = \dfrac{-y\,dx + x\,dy}{x^2 + y^2}$ is a closed form on $\mathbb{C} - \{0\} = \mathbb{R}^2 - \{\mathbf{0}\}$.

Moreover, writing $\mathbf{g}\begin{pmatrix} R\cos t \\ R\sin t \end{pmatrix} = R^n \begin{bmatrix} \cos nt \\ \sin nt \end{bmatrix}$, $0 \leq t \leq 2\pi$, we see that

$$\int_{\partial \overline{B}(\mathbf{0}, R)} \mathbf{g}^*\omega = \int_0^{2\pi} \left(-(\sin nt)(-n \sin nt) + (\cos nt)(n \cos nt)\right) dt = \int_0^{2\pi} n\,dt = 2\pi n,$$

and hence, by Proposition 7.4, we have $\int_{\partial \overline{B}(\mathbf{0}, R)} p^*\omega = 2\pi n$ as well. Now, suppose p had no root in $B(\mathbf{0}, R)$. Then p would actually be a smooth map from all of $\overline{B}(\mathbf{0}, R)$ to $\mathbb{C} - \{0\}$ and we would have

$$2\pi n = \int_{\partial \overline{B}(\mathbf{0}, R)} p^*\omega = \int_{\overline{B}(\mathbf{0}, R)} d(p^*\omega) = \int_{\overline{B}(\mathbf{0}, R)} p^*(d\omega) = 0,$$

which is a contradiction. Therefore, p has at least one root in $B(\mathbf{0}, R)$. The stronger statement of the theorem follows easily by induction on n. ∎

We can actually obtain a stronger, more localized version. We need the following computational result, a more elegant proof of which is suggested in Exercise 8.

---

[9]Recall that complex numbers are of the form $z = x + iy$, $x, y \in \mathbb{R}$. We add complex numbers as vectors in $\mathbb{R}^2$, and we multiply by using the distributive property and the rule $i^2 = -1$: If $z = x + iy$ and $w = u + iv$, then $zw = (xu - yv) + i(xv + yu)$. It is customary to denote the length of the complex number z by $|z|$, and the reader can easily check that $|zw| = |z||w|$. In addition, deMoivre's formula tells us that if $z = r(\cos\theta + i \sin\theta)$, then $z^n = r^n(\cos n\theta + i \sin n\theta)$.

Chapter 8. Differential Forms and Integration on Manifolds

Lemma 7.7 Let $\omega = (-ydx + xdy)/(x^2 + y^2) \in \mathcal{A}^1(\mathbb{C} - \{0\})$, and suppose f and g are smooth maps to $\mathbb{C} - \{0\}$. Then $(fg)^*\omega = f^*\omega + g^*\omega$.

Proof Write $f = u + iv$ and $g = U + iV$. Then $fg = (uU - vV) + i(uV + vU)$, and so, using the product rule and a bit of high school algebra, we obtain

$$(fg)^*\omega = (fg)^*\left(\frac{-ydx + xdy}{x^2 + y^2}\right)$$

$$= \frac{-(uV + vU)d(uU - vV) + (uU - vV)d(uV + vU)}{(uU - vV)^2 + (uV + vU)^2}$$

$$= \frac{-(uV + vU)d(uU - vV) + (uU - vV)d(uV + vU)}{(u^2 + v^2)(U^2 + V^2)}$$

$$= \frac{-(uV + vU)(Udu - Vdv + udU - vdV) + (uU - vV)(Vdu + Udv + udV + vdU)}{(u^2 + v^2)(U^2 + V^2)}$$

$$= \frac{(U^2 + V^2)(-vdu + udv) + (u^2 + v^2)(-VdU + UdV)}{(u^2 + v^2)(U^2 + V^2)}$$

$$= \frac{-vdu + udv}{u^2 + v^2} + \frac{-VdU + UdV}{U^2 + V^2} = f^*\omega + g^*\omega,$$

as required. ∎

Now we have an intriguing application of winding numbers (see Section 3) that gives a two-dimensional analogue of Gauss's law from the preceding section. We make use of the Fundamental Theorem of Algebra.

Proposition 7.8 Let p be a polynomial with complex coefficients. Let $D \subset \mathbb{C}$ be a region so that no root of p lies on $C = \partial D$. Then

$$\frac{1}{2\pi}\int_C p^*\left(\frac{-ydx + xdy}{x^2 + y^2}\right)$$

is equal to the number of roots of p in D.

Proof As usual, let $\omega = (-ydx + xdy)/(x^2 + y^2)$. Using Theorem 7.6, we factor $p(z) = c(z - r_1)(z - r_2) \cdots (z - r_n)$, where $c \neq 0$ and $r_j \in \mathbb{C}$, $j = 1, \ldots, n$, are the roots of p. Let $f_j(z) = z - r_j$. Then we claim that

$$\frac{1}{2\pi}\int_C f_j^*\omega = \begin{cases} 1, & r_j \in D \\ 0, & \text{otherwise} \end{cases}.$$

The former is a consequence of Example 10 on p. 361; the latter follows from Corollary 7.5. Applying Lemma 7.7 repeatedly, we see that $p^*\omega = \sum_{j=1}^n f_j^*\omega$, and so

$$\frac{1}{2\pi}\int_C p^*\omega = \sum_{j=1}^n \frac{1}{2\pi}\int_C f_j^*\omega = \sum_{r_j \in D} 1$$

is equal to the number of roots of p in D. ∎

There are far-reaching generalizations of this result that you may learn about in a differential topology or differential geometry course. An interesting application is the study of how roots of a polynomial vary as we change the polynomial; see Exercise 9.

A *vector field* $\mathbf{v}$ on S^n is a smooth function $\mathbf{v}\colon S^n \to \mathbb{R}^{n+1}$ with the property that $\mathbf{x} \cdot \mathbf{v}(\mathbf{x}) = 0$ for all $\mathbf{x}$. (That is, $\mathbf{v}(\mathbf{x})$ is tangent to the sphere at $\mathbf{x}$.)

▶ **EXAMPLE 3**

There is an obvious nowhere-zero vector field on S^1, the unit circle, which we've seen many times in this chapter:

$$\mathbf{v}\begin{pmatrix} x_1 \\ x_2 \end{pmatrix} = \begin{bmatrix} -x_2 \\ x_1 \end{bmatrix}.$$

Indeed, an analogous formula works on $S^{2m-1} \subset \mathbb{R}^{2m}$:

$$\mathbf{v}\begin{pmatrix} x_1 \\ x_2 \\ \vdots \\ x_{2m-1} \\ x_{2m} \end{pmatrix} = \begin{bmatrix} -x_2 \\ x_1 \\ \vdots \\ -x_{2m} \\ x_{2m-1} \end{bmatrix}.$$

(If we visualize the vector field in the case of the circle as pushing around the circle, in the higher-dimensional case, we imagine pushing in each of the orthogonal $x_1 x_2$-, $x_3 x_4$-, ..., $x_{2m-1} x_{2m}$-planes independently.) ◀

In contrast with the preceding example, however, it is somewhat surprising that there is *no* nowhere-zero vector field on S^n when n is even. The following result is usually affectionately called the Hairy Ball Theorem, as it says that we cannot "comb the hairs" on an even-dimensional sphere.

Theorem 7.9 *Any vector field on the unit sphere S^{2m} must vanish somewhere.*

Proof We proceed by contradiction. Suppose $\mathbf{v}$ were a nowhere-zero vector field on S^{2m}; we may assume (by normalizing) that $\|\mathbf{v}(\mathbf{x})\| = 1$ for all $\mathbf{x} \in S^{2m}$. We now use the vector field to define a homotopy between the identity map $\mathbf{f}\colon S^{2m} \to S^{2m}$ and the antipodal map $\mathbf{g}\colon S^{2m} \to S^{2m}$, $\mathbf{g}(\mathbf{x}) = -\mathbf{x}$. Namely, we follow along the semicircle from $\mathbf{x}$ to $-\mathbf{x}$ in the direction of $\mathbf{v}(\mathbf{x})$, as pictured in Figure 7.3. To be specific, define $\mathbf{H}\colon S^{2m} \times [0,1] \to S^{2m}$ by

$$\mathbf{H}\begin{pmatrix} \mathbf{x} \\ t \end{pmatrix} = (\cos \pi t)\mathbf{x} + (\sin \pi t)\mathbf{v}(\mathbf{x}).$$

Clearly, $\mathbf{H}$ is a smooth function. Now we apply Proposition 7.4, using the form ω defined in Lemma 7.1. In particular, we calculate $\mathbf{g}^*\omega$ explicitly:

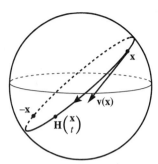

Figure 7.3

$$\mathbf{g}^*\omega = \mathbf{g}^*\left(\sum_{i=1}^{2m+1}(-1)^{i-1}x_i\,dx_1 \wedge \cdots \wedge \widehat{dx_i} \wedge \cdots \wedge dx_{2m+1}\right)$$

$$= \sum_{i=1}^{2m+1}(-1)^{i-1}(-x_i)(-dx_1) \wedge \cdots \wedge \widehat{(-dx_i)} \wedge \cdots \wedge (-dx_{2m+1}) = (-1)^{2m+1}\omega = -\omega.$$

Thus, we have

$$\int_{S^{2m+1}} \omega = \int_{S^{2m+1}} \mathbf{f}^*\omega = \int_{S^{2m+1}} \mathbf{g}^*\omega = -\int_{S^{2m+1}} \omega;$$

since $\int_{S^{2m+1}} \omega \neq 0$, we have arrived at a contradiction. ∎

EXERCISES 8.7

1. Check that the mapping **r** defined in the proof of Corollary 7.3 is in fact smooth.

2. Consider the maps **f** and **g** defined in Example 2 as maps from $[0, 2\pi]$ to $\mathbb{R}^2$ (rather than to S^1). Determine whether they are homotopic.

3. Prove Proposition 7.4.

4. Let $f : \mathbb{C} \to \mathbb{C}$ be given by $f(z) = z^4 - 3z + 9$, and let $\Omega = \{|z| \leq 2\}$. Evaluate $\int_{\partial\Omega} f^*\omega$, where, as usual, $\omega = (-y\,dx + x\,dy)/(x^2 + y^2)$. How many roots does f have in Ω?

5. Show that Corollary 7.3 need not hold on the following spaces:
   (a) S^n,
   (b) the annulus $\{\mathbf{x} \in \mathbb{R}^2 : 1 \leq \|\mathbf{x}\| \leq 2\}$,
   (c) a solid torus,
   (d) B^n (the open unit ball).

6. Prove the following generalization of Theorem 7.2: Let M be any compact, orientable manifold with boundary. Then there is no function $\mathbf{f} : M \to \partial M$ with the property that $\mathbf{f}(\mathbf{x}) = \mathbf{x}$ for all $\mathbf{x} \in \partial M$.

7. As pictured in Figure 7.4, let
$$Z = \{x^2 + y^2 = 1, \ z = 0\} \cup \{x = y = 0\} \cup \{x = z = 0, \ y \geq 1\} \subset \mathbb{R}^3.$$

Suppose ω is a continuously differentiable 1-form on $\mathbb{R}^3 - Z$ satisfying $d\omega = 0$. Suppose, moreover, that $\int_{C_1} \omega = 3$ and $\int_{C_2} \omega = -7$. Calculate $\int_{C_3} \omega$, $\int_{C_4} \omega$, and $\int_{C_5} \omega$. Give your reasoning.

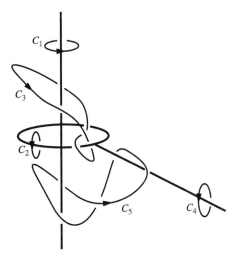

Figure 7.4

8. (a) Let $z = x + iy$. Show that
$$\frac{dz}{z} = \frac{x\,dx + y\,dy}{x^2 + y^2} + i\frac{-y\,dx + x\,dy}{x^2 + y^2}.$$
(b) Let $U \subset \mathbb{C}$ be open, $f, g: U \to \mathbb{C} - \{0\}$ be differentiable, and $\omega = (-y\,dx + x\,dy)/(x^2 + y^2)$. Prove that $(fg)^*\omega = f^*\omega + g^*\omega$. (Hint: What is $(fg)^*(dz/z)$?)

9. Let $\omega = (-y\,dx + x\,dy)/(x^2 + y^2)$.
(a) Suppose $U \subset \mathbb{C}$ is open and $f, g: U \to \mathbb{C} - \{0\}$ are smooth. Let $C \subset U$ be a closed curve and suppose $|g - f| < |f|$ on C. Prove that
$$\int_C g^*\omega = \int_C f^*\omega.$$
(Hint: Use a homotopy similar to that appearing in the proof of Theorem 7.6.)
(b) Let $a_0, a_1, \ldots, a_{n-1} \in \mathbb{C}$ and $p(z) = z^n + a_{n-1}z^{n-1} + \cdots + a_1 z + a_0$. Let $D \subset \mathbb{C}$ be a region so that no root of p lies on $C = \partial D$. Prove that there is $\delta > 0$ so that whenever $|b_j - a_j| < \delta$ for all $j = 0, 1, \ldots, n-1$, the polynomial $P(z) = z^n + b_{n-1}z^{n-1} + \cdots + b_1 z + b_0$ has the same number of roots in D as p.
(c) Deduce from part b that the roots of a polynomial vary continuously with the coefficients.
(Cf. Example 2 on p. 189 and Exercise 6.2.2. See also Exercise 9.4.22 for an interesting application to linear algebra.)

10. Let $\mathbf{f}: S^{2m} \to S^{2m}$ be a smooth map. Prove that there exists $\mathbf{x} \in S^{2m}$ so that either $\mathbf{f}(\mathbf{x}) = \mathbf{x}$ or $\mathbf{f}(\mathbf{x}) = -\mathbf{x}$.

11. Let $n \geq 2$ and $\mathbf{f}: D^n \to \mathbb{R}^n$ be smooth. Suppose $\|\mathbf{f}(\mathbf{x}) - \mathbf{x}\| < 1$ for all $\mathbf{x} \in S^{n-1}$. Prove that there is some $\mathbf{x} \in D^n$ so that $\mathbf{f}(\mathbf{x}) = \mathbf{0}$. (Hint: If not, show that the restriction of the map $\frac{\mathbf{f}}{\|\mathbf{f}\|}: D^n \to S^{n-1}$ to ∂D^n is homotopic to the identity map.)

12. We wish to give a generalization of Proposition 3.3. Suppose $U \subset \mathbb{R}^n$ is an open subset that is star-shaped with respect to the origin.

(a) For any $k = 1, \ldots, n$, given a k-form $\phi = f_I d\mathbf{x}_I$ on U, define the $(k-1)$-form $\mathcal{I}(\phi) =$
$$\left(\int_0^1 t^{k-1} f_I(t\mathbf{x}) dt\right) \sum_{j=1}^k (-1)^{j-1} x_{i_j} dx_{i_1} \wedge \cdots \wedge \widehat{dx_{i_j}} \wedge \cdots \wedge dx_{i_k}.$$
Then make $\mathcal{I}$ linear. Prove that $\phi = d(\mathcal{I}(\phi)) + \mathcal{I}(d\phi)$.

(b) Prove that if ω is a closed k-form on U, then ω is exact.

13. Use the result of Exercise 12 to express each of the following closed forms ω on $\mathbb{R}^3$ in the form $\omega = d\eta$.
  (a) $\omega = (e^x \cos y + z) dx + (2yz^2 - e^x \sin y) dy + (x + 2y^2 z + e^z) dz$.
  (b) $\omega = (2x + y^2) dy \wedge dz + (3y + z) dx \wedge dz + (z - xy) dx \wedge dy$.
  (c) $\omega = xyz \, dx \wedge dy \wedge dz$.

14. Draw an *orientable* surface whose boundary is the boundary curve of the Möbius strip, as pictured in Figure 7.5. (More generally, every simple closed curve in $\mathbb{R}^3$ bounds an orientable surface. Can you see why?)

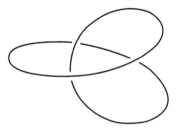
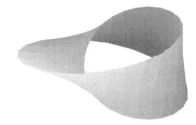

Figure 7.5

15. Find three everywhere linearly independent vector fields on $S^1 \times S^2$.

16. Fill in the details in the following alternative proof of Theorem 7.9, following J. Milnor. Given a (smooth) unit vector field $\mathbf{v}$ on S^n, first extend $\mathbf{v}$ to be a vector field $\mathbf{V}$ on $\mathbb{R}^{n+1}$ by setting
$$\mathbf{V}(\mathbf{x}) = \begin{cases} \|\mathbf{x}\|^2 \mathbf{v}\left(\frac{\mathbf{x}}{\|\mathbf{x}\|}\right), & \mathbf{x} \neq \mathbf{0} \\ \mathbf{0}, & \mathbf{x} = \mathbf{0} \end{cases}.$$

(a) Check that $\mathbf{V}$ is $\mathcal{C}^1$.

(b) Define $\mathbf{f}_t \colon D^{n+1} \to \mathbb{R}^{n+1}$ by $\mathbf{f}_t(\mathbf{x}) = \mathbf{x} + t \mathbf{V}(\mathbf{x})$. Apply the inverse function theorem to prove that for t sufficiently small, $\mathbf{f}_t$ maps the closed unit ball one-to-one and onto the closed ball of radius $\sqrt{1+t^2}$. (Hints: To establish one-to-one, first use the inverse function theorem to show that the function $\mathbf{F} \colon D^{n+1} \times \mathbb{R} \to \mathbb{R}^{n+1} \times \mathbb{R}$ given by $\mathbf{F}\begin{pmatrix} \mathbf{x} \\ t \end{pmatrix} = \begin{bmatrix} \mathbf{f}_t(\mathbf{x}) \\ t \end{bmatrix}$ is locally one-to-one. Now proceed by contradiction: Suppose there were a sequence $t_k \to 0$ and points $\mathbf{x}_k, \mathbf{y}_k \in D^{n+1}$ so that $\mathbf{f}_{t_k}(\mathbf{x}_k) = \mathbf{f}_{t_k}(\mathbf{y}_k)$. Use compactness of D^{n+1} to pass to convergent subsequences $\mathbf{x}_{k_j}$ and $\mathbf{y}_{k_j}$. To establish onto, you will need to use the fact that the only nonempty subset of D^{n+1} that is both open (in D^{n+1}) and closed is D^{n+1} itself.)

(c) Apply the Change of Variables Theorem to see that the volume of $\overline{B}(\mathbf{0}, \sqrt{1+t^2})$ must be a polynomial expression in t.

(d) Deduce that you have arrived at a contradiction when n is even.

CHAPTER

9

EIGENVALUES, EIGENVECTORS, AND APPLICATIONS

We have seen the importance of choosing the appropriate coordinates in doing multiple integration. Now we turn to what is really a much more basic question. Given a linear transformation $T: \mathbb{R}^n \to \mathbb{R}^n$, can we choose appropriate (convenient?) coordinates on $\mathbb{R}^n$ so that the matrix for T (in these coordinates) is as simple as possible, say, diagonal? For this the fundamental tool is eigenvalues and eigenvectors. We then give applications to difference and differential equations and quadratic forms.

▶ 1 LINEAR TRANSFORMATIONS AND CHANGE OF BASIS

In all our previous work, we have referred to the "standard matrix" of a linear transformation. Now we wish to broaden our scope.

Definition Let V be a finite-dimensional vector space and let $T: V \to V$ be a linear transformation. Let $\mathcal{B} = \{\mathbf{v}_1, \ldots, \mathbf{v}_n\}$ be an ordered basis for V. Define numbers a_{ij}, $i = 1, \ldots, n, j = 1, \ldots, n$, by

$$T(\mathbf{v}_j) = a_{1j}\mathbf{v}_1 + a_{2j}\mathbf{v}_2 + \cdots + a_{nj}\mathbf{v}_n, \quad j = 1, \ldots, n.$$

Then we define $A = [a_{ij}]$ to be the *matrix for T with respect to $\mathcal{B}$*, also denoted $[T]_\mathcal{B}$. As before, we have

$$A = \begin{bmatrix} | & | & & | \\ T(\mathbf{v}_1) & T(\mathbf{v}_2) & \cdots & T(\mathbf{v}_n) \\ | & | & & | \end{bmatrix},$$

where now the column vectors are the coordinates of the vectors with respect to the basis $\mathcal{B}$.

We might agree that, generally, the easiest matrices to understand are diagonal. If we think of our examples of projection and reflection in $\mathbb{R}^n$, we obtain some particularly simple diagonal matrices.

EXAMPLE 1

Suppose $V \subset \mathbb{R}^n$ is a subspace. Choose a basis $\{\mathbf{v}_1, \ldots, \mathbf{v}_k\}$ for V and a basis $\{\mathbf{v}_{k+1}, \ldots, \mathbf{v}_n\}$ for $V^\perp$. Then $\mathcal{B} = \{\mathbf{v}_1, \ldots, \mathbf{v}_n\}$ forms a basis for $\mathbb{R}^n$ (why?). Let $T = \text{proj}_V \colon \mathbb{R}^n \to \mathbb{R}^n$ be the linear transformation given by projecting onto V, and let $S \colon \mathbb{R}^n \to \mathbb{R}^n$ be the linear transformation given by reflecting across V. Then we have

$$T(\mathbf{v}_1) = \mathbf{v}_1 \qquad\qquad S(\mathbf{v}_1) = \mathbf{v}_1$$
$$\vdots \qquad\qquad \vdots$$
$$T(\mathbf{v}_k) = \mathbf{v}_k \qquad\qquad S(\mathbf{v}_k) = \mathbf{v}_k$$
$$T(\mathbf{v}_{k+1}) = \mathbf{0} \qquad \text{and} \qquad S(\mathbf{v}_{k+1}) = -\mathbf{v}_{k+1}$$
$$\vdots \qquad\qquad \vdots$$
$$T(\mathbf{v}_n) = \mathbf{0} \qquad\qquad S(\mathbf{v}_n) = -\mathbf{v}_n$$

Then the matrices for T and S with respect to the basis $\mathcal{B}$ are, respectively,

$$B = \left[\begin{array}{c|c} I_k & O \\ \hline O & O \end{array}\right] \quad \text{and} \quad C = \left[\begin{array}{c|c} I_k & O \\ \hline O & -I_{n-k} \end{array}\right].$$

EXAMPLE 2

Let $T \colon \mathbb{R}^2 \to \mathbb{R}^2$ be the linear transformation defined by multiplying by

$$A = \begin{bmatrix} 3 & 1 \\ 2 & 2 \end{bmatrix}.$$

It is rather difficult to understand this function until we discover that if we take

$$\mathbf{v}_1 = \begin{bmatrix} 1 \\ 1 \end{bmatrix} \quad \text{and} \quad \mathbf{v}_2 = \begin{bmatrix} -1 \\ 2 \end{bmatrix},$$

then $T(\mathbf{v}_1) = 4\mathbf{v}_1$ and $T(\mathbf{v}_2) = \mathbf{v}_2$, so that the matrix for T with respect to the ordered basis $\mathcal{B} = \{\mathbf{v}_1, \mathbf{v}_2\}$ is the diagonal matrix

$$B = \begin{bmatrix} 4 & 0 \\ 0 & 1 \end{bmatrix}.$$

Now it is rather straightforward to picture the linear transformation: As we see from Figure 1.1, it stretches the $\mathbf{v}_1$-axis by a factor of 4 and leaves the $\mathbf{v}_2$-axis unchanged. Since we can "pave" the plane by parallelograms formed by $\mathbf{v}_1$ and $\mathbf{v}_2$, we are able to describe the effects of T quite explicitly. We shall soon see how to find $\mathbf{v}_1$ and $\mathbf{v}_2$.

For future reference, let's consider the matrix P with column vectors $\mathbf{v}_1$ and $\mathbf{v}_2$. Since $T(\mathbf{v}_1) = 4\mathbf{v}_1$ and $T(\mathbf{v}_2) = \mathbf{v}_2$, we observe that

$$AP = \begin{bmatrix} 3 & 1 \\ 2 & 2 \end{bmatrix} \begin{bmatrix} 1 & -1 \\ 1 & 2 \end{bmatrix} = \begin{bmatrix} 4 & -1 \\ 4 & 2 \end{bmatrix} = \begin{bmatrix} 1 & -1 \\ 1 & 2 \end{bmatrix} \begin{bmatrix} 4 & 0 \\ 0 & 1 \end{bmatrix} = PB.$$

This might be rewritten as $B = P^{-1}AP$, in the form that will occupy our attention for the rest of this section.

1 Linear Transformations and Change of Basis

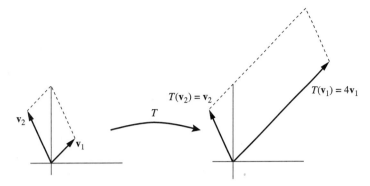

Figure 1.1

It would have been a more honest exercise here to start with the geometric description of T, i.e., its action on the basis vectors $\mathbf{v}_1$ and $\mathbf{v}_2$, and try to find the standard matrix for T. As the reader can check, we have

$$\mathbf{e}_1 = \tfrac{2}{3}\mathbf{v}_1 - \tfrac{1}{3}\mathbf{v}_2$$
$$\mathbf{e}_2 = \tfrac{1}{3}\mathbf{v}_1 + \tfrac{1}{3}\mathbf{v}_2,$$

and so we compute that

$$T(\mathbf{e}_1) = \tfrac{2}{3}T(\mathbf{v}_1) - \tfrac{1}{3}T(\mathbf{v}_2) = \tfrac{8}{3}\mathbf{v}_1 - \tfrac{1}{3}\mathbf{v}_2$$
$$= \begin{bmatrix} 3 \\ 2 \end{bmatrix}, \quad \text{and}$$
$$T(\mathbf{e}_2) = \tfrac{1}{3}T(\mathbf{v}_1) + \tfrac{1}{3}T(\mathbf{v}_2) = \tfrac{4}{3}\mathbf{v}_1 + \tfrac{1}{3}\mathbf{v}_2$$
$$= \begin{bmatrix} 1 \\ 2 \end{bmatrix}.$$

What a relief! ◀

Given a (finite-dimensional) vector space V and an ordered basis $\mathcal{B} = \{\mathbf{v}_1, \ldots, \mathbf{v}_n\}$ for V, we can define a linear transformation

$$C_{\mathcal{B}} \colon V \to \mathbb{R}^n,$$

which assigns to each vector $\mathbf{v}$ its vector of coordinates with respect to the basis $\mathcal{B}$. That is,

$$C_{\mathcal{B}}(c_1\mathbf{v}_1 + c_2\mathbf{v}_2 + \cdots + c_n\mathbf{v}_n) = \begin{bmatrix} c_1 \\ c_2 \\ \vdots \\ c_n \end{bmatrix}.$$

Of course, when $\mathcal{B}$ is the standard basis $\mathcal{E}$ for $\mathbb{R}^n$, this is what you'd expect:

$$C_\mathcal{E}(\mathbf{x}) = \begin{bmatrix} x_1 \\ x_2 \\ \vdots \\ x_n \end{bmatrix}.$$

Suppose $T\colon \mathbb{R}^n \to \mathbb{R}^n$ is a linear transformation and $T(\mathbf{x}) = \mathbf{y}$; to say that A is the standard matrix for T is to say that multiplying A by the coordinate vector of $\mathbf{x}$ (in the standard basis) gives the coordinate vector of $\mathbf{y}$ (in the standard basis). Likewise, suppose $T\colon V \to V$ is a linear transformation, $T(\mathbf{v}) = \mathbf{w}$, and $\mathcal{B}$ is an ordered basis for V. Then let $C_\mathcal{B}(\mathbf{v}) = \mathbf{x}$ be the coordinate vector of $\mathbf{v}$ with respect to the basis $\mathcal{B}$, and let $C_\mathcal{B}(\mathbf{w}) = \mathbf{y}$ be the coordinate vector of $\mathbf{w}$ with respect to the basis $\mathcal{B}$. To say that A is the matrix for T with respect to the basis $\mathcal{B}$ (see the definition on p. 413) is to say $A\mathbf{x} = \mathbf{y}$. (See Figure 1.2.)

Suppose now that we have a linear transformation $T\colon V \to V$ and two ordered bases $\mathcal{B} = \{\mathbf{v}_1, \ldots, \mathbf{v}_n\}$ and $\mathcal{B}' = \{\mathbf{v}'_1, \ldots, \mathbf{v}'_n\}$ for V. (Often in our applications, as the notation suggests, V will be $\mathbb{R}^n$ and $\mathcal{B}$ will be the standard basis $\mathcal{E}$.) Let $A_{\text{old}} = [T]_\mathcal{B}$ be the matrix for T with respect to the "old" basis $\mathcal{B}$, and let $A_{\text{new}} = [T]_{\mathcal{B}'}$ be the matrix for T with respect to the "new" basis $\mathcal{B}'$. The fundamental issue now is to compute A_{new} if we know A_{old}. Define the *change-of-basis matrix* P to be the matrix whose column vectors are the coordinates of the *new* basis vectors with respect to the *old* basis: i.e.,

$$\mathbf{v}'_j = p_{1j}\mathbf{v}_1 + p_{2j}\mathbf{v}_2 + \cdots + p_{nj}\mathbf{v}_n.$$

When $\mathcal{B}$ is the standard basis, we have our usual schematic picture

$$P = \begin{bmatrix} | & | & & | \\ \mathbf{v}'_1 & \mathbf{v}'_2 & \cdots & \mathbf{v}'_n \\ | & | & & | \end{bmatrix}.$$

Note that P must be invertible since we can similarly express each of the old basis vectors as a linear combination of the new basis vectors. (Cf. Proposition 3.4 of Chapter 4.) Then, as the diagram in Figure 1.3 summarizes, we have the following

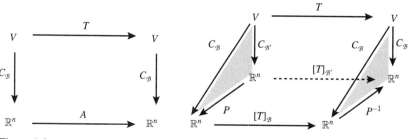

Figure 1.2

Figure 1.3

1 Linear Transformations and Change of Basis

Theorem 1.1 (Change-of-Basis Formula) *Let $T: V \to V$ be a linear transformation, and let $\mathcal{B} = \{\mathbf{v}_1, \ldots, \mathbf{v}_n\}$ and $\mathcal{B}' = \{\mathbf{v}'_1, \ldots, \mathbf{v}'_n\}$ be ordered bases for V. If $[T]_\mathcal{B}$ and $[T]_{\mathcal{B}'}$ are the matrices for T with respect to the respective bases and P is the change-of-basis matrix (whose columns are the coordinates of the new basis vectors with respect the old basis), then we have*

$$[T]_{\mathcal{B}'} = P^{-1}[T]_\mathcal{B} P.$$

Remark Two matrices A and B are called *similar* if $B = P^{-1}AP$ for some invertible matrix P (see Exercise 9). Theorem 1.1 tells us that any two matrices representing a linear map $T: V \to V$ are similar.

Proof Given a vector $\mathbf{v} \in V$, denote by $\mathbf{x}$ and $\mathbf{x}'$, respectively, its coordinate vectors with respect to the bases $\mathcal{B}$ and $\mathcal{B}'$. The important relation here is

$$\mathbf{x} = P\mathbf{x}'.$$

We derive this as follows: Using the equations $\mathbf{v} = \sum_{i=1}^n x_i \mathbf{v}_i$ and

$$\mathbf{v} = \sum_{j=1}^n x'_j \mathbf{v}'_j = \sum_{j=1}^n x'_j \left(\sum_{i=1}^n p_{ij} \mathbf{v}_i \right) = \sum_{i=1}^n \left(\sum_{j=1}^n p_{ij} x'_j \right) \mathbf{v}_i,$$

we deduce from Corollary 4.3.3 that

$$x_i = \sum_{j=1}^n p_{ij} x'_j.$$

(If we think of the old basis as the standard basis for $\mathbb{R}^n$, then this is our familiar fact that multiplying P by $\mathbf{x}'$ takes the appropriate linear combination of the columns of P.)

Likewise, if $T(\mathbf{v}) = \mathbf{w}$, let $\mathbf{y}$ and $\mathbf{y}'$, respectively, denote the coordinate vectors of $\mathbf{w}$ with respect to bases $\mathcal{B}$ and $\mathcal{B}'$. Now compare the equations

$$\mathbf{y}' = [T]_{\mathcal{B}'}\mathbf{x}' \quad \text{and} \quad \mathbf{y} = [T]_\mathcal{B}\mathbf{x},$$

using

$$\mathbf{y} = P\mathbf{y}' \quad \text{and} \quad \mathbf{x} = P\mathbf{x}':$$

On one hand, we have

$$\mathbf{y} = P\mathbf{y}' = P([T]_{\mathcal{B}'}\mathbf{x}') = (P[T]_{\mathcal{B}'})\mathbf{x}',$$

and on the other hand,

$$\mathbf{y} = [T]_\mathcal{B}\mathbf{x} = [T]_\mathcal{B}(P\mathbf{x}') = ([T]_\mathcal{B} P)\mathbf{x}',$$

from which we conclude that

$$[T]_\mathcal{B} P = P[T]_{\mathcal{B}'}; \quad \text{i.e.,} \quad [T]_{\mathcal{B}'} = P^{-1}[T]_\mathcal{B} P. \blacksquare$$

EXAMPLE 3

Let's return to Example 2 as a test case for the change-of-basis formula. (Of course, we've already seen there that it works.) Given the matrix

$$A = [T] = \begin{bmatrix} 3 & 1 \\ 2 & 2 \end{bmatrix}$$

of a linear transformation $T: \mathbb{R}^2 \to \mathbb{R}^2$ with respect to the standard basis, let's calculate its matrix $[T]_{\mathcal{B}'}$ with respect to the new basis $\mathcal{B}' = \{\mathbf{v}_1, \mathbf{v}_2\}$, where

$$\mathbf{v}_1 = \begin{bmatrix} 1 \\ 1 \end{bmatrix} \quad \text{and} \quad \mathbf{v}_2 = \begin{bmatrix} -1 \\ 2 \end{bmatrix}.$$

The change-of-basis matrix is

$$P = \begin{bmatrix} 1 & -1 \\ 1 & 2 \end{bmatrix}, \quad \text{and} \quad P^{-1} = \frac{1}{3}\begin{bmatrix} 2 & 1 \\ -1 & 1 \end{bmatrix},$$

from which the reader should calculate that, indeed,

$$[T]_{\mathcal{B}'} = P^{-1}AP = \frac{1}{3}\begin{bmatrix} 2 & 1 \\ -1 & 1 \end{bmatrix}\begin{bmatrix} 3 & 1 \\ 2 & 2 \end{bmatrix}\begin{bmatrix} 1 & -1 \\ 1 & 2 \end{bmatrix} = \begin{bmatrix} 4 & 0 \\ 0 & 1 \end{bmatrix}.$$

EXAMPLE 4

We wish to calculate the standard matrix for the linear transformation $T = \text{proj}_V$, where $V \subset \mathbb{R}^3$ is the plane $x_1 - 2x_2 + x_3 = 0$. If we choose a basis $\mathcal{B} = \{\mathbf{v}_1, \mathbf{v}_2, \mathbf{v}_3\}$ for $\mathbb{R}^3$ so that $\{\mathbf{v}_1, \mathbf{v}_2\}$ is a basis for V and $\mathbf{v}_3$ is *normal* to the plane, then (see Example 1) we'll have

$$[T]_{\mathcal{B}} = \begin{bmatrix} 1 & 0 & 0 \\ 0 & 1 & 0 \\ 0 & 0 & 0 \end{bmatrix}.$$

So we take

$$\mathbf{v}_1 = \begin{bmatrix} -1 \\ 0 \\ 1 \end{bmatrix}, \quad \mathbf{v}_2 = \begin{bmatrix} 1 \\ 1 \\ 1 \end{bmatrix}, \quad \text{and} \quad \mathbf{v}_3 = \begin{bmatrix} 1 \\ -2 \\ 1 \end{bmatrix}.$$

We wish to know the standard matrix, which means that $\mathcal{B}' = \{\mathbf{e}_1, \mathbf{e}_2, \mathbf{e}_3\}$ should be the standard basis for $\mathbb{R}^3$. Then the *inverse* of the change-of-basis matrix is

$$P^{-1} = \begin{bmatrix} -1 & 1 & 1 \\ 0 & 1 & -2 \\ 1 & 1 & 1 \end{bmatrix},$$

and so

$$P = \begin{bmatrix} -\frac{1}{2} & 0 & \frac{1}{2} \\ \frac{1}{3} & \frac{1}{3} & \frac{1}{3} \\ \frac{1}{6} & -\frac{1}{3} & \frac{1}{6} \end{bmatrix}.$$

Now we use the change-of-basis formula:

$$[T] = [T]_{\mathcal{B}'} = P^{-1}[T]_{\mathcal{B}} P = \begin{bmatrix} -1 & 1 & 1 \\ 0 & 1 & -2 \\ 1 & 1 & 1 \end{bmatrix} \begin{bmatrix} 1 & 0 & 0 \\ 0 & 1 & 0 \\ 0 & 0 & 0 \end{bmatrix} \begin{bmatrix} -\frac{1}{2} & 0 & \frac{1}{2} \\ \frac{1}{3} & \frac{1}{3} & \frac{1}{3} \\ \frac{1}{6} & -\frac{1}{3} & \frac{1}{6} \end{bmatrix}$$

$$= \begin{bmatrix} \frac{5}{6} & \frac{1}{3} & -\frac{1}{6} \\ \frac{1}{3} & \frac{1}{3} & \frac{1}{3} \\ -\frac{1}{6} & \frac{1}{3} & \frac{5}{6} \end{bmatrix}. \blacktriangleleft$$

EXAMPLE 5

Suppose we consider the linear transformation $T: \mathbb{R}^3 \to \mathbb{R}^3$ defined by rotating an angle $2\pi/3$ about the line spanned by $\begin{bmatrix} 1 \\ -1 \\ 1 \end{bmatrix}$. (The angle is measured counterclockwise from a vantage point on the "positive side" of this line.) Once again, the key is to choose a convenient new basis adapted to the geometry of the problem. We choose

$$\mathbf{v}_3 = \begin{bmatrix} 1 \\ -1 \\ 1 \end{bmatrix}$$

along the axis of rotation and $\mathbf{v}_1, \mathbf{v}_2$ to be an orthonormal basis for the plane orthogonal to that axis: e.g.,

$$\mathbf{v}_1 = \frac{1}{\sqrt{2}} \begin{bmatrix} 1 \\ 1 \\ 0 \end{bmatrix} \quad \text{and} \quad \mathbf{v}_2 = \frac{1}{\sqrt{6}} \begin{bmatrix} -1 \\ 1 \\ 2 \end{bmatrix}.$$

Now let's compute:

$$\begin{aligned} T(\mathbf{v}_1) &= -\tfrac{1}{2}\mathbf{v}_1 + \tfrac{\sqrt{3}}{2}\mathbf{v}_2, \\ T(\mathbf{v}_2) &= -\tfrac{\sqrt{3}}{2}\mathbf{v}_1 - \tfrac{1}{2}\mathbf{v}_2, \\ T(\mathbf{v}_3) &= \mathbf{v}_3. \end{aligned}$$

(Now it should be clear why we chose $\mathbf{v}_1, \mathbf{v}_2$ to be orthonormal. We also want $\mathbf{v}_1, \mathbf{v}_2, \mathbf{v}_3$ to form a "right-handed system" so that we're turning in the correct direction, as indicated in Figure 1.4. But there's no need to worry about the length of $\mathbf{v}_3$.) Thus, we have

$$[T]_{\mathcal{B}} = \begin{bmatrix} -\frac{1}{2} & -\frac{\sqrt{3}}{2} & 0 \\ \frac{\sqrt{3}}{2} & -\frac{1}{2} & 0 \\ 0 & 0 & 1 \end{bmatrix}.$$

Next, we take $\mathcal{B}' = \{\mathbf{e}_1, \mathbf{e}_2, \mathbf{e}_3\}$, and the inverse of the change-of-basis matrix is

$$P^{-1} = \begin{bmatrix} \frac{1}{\sqrt{2}} & -\frac{1}{\sqrt{6}} & 1 \\ \frac{1}{\sqrt{2}} & \frac{1}{\sqrt{6}} & -1 \\ 0 & \frac{2}{\sqrt{6}} & 1 \end{bmatrix},$$

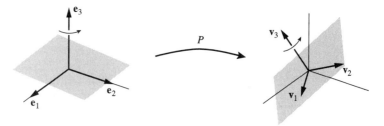

Figure 1.4

so that

$$P = \begin{bmatrix} \frac{1}{\sqrt{2}} & \frac{1}{\sqrt{2}} & 0 \\ -\frac{1}{\sqrt{6}} & \frac{1}{\sqrt{6}} & \frac{2}{\sqrt{6}} \\ \frac{1}{3} & -\frac{1}{3} & \frac{1}{3} \end{bmatrix}.$$

(Exercise 5.5.16 may be helpful here, but, as a last resort, there's always Gaussian elimination.) Once again, we solve for

$$[T] = [T]_{\mathcal{B}'} = P^{-1}[T]_{\mathcal{B}}P = \begin{bmatrix} \frac{1}{\sqrt{2}} & -\frac{1}{\sqrt{6}} & 1 \\ \frac{1}{\sqrt{2}} & \frac{1}{\sqrt{6}} & -1 \\ 0 & \frac{2}{\sqrt{6}} & 1 \end{bmatrix} \begin{bmatrix} -\frac{1}{2} & -\frac{\sqrt{3}}{2} & 0 \\ \frac{\sqrt{3}}{2} & -\frac{1}{2} & 0 \\ 0 & 0 & 1 \end{bmatrix} \begin{bmatrix} \frac{1}{\sqrt{2}} & \frac{1}{\sqrt{2}} & 0 \\ -\frac{1}{\sqrt{6}} & \frac{1}{\sqrt{6}} & \frac{2}{\sqrt{6}} \\ \frac{1}{3} & -\frac{1}{3} & \frac{1}{3} \end{bmatrix}$$

$$= \begin{bmatrix} 0 & -1 & 0 \\ 0 & 0 & -1 \\ 1 & 0 & 0 \end{bmatrix},$$

amazingly enough. In hindsight, then, we should be able to see the effect of T on the standard basis vectors quite plainly. Can you? ◀

Remark Suppose we first rotate $\pi/2$ about the x_3-axis and then rotate $\pi/2$ about the x_1-axis. We leave it to the reader to check that the result is the linear transformation whose matrix we just calculated. This raises a fascinating question: Is the composition of rotations always again a rotation? If so, is there a way of predicting the ultimate axis and angle?

▶ EXERCISES 9.1

*1. Let $\mathbf{v}_1 = \begin{bmatrix} 2 \\ 3 \end{bmatrix}$ and $\mathbf{v}_2 = \begin{bmatrix} 1 \\ 2 \end{bmatrix}$, and consider the basis $\mathcal{B}' = \{\mathbf{v}_1, \mathbf{v}_2\}$ for $\mathbb{R}^2$.

(a) Suppose $T\colon \mathbb{R}^2 \to \mathbb{R}^2$ is a linear transformation whose standard matrix is $[T] = \begin{bmatrix} 1 & 5 \\ 2 & -2 \end{bmatrix}$. Find the matrix for T with respect to the basis $\mathcal{B}'$.

(b) If $S: \mathbb{R}^2 \to \mathbb{R}^2$ is a linear transformation defined by
$$S(\mathbf{v}_1) = 2\mathbf{v}_1 + \mathbf{v}_2$$
$$S(\mathbf{v}_2) = -\mathbf{v}_1 + 3\mathbf{v}_2,$$
then give the standard matrix for S.

2. Derive the result of Exercise 1.4.10a by the change-of-basis formula.

3. Let $T: \mathbb{R}^3 \to \mathbb{R}^3$ be the linear transformation given by reflecting across the plane $-x_1 + x_2 + x_3 = 0$.
(a) Find an orthogonal basis $\{\mathbf{v}_1, \mathbf{v}_2, \mathbf{v}_3\}$ for $\mathbb{R}^3$ so that $\mathbf{v}_1, \mathbf{v}_2$ span the plane and $\mathbf{v}_3$ is orthogonal to it.
(b) Give the matrix representing T with respect to your basis in part a.
(c) Use the change-of-basis theorem to give the matrix representing T with respect to the standard basis.

4. Use the change-of-basis formula to find the standard matrix for projection onto the plane spanned by $\begin{bmatrix} 1 \\ 0 \\ 1 \end{bmatrix}$ and $\begin{bmatrix} 0 \\ 1 \\ -2 \end{bmatrix}$.

*****5.** Let $T: \mathbb{R}^3 \to \mathbb{R}^3$ be the linear transformation given by reflecting across the plane $x_1 - 2x_2 + 2x_3 = 0$. Use the change-of-basis formula to find its standard matrix.

6. Check the result claimed in the remark on p. 420.

7. Let $V \subset \mathbb{R}^3$ be the subspace defined by
$$V = \{\mathbf{x} \in \mathbb{R}^3 : x_1 - x_2 + x_3 = 0\}.$$
Find the standard matrix for each of the following linear transformations:
(a) projection on V,
(b) reflection across V,
(c) rotation of V through angle $\pi/6$ (as viewed from high above).

*****8.** Find the standard matrix for the linear transformation giving projection onto the plane in $\mathbb{R}^4$ spanned by $\begin{bmatrix} 1 \\ 0 \\ 2 \\ 1 \end{bmatrix}$ and $\begin{bmatrix} 0 \\ 1 \\ -1 \\ 1 \end{bmatrix}$.

#**9.** Let A and B be $n \times n$ matrices. We say B is *similar* to A if there is an invertible matrix P so that $B = P^{-1}AP$. (Hint: $B = P^{-1}AP \iff PB = AP$.)
(a) If c is any scalar, show that cI is similar only to itself.
(b) Show that if B is similar to A, then A is similar to B.
(c) Show that $\begin{bmatrix} a & 0 \\ 0 & b \end{bmatrix}$ is similar to $\begin{bmatrix} b & 0 \\ 0 & a \end{bmatrix}$.
(d) Show that for any real numbers a and b, the matrices $\begin{bmatrix} 1 & a \\ 0 & 2 \end{bmatrix}$ and $\begin{bmatrix} 1 & b \\ 0 & 2 \end{bmatrix}$ are similar.
(e) Show that $\begin{bmatrix} 2 & 1 \\ 0 & 2 \end{bmatrix}$ is not similar to $\begin{bmatrix} 2 & 0 \\ 0 & 2 \end{bmatrix}$.
(f) Show that $\begin{bmatrix} 2 & 1 \\ 0 & 2 \end{bmatrix}$ is not diagonalizable, i.e., is not similar to any diagonal matrix.

10. See Exercise 9 for the relevant definition. Prove or give a counterexample:
(a) If B is similar to A, then B^T is similar to A^T.
(b) If B^2 is similar to A^2, then B is similar to A.
(c) If B is similar to A and A is nonsingular, then B is nonsingular.
(d) If B is similar to A and A is symmetric, then B is symmetric.
(e) If B is similar to A, then $\mathbf{N}(B) = \mathbf{N}(A)$.
(f) If B is similar to A, then $\operatorname{rank}(B) = \operatorname{rank}(A)$.

11. See Exercise 9 for the relevant definition. Suppose A and B are $n \times n$ matrices.
(a) Show that if either A or B is nonsingular, then AB and BA are similar.
(b) Must AB and BA be similar in general?

12. *(a) Let $\mathbf{a} = \begin{bmatrix} \sin\phi\cos\theta \\ \sin\phi\sin\theta \\ \cos\phi \end{bmatrix}$, $0 \le \phi < \pi/2$. Prove that the intersection of the circular cylinder $x_1^2 + x_2^2 = 1$ with the plane $\mathbf{a} \cdot \mathbf{x} = 0$ is an ellipse. (Hint: Consider the new basis $\mathbf{v}_1 = \begin{bmatrix} -\sin\theta \\ \cos\theta \\ 0 \end{bmatrix}$, $\mathbf{v}_2 = \begin{bmatrix} -\cos\phi\cos\theta \\ -\cos\phi\sin\theta \\ \sin\phi \end{bmatrix}$, $\mathbf{v}_3 = \mathbf{a}$.)

(b) Describe the projection of the cylindrical region $x_1^2 + x_2^2 = 1$, $-h \le x_3 \le h$ onto the general plane $\mathbf{a} \cdot \mathbf{x} = 0$. (Hint: Special cases are the planes $x_3 = 0$ and $x_1 = 0$.)

13. A cube with vertices at $\begin{bmatrix} \pm 1 \\ \pm 1 \\ \pm 1 \end{bmatrix}$ is rotated about the long diagonal through $\pm \begin{bmatrix} 1 \\ 1 \\ 1 \end{bmatrix}$. Describe the resulting surface and give equation(s) for it.

14. In this exercise we give the general version of the change-of-basis formula for a linear transformation $T: V \to W$.
(a) Suppose $\mathcal{V}$ and $\mathcal{V}'$ are ordered bases for the vector space V and $\mathcal{W}$ and $\mathcal{W}'$ are ordered bases for the vector space W. Let P be the change of basis matrix from $\mathcal{V}$ to $\mathcal{V}'$ and let Q be the change of basis matrix from $\mathcal{W}$ to $\mathcal{W}'$. Suppose $T: V \to W$ is a linear transformation whose matrix with respect to the bases $\mathcal{V}$ and $\mathcal{W}$ is $[T]_{\mathcal{V}}^{\mathcal{W}}$ and whose matrix with respect to the new bases $\mathcal{V}'$ and $\mathcal{W}'$ is $[T]_{\mathcal{V}'}^{\mathcal{W}'}$. Prove that $[T]_{\mathcal{V}'}^{\mathcal{W}'} = Q^{-1}[T]_{\mathcal{V}}^{\mathcal{W}} P$.
(b) Consider the identity transformation $T: V \to V$. Using the basis $\mathcal{V}'$ in the domain and the basis $\mathcal{V}$ in the range, show that the matrix for $[T]_{\mathcal{V}'}^{\mathcal{V}}$ is the change of basis matrix P.

15. (See the discussion on p. 183 and Exercise 4.4.18.) Let A be an $n \times n$ matrix. Prove that the functions $T: \mathbf{R}(A) \to \mathbf{C}(A)$ and $S: \mathbf{C}(A) \to \mathbf{R}(A)$ are inverse functions if and only if $A = QP$, where P is a projection matrix and Q is orthogonal.

▶ 2 EIGENVALUES, EIGENVECTORS, AND DIAGONALIZABILITY

As we shall soon see, it is often necessary in applications to compute (high) powers of a given square matrix. When A is diagonalizable, i.e., there is an invertible matrix P so that $P^{-1}AP = \Lambda$ is diagonal, we have

$$A = P\Lambda P^{-1}, \quad \text{and so}$$

$$A^k = \underbrace{(P\Lambda P^{-1})(P\Lambda P^{-1}) \cdots (P\Lambda P^{-1})}_{k \text{ times}} = P\Lambda^k P^{-1}.$$

Since Λ^k is easy to calculate, we are left with a very computable formula for A^k. We will see a number of applications of this principle in Section 3. We turn first to the matter of finding the diagonal matrix Λ if, in fact, A is diagonalizable. Then we will try to develop some criteria that guarantee diagonalizability.

2.1 The Characteristic Polynomial

Recall that a linear transformation $T\colon V \to V$ is *diagonalizable* if there is an (ordered) basis $\mathcal{B} = \{\mathbf{v}_1, \ldots, \mathbf{v}_n\}$ for V so that the matrix for T with respect to that basis is diagonal. This means precisely that, for some scalars $\lambda_1, \ldots, \lambda_n$, we have

$$T(\mathbf{v}_1) = \lambda_1 \mathbf{v}_1,$$
$$T(\mathbf{v}_2) = \lambda_2 \mathbf{v}_2,$$
$$\vdots$$
$$T(\mathbf{v}_n) = \lambda_n \mathbf{v}_n.$$

Likewise, an $n \times n$ matrix A is diagonalizable if there is a basis $\{\mathbf{v}_1, \ldots, \mathbf{v}_n\}$ for $\mathbb{R}^n$ with the property that $A\mathbf{v}_i = \lambda_i \mathbf{v}_i$ for all $i = 1, \ldots, n$.

This observation leads us to the following

Definition Let $T\colon V \to V$ be a linear transformation. A *nonzero* vector $\mathbf{v} \in V$ is called an *eigenvector* of T if there is a scalar λ so that $T(\mathbf{v}) = \lambda \mathbf{v}$. The scalar λ is called the associated *eigenvalue* of T.

In other words, an eigenvector of a linear transformation T is a (nonzero) vector that is merely stretched (perhaps in the negative direction) by T. The line spanned by the vector is identical to the line spanned by its image under T.

This definition, in turn, leads to a convenient reformulation of diagonalizability:

Proposition 2.1 *The linear transformation $T\colon V \to V$ is diagonalizable if and only if there is a basis for V consisting of eigenvectors of T.*

At this juncture, the obvious question to ask is how we should find eigenvectors. Let's start by observing that, if we include the zero vector, the set of eigenvectors with eigenvalue λ forms a subspace.

Lemma 2.2 *Let $T\colon V \to V$ be a linear transformation, and let λ be any scalar. Then*

$$\mathbf{E}(\lambda) = \{\mathbf{v} \in V : T(\mathbf{v}) = \lambda \mathbf{v}\} = \ker(T - \lambda I)$$

is a subspace of V; $\dim \mathbf{E}(\lambda) > 0$ if and only if λ is an eigenvalue, in which case we call $\mathbf{E}(\lambda)$ the λ-eigenspace.

Proof That $\mathbf{E}(\lambda)$ is a subspace follows immediately once we recognize that it is the kernel (or nullspace) of a linear map. (In the more familiar matrix notation, $\{\mathbf{x} \in \mathbb{R}^n : A\mathbf{x} = \lambda \mathbf{x}\} = \mathbf{N}(A - \lambda I)$.) Now, by definition, λ is an eigenvalue precisely when there is a *nonzero* vector in $\mathbf{E}(\lambda)$. ∎

Chapter 9. Eigenvalues, Eigenvectors, and Applications

We now come to the main computational tool for finding eigenvalues.

Proposition 2.3 *Let A be an $n \times n$ matrix. Then λ is an eigenvalue of A if and only if $\det(A - \lambda I) = 0$.*

Proof From Lemma 2.2 we infer that λ is an eigenvalue if and only if the matrix $A - \lambda I$ is singular. Next we conclude from Theorem 5.5 of Chapter 7 that $A - \lambda I$ is singular precisely when $\det(A - \lambda I) = 0$. Putting the two statements together, we obtain the result. ∎

Once we use this criterion to find the eigenvalues λ, it is an easy matter to find the corresponding eigenvectors merely by finding $\mathbf{N}(A - \lambda I)$.

▶ **EXAMPLE 1**

Let's find the eigenvalues and eigenvectors of the matrix

$$A = \begin{bmatrix} 3 & 1 \\ -3 & 7 \end{bmatrix}.$$

We start by calculating

$$\det(A - tI) = \begin{vmatrix} 3-t & 1 \\ -3 & 7-t \end{vmatrix} = (3-t)(7-t) - (1)(-3) = t^2 - 10t + 24.$$

Since $t^2 - 10t + 24 = (t-4)(t-6) = 0$ when $t = 4$ or $t = 6$, these are our two eigenvalues. We now proceed to find the corresponding eigenspaces:

$\mathbf{E}(4)$: We see that

$$\mathbf{v}_1 = \begin{bmatrix} 1 \\ 1 \end{bmatrix} \text{ is a basis for } \mathbf{N}(A - 4I) = \mathbf{N}\left(\begin{bmatrix} -1 & 1 \\ -3 & 3 \end{bmatrix}\right).$$

$\mathbf{E}(6)$: We see that

$$\mathbf{v}_2 = \begin{bmatrix} 1 \\ 3 \end{bmatrix} \text{ is a basis for } \mathbf{N}(A - 6I) = \mathbf{N}\left(\begin{bmatrix} -3 & 1 \\ -3 & 1 \end{bmatrix}\right).$$

Since we observe that the $\{\mathbf{v}_1, \mathbf{v}_2\}$ is linearly independent, the matrix A is diagonalizable. Indeed, as the reader can check, if we take $P = \begin{bmatrix} 1 & 1 \\ 1 & 3 \end{bmatrix}$, then

$$P^{-1}AP = \begin{bmatrix} \frac{3}{2} & -\frac{1}{2} \\ -\frac{1}{2} & \frac{1}{2} \end{bmatrix} \begin{bmatrix} 3 & 1 \\ -3 & 7 \end{bmatrix} \begin{bmatrix} 1 & 1 \\ 1 & 3 \end{bmatrix} = \begin{bmatrix} 4 & 0 \\ 0 & 6 \end{bmatrix},$$

as should be the case. ◀

▶ EXAMPLE 2

Let's find the eigenvalues and eigenvectors of the matrix
$$A = \begin{bmatrix} 1 & 2 & 1 \\ 0 & 1 & 0 \\ 1 & 3 & 1 \end{bmatrix}.$$

We begin by computing
$$\det(A - tI) = \begin{vmatrix} 1-t & 2 & 1 \\ 0 & 1-t & 0 \\ 1 & 3 & 1-t \end{vmatrix}$$

(expanding in cofactors along the second row)
$$= (1-t)\big((1-t)(1-t) - 1\big) = (1-t)(t^2 - 2t) = -t(t-1)(t-2).$$

Thus, the eigenvalues of A are 0, 1, and 2. We next find the respective eigenspaces:

E(0): We see that
$$\mathbf{v}_1 = \begin{bmatrix} -1 \\ 0 \\ 1 \end{bmatrix} \text{ is a basis for } \mathbf{N}(A - 0I) = \mathbf{N}\left(\begin{bmatrix} 1 & 2 & 1 \\ 0 & 1 & 0 \\ 1 & 3 & 1 \end{bmatrix}\right) = \mathbf{N}\left(\begin{bmatrix} 1 & 0 & 1 \\ 0 & 1 & 0 \\ 0 & 0 & 0 \end{bmatrix}\right).$$

E(1): We see that
$$\mathbf{v}_2 = \begin{bmatrix} 3 \\ -1 \\ 2 \end{bmatrix} \text{ is a basis for } \mathbf{N}(A - 1I) = \mathbf{N}\left(\begin{bmatrix} 0 & 2 & 1 \\ 0 & 0 & 0 \\ 1 & 3 & 0 \end{bmatrix}\right) = \mathbf{N}\left(\begin{bmatrix} 1 & 0 & -\frac{3}{2} \\ 0 & 1 & \frac{1}{2} \\ 0 & 0 & 0 \end{bmatrix}\right).$$

E(2): We see that
$$\mathbf{v}_3 = \begin{bmatrix} 1 \\ 0 \\ 1 \end{bmatrix} \text{ is a basis for } \mathbf{N}(A - 2I) = \mathbf{N}\left(\begin{bmatrix} -1 & 2 & 1 \\ 0 & -1 & 0 \\ 1 & 3 & -1 \end{bmatrix}\right) = \mathbf{N}\left(\begin{bmatrix} 1 & 0 & -1 \\ 0 & 1 & 0 \\ 0 & 0 & 0 \end{bmatrix}\right).$$

Once again, A is diagonalizable: As the reader can check, $\{\mathbf{v}_1, \mathbf{v}_2, \mathbf{v}_3\}$ is linearly independent and therefore gives a basis for $\mathbb{R}^3$. Just to be sure, we let
$$P = \begin{bmatrix} -1 & 3 & 1 \\ 0 & -1 & 0 \\ 1 & 2 & 1 \end{bmatrix};$$

then
$$P^{-1}AP = \begin{bmatrix} -\frac{1}{2} & -\frac{1}{2} & \frac{1}{2} \\ 0 & -1 & 0 \\ \frac{1}{2} & \frac{5}{2} & \frac{1}{2} \end{bmatrix} \begin{bmatrix} 1 & 2 & 1 \\ 0 & 1 & 0 \\ 1 & 3 & 1 \end{bmatrix} \begin{bmatrix} -1 & 3 & 1 \\ 0 & -1 & 0 \\ 1 & 2 & 1 \end{bmatrix} = \begin{bmatrix} 0 & 0 & 0 \\ 0 & 1 & 0 \\ 0 & 0 & 2 \end{bmatrix},$$

as we expected. ◀

Remark There is a built-in check here for the eigenvalues. If λ is truly to be an eigenvalue of A, we must find a nonzero vector in $\mathbf{N}(A - \lambda I)$. If we do not, then λ cannot be an eigenvalue.

▶ EXAMPLE 3

Let's find the eigenvalues and eigenvectors of the matrix

$$A = \begin{bmatrix} 0 & 1 \\ -1 & 0 \end{bmatrix}.$$

As usual, we calculate

$$\det(A - tI) = \begin{vmatrix} -t & 1 \\ -1 & -t \end{vmatrix} = t^2 + 1.$$

Since $t^2 + 1 \geq 1$ for all real numbers t, there is no *real* number λ so that $\det(A - \lambda I) = 0$. Since our scalars are allowed only to be real numbers, this matrix has no eigenvalue. On the other hand, as one might see in a more advanced course, it is often convenient to allow complex numbers as scalars. ◀

It is evident that we are going to find the eigenvalues of a matrix A by finding the (real) roots of the polynomial $\det(A - tI)$. This leads us to our next

Definition Let A be a square matrix. Then $p(t) = p_A(t) = \det(A - tI)$ is called the *characteristic polynomial* of A.[1]

We can restate Proposition 2.3 by saying that the eigenvalues of A are the real roots of the characteristic polynomial $p_A(t)$. It is comforting to observe that similar matrices have the same characteristic polynomial, and hence it makes sense to refer to the characteristic polynomial of a linear map $T: V \to V$.

Lemma 2.4 *If $B = P^{-1}AP$ for some invertible matrix P, then $p_A(t) = p_B(t)$.*

Proof We have

$$p_B(t) = \det(B - tI) = \det(P^{-1}AP - tI)$$
$$= \det\left(P^{-1}(A - tI)P\right) = \det(A - tI) = p_A(t),$$

by virtue of the product rule for determinants, Proposition 5.7 of Chapter 7. ■

As a consequence, if V is a finite-dimensional vector space and $T: V \to V$ is a linear transformation, then we can define the characteristic polynomial of T to be that of the matrix A for T with respect to *any* basis for V. By Lemma 2.4 we'll get the same answer no matter what basis we choose.

---

[1] That the characteristic polynomial of an $n \times n$ matrix is in fact a polynomial of degree n seems pretty evident from examples; but the fastidious reader can establish this by expanding in cofactors.

Remark In order to determine the eigenvalues of a matrix, we must find the roots of its characteristic polynomial. In real-world applications (where the matrices tend to get quite large), one might solve this numerically (e.g., using Newton's method). However, there are more sophisticated methods for finding the eigenvalues without even calculating the characteristic polynomial; a powerful such method is based on the Gram-Schmidt process. The interested reader should consult Strang or Wilkinson for more details.

For the lion's share of the matrices that we shall encounter here, the eigenvalues will be integers, and so we take this opportunity to remind you of a trick from high school algebra.

Proposition 2.5 (Rational Roots Test) Let $p(t) = a_n t^n + a_{n-1} t^{n-1} + \cdots + a_1 t + a_0$ be a polynomial with integer coefficients. If $t = r/s$ is a rational root (in lowest terms) of $p(t)$, then r must be a factor of a_0 and s must be a factor of a_n.

Proof You can find a proof in most abstract algebra texts, but, for obvious reasons, we recommend *Abstract Algebra: A Geometric Approach*, by someone named T. Shifrin, p. 105. ∎

In particular, when the leading coefficient a_n is ± 1, as is always the case with the characteristic polynomial, any rational root must in fact be an integer that divides a_0. So, in practice, we test the various factors of a_0 (being careful to try both positive and negative). Once we find one root r, we can divide $p(t)$ by $t - r$ to obtain a polynomial of smaller degree.

▶ **EXAMPLE 4**

The characteristic polynomial of the matrix

$$A = \begin{bmatrix} 4 & -3 & 3 \\ 0 & 1 & 4 \\ 2 & -2 & 1 \end{bmatrix}$$

is $p(t) = -t^3 + 6t^2 - 11t + 6$. The factors of 6 are $\pm 1, \pm 2, \pm 3$, and ± 6. Since $p(1) = 0$, we know that 1 is a root (so we were lucky). Now,

$$\frac{-p(t)}{t-1} = t^2 - 5t + 6 = (t-2)(t-3),$$

and we have succeeded in finding all three eigenvalues of A. They are 1, 2, and 3. ◀

Remark It might be nice to have a few shortcuts for calculating the characteristic polynomial of small matrices. For 2×2 matrices, it's quite easy:

$$\begin{vmatrix} a-t & b \\ c & d-t \end{vmatrix} = (a-t)(d-t) - bc$$

$$= t^2 - \boxed{(a+d)}\, t + \boxed{(ad-bc)} = t^2 - \boxed{\operatorname{tr} A}\, t + \boxed{\det A}.$$

(Recall that the *trace* of a matrix is the sum of its diagonal entries. The trace of A is denoted $\operatorname{tr} A$.) For 3×3 matrices, similarly,

$$\begin{vmatrix} a_{11}-t & a_{12} & a_{13} \\ a_{21} & a_{22}-t & a_{23} \\ a_{31} & a_{32} & a_{33}-t \end{vmatrix} = -t^3 + \boxed{\operatorname{tr} A}\, t^2 - \boxed{(C_{11}+C_{22}+C_{33})}\, t + \boxed{\det A},$$

where C_{ii} is the ii^{th} cofactor, the determinant of the 2×2 submatrix formed by deleting the ith row and column from A.

In general, the characteristic polynomial $p(t)$ of an $n \times n$ matrix A is always of the form

$$p(t) = (-1)^n t^n + (-1)^{n-1} \boxed{\operatorname{tr} A}\, t^{n-1} + \cdots + \boxed{\det A}.$$

Note that the constant term is always $\det A$ because $p(0) = \det(A - 0I) = \det A$.

In the long run, these formulas notwithstanding, it's sometimes best to calculate the characteristic polynomial of 3×3 matrices by expansion in cofactors. If one is both attentive and fortunate, this may save the trouble of factoring the polynomial.

▶ **EXAMPLE 5**

Let's find the characteristic polynomial of

$$A = \begin{bmatrix} 2 & 0 & 0 \\ 1 & 2 & 1 \\ 0 & 1 & 2 \end{bmatrix}.$$

We calculate the determinant by expanding in cofactors along the first row:

$$\begin{vmatrix} 2-t & 0 & 0 \\ 1 & 2-t & 1 \\ 0 & 1 & 2-t \end{vmatrix} = (2-t)\begin{vmatrix} 2-t & 1 \\ 1 & 2-t \end{vmatrix}$$

$$= (2-t)\bigl((2-t)^2 - 1\bigr) = (2-t)(t^2 - 4t + 3)$$
$$= (2-t)(t-3)(t-1).$$

But that was too easy. Let's try the characteristic polynomial of

$$B = \begin{bmatrix} 2 & 0 & 1 \\ 1 & 3 & 1 \\ 1 & 1 & 2 \end{bmatrix}.$$

Again, we expand in cofactors along the first row:

$$\begin{vmatrix} 2-t & 0 & 1 \\ 1 & 3-t & 1 \\ 1 & 1 & 2-t \end{vmatrix} = (2-t)\begin{vmatrix} 3-t & 1 \\ 1 & 2-t \end{vmatrix} + 1\begin{vmatrix} 1 & 3-t \\ 1 & 1 \end{vmatrix}$$

$$= (2-t)\bigl((3-t)(2-t) - 1\bigr) + \bigl(1 - (3-t)\bigr)$$

$$= (2-t)(t^2 - 5t + 5) - (2-t) = (2-t)(t^2 - 5t + 4)$$
$$= (2-t)(t-1)(t-4).$$

OK, perhaps we were a bit lucky there, too. ◂

2.2 Diagonalizability

Judging by the foregoing examples, it seems to be the case that when an $n \times n$ matrix (or linear transformation) has n *distinct* eigenvalues, the corresponding eigenvectors form a linearly independent set and will therefore give a "diagonalizing basis." Let's begin by proving a slightly stronger statement.

Theorem 2.6 *Let $T: V \to V$ be a linear transformation. Let $\lambda_1, \ldots, \lambda_k$ be k distinct scalars. Suppose $\mathbf{v}_1, \ldots, \mathbf{v}_k$ are eigenvectors of T with respective eigenvalues $\lambda_1, \ldots, \lambda_k$. Then $\{\mathbf{v}_1, \ldots, \mathbf{v}_k\}$ is a linearly independent set of vectors.*

Proof Let m be the largest number between 1 and k (inclusive) so that $\{\mathbf{v}_1, \ldots, \mathbf{v}_m\}$ is linearly independent. We want to see that $m = k$. By way of contradiction, suppose $m < k$. Then we know that $\{\mathbf{v}_1, \ldots, \mathbf{v}_m\}$ is linearly independent and $\{\mathbf{v}_1, \ldots, \mathbf{v}_m, \mathbf{v}_{m+1}\}$ is linearly dependent. It follows from Proposition 3.2 of Chapter 4 that $\mathbf{v}_{m+1} = c_1 \mathbf{v}_1 + \cdots + c_m \mathbf{v}_m$ for some scalars $c_1, \ldots, c_m$. Then (using repeatedly the fact that $T(\mathbf{v}_i) = \lambda_i \mathbf{v}_i$)

$$\mathbf{0} = (T - \lambda_{m+1}I)\mathbf{v}_{m+1} = (T - \lambda_{m+1}I)(c_1\mathbf{v}_1 + \cdots + c_m\mathbf{v}_m)$$
$$= c_1(\lambda_1 - \lambda_{m+1})\mathbf{v}_1 + \cdots + c_m(\lambda_m - \lambda_{m+1})\mathbf{v}_m.$$

Since $\lambda_i - \lambda_{m+1} \neq 0$ for $i = 1, \ldots, m$, and since $\{\mathbf{v}_1, \ldots, \mathbf{v}_m\}$ is linearly independent, the only possibility is that $c_1 = \cdots = c_m = 0$, contradicting the fact that $\mathbf{v}_{m+1} \neq \mathbf{0}$ (by the very definition of eigenvector). Thus, it cannot happen that $m < k$, and the proof is complete. ∎

We now arrive at our first result that gives a sufficient condition for a linear transformation to be diagonalizable.

Corollary 2.7 *Suppose V is an n-dimensional vector space and $T: V \to V$ has n distinct (real) eigenvalues. Then T is diagonalizable.*

Proof The set of the n corresponding eigenvectors will be linearly independent and will hence give a basis for V. The matrix for T with respect to a basis of eigenvectors is always diagonal. ∎

Remark Of course, there are many diagonalizable (indeed, diagonal) matrices with repeated eigenvalues. Certainly the identity matrix and the matrix

$$\begin{bmatrix} 2 & 0 & 0 \\ 0 & 3 & 0 \\ 0 & 0 & 2 \end{bmatrix}$$

are diagonal, and yet they fail to have distinct eigenvalues.

We spend the rest of this section discussing the two ways the hypotheses of Corollary 2.7 can fail: The characteristic polynomial may have complex roots or it may have repeated roots.

EXAMPLE 6

Consider the matrix

$$A = \begin{bmatrix} \frac{1}{\sqrt{2}} & -\frac{1}{\sqrt{2}} \\ \frac{1}{\sqrt{2}} & \frac{1}{\sqrt{2}} \end{bmatrix}.$$

The reader may well recall from Chapter 1 that multiplying by A gives a rotation of the plane through an angle of $\pi/4$. Now, what are the eigenvalues of A? The characteristic polynomial is

$$p(t) = t^2 - (\operatorname{tr} A)t + \det A = t^2 - \sqrt{2}t + 1,$$

whose roots (by the quadratic formula) are

$$\lambda = \frac{\sqrt{2} \pm \sqrt{-2}}{2} = \frac{1 \pm i}{\sqrt{2}}.$$

After a bit of thought, it should come as no surprise that A has no (real) eigenvector, as there can be no line through the origin that is unchanged after a rotation. ◄

We have seen that when the characteristic polynomial has distinct (real) roots, we get a 1-dimensional eigenspace for each. What happens if the characteristic polynomial has some repeated roots?

EXAMPLE 7

Consider the matrix

$$A = \begin{bmatrix} 1 & 1 \\ -1 & 3 \end{bmatrix}.$$

Its characteristic polynomial is $p(t) = t^2 - 4t + 4 = (t-2)^2$, so 2 is a repeated eigenvalue. Now let's find the corresponding eigenvectors:

$$\mathbf{N}(A - 2I) = \mathbf{N}\left(\begin{bmatrix} -1 & 1 \\ -1 & 1 \end{bmatrix}\right) = \mathbf{N}\left(\begin{bmatrix} 1 & -1 \\ 0 & 0 \end{bmatrix}\right)$$

is 1-dimensional, with basis

$$\left\{ \begin{bmatrix} 1 \\ 1 \end{bmatrix} \right\}.$$

It follows that A cannot be diagonalized. (See also Exercise 16.) ◄

▶ **EXAMPLE 8**

Both the matrices

$$A = \left[\begin{array}{cc|cc} 2 & 0 & & \\ 0 & 2 & \multicolumn{2}{c}{O} \\ \hline & & 3 & 1 \\ \multicolumn{2}{c|}{O} & 0 & 3 \end{array}\right] \quad \text{and} \quad B = \left[\begin{array}{cc|cc} 2 & 1 & & \\ 0 & 2 & \multicolumn{2}{c}{O} \\ \hline & & 3 & 0 \\ \multicolumn{2}{c|}{O} & 0 & 3 \end{array}\right]$$

have the characteristic polynomial $p(t) = (t-2)^2(t-3)^2$ (why?). For A, there are two linearly independent eigenvectors with eigenvalue 2 but only one linearly independent eigenvector with eigenvalue 3. For B, there are two linearly independent eigenvectors with eigenvalue 3 but only one linearly independent eigenvector with eigenvalue 2. As a result, neither can be diagonalized. ◀

It would be convenient to have a bit of terminology here.

Definition Let λ be an eigenvalue of a linear transformation. The *algebraic multiplicity* of λ is its multiplicity as a root of the characteristic polynomial $p(t)$, i.e., the highest power of $t - \lambda$ dividing $p(t)$. The *geometric multiplicity* of λ is the dimension of the λ-eigenspace $\mathbf{E}(\lambda)$.

▶ **EXAMPLE 9**

For the matrices in Example 8, both the eigenvalues 2 and 3 have algebraic multiplicity 2. For matrix A, the eigenvalue 2 has geometric multiplicity 2 and the eigenvalue 3 has geometric multiplicity 1; for matrix B, the eigenvalue 2 has geometric multiplicity 1 and the eigenvalue 3 has geometric multiplicity 2. ◀

From the examples we've seen, it seems quite plausible that the geometric multiplicity of an eigenvalue can be no larger than its algebraic multiplicity, but we stop to give a proof.

Proposition 2.8 *Let λ be an eigenvalue of algebraic multiplicity m and geometric multiplicity d. Then $1 \leq d \leq m$.*

Proof Suppose λ is an eigenvalue of the linear transformation T. Then $d = \dim \mathbf{E}(\lambda) \geq 1$ by definition. Now, choose a basis $\{\mathbf{v}_1, \ldots, \mathbf{v}_d\}$ for $\mathbf{E}(\lambda)$ and extend it to a basis $\mathcal{B} = \{\mathbf{v}_1, \ldots, \mathbf{v}_n\}$ for V. Then the matrix for T with respect to the basis $\mathcal{B}$ is of the form

$$A = \left[\begin{array}{c|c} \lambda I_d & B \\ \hline O & C \end{array}\right],$$

and so, by Exercise 7.5.10, the characteristic polynomial

$$p_A(t) = \det(A - tI) = \det\big((\lambda - t)I_d\big) \det(C - tI) = (\lambda - t)^d \det(C - tI).$$

Since the characteristic polynomial does not depend on the basis and since $(t - \lambda)^m$ is the largest power of $t - \lambda$ dividing the characteristic polynomial, it follows that $d \leq m$. ∎

We are now able to give a necessary and sufficient criterion for a linear transformation to be diagonalizable. Based on our experience with examples, it should come as no great surprise.

Theorem 2.9 *Let $T: V \to V$ be a linear transformation. Let its distinct eigenvalues be $\lambda_1, \ldots, \lambda_k$ and assume these are all real numbers. Then T is diagonalizable if and only if the geometric multiplicity, d_i, of each λ_i equals its algebraic multiplicity, m_i.*

Proof Let V be an n-dimensional vector space. Then the characteristic polynomial of T has degree n, and we have

$$p(t) = \pm(t - \lambda_1)^{m_1}(t - \lambda_2)^{m_2} \cdots (t - \lambda_k)^{m_k};$$

therefore,

$$n = \sum_{i=1}^{k} m_i.$$

Now, suppose T is diagonalizable. Then there is a basis $\mathcal{B}$ consisting of eigenvectors. At most, d_i of these basis vectors lie in $\mathbf{E}(\lambda_i)$, and so $n \leq \sum_{i=1}^{k} d_i$. On the other hand, by Proposition 2.8, $d_i \leq m_i$ for $i = 1, \ldots, k$. Putting these together, we have

$$n \leq \sum_{i=1}^{k} d_i \leq \sum_{i=1}^{k} m_i = n.$$

Thus, we must have equality at every stage here, which implies that $d_i = m_i$ for all $i = 1, \ldots, k$.

Conversely, suppose $d_i = m_i$ for $i = 1, \ldots, k$. If we choose a basis $\mathcal{B}_i$ for each eigenspace $\mathbf{E}(\lambda_i)$ and let $\mathcal{B} = \mathcal{B}_1 \cup \cdots \cup \mathcal{B}_k$, then we assert that $\mathcal{B}$ is a basis for V. There are n vectors in $\mathcal{B}$, so we need only check that the set of vectors is linearly independent. This is a generalization of the argument of Theorem 2.6, and we leave it to Exercise 25. ∎

▶ **EXAMPLE 10**

The matrices

$$A = \begin{bmatrix} -1 & 4 & 2 \\ -1 & 3 & 1 \\ -1 & 2 & 2 \end{bmatrix} \quad \text{and} \quad B = \begin{bmatrix} 0 & 3 & 1 \\ -1 & 3 & 1 \\ 0 & 1 & 1 \end{bmatrix}$$

both have characteristic polynomial $p(t) = -(t-1)^2(t-2)$. That is, the eigenvalue 1 has algebraic multiplicity 2 and the eigenvalue 2 has algebraic multiplicity 1. To decide whether the matrices are diagonalizable, we need to know the geometric multiplicity of the eigenvalue 1. Well,

$$A - I = \begin{bmatrix} -2 & 4 & 2 \\ -1 & 2 & 1 \\ -1 & 2 & 1 \end{bmatrix} \rightsquigarrow \begin{bmatrix} 1 & -2 & -1 \\ 0 & 0 & 0 \\ 0 & 0 & 0 \end{bmatrix}$$

has rank 1 and so $\dim \mathbf{E}_A(1) = 2$. We infer from Theorem 2.9 that A is diagonalizable. Indeed, as the reader can check, a diagonalizing basis is

$$\left\{ \begin{bmatrix} 1 \\ 0 \\ 1 \end{bmatrix}, \begin{bmatrix} 1 \\ 1 \\ -1 \end{bmatrix}, \begin{bmatrix} 2 \\ 1 \\ 1 \end{bmatrix} \right\}.$$

On the other hand,

$$B - I = \begin{bmatrix} -1 & 3 & 1 \\ -1 & 3 & 1 \\ 0 & 1 & 0 \end{bmatrix} \rightsquigarrow \begin{bmatrix} 1 & 0 & -1 \\ 0 & 1 & 0 \\ 0 & 0 & 0 \end{bmatrix}$$

has rank 2 and so $\dim \mathbf{E}_B(1) = 1$. Since the eigenvalue 1 has geometric multiplicity 1, it follows from Theorem 2.9 that B is *not* diagonalizable. ◂

In the next section we will see the power of diagonalizing matrices in several applications.

▶ EXERCISES 9.2

1. Find the eigenvalues and eigenvectors of the following matrices.

*(a) $\begin{bmatrix} 1 & 5 \\ 2 & 4 \end{bmatrix}$

(b) $\begin{bmatrix} 0 & 1 \\ 1 & 0 \end{bmatrix}$

(c) $\begin{bmatrix} 10 & -6 \\ 18 & -11 \end{bmatrix}$

(d) $\begin{bmatrix} 1 & 3 \\ 3 & 1 \end{bmatrix}$

*(e) $\begin{bmatrix} 1 & 1 \\ -1 & 3 \end{bmatrix}$

*(f) $\begin{bmatrix} -1 & 1 & 2 \\ 1 & 2 & 1 \\ 2 & 1 & -1 \end{bmatrix}$

(g) $\begin{bmatrix} 1 & 0 & 0 \\ -2 & 1 & 2 \\ -2 & 0 & 3 \end{bmatrix}$

(h) $\begin{bmatrix} 1 & -1 & 2 \\ 0 & 1 & 0 \\ 0 & -2 & 3 \end{bmatrix}$

*(i) $\begin{bmatrix} 2 & 0 & 1 \\ 0 & 1 & 2 \\ 0 & 0 & 1 \end{bmatrix}$

(j) $\begin{bmatrix} 1 & -2 & 2 \\ -1 & 0 & -1 \\ 0 & 2 & -1 \end{bmatrix}$

(k) $\begin{bmatrix} 3 & 1 & 0 \\ 0 & 1 & 2 \\ 0 & 1 & 2 \end{bmatrix}$

*(l) $\begin{bmatrix} 1 & -6 & 4 \\ -2 & -4 & 5 \\ -2 & -6 & 7 \end{bmatrix}$

(m) $\begin{bmatrix} 3 & 2 & -2 \\ 2 & 2 & -1 \\ 2 & 1 & 0 \end{bmatrix}$

(n) $\begin{bmatrix} 1 & 0 & 0 & 1 \\ 0 & 1 & 1 & 1 \\ 0 & 0 & 2 & 0 \\ 0 & 0 & 0 & 2 \end{bmatrix}$

2. Prove that 0 is an eigenvalue of A if and only if A is singular.

3. Prove that the eigenvalues of an upper (or lower) triangular matrix are its diagonal entries.

4. What are the eigenvalues and eigenvectors of a projection? A reflection?

5. Suppose A is nonsingular. Prove that the eigenvalues of A^{-1} are the reciprocals of the eigenvalues of A.

6. Suppose $\mathbf{x}$ is an eigenvector of A with corresponding eigenvalue λ.
  (a) Prove that for any $n \in \mathbb{N}$, $\mathbf{x}$ is an eigenvector of A^n with corresponding eigenvalue λ^n.
  (b) Prove or give a counterexample: $\mathbf{x}$ is an eigenvector of $A + I$.
  (c) If $\mathbf{x}$ is an eigenvector of B with corresponding eigenvalue μ, prove or give a counterexample: $\mathbf{x}$ is an eigenvector of $A + B$ with corresponding eigenvalue $\lambda + \mu$.
  (d) Prove or give a counterexample: If λ is an eigenvalue of A and μ is an eigenvalue of B, then $\lambda + \mu$ is an eigenvalue of $A + B$.

7. Prove or give a counterexample: If A and B have the same characteristic polynomial, then there is an invertible matrix P so that $B = P^{-1}AP$.

♯8. Suppose A is a square matrix. Suppose $\mathbf{x}$ is an eigenvector of A with corresponding eigenvalue λ and $\mathbf{y}$ is an eigenvector of A^T with corresponding eigenvalue μ. Prove that if $\lambda \ne \mu$, then $\mathbf{x} \cdot \mathbf{y} = 0$.

9. Prove or give a counterexample:
  (a) A and A^T have the same eigenvalues.
  (b) A and A^T have the same eigenvectors.

10. Prove that the product of the roots of the characteristic polynomial of A is equal to $\det A$. (Hint: If $\lambda_1, \ldots, \lambda_n$ are the roots, show that $p(t) = \pm(t - \lambda_1)(t - \lambda_2)\ldots(t - \lambda_n)$.)

11. Let A and B be $n \times n$ matrices.
  (a) Suppose A (or B) is nonsingular. Prove that the characteristic polynomials of AB and BA are equal.
  (b) (More challenging) Prove the result of part a when both A and B are singular.

*12. Decide whether each of the matrices in Exercise 1 is diagonalizable. Give your reasoning.

13. Prove or give a counterexample.
  (a) If A is an $n \times n$ matrix with n distinct (real) eigenvalues, then A is diagonalizable.
  (b) If A is diagonalizable and $AB = BA$, then B is diagonalizable.
  (c) If there is an invertible matrix P so that $A = P^{-1}BP$, then A and B have the same eigenvalues.
  (d) If A and B have the same eigenvalues, then there is an invertible matrix P so that $A = P^{-1}BP$.
  (e) There is no real 2×2 matrix A satisfying $A^2 = -I$.
  (f) If A and B are diagonalizable and have the same eigenvalues (with the same algebraic multiplicities), then there is an invertible matrix P so that $A = P^{-1}BP$.

14. Suppose A is a 2×2 matrix whose eigenvalues are integers. If $\det A = 120$, explain why A must be diagonalizable.

15. Is the linear transformation $T: \mathcal{M}_{n \times n} \to \mathcal{M}_{n \times n}$ defined by $T(X) = X^\mathsf{T}$ diagonalizable? (Hint: Consider the equation $X^\mathsf{T} = \lambda X$. What are the corresponding eigenspaces? Exercise 1.4.36 may also be relevant.)

*16. Let $A = \begin{bmatrix} 1 & 1 \\ -1 & 3 \end{bmatrix}$. We saw in Example 7 that A has repeated eigenvalue 2 and $\mathbf{v}_1 = \begin{bmatrix} 1 \\ 1 \end{bmatrix}$ spans $\mathbf{E}(2)$.
  (a) Calculate $(A - 2I)^2$.
  (b) Solve $(A - 2I)\mathbf{v}_2 = \mathbf{v}_1$ for $\mathbf{v}_2$. Explain how we know a priori that this equation has a solution.
  (c) Give the matrix for A with respect to the basis $\{\mathbf{v}_1, \mathbf{v}_2\}$.
  This is the closest to diagonal one can get and is called the *Jordan canonical form* of A.

17. Prove that if λ is an eigenvalue of A with geometric multiplicity d, then λ is an eigenvalue of A^T with geometric multiplicity d. (Hint: Use Theorem 4.5 of Chapter 4.)

18. Suppose A is an $n \times n$ matrix with the property that $A^2 = A$.
    (a) Show that if λ is an eigenvalue of A, then $\lambda = 0$ or $\lambda = 1$.
    (b) Prove that A is diagonalizable. (Hint: See Exercise 4.4.16.)

19. Suppose A is an $n \times n$ matrix with the property that $A^2 = I$.
    (a) Show that if λ is an eigenvalue of A, then $\lambda = 1$ or $\lambda = -1$.
    (b) Prove that
    $$\mathbf{E}(1) = \{\mathbf{x} \in \mathbb{R}^n : \mathbf{x} = \tfrac{1}{2}(\mathbf{u} + A\mathbf{u}) \text{ for some } \mathbf{u} \in \mathbb{R}^n\} \quad \text{and}$$
    $$\mathbf{E}(-1) = \{\mathbf{x} \in \mathbb{R}^n : \mathbf{x} = \tfrac{1}{2}(\mathbf{u} - A\mathbf{u}) \text{ for some } \mathbf{u} \in \mathbb{R}^n\}.$$
    (c) Prove that $\mathbf{E}(1) + \mathbf{E}(-1) = \mathbb{R}^n$ and deduce that A is diagonalizable. (For an application, see Exercise 15.)

20. Let A be an orthogonal 3×3 matrix.
    (a) Prove that the characteristic polynomial p_A has a real root.
    (b) Prove that $\|A\mathbf{x}\| = \|\mathbf{x}\|$ for all $\mathbf{x} \in \mathbb{R}^3$ and deduce that the only (real) eigenvalues of A can be 1 and -1.
    (c) Prove that if $\det A = 1$, then 1 must be an eigenvalue of A.
    (d) Prove that if $\det A = 1$ and $A \neq I$, then $\mu_A \colon \mathbb{R}^3 \to \mathbb{R}^3$ is given by rotation through some angle θ about some axis. (Hint: First show $\dim \mathbf{E}(1) = 1$. Then show that μ_A maps $\mathbf{E}(1)^\perp$ to itself and use Exercise 1.4.34.)
    (e) (Cf. the remark on p. 420.) Prove that the composition of rotations in $\mathbb{R}^3$ is again a rotation.

21. Consider the linear map $T \colon \mathbb{R}^3 \to \mathbb{R}^3$ whose standard matrix is the matrix
    $$C = \begin{bmatrix} \frac{1}{6} & \frac{1}{3} + \frac{\sqrt{6}}{6} & \frac{1}{6} - \frac{\sqrt{6}}{3} \\ \frac{1}{3} - \frac{\sqrt{6}}{6} & \frac{2}{3} & \frac{1}{3} + \frac{\sqrt{6}}{6} \\ \frac{1}{6} + \frac{\sqrt{6}}{3} & \frac{1}{3} - \frac{\sqrt{6}}{6} & \frac{1}{6} \end{bmatrix}$$
    given on p. 28. Show that T is indeed a rotation. Find the axis and angle of rotation.

22. Let A be an $n \times n$ matrix all of whose eigenvalues are real numbers. Prove that there is a basis for $\mathbb{R}^n$ with respect to which the matrix for A becomes upper triangular. (Hint: Consider a basis $\{\mathbf{v}_1, \mathbf{v}'_2, \ldots, \mathbf{v}'_n\}$, where $\mathbf{v}_1$ is an eigenvector.)

♯23. Suppose $T \colon V \to V$ is a linear transformation. Suppose T is diagonalizable (i.e., there is a basis for V consisting of eigenvectors of T). Suppose, moreover, that there is a subspace $W \subset V$ with the property that $T(W) \subset W$. Prove that there is a basis for W consisting of eigenvectors of T. (Hint: Using Exercise 4.3.18, concoct a basis for V by starting with a basis for W. Consider the matrix for T with respect to this basis; what is its characteristic polynomial?)

24. Suppose A and B are $n \times n$ matrices.
    (a) Suppose both A and B are diagonalizable and that they have the same eigenvectors. Prove that $AB = BA$.
    (b) Suppose A has n distinct eigenvalues and $AB = BA$. Prove that every eigenvector of A is also an eigenvector of B. Conclude that B is diagonalizable. (Query: Need every eigenvector of B be an eigenvector of A?)
    (c) Suppose A and B are diagonalizable and $AB = BA$. Prove that A and B are simultaneously diagonalizable; i.e., there is a nonsingular matrix P so that both $P^{-1}AP$ and $P^{-1}BP$ are diagonal.

(Hint: If $\mathbf{E}(\lambda)$ is the λ-eigenspace for A, show that if $\mathbf{v} \in \mathbf{E}(\lambda)$, then $B\mathbf{v} \in \mathbf{E}(\lambda)$. Now use Exercise 23.)

25. (a) Let $\lambda \neq \mu$ be eigenvalues of a linear transformation. Suppose $\{\mathbf{v}_1, \ldots, \mathbf{v}_k\} \subset \mathbf{E}(\lambda)$ is linearly independent and $\{\mathbf{w}_1, \ldots, \mathbf{w}_\ell\} \subset \mathbf{E}(\mu)$ is linearly independent. Prove that $\{\mathbf{v}_1, \ldots, \mathbf{v}_k, \mathbf{w}_1, \ldots, \mathbf{w}_\ell\}$ is linearly independent.
(b) More generally, if $\lambda_1, \ldots, \lambda_k$ are distinct and $\{\mathbf{v}_1^{(i)}, \ldots, \mathbf{v}_{d_i}^{(i)}\} \subset \mathbf{E}(\lambda_i)$ is linearly independent for $i = 1, \ldots, k$, prove that $\{\mathbf{v}_j^{(i)} : i = 1, \ldots, k, j = 1, \ldots, d_i\}$ is linearly independent.

3 DIFFERENCE EQUATIONS AND ORDINARY DIFFERENTIAL EQUATIONS

Suppose A is a diagonalizable matrix. Then there is a nonsingular matrix P so that

$$P^{-1}AP = \Lambda = \begin{bmatrix} \lambda_1 & & & \\ & \lambda_2 & & \\ & & \ddots & \\ & & & \lambda_n \end{bmatrix},$$

where the diagonal entries of Λ are the eigenvalues $\lambda_1, \ldots, \lambda_n$ of A. Then it is easy to use this to calculate the powers of A:

$$A = P\Lambda P^{-1}$$
$$A^2 = (P\Lambda P^{-1})^2 = (P\Lambda P^{-1})(P\Lambda P^{-1}) = P\Lambda(P^{-1}P)\Lambda P^{-1} = P\Lambda^2 P^{-1}$$
$$A^3 = A^2 A = (P\Lambda^2 P^{-1})(P\Lambda P^{-1}) = P\Lambda^2(P^{-1}P)\Lambda P^{-1} = P\Lambda^3 P^{-1}$$
$$\vdots$$
$$A^k = P\Lambda^k P^{-1}.$$

We now show how linear algebra can be applied to solve some simple difference equations and systems of differential equations. Both arise very naturally in modeling economic, physical, and biological problems. For the most basic example, we need only take "exponential growth." When we model a *discrete* growth process and stipulate that population doubles each year, then a_k, the population after k years, obeys the law: $a_{k+1} = 2a_k$. When we model a *continuous* growth process, we stipulate that the rate of change of the population $x(t)$ is proportional to the population at that instant, giving the differential equation $\dot{x}(t) = kx(t)$.

3.1 Difference Equations

EXAMPLE 1

(A cat/mouse population problem) Suppose the cat population at month k is c_k and the mouse population at month k is m_k, and let $\mathbf{x}_k = \begin{bmatrix} c_k \\ m_k \end{bmatrix}$ denote the population vector at month k. Suppose

$$\mathbf{x}_{k+1} = A\mathbf{x}_k, \quad \text{where} \quad A = \begin{bmatrix} 0.7 & 0.2 \\ -0.6 & 1.4 \end{bmatrix},$$

3 Difference Equations and Ordinary Differential Equations

and an initial population vector $\mathbf{x}_0$ is given. Then the population vector $\mathbf{x}_k$ can be computed from

$$\mathbf{x}_k = A^k \mathbf{x}_0,$$

so we want to compute A^k by diagonalizing the matrix A.

Since the characteristic polynomial of A is $p(t) = t^2 - 2.1t + 1.1 = (t-1)(t-1.1)$, we see that the eigenvalues of A are 1 and 1.1. The corresponding eigenvectors are

$$\mathbf{v}_1 = \begin{bmatrix} 2 \\ 3 \end{bmatrix} \quad \text{and} \quad \mathbf{v}_2 = \begin{bmatrix} 1 \\ 2 \end{bmatrix},$$

and so we form the change-of-basis matrix

$$P = \begin{bmatrix} 2 & 1 \\ 3 & 2 \end{bmatrix}.$$

Then we have

$$A = P \Lambda P^{-1}, \quad \text{where} \quad \Lambda = \begin{bmatrix} 1 & 0 \\ 0 & 1.1 \end{bmatrix},$$

and so

$$A^k = P \Lambda^k P^{-1} = \begin{bmatrix} 2 & 1 \\ 3 & 2 \end{bmatrix} \begin{bmatrix} 1 & 0 \\ 0 & (1.1)^k \end{bmatrix} \begin{bmatrix} 2 & -1 \\ -3 & 2 \end{bmatrix}.$$

In particular, if $\mathbf{x}_0 = \begin{bmatrix} c_0 \\ m_0 \end{bmatrix}$ is the original population vector, we have

$$\mathbf{x}_k = \begin{bmatrix} c_k \\ m_k \end{bmatrix} = \begin{bmatrix} 2 & 1 \\ 3 & 2 \end{bmatrix} \begin{bmatrix} 1 & 0 \\ 0 & (1.1)^k \end{bmatrix} \begin{bmatrix} 2 & -1 \\ -3 & 2 \end{bmatrix} \begin{bmatrix} c_0 \\ m_0 \end{bmatrix}$$

$$= \begin{bmatrix} 2 & 1 \\ 3 & 2 \end{bmatrix} \begin{bmatrix} 1 & 0 \\ 0 & (1.1)^k \end{bmatrix} \begin{bmatrix} 2c_0 - m_0 \\ -3c_0 + 2m_0 \end{bmatrix}$$

$$= \begin{bmatrix} 2 & 1 \\ 3 & 2 \end{bmatrix} \begin{bmatrix} 2c_0 - m_0 \\ (1.1)^k(-3c_0 + 2m_0) \end{bmatrix}$$

$$= (2c_0 - m_0) \begin{bmatrix} 2 \\ 3 \end{bmatrix} + (-3c_0 + 2m_0)(1.1)^k \begin{bmatrix} 1 \\ 2 \end{bmatrix}.$$

We can now see what happens as time passes. If $3c_0 = 2m_0$, the second term drops out and the population vector stays constant. If $3c_0 < 2m_0$, the first term is still constant, and the second term increases exponentially; but note that the contribution to the mouse population is double the contribution to the cat population. And if $3c_0 > 2m_0$, we see that the population vector decreases exponentially, the mouse population being the first to disappear (why?). ◀

The story for a general diagonalizable matrix A is the same. The column vectors of P are the eigenvectors $\mathbf{v}_1, \ldots, \mathbf{v}_n$, and the entries of Λ^k are $\lambda_1^k, \ldots, \lambda_n^k$, and so, letting

$$P^{-1}\mathbf{x}_0 = \begin{bmatrix} c_1 \\ c_2 \\ \vdots \\ c_n \end{bmatrix},$$

we have

$$(*) \quad A^k \mathbf{x}_0 = P\Lambda^k(P^{-1}\mathbf{x}_0) = \begin{bmatrix} | & | & & | \\ \mathbf{v}_1 & \mathbf{v}_2 & \cdots & \mathbf{v}_n \\ | & | & & | \end{bmatrix} \begin{bmatrix} \lambda_1^k & & & \\ & \lambda_2^k & & \\ & & \ddots & \\ & & & \lambda_n^k \end{bmatrix} \begin{bmatrix} c_1 \\ c_2 \\ \vdots \\ c_n \end{bmatrix}$$

$$= c_1 \lambda_1^k \mathbf{v}_1 + c_2 \lambda_2^k \mathbf{v}_2 + \cdots + c_n \lambda_n^k \mathbf{v}_n.$$

This formula will have all the information we need, and we will see physical interpretations of analogous formulas when we discuss systems of differential equations shortly.

▶ **EXAMPLE 2**

(**The Fibonacci Sequence**) The renowned Fibonacci sequence,

$$1, \ 1, \ 2, \ 3, \ 5, \ 8, \ 13, \ 21, \ 34, \ 55, \ 89, \ 144, \ \ldots,$$

is obtained by letting each number (starting with the third) be the sum of the preceding two: If we let a_k denote the kth number in the sequence, then

$$a_{k+1} = a_k + a_{k-1}, \quad a_0 = a_1 = 1.$$

Thus, if we define $\mathbf{x}_k = \begin{bmatrix} a_k \\ a_{k+1} \end{bmatrix}$, $k \geq 0$, then we can encode the pattern of the sequence in the matrix equation

$$\begin{bmatrix} a_k \\ a_{k+1} \end{bmatrix} = \begin{bmatrix} 0 & 1 \\ 1 & 1 \end{bmatrix} \begin{bmatrix} a_{k-1} \\ a_k \end{bmatrix}, \quad k \geq 1.$$

In other words, setting

$$A = \begin{bmatrix} 0 & 1 \\ 1 & 1 \end{bmatrix} \quad \text{and} \quad \mathbf{x}_k = \begin{bmatrix} a_k \\ a_{k+1} \end{bmatrix},$$

we have

$$\mathbf{x}_{k+1} = A\mathbf{x}_k \quad \text{for all } k \geq 0, \quad \text{with} \quad \mathbf{x}_0 = \begin{bmatrix} 1 \\ 1 \end{bmatrix}.$$

Once again, by computing the powers of the matrix A, we can calculate $\mathbf{x}_k = A^k \mathbf{x}_0$, and hence the kth term in the Fibonacci sequence.

The characteristic polynomial of A is $p(t) = t^2 - t - 1$, and so the eigenvalues are

$$\lambda_1 = \frac{1 + \sqrt{5}}{2} \quad \text{and} \quad \lambda_2 = \frac{1 - \sqrt{5}}{2}.$$

The corresponding eigenvectors are
$$\mathbf{v}_1 = \begin{bmatrix} 1 \\ \lambda_1 \end{bmatrix} \quad \text{and} \quad \mathbf{v}_2 = \begin{bmatrix} 1 \\ \lambda_2 \end{bmatrix}.$$

Then
$$P = \begin{bmatrix} 1 & 1 \\ \lambda_1 & \lambda_2 \end{bmatrix} \quad \text{and} \quad P^{-1} = \frac{1}{\sqrt{5}} \begin{bmatrix} -\lambda_2 & 1 \\ \lambda_1 & -1 \end{bmatrix},$$

so we have
$$\begin{bmatrix} c_1 \\ c_2 \end{bmatrix} = P^{-1} \begin{bmatrix} 1 \\ 1 \end{bmatrix} = \frac{1}{\sqrt{5}} \begin{bmatrix} 1-\lambda_2 \\ \lambda_1 - 1 \end{bmatrix} = \frac{1}{\sqrt{5}} \begin{bmatrix} \lambda_1 \\ -\lambda_2 \end{bmatrix}.$$

Now we use the formula (∗) above to calculate
$$\mathbf{x}_k = A^k \mathbf{x}_0 = c_1 \lambda_1^k \mathbf{v}_1 + c_2 \lambda_2^k \mathbf{v}_2$$
$$= \frac{\lambda_1}{\sqrt{5}} \lambda_1^k \begin{bmatrix} 1 \\ \lambda_1 \end{bmatrix} - \frac{\lambda_2}{\sqrt{5}} \lambda_2^k \begin{bmatrix} 1 \\ \lambda_2 \end{bmatrix}.$$

In particular, reading off the first coordinate of this vector, we find that the kth number in the Fibonacci sequence is
$$a_k = \frac{1}{\sqrt{5}} \left(\lambda_1^{k+1} - \lambda_2^{k+1} \right) = \frac{1}{\sqrt{5}} \left(\left(\tfrac{1+\sqrt{5}}{2} \right)^{k+1} - \left(\tfrac{1-\sqrt{5}}{2} \right)^{k+1} \right).$$

It's far from obvious (at least to the author) that each such number is an integer. We would be remiss if we didn't point out one of the classic facts about the Fibonacci sequence: If we take the ratio of successive terms, we get
$$\frac{a_{k+1}}{a_k} = \frac{\frac{1}{\sqrt{5}} \left(\lambda_1^{k+2} - \lambda_2^{k+2} \right)}{\frac{1}{\sqrt{5}} \left(\lambda_1^{k+1} - \lambda_2^{k+1} \right)}.$$

Now, $|\lambda_2| \approx .618$, so $\lim_{k \to \infty} \lambda_2^k = 0$ and we have
$$\lim_{k \to \infty} \frac{a_{k+1}}{a_k} = \lambda_1 \approx 1.618.$$

This is the famed *golden ratio*. ◀

3.2 Systems of Differential Equations

Another powerful application of linear algebra comes from the study of systems of ordinary differential equations (ODE's). For example, we have the constant-coefficient system of linear ordinary differential equations:
$$\dot{x}_1(t) = a_{11} x_1(t) + a_{12} x_2(t),$$
$$\dot{x}_2(t) = a_{21} x_1(t) + a_{22} x_2(t).$$

Here, and throughout this section, we use a dot to represent differentiation with respect to t (time).

440 ▶ Chapter 9. Eigenvalues, Eigenvectors, and Applications

The main problem we address in this section is the following: Given an $n \times n$ (constant) matrix A and a vector $\mathbf{x}_0 \in \mathbb{R}^n$, we wish to find all differentiable vector-valued functions $\mathbf{x}(t)$ so that

$$\dot{\mathbf{x}}(t) = A\mathbf{x}(t), \qquad \mathbf{x}(0) = \mathbf{x}_0.$$

(The vector $\mathbf{x}_0$ is called the *initial value* of the solution $\mathbf{x}(t)$.)

▶ **EXAMPLE 3**

Suppose $n = 1$, so that $A = [a]$ for some real number a. Then we have simply the ordinary differential equation

$$\dot{x}(t) = ax(t), \qquad x(0) = x_0.$$

The trick of separating variables that the reader most likely learned in her integral calculus course leads to the solution $x(t) = x_0 e^{at}$. As we can easily check, $\dot{x}(t) = ax(t)$, so we have in fact found a solution. Do we know there can be no more? Suppose $y(t)$ were any solution of the original problem. Then the function $z(t) = y(t)e^{-at}$ satisfies the equation

$$\dot{z}(t) = \dot{y}(t)e^{-at} + y(t)\left(-ae^{-at}\right) = (ay(t))\,e^{-at} + y(t)\left(-ae^{-at}\right) = 0,$$

and so $z(t)$ must be a constant function. Since $z(0) = y(0) = x_0$, we see that $y(t) = x_0 e^{at}$. The original differential equation (with its initial condition) has a unique solution. ◀

▶ **EXAMPLE 4**

Consider perhaps the simplest possible 2×2 example:

$$\begin{aligned} \dot{x}_1(t) &= ax_1(t) \\ \dot{x}_2(t) &= bx_2(t) \end{aligned}$$

with the initial conditions $x_1(0) = (x_1)_0$, $x_2(0) = (x_2)_0$. In matrix notation, this is the ODE

$$\dot{\mathbf{x}}(t) = A\mathbf{x}(t), \quad \mathbf{x}(0) = \mathbf{x}_0, \qquad \text{where}$$

$$A = \begin{bmatrix} a & 0 \\ 0 & b \end{bmatrix}, \quad \mathbf{x}(t) = \begin{bmatrix} x_1(t) \\ x_2(t) \end{bmatrix}, \quad \text{and} \quad \mathbf{x}_0 = \begin{bmatrix} (x_1)_0 \\ (x_2)_0 \end{bmatrix}.$$

Since $x_1(t)$ and $x_2(t)$ appear completely independently in these equations, we infer from Example 3 that the unique solution of this system of equations will be

$$x_1(t) = (x_1)_0 e^{at}, \qquad x_2(t) = (x_2)_0 e^{bt}.$$

In vector notation, we have

$$\mathbf{x}(t) = \begin{bmatrix} x_1(t) \\ x_2(t) \end{bmatrix} = \begin{bmatrix} e^{at} & 0 \\ 0 & e^{bt} \end{bmatrix} \mathbf{x}_0 = E(t)\mathbf{x}_0,$$

where $E(t)$ is the diagonal 2×2 matrix with entries e^{at} and e^{bt}. This result is easily generalized to the case of a diagonal $n \times n$ matrix. ◀

3 Difference Equations and Ordinary Differential Equations ◂ 441

Recall that for any real number x, we have the Taylor series expansion

(†)
$$e^x = \sum_{k=0}^{\infty} \frac{x^k}{k!} = 1 + x + \frac{1}{2}x^2 + \frac{1}{6}x^3 + \cdots + \frac{1}{k!}x^k + \cdots.$$

Now, given an $n \times n$ matrix A, we define a new $n \times n$ matrix e^A, called the *exponential of A*, by

$$e^A = I + A + \frac{1}{2}A^2 + \frac{1}{6}A^3 + \cdots + \frac{1}{k!}A^k + \cdots = \sum_{k=0}^{\infty} \frac{A^k}{k!}.$$

That the series converges is immediate from Proposition 1.1 of Chapter 6. In general, however, trying to evaluate this series directly is extremely difficult because the coefficients of A^k are not easily expressed in terms of the coefficients of A. However, when A is a diagonalizable matrix, it is easy to compute e^A: There is an invertible matrix P so that $\Lambda = P^{-1}AP$ is diagonal. Thus, $A = P\Lambda P^{-1}$ and $A^k = P\Lambda^k P^{-1}$ for all $k \in \mathbb{N}$, and so

$$e^A = \sum_{k=0}^{\infty} \frac{A^k}{k!} = \sum_{k=0}^{\infty} \frac{P\Lambda^k P^{-1}}{k!} = P\left(\sum_{k=0}^{\infty} \frac{\Lambda^k}{k!}\right) P^{-1} = P e^\Lambda P^{-1}.$$

▶ **EXAMPLE 5**

Let $A = \begin{bmatrix} 2 & 0 \\ 3 & -1 \end{bmatrix}$. Then $A = P\Lambda P^{-1}$, where

$$\Lambda = \begin{bmatrix} 2 & \\ & -1 \end{bmatrix} \quad \text{and} \quad P = \begin{bmatrix} 1 & 0 \\ 1 & 1 \end{bmatrix}.$$

Then we have

$$e^{t\Lambda} = \begin{bmatrix} e^{2t} & \\ & e^{-t} \end{bmatrix} \quad \text{and} \quad e^{tA} = Pe^{t\Lambda}P^{-1} = \begin{bmatrix} e^{2t} & 0 \\ e^{2t} - e^{-t} & e^{-t} \end{bmatrix}. \quad \blacktriangleleft$$

The result of Example 4 generalizes to the $n \times n$ case. Indeed, whenever we can solve a problem for diagonal matrices, we can solve it for diagonalizable matrices by making the appropriate change of basis. So we should not be surprised by the following result.

Proposition 3.1 *Let A be a diagonalizable $n \times n$ matrix. The general solution of initial value problem*

(‡)
$$\dot{\mathbf{x}}(t) = A\mathbf{x}(t), \quad \mathbf{x}(0) = \mathbf{x}_0$$

is given by $\mathbf{x}(t) = e^{tA}\mathbf{x}_0$.

Proof As above, since A is diagonalizable, there are an invertible matrix P and a diagonal matrix Λ so that $A = P\Lambda P^{-1}$ and $e^{tA} = Pe^{t\Lambda}P^{-1}$. Since the derivative of the diagonal matrix

$$e^{t\Lambda} = \begin{bmatrix} e^{t\lambda_1} & & & \\ & e^{t\lambda_2} & & \\ & & \ddots & \\ & & & e^{t\lambda_n} \end{bmatrix}$$

is obviously

$$\begin{bmatrix} \lambda_1 e^{t\lambda_1} & & & \\ & \lambda_2 e^{t\lambda_2} & & \\ & & \ddots & \\ & & & \lambda_n e^{t\lambda_n} \end{bmatrix} = \Lambda e^{t\Lambda},$$

then we have

$$(e^{tA})^{\bullet} = (Pe^{t\Lambda}P^{-1})^{\bullet} = P(e^{t\Lambda})^{\bullet}P^{-1}$$
$$= P(\Lambda e^{t\Lambda})P^{-1}$$
$$= (P\Lambda P^{-1})(Pe^{t\Lambda}P^{-1}) = Ae^{tA}.$$

We begin by checking that $\mathbf{x}(t) = e^{tA}\mathbf{x}_0$ is indeed a solution:

$$\dot{\mathbf{x}}(t) = (e^{tA}\mathbf{x}_0)^{\bullet} = (Ae^{tA})\mathbf{x}_0 = A(e^{tA}\mathbf{x}_0) = A\mathbf{x}(t),$$

as required.

Now suppose that $\mathbf{y}(t)$ is a solution of the equation ($\ddagger$), and consider the vector function $\mathbf{z}(t) = e^{-tA}\mathbf{y}(t)$. Then by the product rule, we have

$$\dot{\mathbf{z}}(t) = (e^{-tA})^{\bullet}\mathbf{y}(t) + e^{-tA}\dot{\mathbf{y}}(t)$$
$$= -Ae^{-tA}\mathbf{y}(t) + e^{-tA}(A\mathbf{y}(t)) = (-Ae^{-tA} + e^{-tA}A)\mathbf{y}(t) = \mathbf{0},$$

as $Ae^{-tA} = e^{-tA}A$. This implies that $\mathbf{z}(t)$ must be a constant vector, and so

$$\mathbf{z}(t) = \mathbf{z}(0) = \mathbf{y}(0) = \mathbf{x}_0,$$

whence $\mathbf{y}(t) = e^{tA}\mathbf{z}(t) = e^{tA}\mathbf{x}_0$ for all t, as required. ∎

Remark A more sophisticated interpretation of this result is the following: If we view the system ($\ddagger$) of ODE's in a coordinate system derived from the eigenvectors of the matrix A, then the system is uncoupled.

▶ **EXAMPLE 6**

Continuing Example 5, we see that the general solution of the system $\dot{\mathbf{x}}(t) = A\mathbf{x}(t)$ has the form

$$\mathbf{x}(t) = \begin{bmatrix} x_1(t) \\ x_2(t) \end{bmatrix} = e^{tA}\begin{bmatrix} c_1 \\ c_2 \end{bmatrix} \text{ for appropriate constants } c_1 \text{ and } c_2$$

$$= \begin{bmatrix} c_1 e^{2t} \\ c_1 e^{2t} - c_1 e^{-t} + c_2 e^{-t} \end{bmatrix} = c_1 e^{2t}\begin{bmatrix} 1 \\ 1 \end{bmatrix} + (c_2 - c_1)e^{-t}\begin{bmatrix} 0 \\ 1 \end{bmatrix}.$$

Of course, this is the expression we get when we write

$$\mathbf{x}(t) = e^{tA}\begin{bmatrix} c_1 \\ c_2 \end{bmatrix} = P\left(e^{t\Lambda}P^{-1}\begin{bmatrix} c_1 \\ c_2 \end{bmatrix}\right)$$

and obtain the familiar linear combination of the columns of P (which are the eigenvectors of A). If, in particular, we wish to study the long-term behavior of the solution, we observe that $\lim_{t\to\infty} e^{-t} = 0$ and $\lim_{t\to\infty} e^{2t} = \infty$, so that $\mathbf{x}(t)$ behaves like $c_1 e^{2t}\begin{bmatrix} 1 \\ 1 \end{bmatrix}$ as $t \to \infty$. In general, this type of analysis of diagonalizable systems is called *normal mode analysis*, and the vector functions

$$e^{2t}\begin{bmatrix} 1 \\ 1 \end{bmatrix} \quad \text{and} \quad e^{-t}\begin{bmatrix} 0 \\ 1 \end{bmatrix}$$

corresponding to the eigenvectors are called the *normal modes* of the system. ◂

To emphasize the analogy with the solution of difference equations earlier and the formula (∗) on p. 438, we rephrase Proposition 3.1 to highlight the normal modes.

Corollary 3.2 *Suppose A is diagonalizable, with eigenvalues $\lambda_1, \ldots, \lambda_n$ and corresponding eigenvectors $\mathbf{v}_1, \ldots, \mathbf{v}_n$, and write $A = P\Lambda P^{-1}$, as usual. Then the solution of the initial value problem*

$$\dot{\mathbf{x}}(t) = A\mathbf{x}(t), \quad \mathbf{x}(0) = \mathbf{x}_0$$

is

(††)
$$\mathbf{x}(t) = e^{tA}\mathbf{x}_0 = Pe^{t\Lambda}(P^{-1}\mathbf{x}_0)$$

$$= \begin{bmatrix} | & | & & | \\ \mathbf{v}_1 & \mathbf{v}_2 & \cdots & \mathbf{v}_n \\ | & | & & | \end{bmatrix}\begin{bmatrix} e^{\lambda_1 t} & & & \\ & e^{\lambda_2 t} & & \\ & & \ddots & \\ & & & e^{\lambda_n t} \end{bmatrix}\begin{bmatrix} c_1 \\ c_2 \\ \vdots \\ c_n \end{bmatrix}$$

$$= c_1 e^{\lambda_1 t}\mathbf{v}_1 + c_2 e^{\lambda_2 t}\mathbf{v}_2 + \cdots + c_n e^{\lambda_n t}\mathbf{v}_n,$$

where

$$P^{-1}\mathbf{x}_0 = \begin{bmatrix} c_1 \\ c_2 \\ \vdots \\ c_n \end{bmatrix}.$$

Note that the general solution is a linear combination of the normal modes $e^{\lambda_1 t}\mathbf{v}_1, \ldots, e^{\lambda_n t}\mathbf{v}_n$.

Even when A is not diagonalizable, we may differentiate the exponential series term-by-term[2] to obtain

$$(e^{tA})^{\bullet} = \left(I + tA + \frac{t^2}{2!}A^2 + \frac{t^3}{3!}A^3 + \cdots + \frac{t^k}{k!}A^k + \frac{t^{k+1}}{(k+1)!}A^{k+1} + \cdots\right)^{\bullet}$$

$$= A + tA^2 + \frac{t^2}{2!}A^3 + \cdots + \frac{t^{k-1}}{(k-1)!}A^k + \frac{t^k}{k!}A^{k+1} + \cdots$$

$$= A\left(I + tA + \frac{t^2}{2!}A^2 + \cdots + \frac{t^{k-1}}{(k-1)!}A^{k-1} + \frac{t^k}{k!}A^k + \cdots\right) = Ae^{tA}.$$

Thus, we have

Theorem 3.3 *Suppose A is an $n \times n$ matrix. Then the unique solution of the initial value problem*

$$\dot{\mathbf{x}}(t) = A\mathbf{x}(t), \qquad \mathbf{x}(0) = \mathbf{x}_0$$

is $\mathbf{x}(t) = e^{tA}\mathbf{x}_0$.

▶ **EXAMPLE 7**

Consider the differential equation $\dot{\mathbf{x}}(t) = A\mathbf{x}(t)$ when

$$A = \begin{bmatrix} 0 & -1 \\ 1 & 0 \end{bmatrix}.$$

The unsophisticated (but tricky) approach is to write this system out explicitly:

$$\dot{x}_1(t) = -x_2(t)$$
$$\dot{x}_2(t) = x_1(t)$$

and differentiate again, obtaining

(∗∗)
$$\ddot{x}_1(t) = -\dot{x}_2(t) = -x_1(t)$$
$$\ddot{x}_2(t) = \dot{x}_1(t) = -x_2(t).$$

That is, our vector function $\mathbf{x}(t)$ satisfies the second-order differential equation

$$\ddot{\mathbf{x}}(t) = -\mathbf{x}(t).$$

Now, the equations (∗∗) have the "obvious" solutions

$$x_1(t) = a_1 \cos t + b_1 \sin t \qquad \text{and} \qquad x_2(t) = a_2 \cos t + b_2 \sin t$$

for some constants $a_1, a_2, b_1,$ and b_2 (although it is far from obvious that these are the *only* solutions). Some information was lost in the process; in particular, since $\dot{x}_1 = x_2$, the constants must satisfy the equations

$$a_2 = -b_1 \qquad \text{and} \qquad b_2 = a_1.$$

---
[2] See Spivak, *Calculus*, 3 ed., chap. 24, for the proof in the real case; Proposition 1.1 of Chapter 6 applies to show that it works for the matrix case as well.

That is, the vector function

$$\mathbf{x}(t) = \begin{bmatrix} x_1(t) \\ x_2(t) \end{bmatrix} = \begin{bmatrix} a\cos t - b\sin t \\ a\sin t + b\cos t \end{bmatrix} = \begin{bmatrix} \cos t & -\sin t \\ \sin t & \cos t \end{bmatrix} \begin{bmatrix} a \\ b \end{bmatrix}$$

gives a solution of the original differential equation.

On the other hand, Theorem 3.3 tells us that the general solution should be of the form

$$\mathbf{x}(t) = e^{tA}\mathbf{x}_0,$$

and so we suspect that

$$e^{\left(t\begin{bmatrix} 0 & -1 \\ 1 & 0 \end{bmatrix}\right)} = \begin{bmatrix} \cos t & -\sin t \\ \sin t & \cos t \end{bmatrix}$$

should hold. Well,

$$e^{tA} = I + tA + \frac{t^2}{2!}A^2 + \frac{t^3}{3!}A^3 + \frac{t^4}{4!}A^4 + \cdots$$

$$= \begin{bmatrix} 1 & 0 \\ 0 & 1 \end{bmatrix} + t\begin{bmatrix} 0 & -1 \\ 1 & 0 \end{bmatrix} + \frac{t^2}{2!}\begin{bmatrix} -1 & 0 \\ 0 & -1 \end{bmatrix} + \frac{t^3}{3!}\begin{bmatrix} 0 & 1 \\ -1 & 0 \end{bmatrix} + \frac{t^4}{4!}\begin{bmatrix} 1 & 0 \\ 0 & 1 \end{bmatrix} + \cdots$$

$$= \begin{bmatrix} 1 - \frac{t^2}{2!} + \frac{t^4}{4!} + \cdots & -t + \frac{t^3}{3!} - \frac{t^5}{5!} + \cdots \\ t - \frac{t^3}{3!} + \frac{t^5}{5!} + \cdots & 1 - \frac{t^2}{2!} + \frac{t^4}{4!} + \cdots \end{bmatrix}.$$

Since the power series expansions (Taylor series) for sine and cosine are, indeed,

$$\sin t = t - \frac{1}{3!}t^3 + \frac{1}{5!}t^5 + \cdots + (-1)^k \frac{1}{(2k+1)!}t^{2k+1} + \cdots$$

$$\cos t = 1 - \frac{1}{2!}t^2 + \frac{1}{4!}t^4 + \cdots + (-1)^k \frac{1}{(2k)!}t^{2k} + \cdots,$$

the formulas agree. (Another approach to computing e^{tA} is to diagonalize A over the complex numbers, but we don't stop to do this here.[3]) ◀

▶ **EXAMPLE 8**

Let's now the consider the case of a nondiagonalizable matrix, e.g.,

$$A = \begin{bmatrix} 2 & 1 \\ 0 & 2 \end{bmatrix}.$$

The system

$$\dot{x}_1(t) = 2x_1 + x_2$$
$$\dot{x}_2(t) = 2x_2$$

---

[3] But we must remind you of the famous formula, usually attributed to Euler: $e^{it} = \cos t + i\sin t$.

is already partially uncoupled, so we know that $x_2(t)$ must take the form $x_2(t) = ce^{2t}$ for some constant c. Now, in order to find $x_1(t)$, we must solve the inhomogeneous ODE

$$\dot{x}_1(t) = 2x_1(t) + ce^{2t}.$$

In elementary differential equations courses, one is taught to look for a solution of the form

$$x_1(t) = ae^{2t} + bte^{2t};$$

in this case,

$$\dot{x}_1(t) = (2a+b)e^{2t} + (2b)te^{2t} = 2x_1(t) + be^{2t},$$

and so taking $b = c$ gives the desired solution of our equation. That is, the solution of the system is the vector function

$$\mathbf{x}(t) = \begin{bmatrix} ae^{2t} + cte^{2t} \\ ce^{2t} \end{bmatrix} = \begin{bmatrix} e^{2t} & te^{2t} \\ 0 & e^{2t} \end{bmatrix} \begin{bmatrix} a \\ c \end{bmatrix}.$$

The explanation of the trick is quite simple. Let's calculate the matrix exponential e^{tA} by writing

$$A = \begin{bmatrix} 2 & 0 \\ 0 & 2 \end{bmatrix} + \begin{bmatrix} 0 & 1 \\ 0 & 0 \end{bmatrix} = 2I + B, \quad \text{where} \quad B = \begin{bmatrix} 0 & 1 \\ 0 & 0 \end{bmatrix}.$$

The powers of A are easy to compute because $B^2 = 0$: By the binomial theorem,

$$(2I + B)^k = 2^k I + k2^{k-1} B,$$

and so

$$e^{tA} = \sum_{k=0}^{\infty} \frac{t^k}{k!} A^k = \sum_{k=0}^{\infty} \frac{t^k}{k!} (2^k I + k2^{k-1} B)$$

$$= \sum_{k=0}^{\infty} \frac{(2t)^k}{k!} I + \sum_{k=0}^{\infty} \frac{t^k}{k!} (k2^{k-1}) B$$

$$= e^{2t} I + t \sum_{k=1}^{\infty} \frac{(2t)^{k-1}}{(k-1)!} B = e^{2t} I + t \sum_{k=0}^{\infty} \frac{(2t)^k}{k!} B$$

$$= e^{2t} I + te^{2t} B = \begin{bmatrix} e^{2t} & te^{2t} \\ 0 & e^{2t} \end{bmatrix}.$$

A similar phenomenon occurs more generally (see Exercise 14). ◀

Let's consider the general n^{th} order linear ODE with constant coefficients:

(★) $\quad y^{(n)}(t) + a_{n-1} y^{(n-1)}(t) + \cdots + a_2 \ddot{y}(t) + a_1 \dot{y}(t) + a_0 y(t) = 0.$

Here $a_0, a_1, \ldots, a_{n-1}$ are scalars, and $y(t)$ is assumed to be $\mathcal{C}^n$; $y^{(k)}$ denotes its kth derivative. We can use the power of Theorem 3.3 to derive the following general result.

Theorem 3.4 Let $n \in \mathbb{N}$. The set of solutions of the nth order ODE $(\star)$ is an n-dimensional subspace of $\mathcal{C}^\infty(\mathbb{R})$, the vector space[4] of smooth functions. In particular, the initial value problem

$$y^{(n)}(t) + a_{n-1} y^{(n-1)}(t) + \cdots + a_2 \ddot{y}(t) + a_1 \dot{y}(t) + a_0 y(t) = 0$$
$$y(0) = c_0, \quad \dot{y}(0) = c_1, \quad \ddot{y}(0) = c_2, \quad \ldots, \quad y^{(n-1)}(0) = c_{n-1}$$

has a unique solution.

Proof The trick is to concoct a way to apply Theorem 3.3. We introduce the vector function $\mathbf{x}(t)$ defined by

$$\mathbf{x}(t) = \begin{bmatrix} y(t) \\ \dot{y}(t) \\ \ddot{y}(t) \\ \vdots \\ y^{(n-1)}(t) \end{bmatrix},$$

and observe that it satisfies the *first*-order system of ODE's

$$\dot{\mathbf{x}}(t) = \begin{bmatrix} \dot{y}(t) \\ \ddot{y}(t) \\ \dddot{y}(t) \\ \vdots \\ y^{(n)}(t) \end{bmatrix} = \begin{bmatrix} 0 & 1 & 0 & \cdots & 0 \\ 0 & 0 & 1 & \cdots & 0 \\ 0 & 0 & 0 & \ddots & 0 \\ \vdots & \vdots & \vdots & \ddots & \vdots \\ -a_0 & -a_1 & -a_2 & \cdots & -a_{n-1} \end{bmatrix} \begin{bmatrix} y(t) \\ \dot{y}(t) \\ \ddot{y}(t) \\ \vdots \\ y^{(n-1)}(t) \end{bmatrix}$$
$$= A\mathbf{x}(t),$$

where A is the obvious matrix of coefficients. We infer from Theorem 3.3 that the general solution is $\mathbf{x}(t) = e^{tA} \mathbf{x}_0$, so

$$\begin{bmatrix} y(t) \\ \dot{y}(t) \\ \ddot{y}(t) \\ \vdots \\ y^{(n-1)}(t) \end{bmatrix} = e^{tA} \begin{bmatrix} c_0 \\ c_1 \\ c_2 \\ \vdots \\ c_{n-1} \end{bmatrix} = c_0 \mathbf{v}_1(t) + c_1 \mathbf{v}_2(t) + \cdots + c_{n-1} \mathbf{v}_n(t),$$

where $\mathbf{v}_j(t)$ are the columns of e^{tA}. In particular, if we let $q_1(t), \ldots, q_n(t)$ denote the first entries of the vector functions $\mathbf{v}_1(t), \ldots, \mathbf{v}_n(t)$, respectively, we see that

$$y(t) = c_0 q_1(t) + c_1 q_2(t) + \cdots + c_{n-1} q_n(t);$$

---

[4] See Section 3.1 of Chapter 4.

that is, the functions $q_1, \ldots, q_n$ span the vector space of solutions of the differential equation $(\star)$. Note that these functions are C^∞ since the entries of e^{tA} are. Last, we claim that these functions are linearly independent. Suppose that for some scalars $c_0, c_1, \ldots, c_{n-1}$ we have

$$y(t) = c_0 q_1(t) + c_1 q_2(t) + \cdots + c_{n-1} q_n(t) = 0;$$

then, differentiating, we have the same linear relation among the kth derivatives of $q_1, \ldots, q_n$, $k = 1, \ldots, n-1$, and so we have

$$\mathbf{0} = \begin{bmatrix} y(t) \\ \dot{y}(t) \\ \ddot{y}(t) \\ \vdots \\ y^{(n-1)}(t) \end{bmatrix} = e^{tA} \begin{bmatrix} c_0 \\ c_1 \\ c_2 \\ \vdots \\ c_{n-1} \end{bmatrix}.$$

Since e^{tA} is an invertible matrix (see Exercise 17), we infer that $c_0 = c_1 = \cdots = c_{n-1} = 0$, and so $\{q_1, \ldots, q_n\}$ is linearly independent. ∎

▶ **EXAMPLE 9**

Let

$$A = \begin{bmatrix} -3 & 2 \\ 2 & -3 \end{bmatrix},$$

and consider the *second*-order system of ODE's

$$\ddot{\mathbf{x}}(t) = A\mathbf{x}(t), \quad \mathbf{x}(0) = \mathbf{x}_0, \quad \dot{\mathbf{x}}(0) = \dot{\mathbf{x}}_0.$$

The experience we gained in Example 7 suggests that if we can uncouple this system (by finding eigenvalues and eigenvectors), we should expect to find normal modes that are sinusoidal in nature.

The characteristic polynomial of A is $p(t) = t^2 + 6t + 5$, and so its eigenvalues are $\lambda_1 = -1$ and $\lambda_2 = -5$, with corresponding eigenvectors

$$\mathbf{v}_1 = \begin{bmatrix} 1 \\ 1 \end{bmatrix} \quad \text{and} \quad \mathbf{v}_2 = \begin{bmatrix} 1 \\ -1 \end{bmatrix}.$$

(Note, as a check, that because A is symmetric, the eigenvectors are orthogonal. See Exercise 9.2.8.) As usual, we write $P^{-1}AP = \Lambda$, where

$$\Lambda = \begin{bmatrix} -1 & \\ & -5 \end{bmatrix} \quad \text{and} \quad P = \begin{bmatrix} 1 & 1 \\ 1 & -1 \end{bmatrix}.$$

Let's make the "uncoupling" change of coordinates $\mathbf{y} = P^{-1}\mathbf{x}$; i.e.,

$$\mathbf{y} = \begin{bmatrix} y_1 \\ y_2 \end{bmatrix} = \frac{1}{2} \begin{bmatrix} 1 & 1 \\ 1 & -1 \end{bmatrix} \begin{bmatrix} x_1 \\ x_2 \end{bmatrix}.$$

Then the system of differential equations becomes

$$\ddot{\mathbf{y}}(t) = P^{-1}\ddot{\mathbf{x}}(t) = P^{-1}A\mathbf{x} = \Lambda P^{-1}\mathbf{x} = \Lambda \mathbf{y};$$

i.e.,
$$\ddot{y}_1(t) = -y_1$$
$$\ddot{y}_2(t) = -5y_2,$$

whose general solution is
$$y_1(t) = a_1 \cos t + b_1 \sin t$$
$$y_2(t) = a_2 \cos \sqrt{5}t + b_2 \sin \sqrt{5}t.$$

This means that in the original coordinates, we have $\mathbf{x} = P\mathbf{y}$; i.e.,

$$\mathbf{x} = \begin{bmatrix} x_1 \\ x_2 \end{bmatrix} = \begin{bmatrix} 1 & 1 \\ 1 & -1 \end{bmatrix} \begin{bmatrix} a_1 \cos t + b_1 \sin t \\ a_2 \cos \sqrt{5}t + b_2 \sin \sqrt{5}t \end{bmatrix}$$

$$= (a_1 \cos t + b_1 \sin t) \begin{bmatrix} 1 \\ 1 \end{bmatrix} + (a_2 \cos \sqrt{5}t + b_2 \sin \sqrt{5}t) \begin{bmatrix} 1 \\ -1 \end{bmatrix}.$$

The four constants can be determined from the initial conditions $\mathbf{x}_0$ and $\dot{\mathbf{x}}_0$. In particular, if we start with

$$\mathbf{x}_0 = \begin{bmatrix} 1 \\ 0 \end{bmatrix} \quad \text{and} \quad \dot{\mathbf{x}}_0 = \begin{bmatrix} 0 \\ 0 \end{bmatrix},$$

then $a_1 = a_2 = \frac{1}{2}$ and $b_1 = b_2 = 0$. Note that the form of our solution looks very much like the normal mode decomposition of the solution (††) of the first-order system earlier.

A physical system that leads to this differential equation is the following. Hooke's Law says that a spring with spring constant k exerts a restoring force $F = -kx$ on a mass m that is displaced x units from its equilibrium position (corresponding to the "natural length" of the spring). Now imagine a system, as pictured in Figure 3.1, consisting of two masses (m_1 and m_2) connected to each other and to walls by three springs (with spring constants k_1, k_2, and k_3). Denote by x_1 and x_2 the displacement of masses m_1 and m_2, respectively, from equilibrium position. Hooke's Law, as stated above, and Newton's second law of motion ("force = mass $\times$ acceleration") give us the following system of equations:

$$m_1 \ddot{x}_1(t) = -k_1 x_1 + k_2(x_2 - x_1) = -(k_1 + k_2)x_1 + k_2 x_2$$
$$m_2 \ddot{x}_2(t) = k_2(x_1 - x_2) - k_3 x_2 = k_2 x_1 - (k_2 + k_3)x_2.$$

Setting $m_1 = m_2 = 1$, $k_1 = k_3 = 1$, and $k_2 = 2$ gives the system of differential equations with which we began. Here the normal modes correspond to sinusoidal motion with $x_1 = x_2$ (so we observe the masses moving "in parallel," the middle spring staying at its natural length) and frequency 1 and sinusoidal motion with $x_1 = -x_2$ (so we observe the masses moving "in antiparallel," the middle spring compressing symmetrically) and frequency $\sqrt{5}$. ◀

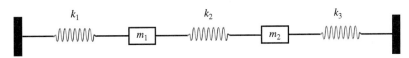

Figure 3.1

3.3 Flows and the Divergence Theorem

Let $U \subset \mathbb{R}^n$ be an open subset. Let $\mathbf{F}: U \to \mathbb{R}^n$ be a vector field on U. So far we have dealt with vector fields of the form $\mathbf{F}(\mathbf{x}) = A\mathbf{x}$, where A is an $n \times n$ matrix. But, more generally, we can try to solve the system of differential equations

$$\dot{\mathbf{x}}(t) = \mathbf{F}(\mathbf{x}(t)), \quad \mathbf{x}(0) = \mathbf{x}_0.$$

We will write the solution of this initial value problem as $\boldsymbol{\phi}_t(\mathbf{x}_0)$, indicating its functional dependence on both time and the initial value. The function $\boldsymbol{\phi}$ is called the *flow* of the vector field $\mathbf{F}$. Note that $\boldsymbol{\phi}_0(\mathbf{x}) = \mathbf{x}$ for all $\mathbf{x} \in U$.

▶ **EXAMPLE 10**

a. The flow of the vector field $F(x) = x$ on $\mathbb{R}$ is $\phi_t(x) = e^t x$.

b. The flow of the vector field $\mathbf{F}\begin{pmatrix} x \\ y \end{pmatrix} = \begin{bmatrix} -y \\ x \end{bmatrix}$ on $\mathbb{R}^2$ is

$$\boldsymbol{\phi}_t\begin{pmatrix} x \\ y \end{pmatrix} = \begin{bmatrix} \cos t & -\sin t \\ \sin t & \cos t \end{bmatrix}\begin{bmatrix} x \\ y \end{bmatrix};$$

i.e., the flow lines are circles centered at the origin.

c. Let $A = \begin{bmatrix} 2 & 1 \\ 1 & -2 \end{bmatrix}$. The flow of the vector field $\mathbf{F}(\mathbf{x}) = A\mathbf{x}$ on $\mathbb{R}^2$ is

$$\boldsymbol{\phi}_t(\mathbf{x}) = e^{tA}\mathbf{x} = Pe^{t\Lambda}(P^{-1}\mathbf{x}) = \frac{1}{6}\left((5x_1 + x_2)e^{3t}\begin{bmatrix} 1 \\ 1 \end{bmatrix} + (-x_1 + x_2)e^{-3t}\begin{bmatrix} -1 \\ 5 \end{bmatrix}\right),$$

where

$$\Lambda = \begin{bmatrix} 3 & \\ & -3 \end{bmatrix} \quad \text{and} \quad P = \begin{bmatrix} 1 & -1 \\ 1 & 5 \end{bmatrix}.$$

d. The flow of the general linear differential equation $\dot{\mathbf{x}}(t) = A\mathbf{x}(t)$ is given by $\boldsymbol{\phi}_t(\mathbf{x}) = e^{tA}\mathbf{x}$. Finding an explicit formula for the flow of a nonlinear differential equation may be somewhat difficult. ◀

It is proved in more advanced courses that if $\mathbf{F}$ is a smooth vector field on an open set $U \subset \mathbb{R}^n$, then for any $\mathbf{x} \in U$, there are a neighborhood V of $\mathbf{x}$ and $\varepsilon > 0$ so that for any $\mathbf{y} \in V$ the flow starting at $\mathbf{y}$, $\boldsymbol{\phi}_t(\mathbf{y})$, is defined for all $|t| < \varepsilon$. Moreover, the function $\boldsymbol{\phi}: V \times (-\varepsilon, \varepsilon) \to \mathbb{R}^n$, $\boldsymbol{\phi}\begin{pmatrix} \mathbf{y} \\ t \end{pmatrix} = \boldsymbol{\phi}_t(\mathbf{y})$, is smooth. We now want to give another interpretation of divergence of the vector field $\mathbf{F}$, first discussed in Section 6 of Chapter 8. It is a natural generalization of the elementary observation that the derivative of the area of a circle with respect to its radius is the circumference.

First we need to extend the definition of divergence to n dimensions: If $\mathbf{F} = \begin{bmatrix} F_1 \\ \vdots \\ F_n \end{bmatrix}$ is a smooth vector field on $\mathbb{R}^n$, we set

$$\operatorname{div} \mathbf{F} = \frac{\partial F_1}{\partial x_1} + \frac{\partial F_2}{\partial x_2} + \cdots + \frac{\partial F_n}{\partial x_n}.$$

Proposition 3.5 *Let $\mathbf{F}$ be a smooth vector field on $U \subset \mathbb{R}^n$, let $\boldsymbol{\phi}_t$ denote the flow of $\mathbf{F}$, and let $\Omega \subset U$ be a compact region with piecewise smooth boundary. Let $\mathcal{V}(t) = \operatorname{vol}(\boldsymbol{\phi}_t(\Omega))$. Then $\dot{\mathcal{V}}(0) = \int_\Omega \operatorname{div} \mathbf{F}\, dV.$*

Remark Using (the obvious generalization of) the Divergence Theorem, Theorem 6.2 of Chapter 8, we have the intuitively appealing result that $\dot{\mathcal{V}}(0) = \int_{\partial \Omega} \mathbf{F} \cdot \mathbf{n}\, dS$. That is, what causes net increase in the volume of the region is flow across its boundary.

► **EXAMPLE 11**

a. In Figure 3.2, we see the flow of the unit square under the vector field $\mathbf{F}\begin{pmatrix} x \\ y \end{pmatrix} = \begin{bmatrix} 2x + y \\ 5x - 2y \end{bmatrix}$. Note that area is preserved under the flow, as $\operatorname{div} \mathbf{F} = 0$.

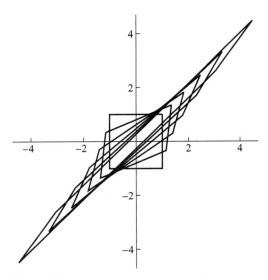

Figure 3.2

b. In Figure 3.3 (with thanks to John Polking's MATLAB software pplane5), we see the flow of certain regions Ω. In (a), the region expands (as $\operatorname{div} \mathbf{F} > 0$), whereas in (b) the region maintains its area (as $\operatorname{div} \mathbf{F} = 0$). ◄

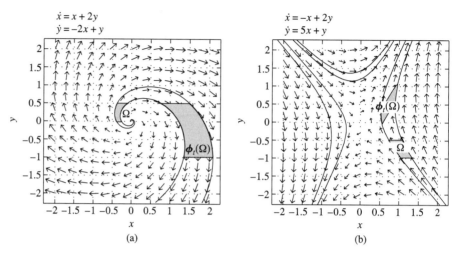

Figure 3.3

Proof We have
$$\mathcal{V}(t) = \int_\Omega \boldsymbol{\phi}_t^*(dx_1 \wedge \cdots \wedge dx_n) = \int_\Omega d(\boldsymbol{\phi}_t)_1 \wedge \cdots \wedge d(\boldsymbol{\phi}_t)_n.$$

By Exercise 7.2.20, we have
$$\dot{\mathcal{V}}(0) = \int_\Omega \frac{\partial}{\partial t}\Big|_{t=0} d(\boldsymbol{\phi}_t)_1 \wedge \cdots \wedge d(\boldsymbol{\phi}_t)_n.$$

Now the fact that mixed partials are equal tells us that $\frac{\partial^2 \boldsymbol{\phi}_t}{\partial t \partial x_i} = \frac{\partial^2 \boldsymbol{\phi}_t}{\partial x_i \partial t}$, and so $\frac{\partial}{\partial t}(d\boldsymbol{\phi}_t) = d\left(\frac{\partial \boldsymbol{\phi}_t}{\partial t}\right) = d(\dot{\boldsymbol{\phi}}_t)$. Moreover, $\dot{\boldsymbol{\phi}}_0(\mathbf{x}) = \mathbf{F}(\mathbf{x})$ (since $\dot{\boldsymbol{\phi}}_t(\mathbf{x}) = \mathbf{F}(\boldsymbol{\phi}_t(\mathbf{x}))$), and $\boldsymbol{\phi}_0(\mathbf{x}) = \mathbf{x}$, so the latter integral can be rewritten

$$\dot{\mathcal{V}}(0) = \int_\Omega \big(dF_1 \wedge dx_2 \wedge \cdots \wedge dx_n + dx_1 \wedge dF_2 \wedge dx_3 \wedge \cdots \wedge dx_n$$
$$+ \cdots + dx_1 \wedge \cdots \wedge dx_{n-1} \wedge dF_n \big)$$
$$= \int_\Omega \operatorname{div} \mathbf{F} \, dx_1 \wedge \cdots \wedge dx_n,$$

as required. ∎

▶ EXERCISES 9.3

1. Let $A = \begin{bmatrix} 2 & 5 \\ 1 & -2 \end{bmatrix}$. Calculate A^k for all $k \geq 1$.

*2. Suppose each of two tubs contains two bottles of beer; two are Budweiser and two are Beck's. Each minute, Fraternity Freddy picks a bottle of beer at random from each tub and replaces it in the other tub. After a long time, what portion of the time will there be exactly one bottle of Beck's in the first tub? At least one bottle of Beck's? (Hint: Let $\mathbf{x}_k$ be the vector whose entries are, respectively, the probabilities that there are two Beck's, one of each, or two Buds in the first tub.)

*3. Gambling Gus has $200 and plays a game where he must continue playing until he has either lost all his money or doubled it. In each game, he has a 2/5 chance of winning $100 and a 3/5 chance of losing $100. What is the probability that he eventually loses all his money? (Warning: Calculator or computer suggested.)

*4. If $a_0 = 2$, $a_1 = 3$, and $a_{k+1} = 3a_k - 2a_{k-1}$, for all $k \geq 1$, use methods of linear algebra to determine the formula for a_k.

5. If $a_0 = a_1 = 1$ and $a_{k+1} = a_k + 6a_{k-1}$ for all $k \geq 1$, use methods of linear algebra to determine the formula for a_k.

6. Suppose $a_0 = 0$, $a_1 = 1$, and $a_{k+1} = 3a_k + 4a_{k-1}$ for all $k \geq 1$. Use methods of linear algebra to find an explicit formula for a_k.

7. If $a_0 = 0$, $a_1 = 1$, and $a_{k+1} = 4a_k - 4a_{k-1}$ for all $k \geq 1$, use methods of linear algebra to determine the formula for a_k. (Hint: The matrix will not be diagonalizable, but you can get close if you stare at Exercise 9.2.16.)

*8. If $a_0 = 0$, $a_1 = a_2 = 1$, and $a_{k+1} = 2a_k + a_{k-1} - 2a_{k-2}$ for $k \geq 2$, use methods of linear algebra to determine the formula for a_k.

9. Consider the cat/mouse population problem studied in Example 1. Solve the following versions, including an investigation of the dependence on the original populations.

(a) $c_{k+1} = 0.7c_k + 0.1m_k$
$m_{k+1} = -0.2c_k + m_k$

(c) $c_{k+1} = 1.1c_k + 0.3m_k$
$m_{k+1} = 0.1c_k + 0.9m_k$

*(b) $c_{k+1} = 1.3c_k + 0.2m_k$
$m_{k+1} = -0.1c_k + m_k$

What conclusions do you draw?

10. Check that if A is an $n \times n$ matrix and the $n \times n$ differentiable matrix function $E(t)$ satisfies $\dot{E}(t) = AE(t)$ and $E(0) = I$, then $E(t) = e^{tA}$ for all $t \in \mathbb{R}$.

11. Calculate e^{tA} and use your answer to solve $\dot{\mathbf{x}}(t) = A\mathbf{x}(t)$, $\mathbf{x}(0) = \mathbf{x}_0$.

*(a) $A = \begin{bmatrix} 1 & 5 \\ 2 & 4 \end{bmatrix}$, $\mathbf{x}_0 = \begin{bmatrix} 6 \\ -1 \end{bmatrix}$

*(d) $A = \begin{bmatrix} 1 & 1 \\ -1 & 3 \end{bmatrix}$, $\mathbf{x}_0 = \begin{bmatrix} 2 \\ -1 \end{bmatrix}$

(b) $A = \begin{bmatrix} 0 & 1 \\ 1 & 0 \end{bmatrix}$, $\mathbf{x}_0 = \begin{bmatrix} 1 \\ 3 \end{bmatrix}$

*(e) $A = \begin{bmatrix} -1 & 1 & 2 \\ 1 & 2 & 1 \\ 2 & 1 & -1 \end{bmatrix}$, $\mathbf{x}_0 = \begin{bmatrix} 2 \\ 0 \\ 4 \end{bmatrix}$

(c) $A = \begin{bmatrix} 1 & 3 \\ 3 & 1 \end{bmatrix}$, $\mathbf{x}_0 = \begin{bmatrix} 5 \\ 1 \end{bmatrix}$

(f) $A = \begin{bmatrix} 1 & -2 & 2 \\ -1 & 0 & -1 \\ 0 & 2 & -1 \end{bmatrix}$, $\mathbf{x}_0 = \begin{bmatrix} 3 \\ -1 \\ -4 \end{bmatrix}$

12. Solve $\ddot{\mathbf{x}} = A\mathbf{x}$, $\mathbf{x}(0) = \mathbf{x}_0$, $\dot{\mathbf{x}}(0) = \dot{\mathbf{x}}_0$.

*(a) $A = \begin{bmatrix} 1 & 5 \\ 2 & 4 \end{bmatrix}$, $\mathbf{x}_0 = \begin{bmatrix} 7 \\ 0 \end{bmatrix}$, $\dot{\mathbf{x}}_0 = \begin{bmatrix} -5 \\ 2 \end{bmatrix}$

(b) $A = \begin{bmatrix} 0 & 1 \\ 1 & 0 \end{bmatrix}$, $\mathbf{x}_0 = \begin{bmatrix} -2 \\ 2 \end{bmatrix}$, $\dot{\mathbf{x}}_0 = \begin{bmatrix} 1 \\ 3 \end{bmatrix}$

(c) $A = \begin{bmatrix} 1 & 3 \\ 3 & 1 \end{bmatrix}$, $\mathbf{x}_0 = \begin{bmatrix} -2 \\ 4 \end{bmatrix}$, $\dot{\mathbf{x}}_0 = \begin{bmatrix} 2 - 3\sqrt{2} \\ 2 + 3\sqrt{2} \end{bmatrix}$

*(d) $A = \begin{bmatrix} 0 & 1 \\ 0 & 0 \end{bmatrix}$, $\mathbf{x}_0 = \begin{bmatrix} 1 \\ 2 \end{bmatrix}$, $\dot{\mathbf{x}}_0 = \begin{bmatrix} 2 \\ 1 \end{bmatrix}$

13. Find the motion of the two-mass, three-spring system in Example 9 when
    (a) $m_1 = m_2 = 1$ and $k_1 = k_3 = 1$, $k_2 = 3$,
    (b) $m_1 = m_2 = 1$ and $k_1 = 1$, $k_2 = 2$, $k_3 = 4$,
    *(c) $m_1 = 1$, $m_2 = 2$, $k_1 = 1$, and $k_2 = k_3 = 2$.

*14. Let
$$J = \begin{bmatrix} 2 & 1 & \\ & 2 & 1 \\ & & 2 \end{bmatrix}.$$
Calculate e^{tJ}.

*15. By mimicking the proof of Theorem 3.4, convert the following second-order differential equations into first-order systems and use matrix exponentials to solve:
    (a) $\ddot{y}(t) - \dot{y}(t) - 2y(t) = 0$, $y(0) = -1$, $\dot{y}(0) = 4$,
    (b) $\ddot{y}(t) - 2\dot{y}(t) + y(t) = 0$, $y(0) = 1$, $\dot{y}(0) = 2$.

16. Let $a, b \in \mathbb{R}$. Convert the constant coefficient second-order differential equation
$$\ddot{y}(t) + a\dot{y}(t) + by(t) = 0$$
into a first-order system by letting $\mathbf{x}(t) = \begin{bmatrix} y(t) \\ \dot{y}(t) \end{bmatrix}$. Considering separately the cases $a^2 - 4b \neq 0$ and $a^2 - 4b = 0$, use matrix exponentials to find the general solution.

17. (a) Prove that for any square matrix A, $(e^A)^{-1} = e^{-A}$. (Hint: Show $(e^{tA})^{-1} = e^{-tA}$ for all $t \in \mathbb{R}$.)
    (b) Prove that if A is skew-symmetric (i.e., $A^\top = -A$), then e^A is an orthogonal matrix.
    (c) Prove that when the eigenvalues of A are real, $\det(e^A) = e^{\operatorname{tr} A}$. (Hint: Prove the result when A is diagonalizable and then use continuity to establish it in general. Alternatively, apply Exercise 9.2.22.)

18. Consider the mapping $\exp\colon \mathcal{M}_{n \times n} \to \mathcal{M}_{n \times n}$ given by $\exp(A) = e^A$. By Exercise 17, e^A is always invertible.
    (a) Use the Inverse Function Theorem to show that for every matrix B sufficiently close to I, there is a unique A sufficiently close to O so that $e^A = B$.
    (b) Can the matrices $\begin{bmatrix} -1 & \\ & -1 \end{bmatrix}$ and $\begin{bmatrix} -2 & \\ & 1 \end{bmatrix}$ be written in the form e^A for some A?

19. Use Proposition 3.5 to deduce that the derivative with respect to r of the volume of a ball of radius r (in $\mathbb{R}^n$) is the volume (surface area) of the sphere of radius r.

20. It can be proved using (a generalization of) the Contraction Mapping Principle, Theorem 1.2 of Chapter 6, that when $\mathbf{F}$ is a smooth vector field, given $\mathbf{a}$, there are $\delta, \varepsilon > 0$ so that the differential equation $\dot{\mathbf{x}}(t) = \mathbf{F}(\mathbf{x}(t))$, $\mathbf{x}(0) = \mathbf{x}_0$, has a *unique* solution for all $\mathbf{x}_0 \in B(\mathbf{a}, \delta)$ and defined for all $|t| < \varepsilon$.
    (a) Assuming this result, prove that whenever $|s|$, $|t|$, and $|s+t| < \varepsilon$, we have $\boldsymbol{\phi}_{s+t} = \boldsymbol{\phi}_s \circ \boldsymbol{\phi}_t$. (Hint: Fix $t = t_0$ and vary s.)
    (b) Deduce that $\boldsymbol{\phi}_{-t} = (\boldsymbol{\phi}_t)^{-1}$.
    (c) By considering the example $F(x) = \sqrt{|x|}$, show that uniqueness may fail when the vector field isn't smooth. Indeed, show that the initial value problem $\dot{x}(t) = \sqrt{|x(t)|}$, $x(0) = 0$, has infinitely many solutions.

21. Generalizing Proposition 3.5 somewhat, prove that $\dot{V}(t) = \int_{\phi_t(\Omega)} \text{div } \mathbf{F} dV$. (Hint: Use Exercise 20 and the proposition as stated.)

22. (a) Show that the space-derivative of the flow ϕ_t satisfies the *first variation equation*
$$(D\phi_t(\mathbf{x}))^{\cdot} = D\mathbf{F}(\phi_t(\mathbf{x}))D\phi_t(\mathbf{x}), \qquad D\phi_0(\mathbf{x}) = I.$$
(b) For fixed $\mathbf{x}$, let $J(t) = \det(D\phi_t(\mathbf{x}))$. Using Exercise 7.5.23, show that
$$\dot{J}(t) = \text{div } \mathbf{F}(\phi_t(\mathbf{x}))J(t).$$
Deduce that $J(t) = e^{\int_0^t \text{div } \mathbf{F}(\phi_s(\mathbf{x}))ds}$.

4 THE SPECTRAL THEOREM

We now turn to the study of a large class of diagonalizable matrices, the symmetric matrices. Recall that a square matrix A is symmetric when $A = A^\top$. To begin our exploration, let's start with a general *symmetric* 2×2 matrix
$$A = \begin{bmatrix} a & b \\ b & c \end{bmatrix},$$
whose characteristic polynomial is $p(t) = t^2 - (a+c)t + (ac - b^2)$. By the quadratic formula, its eigenvalues are
$$\lambda = \frac{(a+c) \pm \sqrt{(a+c)^2 - 4(ac - b^2)}}{2} = \frac{(a+c) \pm \sqrt{(a-c)^2 + 4b^2}}{2}.$$
Only when A is diagonal are the eigenvalues not distinct. Thus, A is diagonalizable. Moreover, the corresponding eigenvectors are
$$\mathbf{v}_1 = \begin{bmatrix} b \\ \lambda_1 - a \end{bmatrix} \quad \text{and} \quad \mathbf{v}_2 = \begin{bmatrix} \lambda_2 - c \\ b \end{bmatrix};$$
note that
$$\mathbf{v}_1 \cdot \mathbf{v}_2 = b(\lambda_2 - c) + (\lambda_1 - a)b = b(\lambda_1 + \lambda_2 - a - c) = 0,$$
and so the eigenvectors are orthogonal. Since there is an orthogonal basis for $\mathbb{R}^2$ consisting of eigenvectors of A, we of course have an orthonormal basis for $\mathbb{R}^2$ consisting of eigenvectors of A. That is, by an appropriate rotation of the usual basis, we obtain a diagonalizing basis for A.

▶ **EXAMPLE 1**

The eigenvalues of
$$A = \begin{bmatrix} 1 & 2 \\ 2 & -2 \end{bmatrix}$$

are $\lambda_1 = 2$ and $\lambda_2 = -3$, with corresponding eigenvectors

$$\mathbf{v}_1 = \begin{bmatrix} 2 \\ 1 \end{bmatrix} \quad \text{and} \quad \mathbf{v}_2 = \begin{bmatrix} -1 \\ 2 \end{bmatrix}.$$

By normalizing the vectors, we obtain an orthonormal basis

$$\mathbf{q}_1 = \frac{1}{\sqrt{5}} \begin{bmatrix} 2 \\ 1 \end{bmatrix}, \quad \mathbf{q}_2 = \frac{1}{\sqrt{5}} \begin{bmatrix} -1 \\ 2 \end{bmatrix}.$$

See Figure 4.1. ◀

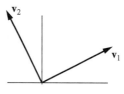

Figure 4.1

From Proposition 4.5 of Chapter 1 we recall that for all $\mathbf{x}, \mathbf{y} \in \mathbb{R}^n$ and $n \times n$ matrices A we have

$$A\mathbf{x} \cdot \mathbf{y} = \mathbf{x} \cdot A^\mathsf{T}\mathbf{y}.$$

In particular, when A is symmetric,

$$\boxed{A\mathbf{x} \cdot \mathbf{y} = \mathbf{x} \cdot A\mathbf{y}.}$$

More generally, we say a linear map $T \colon \mathbb{R}^n \to \mathbb{R}^n$ is *symmetric* if $T(\mathbf{x}) \cdot \mathbf{y} = \mathbf{x} \cdot T(\mathbf{y})$ for all $\mathbf{x}, \mathbf{y} \in \mathbb{R}^n$. It is easy to see that the matrix for a symmetric linear map with respect to *any* orthonormal basis is symmetric.

In general, we have the following important result. Its name comes from the word *spectrum*, associated with the physical concept of decomposing light into its components of different colors.

Theorem 4.1 (Spectral Theorem) *Let $T \colon \mathbb{R}^n \to \mathbb{R}^n$ be a symmetric linear map. Then*

1. *The eigenvalues of T are real.*

2. *There is an orthonormal basis for $\mathbb{R}^n$ consisting of eigenvectors of T. That is, if A is the standard matrix for T, then there is an orthogonal matrix Q so that $Q^{-1}AQ = \Lambda$ is diagonal.*

Proof We proceed by induction on n. The case $n = 1$ is automatic. Now assume that the result is true for all symmetric linear maps $T' \colon \mathbb{R}^{n-1} \to \mathbb{R}^{n-1}$. Given a symmetric linear map $T \colon \mathbb{R}^n \to \mathbb{R}^n$, we begin by proving that it has a real eigenvalue. We choose to use calculus to prove this, but for a purely linear-algebraic proof, see Exercise 16. Consider the function

$$f: \mathbb{R}^n \to \mathbb{R}, \quad f(\mathbf{x}) = A\mathbf{x} \cdot \mathbf{x} = \mathbf{x}^T A\mathbf{x}.$$

By compactness of the unit sphere, f has a maximum subject to the constraint $g(\mathbf{x}) = \|\mathbf{x}\|^2 = 1$. Applying the method of Lagrange multipliers, we infer that there is a unit vector $\mathbf{v}$ so that $Df(\mathbf{v}) = \lambda Dg(\mathbf{v})$ for some scalar λ. By Exercise 3.2.14, this means

$$A\mathbf{v} = \lambda \mathbf{v},$$

and so we've found an eigenvector of A; the Lagrange multiplier is the corresponding eigenvalue. (Incidentally, this was derived at the end of Section 4 of Chapter 5.)

By what we've just established, T has a real eigenvalue λ_1 and a corresponding eigenvector $\mathbf{v}_1$ of length 1. Note that if $\mathbf{w} \cdot \mathbf{v}_1 = 0$, then $T(\mathbf{w}) \cdot \mathbf{v}_1 = \mathbf{w} \cdot T(\mathbf{v}_1) = \lambda_1 \mathbf{w} \cdot \mathbf{v}_1 = 0$, so that $T(\mathbf{w}) \in W$ whenever $\mathbf{w} \in W$. Let $W = \bigl(\text{Span}(\mathbf{v}_1)\bigr)^\perp \subset \mathbb{R}^n$, and let $T' = T|_W$ be the restriction of T to W. Since $\dim W = n - 1$, it follows from our induction hypothesis that there is an orthonormal basis $\{\mathbf{v}_2, \ldots, \mathbf{v}_n\}$ for W consisting of eigenvectors of T'. Then $\{\mathbf{v}_1, \mathbf{v}_2, \ldots, \mathbf{v}_n\}$ is the requisite orthonormal basis for $\mathbb{R}^n$ since $T(\mathbf{v}_1) = \lambda_1 \mathbf{v}_1$ and $T(\mathbf{v}_i) = T'(\mathbf{v}_i) = \lambda_i \mathbf{v}_i$ for $i \geq 2$. ∎

▶ **EXAMPLE 2**

Consider the symmetric matrix

$$A = \begin{bmatrix} 0 & 1 & 1 \\ 1 & 1 & 0 \\ 1 & 0 & 1 \end{bmatrix}.$$

Its characteristic polynomial is $p(t) = -t^3 + 2t^2 + t - 2 = -(t+1)(t-1)(t-2)$, so the eigenvalues of A are -1, 1, and 2. As the reader can check, the corresponding eigenvectors are

$$\mathbf{v}_1 = \begin{bmatrix} -2 \\ 1 \\ 1 \end{bmatrix}, \quad \mathbf{v}_2 = \begin{bmatrix} 0 \\ -1 \\ 1 \end{bmatrix}, \quad \text{and} \quad \mathbf{v}_3 = \begin{bmatrix} 1 \\ 1 \\ 1 \end{bmatrix}.$$

Note that these three vectors form an orthogonal basis for $\mathbb{R}^3$, and we can easily obtain an orthonormal basis by normalizing:

$$\mathbf{q}_1 = \frac{1}{\sqrt{6}} \begin{bmatrix} -2 \\ 1 \\ 1 \end{bmatrix}, \quad \mathbf{q}_2 = \frac{1}{\sqrt{2}} \begin{bmatrix} 0 \\ -1 \\ 1 \end{bmatrix}, \quad \text{and} \quad \mathbf{q}_3 = \frac{1}{\sqrt{3}} \begin{bmatrix} 1 \\ 1 \\ 1 \end{bmatrix}.$$

The orthogonal diagonalizing matrix Q is therefore

$$Q = \begin{bmatrix} -\frac{2}{\sqrt{6}} & 0 & \frac{1}{\sqrt{3}} \\ \frac{1}{\sqrt{6}} & -\frac{1}{\sqrt{2}} & \frac{1}{\sqrt{3}} \\ \frac{1}{\sqrt{6}} & \frac{1}{\sqrt{2}} & \frac{1}{\sqrt{3}} \end{bmatrix}. \quad \blacktriangleleft$$

EXAMPLE 3

Consider the symmetric matrix

$$A = \begin{bmatrix} 5 & -4 & -2 \\ -4 & 5 & -2 \\ -2 & -2 & 8 \end{bmatrix}.$$

Its characteristic polynomial is $p(t) = -t^3 + 18t^2 - 81t = -t(t-9)^2$, so the eigenvalues of A are $0, 9,$ and 9. It is easy to check that

$$\mathbf{v}_1 = \begin{bmatrix} 2 \\ 2 \\ 1 \end{bmatrix}$$

gives a basis for $\mathbf{E}(0) = \mathbf{N}(A)$. As for $\mathbf{E}(9)$, we find

$$A - 9I = \begin{bmatrix} -4 & -4 & -2 \\ -4 & -4 & -2 \\ -2 & -2 & -1 \end{bmatrix},$$

which has rank 1, and so, as the spectral theorem guarantees, $\mathbf{E}(9)$ is 2-dimensional, with basis

$$\mathbf{v}_2 = \begin{bmatrix} -1 \\ 1 \\ 0 \end{bmatrix} \quad \text{and} \quad \mathbf{v}_3 = \begin{bmatrix} -1 \\ 0 \\ 2 \end{bmatrix}.$$

If we want an orthogonal (or orthonormal) basis, we must use the Gram-Schmidt process, Theorem 5.3 of Chapter 5: We take $\mathbf{w}_2 = \mathbf{v}_2$ and let

$$\mathbf{w}_3 = \mathbf{v}_3 - \text{proj}_{\mathbf{w}_2} \mathbf{v}_3 = \begin{bmatrix} -1 \\ 0 \\ 2 \end{bmatrix} - \frac{1}{2}\begin{bmatrix} -1 \\ 1 \\ 0 \end{bmatrix} = \begin{bmatrix} -\frac{1}{2} \\ -\frac{1}{2} \\ 2 \end{bmatrix}.$$

It is convenient to eschew fractions, and so we let

$$\mathbf{w}'_3 = 2\mathbf{w}_3 = \begin{bmatrix} -1 \\ -1 \\ 4 \end{bmatrix}.$$

As a check, note that $\mathbf{v}_1, \mathbf{w}_2, \mathbf{w}'_3$ do in fact form an orthogonal basis. As before, if we want the orthogonal diagonalizing matrix Q, we take

$$\mathbf{q}_1 = \frac{1}{3}\begin{bmatrix} 2 \\ 2 \\ 1 \end{bmatrix}, \quad \mathbf{q}_2 = \frac{1}{\sqrt{2}}\begin{bmatrix} -1 \\ 1 \\ 0 \end{bmatrix}, \quad \text{and} \quad \mathbf{q}_3 = \frac{1}{3\sqrt{2}}\begin{bmatrix} -1 \\ -1 \\ 4 \end{bmatrix},$$

whence

$$Q = \begin{bmatrix} \frac{2}{3} & -\frac{1}{\sqrt{2}} & -\frac{1}{3\sqrt{2}} \\ \frac{2}{3} & \frac{1}{\sqrt{2}} & -\frac{1}{3\sqrt{2}} \\ \frac{1}{3} & 0 & \frac{4}{3\sqrt{2}} \end{bmatrix}.$$

We reiterate that repeated eigenvalues cause no problem with symmetric matrices.

We conclude this discussion with a comparison to our study of projections in Chapter 5. Note that if we write out $A = Q\Lambda Q^{-1} = Q\Lambda Q^\top$, we see that

$$A = \begin{bmatrix} | & | & & | \\ \mathbf{q}_1 & \mathbf{q}_2 & \cdots & \mathbf{q}_n \\ | & | & & | \end{bmatrix} \begin{bmatrix} \lambda_1 & & & \\ & \lambda_2 & & \\ & & \ddots & \\ & & & \lambda_n \end{bmatrix} \begin{bmatrix} - & \mathbf{q}_1^\top & - \\ - & \mathbf{q}_2^\top & - \\ & \vdots & \\ - & \mathbf{q}_n^\top & - \end{bmatrix}$$

$$= \sum_{i=1}^{n} \lambda_i \mathbf{q}_i \mathbf{q}_i^\top.$$

This is the so-called *spectral decomposition* of A: Multiplying by a symmetric matrix A is the same as taking a weighted sum (weighted by the eigenvalues) of projections onto the respective eigenspaces. This is, indeed, a beautiful result with many applications in higher mathematics and physics.

4.1 Conics and Quadric Surfaces

We now use the Spectral Theorem to analyze the equations of conic sections and quadric surfaces.

▶ **EXAMPLE 4**

Suppose we are given the quadratic equation

$$x_1^2 + 4x_1 x_2 - 2x_2^2 = 6$$

to graph. Then we notice that we can write the quadratic expression

$$x_1^2 + 4x_1 x_2 - 2x_2^2 = \begin{bmatrix} x_1 & x_2 \end{bmatrix} \begin{bmatrix} 1 & 2 \\ 2 & -2 \end{bmatrix} \begin{bmatrix} x_1 \\ x_2 \end{bmatrix} = \mathbf{x}^\top A \mathbf{x},$$

where

$$A = \begin{bmatrix} 1 & 2 \\ 2 & -2 \end{bmatrix}$$

is the symmetric matrix we analyzed in Example 1 above. Thus, we know that

$$A = Q\Lambda Q^\top, \quad \text{where} \quad Q = \frac{1}{\sqrt{5}} \begin{bmatrix} 2 & -1 \\ 1 & 2 \end{bmatrix} \quad \text{and} \quad \Lambda = \begin{bmatrix} 2 & 0 \\ 0 & -3 \end{bmatrix}.$$

So, if we make the substitution $\mathbf{y} = Q^\top \mathbf{x}$, then we have

$$\mathbf{x}^\top A \mathbf{x} = \mathbf{x}^\top (Q\Lambda Q^\top) \mathbf{x} = (Q^\top \mathbf{x})^\top \Lambda (Q^\top \mathbf{x}) = \mathbf{y}^\top \Lambda \mathbf{y} = 2y_1^2 - 3y_2^2.$$

Note that the conic is much easier to understand in the $y_1 y_2$-coordinates. Indeed, we recognize that the equation $2y_1^2 - 3y_2^2 = 6$ can be written in the form

$$\frac{y_1^2}{3} - \frac{y_2^2}{2} = 1,$$

from which we see that this is a hyperbola with asymptotes $y_2 = \pm\sqrt{\frac{2}{3}} y_1$, as pictured in Figure 4.2. Now recall that the $y_1 y_2$-coordinates are the coordinates with respect to the basis formed by the column vectors of Q. Thus, if we want to sketch the picture in the original x_1x_2-coordinates, we first draw in the basis vectors $\mathbf{q}_1$ and $\mathbf{q}_2$, and these establish the y_1- and y_2-axes, respectively, as shown in Figure 4.3. ◀

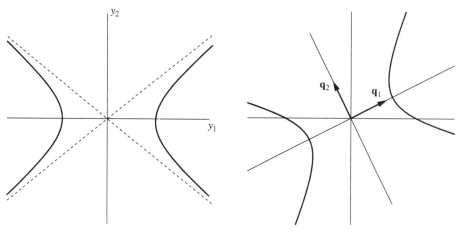

Figure 4.2   **Figure 4.3**

It's worth recalling that the equation

$$\frac{x_1^2}{a^2} + \frac{x_2^2}{b^2} = 1$$

represents an ellipse (with semiaxes a and b), whereas the equation

$$\frac{x_1^2}{a^2} - \frac{x_2^2}{b^2} = 1$$

represents a hyperbola with vertices $\begin{bmatrix} \pm a \\ 0 \end{bmatrix}$ and asymptotes $x_2 = \pm \frac{b}{a} x_1$.

Quadric surfaces include those shown in Figure 4.4: ellipsoids, cylinders, and hyperboloids of 1 and 2 sheets. There are also paraboloids (both elliptic and hyperbolic), but we come to these a bit later. We turn to another example.

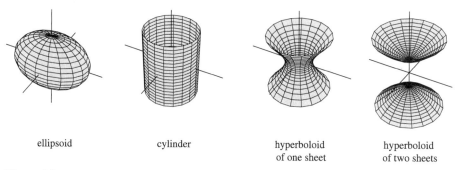

ellipsoid   cylinder   hyperboloid of one sheet   hyperboloid of two sheets

Figure 4.4

EXAMPLE 5

Consider the surface defined by the equation

$$2x_1x_2 + 2x_1x_3 + x_2^2 + x_3^2 = 2.$$

We observe that if

$$A = \begin{bmatrix} 0 & 1 & 1 \\ 1 & 1 & 0 \\ 1 & 0 & 1 \end{bmatrix}$$

is the symmetric matrix from Example 2, then

$$\mathbf{x}^T A \mathbf{x} = 2x_1x_2 + 2x_1x_3 + x_2^2 + x_3^2,$$

and so we use the diagonalization and the substitution $\mathbf{y} = Q^T\mathbf{x}$, as before, to write

$$\mathbf{x}^T A \mathbf{x} = \mathbf{y}^T \Lambda \mathbf{y}, \quad \text{where} \quad \Lambda = \begin{bmatrix} -1 & 0 & 0 \\ 0 & 1 & 0 \\ 0 & 0 & 2 \end{bmatrix};$$

that is, in terms of the coordinates $\mathbf{y} = \begin{bmatrix} y_1 \\ y_2 \\ y_3 \end{bmatrix}$, we have

$$2x_1x_2 + 2x_1x_3 + x_2^2 + x_3^2 = -y_1^2 + y_2^2 + 2y_3^2,$$

and the graph of $-y_1^2 + y_2^2 + 2y_3^2 = 2$ is the hyperboloid of one sheet shown in Figure 4.5. This is the picture with respect to the "new basis" $\{\mathbf{q}_1, \mathbf{q}_2, \mathbf{q}_3\}$ (given in the solution of Example 2). The picture with respect to the standard basis, then, is as shown in Figure 4.6. (This figure is obtained by multiplying by the matrix Q. Why?) ◄

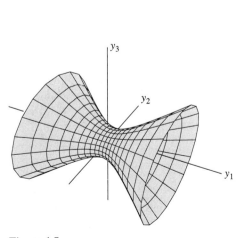

Figure 4.5

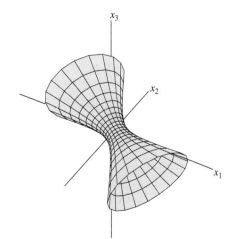

Figure 4.6

The alert reader may have noticed that we're lacking certain curves and surfaces. If there are linear terms present along with the quadratic, we must adjust accordingly. For example, we recognize that

$$x_1^2 + 2x_2^2 = 1$$

is the equation of an ellipse centered at the origin. Correspondingly, by completing the square, we see that

$$x_1^2 + 2x_1 + 2x_2^2 - 3x_2 = \tfrac{13}{2}$$

is the equation of a congruent ellipse centered at $\begin{bmatrix} -1 \\ 3/4 \end{bmatrix}$. However, the linear terms become all important when the symmetric matrix defining the quadratic terms is singular. For example,

$$x_1^2 - x_1 = 1$$

defines a pair of lines, whereas

$$x_1^2 - x_2 = 1$$

defines a parabola.

▶ **EXAMPLE 6**

We wish to sketch the surface

$$5x_1^2 - 8x_1x_2 - 4x_1x_3 + 5x_2^2 - 4x_2x_3 + 8x_3^2 + 2x_1 + 2x_2 + x_3 = 9.$$

No, we did not pull this mess out of a hat. The quadratic terms came, as might be predicted, from Example 3. Thus, we make the change of coordinates given by $\mathbf{y} = Q^T\mathbf{x}$, with

$$Q = \begin{bmatrix} \tfrac{2}{3} & -\tfrac{1}{\sqrt{2}} & -\tfrac{1}{3\sqrt{2}} \\ \tfrac{2}{3} & \tfrac{1}{\sqrt{2}} & -\tfrac{1}{3\sqrt{2}} \\ \tfrac{1}{3} & 0 & \tfrac{4}{3\sqrt{2}} \end{bmatrix}.$$

Since $\mathbf{x} = Q\mathbf{y}$, we have

$$2x_1 + 2x_2 + x_3 = \begin{bmatrix} 2 & 2 & 1 \end{bmatrix} Q\mathbf{y} = \begin{bmatrix} 2 & 2 & 1 \end{bmatrix} \begin{bmatrix} \tfrac{2}{3} & -\tfrac{1}{\sqrt{2}} & -\tfrac{1}{3\sqrt{2}} \\ \tfrac{2}{3} & \tfrac{1}{\sqrt{2}} & -\tfrac{1}{3\sqrt{2}} \\ \tfrac{1}{3} & 0 & \tfrac{4}{3\sqrt{2}} \end{bmatrix} \begin{bmatrix} y_1 \\ y_2 \\ y_3 \end{bmatrix} = 3y_1,$$

and so our given equation becomes, in the $y_1y_2y_3$-coordinates,

$$9y_2^2 + 9y_3^2 + 3y_1 = 9.$$

Rewriting this a bit, we have

$$y_1 = 3(1 - y_2^2 - y_3^2),$$

which we recognize as a (circular) paraboloid, shown in Figure 4.7. The sketch of the surface in our original $x_1x_2x_3$-coordinates is then shown in Figure 4.8. ◀

4 The Spectral Theorem ◀ **463**

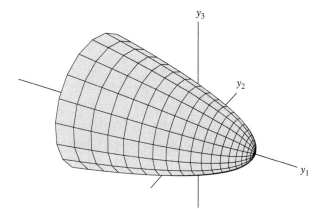

Figure 4.7

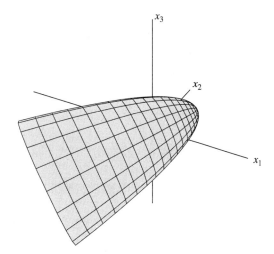

Figure 4.8

▶ EXERCISES 9.4

1. Find orthogonal matrices that diagonalize each of the following symmetric matrices:

*(a) $\begin{bmatrix} 6 & 2 \\ 2 & 9 \end{bmatrix}$

*(c) $\begin{bmatrix} 2 & 2 & -2 \\ 2 & -1 & -1 \\ -2 & -1 & -1 \end{bmatrix}$

(e) $\begin{bmatrix} 1 & -2 & 2 \\ -2 & 1 & 2 \\ 2 & 2 & 1 \end{bmatrix}$

(b) $\begin{bmatrix} 2 & 0 & 0 \\ 0 & 1 & -1 \\ 0 & -1 & 1 \end{bmatrix}$

*(d) $\begin{bmatrix} 3 & 2 & 2 \\ 2 & 2 & 0 \\ 2 & 0 & 4 \end{bmatrix}$

(f) $\begin{bmatrix} 1 & 0 & 1 & 0 \\ 0 & 1 & 0 & 1 \\ 1 & 0 & 1 & 0 \\ 0 & 1 & 0 & 1 \end{bmatrix}$

*2. Suppose A is a symmetric matrix with eigenvalues 2 and 5. If the vectors $\begin{bmatrix} 1 \\ 1 \\ -1 \end{bmatrix}$ and $\begin{bmatrix} 1 \\ -1 \\ 0 \end{bmatrix}$ span the 5-eigenspace, what is $A \begin{bmatrix} 1 \\ 1 \\ 2 \end{bmatrix}$? Give your reasoning.

3. A symmetric matrix A has eigenvalues 1 and 2. Find A if $\begin{bmatrix} 1 \\ 1 \\ 1 \end{bmatrix}$ spans $\mathbf{E}(2)$.

4. Suppose A is symmetric, $A \begin{bmatrix} 1 \\ 1 \end{bmatrix} = \begin{bmatrix} 2 \\ 2 \end{bmatrix}$, and $\det A = 6$. Give the matrix A. Explain your reasoning clearly. (Hint: What are the eigenvalues of A?)

*5. Prove that if λ is the only eigenvalue of a symmetric matrix A, then $A = \lambda I$.

6. Decide (as efficiently as possible) which of the following matrices are diagonalizable. Give your reasoning.

$$A = \begin{bmatrix} 5 & 0 & 2 \\ 0 & 5 & 0 \\ 0 & 0 & 5 \end{bmatrix}, \quad B = \begin{bmatrix} 5 & 0 & 2 \\ 0 & 5 & 0 \\ 2 & 0 & 5 \end{bmatrix},$$

$$C = \begin{bmatrix} 1 & 2 & 4 \\ 0 & 2 & 2 \\ 0 & 0 & 3 \end{bmatrix}, \quad D = \begin{bmatrix} 1 & 2 & 4 \\ 0 & 2 & 2 \\ 0 & 0 & 1 \end{bmatrix}.$$

7. Suppose A is a diagonalizable matrix whose eigenspaces are orthogonal. Prove that A is symmetric.

8. Suppose A is a symmetric $n \times n$ matrix. Using the spectral theorem, prove that if $A\mathbf{x} \cdot \mathbf{x} = 0$ for every vector $\mathbf{x} \in \mathbb{R}^n$, then $A = O$.

9. Apply the spectral theorem to prove that any symmetric matrix A satisfying $A^2 = A$ is in fact a projection matrix.

10. Suppose T is a symmetric linear map satisfying $[T]^4 = I$. Use the spectral theorem to give a complete description of $T : \mathbb{R}^n \to \mathbb{R}^n$. (Hint: For starters, what are the potential eigenvalues of T?)

11. Let A be an $m \times n$ matrix. Show that $\|A\| = \sqrt{\lambda}$, where λ is the largest eigenvalue of the symmetric matrix $A^\mathsf{T} A$.

12. We say a symmetric matrix A is *positive definite* if $A\mathbf{x} \cdot \mathbf{x} > 0$ for all $\mathbf{x} \neq \mathbf{0}$, *negative definite* if $A\mathbf{x} \cdot \mathbf{x} < 0$ for all $\mathbf{x} \neq \mathbf{0}$, and *positive* (resp., *negative*) *semidefinite* if $A\mathbf{x} \cdot \mathbf{x} \geq 0$ (resp., ≤ 0) for all $\mathbf{x}$.

(a) Prove that if A and B are positive (negative) definite, then so is $A + B$.

(b) Prove that A is positive (resp., negative) definite if and only if all its eigenvalues are positive (resp., negative).

(c) Prove that A is positive (resp., negative) semidefinite if and only if all its eigenvalues are nonnegative (resp., nonpositive).

(d) Prove that if C is any $m \times n$ matrix of rank n, then $A = C^\mathsf{T} C$ has positive eigenvalues.

(e) Prove or give a counterexample: If A and B are positive definite, then so is AB. What about $AB + BA$?

13. Let A be an $n \times n$ matrix. Prove that A is nonsingular if and only if every eigenvalue of $A^\mathsf{T} A$ is positive.

14. Prove that if A is a positive semidefinite (symmetric) matrix, then there is a unique positive semidefinite (symmetric) matrix B with $B^2 = A$.

15. Suppose A and B are symmetric and $AB = BA$. Prove there is an orthogonal matrix Q so that both $Q^{-1}AQ$ and $Q^{-1}BQ$ are diagonal. (Hint: Let λ be an eigenvalue of A. Use the Spectral Theorem to show that there is an orthonormal basis for $\mathbf{E}(\lambda)$ consisting of eigenvectors of B.)

16. Prove, using only methods of linear algebra, that the eigenvalues of a symmetric matrix are real. (Hints: Let $\lambda = a + bi$ be a putative complex eigenvalue of A, and consider the *real* matrix
$$B = (A - (a+bi)I)(A - (a-bi)I) = A^2 - 2aA + (a^2 + b^2)I = (A - aI)^2 + b^2 I.$$
Show that B is singular, and that if $\mathbf{v} \in \mathbf{N}(B)$ is a nonzero vector, then $(A - aI)\mathbf{v} = \mathbf{0}$ and $b = 0$.)

17. If A is a positive definite symmetric $n \times n$ matrix, what is the volume of the n-dimensional ellipsoid $\{\mathbf{x} \in \mathbb{R}^n : A\mathbf{x} \cdot \mathbf{x} \le 1\}$? (See also Exercise 7.6.3.)

18. Sketch the following conic sections, giving axes of symmetry and asymptotes (if any).
    (a) $6x_1 x_2 - 8x_2^2 = 9$
    *(b) $3x_1^2 - 2x_1 x_2 + 3x_2^2 = 4$
    *(c) $16x_1^2 + 24x_1 x_2 + 9x_2^2 - 3x_1 + 4x_2 = 5$
    (d) $10x_1^2 + 6x_1 x_2 + 2x_2^2 = 11$
    (e) $7x_1^2 + 12x_1 x_2 - 2x_2^2 - 2x_1 + 4x_2 = 6$

19. Sketch the following quadric surfaces.
    *(a) $3x_1^2 + 2x_1 x_2 + 2x_1 x_3 + 4x_2 x_3 = 4$
    (b) $4x_1^2 - 2x_1 x_2 - 2x_1 x_3 + 3x_2^2 + 4x_2 x_3 + 3x_3^2 = 6$
    (c) $-x_1^2 + 2x_2^2 - x_3^2 - 4x_1 x_2 - 10x_1 x_3 + 4x_2 x_3 = 6$
    *(d) $2x_1^2 + 2x_1 x_2 + 2x_1 x_3 + 2x_2 x_3 - x_1 + x_2 + x_3 = 1$
    (e) $3x_1^2 + 4x_1 x_2 + 8x_1 x_3 + 4x_2 x_3 + 3x_3^2 = 8$
    (f) $3x_1^2 + 2x_1 x_3 - x_2^2 + 3x_3^2 + 2x_2 = 0$

20. Let $a, b, c \in \mathbb{R}$, and let $\mathcal{Q}(\mathbf{x}) = ax_1^2 + 2bx_1 x_2 + cx_2^2$.
(a) The Spectral Theorem tells us that there exists an orthonormal basis for $\mathbb{R}^2$ with respect to whose coordinates y_1, y_2 we have
$$\mathcal{Q}(\mathbf{x}) = \tilde{\mathcal{Q}}(\mathbf{y}) = \lambda y_1^2 + \mu y_2^2.$$
In high school analytic geometry, one derives the formula
$$\cot 2\alpha = \frac{a-c}{2b}$$
for the angle α through which we must rotate the $x_1 x_2$-axes to get the appropriate $y_1 y_2$-axes. Derive this by using eigenvalues and eigenvectors, and determine the type (ellipse, hyperbola, etc.) of the conic section $\mathcal{Q}(\mathbf{x}) = 1$ from a, b, and c. (Hint: Use the characteristic polynomial to eliminate λ^2 in your computation of $\tan 2\alpha$.)
(b) Use the formula for $\tilde{\mathcal{Q}}$ above to find the maximum and minimum of $\mathcal{Q}$ on the unit circle $\|\mathbf{x}\| = 1$.

21. In this exercise we consider the nature of the restriction of a quadratic form to a hyperplane. Let A be a symmetric $n \times n$ matrix.
(a) Show that the quadratic form $\mathcal{Q}(\mathbf{x}) = \mathbf{x}^\mathsf{T} A \mathbf{x}$ on $\mathbb{R}^n$ is positive definite when restricted to the subspace $x_n = 0$ if and only if all the roots of

$$\begin{vmatrix} & & & 0 \\ & A - tI & & \vdots \\ & & & 0 \\ & & & 1 \\ \hline 0 & \cdots & 0 & 1 & 0 \end{vmatrix} = 0$$

are positive.

(b) Use the change-of-basis theorem to prove that the restriction to the subspace $\mathbf{b} \cdot \mathbf{x} = 0$ is positive definite if and only if all the roots of

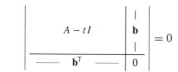

are positive.

(c) Use this result to give a bordered Hessian test for the point $\mathbf{a}$ to be a constrained maximum (minimum) of the function f subject to the constraint $g = c$. (See Exercises 5.4.34 and 5.4.32b.)

(d) What is the analogous result for an arbitrary subspace?

22. We saw in Section 3 of Chapter 5 that we can write a symmetric $n \times n$ matrix A in the form $A = LDL^\mathsf{T}$ (where L is lower triangular with diagonal entries 1 and D is diagonal); we saw in this section that we can write $A = Q\Lambda Q^\mathsf{T}$ for some orthogonal matrix Q. Although the diagonal entries of D obviously need not be the eigenvalues of A, the point of this exercise is to see that the signs of these numbers must agree. That is, the number of positive entries in D equals the number of positive eigenvalues of A, the number of negative entries in D equals the number of negative eigenvalues of A, and the number of zero (diagonal) entries in D equals the number of zero eigenvalues.

(a) Assume first that A is nonsingular. Consider the "straight line path" joining I and L (stick a parameter s in front of the nondiagonal entries of L and let s vary from 0 to 1). We then obtain a path in $\mathcal{M}_{n \times n}$ joining D and A. Show that all the matrices in this path are nonsingular and, applying Exercise 8.7.9, show that the number of positive eigenvalues of D equals the number of positive eigenvalues of A. Deduce the result in this case.

(b) In general, prove that the number of zero diagonal entries in D is equal to $\dim \mathbf{N}(A) = \dim \mathbf{E}(0)$. By considering the matrix $A + \varepsilon I$ for $\varepsilon > 0$ sufficiently small, use part a to deduce the result.

Remark Comparing Proposition 3.5 of Chapter 5 with Exercise 12 above, we can easily derive the result of this exercise when A is either positive or negative definite. But the indefinite case is more subtle.

GLOSSARY OF NOTATIONS AND RESULTS FROM SINGLE-VARIABLE CALCULUS

▶ Notations

Notation	Definition	Discussion/Page reference
$\in$	is an element of	$x \in X$ means that x belongs to the set X
$\subset$	subset	$X \subset Y$ means that every element x of X belongs to Y as well; two sets X and Y are equal if $X \subset Y$ and $Y \subset X$
$\subsetneq$	proper subset	$X \subsetneq Y$ means that $X \subset Y$ and $X \neq Y$
$\implies$	implies	$P \implies Q$ means that whenever P is true, Q must be as well
$\iff$	if and only if	$P \iff Q$ means $P \implies Q$ and $Q \implies P$
$\leadsto$	gives by row operations	See p. 130
$\binom{n}{k}$	binomial coefficient	171
$\dfrac{\partial f}{\partial x_j}$	partial derivative of f with respect to x_j	82
$\dfrac{\partial^2 f}{\partial x_j \partial x_i}$	second-order partial derivative	120
$\nabla^2 f$	Laplacian of f	122
∇f	gradient of f	104
$\int_R f\,dA,\ \int_R f\,dV$	integral of f over R	268
$\wedge$	wedge product	338
∂R	boundary of R	358
$\mathbf{A}_i$	i^{th} row vector of the matrix A	28
$\mathbf{a}_j$	j^{th} column vector of the matrix A	28
A^{-1}	inverse of the matrix A	34
$A^\top$	transpose of the matrix A	36
A_θ	matrix giving rotation through angle θ	27

Notation	Definition	Discussion/Page reference
$\overrightarrow{AB}$	vector corresponding to the directed line segment from A to B	1
AB	product of the matrices A and B	31
$A\mathbf{x}$	product of the matrix A and the vector $\mathbf{x}$	28
A_{ij}	$(n-1) \times (n-1)$ matrix obtained by deleting the ith row and the jth column from the $n \times n$ matrix A	315
$B(\mathbf{a}, \delta)$	ball	65
$\overline{B}(\mathbf{a}, \delta)$	closed ball	70
$\mathcal{B}$	basis	413
$\mathcal{C}^1, \mathcal{C}^k, \mathcal{C}^\infty$	continuously differentiable, smooth functions	93, 120, 120, 167
$C(A)$	column space of the matrix A	171
$C_\mathcal{B}$	coordinates with respect to a basis $\mathcal{B}$	415
c_{ij}	ij^{th} cofactor	315
curl $\mathbf{F}$	curl of vector field $\mathbf{F}$	393
d	(exterior) derivative	340
$D\mathbf{f}(\mathbf{a})$	derivative of $\mathbf{f}$ at $\mathbf{a}$	87
$D_\mathbf{v}\mathbf{f}(\mathbf{a})$	directional derivative of $\mathbf{f}$ at $\mathbf{a}$ in direction $\mathbf{v}$	83
det A	determinant of the square matrix A	309
div $\mathbf{F}$	divergence of vector field $\mathbf{F}$	395
$\{\mathbf{e}_1, \ldots, \mathbf{e}_n\}$	standard basis for $\mathbb{R}^n$	19, 162
$\mathbf{E}(\lambda)$	λ-eigenspace	423
e^A	exponential of the square matrix A	441
$f\begin{pmatrix} x_1 \\ \vdots \\ x_n \end{pmatrix}$	function of a vector variable	60
$\tilde{f}$	extension of f by 0	272
$\overline{\mathbf{f}}$	average value of $\mathbf{f}$	300
$\mathbf{g}^*\omega$	pullback of ω by $\mathbf{g}$	342
graph(f)	graph of the function f	57
Hess(f)	Hessian matrix of f	209
$\mathcal{H}_{f,\mathbf{a}}$	quadratic form associated to Hessian of f	209
I	identity matrix, moment of inertia	303
I_n	$n \times n$ identity matrix	34
image(T)	image of a linear transformation T	172
$\kappa(s)$	curvature	115
ker(T)	kernel of a linear transformation T	172
$\lim_{\mathbf{x} \to \mathbf{a}} \mathbf{f}(\mathbf{x})$	limit of $\mathbf{f}(\mathbf{x})$ as $\mathbf{x}$ approaches $\mathbf{a}$	72
$\ell(\mathbf{g})$	arclength of $\mathbf{g}$	112
$L(f, \mathcal{P})$	lower sum of f with respect to partition $\mathcal{P}$	267

Glossary

Notation	Definition	Discussion/Page reference
$\Lambda^k(\mathbb{R}^n)^*$	vector space of alternating multilinear functions from $(\mathbb{R}^n)^k$ to $\mathbb{R}$	337
$\mathcal{M}_{m\times n}$	vector space of $m \times n$ matrices	167
μ_A	linear transformation defined by multiplication by A	28
$\mathbf{n}$	outward-pointing unit normal	364
$\mathbf{N}(A)$	nullspace of the matrix A	172
$\mathbf{N}(s)$	principal normal vector	115
ω	differential form	339
$\star\omega$	star operator	346
Ω	region	273
$\mathcal{P}$	plane, parallelogram, or partition	16, 43, 267
$\mathcal{P}_k$	vector space of polynomials of degree $\leq k$	168
$p_A(t)$	characteristic polynomial of the matrix A	426
$\text{proj}_\mathbf{y}\mathbf{x}$	projection of $\mathbf{x}$ onto $\mathbf{y}$	10
$\text{proj}_V\mathbf{b}$	projection of $\mathbf{b}$ onto the subspace V	226
$\mathcal{Q}$	quadratic form	210
r, θ	polar coordinates	288
r, θ, z	cylindrical coordinates	292
ρ, ϕ, θ	spherical coordinates	294
$\mathbb{R}^n$	(real) n-dimensional space	1
$(\mathbb{R}^n)^*$	vector space of linear maps from $\mathbb{R}^n$ to $\mathbb{R}$	335
R	rectangle	65
$\mathbf{R}(A)$	row space of the matrix A	171
$\rho(\mathbf{x})$	rotation of $\mathbf{x} \in \mathbb{R}^2$ through angle $\pi/2$	14
S^{n-1}	unit sphere	403
$\text{Span}(\mathbf{v}_1,\ldots,\mathbf{v}_k)$	span of $\mathbf{v}_1,\ldots,\mathbf{v}_k$	19
$\sup S$	least upper bound (supremum) of S	69
$\overline{S}$	closure of the subset S	70
T	linear map (or transformation)	24
$[T]$	standard matrix of linear map T	24
$\|T\|$	norm of the linear map T	97
$\|T\|_\square$	cubical norm of the linear map T	324
$\mathbf{T}(s)$	unit tangent vector	115
$\text{tr}\,A$	trace of the matrix A	41
$U(f,\mathcal{P})$	upper sum of f with respect to partition $\mathcal{P}$	267
$U+V$	sum of the subspaces U and V	23
$U\cap V$	intersection of the subspaces U and V	23
$V^\perp$	orthogonal complement of subspace V	22
$\mathbf{x}\times\mathbf{y}$	cross product of the vectors $\mathbf{x},\mathbf{y}\in\mathbb{R}^3$	48
$\|\mathbf{x}\|$	length of the vector $\mathbf{x}$	2
$\mathbf{x}_k$	sequence	66

Notation	Definition	Discussion/Page reference
$\mathbf{x}_{k_j}$	subsequence	70
$\bar{\mathbf{x}}$	least squares solution	227
$\mathbf{x} \cdot \mathbf{y}$	dot product of the vectors $\mathbf{x}$ and $\mathbf{y}$	8
$\langle \mathbf{x}, \mathbf{y} \rangle$	inner product of the vectors $\mathbf{x}$ and $\mathbf{y}$	238
$\mathbf{x}^\|, \mathbf{x}^\perp$	components of $\mathbf{x}$ parallel to and orthogonal to another vector	9
$\mathbf{0}$	zero vector	1
O	zero matrix	30

▶ RESULTS FROM SINGLE-VARIABLE CALCULUS

Intermediate Value Theorem: Let $f: [a, b] \to \mathbb{R}$ be continuous. Then for any y between $f(a)$ and $f(b)$, there is $x \in [a, b]$ with $f(x) = y$.

Rolle's Theorem: Suppose $f: [a, b] \to \mathbb{R}$ is continuous, f is differentiable on (a, b), and $f(a) = f(b)$. Then there is $c \in (a, b)$ so that $f'(c) = 0$. *Proof:* By the maximum value theorem, Theorem 1.2 of Chapter 5, f takes on its maximum and minimum values on $[a, b]$. If f is constant on $[a, b]$, then $f'(c) = 0$ for all $c \in (a, b)$. If not, say $f(x) > f(a)$ for some $x \in (a, b)$, in which case f takes on a global maximum at some $c \in (a, b)$. Then $f'(c) = 0$ (by Lemma 2.1 of Chapter 5). Alternatively, $f(x) < f(a)$ for some $x \in (a, b)$, in which case f takes on a global minimum at some $c \in (a, b)$. Then in this case, as well, $f'(c) = 0$.

Mean Value Theorem: Suppose $f: [a, b] \to \mathbb{R}$ is continuous and f is differentiable on (a, b). Then there is $c \in (a, b)$ so that $f(b) - f(a) = f'(c)(b - a)$.

Fundamental Theorem of Calculus, Part I: Suppose f is continuous on $[a, b]$ and we set $F(x) = \int_a^x f(t)dt$. Then $F'(x) = f(x)$ for all $x \in (a, b)$.

Fundamental Theorem of Calculus, Part II: Suppose f is integrable on $[a, b]$ and $f = F'$. Then $\int_a^b f(x)dx = F(b) - F(a)$.

Basic Differentiation Formulas:

product rule: $(fg)' = f'g + fg'$

quotient rule: $(f/g)' = (f'g - fg')/g^2$

chain rule: $(f \circ g)' = (f' \circ g)g'$

Note: We use log to denote the natural logarithm (ln).

Function	Derivative
x^n	nx^{n-1}
e^x	e^x
$\log x$	$1/x$
$\sin x$	$\cos x$
$\cos x$	$-\sin x$
$\tan x$	$\sec^2 x$
$\sec x$	$\sec x \tan x$
$\cot x$	$-\csc^2 x$
$\csc x$	$-\csc x \cot x$
$\arcsin x$	$1/\sqrt{1-x^2}$
$\arctan x$	$1/(1+x^2)$

Basic Trigonometric Formulas:

$$\sin^2 \theta + \cos^2 \theta = 1 \qquad \tan^2 \theta + 1 = \sec^2 \theta \qquad \cot^2 \theta + 1 = \csc^2 \theta$$

$$\cos 2\theta = \cos^2 \theta - \sin^2 \theta = 2\cos^2 \theta - 1 = 1 - 2\sin^2 \theta$$

$$\sin 2\theta = 2 \sin \theta \cos \theta$$

law of cosines: $c^2 = a^2 + b^2 - 2ab \cos \gamma$

law of sines: $\dfrac{\sin \alpha}{a} = \dfrac{\sin \beta}{b} = \dfrac{\sin \gamma}{c}$

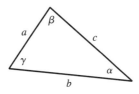

Basic Integration Formulas:

integration by parts: $\displaystyle\int f'(x)g(x)dx = f(x)g(x) - \int f(x)g'(x)dx$

integration by substitution: $\displaystyle\int f(g(x))g'(x)dx = F(g(x))$, where $F(u) = \int f(u)du$

Miscellaneous Integration Formulas:

$\displaystyle\int x^n dx = \dfrac{x^{n+1}}{n+1}, \quad n \neq -1$

$\displaystyle\int \dfrac{dx}{x} = \log |x|$

$\displaystyle\int \cos x \, dx = \sin x$

$\displaystyle\int \sin^2 x \, dx = \dfrac{1}{2}(x - \sin x \cos x)$

$\displaystyle\int \tan^2 x \, dx = \tan x - x$

$\displaystyle\int e^x dx = e^x$

$\displaystyle\int \sin x \, dx = -\cos x$

$\displaystyle\int \tan x \, dx = -\log |\cos x|$

$\displaystyle\int \cos^2 x \, dx = \dfrac{1}{2}(x + \sin x \cos x)$

$\displaystyle\int \sec x \, dx = \log |\sec x + \tan x|$

$$\int \sec^2 x\,dx = \tan x$$

$$\int \sec^3 x\,dx = \frac{1}{2}(\sec x \tan x + \log|\sec x + \tan x|)$$

$$\int \sin^3 x\,dx = -\cos x + \frac{1}{3}\cos^3 x$$

$$\int \cos^3 x\,dx = \sin x - \frac{1}{3}\sin^3 x$$

$$\int \tan^3 x\,dx = \frac{1}{2}\tan^2 x + \log|\cos x|$$

$$\int \frac{dx}{\sqrt{a^2 - x^2}} = \arcsin \frac{x}{a}$$

$$\int \frac{dx}{a^2 + x^2} = \frac{1}{a}\arctan \frac{x}{a}$$

$$\int \frac{dx}{a^2 - x^2} = \frac{1}{2}\log\left|\frac{x+a}{x-a}\right|$$

$$\int \sqrt{x^2 \pm a^2}\,dx = \frac{x}{2}\sqrt{x^2 \pm a^2} \pm \frac{a^2}{2}\log\left|x + \sqrt{x^2 \pm a^2}\right|$$

$$\int \frac{dx}{\sqrt{x^2 \pm a^2}} = \log\left|x + \sqrt{x^2 \pm a^2}\right|$$

$$\int \sqrt{a^2 - x^2}\,dx = \frac{x}{2}\sqrt{a^2 - x^2} + \frac{a^2}{2}\arcsin \frac{x}{a}$$

$$\int \log x\,dx = x\log x - x$$

$$\int e^{ax} \sin bx\,dx = \frac{e^{ax}}{a^2 + b^2}(a\sin bx - b\cos bx)$$

$$\int e^{ax} \cos bx\,dx = \frac{e^{ax}}{a^2 + b^2}(a\cos bx + b\sin bx)$$

▶ GREEK ALPHABET

alpha	α	iota	ι	rho	ρ
beta	β	kappa	κ	sigma	σ Σ
gamma	γ Γ	lambda	λ Λ	tau	τ
delta	δ Δ	mu	μ	upsilon	υ Υ
epsilon	$\epsilon\ (\varepsilon)$	nu	ν	phi	$\phi\ (\varphi)$ Φ
zeta	ζ	xi	ξ Ξ	chi	χ
eta	η	omicron	o	psi	ψ Ψ
theta	θ Θ	pi	π Π	omega	ω Ω

FOR FURTHER READING

Adams, Malcolm, and Theodore Shifrin. *Linear Algebra: A Geometric Approach*. New York: Freeman, 2002. Includes a few advanced topics in linear algebra that we did not have time to discuss in this text, e.g., complex eigenvalues, Jordan canonical form, and computer graphics.

Apostol, Tom M. *Calculus* (2 vols.), 2nd ed. Waltham, Mass.: Blaisdell, 1967. Although the first volume is needed for rudimentary vector algebra, the second volume includes linear algebra, multivariable calculus (although only treating the "classic" versions of Stokes's Theorem), and an introduction to probability theory and numerical analysis.

Bamberg, Paul, and Shlomo Sternberg. *A Course in Mathematics for Students of Physics* (2 vols.). Cambridge: Cambridge University Press, 1988. This book includes much of the mathematics of our course, as well as a volume's worth of interesting physics (using differential forms).

Edwards, Jr., C. H. *Advanced Calculus of Several Variables*. New York: Dover, 1994 (originally published by Academic Press, 1973). This very well-written book parallels ours for students who have already had standard courses in linear algebra and multivariable calculus. Of particular note is the last chapter, on the calculus of variations.

Friedberg, Stephen H., Arnold J. Insel, and Lawrence E. Spence. *Linear Algebra*, 3rd ed. Upper Saddle River, N.J.: Prentice Hall, 1997. A well-written, somewhat more advanced book, concentrating on the theoretical aspects of linear algebra.

Hubbard, John H., and Barbara Burke Hubbard. *Vector Calculus, Linear Algebra, and Differential Forms: A Unified Approach*, 2nd ed. Upper Saddle River, N.J.: Prentice Hall, 2002. Very similar in spirit to our text, this book is wonderfully idiosyncratic and includes Lebesgue integration, Kantarovich's Theorem, and the exterior derivative from a nonstandard definition. It also treats the Taylor polynomial in several variables.

Spivak, Michael. *Calculus*, 3rd ed. Houston, Tex.: Publish or Perish, 1994. The beautiful, ultimate source for single-variable calculus "done right."

Spivak, Michael. *Calculus on Manifolds: A Modern Approach to Classical Theorems of Advanced Calculus*. Boulder, Col.: Westview, 1965. A very terse and sophisticated version of this text, intended to introduce students who've seen linear algebra and multivariable calculus to the rigorous approach and to a more formal treatment of Stokes's Theorem.

Strang, Gilbert. *Linear Algebra and Its Applications*, 3rd ed. Philadelphia: Saunders, 1988. A classic text, with far more depth on applications.

More Advanced Reading

Do Carmo, Manfredo P., *Differential Geometry of Curves and Surfaces*. Englewood Cliffs, N.J.: Prentice Hall, 1976. A sophisticated approach using much of the material of this text.

Flanders, Harley. *Differential Forms with Applications to the Physical Sciences*. New York: Dover, 1989 (originally published by Academic Press, 1963). A short, sophisticated treatment of differential forms with applications to physics, topology, differential geometry, and partial differential equations.

Guillemin, Victor, and Alan Pollack. *Differential Topology*, Englewood Cliffs, N.J.: Prentice Hall, 1974. The perfect follow-up to our introduction to manifolds and the material of Chapter 8, Section 7.

Munkres, James. *Topology*, 2nd ed. Upper Saddle River, N.J.: Prentice Hall, 2000. A classic, extremely well-written text on point-set topology, to follow up on our discussion of open and closed sets, compactness, maximum value theorem, etc.

Shifrin, Theodore. *Abstract Algebra: A Geometric Approach*. Upper Saddle River, N.J.: Prentice Hall, 1996. A first course in abstract algebra that will be accessible to anyone who's enjoyed this course.

Wilkinson, J. M. *The Algebraic Eigenvalue Problem*. New York: Oxford University Press, 1965. An advanced book that includes a proof of the algorithm based on the Gram-Schmidt process to calculate eigenvalues and eigenvectors numerically.

ANSWERS TO SELECTED EXERCISES

1.1.2. $\begin{bmatrix} 4 \\ 3 \\ 7 \end{bmatrix}, \begin{bmatrix} 0 \\ 5 \\ -1 \end{bmatrix}, \begin{bmatrix} 2 \\ -1 \\ 3 \end{bmatrix}$

1.1.6 $c = 5/6, 5:1$

1.2.1 c. $-25, \theta = \arccos(-5/13);$ **f.** $2, \theta = \arccos(1/5)$

1.2.2 c. $-\dfrac{5}{13}\begin{bmatrix} 7 \\ -4 \end{bmatrix}, -\dfrac{5}{13}\begin{bmatrix} 1 \\ 8 \end{bmatrix};$ **f.** $\begin{bmatrix} -1 \\ 0 \\ 1 \end{bmatrix}, \dfrac{1}{25}\begin{bmatrix} 3 \\ -4 \\ 5 \end{bmatrix}$

1.2.3 $\arccos\sqrt{2/3} \approx .62$ radians $\approx 35.3°$

1.2.8 $\pi/6$

1.2.12 Let $\mathbf{x} = \overrightarrow{CA}$ and $\mathbf{y} = \overrightarrow{CB}$. Then $\overrightarrow{AB} = \mathbf{y} - \mathbf{x}$, and $\|\overrightarrow{AB}\|^2 = \|\mathbf{y} - \mathbf{x}\|^2 = \|\mathbf{y}\|^2 - 2\mathbf{y} \cdot \mathbf{x} + \|\mathbf{x}\|^2 = a^2 - 2ab\cos\theta + b^2$.

1.3.1 b., e., g., h. yes; **a., c., d., f., i.** no

1.3.2 The argument is valid only if there is *some* vector in the subspace. The first criterion is equivalent to the subspace's being nonempty.

1.3.8 If $\mathbf{v} \in V \cap V^\perp$, then $\mathbf{v} \cdot \mathbf{v} = 0$, so $\mathbf{v} = \mathbf{0}$.

1.4.1 b. $\begin{bmatrix} 0 & 3 \\ 2 & 5 \end{bmatrix};$ **f.** $\begin{bmatrix} 5 & 8 \\ 13 & 20 \end{bmatrix};$ **g.** $\begin{bmatrix} 1 & 4 & 5 \\ 3 & 10 & 11 \end{bmatrix};$ **h.** not defined; **k.** $\begin{bmatrix} 4 & 4 \\ 5 & 6 \end{bmatrix};$ **l.** $\begin{bmatrix} 0 & 1 & 2 \\ 1 & 2 & 1 \\ 2 & 7 & 8 \end{bmatrix}$

1.4.5 b. $A = \begin{bmatrix} \frac{1}{2} & \frac{1}{2} \\ \frac{1}{2} & \frac{1}{2} \end{bmatrix};$ **e.** $A = \begin{bmatrix} -\frac{2}{5} & -\frac{4}{5} \\ \frac{1}{5} & \frac{2}{5} \end{bmatrix}$

1.4.9 b. Either $A = \begin{bmatrix} 0 & \beta \\ 0 & 0 \end{bmatrix}$ or $\begin{bmatrix} 0 & 0 \\ \beta & 0 \end{bmatrix}$ for some real number β or $A = \alpha\begin{bmatrix} 1 & \beta \\ -1/\beta & -1 \end{bmatrix}$, α any real number, $\beta \neq 0$.

1.4.13 a. $S: \begin{bmatrix} 0 & -1 \\ -1 & 0 \end{bmatrix}, T: \begin{bmatrix} 0 & -1 \\ 1 & 0 \end{bmatrix};$ **b.** $\begin{bmatrix} 1 & 0 \\ 0 & -1 \end{bmatrix};$ **c.** $\begin{bmatrix} -1 & 0 \\ 0 & 1 \end{bmatrix}$

1.4.17 a. $BA^2B^{-1};$ **b.** $BA^nB^{-1};$ **c.** assuming A invertible, $BA^{-1}B^{-1}$

1.4.19 Since $A\mathbf{x} = 7\mathbf{x}$, we have $\mathbf{x} = (A^{-1}A)\mathbf{x} = A^{-1}(A\mathbf{x}) = A^{-1}(7\mathbf{x}) = 7(A^{-1}\mathbf{x})$, and so $A^{-1}\mathbf{x} = \frac{1}{7}\mathbf{x}$.

1.4.23 b. $\begin{bmatrix} 0 & 0 \\ 5 & 5 \end{bmatrix}$; **e.** $\begin{bmatrix} 1 & 5 & 7 \\ 2 & 8 & 10 \end{bmatrix}$; **g.** $\begin{bmatrix} 1 & 3 \\ 4 & 10 \\ 5 & 11 \end{bmatrix}$; **j.** $\begin{bmatrix} 6 & 4 \\ 4 & 5 \end{bmatrix}$; **k.** $\begin{bmatrix} 1 & 2 & 1 \\ 2 & 5 & 4 \\ 1 & 4 & 5 \end{bmatrix}$

1.4.24 $(AB)^T = B^T A^T = BA$; thus, $(AB)^T = AB$ if and only if $BA = AB$.

1.4.27 A hint: It suffices to see why $A_\theta \mathbf{x} \cdot \mathbf{y} = \mathbf{x} \cdot A_\theta^{-1} \mathbf{y}$. Since rotation doesn't change the length of vectors, we only need to see that the angle between $A_\theta \mathbf{x}$ and $\mathbf{y}$ is the same as the angle between $\mathbf{x}$ and $A_\theta^{-1} \mathbf{y}$.

1.4.31 a. $(A^{-1})^T$; **b.** switch the second and third columns of A^{-1}; **c.** multiply the third column of A^{-1} by $1/2$.

1.4.32 Since $(A^T A)\mathbf{x} = \mathbf{0}$, we have $(A^T A)\mathbf{x} \cdot \mathbf{x} = 0$. By Proposition 4.5, $A\mathbf{x} \cdot A\mathbf{x} = 0$, and so $\|A\mathbf{x}\| = 0$. This means that $A\mathbf{x} = \mathbf{0}$.

1.4.34 d. By part c, every orthogonal matrix can be written either as A_θ or as $A_\theta \begin{bmatrix} 1 & 0 \\ 0 & -1 \end{bmatrix}$.

1.4.35 b. If A is orthogonal, then $A^{-1} = A^T$, and so $(A^{-1})^T(A^{-1}) = (A^T)^T A^T = AA^T = I$ by Exercise 34e.

1.5.5 a. $\begin{bmatrix} 2 \\ -2 \\ 2 \end{bmatrix}$

1.5.6 a. $\sqrt{3}$

1.5.7 a. $x_1 - x_2 + x_3 = 0$; **d.** $3x_1 + 4x_2 + 5x_3 = 12$

1.5.8 $x_1 + x_2 - x_3 = -2$

1.5.11 $1/\sqrt{6}$

1.5.12 2

1.5.15 $\begin{bmatrix} 0 & -c & b \\ c & 0 & -a \\ -b & a & 0 \end{bmatrix}$

2.1.1 b. $\mathbf{f}(t) = \begin{bmatrix} -1 \\ 2 \end{bmatrix} + t \begin{bmatrix} 3 \\ 1 \end{bmatrix}$; **e.** $\mathbf{f}(t) = \begin{bmatrix} 1 \\ 1 \\ 0 \\ -1 \end{bmatrix} + t \begin{bmatrix} 1 \\ -2 \\ 3 \\ -1 \end{bmatrix}$

2.1.5 a. $\begin{bmatrix} (a+b)\cos\theta - b\cos\left((a+b)\theta/b\right) \\ (a+b)\sin\theta - b\sin\left((a+b)\theta/b\right) \end{bmatrix}$

2.1.7 a. $x = \cos\theta - \log(\csc\theta + \cot\theta)$, $y = \sin\theta$;
b. $x = t - (e^{2t} - 1)/(e^{2t} + 1)$, $y = 2e^t/(e^{2t} + 1)$

2.1.11 b. $(x^2 + y^2 + z^2)^2 - 10(x^2 + y^2) + 6z^2 + 9 = 0$

2.2.1 a., k. neither, **c., e., f., h., i., l.** open, **b., d., g., j.** closed, **m.** both

2.2.5 If $\mathbf{y} \notin \overline{B}(\mathbf{a}, r)$, then $\|\mathbf{y} - \mathbf{a}\| = s > r$. Let $\delta = s - r$. By the triangle inequality, for every point $\mathbf{z} \in B(\mathbf{y}, \delta)$ we have $\|\mathbf{y} - \mathbf{a}\| \leq \|\mathbf{y} - \mathbf{z}\| + \|\mathbf{z} - \mathbf{a}\|$, so $\|\mathbf{z} - \mathbf{a}\| \geq \|\mathbf{y} - \mathbf{a}\| - \|\mathbf{y} - \mathbf{z}\| > s - \delta = r$. Therefore, $B(\mathbf{y}, \delta) \subset \mathbb{R}^n - \overline{B}(\mathbf{a}, r)$, and so, by Proposition 2.1, $\overline{B}(\mathbf{a}, r)$ is closed.

2.2.10 a. If $I_k = [a_k, b_k]$, let $x = \sup\{a_k\}$. The set of left-hand endpoints is bounded above (e.g., by b_1), and so the least upper bound exists. We have $a_k \le x$ for all k automatically. Now, if $x > b_j$ for some j, then since $I_k \subset I_j$ for all $k > j$, this means that b_j is an upper bound of the set as well, contradicting the fact that x is the *least* upper bound. **b.** Take $I_k = (0, 1/k)$.

2.2.13 Choose $\varepsilon = 1$. Then there is $K \in \mathbb{N}$ so that for all $k > K$ we have $\|\mathbf{x}_k - \mathbf{x}_{K+1}\| < 1$, so $\|\mathbf{x}_k\| < 1 + \|\mathbf{x}_{K+1}\|$. Therefore, for all $j \in \mathbb{N}$, we have $\|\mathbf{x}_j\| \le \min\left(\|\mathbf{x}_1\|, \|\mathbf{x}_2\|, \ldots, \|\mathbf{x}_K\|, \|\mathbf{x}_{K+1}\| + 1\right)$.

2.3.8 a., b., c., d., e., g., h. yes; f., i., j. no

2.3.10 a. 2; **d.** $\sqrt{2}$

3.1.1 a. $\frac{\partial f}{\partial x} = 3x^2 + 3y^2$, $\frac{\partial f}{\partial y} = 6xy - 2$; **c.** $\frac{\partial f}{\partial x} = -y/(x^2+y^2)$, $\frac{\partial f}{\partial y} = x/(x^2+y^2)$

3.1.2 a. 3; **b.** $3/\sqrt{2}$

3.1.3 a. $\dfrac{1}{\sqrt{29}}\begin{bmatrix}5\\2\end{bmatrix}$

3.1.11 $T(\mathbf{v})$

3.2.1 a. $z = e^{-2}(2x - y + 5)$; **f.** $w = -x + y + z + 3$

3.2.2 a. 0; **e.** 1

3.2.3 a. $\begin{bmatrix} y & x \\ 2x & 2y \end{bmatrix}$; **b.** $\begin{bmatrix} -\sin t \\ \cos t \\ e^t \end{bmatrix}$; **c.** $\begin{bmatrix} \cos t & -s\sin t \\ \sin t & s\cos t \end{bmatrix}$; **d.** $\begin{bmatrix} yz & xz & xy \\ 1 & 1 & 2z \end{bmatrix}$; **e.** $\begin{bmatrix} \cos y & -x\sin y \\ \sin y & x\cos y \\ 0 & 1 \end{bmatrix}$

3.2.4 We get the approximate answer 34 taking $a = 240$ and $b = 6$ and the approximate answer 34.14 taking $a = 210$ and $b = 7$. My calculator gives me the "correct answer" 34.46 to two decimals.

3.3.1 -1

3.3.2 $D(\mathbf{f} \circ \mathbf{g})(\mathbf{0}) = D\mathbf{f}\begin{pmatrix}0\\1\\1\end{pmatrix} D\mathbf{g}\begin{pmatrix}0\\0\\0\end{pmatrix} = \begin{bmatrix} -1 & -1 & 1 \\ 6e^3 & e^3 & -e^3 \\ 6 & 1 & -1 \end{bmatrix}$;

$D(\mathbf{g}\circ\mathbf{f})(\mathbf{0}) = D\mathbf{g}\begin{pmatrix}0\\1\\0\end{pmatrix} D\mathbf{f}\begin{pmatrix}0\\0\end{pmatrix} = \begin{bmatrix} -2 & 9 \\ -1 & 2 \end{bmatrix}$

3.3.5 a. 266 mph; **b.** approx. 187 mph

3.3.6 -2.5 atm/min

3.3.14 $F'(t) = h(v(t))v'(t) - h(u(t))u'(t)$

3.3.16 $\left(\dfrac{\partial f}{\partial x}\right)^2 + \left(\dfrac{\partial f}{\partial y}\right)^2$ (evaluated at $\begin{bmatrix} r\cos\theta \\ r\sin\theta \end{bmatrix}$)

3.4.1 a. $x + 4y = 9$

3.4.2 b. $4x + y + 14z = 16$

3.4.4 a. $\begin{bmatrix} 3 \\ -4 \\ 25 \end{bmatrix}$; **b.** $-\arctan 4$

3.4.12 a. $c = -5/4$ or $c = 1$; **b.** $c = (\sqrt{17} - 1)/8$

3.5.7 a. $\sqrt{3}(e^b - e^a)$; **c.** 7

3.5.8 a. $\kappa = 1$; **b.** $\kappa = \sqrt{2}/(e^t + e^{-t})^2$

3.5.12 $\kappa = a/(a^2 + b^2)$, $\tau = b/(a^2 + b^2)$

3.6.5 $\phi(x) = \frac{1}{2}(h(x) + \frac{1}{c}\int_0^x k(u)\,du)$, $\psi(x) = \frac{1}{2}(h(x) - \frac{1}{c}\int_0^x k(u)\,du)$

3.6.9 $F\begin{pmatrix} r \\ \theta \end{pmatrix} = c\log r + k$ for some constants c and k.

4.1.2 **b., c., d., f., g.** are in echelon form; **c.** and **g.** are in reduced echelon form.

4.1.3 b. $\begin{bmatrix} 1 & -1 & 2 \\ 0 & 0 & 0 \\ 0 & 0 & 0 \end{bmatrix}$, $\mathbf{x} = x_2 \begin{bmatrix} 1 \\ 1 \\ 0 \end{bmatrix} + x_3 \begin{bmatrix} -2 \\ 0 \\ 1 \end{bmatrix}$; **e.** $\begin{bmatrix} 1 & 0 & 0 & -1 \\ 0 & 1 & 0 & 1 \\ 0 & 0 & 1 & 1 \\ 0 & 0 & 0 & 0 \end{bmatrix}$, $\mathbf{x} = x_4 \begin{bmatrix} 1 \\ -1 \\ -1 \\ 1 \end{bmatrix}$;

f. $\begin{bmatrix} 1 & 0 & 2 & 0 & -1 \\ 0 & 1 & -1 & 0 & 2 \\ 0 & 0 & 0 & 1 & 4 \\ 0 & 0 & 0 & 0 & 0 \end{bmatrix}$, $\mathbf{x} = x_3 \begin{bmatrix} -2 \\ 1 \\ 1 \\ 0 \\ 0 \end{bmatrix} + x_5 \begin{bmatrix} 1 \\ -2 \\ 0 \\ -4 \\ 1 \end{bmatrix}$

4.1.4 a. $\mathbf{x} = \begin{bmatrix} 2 \\ -1 \\ 0 \end{bmatrix} + x_3 \begin{bmatrix} 1 \\ -1 \\ 1 \end{bmatrix}$

4.1.5 $\mathbf{x} = \begin{bmatrix} 1/\sqrt{2} \\ 1/\sqrt{2} \\ 0 \end{bmatrix}$

4.1.6 a. $\begin{bmatrix} 1 \\ -1 \\ -1 \\ 1 \end{bmatrix}$

4.1.7 center $\begin{bmatrix} -1 \\ 2 \end{bmatrix}$, radius 5

4.1.8 $\mathbf{b} = \mathbf{v}_1 - \mathbf{v}_2 + \mathbf{v}_3$

4.1.9 b. yes; **a., c.** no

4.1.10 d. yes; **a., b., c.** no

4.1.11 b. $2b_1 + b_2 - b_3 = 0$

4.1.13 a. By Proposition 1.4, $A \begin{bmatrix} 1 \\ 1 \\ 0 \end{bmatrix} = \mathbf{0}$, but since $\begin{bmatrix} 1 \\ 0 \\ 1 \end{bmatrix} \cdot \begin{bmatrix} 1 \\ 1 \\ 0 \end{bmatrix} \neq 0$, this is impossible.

4.1.14 a. 0, 3; **b.** for $\alpha = 0$, $\mathbf{b}$ must satisfy $b_2 = 0$; for $\alpha = 3$, $\mathbf{b}$ must satisfy $b_2 = 3b_1$.

4.1.18 a. none, as $A\mathbf{x} = \mathbf{0}$ is always consistent; **b.** take $r = m = n$; **e.** take $r < n$

4.2.1 b. $\begin{bmatrix} 1 & 0 & 0 \\ 0 & 1 & 0 \\ -3 & 0 & 1 \end{bmatrix} \begin{bmatrix} 1 & 0 & 0 \\ 1 & 1 & 0 \\ 0 & 0 & 1 \end{bmatrix} \begin{bmatrix} \frac{1}{2} & 0 & 0 \\ 0 & 1 & 0 \\ 0 & 0 & 1 \end{bmatrix} = \begin{bmatrix} \frac{1}{2} & 0 & 0 \\ \frac{1}{2} & 1 & 0 \\ -\frac{3}{2} & 0 & 1 \end{bmatrix}$, $\frac{1}{2}b_1 + b_2 = -\frac{3}{2}b_1 + b_3 = 0$;

e. $\begin{bmatrix} 1 & 0 & 0 & 0 \\ 0 & 1 & 0 & 0 \\ 0 & 0 & 1 & 0 \\ 0 & 0 & -1 & 1 \end{bmatrix} \begin{bmatrix} 1 & 0 & 0 & 0 \\ 0 & 1 & 0 & 0 \\ 0 & 0 & 1 & 0 \\ 0 & -1 & 0 & 1 \end{bmatrix} \begin{bmatrix} 1 & 0 & 0 & 0 \\ 0 & 1 & 0 & 0 \\ 0 & -2 & 1 & 0 \\ 0 & 0 & 0 & 1 \end{bmatrix} \begin{bmatrix} 1 & 0 & 0 & 0 \\ 0 & 1 & 0 & 0 \\ 0 & 0 & 1 & 0 \\ -1 & 0 & 0 & 1 \end{bmatrix} \begin{bmatrix} 1 & 0 & 0 & 0 \\ 0 & 1 & 0 & 0 \\ -1 & 0 & 1 & 0 \\ 0 & 0 & 0 & 1 \end{bmatrix} \cdots$

$\cdots \begin{bmatrix} 1 & 0 & 0 & 0 \\ -1 & 1 & 0 & 0 \\ 0 & 0 & 1 & 0 \\ 0 & 0 & 0 & 1 \end{bmatrix} = \begin{bmatrix} 1 & 0 & 0 & 0 \\ -1 & 1 & 0 & 0 \\ 1 & -2 & 1 & 0 \\ -1 & 1 & -1 & 1 \end{bmatrix}$, $-b_1 + b_2 - b_3 + b_4 = 0$;

f. $\begin{bmatrix} 1 & 0 & 0 & 0 \\ 0 & 1 & 0 & 0 \\ 0 & 0 & -1 & 0 \\ 0 & 0 & 0 & 1 \end{bmatrix} \begin{bmatrix} 1 & 0 & 0 & 0 \\ 0 & -1 & 0 & 0 \\ 0 & 0 & 1 & 0 \\ 0 & 0 & 0 & 1 \end{bmatrix} \begin{bmatrix} 1 & 0 & 0 & 0 \\ 0 & 1 & 0 & 0 \\ 0 & 0 & 1 & 0 \\ 0 & 0 & -2 & 1 \end{bmatrix} \begin{bmatrix} 1 & 0 & 0 & 0 \\ 0 & 1 & 0 & 0 \\ 0 & 0 & 1 & 0 \\ 0 & -7 & 0 & 1 \end{bmatrix} \cdots$

$\cdots \begin{bmatrix} 1 & 0 & 0 & 0 \\ 0 & 1 & 0 & 0 \\ 0 & -3 & 1 & 0 \\ 0 & 0 & 0 & 1 \end{bmatrix} \begin{bmatrix} 1 & 0 & 0 & 0 \\ 0 & 1 & 0 & 0 \\ 0 & 0 & 1 & 0 \\ -2 & 0 & 0 & 1 \end{bmatrix} \begin{bmatrix} 1 & 0 & 0 & 0 \\ 0 & 1 & 0 & 0 \\ -1 & 0 & 1 & 0 \\ 0 & 0 & 0 & 1 \end{bmatrix} \begin{bmatrix} 1 & 0 & 0 & 0 \\ 1 & 1 & 0 & 0 \\ 0 & 0 & 1 & 0 \\ 0 & 0 & 0 & 1 \end{bmatrix} =$

$\begin{bmatrix} 1 & 0 & 0 & 0 \\ -1 & -1 & 0 & 0 \\ 4 & 3 & -1 & 0 \\ -1 & -1 & -2 & 1 \end{bmatrix}$, $-b_1 - b_2 - 2b_3 + b_4 = 0$.

4.2.2 c. $A^{-1} = \begin{bmatrix} -1 & 3 & -2 \\ -1 & 2 & -1 \\ 2 & -3 & 2 \end{bmatrix}$; **e.** $A^{-1} = \begin{bmatrix} -1 & 2 & 1 \\ 5 & -8 & -6 \\ -3 & 5 & 4 \end{bmatrix}$

4.2.3 b. $A^{-1} = \begin{bmatrix} -2 & 0 & 1 \\ 9 & -1 & -3 \\ -6 & 1 & 2 \end{bmatrix}$, $\mathbf{x} = \begin{bmatrix} 0 \\ 2 \\ -1 \end{bmatrix}$; **d.** $A^{-1} = \begin{bmatrix} 1 & -1 & 0 & 0 \\ 0 & 1 & -3 & 2 \\ 0 & 0 & 4 & -3 \\ 0 & 0 & -1 & 1 \end{bmatrix}$, $\mathbf{x} = \begin{bmatrix} 2 \\ -1 \\ 1 \\ 0 \end{bmatrix}$

4.3.2 a., b., d., e. yes; **c., f.** no

4.3.12 d. yes; **a., b., c.** no

4.3.13 a. $\left\{ \begin{bmatrix} 1 \\ 0 \\ 1 \end{bmatrix}, \begin{bmatrix} 0 \\ 1 \\ 1 \end{bmatrix} \right\}$, dim 2

4.3.14 b. $\begin{bmatrix} 0 \\ 2 \\ -1 \end{bmatrix}$; **d.** $\begin{bmatrix} 2 \\ -1 \\ 1 \\ 0 \end{bmatrix}$

4.3.21 a. A hint: Use the definition of $U + V$ to show that the vectors span. To establish linear independence, suppose $c_1\mathbf{u}_1 + c_2\mathbf{u}_2 + \cdots + c_k\mathbf{u}_k + d_1\mathbf{v}_1 + \cdots + d_\ell\mathbf{v}_\ell = \mathbf{0}$. Then what can you say about the vector $c_1\mathbf{u}_1 + c_2\mathbf{u}_2 + \cdots + c_k\mathbf{u}_k = -(d_1\mathbf{v}_1 + \cdots + d_\ell\mathbf{v}_\ell)$?

4.3.23 a., c., e. yes; **b., d., f.** no

4.4.1 Let's show that $\mathbf{R}(B) \subset \mathbf{R}(A)$ if B is obtained by performing any row operation on A. Obviously, a row interchange doesn't affect the span. If $\mathbf{B}_i = c\mathbf{A}_i$ and all the other rows are the same, $c_1\mathbf{B}_1 + \cdots + c_i\mathbf{B}_i + \cdots + c_m\mathbf{B}_m = c_1\mathbf{A}_1 + \cdots + (c_ic)\mathbf{A}_i + \cdots + c_m\mathbf{A}_m$, so any vector in $\mathbf{R}(B)$ is in $\mathbf{R}(A)$. If $\mathbf{B}_i = \mathbf{A}_i + c\mathbf{A}_j$ and all the other rows are the same, then $c_1\mathbf{B}_1 + \cdots + c_i\mathbf{B}_i + \cdots + c_m\mathbf{B}_m = c_1\mathbf{A}_1 + \cdots + c_i(\mathbf{A}_i + c\mathbf{A}_j) + \cdots + c_m\mathbf{A}_m = c_1\mathbf{A}_1 + \cdots + c_i\mathbf{A}_i + \cdots + (c_j + cc_i)\mathbf{A}_j + \cdots + c_m\mathbf{A}_m$, so once again any vector in $\mathbf{R}(B)$ is in $\mathbf{R}(A)$.

To see that $\mathbf{R}(A) \subset \mathbf{R}(B)$, we observe that the matrix A is obtained from B by performing the (inverse) row operation (this is why we need $c \neq 0$ for the second type of row operation). Since $\mathbf{R}(B) \subset \mathbf{R}(A)$ and $\mathbf{R}(A) \subset \mathbf{R}(B)$, we have $\mathbf{R}(A) = \mathbf{R}(B)$.

4.4.3 f. $\mathbf{R}(A): \left\{ \begin{bmatrix} 1 \\ 0 \\ -1 \\ 2 \\ 0 \\ 1 \end{bmatrix}, \begin{bmatrix} 0 \\ 1 \\ 1 \\ 3 \\ 0 \\ -2 \end{bmatrix}, \begin{bmatrix} 0 \\ 0 \\ 0 \\ 0 \\ 1 \\ -1 \end{bmatrix} \right\}$, $\mathbf{C}(A): \left\{ \begin{bmatrix} 1 \\ 0 \\ -1 \\ 0 \end{bmatrix}, \begin{bmatrix} 1 \\ 1 \\ 2 \\ 4 \end{bmatrix}, \begin{bmatrix} 0 \\ -2 \\ 1 \\ -1 \end{bmatrix} \right\}$,

$\mathbf{N}(A): \left\{ \begin{bmatrix} 1 \\ -1 \\ 1 \\ 0 \\ 0 \\ 0 \end{bmatrix}, \begin{bmatrix} -2 \\ -3 \\ 0 \\ 1 \\ 0 \\ 0 \end{bmatrix}, \begin{bmatrix} -1 \\ 2 \\ 0 \\ 0 \\ 1 \\ 1 \end{bmatrix} \right\}$, $\mathbf{N}(A^\mathsf{T}): \left\{ \begin{bmatrix} 1 \\ 1 \\ 1 \\ -1 \end{bmatrix} \right\}$

4.4.4 a. $X = \begin{bmatrix} 2 & -1 & 0 \\ 3 & 0 & 1 \end{bmatrix}$, $Y = \begin{bmatrix} 1 & 2 \\ 3 & 6 \end{bmatrix}$

4.4.5 b. $\begin{bmatrix} 1 & 0 & -1 & -1 \\ 0 & 1 & 0 & 0 \\ 0 & 1 & 0 & 0 \end{bmatrix}$; **c.** $\begin{bmatrix} 2 & -1 & 0 \\ 0 & 0 & 0 \\ 2 & -1 & 0 \end{bmatrix}$; **e.** $\begin{bmatrix} 2 & 0 & 1 \\ 0 & 2 & 1 \\ 2 & 2 & 2 \end{bmatrix}$

4.4.8 a. $\left\{ \begin{bmatrix} -3 \\ -2 \\ 1 \\ 0 \end{bmatrix}, \begin{bmatrix} -4 \\ 5 \\ 0 \\ 1 \end{bmatrix} \right\}$; **b.** $\left\{ \begin{bmatrix} 1 \\ 0 \\ 3 \\ 4 \end{bmatrix}, \begin{bmatrix} 0 \\ 1 \\ 2 \\ -5 \end{bmatrix} \right\}$

4.4.10 Since U is a matrix in echelon form, its *last* $m - r$ rows are $\mathbf{0}$. When we consider the matrix product $A = BU$, we see that every column of A is a linear combinations of the *first* r columns of B; hence, these r column vectors span $\mathbf{C}(A)$. Since $\dim \mathbf{C}(A) = r$, these column vectors must give a basis (see Proposition 3.9).

4.4.11 b. $\begin{bmatrix} \frac{1}{3}b_1 \\ \frac{1}{3}b_1 + \frac{1}{2}b_2 \\ \frac{1}{3}b_1 - \frac{1}{2}b_2 \end{bmatrix}$

4.5.3 a. $\frac{\partial f}{\partial z}(\mathbf{a}) = 2 \neq 0$; **b.** $\frac{\partial \phi}{\partial x}\begin{pmatrix}1\\-1\end{pmatrix} = -\frac{1}{2}$, $\frac{\partial \phi}{\partial y}\begin{pmatrix}1\\-1\end{pmatrix} = 1$; **c.** $x - 2y + 2z = 3$

4.5.6 It is a smooth surface away from the curve $\mathbf{g}(t) = \begin{bmatrix} t \\ t^2 \\ t^3 \end{bmatrix}$. Indeed, M is the collection of all the tangent lines to this curve; this surface has a cuspidal edge along the curve, as one can perhaps see from the picture below:

4.5.11 a. Let $\mathbf{F}(\mathbf{x}) = \begin{bmatrix} x_1^2 + x_2^2 + x_3^2 + x_4^2 - 1 \\ x_1 x_2 - x_3 x_4 \end{bmatrix}$. Then $M = \mathbf{F}^{-1}(\{\mathbf{0}\})$. Now $D\mathbf{F}(\mathbf{x}) = \begin{bmatrix} 2x_1 & 2x_2 & 2x_3 & 2x_4 \\ x_2 & x_1 & -x_4 & -x_3 \end{bmatrix}$ has rank < 2 only at the origin.
b. $x_1 = 1, x_2 = 0$; $x_1 - x_2 = x_4 - x_3 = 1$.

5.1.1 a., b., h., k., l. are compact

5.1.4 a. 2

5.1.9 First, S is closed: Any convergent sequence of points in S has a subsequence converging to a point of S and hence must converge itself to that point of S (Exercise 2.2.6). Next, S is bounded: If not, we could take $\mathbf{x}_k \in S$ with $\|\mathbf{x}_k\| > k$; then $\{\mathbf{x}_k\}$ would have no convergent subsequence.

5.2.1 a. $\begin{bmatrix} -3/2 \\ 1 \end{bmatrix}$; **g.** $\begin{bmatrix} 1 \\ -1 \end{bmatrix}, \begin{bmatrix} -1 \\ 1 \end{bmatrix}$; **i.** $\mathbf{0}$

5.2.3 length and width $2r/\sqrt{3}$, height $r/\sqrt{3}$

5.2.4 max 5, min -1

5.2.6 $2' \times 2' \times 1'$

5.2.10 Bend up $4''$ on either side at an angle of $\pi/3$.

5.3.1 a. saddle point; **g.** $\begin{bmatrix} 1 \\ -1 \end{bmatrix}$ is a local maximum point, $\begin{bmatrix} -1 \\ 1 \end{bmatrix}$ is a local minimum point; **i.** saddle point

5.3.3 We see two mountain peaks (global maxima) joined by two ridges (two saddle points) with a deep valley (global minimum) between them.

5.3.6 c. $A = \begin{bmatrix} 1 & & \\ 1 & 1 & \\ -1 & -2 & 1 \end{bmatrix} \begin{bmatrix} 2 & & \\ & -3 & \\ & & 11 \end{bmatrix} \begin{bmatrix} 1 & 1 & -1 \\ & 1 & -2 \\ & & 1 \end{bmatrix}$, Q is indefinite.

5.3.8 b. $2x_1x_2 = (\frac{1}{2}x_1 + x_2)^2 - (-\frac{1}{2}x_1 + x_2)^2$

5.4.2 $0, 49/8$

5.4.13 $x^2/18 + y^2/2 = 1$

5.4.15 semimajor axis 1, semiminor axis $1/\sqrt{6}$

5.4.19 $(\delta/\sqrt{n})^n$

5.4.25 $\dfrac{1}{3}\begin{bmatrix} 1 \\ 5 \\ 4 \end{bmatrix}$

5.4.26 a. $\begin{bmatrix} -2 \\ 4 \\ 2 \end{bmatrix}$

5.4.27 $\pm \begin{bmatrix} 1 \\ 0 \\ 0 \end{bmatrix}, \pm \begin{bmatrix} 0 \\ 1 \\ 0 \end{bmatrix}$

5.4.29 a. $\mathbf{x} = \pm \begin{bmatrix} 2/\sqrt{5} \\ 1/\sqrt{5} \end{bmatrix}, \lambda = 2; \mathbf{x} = \pm \begin{bmatrix} 1/\sqrt{5} \\ -2/\sqrt{5} \end{bmatrix}, \lambda = -3$

5.4.30 a. $\frac{1+\sqrt{5}}{2} \approx 1.62$

5.5.1 b. $\begin{bmatrix} -1 \\ 0 \\ 1 \\ 3 \end{bmatrix}$

5.5.4 a. $\bar{\mathbf{x}} = \dfrac{1}{14}\begin{bmatrix} 1 \\ 4 \end{bmatrix}$; **b.** $\dfrac{1}{14}\begin{bmatrix} 5 \\ 6 \\ -3 \end{bmatrix}$.

5.5.7 a. $a = 9/4$; **c.** $a = \frac{1}{4}, b = \frac{29}{20}, c = \frac{23}{20}$.

5.5.12 c. $\mathbf{q}_1 = \dfrac{1}{\sqrt{2}}\begin{bmatrix} 1 \\ 0 \\ 1 \\ 0 \end{bmatrix}, \mathbf{q}_2 = \dfrac{1}{2}\begin{bmatrix} 1 \\ 1 \\ -1 \\ 1 \end{bmatrix}, \mathbf{q}_3 = \dfrac{1}{\sqrt{2}}\begin{bmatrix} 0 \\ 1 \\ 0 \\ -1 \end{bmatrix}$

5.5.13 a. $\mathbf{w}_1 = \begin{bmatrix} 1 \\ -1 \\ 0 \\ 2 \end{bmatrix}, \mathbf{w}_2 = \dfrac{1}{2}\begin{bmatrix} 1 \\ 1 \\ 2 \\ 0 \end{bmatrix}$; **b.** $\mathbf{p} = \begin{bmatrix} 1 \\ -1 \\ 0 \\ 2 \end{bmatrix}$; **c.** $\bar{\mathbf{x}} = \begin{bmatrix} 1 \\ 0 \end{bmatrix}$

5.5.15 b. $\begin{bmatrix} \frac{1}{3}b_1 \\ \frac{1}{3}b_1 + \frac{1}{2}b_2 \\ \frac{1}{3}b_1 - \frac{1}{2}b_2 \end{bmatrix}$

5.5.16 b. The ith row of A^{-1} is $\mathbf{a}_i^T / \|\mathbf{a}_i\|^2$.

5.5.19 b. $f(t) = t^2 - t + \frac{1}{6}$ gives a basis.

5.5.20 Hint: Use the addition formulas for sin and cos to derive the formulas
$$\sin kt \sin \ell t = \tfrac{1}{2}\big(\cos(k-\ell)t - \cos(k+\ell)t\big)$$
$$\sin kt \cos \ell t = \tfrac{1}{2}\big(\sin(k+\ell)t + \sin(k-\ell)t\big).$$

6.1.3 $\mathbf{x}_k - \mathbf{x} = \big(\mathbf{x}_0 + \sum_{j=1}^{k}(\mathbf{x}_j - \mathbf{x}_{j-1})\big) - \big(\mathbf{x}_0 + \sum_{j=1}^{\infty}(\mathbf{x}_j - \mathbf{x}_{j-1})\big) = -\sum_{j=k+1}^{\infty}(\mathbf{x}_j - \mathbf{x}_{j-1})$, so
$\|\mathbf{x}_k - \mathbf{x}\| \le \sum_{j=k+1}^{\infty} \|\mathbf{x}_j - \mathbf{x}_{j-1}\| \le \big(\sum_{j=k+1}^{\infty} c^{j-1}\big)\|\mathbf{x}_1 - \mathbf{x}_0\| = \frac{c^k}{1-c}\|\mathbf{x}_1 - \mathbf{x}_0\|$.

6.1.9 a. $[1, 2]$; $x_0 = 1$, $x_1 = 1.5$, $x_2 = 1.41667$; **c.** $[0.26, 0.79]$, $x_0 = 0.785398$, $x_1 = 0.523599$, $x_2 = 0.514961$

6.1.11 a. $B\left(\begin{bmatrix}1\\1/4\end{bmatrix}, 1/4\right)$; to three decimals, the root is $\begin{bmatrix}.984\\.254\end{bmatrix}$

6.2.1 a. any $\mathbf{x}_0 \ne \mathbf{0}$, $D\mathbf{g}(\mathbf{f}(\mathbf{x}_0)) = \frac{1}{x_0^2 + y_0^2}\begin{bmatrix}x_0 & y_0\\-y_0 & x_0\end{bmatrix}$

6.2.3 b. $D\phi\begin{pmatrix}1\\2\end{pmatrix} = \begin{bmatrix}0 & 0\end{bmatrix}$

6.2.4 $\begin{bmatrix}-1\\1\\1\end{bmatrix} + s\begin{bmatrix}-1\\3\\1\end{bmatrix}$

6.2.6 the opposite

6.2.9 a. $\alpha = 1/T$, $\beta = 1/p$

6.3.1 If X were a 1-dimensional manifold, in a neighborhood of $\mathbf{0}$ it would have to be a graph of the form $y = f(x)$ or $x = f(y)$ for some smooth function f. It is clearly not a graph over the y-axis, and $f(x) = |x|$ is far from differentiable. The so-called parametrization has 0 derivative at $t = 0$.

6.3.5 a. zero set of $\mathbf{F}\begin{pmatrix}x\\y\\z\end{pmatrix} = \begin{bmatrix}x - y^2\\z - y^4\end{bmatrix}$ and graph of $\mathbf{f}(y) = \begin{bmatrix}y^2\\y^4\end{bmatrix}$.

6.3.7 c. zero set of $F\begin{pmatrix}x\\y\\z\end{pmatrix} = y - x\tan z$ and graph of $f\begin{pmatrix}x\\z\end{pmatrix} = x\tan z$ away from $z = (2n+1)\pi/2$, $n \in \mathbb{Z}$; near such points, use $F\begin{pmatrix}x\\y\\z\end{pmatrix} = x - y\cot z$ and $f\begin{pmatrix}y\\z\end{pmatrix} = y\cot z$.

7.1.1 Let $\mathcal{P}_1 = \{0 = x_0 < x_1 = 1\}$ be the trivial partition of $[0, 1]$ and let $\mathcal{P}_2 = \{0 = y_0 < y_1 < y_2 < y_3 = 1\}$ be a partition of $[0, 1]$ with the properties that $y_1 \le \frac{1}{2} < y_2$ and $y_2 - y_1 < \varepsilon$; set $\mathcal{P} = \mathcal{P}_1 \times \mathcal{P}_2$. Then for $j = 1, 3$, we have $m_{1j} = M_{1j}$, whereas $m_{12} = 0$ and $M_{12} = 1$. Then
$$U(f, \mathcal{P}) - L(f, \mathcal{P}) = (M_{12} - m_{12})(y_2 - y_1) = y_2 - y_1 < \varepsilon,$$
and so, by the Convenient Criterion, Proposition 1.3, we infer that f is integrable. Now, for our

Answers to Selected Exercises ◀ **483**

particular partition $\mathcal{P}$, we have
$$L(f, \mathcal{P}) = 1 - y_2 < \tfrac{1}{2} \le 1 - y_1 = U(f, \mathcal{P});$$
thus, $1/2$ is the only number that can lie between all lower and upper sums, and therefore $\int_R f\, dA = 1/2$.

7.1.8 Hint: Let R'' be the intersection of R and R'. Then show that $\int_R \tilde{f}\, dV = \int_{R''} \tilde{f}''\, dV = \int_{R'} \tilde{f}'\, dV$.

7.2.1 a. $e - 1$; **c.** $\log(8/3\sqrt{3})$

7.2.2 b. $\displaystyle\int_0^2 \int_{y/2}^1 f\begin{pmatrix}x\\y\end{pmatrix} dx\, dy$; **f.** $\displaystyle\int_0^1 \int_{-\sqrt{y}}^{\sqrt{y}} f\begin{pmatrix}x\\y\end{pmatrix} dx\, dy + \int_1^4 \int_{y-2}^{\sqrt{y}} f\begin{pmatrix}x\\y\end{pmatrix} dx\, dy$

7.2.3 c. $\tfrac{1}{2}(\tfrac{1}{2}\log 2 - 1 + \pi/4)$

7.2.8 a. $\tfrac{2}{3}\log 3$

7.2.11 $16/3$

7.2.12 a. $\displaystyle\int_0^1 \int_0^{1-z} \int_0^{1-x-z} f\begin{pmatrix}x\\y\\z\end{pmatrix} dy\, dx\, dz$;

d. $\displaystyle\int_0^1 \int_0^x \int_0^{1-x^2} f\begin{pmatrix}x\\y\\z\end{pmatrix} dy\, dz\, dx + \int_0^1 \int_x^{1+x-x^2} \int_{z-x}^{1-x^2} f\begin{pmatrix}x\\y\\z\end{pmatrix} dy\, dz\, dx$

7.2.13 $abc/6$

7.2.15 $1/8$

7.2.24 a. $(2 + \pi)/8x^3$

7.3.5 $3\pi/4$

7.3.6 $(\sqrt{2} - 1)/12$

7.3.9 $\pi(e - 1)/4$

7.3.12 $8/9$

7.3.17 a. $\displaystyle\int_0^{2\pi} \int_0^a \int_{hr/a}^h r\, dz\, dr\, d\theta$; **b.** $\displaystyle\int_0^{2\pi} \int_0^{\arctan(a/h)} \int_0^{h\sec\phi} \rho^2 \sin\phi\, d\rho\, d\phi\, d\theta$

7.3.20 $\tfrac{1}{3}\displaystyle\int_S \rho^2\, dV = 4\pi/15$

7.3.22 $625\pi/32$

7.4.1 $\tfrac{2}{3}a$

7.4.3 $\tfrac{32}{9\pi}a$

7.4.4 $6a/5$

7.4.7 mass $= 2(\sqrt{3} - \tfrac{\pi}{3})$, $\bar{x} = \dfrac{\sqrt{3}}{2(\sqrt{3} - \tfrac{\pi}{3})}$, $\bar{y} = 0$ by symmetry.

7.4.11 $\dfrac{1}{4}\begin{bmatrix} a \\ b \\ c \end{bmatrix}$

7.4.12 $I = \tfrac{3}{5}ma^2$

7.4.16 a. $k = 1$; **b.** $k = 1/2$; **c.** $k = 3/10$

7.5.1 b. -4; **d.** 6

7.5.11 $\det A = \pm 1$.

7.5.13 a. $-4/3$; **b.** $\begin{bmatrix} 2 & 0 & -1 \\ -\tfrac{4}{3} & \tfrac{1}{3} & \tfrac{2}{3} \\ \tfrac{5}{3} & -\tfrac{2}{3} & -\tfrac{1}{3} \end{bmatrix}$

7.5.14 -3; $\begin{bmatrix} -1 & 0 & 1 \\ 2 & 1 & -2 \\ -\tfrac{4}{3} & -\tfrac{2}{3} & \tfrac{5}{3} \end{bmatrix}$

7.6.5 $\tfrac{1}{14}(3 - \log 5)$

7.6.7 $\tfrac{1}{2}\sin 1$

7.6.10 $\tfrac{1}{2}$

7.6.13 $(\pi\sqrt{3} - 3)/9$

7.6.14 This is a solid torus with core radius a and little radius b. Its volume is $2\pi^2 ab^2$.

7.6.17 $2\pi^2 a^5/5$

8.2.4 a. $-5\, dx \wedge dy + 10\, dy \wedge dz$; **c.** $11\, dx \wedge dy \wedge dz$

8.2.6 a. $-xe^{xy}\, dx \wedge dy$; **b.** $2(x\, dx \wedge dy + z\, dz \wedge dx + y\, dy \wedge dz)$; **c.** $2(x + y + z)\, dx \wedge dy \wedge dz$; **d.** $x_2\, dx_1 \wedge dx_3 \wedge dx_4 + x_1\, dx_2 \wedge dx_3 \wedge dx_4$

8.2.7 a., c., f. no (although for f. we might think the opposite); **b.** $f\begin{pmatrix} x \\ y \end{pmatrix} = x^2 y$;

d. $f\begin{pmatrix} x \\ y \\ z \end{pmatrix} = \tfrac{1}{3}x^3 + xyz + \sin y + \tfrac{1}{2}z^2$; **e.** $f\begin{pmatrix} x \\ y \end{pmatrix} = \log\sqrt{x^2 + y^2}$

8.2.11 b. $18\, dv$, **e.** $\sin^2 u\, du \wedge dv$

8.3.1 a., b., e. 1; **c., d., f.** -1

8.3.2 a. $5/6$; **b.** $13/15$; **c.** $-5/6$; **d.** $-5/6$; **e.** $5/3 - \pi/4$; **f.** $-5/3 + \pi/4$

8.3.10 c. $\pi a^4/2$; **d.** $32/3$

8.3.12 πab

8.3.14 $\pi(2a^2 + b^2)$

8.4.4 16

8.4.8 $2\pi ah$

8.4.10 $\begin{bmatrix} u \\ v \end{bmatrix} = \dfrac{1}{1-z}\begin{bmatrix} x \\ y \end{bmatrix}$, $\mathbf{g}\begin{pmatrix} u \\ v \end{pmatrix} = \begin{bmatrix} x \\ y \\ z \end{bmatrix} = \dfrac{1}{1+u^2+v^2}\begin{bmatrix} 2u \\ 2v \\ u^2+v^2-1 \end{bmatrix}$

8.4.14 $\frac{2}{3}ma^2 = \frac{8}{3}\delta\pi a^4$

8.4.15 a. $4\pi a^3$; **b.** $4\pi a^2 h$; **c.** $6\pi a^2 h$; **d.** 24

8.4.17 $88\pi/3$

8.4.18 a., c., d. 4π; **b.** $4\pi h/\sqrt{a^2+h^2}$

8.4.22 a., b. 0; **c.** π^2

8.5.1 Since the outward-pointing normal to $\partial\mathbb{R}_+^k$ is $-\mathbf{e}_k$, we must decide whether $\{\underbrace{-\mathbf{e}_k, \quad \mathbf{e}_1,\ldots,\mathbf{e}_{k-1}}_{\text{standard positive basis for } \mathbb{R}^{k-1}}\}$ is a positively-oriented basis for $\mathbb{R}^k$. We need $k-1$ exchanges and one change of sign to obtain $\{\mathbf{e}_1,\ldots,\mathbf{e}_k\}$. This is k sign changes in all, and hence the standard positive basis for $\mathbb{R}^{k-1}$ gives the correct orientation precisely when $(-1)^k = +1$.

8.5.3 $2\pi a(a+b)$

8.5.6 πa^4

8.5.12 a. $-8\pi/15$

8.5.18 $\pi/2\sqrt{3}$

8.6.7 c., d., e., g., h. div $= 0$; **a., f., g., h.** curl $= \mathbf{0}$

9.1.1 a. $[T]_{B'} = \begin{bmatrix} 36 & 24 \\ -55 & -37 \end{bmatrix}$; **b.** $[S]_{\mathcal{E}} = \begin{bmatrix} 7 & -3 \\ 7 & -2 \end{bmatrix}$

9.1.5 $\dfrac{1}{9}\begin{bmatrix} 7 & 4 & -4 \\ 4 & 1 & 8 \\ -4 & 8 & 1 \end{bmatrix}$

9.1.8 $\dfrac{1}{17}\begin{bmatrix} 3 & 1 & 5 & 4 \\ 1 & 6 & -4 & 7 \\ 5 & -4 & 14 & 1 \\ 4 & 7 & 1 & 11 \end{bmatrix}$

9.1.12 a. With respect to the "new" coordinates y_1, y_2, y_3, the equation of the curve of intersection is $y_1^2 + \cos^2\phi\, y_2^2 = 1$, $y_3 = 0$.

9.2.1 a. eigenvalues $-1, 6$; eigenvectors $\begin{bmatrix} -5 \\ 2 \end{bmatrix}, \begin{bmatrix} 1 \\ 1 \end{bmatrix}$;

e. eigenvalues $2, 2$; only eigenvector $\begin{bmatrix} 1 \\ 1 \end{bmatrix}$;

f. eigenvalues $-3, 0, 3$; eigenvectors $\begin{bmatrix} -1 \\ 0 \\ 1 \end{bmatrix}, \begin{bmatrix} 1 \\ -1 \\ 1 \end{bmatrix}, \begin{bmatrix} 1 \\ 2 \\ 1 \end{bmatrix}$;

i. eigenvalues 1, 1, 2; eigenvectors $\begin{bmatrix} 0 \\ 1 \\ 0 \end{bmatrix}, \begin{bmatrix} 1 \\ 0 \\ 0 \end{bmatrix}$;

l. eigenvalues $-1, 2, 3$; eigenvectors $\begin{bmatrix} 1 \\ 1 \\ 1 \end{bmatrix}, \begin{bmatrix} 2 \\ 1 \\ 2 \end{bmatrix}, \begin{bmatrix} -1 \\ 1 \\ 1 \end{bmatrix}$

9.2.12 a., f., l. diagonalizable; **e., i.** not diagonalizable

9.2.16 a. O; **b.** $\begin{bmatrix} 2 \\ 3 \end{bmatrix}$; $C(A - 2I) = N(A - 2I)$ because of part a; **c.** $\begin{bmatrix} 2 & 1 \\ 0 & 2 \end{bmatrix}$

9.3.2 $2/3$; $5/6$

9.3.3 $9/13$

9.3.4 $a_k = 2^k + 1$

9.3.8 $a_k = \frac{1}{3}\left(2^k + (-1)^{k+1}\right)$

9.3.9 b. $\mathbf{x}_k = \begin{bmatrix} c_k \\ m_k \end{bmatrix} = (c_0 + 2m_0)(1.1)^k \begin{bmatrix} -1 \\ 1 \end{bmatrix} + (c_0 + m_0)(1.2)^k \begin{bmatrix} 2 \\ -1 \end{bmatrix}$, so—no matter what the original cat/mouse populations—the cats proliferate and the mice die out.

9.3.11 a. $-e^{-t}\begin{bmatrix} -5 \\ 2 \end{bmatrix} + e^{6t}\begin{bmatrix} 1 \\ 1 \end{bmatrix}$; **d.** $e^{2t}\left((2-3t)\begin{bmatrix} 1 \\ 1 \end{bmatrix} - 3\begin{bmatrix} 0 \\ 1 \end{bmatrix}\right)$;

e. $e^{-3t}\begin{bmatrix} -1 \\ 0 \\ 1 \end{bmatrix} + 2\begin{bmatrix} 1 \\ -1 \\ 1 \end{bmatrix} + e^{3t}\begin{bmatrix} 1 \\ 2 \\ 1 \end{bmatrix}$

9.3.12 a. $(-\cos t + \sin t)\begin{bmatrix} -5 \\ 2 \end{bmatrix} + (e^{\sqrt{6}t} + e^{-\sqrt{6}t})\begin{bmatrix} 1 \\ 1 \end{bmatrix}$; **d.** $\begin{bmatrix} \frac{1}{6}t^3 + t^2 + 2t + 1 \\ t + 2 \end{bmatrix}$

9.3.13 c. normal modes $\cos t \begin{bmatrix} 1 \\ 1 \end{bmatrix}$, $\sin t \begin{bmatrix} 1 \\ 1 \end{bmatrix}$, $\cos 2t \begin{bmatrix} 2 \\ -1 \end{bmatrix}$, and $\sin 2t \begin{bmatrix} 2 \\ -1 \end{bmatrix}$

9.3.14 $\begin{bmatrix} e^{2t} & te^{2t} & \frac{1}{2}t^2 e^{2t} \\ 0 & e^{2t} & te^{2t} \\ 0 & 0 & e^{2t} \end{bmatrix}$

9.3.15 a. $y(t) = e^{2t} - 2e^{-t}$; **b.** $y(t) = e^t + te^t$

9.4.1 a. $\dfrac{1}{\sqrt{5}}\begin{bmatrix} -2 & 1 \\ 1 & 2 \end{bmatrix}$; **c.** $\begin{bmatrix} \frac{2}{\sqrt{6}} & 0 & \frac{1}{\sqrt{3}} \\ \frac{1}{\sqrt{6}} & \frac{1}{\sqrt{2}} & -\frac{1}{\sqrt{3}} \\ -\frac{1}{\sqrt{6}} & \frac{1}{\sqrt{2}} & \frac{1}{\sqrt{3}} \end{bmatrix}$; **d.** $\dfrac{1}{3}\begin{bmatrix} -2 & 1 & 2 \\ 2 & 2 & 1 \\ 1 & -2 & 2 \end{bmatrix}$

9.4.2 $\begin{bmatrix} 2 \\ 2 \\ 4 \end{bmatrix}$

9.4.5 There is an orthogonal matrix Q so that $Q^{-1}AQ = \Lambda = \lambda I$. But then $A = Q(\lambda I)Q^{-1} = \lambda I$.

9.4.18 b. ellipse $y_1^2 + 2y_2^2 = 2$, where $\mathbf{y} = \dfrac{1}{\sqrt{2}}\begin{bmatrix} 1 & -1 \\ 1 & 1 \end{bmatrix}^T \mathbf{x}$;

c. parabola $y_1 = 5y_2^2 - 1$, where $\mathbf{y} = \dfrac{1}{5}\begin{bmatrix} 3 & -4 \\ 4 & 3 \end{bmatrix}^T \mathbf{x}$.

9.4.19 a. hyperboloid of 1 sheet $-2y_1^2 + y_2^2 + 4y_3^2 = 4$, where $\mathbf{y} = \begin{bmatrix} 0 & -\frac{1}{\sqrt{3}} & \frac{2}{\sqrt{6}} \\ \frac{1}{\sqrt{2}} & \frac{1}{\sqrt{3}} & \frac{1}{\sqrt{6}} \\ -\frac{1}{\sqrt{2}} & \frac{1}{\sqrt{3}} & \frac{1}{\sqrt{6}} \end{bmatrix}^T \mathbf{x}$;

d. hyperbolic paraboloid (saddle surface) $-y_1^2 + 3y_3^2 + \sqrt{3}y_2 = 1$, where

$\mathbf{y} = \begin{bmatrix} 0 & -\frac{1}{\sqrt{3}} & \frac{2}{\sqrt{6}} \\ \frac{1}{\sqrt{2}} & \frac{1}{\sqrt{3}} & \frac{1}{\sqrt{6}} \\ -\frac{1}{\sqrt{2}} & \frac{1}{\sqrt{3}} & \frac{1}{\sqrt{6}} \end{bmatrix}^T \mathbf{x}$.

INDEX

acceleration, 109
Ampère's law, 398
angle, 11, 239
arclength, 112
arclength-parametrized, 114
area, 43
   signed, 44
area form, 370
augmented matrix, 130
average value, 298, 299
   weighted, 301

ball, 65
   closed, 70
basis, 161
   change of, 416
   orthogonal, 233, 236
   orthonormal, 236, 456
   standard, 162
binomial coefficient, 171
binormal, 117
boundary orientation, 382
boundary point, 380
bounded, 197, 199
Brouwer Fixed Point Theorem, 404
bump function, 381

$\mathcal{C}^k, \mathcal{C}^\infty$, 120, 167–168
catenoid, 124
Cauchy sequence, 71, 249
Cauchy-Schwarz inequality, 11, 15, 239
Cavalieri's principle, 324
center of mass, 300
centroid, 300
Change of Variables Theorem, 326
Change-of-Basis Formula, 417
change-of-basis matrix, 416
characteristic polynomial, 426–428
checkerboard, 315
circulation, 394
closed, 69, 347
closed ball, 70

closure, 70
cofactor, 315
column space, 171
   basis for, 177
column vector(s), 28
   linear combination of, 136
compact, 197
complement
   orthogonal, 22
conic section, 108, 459–460
connected, 103, 352, 361
   simply, *see* simply connected
conservative, 352
consistent, 138, 140, 172
constraint equation, 139, 140, 150, 179, 182
continuity, 75
   properties of, 76–78
   uniform, 200, 271, 287
contour curves, 57
contraction mapping, 245
convergent, 67
coordinate chart, 380
coordinates, 163, 233, 413, 415, 416
cos, power series of, 445
Cramer's Rule, 318
critical point, 203
cross product, 48
cubic
   cuspidal, 55
   nodal, 55, 187
   twisted, 56
cubical norm, 324
curl, 393
curvature, 115
cuspidal cubic, 55
cycloid, 56, 187
cylinder, 460
cylindrical coordinates, 291–294, 329

d, 341
deMoivre's formula, 407

determinant, 46, 309, 314
diagonalizable, 423, 429, 432, 455
   simultaneously, 435, 465
difference equation, 436
differentiable, 87
   continuously, 93
differential equations, system of, 439, 441
differential form, 339
   closed, 347
   exact, 347, 386
differentiating under the integral sign, 287
dimension, 165
directional derivative, 83
distributive property, 8, 34
divergence, 395, 450
Divergence Theorem, 395
domain, 24
dot product, 8

echelon form, 131
eigenspace, 423
eigenvalue, 222, 423
eigenvector, 222, 423
elementary matrix, 147–148, 312
elementary operations, 128
   column, 309
   row, 130
ellipse, 105–106, 108, 460
ellipsoid, 460
envelope, 261
epicycloid, 62
Euler, 103, 445
exact, 347
exponential, power series of, 441
exterior derivative, 341
extremum, 202

Faraday's law, 398
Fibonacci Sequence, 438
finite-dimensional, 168
first variation equation, 455

fixed point, 79, 245
flow, 450
flow line, 401
flux, 373, 395
force
    central, 110
    conservative, 352
free variable, 131
Frenet formulas, 118
frontier, 71, 273
Fubini's Theorem, 279, 281
function, 24
function space, 167
Fundamental Theorem of Algebra, 406

$GL(n)$, 249, 321
Gauss's law, 396, 398
Gauss's Theorem, 395
Gaussian
    elimination, 132, 152
    integral, 287, 291
general solution, 132
    standard form of, 132
global maximum (minimum), 202
golden ratio, 439
gradient, 104, 192, 395
Gram-Schmidt Process, 236, 427
gravitation, 304, 396–398
Green's Formulas, 400
Green's Theorem, 358–362

harmonic function, 122, 400
    maximum principle for, 401
    mean value property of, 400
heat equation, 125, 403
helicoid, 124, 368
helix, 114
Heron's formula, 52
Hessian, 209
    bordered, 224, 466
homogeneous function, 103, 171
homogeneous system, 140
homotopic, 405
Hooke's Law, 449
hyperbola, 459
hyperboloid, 460
hyperplane, 18
hypocycloid, 62, 187

identity matrix, 34
image, 24, 79, 170, 172, 183
implicit differentiation, 190
improper integral, 291, 297
inconsistent system, 138, 227, 230
infinite-dimensional, 168
inhomogeneous system, 140
initial value, 440
inner product space, 238
integrable, 268
integral, 268
    Gaussian, 287, 291
interior, 71
intersection, 23
inverse matrix, 34
    formula for, 154, 319
    right (left), 154, 156
invertible, 34
iterated integral, 277–284

Jacobian matrix, 88
Jordan canonical form, 434

k-form, 339
k-tuple, increasing, 337
kernel, 170, 172
kinetic energy, 109, 302, 350, 352
knot, 116

Lagrange interpolation, 239
Lagrange multiplier, 218, 457
Laplacian, 122, 125, 346, 391
leading entry, 131
least squares line, 230
least squares solution, 226, 227
least upper bound property, 68–69
length, 2, 239
level curves, 57
level set, 78, 192
limit, 72
    properties of, 74–75
line, 16
    of regression, 230
linear combination, 19
    trivial, 158
linear map, 24
    matrix for, 413, 416
    standard matrix for, 24
    symmetric, 456
linear transformation, *see* linear map

linearly dependent, 158
linearly independent, 158
local maximum (minimum), 202
lower sum, 267

manifold, 192, 262
    with boundary, 380
matrix
    addition of, 29
    change-of-basis, 416
    diagonal, 29
    exponential, 441
    identity, 34
    nonsingular, 142, 152, 163, 312, 464
    orthogonal, 42, 266, 322, 456
    permutation, 42, 320
    positive (negative) definite, 464
    powers, 32
    product, 31, 147
        not commutative, 32
    singular, 142, 433
    skew-symmetric, 36
    square, 29
    symmetric, 36, 455
    upper (lower) triangular, 29, 314
    zero, 30
matrix multiplication
    associative property of, 34, 36, 37, 152, 165
    block, 39, 156
    distributive property of, 34
maximum, 202
maximum principle, 401
Maximum Value Theorem, 199
Maxwell's equations, 398–400
Mean Value Inequality, 248
Mean Value Theorem, 94, 121, 208, 328, 470
    for integrals, 274
measure zero, 275
median, 5
minimal surface, 124, 392
minimum, 202
moment of inertia, 303, 308
monkey saddle, 204
multiplicity
    algebraic, 431
    geometric, 431

negative definite, 210, 464
neighborhood, 65
Newton
  law of gravitation, 304, 396
  second law of motion, 109, 449
Newton's method, 247, 250
  n-dimensional, 250
nodal cubic, 55
nonsingular, 142, 152, 163, 312, 464
norm, 97, 199, 248, 324
normal equations, 227
normal mode, 443, 449
normal vector, 18, 49, 105
  outward-pointing, 370
nullity, 181
Nullity-Rank Theorem, 170, 181
nullspace, 172
  basis for, 177

open, 65
open rectangle, 65
orientable, 370, 381
orientation, 370, 380
orthogonal, 9
  basis, 233, 236
  complement, 22, 181–182, 226
  matrix, see matrix, orthogonal
  set, 233
  subspaces, 21
orthonormal basis, 236, 456
oscillation, 276

Pappus's Theorem, 308
parabola, 108, 462
paraboloid, 462
parallel, 3, 160
parallel axis theorem, 308
parametrization, 53
parametrized k-dimensional
  manifold, 345
parametrized curve, 53
partial derivative, 81
partial differential equation, 122
particular solution, 132
partition, 267
partition of unity, 381
path-independent, 352
permutation, 320
permutation matrix, 42, 320
piecewise-$\mathcal{C}^1$, 348

pivot, 131
  column, 131
  variable, 131
plane, 17
  affine, 21, 49
planimeter, 366
polar coordinates, 60, 288–291, 329
positive definite, 210, 464
potential energy, 352
potential function, 352–357
preimage, 76
principal normal vector, 115
product, 45
  wedge, 338
product rule, 312
projection, 10, 226, 234, 459
projection matrix, 228, 414
pullback, 342
Pythagorean Theorem, 2, 9, 51, 227
Pythagorean triple, 61

quadratic form, 209–210, 214–215,
  465–466
quadric surface, 460–462

range, 24
rank, 140, 180–181
Rational Roots Test, 427
rectangle, 65, 267, 268
reduced echelon form, 131
refinement, 269
  common, 269
reflection matrix, 39, 414
region, 273
regression, line of, 230
rotation matrix, 27, 39, 419–420
row space, 171
  basis for, 177
row vector, 28–29

saddle point, 203
scalar multiplication, 2
sequence, 66
  Cauchy, 71, 249
  convergent, 67
shear, 26, 314
sign, 320
similar, 417, 421
simply connected, 361, 406
sin, power series of, 445
singular, 142, 433

skew-commutative, 339
smooth, 120
span, 19, 20
spectral decomposition, 459
Spectral Theorem, 456
speed, 109
sphere, 78, 197, 409
spherical coordinates, 294–296, 329
squeeze principle, 79
standard basis, 19, 162
standard matrix, 24
star operator, 346, 395
star-shaped, 355, 411
stereographic projection, 376
Stokes's Theorem, 383–389
  classical, 394
subsequence, 70
  convergent, 197
subspace, 16
  fundamental, 171–183
  trivial, 16, 164
subspaces, sum of, 23
surface area, 375
symmetric, 36, 455

tangent plane, 87, 105, 188, 191
tangent space, 219, 264
Taylor polynomial, 210
torsion, 118
torus, 367
trace, 41, 428
tractrix, 63
transpose, 36
  determinant of, 313
triangle inequality, 12
trivial solution, 141, 142
trivial subspace, 16, 164
twisted cubic, 56

uniformly continuous, 200, 271, 287
unique solution, 141, 142, 152
unit tangent vector, 115
unit vector, 2
upper sum, 267

variable
  free, 131
  pivot, 131
vector
  addition, 3
  column, 28

row, 28–29
subtraction, 4
zero, 1
vector field, 348, 393–395, 409
vector space, 167

velocity, 109
volume, 268, 314
  signed, 314
volume form, 381
volume zero, 271

wave equation, 122, 403
wedge product, 338
weighted average, 301
winding number, 362
work, 350